Dan Shen (Salvia miltiorrhiza) in Medicine

Xijun Yan
Editor

Dan Shen (Salvia miltiorrhiza) in Medicine

Volume 2. Pharmacology and Quality Control

人民卫生出版社
PMPH PEOPLE'S MEDICAL PUBLISHING HOUSE

Springer

Editor
Xijun Yan
Tianjin
China

ISBN 978-94-017-9462-6 ISBN 978-94-017-9463-3 (eBook)
DOI 10.1007/978-94-017-9463-3
Springer Dordrecht Heidelberg New York London

Library of Congress Control Number: 2014950646

Printed on acid-free paper

Springer is part of Springer Science+Business Media (www.springer.com)

Preface to Dan Shen (Salvia miltiorrhiza) in Medicine

In the 2008 press conference on the publication of the Chinese edition of the *Dan Shen* (*Salvia miltiorrhiza*) *in Medicine*, several volume editors suggested that the book should be translated into English and distributed internationally. They all believed that the medical communities are enthusiastic about TCM research, and that among the studies on single herbs, the study of Danshen has taken the lead. Therefore, it was a worthy undertaking to introduce the study conducted by the Chinese people over the past 1,000 years, and especially in the past 30 years, to the world. Meanwhile, I was asked unanimously to be its editor-in chief. After several years of hard work by nearly 100 professors and research scientists, the translation is finally complete.

The English edition of *Dan Shen* (*Salvia miltiorrhiza*) *in Medicine* is based on its Chinese edition. Modifications include changing the five-volume format to one volume and deleting some duplicated portions in the Chinese edition. Since the chapters in each volume of the Chinese edition were written by many individuals, details such as biological properties and ancient literature reviews were repeated many times, and the duplications were deleted in the English edition. Also, the various names of Danshen were unified. The appendix in the fifth volume, the prescriptions or formulas in ancient China, and the chapter about information management in the fourth volume were also deleted. The introduction to the production region, common names, and phytochemical components had appeared in the first three volumes, while this time only their first appearance was preserved. Some typos and oversights were corrected after consulting with volume editors. New progress in Danshen research was included in this book, such as the development of Salvianolate Lyophilized Injection, which finally came on the market in 2011 after 8 years of strict examination, and it was a landmark event in the development of TCM injections. It is unfortunate that we could not include the data on Qishenyiqi Dripping Pills, as the papers have not been published yet. The drug, developed by academician Boli Zhang, passed large-scale, evidence-based medicine clinical research trials in 2010, the first for a TCM drug, and won the 2011 National Science and Technology Progress Award.

In principle, the English edition of *Dan Shen* (*Salvia miltiorrhiza*) *in Medicine* is the translation of the Chinese edition, thus preserving the latter's framework. Because the Chinese edition was written by more than 100 scholars and published in five volumes, the styles and layouts were not identical. For example, some references were listed at the end of the chapter, while some were listed at the end of the section. The English edition did not change the style.

Dan Shen (*Salvia miltiorrhiza*) *in Medicine* has amassed Danshen research results since the times of ancient China—it is not only a magnificent historical scroll, but also a huge work which shines the light of modern science and technology.

I sincerely thank academicians Yongyan Wang and Boli Zhang. They have given me so much substantive guidance and encouragement despite their busy schedules. Without their help, it would have been impossible to finish the work. I also want to thank every author and volume editor who has participated in the writing and editing of both editions of this book; they have solved various problems that arose during the writing and translating processes. Last but not least, I want to thank the comrades working in the office of *Dan Shen* (*Salvia miltiorrhiza*) *in Medicine*, who have worked patiently and diligently over the past 15 years, collecting and organizing data and information.

The publication of the English edition of *Dan Shen* (*Salvia miltiorrhiza*) *in Medicine* is a testimony of our sincere desire for the communication and discussion of TCM among international communities. We earnestly welcome suggestions and criticism from our colleagues around the world.

Xijun Yan

Editing Committee

Volume 1. Biology and Chemistry

Chief Editors
Luqi Huang
Lianniang Li
Houwei Luo

Associate Chief Editors
Deyou Qiu
Zongsuo Liang
Min Chen

Editors
Xuefeng Feng
Wenting Liu
Xirong He
Jingyuan Song
Deyou Qiu
Min Chen
Jinda Hao
Lanping Guo
Guanghon Cui
Zongsuo Liang
Luqi Huang
Houwei Luo
Lianniang Li

Volume 2. Pharmacology and Quality Control

Chief Editors
Juntian Zhang
Guanhua Du
Dean Guo

Associate Chief Editors
Xiumei Gao
Shen Ji
Ming Zhu
Rongxia Liu
Haibin Qu

Editors
Lianhua Fang
Yitao Wang
Xiaoying Wang
Yuehua Wang
Dongxia Wang
Xiaoming Wang

Rongxia Liu
Shunhang Liu
Aihua Liu
Xuesong Liu
Jianghao Sun
Guoqing Wu
Ling Tong
Yongjiang Wu
Wanying Wu
Xiaoqian Zhang
Xuemin Zhang
Jinlan Zhang
Xian Zhang
Yunfei Li
Manling Li
Min Yang
Yuewu Yang
Bilian Chen
Shen Ji
Jianping Lin
Dean Guo
Guoqiang Fan
Xiaohui Fan
Qiang Zheng
Zhangzhao Jin
Ming Zhu
Jun Gao
Hongfang Cui
Fenglan Cao
Rixin Liang
Shunnan Zhang
Qing Gong
Haiou Dong
Jianping Han
Haibin Qu

Volume 3. Clinical Research

Chief Editors
Guoguang Zhu
Xuewen Zhang
Xinde Shi

Associate Chief Editors
Yi Zheng
Ruizhi Luo
Naifeng Wu

Contents

About the Chief Editor

Dr. Xijun Yan was born in 1953 in Zhenyuan, Gansu province. Dr. Yan is a chief pharmacist and enjoys the Special Expert's Allowance from the State Council of China. At present, he is a President of Tianjin Tasly Group; serves as a member of traditional Chinese medicine (TCM) Standardization Technical Committee; the deputy director of Science Popularization Committee of Chinese Pharmaceutical Association; a member of Expert Committee of The National Pharmaceutical Industry Policy Research Project of Chinese Pharmaceutical Association; the vice president of Tianjin Pharmaceutical Association. Dr. Yan is the first person who proposed the concept of "modern TMC drugs" and the new mode of TCM R&D, and has engaged for a considerable length of time in research and industrial development of modern Chinese medicine. Dr. Yan has actively undertaken key national projects on the research and development of important new drugs, supervised or participated in 38 projects in the National Key Science and Technology Projects in the 9th and 10th "Five Year Plans"; National Level Promotion Program of Scientific and Technological Achievements; National High-tech Industrialization Demonstration Project, etc. Dr. Yan has 69 patents and more than 50 publications, including The Standards of Diagnosis, Efficacy, and Medication in TCM's Heart Disorder (one of the editors-in-chief); How TCM Drugs Enter the EC Market (associate editor-in-chief); The Ideas and Methods of Modernization of TCM Research (one of the editors-in-chief); The Reflection and Practice of the Modernization of TCM Preparations; On the Trends of Modern TCM Industry, etc. Dr. Yan has been awarded honorary titles like National Model Worker; National Outstanding Scientific and Technological Worker; National Outstanding Pharmaceutical Entrepreneur; National Health Industry and Enterprise Advanced Individual; National Model Worker in Chinese Medicine System. Dr. Yan has received numerous awards, including the Third National Prize for Scientific Advancement, the Second Prize for Scientific Advancement by the People's Liberation Army, and the First Prize for Scientific Advancement by Tianjin Municipal Government.

Introduction by Chief Editor

After the hard efforts of an entire decade, by nearly a 100 experts from home and abroad, the *Dan Shen (Salvia miltiorrhiza) in Medicine* is finally going to press.

Danshen (*Salvia Miltiorrhiza*) has a time-honored history of research, development, and application in China's traditional medicine, and is held in very high esteem by the medical community. We hope that the five-volume series now made available to the readership will make a worthwhile contribution to its research and application.

Salvia Carries the Dream of Internationalization of Traditional Chinese Medicine

In 1996, China began to formulate and implement a strategy for the modernization of traditional Chinese medicine (TCM). However, the question of how in practice to systematically develop traditional medicine, passed on from generation to generation over several millennia, remained a real and persistent challenge to the academic and industrial dimensions of TCM. In 1998, a discussion among several TCM experts from home and abroad gave us the idea of compiling the *Dan Shen (Salvia miltiorrhiza) in Medicine*. The experts discussed not only the present state and the future prospects of modernizing Chinese medicine itself, but also the growth of Chinese medicine industry and the issues in developing big brand modern TCM drugs. The discussion was especially centered on the utilization of Danshen and the topics ranged from its collection and processing in ancient times to the widespread application of various Danshen preparation, from its effective components to its pharmaceutical action, from its cultivation and plantation to its modern industry chain, from its compound prescriptions and to modern Compound Danshen Dripping Pill (Dantonic™), from its R&D to the rapid growth of a modern TCM enterprise—the Tasly Group. Is it possible that a unique industrial technology and economy grows out from a single TCM drug and a series of research activities focused on the drug? Is it possible that the unique industrial technology and economy stimulates new lines of thinking and novel approaches to the modernization and internationalization of TCM? Is it possible the unique industrial technology and economy pushes forward the systematic project of TCM research and development? These and several other questions roused a profound interest among experts in deepening the research on Danshen, and the compilation of the *Dan Shen* (*Salvia miltiorrhiza*) *in Medicine* was originated from this initial driving force.

For thousands of years, TCM has made enormous contributions to the health and multiplication of the Chinese people. But why is it so hard for TCM to be accepted and acknowledged in the international community? Why do some people still have doubts about the scientific nature of TCM? To be able to continually promote the modernization and internationalization of TCM, these questions must have affirmative answers. In order to introduce Chinese medicine to the rest of the world and let the international community understand, accept, and use TCM, we must resort to modern technology to re-develop TCM. We also have to give TCM a fresh interpretation, using the standardized scientific and digitized languages. It is indeed necessary to select a certain Chinese medicine for an exploratory trial, and this Chinese medicine must meet a number of requirements. First, it should have both a long history of inheritance and deep accumulation of clinical knowledge; second, it has been systemically studied with modern means of science and technology, and its effective substances and the mechanism of action have been elucidated, relatively speaking; third, it has made a comparatively great contribution to human health, especially in terms of satisfactory effects in the treatment of serious diseases; fourth, its industrialization has been successful, having representative name brand products; fifth, it has sufficient resources to ensure sustainable industrialization; and sixth, it should be conducive to a progressive growth of Chinese medical research. We believe that Danshen meets all these requirements, and it could showcase a wealth of innovative achievements and profound knowledge, and become a model in promoting the modernization and internationalization of TCM.

Hundreds of Scientists Involved in the Work

It is with this underlying ambition in mind that we started data collection, collation, and *compilation of the this book. Nearly 100 experts* from home and abroad have participated in this huge project. The expert team includes both world renowned senior scientists and young scholars with outstanding achievements. Some of them are from prestigious research institutions and universities, others from industrial regulatory bodies, and still others from the frontline of industrial development. Their expertise covers multiple research areas, including medicinal botany, phytochemistry, pharmaceutical analytics, pharmacology, toxicology, medical preparation, medicine reviewing, TCM, and integrative medicine. We have been particularly fortunate in that our research and compilation work has received strong support and guidance from the academicians Yongyan Wang and Boli Zhang, as well as from several relevant leaders and experts. Each writer adhered to the mission of "being responsible for both past and future generations" and followed stringent scientific spirit and serious academic attitude, searching extensively, and studying strenuously and carefully. They referenced nearly 50,000 pieces of literature, including books, research articles, trial reports, and others, directly quoted 7,235 references, thus laying a solid literature foundation for the work.

Botany of Danshen documents Danshen herbal research, medical source survey, identification of medicinal characteristics, distribution and ecological environment, biological features, in vitro culture, and genetic breeding; describes and reviews research on Danshen germplasm resources and genetic diversity. The detailed research provides guidance for scientific and large-scale cultivation of Danshen, and contributes to the securing of resources for its further development and industrialization.

The phytochemical research of Danshen originated in the 1930s. Centered on the extraction and isolation of the effective components from Danshen, researchers have utilized various techniques and spectral analytical methods, including the 2D nuclear magnetic resonance. These developments are all accounted for in *Phytochemistry of Danshen*. The volume also systematically describes the chemical structure of Danshen's liposoluble ingredients (tanshinones) and water-soluble ingredients (salvianolic acids), methods for extraction and isolation, spectral characteristics, the chemical and physical properties, the biosynthesis pathways, and chemical synthesis. The chemical ingredients of other plants in the genus *Salvia* are also described.

Since the 1930s, domestic and international experts have carried out in-depth, or even spectacular, intensive pharmacological research into the effective ingredients of Danshen. They have analyzed its pharmacological action and mechanisms in the cardio-cerebrovascular system, the nervous system, the digestive system, and its anti-bacterial and anti-inflammatory effects. In the *Pharmacology of Danshen*, the liposoluble and water-soluble ingredients of Danshen are described in detail for their pharmaceutical functions. The lipo-soluble ingredients have the functions of anti-bacteria and regulation of the endocrine, while the water-soluble ingredients have the functions of anti-oxidation, anti-ischemia, and inhibiting the expression of cell adhesion molecules. The multi-target effects of Danshen and its preparations in the treatment of microcirculatory dysfunction have been recognized by the international pharmacological scholars. The volume reviews and summarizes the achievements of these researches in considerable detail, revealing the leaps and bounds of modern pharmacology in recent years.

Quality Control of Danshen is based on the research into the pharmacologically effective components of Danshen. The volume introduces a QC system which combines fingerprinting technology with multi-indicator analysis. In fact, this QC system is a full range of Danshen quality control system, integrating modern TCM chemistry, pharmacological pharmacodynamic, and pharmacokinetics, covering Danshen identification, content determination, fingerprinting spectra, in vivo metabolic processes of major ingredients, and the effect of the preparation process on the quality of drugs, etc. The volume shows comprehensively and systematically the present state of Danshen quality control, highlights the most recent achievements.

Clinical Research of Danshen records 1,261 carefully selected Danshen-containing prescriptions from over 2,000 medical works since the Qin and Han dynasties. It reports the summarized results of mathematical and statistical analysis, showing how Danshen-containing drugs were processed, with which herbs Danshen was combined, and what kinds of preparations were used. The changes in the frequency of other herbs combined with Danshen, and the indications of these prescriptions have been compared. The ancient Chinese pointed out explicitly a long time ago that the basic functions of Danshen is to invigorate blood and dissolve stasis, and clear blood vessels. They also summarized that "the functions of a single Danshen is equal to those of Four Substances." In the field of modern medicine, Danshen enjoys an even wider application. The authors of the volume have carried out an exhaustive documentation and summarization of the clinical applications Danshen, especially the experience from famous TCM doctors. In this volume, the methods of evidence-based medicine have been used to summarize the functions of modern TCM drugs, Compound Danshen Dripping Pill (Dantonic™) in the prevention and treatment of coronary heart disease and its multiple risk factors, and proposes for the first time the multiple-target mechanism of Dantonic™'s effects on the treatment of coronary heart disease and angina. In addition to their wide application in the treatment of cardio-cerebral vascular diseases like coronary heart disease and ischemic stroke, Danshen preparations are used to treat other diseases, and it has proved to be effective for chronic hepatitis, chronic nephritis, chronic kidney failure, type II diabetes, blood diseases, infectious diseases, and skin diseases. All of these are systematically described in the volume. Such an integration will doubtlessly be beneficial to clinical workers in the fields concerned.

Looking at the history, it is safe to say that Danshen has been one of the most extensively used drug in TCM since ancient times, and it is one of the TCM herbs which were studied with modern techniques the earliest and most thoroughly. Through the research on Danshen in botany, biology, chemistry, pharmacology, and clinical trial, we now have comprehensive understanding of its effective components, clinical effects, and indications. The study on the effective components of Danshen serves as the foundation for the study on its pharmacological effects, and based on the pharmacological study, a batch of widely applied and effective modern Danshen preparations have been developed. Danshen research is a model of multidisciplinary, multi-domain, multifaceted integration and cooperation, and will surely be an important milestone in the process of modernization of TCM.

New Challenges and Opportunities in the Era of Health Caring

Since the 1970s, the spectrum of human diseases has shifted from the infectious to lifestyle diseases, geriatrics, and degenerative diseases. This has brought about significant changes in medical modes and treatment philosophies; from the biological medicine mode to a "biological-psychological-social" mode; from the purely passive disease treatment to the combination of "prevention–healthcare–treatment–rehabilitation." Also, people are paying more attention to a timely adjustment of, and recovery from, sub-health states. These trends suggest that the society has entered into an era of "enlarged health." People hope to attain the objective of enjoying a longer and healthier life by means of full care and comprehensive protection. To put it simply, we now strive to have a life journey which consists of "eugenic birth, longevity life, delayed aging, and peaceful departure." These changes have posed new challenges to medical research, and guided medical R&D into a new age.

Started from searching for the chemical basis of life substances in the 1780s, the drug development mode of modern medicine has gradually evolved into looking for chemical compounds. The trends of contemporary drug development show that a new drug can only be successfully developed after screening thousands of compounds, and the cost of R&D for listed drugs is increasing annually, while the speed of new drug development is slowing down, which leads to a vicious circle characterized by high input, high risk, and low output. There is an urgent need for a fresh drug R&D mode.

Very different from the western mode, the TCM mode started from a holistic concept, based on individualized diagnosis and treatment and knowledge accumulation, used the resources of plants, animals, and minerals, and developed preparations of pills, powders, pastes, etc. These TCM preparations, tested through clinical application over thousands of years, contained a variety of pharmacodynamic active substances and their action mechanisms integrated antagonistic, supplement, and regulation as one. Thus, our ancestors have left us a drug resource treasure, which is our unique advantage. This advantage is also a more economical and more efficient path to the discovery of new drugs.

The promotion of TCM modernization strategy, the implementation of policies such as the Scientific and Technological Action of TCM Modernization (Outline), and the Development Outline for TCM Modernization has enabled the construction of a system for scientific innovation of Chinese medicine. Contemporary high-tech innovations are increasingly used in the field of TCM, engendering a diversified research and development of TCM. Although the most varieties and the most widely used TCM drugs are the traditional preparations, modern TCM drugs have gradually grown up, and the exploration of chemical TCM, biological TCM, and metabolic TCM is receiving increasing attention. Modern TCM drugs are drugs based on TCM formulas and prescriptions, but manufactured using modern technology to extract effective components, using fingerprinting and chromatographic techniques for qualitative and quantitative quality control, using online data collection, analysis and feedback function of information technology to adapt to the new industrialized production requirements. Chemical TCM drugs are drugs obtained using chemical methods to isolate from TCM or herbal drugs single effective ingredients which cannot be totally synthesized, and use them as the lead compounds to generate monomer compounds drugs with defined structures after in-depth study of chemical and biological activities, and by structural modification and transformation.

The priorities in the exploration of biological TCM drugs are the plantation of medicinal plants, extraction of effective components, and manufacture of the preparations. In addition, biotechniques could be used to discover or invent new effective components to improve the effectiveness of TCM drugs and the recovery of the effective components, as well as to reduce toxic and side effects. The so-called metabolic TCM drugs are to search for new active substances (effective components, effective fractions, or new lead compounds) from the metabolic process of TCM drugs in human or animal bodies, and use them for new drug development. It can be predicted that new techniques and new fields of Chinese medicine R&D will keep springing up. The key is to encourage diversified innovation and exploration based on different technical routes.

The aim of *Encyclopedia of Danshen* is not just to focus on a single TCM herb, rather, it is to use the rich TCM drugs and prescriptions as a medical resource treasury, to combine the innovative thinking in TCM with modern medical techniques, and to explore new modes for pharmaceutical R&D.

Cooperation to Develop a Blockbuster of Traditional Chinese Medicine

Historically, the typical mode of practicing TCM has been to run a mom-and-pop style operation, composed of a storefront and a backroom workshop. The introduction of modern industrialization has changed the modes of both innovation and industrial organization of TCM. Multidisciplinary cooperation and industry—academia—research integration have become the best approach to TCM pharmaceutical R&D.

China has a large number of TCM preparations, but few of them have market sales in excess of 500 million Yuan, and even fewer with market sales over 1 billion Yuan. For TCM drugs to meet the standards of modernization and internationalization, for TCM enterprises to become large industry, characteristic industry, or even competitive industry, name brand products and large-scale production are essential. Analyzing the knowledge economy, industry economy, and technology economy of TCM from the point of view of economics, we can see both the huge potential of TCM and the gap between TCM industry and modern pharmaceutical industry.

Starting with a single TCM herb, Danshen, bringing together the elite researchers at home and abroad, through a comprehensive and systematic study, *Dan Shen (Salvia miltiorrhiza) in Medicine* has demonstrated the quality, effectiveness, safety, and toxicological and pharmacokinetic characteristics of Danshen-containing drugs. Moreover, it has demonstrated the scientific nature of TCM with detailed and accurate research results, which will unquestionably contribute to the understanding, acceptance, and utilization of TCM drugs by more people, and this is in turn laying a scientific foundation for the expansion and further strengthening of the TCM industry.

The comprehensive and systemic study of Danshen shows that a "blockbuster" type of TCM drug can be developed. The industrialization of TCM has a character of close connections among multiple industries, so the key is to actively exert the gathering and coordinating role of TCM industrial chain to promote the technological transformation, standards upgrades, and structural optimization in the entire industry, and realize the economic value of TCM in various links of the chain, including TCM agriculture, TCM industry, TCM commerce, and TCM knowledge industry.

A single TCM herb can be used in a variety of prescriptions to benefit the health of many patients; a TCM industrial chain can also boost the economy of certain areas. I believe that the *Encyclopedia* can provide inspiration for the development of TCM resources and TCM industry.

Inheritance and Innovation

A comprehensive summary of Danshen research over thousands of years is both a grand event to revitalize TCM and a difficult task. This series contains not only the original research on Danshen by ancient people and contemporary scientists, but is also the fruit of strenuous labor and a large amount of pioneering work on the part of each author. To summarize Danshen research is to shoulder not only the burden of inheriting previous experience, but also the burden of opening the future and deepening innovation.

As the titled Encyclopedia indicated, this series not only strives to be "extensive and comprehensive," but also attempts to be "excellent and profound." Therefore, in terms of compilation style, we not only try to maintain uniformity in style and systematicness in content, but also take into consideration the uniqueness of each volume, so that the characteristics and progress in each field can be reflected. The book is based on a single herb, Danshen, thoroughly going through the literature from ancient till modern times, with the purpose of reflecting from a certain angle the progress and development of TCM through generations. We divide the series into volumes in terms of botany, phytochemistry, pharmacology, quality control, and clinical research, in order to reflect the entire picture of Danshen research on one hand, and extract the essence of the research on the other, so as to reach to goal of integrating the practicality, comprehensiveness, and prospectiveness in one book. We adopted the approach of "breaking down the institution barriers," bringing together nearly 100 experts from the forefronts of their relevant fields. We advocated brainstorming and free debate, as well as the accumulation of collective knowledge and wisdom to develop the Encyclopedia into a genuine milestone in Danshen research.

Memories pass through thousands of years, and science is an endless frontier. The modernization and internationalization of TCM is an ongoing process, science is being advanced, technology is being innovated, cross-disciplinary achievements are being integrated, and in-depth research into Danshen is being pushed forward in multiple sectors. There is every reason to believe that this research will continue to be enriched and replenished with an ever-increasing number of new and even more in-depth fruits. The modernization of Chinese medicine is of course more than just producing a collection of academic and technical research papers and awards. It also calls for more TCM products to drive the relevant industries, even new cultural and humanitarian philosophies, which in turn pose new themes and topics to Danshen research for our continuously to explore and innovate. With further research and accumulation of findings, we will no doubt be able to come up with additional volumes or a sequel to this book, bringing Danshen research to ever deeper and higher levels.

The publication of this book is first of all made possible by the creative achievements of our predecessors in various ages, and by the long history and timeless essence of Chinese culture and Chinese medicine. Our most deeply felt gratitude, then, goes to our ancestors for this precious scientific and cultural heritage.

In the process of compilation, the associate editors of the book, leaders of various institutions and agencies, colleagues and friends from home and abroad, who are enthusiastic about TCM, have given strong support and great help. The chief editors, associate editors, and editors of individual volume, as well as colleagues of the *Encyclopedia* compilation office, have made enormously strenuous efforts toward the publication of the book, and it is to them that I hereby extend my sincerest gratitude. A special mention is due to these experts who participated in this project. They themselves are accomplished experts on Danshen research. They bear the heavy research work of their own, and shared the work of writing this book. Thanks go to the experts for their research accomplishment and hard work. The academicians Yongyan Wang and Boli Zhang sacrificed their precious time to review the manuscript and to author the Prefaces. All of our editors and compilers extend their deepest gratitude to them.

In terms of compilation style, we applied uniform requirements while leaving relatively great room for each volume in order to allow the chief editors of each volume to exercise judgment according to discipline-specific features and the particular status of research. Therefore, it was impossible to achieve perfect coherence in the writing style and length. The research literature on Danshen transcends several millennia and country borders, and covers multiple disciplines. Despite arduous research and diligent search efforts on the part of the various editors and compilers, errors and mistakes may still exist. Collective authoring of a five-volume encyclopedic series calls for enormous amount of work. Therefore, even repeated elaboration and modification by multiple experts may still leave some room for improvement. We sincerely hope that friends and colleagues in Chinese medicine research and industry, as well as the hopefully vast readership at large will offer their criticism and comments.

Xijun Yan

Part I
Pharmacology

1 Overview of Modern Research on Danshen

Guanhua Du and Juntian Zhang

Danshen (*Salvia miltiorrhiza*) is one of the most widely and earliest used herbal drugs in traditional Chinese medicine. A comprehensive knowledge about its clinical therapeutic effects and indications has been acquired over long-term clinical application. It is also one of the main Chinese herbal drugs that have been studied with modern medical technology. Thanks to the joint efforts of chemists, pharmacologists, and medical researchers we now have almost a thorough knowledge and understanding of Danshen regarding its biological, chemical, and pharmacological properties and clinical therapeutic effects. Besides these fundamental achievements, many clinical preparations with Danshen as the principal active component have been developed, which play an important role in the prevention and treatment of diseases. In this chapter, the current state of research on Danshen is reviewed, in which much data are drawn from the findings in our own laboratory.

G. Du (✉) · J. Zhang
Institute of Materia Medica, Chinese Academy of Medical Science and Peking Union Medical College, Beijing, China
e-mail: dugh@imm.ac.cn

1.1 General Situation of Application and Research of Danshen

Danshen (*Radix Salviae miltiorrhiza*) is the dry root and rhizome of Danshen (*S. miltiorrhiza* Bunge), a labiate plant. According to the existing literature, Danshen was first recorded in *Shen Nong's Classic of the Materia Medica* and was then recorded in the other ancient monographs about materia medica such as *Wu Pu's Materia Medica* and *Miscellaneous Records of Famous Physicians*. According to *Shen Nong's Classic of the Materia Medica*: "Danshen can be used to treat the pathogenic factors in heart and abdomen, treat the syndromes of increased bowel sounds, chills and fever, addiction and abdominal mass, eliminate restlessness and invigorate qi." Based on this, subsequent medical researchers have described Danshen in their works. In *The Grand Compendium of Materia Medica*, the descriptions of the functions of Danshen have been extended to many aspects: "eliminating the pathogenic factors in heart and abdomen," "replenishing the heart, resuscitation, and relieving mental stress," "quieting vital fetus and removing the dead fetus, arresting hemorrhage, mitigating leukorrhea and regulating the irregular channels of women," "promoting blood circulation, regulating pericardium, invigorating qi and nourishing the blood." Thereafter, there are many new findings about the application of Danshen after long-term medical practice (see Table 1.1).

X. Yan (ed.), *Dan Shen (Salvia miltiorrhiza) in Medicine*,
DOI 10.1007/978-94-017-9463-3_1,

Table 1.1 The Functions and indications of Danshen recorded in the works of materia medica at different times

Time	Works[a]	Functions and indications							
		Relieving mental stress	Removing pathogenic heat from blood	Nourishing the blood and invigorating qi	Quickening the blood and transforming stasis	Promoting circulation of qi and to relieve pain	Eliminating numbness and strengthening bone	Regulating menstruation and stopping metrorrhagia	Resolving toxin and evacuating pus
200 BC	Shennong Bencao Jing	+	+	+	+	+			
220–450	Mingyi Bielu		+	+		+	+		
494	Bencaojing Jizhu		+	+	+	+	+		
1612	Yao Xing Lun	+		+		+	+		
1565	Bencao Mengquan	+	+	+	+	+	+	+	+
1578	Bencao Gangmu	+	+	+	+	+	+	+	+
1757	Bencao Congxin	+	+	+	+	+	+	+	+

[a] PMPH standard translation
Shennong Bencao Jing: *Shen Nong's Classic of the Materia Medica*
Wu Pu Bencao: *Wu Pu's Materia Medica*
Mingyi Bielu: *Miscellaneous Records of Famous Physicians*
Bencaojing Jizhu: *Collective Commentaries on the Classic of Materia Medica*
Yao Xing Lun: *Treatise on Medicinal Properties*
Rihuazi Bencao: *Ri Hua-zi's Materia Medica*
Bencao Mengquan: *Enlightening Primer of Materia Medica*
Bencao Gangmu: *The Grand Compendium of Materia Medica*
Bencao Congxin: *Thoroughly Revised Materia Medica*

It is recorded in *Chinese Pharmacopoeia* that the actions of Danshen include "dispelling stasis, assuaging pain, promoting blood circulation, restoring menstrual flow, clearing away the heart-fire, and relieving restlessness," which indicates that Danshen has multiple functions.

Research on Danshen using modern scientific methods has been ongoing since the 1930s. In the last few decades, various aspects of Danshen have been studied by chemists, pharmacists, pharmacologists, and clinical workers. A great deal of work has been carried out on the identification and isolation of the active components of Danshen and their pharmacological actions, and a great variety of Danshen preparations have been developed for clinical application. As for the latter, Danshen preparations not only can be used to treat cardiovascular and cerebrovascular diseases, such as coronary heart disease and ischemic stroke, but also show therapeutic effects on many other diseases, such as chronic liver diseases, chronic nephritis, chronic renal failure, type 2 diabetes, blood diseases, infectious diseases, skin diseases, etc., which have been proved by clinical observations and studies. Therefore, we have a more complete and thorough understanding about Danshen [1–3].

1.2 The Research on the Chemical Components of Danshen

1.2.1 Overview

Research on the chemical components of Danshen began in the 1930s, when Japanese scholars first isolated three liposoluble components from Danshen, which were named tanshinones I, II, and III. Since then, Chinese scholars continued on this path demonstrating that tanshinone II was a mixture of two components, which were then named tanshinone II_A and II_B. They also illuminated their chemical structures and discovered a great deal of new compounds.

As mentioned above, Danshen's chemical components can be either hydrosoluble or liposoluble. The early research on the active components of Danshen was mainly focused on the liposoluble components, and more than 40 compounds have been identified. These compounds can be divided into two classes: tanshinone (orthoquinoid form) and royleanone (ortho-hydroxyl paraquinoid form). Most of the tanshinone compounds belong to diterpenoids, and the majority of the latter are diterpene quinones.

Although research on the hydrosoluble components of Danshen was conducted later, it progressed quickly. Since the discovery of Tanshinol, a hydrosoluble component of Danshen, was reported in the early 1980s, a series of hydrosoluble compounds have been isolated and these compounds have been developed into many preparations for clinical application.

1.2.2 Research on Liposoluble Components of Danshen

Since the discovery of tanshinone, the liposoluble components of Danshen have been studied by many researchers and more than 40 types of compounds have been identified, including tanshinone I, tanshinone II_A, tanshinone II_B, tanshinone III, isotanshinone I, isotanshinone II_A, cryptotanshinone, isocryptotanshinone, hydroxytanshinone II_A, dihydrotanshinone I, L-dihydrotanshinone I, neotanshinone A, neotanshinone B, and neotanshinone C. After a few decades' work, comprehensive knowledge about the liposoluble components of Danshen has been obtained, and important progress has been acquired in the research on the biological activities of various components of Danshen.

In recent years, due to the application of new techniques and the discovery of new biological activities, research on the liposoluble components of Danshen has received further attention. New compounds have been discovered continuously, including some new tanshinone compounds with diterpene quinone structures [4] and other types of compounds such as neocryptotanshinone II [5], cyclic phenyllactamide (2,10,11-trihydroxy-8-methoxy-1,6,7,8-tetrahydro-2H-benzo[e]azecine-3,5-dione) [6], oleoyl neocryptotanshinone, oleoyl danshenxinkun A [7], etc.

The research on tanshinones not only resulted in the discovery of new compounds, but also demonstrated the pharmacological actions of these compounds. These studies in turn led to the development of a number of new drugs for clinical application with excellent effects.

1.2.3 Research on the Hydrosoluble Chemical Components of Danshen

All the hydrosoluble components of Danshen have the structure of phenolic acid. Tanshinol was the first discovered of these, with the chemical name of (3,4-hydroxybenzyl) lactic acid. Thereafter, a series of Phenolic acid compounds have been found, named as salvianolic acids in alphabetic order. The hydrosoluble components mainly include salvianolic acids A, B, C, D, E, G, H, I, J, tetramethyl salvianolic acid F, isosalvianolic acid C, rosmarinic acid, lithospermic acid, etc. Most salvianolic acids are formed through the combination of Tanshinol and other organic acids. For example, rosmarinic acid is formed through the combination of Tanshinol and one molecule of caffeic acid; salvianolic acid A is formed through the condensation of Tanshinol and two molecules of caffeic acid, salvianolic acid B is formed through the condensation of three molecules of Tanshinol and one molecule of caffeic acid, and salvianolic acid C is formed through the condensation of two molecules of Tanshinol.

It has been proven by experimental studies that the hydrosoluble components of Danshen have multiple pharmacological activities, for example, they have strong antioxidation activity. These discoveries have attracted researchers to conduct more investigations, and many drugs for clinical application have been developed [8, 9].

1.2.4 Research Methods for the Chemical Components

The research methods for the chemical components of Danshen have been developed with advances in science and technology. In recent years, supercritical extraction, high-speed counter-current chromatography, and high performance liquid chromatography (HPLC) have been used to extract the chemical components of Danshen, which provide a technical base for the investigation and application of the active components of Danshen [10–13].

On the other hand, some methods with high specificity and sensitivity have been developed for the analysis and detection of the components, which provide new means for the detection of the active components and quality control of preparations. The development and the application of these new technologies are the bases of Danshen's quality control and safe usage [14, 15].

1.3 Research on the Active Components of Danshen and Their Pharmacological Actions

The chemical components of Danshen are mainly liposoluble and hydrosoluble components, and it has been proven through long-term studies that both types of components have significant pharmacological actions. Therefore, both are the active components of Danshen.

However, the two components have significantly different pharmacological actions. The hydrosoluble one mainly has antibacterial and endocrine regulation actions, while the liposoluble one has significant antioxidation actions. The multi-functionality is the basis for the broad indications of Danshen.

As for specific compounds, their characteristics and the strengths of their pharmacological actions are obviously different, and these differences provide important information for the identification of the leading compounds and the development of highly effective drugs.

1.3.1 Pharmacological Actions of the Liposoluble Components of Danshen

The first known pharmacological actions of the liposoluble compounds of Danshen were mainly the antibacterial actions. It was demonstrated later by experiments that tanshinones not only have antibacterial properties, but also strong pharmacological effects on the cardiovascular and nervous systems [16].

1.3.1.1 Antibacterial Actions

Tanshinone has strong inhibitory effects on a variety of Gram-positive bacteria and relatively weak inhibitory effects on Gram-negative bacteria. It has been shown that during the metabolic process, tanshinones can produce free radicals which can damage bacterial constituents such as DNA, leading to the death of bacteria. The treatment of acne with tanshinone preparations is based on the powerful antibacterial actions of tanshinone [17].

1.3.1.2 Actions on the Cardiovascular System

Tanshinones have extensive effects on the cardiovascular system, including the protection of endothelial cells, anti-myocardial-ischemia, improvement of blood Metabolism, and protection of cardiac muscles. Study on the mechanisms of action showed that tanshinones can protect endothelial cells against ischemia-reperfusion injury, but this action is not significant against myocardial cells cultured in vitro. Tanshinones can suppress the oxidation of low-density lipoproteins, inhibit the activities of lipid metabolizing enzymes, improve the metabolic process of lipids, and prevent and treat cardiovascular diseases [18–22].

1.3.1.3 Anti-inflammatory Actions

It has been shown by in vivo and in vitro experiments that tanshinones have the function of inhibiting inflammation, which may be related to the other functions of tanshinones, namely suppressing the production of cell factors in inflammatory cells and inhibiting the metabolism of arachidonic acids [23–26].

1.3.1.4 Induction of Apoptosis and the Actions on Tumors

It has been shown by in vitro experiments that tanshinone II_A and tanshinone I can induce apoptosis, which may be caused by the release of cytochrome C and the activation of caspase-3, both processes induced by tanshinones. There was a report stating that Danshen may facilitate the growth and metastasis of tumors. However, experiments have demonstrated that sodium tanshinone II_A sulfonate does not promote the growth and metastasis of tumor cells, which revises the idea that Danshen facilitates carcinomatous metastasis [27–29].

1.3.1.5 Protection of Nervous System

It has been found that tanshinones and tanshinone II_A have significant inhibitory effects on brain cell damage caused by cerebral ischemia, indicating it has protective actions on neurocytes. In the localized cerebral ischemia experiment in rats, tanshinones could reduce the ischemic area of brain tissue and mitigate the symptoms caused by cerebral ischemia, which proved the protective actions of tanshinone on the brain. It has been proved by in vitro experiments that tanshinones can improve the activity of Na^+–K^+–ATPase in the microsomes of the brain of rats, which provides experimental evidence for the pharmacological mechanisms of tanshinones [29–31]. Furthermore, tanshinones have antioxidation actions and may be involved in protecting the brain against ischemic injury.

1.3.1.6 Influence on Drug Metabolism

Tanshinones can suppress the activity of drug metabolizing enzymes in the liver and tanshinone II_A is the principal active component involved in this action. Experimental results have shown that tanshinone II_A can selectively inhibit human cytochrome P450 1A2 with an inhibition constant K(i) of 7.2 ± 0.7 nM, indicating that it has a strong competitive inhibition effect. Compared with tanshinone II_A, the selectivities of tanshinone I and cryptotanshinone are lower. These actions may influence the effects of combined medications and are important for clinical application [32].

1.3.2 Pharmacological Actions of the Hydrosoluble Components of Danshen

1.3.2.1 Tanshinol

The hydrosoluble components of Danshen are the primary subjects of recent research on Danshen. The first isolated hydrosoluble component was Tanshinol [β-(3,4-dihydroxy phenyl)sodium lactate]. After thorough research, it has been proved that Tanshinol functions in dilating mesenteric arterioles, anti-platelet-aggregation, anti-thrombosis, facilitating fibrin degradation, and anti-myocardial-ischemia. Tanshinol has been used in clinical practice. Experiments using cultured CHO cells and fibroblasts have shown that Tanshinol inhibits the synthesis of cholesterol in these cells, reduces the content of malonaldehyde (MDA) in the oxidized lipoproteins, and significantly decreases the cytotoxicity of oxidized lipoproteins, indicating that Tanshinol can be used to prevent and treat arteriosclerosis. Recent research shows that Tanshinol has inhibiting actions on the allergic reactions caused by antigens [33, 34].

1.3.2.2 Salvianolic Acids

Salvianolic acids are a class of organic acids containing phenolic hydroxyls and are important members of the hydrosoluble components of Danshen. There are a great variety of salvianolic acids, which have similar basic structures and similar actions.

Antioxidation Actions

Salvianolic acids have strong antioxidant actions. It has been shown by in vivo and in vitro experiments that salvianolic acids can scavenge oxygen free radicals, inhibit the peroxide reactions of lipids, and these actions have been proved by different researchers under different experimental conditions [35, 36]. It has been demonstrated that the antioxidant actions are not the main mechanism for their pharmacological effects.

Influence on the Activity of ATPase

Experimental results have shown that at a certain concentration (<1 μmol/L), salvianolic acids can increase the activity of Na^+–K^+–ATPase, but higher concentrations lead to the reverse effect. The actions of salvianolic acids on Ca^{2+}–Mg^{2+} ATPase are similar to those on Na^+–K^+–ATPase, but the effects on the activities of enzymes in different tissues are not consistent. Application of Salvianolic acids [3–10 mg/kg, intraperitoneal injection (ip)] for 1 week can lower the enhancing effects on the expression of the genes of Na^+–K^+–ATPase in the cardiac muscles of rats with spontaneous hypertension, indicating that they have regulatory actions on the expression of the genes encoding Na^+–K^+–ATPase [37].

Actions on Cardiovascular Diseases

Salvianolic acids have protective actions on the in vitro perfused rat heart against the injury caused by ischemia-reperfusion, and can decrease the incidence rate of ventricular fibrillation, alleviate the level of destruction of the ultrastructure of myocardial cells, and reduce the content of MDA in tissues, so they have obvious protective actions.

It has been reported that salvianolic acids have a significant influence on the functions of platelets. It has been shown by in vitro and in vivo experiments that artificial semisynthetic acetylated salvianolic acid A has significant inhibiting effects on the aggregation of platelets caused by multi-factors, and at the same time also inhibits the release of serotonin (5-HT) by platelets induced by collagens [38].

Recent research shows that salvianolic acids can influence the metabolism of lipids in blood and may have inhibiting actions on the generation of arteriosclerosis. In vitro experiments show that salvianolic acids have significant inhibiting actions on the activities of aldose reductase in the crystalline lens of rat, which agrees with the in vivo experimental results. This action may be relevant to the treatment of diabetes [39–41].

Protection Against Cerebral Ischemic Injury and Influence on the Functions of Learning and Memory

Using cerebral ischemia experimental animal models (mice and rats), researchers have shown that salvianolic acids [3–10 mg/kg, intravenous injection(iv)] protect the brains against cerebral ischemic and ischemia-reperfusion injury, and can reduce the area of ischemic brain, decrease the content of MDA, alleviate ethological disorders caused by cerebral ischemia, and significantly improve disorders of memory function. These actions may be related to their antioxidation actions. In addition, salvianolic acids can mitigate memory functions disorders in mice caused by anisodine or scopolamine, and improve the performance of animals, but the mechanisms of these actions remain unknown. The hydrosoluble components of Danshen can also improve the indices of senile dementia, indicating that salvianolic acid can be used to prevent senescence [42].

The research conducted in our laboratory has shown that 5 days' administration of salvianolic acids (3–10 mg/kg, ip, qd) can change the content of transmitters in the brains of normal rats and decrease the content of the metabolites of dopamine (DA) and 5-HT, but has no significant influence on the contents of these transmitters themselves. It has been shown by recent studies that salvianolic acids have some effects on the animal models of senescence and senile dementia. Studies on the mechanisms of action have shown that salvianolic acids can not only protect neurocytes through antioxidation actions, but also have significant inhibiting actions on the aggregation of β amyloids [43–49].

Protective Effects on Endothelial Cells

Experimental results have shown that salvianolic acids can protect blood vessel endothelial cells against oxidative stress injuries and inhibit the adhesion of neutrophils and endothelial cells, and have protective actions on endothelial cells against injury by oxidized cholesterols. Therefore, salvianolic acids may exert their cell protective actions through antioxidation, but other actions cannot be excluded [50–52].

Effects on Hepatic and Renal Function

The magnesium salts of salvianolic acid B can improve renal function and increase the excretion of prostaglandin E2 from urine, and have inhibiting actions on the activity of cyclooxygenase in the metabolic process of arachidonic acid. Japanese researchers have performed systemic research on the magnesium salts of salvianolic acid B and found that the magnesium salts of salvianolic acid B can resist the hepatic injury caused by D-galactose in rats, significantly decrease the activities of serum alanine aminotransferase and aspartate aminotransferase, and alleviate the degree of the necrosis of liver cells [38, 53–56].

Inhibiting Effects on the Formation of Galactose Cataracts in Rats

Salvianolic acid A has inhibiting actions on the formation of cataracts caused by galactose in rats, and can reduce the contents of the peroxidized products of lipids in crystalline lens, strengthen the antioxidant ability of crystalline lens, and have inhibiting actions on the oxidizing formation of cataracts. Crystalline lens cultured in vitro may form cataracts when hydrogen peroxide is present. After adding salvianolic acid A into the culture fluid, the formation of cataracts is significantly inhibited, and the contents of the peroxidized products of lipids in the crystalline lens are reduced [57].

Drug Interactions

It has been shown by the above studies that salvianolic acids are hydrosoluble components of Danshen with a great variety of biological activities and play an important role in its functions. It has been proved that Danshen extract inhibits the activity of the drug metabolizing enzymes in liver, and may have effects on the metabolism of drugs when combined with other drugs. Since salvianolic acids have anticoagulation and anti-platelet actions, it is necessary to pay special attention to hemorrhage when they are used together with anticoagulant drugs such as Warfarin [58–61].

1.4 Development of the Preparations of Danshen

Preparations are required for both research and clinical application of Danshen. As a traditional Chinese medicine, besides herbal pieces prepared for decoction, Danshen has a great variety of preparation types. The preparations for clinical application need to be approved by the authoritative department of the government before they can be used as a new drug, but the preparations used for experimental studies could have multiple forms, according to the conditions and objectives.

1.4.1 Danshen Preparations Used for Research Purposes

Danshen preparations for research use have many types and can be divided into extracts and pure compounds. A pure compound is the chemical component contained in Danshen, and many such compounds have been identified so far. Among the commonly used compounds are tanshinone I, tanshinone II, salvianolic acid A, and salvianolic acid B. Research has shown that these chemicals have good pharmacological activities and that they are the active components of Danshen. However, since Danshen contains many compounds it is difficult to obtain all of these substances for comprehensive studies. At present, only a small percentage of these compounds have been studied and greater effort is required to illuminate every compound's pharmacological activities.

The Danshen extract used for investigation is a complicated system. Besides variations in the sources of raw materials, the extractants themselves differ quite a lot, which include those labeled as "standard" extractants, because there are no criteria for "standard". This situation has caused much difficulty in the study of Danshen. That being said, using various extracts to investigate Danshen can yield data from many different aspects, which might be beneficial to research.

The commonly used extracts include the following types: (1) Decoction of Danshen: It can be used directly for animal experiments, and the principal objective of using this type of extract is to observe the actions of Danshen when used in the traditional way. (2) Water extract of

Table 1.2 Danshen preparations currently used in clinical practice

Forms	Products on the market[a]
Injections	Danshen injectable powder; Danshen glucose injection; sodium tanshinone tablet II_A sulfonate injection; Danshen injection; compound Danshen injection; Rosewood and Danshen injection; Danshen polyphenols for injection; Tanshinol for injection; total Danshen phenolic acid for injection
Tablets	Danshen tablets; Danshen heart-comforting tablets; tanshinone tablets; Compound Danshen tablet; No. 2 coronary artery muscle disease tablets; ginseng royal jelly Danshen tablets; Yizhi compound Danshen tablets; salvianolate B tablets, compound Danshen buccal tablets
Capsules and dripping pills	Danshen heart-comforting capsules; tanshinone capsules; dripping pills; compound Danshen preparations soft capsules; Danshen dripping pills for coronary artery muscle disease; Yizhi Danshen capsules for coronary artery muscle disease; Danshen dripping pills; tanshinone (capsules)
Granules and medicinal granules	Aoxing Danshen granules; Danshen medicinal granules; compound Danshen granules; Notoginseng Danshen granules; compound Danshen oral solution; Mune Danshen oral solution
Others	Danshen polyphenols salts; Danshen alcohol extract concentrated solution; Danshen alcohol extract solution; Danshen extracts; tanshinone I; tanshinone II_A; tanshinone II_B

[a] Due to inappropriate formula and medicament forms, the production of compound nitroglycerol bag (Xin Fu Kang) has been stopped. Some other preparations, in which Danshen is not the main component, are not listed

Danshen: After extracting Danshen with water, materials like proteins are precipitated with alcohol or other solvents. The extracts obtained using this method are relatively pure and are mainly hydrosoluble components. (3) Alcohol extract: This type of extract mainly contains liposoluble components, most of which are tanshinones. In addition, further purified tanshinones and salvianolic acids are often used in research. Because the chemical compositions of these extracts differ a lot due to different extraction methods, the results obtained with them are difficult to compare, a point that needs special attention when analyzing the literature.

1.4.2 Clinically Used Danshen Preparations

There are many types of Danshen preparations for clinical application. Except for the herbal pieces prepared for decoction, all other preparations have corresponding quality standards, so the results obtained from their applications are of high statistical value. In recent years, more and more Danshen preparations have been approved by the government for production, and as a result, the application of Danshen has become more and more popular. Right now, common Danshen preparations used in the clinical practice are mainly compound preparations with Danshen as the main component, including injections, tablets, capsules, dripping pills, granules, and medicinal granules. As Table 1.2 shows, new preparations are continuously being developed.

1.5 Research on the Pharmacological Actions of Danshen

Most studies, except for the studies on the above-mentioned definite chemical components, use the crude extracts of Danshen. In order to give a full picture of Danshen, the results obtained with these crude extracts are included here for reference.

1.5.1 Circulatory System

Myocardial cell cultures from newborn rats were employed to test the effects of Danshen, and it was found that Danshen can improve their electrophoretic performance. Danshen also affects the action potentials and physiological activities of the papillary muscles in the ventricles of guinea pigs. It can increase excitability, decrease autorhythmicity, and prolong the time course of the action potential [62].

It has been shown by animal experiments that Danshen has significant protective actions on the hearts of rats and rabbits against ischemia-reperfusion injury. Whether the local ischemia-reperfusion was caused by the ligation of coronary artery muscle disease, or by cardiac arrest-reperfusion caused by using cold arresting solution, or by myocardial infarction-reperfusion caused by acute hypercholesterolemia stage, Danshen showed significant protective actions in all cases. Danshen also has preventive actions on simple ischemic myocardial damage due to ligation of coronary arteries [63].

The hydrosoluble components of Danshen have preventive and curative actions on the ventricular fibrillation of rats caused by isoprenaline, protect against abnormal enhancement of the electric activities of myocardial cells induced by isoprenaline, and prevent against increased permeability of the myocardial cell membrane caused by radiation [64, 65].

In in vivo experiments, when there is stenosis of the descending anterior branch of the coronary artery muscle in anesthetized dogs, the cross-sectional area of the coronary artery muscle was reduced by 87 % and the indices for function (average blood flow in left ventricle and maximal rate of decrease of left ventricle pressure) significantly decreased. After Danshen was injected into the left atrium, the function of the left ventricle improved. Danshen can decrease the contents of angiotensin II and atrial natriuretic polypeptins in the plasma of rabbits, which is related to the dilation of the coronary artery muscle, and this action is related with increased flow rate in the coronary artery muscle. The protective actions of Danshen on ischemic cardiac muscles not only depend on its dilating actions on coronary artery muscle but are also associated with the alleviation of inflammatory infiltration in the ischemic area, inhibition of the activation of white blood cells, and reduction of the production of free radicals. Furthermore, Danshen can inhibit the elevation of prostacyclin (PGI2) and inositol triphosphate (IP3) and protect cardiac muscles [66, 67].

The anticoagulation actions of Danshen have been experimentally proven. In bovine endothelial cell cultures, Danshen and Tanshinol can enhance the anticoagulation and fibrinolytic functions. It can also improve disorders in the coagulation and fibrinolytic functions of endothelial cells caused by endotoxins. Danshen has inhibiting actions on the formation of atherosclerosis as well [68].

1.5.2 Liver Diseases

In animal experiments, Danshen has obvious protective effects on liver injuries caused by multiple factors such as lipid peroxidation and carbon tetrachloride. Danshen has protective actions on the liver and can facilitate the recovery of liver function, alleviate injury to the morphology of hepatic cells, and inhibit the development of liver cirrhosis.

In in vitro experiments, rat liver perfused with 0.12 % carbon tetrachloride for 60 min led to the degeneration and disarrangement of live cells, atrophy of the hepatic cord, obscure the structure of hepatic lobules, and a significant increase of Glutamate Pyruvate Transaminase (GPT) in the effluent liquid. After perfusion with a solution containing 0.15 % Danshen, the degree of injury to the liver was significantly alleviated, and there was no significant increase in GPT [69–71].

Danshen can significantly increase the survival rate and hepatic coefficient in rats with acute hepatic injury caused by D-galactose amine. The degree of atrophy of animal livers is significantly alleviated, the level of GPT in serum is significantly decreased, and the level of fibronectin in plasma is significantly increased. It has been shown that the release of GPT and the production of MDA increase, while urea and protein synthesis decreases in the hepatic cells of primary culture injured by carbon tetrachloride. After adding Danshen to the culture fluid, the changes caused by carbon tetrachloride could be inhibited, and the morphological changes of hepatic cells were alleviated [72–74].

It is generally accepted that the mechanisms of Danshen's protective actions on liver are concerned with the following aspects: (1) The inhibitory actions on calcium channels; (2) Elevation of the level of Fibronectin in plasma; (3) Improvement of liver microcirculation and increase in liver blood flow; (4) Inhibition of lipid peroxidation in hepatic cells; (5) Induction of cytochrome P450 (this action may be related to the protective actions of the liver) [75, 76].

1.5.3 Kidney Diseases

The protective effects of Danshen have been demonstrated in the animal renal function impairment model caused by different nephrotoxic materials, such as gentamicin, ciclosporin A, adenine, and grass carp gallbladder. The results showed that Danshen can improve the function of kidneys with renal failure and increase the filterability and renal blood flow of kidneys. After administration of grass carp gallbladder to rats, the urinary volume decreased, the concentration of blood creatinine increased, the creatinine clearance significantly decreased, and there was significant change in the morphology of the kidneys. After administration of Danshen (0.25 ml/100 g, ip, bid) for 2 days, the indices of renal impairment were significantly improved, the urinary volume and creatinine clearance increased, and the concentration of blood creatinine decreased. The morphology of kidney also improved significantly [77, 78].

If Danshen is used with gentamicin, it can significantly lower the toxicity of gentamicin to kidney in rats, and can alleviate the toxic effects of gentamicin and protect renal function. Research on the mechanisms has shown that Danshen has significant antagonistic action on the inhibition of Na^{+}–K^{+}–ATPase by gentamicin, which may be one of the basic mechanisms for the improvement of renal Functions [79, 80].

1.5.4 Respiratory Diseases

Danshen has therapeutic effects on respiratory diseases. In dogs, acute hemorrhage can cause necrotic pancreatitis, which leads to changes in pulmonary function and pulmonary morphology, pulmonary edema, serum amylase, lipase and lactate dehydrogenase, congestion of pulmonary capillaries, formation of hyaloid membranes in the alveoli of the lung, thickening of the alveolar walsl, dilation of some alveoli, compensatory emphysema, appearance of many neutrophils and monocytes in the pulmonary interstitial substance and alveoli of the lung, and necrosis and exfoliation of bronchiolar epithelia. Continuous intravenous dripping of Danshen (5 g/kg) significantly inhibited these changes in the function and morphology of the lung caused by necrotic pancreatitis [81].

Under hypoxic conditions, compound Danshen injection can improve the pulmonary hypertension of rabbits caused by hypoxia, decrease pulmonary vascular resistance, ameliorate cardiac functions, increase cardiac output and stroke volume, elevate arterial PO_2, and improve cardiopulmonary functions. Moreover, Danshen injection can alleviate radiation injury to animal lung tissue and facilitate the restoration of injured tissues [82].

The protective actions of Danshen on pulmonary functions may be the result of its actions in lowering capillary permeability and improving microcirculation. As for lung injury caused by inflammation, Danshen can prevent the migration and aggregation of white blood cells, inhibit the release of free radicals by white blood cells, eliminate the free radicals produced in inflammation, avoid the release of lysosomes and oxidation metabolites, alleviate tissue damage, and control the development of inflammation [83]. It has been reported that Danshen has significant inhibiting actions on elastase, which is also a possible mechanism for the therapeutic actions of Danshen on emphysema [53].

1.5.5 Cancers

Danshen has been used to treat various tumors in the clinical practice, and significant therapeutic effects have been reported. As for whether Danshen can accelerate the metastasis of tumor cells, it has been reported that sodium tanshinone II_A

sulfonate and the aqueous extracts of Danshen do not facilitate the growth and metastasis of cancer cells.

There has been no definite conclusion about the therapeutic actions of tumors, but Danshen has been used in combined treatments for a great variety of tumors, and it has synergistic effects on anti-cancer drugs [84]. In addition, it has been demonstrated that Danshen protects against injury caused by chemotherapy and radiation therapy, which may be one of the mechanisms of the enhancing action of Danshen on the therapeutic effects of anti-cancer drugs [80, 81, 85].

1.5.6 Effects on Immunologic Functions

Danshen and compound Danshen preparations can significantly enhance the immunologic functions of normal mice. It has been shown that Tanshinol can significantly inhibit the proliferation of splenic cells in vitro and suppress the formation of hemolytic spots. The administration of Tanshinol to animals can significantly inhibit the formation of hemolytic spots in the splenic cells during first and secondary sensitization by sheep red blood cells. Moreover, it has inhibiting actions on the delayed hypersensitivity reaction of skin of mice caused by dinitrochlorobenzol. Therefore, Danshen has influences on immunologic functions [83, 84, 86].

1.5.7 Others

1.5.7.1 Antioxidation Actions

Almost all Danshen preparations have reported powerful antioxidation actions. The preparations could scavenge the free radicals generated from various sources. These actions may play a role in the various pharmacological actions of Danshen [87, 88].

1.5.7.2 Effects on the Functions of Biomembranes

It has been found Danshen can influence the conformation of lipids in synthetic membranes, regulate the physical conditions of lipoidal double-layer membranes, change the lipoidal environment of membrane proteins, and improve the functions and metabolic process of the membrane. These actions can improve cell functions in some diseases, such as coronary heart disease [89].

1.5.7.3 Analgesic Actions

The Analgesic actions of Danshen have been reported, but the research data are insufficient to demonstrate that Danshen could inhibit the pain directly. Using extracellular microelectrodes and brain stereotaxic technique, it has been shown that Danshen can inhibit the electric discharge of visceralgia by nucleus posterior thalami, indicating it has inhibiting actions on the nervous system, but the clinical value of this activity has not been evaluated.

1.5.7.4 Anti-senescence

It has been reported that Danshen can enhance the activity of superoxide dismutase in the plasma and tissues of aged mice. It is suggested that Danshen may have actions of postponing senescence [90].

Besides the actions mentioned above, Danshen has other actions as well. For example, Danshen can inhibit inflammation, improve the functions of the stomach and the barrier actions of the stomach mucous membrane, influence the metabolism of arachidonic acid and prostaglandin, be used for surgery, and influence the sense of hearing, etc. [91–94].

1.6 Clinical Applications of Danshen

1.6.1 Treatment of Cardiovascular and cerebrovascular Diseases

The main indication of Danshen is for cardiovascular disease, and it has definite therapeutic effects on a great variety of cardiovascular diseases. It has been used to treat pulmonary heart disease, viral myocarditis, hypertension and open heart surgery, and good effects ($P < 0.005$) have been obtained [65, 95–98]. It also has good therapeutic actions on ischemic brain injury, and the dose required is lower than that required for cardiovascular disease.

1.6.2 Treatment of Hepatic and Renal Diseases

Combined with Polyporus umbellatus polysaccharide and other Western medicines, Danshen has been used to treat many refractory hepatic diseases, such as chronic hepatitis B, serious viral hepatitis, late-stage liver cirrhosis, and liver cancer and good therapeutic effects have been obtained. Danshen also has therapeutic effects on chronic renal insufficiency, chronic renal failure, and acute renal failure. Furthermore, Danshen can accelerate the recovery of renal functions and enhance the therapeutic effects of other drugs [97, 98].

1.6.3 Treatment of Respiratory Diseases

It has been found in recent years that Danshen can improve pulmonary function and can be used to treat bronchial asthma, bronchitis, pulmonary heart disease, and pneumonia. After administration of Danshen, the length of fever was decreased compared to the control group, which may be the result of Danshen's actions of accelerating microcirculation blood flow and the delivery speed of drugs and enhancing the action of antibiotics. When Danshen is used as an auxiliary therapy for pneumonia treatment, it can control the development of symptoms, shorten the course of the disease, and increase therapeutic effects.

1.6.4 Actions on Tumors

Danshen can be used as an auxiliary therapy for tumors. A combination of Danshen and hydroxyl camptothecin can be used to treat cancer of the digestive system. It has some therapeutic effects on liver and stomach cancer, and can significantly prolong the survival time of patients and reduce the size of tumors. Certain therapeutic effects were also observed when Danshen was combined with homoharringtonine to treat promyelocytic leukemia and combined with chemotherapy to treat refractory leukemia. It has been proved the anti-tumor actions of Danshen, used alone.

1.6.5 Others

1.6.5.1 Used in Surgery

Good therapeutic effects have been obtained when Danshen is used to treat adhesive ileus. Postoperative injection of compound Danshen injection into the abdominal cavity can significantly inhibit postoperative intestinal adhesion, and the therapeutic effects are significantly better than those of the general treatment group. Danshen can significantly increase the efficiency of peritoneal dialysis and eliminate oxygen free radicals in the blood in extracorporeal circulation, and Danshen has significant effects on recovery after bone injuries.

1.6.5.2 Effects on Infectious Diseases

In recent years, Danshen has been used to treat epidemic hemorrhagic fever, epidemic meningitis, viral myocarditis, and pneumonia in children. Danshen injection has been used to treat various inflammatory diseases, such as mastitis, and good clinical effects have been obtained. As for skin diseases, tanshinone has been widely used to treat acne and has become an effective drug for the treatment of acne.

Danshen has a broad spectrum of actions and these actions have been shown in the clinical practice. For example, it has been used to treat keratitis (ophthalmologic disease), periodontal tissue diseases (dental disease), hematoma of ectopic pregnancy (gynecologic disease), etc. Other diseases such as pulmonary encephalopathy, pediatric allergic purpura, diabetes, bone fractures, and nerve deafness have also been treated with Danshen, with good therapeutic effects observed. It has been shown that Danshen even has inhibiting actions on HIV. However, continued accumulation of experience and in-depth research are still required to confirm the therapeutic effects of Danshen on these diseases [99].

1.6.6 Adverse Reactions of Danshen

Danshen has little toxicity and no obvious adverse reactions have been reported for regular usage. There was one case report of anaphylactic shock with an intravenous drip of compound Danshen injection [100], and there was one case report of bradycardia and hypotensive shock caused by high concentrations of Danshen. Since Danshen can improve cardiac function, it may lead to false negative results when dipyridamole ECG examination is conducted on cardiac patients. Therefore, in order to obtain reliable results, a doctor should ask whether the patient has been using Danshen during the clinical examination. It has also been reported that Danshen preparation may cause dyskinesia, so it should be used with caution in children [101, 102].

1.7 Summary

Guanhua Du and Juntian Zhang

Based on historical records and modern studies, we can conclude that Danshen is a traditional herbal drug with long history of application and various Functions. It can be used for the treatment of many diseases, and is an important drug commonly used in the clinical practice.

Tanshinones are the main liposoluble components of Danshen, which can dilate blood vessels, improve microcirculation and kill bacteria, and have immunoregulatory, hormone-like, and antioxidative functions.

The hydrosoluble components of Danshen, such as salvianolic acids, play an important role in the actions of Danshen. Their most fundamental function is antioxidation and they also have many other functions. The combination of these functions makes Danshen have good effects in the treatment of various diseases.

To sum up, the primary functions of Danshen are circulation improvement, tissue protection, and antioxidation. This conclusion agrees with the functions of "promoting the circulation of qi, invigorating qi, promoting blood circulation, Nourishing the blood" and "eliminating the pathogenic factors in heart and abdomen and removing numbness" of Danshen in traditional Chinese medical theory. Danshen can regulate the functions of the body and facilitate the recovery of the functions of injured tissues, so it can be used as an important adjunctive therapy.

Danshen contains a great variety of active components, which have multiple pharmacological actions. However, few studies are focused on the mechanisms of these actions by the active components. A great deal of work is needed on the applied research on the treatment of many diseases with the active components of Danshen. Only after systematic scientific studies can the full potential of Danshen be realized.

References

1. Guo J, et al. Study and clinical application of Danshen. Beijing: The Medicine Science and Technology Press of China; 1992.
2. Xu R, et al. Biology and application of Danshen. Beijing: Science Press; 1990.
3. Institute of Materia Medica, Chinese Academy of Medical Sciences. Danshen–modern study on Chinese herbal medicine, vol. 2. Beijing: Combination Press of Beijing, Medical University and Peking Union Medical College; 1996. p. 472.
4. Lin H, et al. Chin Pharm J. 1993;45:615.
5. Choi JS, Kang HS, Jung HA, et al. Fitoterapia. 2001;72(1):30.
6. Lin HC, Chang WL. Phytochem. 2000;53(8):951.
7. Lin HC, Ding HY, Chang WL. J Nat Prod. 2001;64(5):648.
8. Ai CB, Li LN. J Nat Prod. 1988;51:145.
9. Du G, Zhang J. Basic and Clinical Medicine, 2000, 20(5):394.
10. Tian G, Zhang Y, Zhang T, et al. J Chromatogr A. 2000;904(1):107.
11. Pan X, Niu G, Liu H. J Chromatogr A. 2001;922(1–2):371.
12. Li HB, Chen F. J Chromatogr A. 2001;925(1–2):109.

13. Li YC, Zeng JQ, Liu LM, et al. Chromatogr. 2002;20(1):40.
14. Zhang QW, Zhang Y, Li JP, et al. China J Chin Materia Medica. 2001;26(12):848.
15. Chen B, Zhu M, Xing WX, et al. Acta Pharma Sin. 2001;36(1):55.
16. Goren AC, Topcu G, Oksuz S, et al. Nat Prod Lett. 2002;16(1):47.
17. Lee DS, Lee SH, Noh JG, et al. Biotechnol Biochem. 1999;63(12):2236.
18. Maki T, Kawahara Y, Tanonaka K, et al. Planta Med. 2002;68(12):1103.
19. Takahashi K, Ouyang X, Komatsu K, et al. Biochem Pharmacol. 2002;64(4):745.
20. Niu XL, Ichimori K, Yang X, et al. Free Radic Res. 2000;33(3):305.
21. Ko JS, Ryu SY, Kim YS, et al. Arch Pharm Res. 2002;25(4):446.
22. Park JW, Lee SH, Yang MK, et al. Arch Pharm Res. 2008;31(1):47.
23. Kang BY, Chung SW, Kim SH, et al. Immunopharmacol. 2000;49(3):355.
24. Yagi A, Takeo S. Yakugaku Zasshi. 2003;123 (7):517.
25. Kim SY, Moon TC, Chang HW, et al. Phytother Res. 2002;16(7):616.
26. Ryu SY, Oak MH, Kim KM. Planta Med. 1999;65 (7):654.
27. Sung HJ, Choi SM, Yoon Y, et al. Exp Mol Med. 1999;31(4):174.
28. Yoon Y, Kim YO, Jeon WK, et al. J Ethnopharmacol. 1999.;68(1–3):121.
29. Kim JY, Kim KM, Nan JX, et al. Pharmacol Toxicol. 2003;92(4):195.
30. Wang YL, et al. Chin J Pharmacol Toxicol. 1994;8:19.
31. Lam BY, Lo AC, Sun X, et al. Phytomedicine. 2003;10(4):286.
32. Ueng YF, Kuo YH, Peng HC, et al. Xenobiotica. 2003;33(6):603.
33. Sun X, et al. Chin Tradit Herbal Med. 1991;22:20.
34. Zhang X, et al. Chin J Pharmacol Toxicol. 1991;5:27.
35. Petersen M, Simmonds MS. Phytochem. 2003;62 (2):121.
36. Huang DS, Zhang JT. Acta Pharm Sin. 1992;27:96.
37. Du GH, et al. Chin J Pathophysiol. 1998;14(7):862.
38. Xu WG, Xu LN. Acta Pharm Sin. 1994;29:412.
39. Du GH, et al. Acta Pharm Sin. 1995;30:731.
40. Lee GT, Ha H, Jung M, et al. J Am Soc Nephrol. 2003;14(3):709.
41. Liu YL, Liu GT. Acta Pharm Sin. 2002;37(2):81.
42. Durairajan SS, Yuall Q, Xie L, et al. Neurochem Int. 2008;52(4–5):741.
43. Du GH, et al. Chin Med J. 1997;110:65.
44. Du GH, Qiu Y, Zhang JT. J Asia Nat Prod Res (JANPR). 2000;2:145.
45. Tang MK, Ren DC, Zhang JT, et al. Phytomedicine. 2002;9(5):405.
46. Du GH, et al. Acta Pharma Sin. 1995;30:184.
47. Du GH, Zhang JT. Shandong Med J. 1994;34:8.
48. Ren DC, Du GH, Zhang JT. Chin Pharmacol Bull. 2002;18:275.
49. Zhang JT, Du GH, Tang MK, et al. Anti-senile-dementia actions of salvianolic acid B. Symposiums of Annual Conference of Chinese Academy of Medical Sciences and Peking Union Medical College. Press of Peking Union Medical College; 2002, p. 238.
50. Ren DC, Du GH, Zhang JT. Chin J Pharmacol Toxical. 2003;17(5):333.
51. Ren DC, Du GH, Zhang JT. Jpn J Pharmaco. 2002;90:276.
52. Ren DC, Du GH, Zhang JT. Chin Med J. 2003;116 (4):630.
53. Yokozawa T, et al. Nippon jinzo Gakkai shi. 1993;35:337.
54. Yokozawa T, et al. Chem Pharm Bull. 1991;39:732.
55. Liu P, Hu Y, Liu C, et al. Liver. 2001;21(6):384.
56. Jung M, Lee HC, Ahn CW, et al. Chem Pharm Bull. 2002;50(8):115.
57. Du GH, et al. Acta Pharma Sin. 1995;30:561.
58. Yang XF, Wang NP, Zeng FD. China J Chin Materia Medica. 2002;27(5):325.
59. Chan TY. Ann Pharmacother 2001;35(4):501.
60. Jinping Q, Peillng H, Yawei L, et al. J Pharm Pharmacol. 2003;55(8):1163.
61. Qiao LX, et al. Chin Med Pharmacol. 1992;6:40.
62. Li ZH, et al. Chin J Pathophysiol. 1993;9:17.
63. Bi HC, Zuo Z, Chen X, et al. Xenobiotica. 2008;38 (2):185.
64. FugkBerman A. Prev Cardiol. 2000;3(1):24.
65. Sugiyama A, Zhu BM, Takahara A, et al. Circ J. 2002;66(2):182.
66. Shi X, Xia Z, Fang J. Chin J Integr Tradit W Med. 2000;20(12):896.
67. Wang SP, Zang WJ, Kong SS, et al. Eur J Pharmacol. 2008;579(1–3):283.
68. Shi W, Fu S, Du N. Chin J Integr Tradit W Med. 2000;20(4):269.
69. Zhang WH, Wang JS, Zhou Y, et al. World J Gastroenterol. 2003;9(9):2040.
70. Ye HG, Song YF, et al. Chin J Digestion. 1991;11:87.
71. Yu ZF, Jiang JX, et al. J Yunnan Coll Tradit Chin Med. 1992;15:23.
72. Huang LB, Guisheng X. J of Gannan Med Coll. 1990;10:215.
73. Yu ZF, et al. J Yunnan Coll Tradit Chin Med. 1991;14:41.
74. Deng HJ, et al. China J Chin Materia Medica. 1992;17:233.
75. Qi XG, et al. Chin J Integr Tradit W Med. 1991;11:102.
76. Zeng MD, et al. J Shanghai Second Med University. 1991;11:204.
77. Lu YS. Hubei J Tradit Chin Med. 1990;4:14.
78. Zhang BZ, et al. China J Tradit Chin Med Pharm. 1991;6:26.

79. Yokozawa T, et al. Nephron. 1991;57:78.
80. Chen FM, et al. Chin J Nephrol. 1989;5:66.
81. Zheng SS, et al. J Integr Tradit Chin W Med. 1989;9:158.
82. Zheng XK, et al. J Integr Tradit Chin W Med. 1991;11:733.
83. Du HW, et al. J Integr Tradit Chin W Med. 1990;10:230.
84. YoN Hu. J Xian Med University. 1991;12:139.
85. Fang ZY, I. in R, Yuan BX, et al. Life Sci. 2007;81 (17–18):1339–1345.
86. Fang ZY, Lin R, Yuan BX, et al. J Ethnopharmacol. 2008;115(2):217–222.
87. Yang WD, et al. Chin Pharmacol Bull. 1990;6:118.
88. Liu JR, Chen GF, Shih NH, Kuo PC. Phytomedicine. 2008;15(1–2):23–30.
89. Qi SP, et al. Chin Sci Bull. 1989;7:547.
90. Guo ZX, et al. Chin Tradit Patent Med. 1993;15:27.
91. Wang GZ, Ru X, Ding LH, et al. World J Gastroenterol. 1998;4(2):169.
92. Yang X, Lu Y, Xie D. Bull Hunan Med University. 1999;24(5):465.
93. Zhang LH, Yao CB, Li HQ. World J Gastroenterol. 2001;7(5):672.
94. Wang XF, Zhao MQ. J First Military Med University. 2003;23(9):969.
95. Wang F. J Tradit Chin Med. 2002;22(1):3.
96. Yi LQ, Wang XW. C J Integr Tradit W Med. 2002;22(7):543.
97. Bao HY, Yu HL, Wang L. C J Integr Tradit W Med. 2002;22(1):28.
98. Yu W, Shen F, Ma Z. C J Integr Tradit W Med. 2000;20(2):94.
99. Abd-Elazem IS, Chen HS, Bates RB, et al. Antiviral Res. 2002;55(1):91.
100. Xie XQ, et al. C J Integr Tradit W Med. 1993;13:527.
101. Yao GH, et al. C J Integr Tradit W Med. 1993;13:369.
102. Wang XP, Yang RM. Fur Neurol. 2003;50(3):153.
103. Cao EH, Liu XQ, Wang JJ, et al. Free Radic Biol Med. 1996;20(6):801.
104. Chen YH, Du GH, Zhang JT. Acta Pharmacol Sin. 2000;21:463.

The Pharmacological Actions of Danshen ThemeD Formulas

2

Zhihao Jiang, Yi Wang, Xiumei Gao, Hongcai Shang and Xiaoying Wang

2.1 Composite Danshen Dropping Pill

Danshen has been used for medicinal purposes in China for more than 1,000 years. In clinical applications, Danshen is usually used in combination with other drugs; for the treatment of cardiovascular diseases, Danshen's partner is normally Notoginseng (*Panax pseudoginseng*). Many Chinese patent medicines have been developed mainly based on this combination, such as Compound Danshen Tablet, Composite Danshen Dropping Pill, Cardio Danshen Dropping Pill, Danshen Notoginseng Tablet, etc. Compound Danshen Tablet originated from the classic TCM formula and contains Danshen, Notoginseng, and Borneol. This formula has the function of removing blood stasis and relieving pain, and is used specifically for the treatment of angina pectoris and atherosclerosis. Since it was developed in 1977, the formula has generated many preparations which have been manufactured by more than 100 pharmaceutical companies.

Danshen drug preparations include tablet, dropping pill, injection solution, capsule, granule, sprayer, slow release formulation, soft gel, oral liquid, etc. Because the composition and proportions of these preparations are not exactly the same, their clinical effects are different. Composite Danshen Dropping Pill is an advanced product developed from Compound Danshen Tablet. The preparation is based on a solid molecular dispersion system, which enlarges the surface area of the components. As a result, its dissolution rate, utilization rate, and curative effect are higher than those of the tablets and it is more popular clinically.

2.1.1 Compatibility Studies on Compound Danshen Dropping Pill (CDDP)

2.1.1.1 The Compatibility of Danshen and Notoginseng

Experimental Method: Baseline Geometric Proportion Increasing and Decreasing Design Baseline geometric proportion increasing and decreasing (BGPID) design is a method suitable to the research on combination ratios for the majority of small complex prescriptions of Chinese medicines. The method is guided by TCM theory and professional knowledge. With the CDDP's functions already known and the optimal ratio of the components unclear, this method is particularly useful. Specifically, the detailed combination ratio of the compound of Danshen and Notoginseng is studied and analyzed. The experiment design followed the principle of pharmacopoeia standards for these two

Z. Jiang · Y. Wang · X. Gao (✉) · H. Shang · X. Wang
Tianjin University of Traditional Chinese Medicine, Tianjin, China
e-mail: gxm6986@sina.com

X. Yan (ed.), *Dan Shen (Salvia miltiorrhiza) in Medicine*,
DOI 10.1007/978-94-017-9463-3_2,

medicines. 10/3 (Danshen/Notoginseng) was set as the baseline, with the ratio of the two components decreasing or increasing by 10–30 %. The endpoints were 10/0 and 0/10, respectively, with several variance groups in between (10/1, 10/6, 10/10, 1/10), totaling seven groups. Each group was set as the independent variable, and effect indicators at each time point were set as the dependent variables. The relationship between the variables was analyzed. The key indicators were the major functions of the two medicines. Animal models were used and multi-indicator optimization and multidimensional time series analyses were processed. Moreover, other results from in vitro organ and cellular experiments were integrated, and all data was fed into Intelligent Information Processing methods to finalize the ultimate report.

The Effect of Different Ratios of Danshen and Notoginseng on the Function of Myocardial Ischemic Model and Cardiac Hemodynamics [1–3]

A Myocardial ischemic model from the ligated coronary arteries of dogs was made and integrated with a hemodynamic model. Furthermore, by the baseline method, seven ratio groups (Danshen/Notoginseng: 10/0, 10/1, 10/3, 10/6, 10/0, 1/10, 0/10) were prepared for the pharmacodynamic research. The entire dosage for each group was 1.31 g/kg. Meanwhile, a sham group, model group and positive control group (isosorbide dinitrate) were set to complete the experiment. The indicators used for effect evaluation were epicardial electrocardiogram (Epi-ECG), scale of myocardial ischemia area, cardiac markers, coronary blood flow, oxygen consumption of myocardium, hemodynamic, nitric oxide, endothelin, free radicals, etc.

The results show as the following:

1. Effect on myocardial ischemia: Four groups in which Danshen was the major component (10/6, 10/3, 10/1, 10/0) showed a significant effect on improving pathologic status. Epi-ECG showed that the treatment groups mitigated myocardial ischemia. TTC staining indicated that the ischemic area decreased in treatment groups, especially the 10/6 and 10/3 groups.
2. Effect on cardiac markers: Treatment groups with Danshen as the major component showed significant inhibition of the increase of cardiac Troponin. They showed that Danshen was more effective than Notoginseng on the cardiovascular system, and the result was the same as those obtained with Epi-ECG and myocardial staining methods.
3. Effect on coronary blood flow and oxygen consumption of myocardium: Treatment groups (10/0, 10/3, 10/6) with Danshen as the major component showed better results. Compared with the positive control group (isosorbide dinitrate), which had a fast effect (30 min after medication) but a very short active duration, TCM groups lasted for a longer period of time and acted steadily.
4. Effect on hemodynamics: All TCM groups showed certain effect. Groups 10/6 and 10/3 were much more effective than other groups in improving systole and diastole, cardiac output, and reducing total systemic vascular resistance. However, left ventricle work did not show a significant increase, and the mean atrial pressure and heart rate showed fewer changes.
5. Effect on nitric oxide (NO), endothelin (ET), and free radicals: Group 0/10 acted better than the other groups in improving NO release, keeping SOD active, decreasing ET and Malondialdehyde (MDA).

The Effect of Different Ratios of Danshen and Notoginseng on the Function of Thrombocyte Aggregation and Adhesion [4]

New Zealand Rabbits were medicated orally for 4 days, then the common carotid arteries were exsanguinated. Thrombocyte aggregation was measured via turbidimetric analysis. Six treatments (10/0, 10/1, 10/3, 10/6, 1/10, 0/10) were conducted. Results: Groups 10/3, 10/6, 0/10 showed obvious inhibition of platelet adhesion. Group 0/10 was the best. Groups 10/0, 10/1, and 1/10 had no significant effect.

The Effect of Different Ratios of Danshen and Notoginseng on the Function of CMEC [5]

Under normal culture conditions, groups with different Danshen to Notoginseng ratios could lower the levels of LDH in CMEC. The levels of NO and ET were also decreased in the treated groups. Groups 10/0–1/10 had a positive effect on raising cell activities; group 10/0 showed the best result, while pure Notoginseng (0/10) showed no effect.

Under hypoxic conditions, the results from all groups showed lowered LDH release by CMEC. All groups showed reduced NO release by CMEC, and except for group 10/6, the reductions were statistically significant. In ET secretion, all groups showed reduced secretion, and the results from groups 10/1 and 10/6 were better than those from other groups. Group 10/0 showed the best result in increasing cell viability, while results of group 0/l0 showed a decrease. From groups 10/1 to 10/6, the effect increased with an increasing proportion of Notoginseng . From groups 10/0 to 1/10, the effect was attenuated with a decrease in Danshen.

Nonlinear Systems Model: Multi-objective Optimization [6, 7]

After the multiple effects of compound Danshen were observed, ED-NM-MO triplex method was employed to perform nonlinear curve fitting and multi-objective optimization on the results from the above experiments [8]. Method: based on the data obtained using BGPID, seven indicators including myocardial ischemia, Hemodynamics, etc. were used as targets for analysis. Results: seven pharmacodynamic indicators and the six best ratios of combinations were obtained through the whole process. ED-NM-MO triplex method is suitable for the optimization of compound formulas. Conclusion: Danshen targets blood vessels and is better than Notoginseng regarding vasodilatory activity; Notoginseng focuses on cardiac muscles which leads to a better effect in protecting myocardial ischemia than Danshen. The best ratio range for myocardial ischemia in the combination of Danshen and Notoginseng was 10/3–10/6.

2.1.1.2 Assisting Function of Borneol in CDDP and Research on Its Target

Borneol is a fragrant herb which has an assistive function in compound formulas. *Chinese Pharmacopoeia* (2000 edition) has recorded more than 20 Chinese Patent Medicines containing Borneol. It is used commonly in the treatment of cardiovascular diseases. However, the targets of borneol in most of the formulas are not clear. Because the human body is an organic whole, in normal physiological conditions, the cardiovascular systemular system is regulated by nerves and body fluids. However, in pathological conditions, the regulation becomes chaotic. Therefore, the aim of this research is to clarify the dosage and effect of borneol on myocardial ischemia, and to identify its targets.

Dosage and Effect of Borneol in CDDP [9]

An acute myocardium ischemia rat model was established by the sublingual venous injection of Pituitrin (Pit). The changes in ECG, myocardial enzymes, and oxygen free radicals were recorded, and the relationship between effect and dosage was investigated by using five dosages of borneol in the compound Danshen formula. Result: All five dosages showed an effect on the resistance to the acute myocardium ischemia induced by Pituitrin, improvement of ECG, reduction of MDA, and suppression of myocardial enzymes like CK and LDH. However, no significant difference was found among these five borneol dosages.

Borneol's Effect on the Regulation of Cerebral Neurotransmitters [10, 11]

The effects of borneol on plasma and hypothalamus neurotransmitters and other active substances were observed to investigate the definite targets of Borneol. Result: borneol decreased the levels of hypothalamic monoamine neurotransmitters NE and E at all five dosages. The effects decreased with a decrease in dosage, and could last for 24 h. On the other hand, borneol did not make significant changes in hypothalamus acetylcholine. Borneol increased MAO significantly in the hypothalamus and the effect lasted for more than 24 h compared to the control. In summary, the effect of borneol on **myocardial ischemia** was partly due to the regulation of the central nervous system via balancing the content of monoamines.

Borneol's Effect on the Blood-Brain Barrier (BBB) [12, 13]
A BBB model was constructed through the co-culture of brain microvascular endothelial cells (BMEC) and astrocytes (AS) in vitro. Borneol's permeability in BBB and salvianolic acid B's permeability were observed. RT-PCR was utilized to evaluate the effect of borneol and salvianolic acid B on the expression of *Mdrla* and *Mdrlb* mRNA. Result: The BBB model was successfully constructed, and borneol had no effect on opening the BBB; the combination of borneol and salvianolic acid B could inhibit P-protein expression in BMEC to improve the permeability of other medicines.

2.1.2 Pharmacological Research

CDDP can enlarge coronary arteries, lower blood lipid levels, resist atherosclerosis, inhibit platelet aggregation, ameliorate hemorheological status, scavenge free radicals, and restrain lipid peroxidation (LPO).

2.1.2.1 Protection of Cardiac Muscle [14, 15]

The myocardial ischemic model of rats was built and the function of CDDP was observed. Results: The infarct size (IS)/area at risk (AAR) ratios in the animals of the re-perfusion group and preconditioning groups in early and late periods were 45, 47, 45 %, respectively; arrhythmia incidence rates were higher than normal at 76.46, 55.56, 73.33 %, LDH in plasma was also much higher than the control group. After ischemic preconditioning (IPC), the early group's IS/AAR ratio was reduced by 17 %, and that of the late group by 4 %; Arrhythmia incidence rates in these two groups were 11.76 and 43.75 %, respectively, much lower than the re-perfusion group and preconditioning groups; the release of IPC and LDH also decreased. These results showed that ischemic preadaptation could minimize the ischemic range and lessen the risk of an arrhythmia incident. After CDDP preconditioning, IS/ARR was reduced by 10–17 % in further step and LDH release also decreased. Also, MDA in serum deceased and SOD activities increased.

Gao et al. [16] observed the compound Danshen's effect on proteinkinase C (PKC) expression in IPC. Expression of PKC mRNA and the effect of compound Danshen were studied via immuno-histochemical method and RT-PCR. The objective was to test the hypothesis that the mechanism of strengthening IPC with compound Danshen was by promoting the expression of PKC at protein or mRNA levels. Results showed that PKC, which is normally located in the cytosol, was also found in the membrane and nucleus after IPC or compound Danshen medication. PKC's expression was higher in both the early and late IPC groups than in the false preconditioning group, and the early IPC group was higher than the late group. After compound Danshen medication, PKC expression was much higher than the re-perfusion group and IPC groups, and the compound Danshen-IPC (DIPC) group was much higher than the early IPC group. There was no obvious difference in PKC mRNA levels between normal myocardial tissue, the re-perfusion group, and the false preconditioning group, but IPC groups were higher than the false preconditioning ones, and the late IPC group gave more obvious results. The medication of compound Danshen could promote the expression of PKC mRNA in the early and late IPC groups and re-perfusion group. Conclusion: compound Danshen can activate and translocate PKC. This may be the main mechanism of IPC enforcement. Experiments showed that the mechanism of compound Danshen's protective functions for cardiac muscles could involve multiple pathways aimed at multiple targets.

Myocardial cell death is the worst injury of re-perfusion. In recent years, Zhao et al. [17] used ligation on the left coronary artery muscle to make re-perfusion models of myocardial ischemia in order to study CDDP's function on myocardial cell death. In addition, TUNEL and immune histochemical staining were used to test cells death and changes in *Fas*, *Bcl*-2 mRNA expressions. Results showed that compared to the control group, the re-perfusion group's apoptosis index and the levels of *Fas* and *Bcl*-2 expression increased significantly; both indices decreased in the medication group. Furthermore, when the dosage of medication increased, these two indices decreased further. The conclusion was that CDDP could protect myocardial cells by inhibiting cell death and down-regulating the expression of *Fas* and *Bcl*-2 mRNA.

Xu et al. [18] studied CDDP's effect on *Fas*/*Fas*L expression under hypoxia. The results: in the time between 4.5 and 10.5 h of hypoxia, *Fas*/*Fas*L expressions were much higher than those in the control group, and the two indicators had no obvious difference. The protein levels in the medication group were much lower than those in the hypoxia groups. After 30 min of hypoxia, oxygen was supplied for 4 and 10 h. The levels of *Fas*/*Fas*L were obviously higher than in the control group, but there was no difference between the two oxygen supply groups. In the medication group, the levels of the two proteins were much lower than in the hypoxia- re-oxygenation groups. Conclusion: hypoxia and re-oxygenation made *Fas*/*Fas*L increase and CDDP could reduce these levels to protect cardiac muscles.

Wang et al. [19] studied the mechanism of CDDP's effect on re-perfusion of rat cardiac muscles at whole animal and cell levels. An acute MRI (myocardial reperfusion injury) model was built by loosening ligations of rat LAD (left anterior descending coronary). Myocardial cells with oxygen-glucose deprivation were cultured in vitro, then resupplied with oxygen and glucose. The changes in ECG, lactate Dehydrogenase, creatine kinase leakage, free calcium concentration, SOD, MDA, ET-1, and NO content caused by CDDP were observed. Results: CDDP could improve MRI rats' ECG, lactate Dehydrogenase, creatine kinase leakage, free calcium concentration, SOD, MDA, ET-1, and NO contents. Conclusion: CDDP probably acted on MRI by helping with calcium overload, oxygen free radicals, and endothelial cell injury.

Zhao et al. [20, 21] made hypoxia/re-oxygenation models in rats with the Langendorff method and medicated with CDDP and isosorbide dinitrate to protect cardiac muscles before or after hypoxia. HPLC was used to test the content changes of high-energy phosphate compounds and change in LPO. H-600 Transmission Electron Microscope was used to observe the ultrastructure of cells. Conclusion: CDDP acted well in both occasions because it increased the content of high-energy phosphate compounds and protected cells and myocardial ultrastructure. It was much better than isosorbide dinitrate.

2.1.2.2 Effect on Hyperlipidemia and Atherosclerosis

Using a rat hyperlipidemia model, Li et al. [22] observed CDDP's dosage effect (50, 150, 450 mg/kg). The positive control group was set with medication of Alginic Sodium Diester (25 mg/kg). After 13 days of oral medication, thrombosis index, high shear whole blood viscosity variability and low shear rate, and whole blood reduced viscosity were measured. Results showed an obvious increase in the platelet adhesion rate and thrombosis index in hyperlipidemic rats. CDDP at 150 and 450 mg/kg and the positive control reduced these indices. These data suggest that CDDP could prevent atherosclerosis.

Guan et al. [23] also used hyperlipidemic rats to test many indicators to analyze the effect of CDDP on the function of blood lipids. Results showed that both preventive administration and treatment with CDDP medication could reduce total cholesterol (TC), triglyceride (TG), and low-density lipoprotein (LDL), and increase the ratio of HDL/LDL in the blood of hyperlipidemic rats.

Using adrenaline and high molecular dextran injections, Xu et al. [24] made a rat model of high-viscosity-hyperlipidemia to analyze the functional mechanism of CDDP. CDDP was applied once 90 min before model establishment. 60 min after model establishment, femoral artery samplings were performed and deforming red blood cells were measured. Results: CDDP can ameliorate red blood cell deformation and dosage effects were found in a range from 15 to 75 mg/kg.

Guo et al. [25] studied the effect of CDDP on hemorrheology and red blood cell micro-rheology in hyperlipidemia model dogs. Methods: The model was maintained for 6 months, then medicated with CDDP for 4 months. After that, the contents of triglyceride, cholesterol, and hemorrheology, RBC-EPM, osmotic fragility, membrane fluidity, and Ch/Pl were measured. Results: CDDP reduced the model's whole blood viscosity, osmotic fragility, Ch/Pl, and improved the deformation index, orientation index, RBC-EPM value, and membrane fluidity. Conclusion: CDDP can improve blood lipid disorders and hemorrheology value.

Tian et al. [26] studied the effect of CDDP on rabbits fed with high fat food. Methods: 48 rabbits were divided into five groups: control group, atherosclerosis group, simvastatin group, and large and small CDDP dosage groups. Animals in the control group were fed normally, and animals in other groups were fed with high fat food. Results: CDDP reduced the contents of cholesterol, triglyceride, and LDL, and increased HDL content. In addition, it could attenuate the endangium and reduce the endangium/Intima-media thickness (IMT) ratio. Compared with the control group, the inhibition rate on IMT of the simvastatin group, small, and large CDDP dosage groups were 32, 24, and 41 %, respectively. In conclusion: CDDP can effectively reduce blood lipid levels of the model rabbits and prevent atherosclerosis.

Chen et al. [27] observed CDDP's function and mechanism on rabbit vascular smooth muscle cells (VSMCs) induced by high blood sugar or high insulin. Methods: VSMCs were first cultured, then high blood sugar or high insulin was used to induce an increase in VSMCs. The model group had a lower content of nitric oxide synthase (NOS) and NO than the control group, but he CDDP groups increased NOS and NO significantly, which could be partially blocked by L-Nitro-Arginine Methyl Ester (L-NAME). Conclusion: CDDP can inhibit the proliferation of VSMCs by the mediation of NO.

Shi et al. [28] studied CDDP's effect on atherosclerosis (AS) in rabbits. Results: CDDP could ameliorate blood lipids. In the AS group, the NO content of the CDDP group was much higher than that of the AS group, but the ET level was reversed. Conclusion: CDDP can improve the levels of blood lipids, NO, and ET much better in AS rabbits.

Wang et al. [29] found that CDDP could inhibit the increase of blood lipids and prevent atherosclerosis, and reduce atheromatous plaque caused by arterial stenosis.

2.1.2.3 Effect on Blood System

Feng et al. [30] studied CDDP's effect on ADP, thrombin, and thrombocyte aggregation induced by collagen. Results: CDDP had inhibitive effects in all these aspects, and the effect was dosage-dependent.

Li et al. [31] studied CDDP's effect on coronary heart diseases and angina pectoris by utilized fluorescence polarization immunoassay to measure the fluorescence polarization P of platelet membrane, micro-viscosity, and membrane fluidity (LFU). Results: in CDDP treated rabbits, LFU increased, micro-viscosity decreased, and the difference between the treatment and control was highly significant. Conclusion: CDDP can efficiently improve LFU and lower micro-viscosity.

Zhang et al. [32] analyzed CDDP's effect on insulin resistant dogs. Results: A high fat diet for 6 months resulted in abnormal metabolism in the animals and insulin resistance, as well as excessive platelet activation. CDDP could reduce blood cholesterol significantly, and showed resistance to oxidation. Both aspirin and CDDP could inhibit platelet activation, but CDDP's inhibition of platelet aggregation and the expression of platelet surface adhesion molecules were more significant. Conclusion: CDDP was able to efficiently inhibit insulin resistance and platelet overactivity.

2.2 The Pharmacological Functions of Other Danshen-Containing Prescriptions

Danshen is widely used in a variety of compound preparations (both in traditional formulas as well as in new modern medicines), and its efficacy has been demonstrated through clinical pharmacology research and experimental studies. This section introduces some commonly used Danshen-containing Chinese patent medicines and compound preparations, and analyzes and summarizes the results of the clinical and experimental research. The patent medicines and preparations are: Compound Danshen injection (CDI), Xinmaitong Tablet (XT), Quanxinning injection, Xinkening Capsule, the Danqi Hemiplegia Capsule, Yangxinshe Tablet, Ningxin-anshen Capsule, Bushenyishou Capsule, Compound Dangshen Tablet, Rukuaixin Tablet (RT), Ningshenbuxin Tablet (NT), Huganning Tablet, Compound Xueshuantong Capsules, etc.

2.2.1 Composite Danshen Injection

CDI is also named Xiangdan injection, and its main ingredients are Danshen and rosewood (Lignum Dalbergiae Odoriferae), which are extracted with water and precipitated with ethanol. This medicine has the function of expanding coronary arteries and blood vessels, inhibiting blood clotting, removing free radicals, reducing lipids, enhancing the blood flow of the liver and kidney, improving immune functions, promoting bone fracture, and wound healing, etc.

2.2.1.1 Explanation of the Formula

The main ingredients of CDI are Danshen and rosewood. Rosewood has other names such as Jiangzhenxiang (降真香), Zijiangxiang (紫降香), and Hualimu (花梨母). It has an acrid flavor and warm nature, and its meridian distribution belongs to the liver and heart. The effects of this medicine include activating blood circulation to dissipate blood stasis, and stopping bleeding and pain. It contains naphtha, which is mainly β-bisabolene, *trans*-β-farnesene, *trans*-nerolidol, etc. It can be used for abdominal pain, liver depression and hypochondriac pain, thoracic obstruction and tingling, injuries from falls, and traumatic hemorrhage. Danshen's nature and flavor are slightly cold and bitter, respectively. The combination of the two herbs can expand blood vessels and increase the blood flow of the coronary artery muscle, and can be used for the symptoms like angina, myocardial infarction, etc. It also has the function of dissipating blood stasis to stop pain, activating blood to promote menstruation, and clearing heart heat to repel vexation.

2.2.1.2 Clinical Application

CDI is commonly used clinically for ischemic cardio- and cerebro-vascular diseases (CCVD). As research on the medicine has deepened, its clinical application has expanded to the areas of the CCVD system, respiratory system, urinary system, digestive system, as well as to orthopedic, neonatal, ophthalmology, surgery and other treatment of diseases. It also Functions as an adjuvant therapy for some infectious diseases. Its clinical application has occasional allergic reactions. It should not be used together with anticancer drugs like cyclophosphamide and combined with cytochrome c.

2.2.1.3 Pharmacological Research

Clinical Research

Cardiovascular Disease

The conventional treatments of dilated cardiomyopathy with heart failure include salt and water restrictions and medication with Western drugs such as Digitalis, diuretics, and vasodilators. If these drugs are used in combination with a large dose of CDI, it can improve heart functions, such as reducing the heart rate, increasing left ventricular ejection fraction (LVEF), and eliminating pulmonary rales and lower limb edema. Occasionally CDI can cause intestinal noise and loose stool. The effects will decrease with lower dosages [33].

The combined use of CDI with Shengmai injection and Huangqi injection in the treatment of coronary heart disease in the elderly can ameliorate the symptoms of chest tightness, palpitation, and fatigue, reduce the seizure frequency of angina, and recover the resting electrocardiogram. The combined use of CDI with Shengmai injection is better for patients with hypotension, infirmities, and complications. The combined use of CDI with Huangqi injection is better for patients with qi-deficiency (qi is the traditional Chinese medicine term meaning body energy) after diagnosis [34, 35]. CDI also has certain effects on unstable angina pectoris [36]. Some research showed that Danshen injection's treatment of blood stasis types of coronary heart disease could increase endothelial nitric oxide synthase mRNA, decrease the positive rate of ET-1 mRNA, and improve endothelial function [37].

The combined use of CDI with Propafenone or Shengmai injection shows good effects on the treatment of coronary heart disease, various cardiac arrhythmias like ventricular premature beat, junctional premature beat, a atrial premature beat complicated by myocarditis. It also has the best effect for blood stasis type patients, and can improve the heart function of congestive heart failure patients [38, 40].

CDI is also used in the treatment of pulmonary heart disease in the acute exacerbation stage, acute exacerbation of chronic pulmonary heart disease, refractory heart failure with pulmonary heart disease, etc.

Cerebrovascular Disease

48 h after subarachnoid hemorrhage (SAH), combined with treatments to stop bleeding, reduce intracranial pressure, and treat dehydration, CDI can significantly relieve headaches, reduce cerebral vasospasms and the incidence of recurrent bleeding, and decrease the death rate. For the treatment of hypertensive cerebral hemorrhage, the application of CDI 12 days after bleeding can effectively prevent cerebral edema and then increase cerebral vascular permeability, and showed better effects of relieving vasospasm, improving microcirculation, promoting absorption of the hematoma, and recovering regional cerebral blood flow, and the effects are better than if treated with Nimodipine. Danshene has a two-way adjustment function on the fibrinolytic system and can improve the redistribution of blood flow, prevent re-bleeding and increase in brain hemorrhage, and relieve the degree of disability and sequelae under the condition of maintaining or reducing vascular pressure [41–43].

For the treatment of acute cerebral infarctions, this medicine can reduce the levels of serum lipid peroxide and apolipoprotein B100 (ApoB100) and increase the level of SOD and ApoA1, thereby removing free radicals, regulating the metabolism of Apo, and preventing atherosclerosis. The injection can improve hemorrheology (whole blood viscosity, plasma viscosity, erythrocyte electrophoresis, fibrinogen, platelet adhesion rate), and the degree of the neurologic deficits of patients with cerebral infarction [44, 45].

Respiratory Diseases

For treatment of pneumoconiosis with acute infection, effective antibiotics should be used to improve ventilation function, stop coughing and asthma, regulate low-flow oxygen, expand coronary artery muscle, strengthen the heart, and correct electrolyte imbalance. In addition, CDI and Venoruton injection can be combined with these treatments to promote the improvement of symptoms like cough, asthma, fever and phlegm, and significantly reduce moist and dry rales in the lung. The combination treatment shows good effects on the restoration of complete blood count, and on the improvement of the lung inflammation, as shown by X-ray [46].

For the treatment of bronchopneumonia, the use of CDI can promote the improvement of symptoms such as cough, asthma, fever and phlegm, and significantly clear the moist rales in the lung [47].

The pathogens of lower respiratory tract infections in the elderly are mainly *Streptococcus pneumoniae* and *Haemophilus influenzae*. Antibiotics are usually used for treatment in the early stage. However, antibiotic treatment alone has some shortcomings, such as a longer disease course, more complications, and slower absorption. But combined with CDI, it can shorten the course of the disease, promote absorption of inflammation, and reduce complications [48].

The combined use of CDI and other conventional methods such as anti-tubercular drugs and hormones to treat tuberculous exudative pleurisy can improve the cure rate and lower the relapse rate [49].

The acute exacerbation of chronic pulmonary heart disease can be treated with conventional treatments plus CDI alginic sodium diester. The combined treatments can increase the partial pressure of oxygen in arterial blood (PaO_2) and the percentage of available hemoglobin that is saturated with oxygen (SaO_2), decrease the atrial partial pressure of carbon dioxide ($PaCO_2$), adjust hypoxia and carbon dioxide retention, release elements like ET, eliminate pulmonary artery spasm, reduce pulmonary hypertension, expand the renal artery, promote diuresis, lower blood viscosity and pulmonary blood flow resistance, and so on, in this way achieving therapeutic effects [50].

For the treatment of chronic obstructive pulmonary disease, in addition to the conventional effects such as stopping cough, decreasing spasm, reducing phlegm, etc., the use of CDI can improve vital capacity, maximum voluntary ventilation, and atrial blood gas [51].

For the treatment of pulmonary encephalopathy, respiratory stimulants should be used, and measures of reducing intracranial pressure, supplying oxygen, correcting acid-base and water-electrolyte imbalances, and controlling heart failure should be adopted. Meanwhile, the use of CDI and Angongniuhuang Pill (one pill, twice or three times daily) can reduce the death rate [52].

Forty patients with carcinoma of the esophagus were randomly divided into two groups: control group and CDI group. During anesthesia, the former group was given sodium lactate Ringer's injection and the latter group was given sodium lactate Ringer's containing 10 ml of CDI. The same method of anesthesia was used. Serum superoxide dismutase (SOD), glutathione peroxidase (GSH2PX) and malindialdehyde (MDA) levels were measured before and 1, 2, 4, and 6 h after anesthesia. The ultramicroscopic structure of the lungs of both groups was investigated. Results: Group D's SOD activity was significantly decreased at the 4th and 6th hour after inspiring, the GSH2PX activity of group D significantly increased at the 1st hour after breathing oxygen and the MDA level of group D significantly decreased at the 2nd hour of inspiring. The ultramicroscopic changes of the pulmonary tissue of group C were worse than those of group D at the 4th and 6th hour after inspiring oxygen. Conclusion: CDI can relieve human lung injury caused by breathing pure oxygen, the possible mechanism being that it can eliminate oxygen free radicals and enhance the activity of SOD and GSH2PX [53].

Liver Diseases

The basic treatment for ascites due to cirrhosis is sodium restriction, diuresis, liver protection, plasma transfusion, antibiotics, and so on. The application of CDI and potassium aspartate and magnesium aspartate injection (one 14 day course of treatment) can reduce the levels of alanine aminotransferase and aspartate aminotransferase, and relieve the clinical symptoms. However, the albumin globulin ratio (A/G) has no significant change before and after treatment [54].

30 and 60 days after the treatment of cirrhosis with CDI, nailfold microcirculation and changes in hepatic portal hemodynamics were observed. It showed that there was improvement of microflow speed, Dpv, Vpvx, Vpvm, Dspv, Vspvx, and Qpv, but no significant effect on Qspv and Vspvm. It suggested that this injection could lower the portal vein pressure without affecting blood supply to the liver and could effectively improve the microcirculation of the liver [55].

Earlier and sustained high-dosage applications of compound Danshen preparations could contribute to the stability of the disease and to reduced complications so as to improve long-term survival [56].

CDI was combined with other medications to treat hepatic fibrosis. The results showed reductions in serum hyaluronic acid, laminin, collagen type IV, procollagen type III, MDA, lipid peroxides (LPO), and an increase in SOD activity, indicating that the free radical cleaning function of this injection is the key mechanism of preventing and curing hepatic fibrosis [57].

Diammonium glycyrrhizinate can act as an adrenal cortex hormone, with functions of anti-inflammation, liver cell membrane protection, bilirubin metabolism promotion, and liver function improvement. It has a special effect on the increasing of alanine aminotransferase caused by liver cell injury, but no obvious effect in reducing jaundice for chronic hepatitis B. However, when combined with CDI, the treatment can obviously reduce jaundice and Transaminase. It can also decrease total bilirubin and increase the value of A/G. The rebound of ALT and the symptoms would be reduced after cessation of intravenous infusion. So, this injection can significantly relieve liver cell injury, promote the recovery of liver function, and slow down the development of chronic hepatitis to liver cirrhosis [58, 59].

The combined use of CDI and diammonium glycyrrhizinate in the treatment for fatty liver can obviously reduce the TC, TG, ALT, AST, and GGT and the effect is better than any single application of the two medicines [60].

The combined use of CDI and other conventional methods in the treatment of stasis cholestatic hepatitis can enhance the regeneration of epithelial cells, maintain an unobstructed bile canaliculus, improve the stasis of bile, clear liver inflammation, and dilute bile so as to reduce jaundice [61].

The combination of CDI and interferons in the treatment of chronic hepatitis B has a certain synergistic effect on the inhibition of Hepatitis B virus (HBV) replication and on the protection of liver cells from fibrosis. Whether it can have long-term anti-HBV effects and prevent the formation of liver fibrosis is not exact [62].

Using CDI for pretreatment is the effective way to prevent hematopoietic stem cell transplantation from hepatic veno-occlusive disease (HVOD). CDI can reduce the incidence of abdominal pain, liver expansion and ascites, reduce AST, ALT, PT, APTT and other indices, and prevent the incidence of HVOD. It is presently believed that HVOD is caused by multiple factors, but endothelial cell injury is the foremost critical factor. CDI's function of relieving endothelial cell injury might be its mechanism of HVOD prevention [63].

Many drugs for blood diseases have certain side effects on the liver. The combined use of Chuanxiong injection and CDI can prevent these side effects. Reports showed that the two drugs can reduce liver damage in the treatment of leukemia, and reduce liver damage as well as increase curative effects [64].

Nervous System Disease

Diabetic peripheral neuropathy disease (DPND) is a common complication of diabetes, and the pathogenesis is related to metabolic disturbance, diabetic microangiopathy, nerve growth factor reduction, etc., with the symptoms including pain, numbness, and hypoaesthesia. The combination of CDI and Methycobal or mexiletine has certain clinical effect on DPND. With mexiletine, it can relieve pain, and with Methycobal, it can relieve the symptom of numbness. CDI can reduce the level of Thromboxane A2, arteriolospasm and platelet aggregation, improve microcirculation, accelerate blood flow, reduce aggregation of RBC, and increase blood and oxygen supply to the tissues. So, the addition of CDI to the treatment of DPND can improve the clinical effect [65, 66].

CDI is widely used in the treatment of kidney diseases (chronic and acute glomerulonephritis, nephrotic syndrome, chronic renal failure, diabetes, and so on), and is usually combined with Astragalus injection, Ligustrazine injection, and Shengmai injection. Together they can improve clinical symptoms and increase kidney function.

CDI has an adjuvant therapeutical effect on pediatric diseases such as neonatal hypoxic-ischemic encephalopathy, neonatal sclera edema, infantile pneumonia, bronchopneumonia, allergic purpura, nephrotic syndrome, etc.

CDI can prevent steroid-induced necrosis of the femoral head, and has clinical effects on scapulohumeral periarthritis, osteoarthritis, and other osteopathy. It also has therapeutic effects and auxiliary therapeutic effects on dermatological, surgical, ENT, mental, gynecological, and other diseases.

Experimental Research

It has been reported that CDI could improve the activity of SOD and reduce the content of MDA in rat cardiac tissue. The increased activities of Na^+-ATP, K^+-ATP, and Ca^{2+}-ATP had a protective effect against acute myocardial ischemia reperfusion injury in rats, and the protective mechanism was related to the function of CDI (including improving the microcirculation of cardiac muscle, cleaning free radicals, reducing LPO, resisting the toxicity of free radicals and calcium overload, and maintaining the function of myocardial cell membranes and mitochondria) [67].

Ischemic cerebrovascular disease is caused by the expression of surface adhesion molecules in vascular endothelial cells and white blood cells, which leads to the infiltration of peripheral blood leucocytes in the ischemic area. This movement blocks microcirculation and affects the blood supply to the tissue. On the other hand, the activated white blood cells release many inflammatory mediators and cytokines which can damage local blood vessels, increasing their permeability which leads to tissue edema; damage neurons and neuroglial cells which worsen nervous tissue injury, making more WBC attack the tissue, and causing more damage. Aided by micromanipulation techniques, researchers found that CDI can inhibit the changes in adhesion of Polymorphonuclear leukocytes and cerebral capillary endothelial cells of rats, which confirmed the observation that CDI reduced adhesion of cells (4th, 12th, 24th hour of cerebral ischemia reperfusion) [68].

Mesenchymal stem cells are derived from the mesoderm during the early period of mesenchymal stem cells, and they are non-hematopoietic stem cells found in bone marrow. The effect of CDI on the differentiation of rat bone marrow mesenchymal stem cells (rMSCs) has been studied. The results showed that CDI could induce the differentiation of rMSCs into neuron-like cells. However, in in vitro culture, there was less than 5 % neuroglia cells, which might indicate the process of MSC's differentiation: from pluripotent stem cells, to neural stem cells, to neural precursor cells, and then to mature nerve cells. It was suggested that the antioxidant activity of Danshen is one of the main factors in inducing the differentiation of rMSCs [69].

Spinal cord injury (SCI) model rats were prepared with the improved Allen's method, CDI was injected intraperitoneally, and the locomotion of the rats on a inclined plane was measured using the Basso Beattie and Bresnahan Locomotor rating scales. Results: CDI could promote motor function recovery in rats with acute SCI, and the application of cycloheximide and CDI could significantly reduce the apoptosis of spinal cord cells [70, 71].

The combined use of methylprednisolone and CDI can promote the recovery of motor function in the hind limbs of SCI rats, protect nerve conduction, stabilize mitochondrial structure, and maintain the normal morphology and function of neurons. Light and electron microscopy showed that hydromyelia, axonal degeneration, changes in myelin, and mitochondrial and capillary injuries in SCI rats were reduced with treatment with methylprednisolone and CDI. The treatment also returned synaptic vesicle numbers to normal [72].

2.2.2 Dan-Qi Hemiplegia Capsules

Dan-qi hemiplegia capsule is a new medicine to cure strokes and is suitable for strokes from qi-deficiency blood stasis and for meridians stopped by wind and phlegm. The drug has the function of dissolving cerebral thrombosis, improving blood circulation, increasing blood flow in the brain, and repairing nerve cells. There are 14 main ingredients in the drug, including Danshen, astragalus root, Rhizoma Chuanxiong, leech, calculus bovis artifactus, cornu saigae tataricae, scorpion, etc. The principle of this formula is mainly to enhance qi, activate blood circulation, reduce stasis, unblock the collaterals, repel wind, and open the orifices (all in traditional Chinese medicine terms). Its main indications are dyskinesia and mental disturbance after stroke.

2.2.2.1 Explanation of the Formula

Astragalus root has a sweet flavor and warm nature, and it belongs to the meridian of lung and spleen. A large dosage of this herb can nourish qi, which in traditional Chinese medicine terminology means body energy, unblock the collaterals, repel blood stasis without harming healthy qi, and is prescribed as a chief medicine.

Rhizoma Chuanxiong has an acrid and dispersing nature. It can promote blood circulation by removing blood stasis and is best for repelling wind. It also can free the qi and repel the wind in blood, which makes other herbs move to head.

Leech can remove blood stasis and stop pain, free the blood and meridian, and clear the mind by removing depression. Combined with the chief medicine astragalus root, they can nourish qi, free blood, remove blood stasis, and free the meridian. It is used as a deputy drug.

Calculus bovis artifactus, antelope horn, and scorpion enter the meridian of the liver and belong to comforting liver and repelling wind medicines; they can stop pain and free the meridian, calm the liver and wind, and relieve distortion. The combined use of the three herbs can clear heat and free phlegm. Rhizoma Acori Tatarinowii and other herbs dispel dampness and remove turbidity, functioning mainly in opening the orifices. Combined with calculus bovis, they can **clear heat** and eliminate phlegm. The above are assistant drugs.

So, the entire formula can nourish qi and restore body energy, refresh and clear the mind, and free the blood and meridian.

2.2.2.2 Clinical Application

It is used for hemorrhagic strokes from qi deficiency and blood stasis and has significant effects on the late stage of this disease.

2.2.2.3 The Pharmacological Research

The main pathological changes of the cerebral hemorrhage are the hematoma and the surrounding edema. The hematoma is the fundamental factor in causing cerebral edema. Many animal experiments, clinical research studies, and iconography data such as MRI and CT scans have confirmed that the local blood flow in the brain in the hematoma area is reduced, the affected area is far larger than the hemorrhage area, and the edema area is the same as the area with reduced brain blood flow.

Experimental studies have showed that the hematoma oppresses the surrounding brain tissue, causing large reduction of brain blood flow in areas near and far from the hematoma, resulting in delayed ischemia and hypoxia. This phenomenon did not disappear with the disappearance of the hematoma, suggesting that injury to nerve function is related to the delayed ischemia and hypoxia. So, the measures to prevent injury to nerve function are to prevent ischemia and hypoxia in area. Many animal experiments revealed that herbs such as Danshen, astragalus root, Sichuan lovage root, etc., could reduce hematocrit and fibrinogen blood levels, lower blood viscosity, accelerate blood flow, relieve injury to brain tissue, shorten the recovery time of the Edema area, and greatly protect nerve functions. Dan-qi hemiplegia capsule can improve paralysis of the face, upper and lower limbs, and fingers and toes, decrease blood lipids, and significantly affect recovery from late stage hemorrhagic stroke [73].

2.2.3 Compound Radix Codonopsitis Tablet

This medicine contains the herbs radix codonopsitis, Danshen, angelica, radix glehniae, radix tinosporae, etc. It can promote blood circulation by removing blood stasis, nourishing qi, and calming the heart, and is used for angina and chest distress caused by myocardial ischemia. Recent research on the medicine has shed more light on its pharmacological mechanism and scope of indications.

2.2.3.1 Explanation of the Formula

This medicine is a pure Chinese traditional medical preparation and contains mainly radix codonopsitis, Danshen, angelica, radix glehniae, radix tinosporae, etc. Radix codonopsitis can nourish spleen qi and lung qi; Danshen can promote blood circulation by removing blood stasis; angelica can nourish and activate blood; radix glehniae can nourish yin and clear the lung; and radix tinosporae can clear heat and repel toxins. Together, they nourish qi and activate blood.

2.2.3.2 Clinical Application

It is used for the treatment of angina and chest tightness caused by myocardial ischemia cerebrovascular disease, and for injuries to the heart, brain, and lung under the hypoxic environment of high altitudes.

2.2.3.3 Pharmacological Research

Clinical Studies

To study the effects of Compound Radix Codonopsitis Tablet (CRCT) on the protection and improvement of brain functions, healthy people living at an altitude of 3,700 m were asked to take CRCT orally, and then their intelligence, and memory functions were measured in several psychological tests. The results showed that CRCT can significantly improve Verbal IQ, Performance IQ and Full-scale IQ. The results of memory comparison revealed that the indexes of orientation, 100-1, picture memory, imagination, touch, recite and memory quotient are significantly higher than those measured before CRCT administration. Several psychological tests of DX-A also revealed a significant change after administration. The above tests showed that this medicine can reduce hypoxic brain injury and can maintain and improve brain functions for people in hypoxic environments. The comparison of hemodynamic parameters 1 month after the administration revealed that pulse (P), total peripheral resistance (TPR), whole blood viscosity (η), altering latter time (ALT), pulmonary artery wedge pressure (PAWP), and Coronary perfusion pressure (CCP) were all reduced, and stroke volume (SV), effective blood volume (BV) and mean arterial blood pressure (mAP) were all increased (statistically significant). Thus, CRCT can reduce oxygen consumption, improve the circulation system and brain function, and protect from myocardial injury. Other tests on humans also showed similar effects [74, 75].

The main cause of high altitude pulmonary hypertension is hypoxic pulmonary vasoconstriction. Hypoxia causes the worsening of pulmonary vasoconstriction and the tension of pulmonary vascular smooth muscles, which increases the pulmonary artery pressure. Pulmonary vasoconstriction caused by hypoxia is the result of the depolarization of PVSM cell membranes and the transmembrane backflow of calcium ions. Improved heart function can delay the development of pulmonary artery pressure and reduce its injury. CRCT belongs to nourishing medicines, and many studies have confirmed its function of resisting hypoxia, and its mechanism is to reduce anoxic symptoms in an oxygen-deficient environment and to maintain normal physiological and biochemical functions. Long-term use of CRCT can turn changes under adverse conditions back to normal. Research results have shown that long-term use of this medicine can prevent the pulmonary artery pressure for people living above an altitude of 5,000 m [76].

CRCT was administered to people between the age of 50–77, twice daily, five tablets each, continued for 3 months. The results showed that the indexes of picture memory (PM), velocity of mental arithmetic (VMA), step velocity (SV), SD, and AFWL were significantly improved. There was no significant improvement for SDMEMA and FSM were not significantly improved until the 3rd month. The results show that this medicine can significantly improve brain function and delay aging [77].

CRCT has been used to treat angina of qi deficiency and blood stasis style. Patients more than 35 years of age were given CRCT three times daily, three tablets at each time, over 30 days. The results showed that CRCT could greatly relieve the main symptoms such as chest pain and tightness and TCM symptoms such as qi deficiency and blood stasis. ECGs were back to normal or basically normal. The depression segment returned to 0.05 mV, the T wave was reduced by more than 25 % or change from flat to erect, and the atrioventricular block was also reduced. Meanwhile, it was observed that CRCT had no effect on liver and kidney function, as indicated by serum ALT, inosine, and blood urea nitrogen. It could reduce cholesterol levels, but had no influence on triglyceride content [78].

Experimental Studies

An acute decompression hypoxia and confined hypoxic condition at an altitude of 8,000 m was simulated. After 7 h under such a condition, the changes of cytochrome oxidase and succinic dehydrogenase of myocardial cells in mice were tested. CRCT (0.2 ml/each) was orally delivered to mice 1 day before hypoxia, and the same volume of distilled water was delivered to the control group. 1 h before hypoxia, the same dosage of CRCT or water was delivered by abdominal cavity injection. The animals were placed in a decompression chamber, where the pressure was reduced to that at an altitude of 8,000 m, maintained for 7 h, and then returned to pressure at sea level. The mice were executed and the hearts collected to measure cytochrome oxidase and succinic dehydrogenase activities. The results showed that the activities of cytochrome oxidase and succinic dehydrogenase in the hypoxic group were reduced when compared to those of the normal pressure group, and the difference was statistically significant. The levels of the two enzymes in the CRCT treatment group were higher than in the hypoxia group and close to those in the normal pressure group, but statistically insignificant. Acute hypoxia can reduce the activity of cytochrome oxidase and succinic dehydrogenase, which is harmful to body. However, CRCT could increase these activities and the levels were close to normal, suggesting that the medicine can prevent the reduction of enzymatic activities and improve anoxia tolerance of their bodies [79].

2.2.4 Xinkening Capsules

This medicine contains Danshen, Notoginseng, safflower, bezoar, borneol, senso, cornu bubali, and ginseng fibrous root. It can promote blood circulation by removing blood stasis, stop pain, and refresh the mind. It is used for the treatment of CCVD clinically.

2.2.4.1 Explanation of the Formula

Danshen is the chief drug in a prescription, and with the aid of Notoginseng and safflower, it can promote blood circulation and remove blood stasis, free meridians, and stop pain. Bezoar, borneol, senso, and cornu bubali can refresh the mind and dispel filth. Ginseng fibrous root can powerfully nourish original qi, restore vessels, and rescue from desertion, promote fluid production, calm the mind, and protect the body energy by reinforcing healthy qi and dispelling pathogens. The nature of senso is pungent and scurrying, the flavor is warm, and it has the function of opening the orifices and dispelling filth. *Convenient Reader on Materia Medica* says, "senso is good at clearing the mind and repelling harm, and is used to protect the body in an emergency." The major chemical components of senso are steroids, and collectively they are called toad diene lipid. Senso has the cardiotonic effect of digitalis and the boosting pressure effect and central respiratory stimulation effect of adrenaline. In addition, it can stop pain, activate striated muscle, uterus and fallopian tube, relieve coughing and panting, resist inflammation, cancer and radiation, increase white blood cells, etc. Senso is used clinically for respiratory and circulatory failures. Ginseng is also used for the symptoms of asthenia, cool limbs and weak pulse.

2.2.4.2 Clinical Application

This medicine is used for coronary heart disease, angina, chest tightness, palpitations, and dizziness. It can relieve hypertension, and reduce blood lipid.

2.2.4.3 Pharmacological Research

Patients were selected in line with WTO diagnostic criteria: at least 3 months of angina history, angina pectoris at least 5 times per week, symptoms able to be relieved with resting or sublingual nitroglycerin, and no liver, heart, or kidney disorders. After treatment with Xinkening capsules for 4 weeks, the myocardial oxygen consumption index (expressed as rate-pressure product) and the degree of myocardial ischemia decreased significantly. The total incidence of myocardial ischemia within 24 h reduced by 41 %, both the sustained and total time of ST segment depression within 24 h were reduced, and the maximal platelet aggregation (MPAG) was significantly lowered compared to before treatment. Blood sugar and liver and kidney function were not affected. The mechanism of this drug's Functions is unclear, possibly acting by expanding the coronary artery muscle, reducing vascular resistance, increasing blood flow, improving myocardial nutrition and myocardial ischemia, and reducing the area of ischemia, myocardial oxygen consumption. Meanwhile, it can reduce blood lipids, regulate blood viscosity, accelerate blood flow, etc. [80].

2.2.5 Fufang Xueshuantong Capsule (Copound Xue-Shuantong Capsule)

This drug is mainly used for treating vascular diseases by removing blood stasis, and nourishes qi and yin. It has been used for eye problems. Its main ingredients include Danshen, Notoginseng and astragalus root, etc.

2.2.5.1 Explanation of the Formula

The main ingredient in this medicine is Notoginseng, which is aided by Danshen, astragalus root, and figwort. The flavor of Notoginseng is sweet and slight bitterly and the nature is warm, which can remove blood stasis, stop bleeding, and free meridians. The flavor of astragalus root is sweet and the nature is slightly warm, and it belongs to the meridian of lung and spleen. It can nourish qi, free meridians, and repel blood stasis without hurting the healthy qi. The nature of Danshen is slightly cold and the flavor is bitter, which can free meridians, remove blood stasis, and promote blood. The nature of radix scrophulariae is slightly cold and the flavor is bitter, sweet, and salty, which can cool blood and promote yin, clear heat and reduce toxins. The combined use of these herbs can improve blood circulation by removing blood stasis and nourishing qi and yin, without hurting the blood.

2.2.5.2 Clinical Application

This medicine can expand the coronary artery muscle, improve microcirculation, and reduce blood lipid and platelet aggregation, and is used for coronary heart disease, angina, eye problems, and other vascular diseases.

2.2.5.3 Pharmacological Research

Compound Xue-shuantong Capsule (CXSC) can expand blood vessels, improve microcirculation, and reduce lipid and platelet aggregation. It is used for the treatment of cerebrovascular disease, coronary heart disease, angina, and a variety of other vascular diseases of the fundus and traumatic bleeding.

Clinical Application

CXSC stops bleeding and repels stasis simultaneously, activating blood circulation without leaving the stasis. It has an anti-inflammatory analgesic effect. The clinical studies showed that it can shorten the clotting time, promote clot lysis, resist thrombosis, improve microcirculation and increase blood flow, and is used for heart, brain, and eye problems of blood stasis and hemorrhagic style.

CXSC and gingko leaf tablets were used together to treat atherosclerosis-related vertigo. The patients receiving the treatment should meet the following criteria: (1) over the age of 50; (2) eye-base arteriosclerosis was above level II; (3) high blood lipid or blood viscosity; (4) cerebral arteriosclerosis proven by either transcranial Doppler (TCD) or cerebral blood flow diagram; (5) no other diseases considered blood stasis type of vertigo by TCM. The patients were treated for 6 months, receiving three capsules/time, 3 times daily for CXSC, and 40 mg/time and 3 times daily for the gingko leaf tablet [81]. The results showed that the treatment reduced TC, TG and LDL, increased HDL, and maintained the above four indexes around the normal level. The combination of the two drugs gave better results than gingko leaf tablets used alone. The mechanism of these drugs might be related to the function of reducing blood viscosity and thrombosis and increasing peripheral blood flow.

CXSC has been used to treat migraine patients, who were selected based on the diagnostic criteria in the 1995 edition of the "Chinese Medical Association Neural Science." Each patient received two capsules of CXSC, and at 50 and 100 min post administration, color TCD ultrasound technology was used to detect the bilateral middle cerebral artery (MCA), and peak systolic velocity (VS) changes in the anterior cerebral artery (ACA) and posterior cerebral artery (PCA). The results showed that MCA, ACA, and VS of PCA were significantly reduced 50 min after the administration, and MCA, ACA, and VS of PCA continued to decrease 100 min after the administration, which was statistically significant. This medicine can increase peripheral, carotid artery and brain blood flow, improve brain circulation, increase blood hypoxia tolerance and inhibit various vascular diseases. The peak function time is from 50 to 100 min after administration, and this drug only has an effect on the flow velocity of the headache side, but has no obvious effect on the normal flow [82].

CXSC has been used for the treatment of vitreous hemorrhage. Patients were administered 2 capsules each time, three times daily, for 2 months. Patients with diabetes and hypertension were also given hypoglycemic and antihypertensive drugs, respectively. Among the 46 cases, 28 % showed an obvious effect, and 52 % showed improvement. Vitreous hemorrhage results from accumulation of blood in the vitreous chamber, which is caused by rupturing of the retinal and uveal blood vessels. Since the metabolism of the vitreous body is low, it can cause proliferative vitreoretinopathy and obvious reduction of eyesight if not treated in time. Meanwhile, surgical treatment of the disease requires certain equipment and technique, which is not available everywhere. CXSC promotes the absorption of vitreous hemorrhage, and improves vision in a simple and practical way [83].

CXSC has a good effect on curing hyphema. The indications also include the concomitant symptoms such as bleeding under the eyelid skin, injury of the eyelid skin, corneal abrasion, iridodialysis, secondary glaucoma, commotio retinae, vitreous hemorrhage, etc. The clinical effects showed that it could quickly promote the absorption of hematoceles and improve eyesight [84].

Diabetic retinopathy is one of the most common complications of diabetes and its cause is related to retinal ischemia and hypoxia. The total saponins of Notoginseng in this medicine stop the α-receptor of vascular smooth muscle cells to control calcium channels, which can expand blood vessels, reduce blood viscosity, promote blood circulation, and improve tolerance of hypoxia. Clinically, the common treatment for diabetic retinopathy is retinal photocoagulation, combined with oral administration of CXSC, or a single treatment with CXSC. The dosage is 2–3 capsules each time, three times a day, continued for more than 3 months. Meanwhile, hypoglycemic and antihypertensive drugs should be administered. The treatment can improve eyesight by more than 1–2 lines, control fundus diseases, and stop the development of retinal hemorrhage, exudation, and microaneurysm. It has good effects on central retinal vein occlusion, central retinal artery and artery branch occlusion, central serous chorioretinopathy, age-related macular degeneration, optic disk vasculitis, etc., hence improving patients' quality of life [85].

In treating glaucoma of controlled intraocular pressure, this drug shows good effects on glaucomatous optic neuropathy. It can broaden the view and improve the visual function of the patients, but the effects are not stable for elderly patients. The pharmacological foundation of controlling intraocular pressure and preventing deterioration of visual function from glaucoma is to improve microcirculation of the optic nerve and ischemia and anoxic conditions, and enhance the excitability of visual cells [86].

Experimental Studies

CXSC has been tested on rats and rabbits, and results showed that it could expand the mesenteric artery and vein, improve blood circulation, prolong the survival time of tissues under hypoxia, resist thrombosis, promote clot lysis, and shorten the clotting time. The above function is the result of synergy of all herbs. It showed that Notoginseng could increase thrombin levels, shorten the clotting time, and also promote clot dissolution, while diminishing inflammation and stopping itching. The other herbs in this medicine can expand blood vessels, improve blood circulation, resist thrombosis, clear heat, nourish yin, etc. [87].

In a study on the protection of ischemic heart muscle in dogs, CXSC has been shown to enhance cardiac muscle systole, maintain blood pressure stability after acute myocardial infarction, make glucogen and lactic acid contents close to the levels of the normal control group, make the contents of MDA and LDH lower than those of ischemic group, and the activity of SOD higher than those of the ischemic group. These data revealed that CXSC can maintain cardiac pump function, promote blood and oxygen supply, normalize the aerobic oxidation of sugar, regulate cardiac muscle injury, and reduce the leakage of LDH. Myocardial ischemia causes the blockage of aerobic oxidation, the reduction of ATP, and disorder of the ion gradient and membrane stability; meanwhile, the generation of oxygen free radicals would damage the cardiac muscle. CXSC can enhance aerobic oxidation, increase the activity of SOD, maintain ATP generation and membrane stability, reduce the content of MDA, and remove free radicals [88].

Another study has shown that CXSC can improve the ischemic change of epicardial electrocardiograms of myocardial infarction in experimental dogs. CXSC can significantly reduce the degree of myocardial ischemia ($\sum$-ST) and the area of the myocardial ischemia (N-ST). Compared with the ischemic group, the $\sum$-ST and N-ST at the corresponding time points are significantly reduced. So, CXSC's ability of resisting myocardial ischemia is confirmed. The pharmacological pathway of this medicine is possibly that the medicine can activate blood circulation, remove stasis, and promote blood and oxygen supply, which normalizes the pump function of Na^{+}-K^{+} in the membrane of cardiac muscle cells, and promotes the index of cardiac electric activity of the epicardium in ECG [89].

In a study on hypoxia tolerance, it has been shown that 1 h after administration of CXSC, both the atmospheric hypoxia tolerance and the ability of isoproterenol to increase hypoxia tolerance in mice were increased, the survival time of mice was prolonged, and the electrophysiology of myocardial ischemia was maintained. These results suggest that CXSC can reduce myocardial oxygen consumption, improve metabolism and improve anoxia tolerance.

2.2.6 Guanxinning Injection

This medicine contains Danshen and Sichuan lovage root. The two herbs together can improve blood circulation, remove blood stasis, nourish the mind, free meridians, free qi, and stop pain.

2.2.6.1 Explanation of the Formula

The flavor and nature of Danshen is slightly cold and bitter, which can invigorate blood, remove blood stasis, and nourish blood. According to TCM classics, although Danshen by itself is only one herb, it has the functions of four different herbs. The flavor and nature of Sichuan lovage root is pungent and enters the liver and gallbladder meridian, which is the qi medicine in blood, and can promote blood circulation, free qi, and repel depression, wind and pain, which is used for stopping pain, reducing pressure, expanding the coronary artery muscle, increasing coronary artery muscle blood flow and resisting myocardial ischemia. The combination of the two herbs can easily remove blood stasis and promote blood circulation, greatly free qi and meridians, effectively nourish the mind and stop pain, and largely enhance the clinical effects without causing too much dryness.

2.2.6.2 Clinical Application

It is used for CCVD such as coronary heart disease, angina, diabetic nephropathy (DN), gynecological and pediatric diseases.

2.2.6.3 Pharmacological Research

Cardiovascular Effects

The clinical effects of Guanxinning injection (GI) treatment for coronary heart disease and angina have been observed. GI was delivered to coronary heart disease and angina patients at 20 ml/day, with a treatment lasting 15 days. Those with high blood pressure used hypotensive drugs, those with acute cardiac insufficiency used digitalis preparations and diuretics, and those with angina used nitrate medications. The results showed that after administration of the drug, the same level of fatigue did not cause or reduce the frequency of angina, and the ECG ST segment rebounded [90].

Another research group studied the effect of combining conventional therapies, such as a low-salt and low-fat diet and using aspirin and Betaloc, with GI and ginkgo leaf injections. Each administration (intravenous drip) was 10 ml per day continued for 14 days. The results showed that the incidence of angina, ST-T (ECG) and WBV were greatly improved over a single conventional therapy [91].

In treating cough, cyanosis, dyspnea, oliguria, edema of lower extremities, lung noise, and hepatojugular reflux of some pulmonary heart disease positive patients, GI relieved the above symptoms in 60 % of the patients, and the total effective rate was 90 % [92].

Effects on Microvessels

DN is one of the chronic complications of diabetes microvascular disease. Combining conventional therapy (like control of blood pressure, sugar, lipid, weight, and other indexes) with GI showed that fasting plasma glucose, postprandial 2 h glucose, glycosylated hemoglobin, blood urea nitrogen, serum creatinine, blood and urine β2-microglobulin, morning urine microalbumin, urinary protein excretion, fibrinogen, and whole blood viscosity were significantly reduced after two periods of treatment. Meanwhile, the thromboplastin time and thrombin time were prolonged, but plasma viscosity was not changed. The dosage was 20 ml/time, once daily, for a course of 14 days [93].

To sum up, GI can reduce platelet surface activity, inhibit platelet aggregation, improve the activity of fibrinoclase, adjust blood rheology, and improve immune function. Danshen can inhibit α and M receptors, expand blood vessels, reduce blood viscosity, increase renal blood flow and glomerular filtration rate, inhibit immune complex allergic reactions and the form of Thromboxane, resist blood capillary spasm and platelet aggregation, free the blockage of renal blood vessels, protect GBM and promote diuresis, and also has calming, antibacterial, and anti-inflammatory effects. Sichuan lovage root can alleviate the spasmodic smooth muscle of the trachea, inhibit the accumulation of Ca^{2+} in cells, reduce the activity of ospholipase, ATPase, cathepsin, lipoxidase, Cyclooxygenase and TXA_2, clear oxyradicals, resist LPO, maintain the balance of Thromboxane (A_2) and prostacyclin, expand blood vessels, reduce blood viscosity, improve microcirculation, and reduce the blood anoxic symptoms of tissue. The combination of the two herbs can even better change the abnormal hemodynamics and hypercoagulative state of Glomeruli, reduce urinary albumin excretion, and positively affect diabetes and renal diseases. Reduction of pulmonary hypertension and right heart load has a therapeutic effect on coronary heart disease, angina, and pulmonary heart disease.

2.2.7 Bushenyishou Capsule

This capsule contains famous Chinese herbs like ginseng, glossy ganoderma, and Danshen, etc. In TCM, these herbs can nourish qi and kidneys, and this medicine can relieve many clinical symptoms related to kidney qi-deficiency and improve cellular immune function and to increase the quality of life of the patients. Recent research shows that this capsule can resist LPO and stress and regulate immune function, adrenocortical function, and hypogonadism.

2.2.7.1 Explanation of the Formula

In this formula, red ginseng and licorice can nourish qi, and the lung and spleen. Glossy ganoderma, lycium fruit, epimedium and polygonati rhizome can nourish qi, yin, and kidneys. Fleeceflower root and Danshen can nourish and activate blood, and clear and calm the mind. Pearl can calm the liver and subdue yang, tranquilize the heart and repel horror. All together, these herbs nourish kidney and qi.

2.2.7.2 Clinical Application

It is used for clinical symptoms of kidney qi-deficiency (like fatigue, short breath, spontaneous sweating, aching and weakness at the waist and knee, and nocturia), has a therapeutic effect on male ED, and improves the quality of life. It can improve cellular immune function, has a significant antioxidant effect, and can delay aging in experimental animals.

2.2.7.3 Pharmacological Research

Clinical Research

Kidney Qi-deficiency

80 patients with kidney qi-deficiency symptoms (shortness of breath, mental and physical fatigue, spontaneous sweating, weak knees, dizziness, tinnitus, long urination, incomplete urination, nocturia, loss of sex drive) were treated for 1 month. The T achroacyte subgroup (CD3, CD4, and CD8) was observed before and after treatment. The results show that this medicine can improve the clinical symptoms of kidney qi-deficiency and the quality of life of the patients, improving cellular immune function, and increasing serum levels of CD3, CD4, and CD8.

Atypical Organic ED

The patients who scored below 21 on the International Index of Erectile Function and had been sick for 3 months to 2 years were treated with this medicine for 8 weeks. Compared with the control group, the treated group showed a significant improvement. Within the treated group, patients with mild or moderate ED had a better improvement than those with serious ED. It suggests that the level of ED is related to the medical effect; the milder the symptoms are, the better the therapeutic effect of the drug will be [94].

Immune Function of the Elderly

Three groups of people were studied; a normal adult group, normal elderly control group, and normal elderly treatment group. The normal elderly treatment group was given six capsules each time, three times daily for 2 months. No drugs were given to the other two groups. Blood samples were collected to check the Ts function, ConA, and PHA. The results showed that after treatment, PHA returned to normal, ConA increased to close to normal adult levels, and the Ts function also increased, but these changes were statistically insignificant.

Experimental Study

Anti-Stress Effect and Influence on Immune Function

The effects of the medicine on the tolerances of high temperature, cold, hypoxia, swimming endurance, anti-inflammation, the peritoneal macrophage phagocytic function, the production of antibodies, and the impact of delayed hypersensitivity were studied on mice. The results showed that after several administrations, the drug can improve the mice's ability to tolerate high and low temperatures, survive longer in high or freezing temperatures, and that the mice's tolerance of cold temperature was stronger than of high temperature. The drug extended the time of death in mice in occlusion hypoxia at atmospheric pressure, and similarly in the treated with isoproterenol to increase their cardiac load. The drug strengthened the endurance of the mice and prolonged their swimming time. The drug had an inhibitive effect on ear edema in mice induced by croton oil, but the effect was not as good as that of aspirin.

In a test of macrophage cell function in the abdominal cavities of mice, the experimental animals were administered continuously for seven days, injected with 0.2 ml of hepatic glycogen solution on the 5th day, 1 ml of 5 % chicken red blood cells (CRBC) with normal saline solution 1 h after the last administration. The animals were executed 1 h later, and the percentage of macrophage cells was counted. The results showed that this medicine can improve the phagocytic activity of mouse peritoneal macrophages, as both the phagocytic percentage and phagocytic index increased significantly.

Mice were injected with CRBC intraperitoneally, and the post-immune serum was collected and mixed with guinea pig serum (mouse: guinea pig = 1:10) and 10 % CRBC. The hemolysis value was measured after incubation at 37 °C. The drug's effect on of antibody generation was observed and compared. The results showed that this capsule can significantly promote hemolysis of the antibody against CRBC.

A high dose of this capsule could significantly inhibit DNFB-induced delayed hypersensitivity (DTH). The effects of medium and low dosages were not remarkable, but they could inhibit DTH induced by cyclophosphamide [95].

Antioxidation and Neurohumoral Regulation

An anti-aging experiment was carried out on *Drosophila melanogaster* using culture medium containing 0, 0.5, 1, and 5 % of the medicine. The results showed that this medicine could prolong the life of *D. melanogaster*. Compared with the control group, the longest and the average age increased (with no difference between genders). This suggested that this capsule could delay the aging and death of *D. melanogaster* to some extent [96].

The drug's effect on the antioxidation activity of aged mice was studied. Five groups of mice aged 18 months, were fed a solution containing different concentrations of this drug, vitamin E or drinking water, and young control mice were fed drinking water. The administration was once per day, six times per week, and continued for one and a half months. 1 h after the final administration, the eyeballs were excised to collect blood, and the livers and hearts were quickly excised to prepare serum, liver and myocardial extracts. The activity of SOD in serum and liver tissue was measured by the hydroxylamine method, the content of LPO was measured with the thiobarbituric acid method, the content of glutathione peroxidase in red blood cells was measured with the improved Hafeman method, and the content of lipofuscin in heart muscle was measured with fluorescence spectrophotometry. The results showed that this medicine could obviously increase the activity of SOD in serum and liver tissue and glutathione peroxidase in red blood cells. It could reduce the content of LPO and lipofuscin. Since the levels of LPO and lipofuscin are positively related to the degree of aging, the results suggest that long term intake of this capsule can delay aging. This function may be related to the increased activity of SOD and glutathione peroxidase and reduced LPO, which prevents injury to the biomembrane.

The drug's effect on vitamin C levels in the adrenal glands of SD rats was studied by randomly dividing the rats into six groups (half male and female), and feeding them the drug or drinking water, once daily. The administration continued for 1, 7, and 14 days. The results showed that this drug could reduce the content of vitamin C in adrenal glands and excite the adrenal cortex, but a one-time administration showed no obvious effect. The pituitaries of the control and treatment groups were extracted, and the extracts could reduce the content of vitamin C in the adrenal glands. It was suggested that this might be caused by adrenocorticotropic hormone (ACTH) contained in the pituitary that could quickly excite the adrenal cortex. Since this medicine could increase the content of ACTH in pituitary, it could delay aging.

The effects of Bushenyiqi Capsule on T-cell apoptosis: The model and treatment groups (SD rats) were injected 10 mg/kg CORT to make the model. Meanwhile, they both were given the drug once daily, for 14 days. After that, their spleens were collected in germ free conditions and the T-cells were isolated and purified. The whole process was finished and cells were cultured on six-well incubation plate pre-coated with CD3 monoclonal antibody. 14 h later, they were prepared as specimens for electron microscopy. TUNEL labeled apoptotic cells were detected by flow cytometry, and the rate of apoptosis was calculated with area integral methods.

Morphological studies showed that CD3 monoclonal antibody could induce apoptosis of T-cells of any group. The apoptotic cells had typical nucleolus changes; nuclear chromatin formed chromatin clumps which showed high electron density under electron microscopy. The test of TUNEL by flow cytometry and the mark of apoptosed cells showed a significant difference between model and normal control groups on T-cell apoptosis. Results showed that Bushenyishou Capsule could downregulate the inducement of T-cell apoptosis [97].

Another test showed that continuous administration of this capsule for 1 or 2 weeks could obviously increase serum corticosterone, but that a single administration had no obvious effect.

In a clinical study on the effect of this capsule on SOD, MDA, and GSH-Px levels in kidney qi-deficiency patients, the results showed that the levels of SOD and GSH-Px increased and MDA decreased after the treatment. The effects of Bushenyishou capsule were more obvious than those of Guifudihuang Pill. The experiment was conducted as follows: 80 kidney qi-deficiency patients were randomly divided into two equal groups, the treatment group and the control group. The treatment group had Bushenyishou Capsule (two capsules per time, three times per day, 0.26 g per capsule), the control group had Guifudihuang Pill (6 g per time and twice daily), and the treatment cycle for both groups was 30 days.

2.2.8 Huganning Tablet

This medicine is a commonly used Chinese medicine that can clear heat and dampness, relax liver, stop pain, and reduce jaundice and ALT. This is used for acute and chronic hepatitis with the effect of improving blood circulation in the liver, inhibiting liver fibrosis, and increasing immune function. The ingredients of this medicine are stringy stonecrop, giant knotweed rhizome, Danshen, and glossy ganoderma.

2.2.8.1 Explanation of the Formula

The sarmentosin of stringy stonecrop can promote metabolism in liver cells and enhance the liver's detoxification ability. Giant knotweed rhizome can repel wind and dampness, stop pain, stasis, cough and phlegm, reduce lipids in the liver, free fatty acids in serum, ALT, AST, etc. Glossy ganoderma can nourish the qi in five organs, promote body energy and restore the vital qi, inhibit the HBV, regulate immune function and inhibit platelet aggregation, etc. Danshen can nourish and activate blood, free meridians, and repel stasis. The mixture of these can clear heat, repel dampness, protect and free the liver, stop stasis, and stop pain.

2.2.8.2 Clinical Application

This medicine can relieve common symptoms related with hepatitis (fatigue, abdominal distension, hypochondriac pain, dark urine, etc.). Clinical observations showed that this medicine could also cure hepatitis and liver damage caused by alcohol-induced liver disease and fatty liver, etc. It also has an anti-hepatic fibrosis function.

2.2.8.3 Pharmacological Research

Research on the protective effects of Huganning Tablet on liver damage caused by carbon tetrachloride revealed that this medicine could obviously reduce the level of ALT and AST in the serum of poisoned mice. Pathological examination of the liver tissues showed that carbon tetrachloride could cause edema or diffuse edema of the liver cells centered around the central vein, with focal necrosis and patchy necrosis of liver tissue, inflammatory cells, neutrophile granulocyte, and lymphocyte infiltration, surrounding the blood vessels and lymphatics. Tissues from mice treated with Huganning Tablet only showed edema of the liver cells and inflammatory cell infiltration (surrounding the blood vessels) and slight necrosis of liver cells. The results suggest that this medicine could protect the liver tissue of mice poisoned with carbon tetrachloride [98].

2.2.9 Xinmaitong Tablet

This medicine is used for the prevention and treatment of cardiovascular diseases. It can promote blood circulation, remove stasis, free meridians, nourish the heart, improve microcirculation, increase coronary blood flow, reduce myocardial oxygen consumption, lower blood pressure and blood lipids, and soften blood vessels, and is suitable for thromboembolic diseases.

2.2.9.1 Explanation of the Formula

There are 11 main ingredients in XT; they are hairy holly root, cassia seeds, gambir plant, angelica, Danshen, Notoginseng, two-toothed achyranthes root, pueraria root, pagoda tree flower, common self-heal fruit-spike, etc. Danshen and Notoginseng can repel stasis and stop pain. The nature and flavor of two-toothed achyranthes plant is even, bitter and acidic, and it can repel stasis and free meridians, and promote the blood to the lower part of the body. The nature and flavor of hairy holly root is cold and bitter, and can clear heat and toxins, free blood and meridians, and reduce the heat in stasis. The nature and flavor of angelica is warm, sweet and pungent, and can nourish and free blood. The above herbs together can nourish, activate blood, free meridians, and reduce stasis without affecting the body energy. Radix puerariae can clear heat, generate body fluid and stop spasms. Common self-heal fruit-spike is cold and pungent, and can clear heat and improve eye function. Cassia seed has the same function as common self-heal fruit-spike. The flavor and nature of gambir plant is cool and sweet, and can clear heat and calm the liver, repel wind and reduce horror. Pagoda tree flower can cool and stop blood, and clear heat in the liver. The combined use of these can promote blood circulation, remove stasis, free meridians, and nourish the heart.

2.2.9.2 Clinical Application

This medicine can improve microcirculation, increase coronary blood flow, reduce myocardial oxygen consumption, lower blood pressure and blood lipids, soften blood vessels, and is clinically used for hypertension and hyperlipidemia

2.2.9.3 Pharmacological Research

Clinical Research

In a study on reducing myocardial oxygen consumption, XT was used to treat the patients of myocardial infarction, coronary heart disease, myocardial ischemia, and angina, and the myocardial oxygen consumption and blood supply were observed. Rate-pressure product (RPP) was used to represent the myocardial oxygen consumption index, and $\sum$ST and NST were used to express the total reduction of ST part of conventional 12 lead and the lead index of reduction (≥0.25 mm), respectively. This medicine was administered four pills at a time, three times a day. The control group used conventional western drugs, isoamyl nitrite, Compound Danshen Pill and Aspirin Enteric-coated Tablet. The dose of isoamyl nitrite was 10 mg, Compound Danshen Pill was three pills three times a day (orally), and Aspirin Enteric-coated Tablet was 75 mg once and three times daily (orally). The treatment cycle lasted 1 month. The results showed that the level of myocardial ischemia was close between the treatment and control groups, but the RPP of the treatment group was better than the control group. This revealed that this medicine could reduce myocardial oxygen consumption and protect the Cardiac Muscle [99].

Experimental Research

The effect of the medicine on carotid artery thrombosis and clotting time was studied, and the results showed that XT had a strong antithrombotic effect. New Zealand rabbits were divided into an XT group, Compound Danshen group, and saline group. Each group had six rabbits. Different drugs were administered 3 days before the experiment, three times per day for 3 days. On the 4th day, drugs were given only once, then the animals were anesthetized with urethane, a 4 cm segment of carotid artery was selected, and its two ends were clamped, stopping blood flow in the segment. 3 cm of No. 1 silk thread was put into the selected artery, and the end of the thread farther from the heart was floating in the blood. At this time the clamps were loosened and blood began to flow. 2 h later, this 4 cm segment was departed and cut open. The thread was also taken out and put on filter paper to absorb the blood. Then, the threads with thrombus were weighed to figure out the net weight of thrombosis by deducting the weight of the threads. Meanwhile, once again carotid artery sampling and blood clotting time was measured by slide method. Results showed that the thrombosis weights of XT groups with intermediate and high dosages were significantly lighter than those of the saline group, and the time of clotting was also prolonged, which was the same as the Compound Danshen group [100].

2.2.10 Ningxinanshen Capsule

This medicine can clear heat and depression, calm the mind, and comfort the chest and heart which can largely relieve symptoms such as palpitations, insomnia, irritability, hot flashes, night sweats, zygomatic redness, and thirst, which are caused by yin deficiency and vigorous fire. There are 11 main ingredients in this medicine, including Danshen and coptis rhizome, etc.

2.2.10.1 Explanation of the Formula

There are 11 main ingredients in Ningxinanshen Capsule (NC): coptis rhizome, amber, acorus, thin-leaf milkwort root, poria, Danshen, licorice, red jujube, wheat, magnet (calcined), and mother-of-pearl. The nature and flavor of coptis rhizome is bitter and cold, which is used for irritability and insomnia caused by heat of heart. The nature and flavor of amber is even and sweet, and can clear and calm the mind. The nature and flavor of magnet is cold and pungent, and can oppress yang, contain qi, oppress horror, and calm the mind. The nature and flavor of mother-of-pearl is salty, sweet and cold, and can calm liver, oppress yang and calm the mind. Thin-leaf milkwort root and acorus can repel phlegm, fresh the mind, regulate heart and kidney, and nourish heart. Licorice and wheat can improve spleen and calm the mind. Danshen and red jujube can nourish blood and calm the mind. The combination of them can regulate the emotion and spirit, nourish yin and calm the mind.

2.2.10.2 Clinical Application

It is used for menopause syndrome and neurasthenia, etc.

2.2.10.3 Pharmacological Research

Insomnia is a common sickness caused by yin-deficiency with vigorous fire, and it can be treated with NC with good effect. Taking four pills of NC orally each time, three times per day, and continuously taking this medicine for 4 weeks can greatly relieve the symptoms caused by yin-deficiency with vigorous fire. Of 32 patients with the symptom of insomnia and treated with NC, 12.5 % slept normally, accompanied with the disappearance of other symptoms, without recurrence for more than 3 months; 43 % of them had better sleep (time to fall asleep <20 min, sleep time >5 h), and 40 % fell asleep in a shorter amount of time for a prolonged sleep time, with the disappearance of other problems. The depth of sleep increased and the mental state was good after waking up [101].

There were 40 cases which western medicine diagnosed with menopausal syndrome and TCM diagnosed with yin-deficiency and excess of heart fire. They were treated with NC for 30 days (three times per day, four pills each time). 15 % of them had improved menopausal syndrome symptoms. Another 15 % had significant improvement of symptoms, and 67 % of them had their symptoms eased. It was shown that this medicine had better effects on female menopausal syndrome (western medicine), yin-deficiency and excess of heart fire [102].

2.2.11 Ningshenbuxin Tablet

This medicine is used for treating the symptoms of dizziness, tinnitus, palpitation, forgetfulness, insomnia, etc. caused by Kidney yin-deficiency, and has the function of nourishing the blood, liver and kidney and calming the mind.

2.2.11.1 Explanation of the Formula

NT contains 10 main ingredients, including Danshen, schisandra, fresh rehmannia, prepared rehmannia root, prepared privet fruit, eclipta, mother-of-pearl, etc. Danshen can nourish and activate blood, free meridians, repel distress, and calm the mind. The nature and flavor of schisandra is acidic and warm, and can restrain, nourish, and calm mind and body. Prepared privet fruit, prepared rehmannia root and eclipta can **nourish yin** and blood, and supplement and boost liver and kidney. The mixture of prepared privet fruit, schisandra, prepared rehmannia root and eclipta together can **nourish yin**. Cortex albizziae and mother-of-pearl can free qi, repel depression and calm the mind. Acorus is pungent, bitter, warm and aromatic, and can nourish the heart, liver, eyes, ears, and throat. The combination of these can nourish the liver and kidney, calm heart, and mind, and in traditional Chinese terminology is the balance and the mutual promotion of coordination between water and fire so as to calm the heart and mind.

2.2.11.2 Clinical Application

This medicine is suitable for symptoms such as insomnia, forgetfulness, heart palpitations, dizziness, tinnitus, etc., caused by liver and kidney yin-deficiency.

2.2.11.3 Pharmacological Research

Clinical research was performed on the influence of this medicine on insomnia; there were 300 cases diagnosed as insomnia or restlessness, with profuse dreaming and easiness to wake from sleep, tinnitus, dizziness, dry mouth and throat, palpitations, sweating, forgetfulness, weak waist (men), nocturnal emission (men), irregular menstruation (women), red and dry tongue (women), etc. The medicine was administered three times daily, six pills each time, and continued for 30 days. 35 % of patients improved their sleep duration to normal levels or to more than 6 h during the night, and 93 % had their symptoms relieved. The clinical effects were better than those of oryzanol pill and doxepin pill.

According to TCM theory, insomnia is caused by yin-deficiency in the kidney and liver; the yin and yang are in disorder, yin is in short supply, and yang is out of control. Yin-deficiency results in vigorous fire, which goes up and disturbs the mind. Because there is no interaction between the heart and the kidney, the mind is uneasy, resulting in insomnia. NT can nourish the yin of liver and kidney and calm the heart and mind, reaching a balance between water and fire and establishing the interaction between heart and kidney, returning sleep back to normal [103].

2.2.12 Yangxinshi Tablet

This medicine contains 13 ingredients, including astragalus root, codonopsis root, Danshen and pueraria root, etc. It is a Chinese medicine used to treat coronary heart disease. It can improve the blood and oxygen supply to ischemic myocardium, ease the symptoms of myocardial ischemia, and has two-way regulations on abnormal heart rate. According to TCM, this medicine can nourish qi, activate blood, free meridians, and stop pain.

2.2.12.1 Explanation of the Formula

There are 13 main ingredients; astragalus root, codonopsis root, Danshen, pueraria root, epimedium, Chinese hawthorn, rehmannia, angelica, coptis rhizome, corydalis rhizome, glossy ganoderma, ginseng, and licorice. The chief herbs in this formula are astragalus root and codonopsis root, which can powerfully supplement the original qi, and restore the pulse and stem desertion. The chief herbs are aided by the deputy herbs Danshen, radices rehmanniae and angelica, etc., which can activate blood circulation and nourish yin. Chinese hawthorn and Danshen can promote blood circulation and remove stasis. The nature and flavor of glossy ganoderma is even and bitter, not toxic, and it can nourish qi of the heart, activate blood, and promote meridians. Rhizoma corydalis belongs to the heart meridian, and can activate blood, free qi and stop pain. Modern pharmacological studies show that astragalus root can improve heart function and increase cardiac output effect. Danshen and angelica can resist platelet aggregation and improve peripheral microcirculation. Hawthorn can expand the blood vessels and reduce blood lipids. Glossy ganoderma can nourish the body and kidney and free urine. Used together they can activate the blood, stop pain, and nourish qi.

2.2.12.2 Clinical Application

This medicine is used for coronary heart disease, angina, cardiac arrhythmia and heart diseases associated with diabetes, and shows good effects on ischemic cerebro-vascular disease.

2.2.12.3 Pharmacological Research

Clinical Research

Treatment of Coronary Heart Disease and Angina

All patients were diagnosed according to the criteria established by the WHO for angina pectoris. The effective criteria are according to the Guidelines for Clinical Research of cardiovascular system Drugs, published by the Pharmaceutical Council of the Ministry of Health of the People's Republic of China (1993 version). The clinical effect criteria are, Significant effect: The occurrence of angina disappeared, or the incidence is 90 % less at the end of treatment than before. Effect rate: Significantly alleviated the symptoms of angina, or the incidence is 50–89 % less than before treatment. No effect: There was no alleviation of angina or the incidence, or it is 49 % less than before treatment. The effects of ECG, Effect: ST-T returns to normal or close to normal, stress ECG test changes from positive to negative. No effect: no improvement on ST-T, stress ECG still positive.

Yangxinshi Tablet (YT) was used to treat patients for 30 days (three pills per time, three times per day). The clinical results were that 15 cases showed a significant effect, 22 cases showed improvement, and 3 cases showed no effect. The total effective rate was 92.5 %. The ECG results were that a significant effect was found in 10 cases, improvement in 15 cases, and no effect in 15 cases, and the total effective rate was 62.5 %. The hemorrheological study showed that whole blood viscosities at high shear (HBV), low shear (LBV), packed cell volume (PCV), plasma viscosity, and platelet aggregation rate were all reduced. The indexes of nailfold microcirculation showed improvement after treatment [104].

Regarding improvement of the clinical symptoms of CHD and angina, the document showed that this medicine (1.8 g, three times daily, continued for 2 months) could relieve symptoms of chest tightness, heart palpitations, shortness of breath, paroxysmal retrosternal pain, arrhythmia, and reduce the indexes of serum TC and triglycerides, etc. [105].

Isosorbide dinitrate was combined with YT to treat angina for 2 months, and the blood, urine, blood lipids, liver function, kidney function, blood pressure, heart rate and ECG, and the incidence of angina were observed. The results showed that the effects of the combination were much better than when isosorbide dinitrate was used alone [106, 107].

YT and isosorbide dinitrate were used to treat patients for 3, 6, and 10 weeks. The results showed that the longer the treatment took, the better the results were. It showed that the combined use of the above two medicines could have a longer effect on stable angina [108].

YT was also used for diabetic CHD and angina. The results showed that YT could relieve the symptoms of diabetic CHD and angina, and was best for type I and II angina and had significant effects on the symptoms of CHD (headaches, heart palpitations, chest tightness, shortness of breath, and weakness, etc.), and the corresponding ST changed obviously. It could also obviously reduce blood sugar and lipids [109, 110].

YT could be used to treat ventricular premature beat associated with CHD, myocarditis and hypertensive heart diseases. YT was given 4 pills each time, three times per day, and for 30 days as one treatment cycle. The results showed that the average number of hourly premature beats in six of 30 patients was reduced by more than 90 % or totally disappeared after the treatment, and in 20 patients, the reduction was more than 60 % after the treatment [111].

Treatment of Cerebrovascular Diseases

Patients of cerebral infarction were treated 5 h to 16 days after the onset, and the conventional therapies were Danshen injection, Mannitol injection, ATP injection, CoA injection and cerebrolysin vials. Meanwhile, nimodipine and YT (5 pills each time, three times per day) were administered orally. The treatment cycle was 4 weeks. The results showed that the main symptoms, body function, and the degree of disability were improved. The clinical effects were better than in the groups without YT [112].

YT has affirmative effects on cerebral arteriosclerosis and hypercalcinuria. It could treat dizziness, headache, fatigue, sleepiness, forgetfulness, and improve blood lipids, and blood rheology. Published results showed that it could significantly relieve the symptoms mentioned above, and the total effective rate was 87.80 % in a total of 82 patients after four treatment cycles. Side effects on liver and kidney function were observed by checking the blood and urine, with no abnormal changes found [113].

The patients were sent to the hospital within 48 h after the occurrence, and head CTs or MRIs were checked and the blood rheology and brain blood flow were measured before and after the treatment. 100 patients were randomly divided into two groups; the control group had 20 ml of Compound Danshen **injection** (added with 250 ml of NS, intravenous, once daily), the treatment group had Compound Danshen **injection** plus YT (5 pills each time, three times daily) added into their prescription, and both groups were treated for 2 weeks. The results showed that blood supply to the frontal lobe, parietal lobe, temporal lobe, occipital lobe, area centralis, brain stem, the focus areas and the surrounding area increased after treatment, and the indexes of the blood rheology (whole blood viscosity, plasma viscosity, hematocrit value, platelet adhesion rate, erythrocyte sedimentation rate and maximum platelet aggregation rate) improved. The blood flow in each cerebral area (excluding area centralis) and the indexes of the blood rheology of the treatment group of YT were better than the control group [114].

Experimental Research

The Effect of YT on Pharmacokinetic Parameters of Ciclosporin A

Cyclosporine A (CsA) is an immunosuppressive agent and at present is commonly used to prevent organ transplant rejection and in the treatment of immune diseases. CsA can interact with many drugs, and its bioavailability in different individuals is different. It also has liver and kidney toxicity. Long-term consumption of CsA may cause symptoms of arrhythmia and high blood pressure. Clinically, YT was usually used to offset the side effects caused by CsA. In the study of YT's effect on CsA, results showed that YT could cause an increase in CsA concentration in the blood and shorten the elimination phase half-life. The study was done as follows: Rabbits were without food and water for 12 h. The CsA group was given medicine at 30 mg/kg, blood samples (2.0 ml) were collected at intervals of 0.5, 1, 2, 3, 4, 6, 8, 10 and 12 h, and CsA blood concentration was measured by RP-HPLC. After one week, YT (25 mg/kg) was delivered to the rabbits for 3 days, CsA (30 mg/kg) was given 0.5 h after the 4th day of administration, blood samples (2.0 ml), were collected at the same time points of the CsA group, and CsA was again measured by RP-HPLC. The pharmacokinetic parameters were calculated according to pharmacokinetic procedure. The results suggest that the blood concentration of CsA during the combined use of the two medicines should be monitored and the dosage adjusted to prevent the occurrence of side effects from CsA [115].

2.2.13 Rukuaixin Tablet

This medicine is a Chinese medicine which can treat hyperplasia of the breast. It is effective against unilateral or bilateral breast pain and against enlarged and hardens breast bumps. It can also relieve pain, soften the lumps of the breast and partly inhibit hyperplasia. The medicine can be used alone or combined with Western medicines. The main ingredients are tangerine leaf, Danshen, Chinese honeylocust spine, cowherb seed, Sichuan chinaberry, earthworm, etc.

2.2.13.1 Explanation of the Formula

The main ingredients are tangerine leaf, Danshen, Chinese honeylocust spine, cowherb seed, Sichuan chinaberry, earthworm, etc. Tangerine leaf belongs to the liver meridian, and it can free liver and qi, and reduce phlegm, bumps, and toxin. Danshen can activate blood circulation, remove stasis, cool blood, and repel carbuncles. The nature of cowherb seed is even and the flavor is bitter, and it can free the blood and meridian, and reduce bumps and pain. The nature of earthworm is cold and the flavor is salty, and it belongs to the meridians of liver, spleen and lung. It can clear heat, reduce toxins, free meridians and stop spasms. The above four medicines work together to activate blood circulation, remove stasis, and free qi and meridians. Sichuan chinaberry belongs to the liver and stomach meridians, and can repel liver heat, free qi, reduce depression, and stop pain. The nature of Chinese honeylocust spine is warm and the flavor is pungent. It can reduce swelling, draw out pus, remove abscesses, and is suitable for mammary abscesses. The mixture of these herbs can free the liver and meridians, soften hard lumps, dispel nodes, activate blood circulation, and reduce stasis.

2.2.13.2 Clinical Application

It is used for the treatment of breast hyperplasia and related pain, hard lumps and nodes.

2.2.13.3 Pharmacological Research

In numerous clinical study reports, there have been 3 ways of using RT to treat breast hyperplasia: alone, combined with other TCM drugs, and combined with Western drugs.

There is a report showing the use of RT alone to treat 80 cases, with five pills each time, three times per day, for 2 months. The cure rate was 45.0 %, the significant efficacy rate was 41.2 %, and the efficacy rate was 10.0 % [116].

Another report showed that 60 patients of breast hyperplasia were treated with RT alone. After 1 month, the cure rate was 25 % and the significant efficacy rate was 36 %. After 2 months, the cure rate was 46 % and the significant efficacy rate was 25 %. After treatment for 3 months, the cure rate was 65 % and the significant efficacy rate was 25 % [117].

Some reports showed RT combined with antiphlogistic tablets (0.6 g/time) to treat 340 breast hyperplasia patients with 1 month to 3 years of disease history. The results: the cure rate was 32.0 % (128 cases), the significant efficacy rate was 45.0 % (180 cases) and the efficacy rate was 19.25 % (77 cases) [118].

In an integrated TCM and Western medicine treatment of 820 patients, tamoxifen (10 mg per time and twice per day) and RT (five pills each time and three times daily) were used, and the treatment time was 6–18 months (average of 8 months). The cure rate was 93.54 % (767 cases), partial cure rate was 6.46 % (53 cases) and the total efficacy rate was 100 %.

Tamoxifen is an estrogen receptor antagonist, competing with the hormone in the body for the receptor, thus inhibiting hormonal stimulation of the hyperplasia of the glandular tube and the surrounding tissues. Clinically, it can quickly relieve the pain and soften the lumps of the breast. However, once the medication is stopped, the recurrence of the disease is more likely. Also, tamoxifen's efficacy is not satisfactory in moderate to severe hyperplasia. Tamoxifen can also cause side effects such as menstrual disorders, even amenorrhea. Long term intake of RT has few side effects, except for occasional gastrointestinal tract discomfort, so it is better to take RT after meals. The combined use of the two medicines can reduce side effects and improve efficacy [119].

The above summarized the information about Danshen-containing Chinese patent medicines approved by the Chinese government, which have special production sites and are used clinically. Many of them have been studied extensively in recent years. There are more than a hundred formulas in *Chinese Pharmacopoeia* of which Danshen is the main ingredient or is one of many components. They are not included in this article.

Danshen is currently widely used, and its research is deepened. However, some problems need to be pointed out. First, the production processes and technique levels are uneven, which makes quality control of the product difficult. Second, the lack of supervision of the raw material sources and quality is serious. Third, the contraindications of clinical use of Danshen are not very clear, and the studies in this area need improvement. There are some adverse reactions in the clinical application of Danshen which should be taken seriously. So, further research is needed and the data should be reviewed and integrated so that the foundation of Danshen's clinical application is solid.

References

1. Shang HC. A comparative pharmacodynamics research on the different proportion of danshen root with sanchi—establishment of baseline geometric proportion increasing and decreasing design. Dissertations of master postgraduate of 1999, 2002, 4.
2. Shang HC, Zhang BL, Gao XM, et al. A comparative pharmacodynamics research on the different proportion of danshen root with sanchi. The second world seminar of integrated Chinese and Western medicine dissertation abstracts.
3. Shang HC, Zhang BL, Gao XM, et al. Tianjin J Tradit Chin Med. 2002;19(2):43.
4. Liu T, Qin CL, Zhang Y, et al. China J Chin Materia Medica. 2002;27(8):609.
5. Zhang M. Effect of different ratios of salvianolic acid B to salviol IIA in composing formulas on heart cells. Dissertations of master postgraduate of 1999;2002: 6.
6. Wang R, Shang HC, Wang YY, et al. Tianjin J Tradit Chin Med. 2006;23(3):242.
7. Shang HC, Zhang BL, Wang YY, et al. Chin J Exp Tradit Med Formulae. 2003;9(3):1.
8. Shang HC, Zhang BL, Wang YY, et al. Drug Eval. 2005;2(3):209.
9. Wang Y, Gao XM, Zhang BL. Chin Tradit Herbal Drugs. 2004;35(6):672.
10. Liu YF, Zhang JP, Zhang BL. Tianjin J Tradit Chin Med. 2004;21(2):144.
11. Liu YF, Zhang JP, Zhang BL. Chin J Inf Tradit Chin Med. 2004;11(2):122.
12. Fan X. Doctoral graduation dissertation (in grade 2004), 2006;6.
13. Yuan Z, Zhang JP, Liu YF, et al. Tianjin J Tradit Chin Med. 2006;23(3):261.
14. Gao XM, Zhang BL, Shang HC, et al. Composite salviae prescription enhancing the protective effect of ischemic preconditioning on myocardium. In: The second world seminar of integrated Chinese and western medicine dissertation abstracts.
15. Gao XM, Zhang BL, Shang HC, et al. J Clin Rehabilitative Tissue Eng Res. 2003;7(12):1754.
16. Gao XM, Zhang BL, Shang HC, et al. Chin Tradit Herbal Drugs. 2003;34(10):933.
17. Zhao MZ, Wang JR, Wei JP, et al. Chin J Clin Pharmacol., 1999;15(4):288.
18. Xu M, Li QF, Zhang WH, et al. Chin J Pathophysiol. 2003;19(4):499.
19. Wang XL, Gu ZL, Wang DS, et al. Chin J Integ Med Cardio-/Cerebrovascular Dis. 2006;4(9):789.
20. Zhao YJ, Zhu SJ, Shi CN, et al. Chin J Pathophysiol. 2001;17(7):697.
21. Zhao YJ, Zhu SJ, Shi CN, et al. J Harbin Med Univ. 2003;37(4):290.
22. Li J, Lin J, Li Z, et al. Chin Arch Tradit Chin Med. 2002;20:496.
23. Guan H, Zhang WP. Med World. 2007;2:6.
24. Xu ZP, Zhang BL, Wang YM, et al. Chin Tradit Herbal Drugs. 2000;31(4):283.
25. Guo J, Zhang L, Zeng Z, et al. Chin J Hemorheology. 2004;14(1):1.
26. Tian JY, Chen JZ, Gu Y, et al. J Clin Rehabilitative Tissue Eng Res. 2004;8(9):1708.
27. Chen P, Xu XJ. Chin J Integr Tradit W Med Intensive Crit Care. 2007;14(5):313.
28. Shi YF, Li J. Chin J Cardiovasc Rehabil Med. 2002;11(3):226.
29. Wang P, Zhang Y, Fan YM, et al. J Cardiovasc Pulm Dis. 1996; 15(1):52.
30. Feng Q, Wang SC. Chin J Misdiagn. 2006;6(12):2261.
31. Li YP, Liu FZ, Sun JS, et al. Chin J Hemorheology. 2001; 11(2).
32. Zhang L, Liu CL, Jin TT. Chin J Basic Med Tradit Chin Med. 2007;13(12):904.
33. Xu ZX. Tianjin J Tradit Chin Med. 2003;20(2):74.
34. Fan YZ, Yan L, Song XL. Chin J Integr Tradit Chin W Med Intensive Crit Care. 2003;10(3):185.
35. Wei J, Du X, Hou ZX, et al. Tianjin Pharm. 2004;16(2):39.
36. Jia XH, Yu XR. J Shanxi Med Coll Continuing Educ. 2003;13(2):36.
37. Wu SD, Wang J, Chen SC, et al. J Chin Integr Med. 2004;2(2):94.
38. Xu C. J Emer Tradit Chin Med. 2003;12(2):168.
39. Chen BS, Feng LW. J Emer Tradit Chin Med. 2003;12(5):397.
40. Liu FP. China' s Naturopathy. 2003;11(4):43.
41. Li MR, Wang CS, Cai YM. J Chin Med Mater. 2003;26(8):615.
42. Qiu WQ, Teng AL. Shanxi J Tradit Chin Med. 2004;20(1):21.
43. Chai XT, Cui HL. Chin J Primary Med Pharm. 2003;10(12):1247.
44. Hua K. Clin J TCM. 2004;1(4):302.
45. Liu JB, Shi XD. Clin J Anhui Tradit Chin Med, 2003; 15(4):279.
46. Cao XF, Xie XW. J Occup Health Damage. 2003;18(3):193.
47. Wu MH. Chin Med Factory Mine. 2004;17(3):241.
48. Li W, Lai Y. J Guiyang Med Coll. 2003;28(3):259.
49. Yu YZ. Zhejiang J Integr Tradit Chin W Med. 2003;16(7):1023.
50. Xiong XD. Chin J Rural Med Pharm. 2004;11(5):43.
51. Pu MZ. Jilin J Tradit Chin Med. 2003;23(2):14.
52. Wang P. Chin Commun Phys. 2002;(4):42.
53. Yu YG, Fan XJ. Med J QILU. 2003;18(3):260.
54. Zhou BH, Wang XC, Zhao QF, et al. J Jilin Mil Med Coll Fourth Mil Med Univ. 2003;25(1):40.
55. Liu LX, Zhang DP, Liu SY, et al. J Chin Microcirc. 2002;6(5):314.
56. Ye F, Mei CJ, Hao LP, et al. Chin J Clin. 2003;31(3):28.
57. Li HH, Zhou X, Yang RY, et al. J Hum. Coll Tradit Chin Med. 2003;23(2):46.

58. Yang BH, Wang G. J Qiqihar Med Coll. 2003;24 (6):641.
59. Yang Y. Chin J Pract Tradit Chin W Med. 2003;16 (13):1920.
60. Liu XJ, Wu QM, Xie GJ. Mod J Integr Tradit Chin W Med. 2003;12(21):2298.
61. Qiu GL. Heilongjiang J Tradit Chin Med. 2002;6:16.
62. Liu QR, Wang GX, Li JL. J Prac Tradit Chinese Med. 2003; 19(11):591.
63. Zhang GL, Song YP, Wei XD, et al. Leukemia-Lymphoma. 2003;12(3):146.
64. Hu ZY, Liu JX, Wang BH, et al. J Fourth Mil Med Univ. 2003;24(24):2304.
65. Xue MG, Qi WX. J Binzhou Med Univ. 2004;27 (1):71.
66. Peng Z, Yu Y. Mod J Integr Tradit Chin W Med, 2004;13(13):1730.
67. Qi GS, Pei SX. J Binzhou Med Univ, 2004;Z 4/ (4):318.
68. Huang W, Chen ME, Zhou HD. West China J Pharm Sci. 2002;17(5):323.
69. Huang H, Tang YA, Zhang C. J Chin Med Mater. 2004;27(8):585.
70. Liu DM, Liu ZX, Wang WM. Pharm Clin Chin Mater Medica. 2002;18(2):6.
71. Jiang ZL, Mao ZB, Wu XT, et al. Chin J Clin Rehabilit. 2003;7(23):3182.
72. Song Y, Liu DM. Jiangxi Sci. 2003;21(4):311.
73. Tang Q, Xing YL, Wang WH, et al. Chin J Tradit Med Sci Technol. 2003;10 (2):69.
74. Ma Y, Zhang XZ, Wang W, et al. J High Altitude Med. 1997;7(4):11.
75. Wang W, Zhu YA, Zhang XZ, et al. Med J Natl Defending Forces Northwest China. 2002;23 (5):343.
76. Ma Y, Zhang XZ, Wang KZ, et al. J High Altitude Med. 1997;7(3):43.
77. Ding ZY, Liu ZY, Jiu JP, et al. Chin J Appl Physiol. 1996;12(1):94.
78. Yan SF, Wu M. Occup Health Emer Rescue. 2004;22(2):110.
79. Xue ZS, Peng HF. Occup Health Emer Rescue, 1994;12(2):113.
80. Zhang HY. Chin J Pract Tradit Chin W Med. 2003;3 (10):1351.
81. Chen XH. Hubei J Tradit Chin Med. 2002;24(5):27.
82. Deng BZ. Guangdong Med J. 2004;25(8):984.
83. Li DH. Guangdong Med J. 2003;24(9):1007.
84. Wu XY. Guangdong Med J. 2003;24(2):203.
85. Chen C. Clin J Integr Tradit Chin W Med. 2002;2 (5):34.
86. Zhang Q, Chen YH. J Clin Ophthalmol. 2003;11 (1):81.
87. Zhang JH, Huang XL, Huang HB, et al. Chin Pharm J. 2004;39(5):350.
88. Tan HM, Chen JP, Wu WH, et al. J First Mili Med Univ. 2000;20(3):241.
89. Zhou L, Wu WK, Chen GP, et al. Shenzhen J Integr Tradit Chin W Med, 1999;9(2):18.
90. Miao HX, Zhang LJ. Harbin Med. 2004;24(2):69.
91. Zhang Y. J Changzhi Med Coll. 2004;18(3):179.
92. Huang Q, Wang HW, Zhu HY. J Pract Tradit Chin Med. 2004;20(7):393.
93. Han L, Li HM. Cent Plains Med J. 2003;30(22):46.
94. He SF, Chen JM, Li CF. Actoandrologica Sinica. 2001;3:207.
95. Wu ZQ, Chen SS, Hu GR, et al. Pharmacol Clin Chin Materia Medica, 1997;13(3):41.
96. Chen SS, Meng HM, Liu JB, et al. Pharmacol Clin Chin Materia Medica. 1997;13(6):40.
97. Shen ZY, Zheng Z, Guo WM, et al. Chin J Integr Tradit W Med. 2002;22(3):178.
98. Miao S, Liu FG, Su FQ, et al. Drug research, 2002;11(12):25.
99. Zheng XK. J Henan Univ Tradit Chin Med. 2003;18 (4):57.
100. Chen JS, Sun SL, Xu H. Chin J Thromb Hemost. 1998;5(2):76.
101. Qian HL, Zhou FG, Feng WM, et al. Zhejiang Clin Med J. 2004;6(3):226.
102. Mao YP. J Zhejiang Univ Tradit Chin Med. 2003;27 (1):46.
103. Zhao QC. Forum Tradit Chin Med. 2001;16(2):38.
104. Mou NZ, Fan LR. Shandong J Tradit Chin Med. 1997;16(9):398.
105. Ding XS, Zhou XM, Wang JH. J Qiqihaer Med Univ. 2002;23(1):14.
106. Li P, Wang JC. Zhejiang J Integr Tradit Chin W Med. 2002;12(1):39.
107. Qin TE. Acta Med Sin. 2004;17(2):210.
108. Ma JH, He XN, Sun XM. J Jinlin Univ (Med). 2003;29(3):272.
109. Liang W, Gao Y. Tianjin J Tradit Chin Med. 2002;19(2):17.
110. Wu YJ, Wang JS, Yin XJ. Chin J Tradit Med Sci Technol. 2003;10 (2):119.
111. Yao SM, Lv YG. Shandong J Tradit Chin Med. 2000;19(7):443.
112. Chu QX, Wang DX. Pract J Cardiac Cerebral Pneumal Vasc Dis. 2003;11(6):353.
113. Zhao M, Meng LB, Jiang X. Chin J Cardiovasc Rehabil Med. 1999;8(4):174.
114. Lai SR, Wang DX. Chin J Integr Med Cardio-/ Cerebrovasc Dis. 2003;7:387.
115. Liang DS, Yuan JH, Zhu XM et al. Chin Acad Med Mag Organ. 2002;(4):52.
116. Lin YZ, Weng Q, Huang MH. Fujian J Tradit Chin Med. 2003;34(6):41.
117. Chen HL, Chen MX, Lin T, et al. J Qiqihaer Med Univ. 2003;24(4):392.
118. Chen YH, Tan HP, Ma BS. Lishizhen Med Materia Medica Res. 2004;15(10):684.
119. Li J, Wang XM, Liu YH, et al. J Shanxi Tradit Chin Med. 2003;24(5):419.

3 Active Constituents in Danshen and Their Pharmacological Actions

Ailin Liu, Yitao Wang, Guanhua Du, Xiuying Yang, Juntian Zhang, Minke Tang, Ji Chen, Yonghong Chen, Zhiwei Qu, Jie Wang, Xiaoying Wang, Yan Sun, Ping Chen and Chuan Li

Early in the 1930s, studies were initiated on the material basis of pharmacological activity in Danshen. Initially, major liposoluble constituents, e.g., tanshinone I, cryptotanshinone, tanshinone II_A, and tanshinone II_B, were isolated from the root of Danshen. In the 1940s, Wang Xu, et al. proved that tanshinone I structurally belongs to the tetracyclic compounds of diterpene quinone. In the 1970s, a large number of studies in China were conducted on the chemistry and pharmacology of lipo- and water-soluble constituents in Danshen. In recent years, with extensive application of new techniques and improvements in the techniques of compound isolation and analysis, more and more chemical constituents have been identified from Danshen. At present, more than 100 chemical constituents have been isolated and identified from Danshen, of which the pharmacological activities of most have been studied. In this chapter, based on the dissolubility of chemical constituents, active constituents in Danshen are divided into liposoluble constituents, water-soluble constituents, and other constituents, and their physicochemical properties and pharmacological activities are reviewed with the hope of providing a reference for further studies and development on Danshen.

3.1 Major Constituents and Their Pharmacological Actions

Ailin Liu, Yitao Wang and Guanhua Du

3.1.1 Liposoluble Constituents

Most of Danshen's liposoluble constituents belong to compounds of terpene quinone, and the rest are compounds of other types. At present, more than 50 liposoluble constituents have been isolated and they are briefly introduced below.

1. Tanshinone I: melting point (MP) 233–234 °C, $C_{18}H_{12}O_3$ (276.3). It has an antibacterial effect and estrogen-like activity.
2. Tanshinone II_A: MP 198–200 °C, $C_{19}H_{18}O_3$ (294.4). It has antibacterial, anti-arterial atherosclerosis, antithrombosis, blood vessel dilation, and myocardial preservation functions. It is cytotoxic to human cancer cell lines HeLa, HepG2, and OVCAR-3 with an IC_{50} of 8.5, 26.5, and 9.52 μM, respectively. It can prevent blood clotting and restore blood circulation. Research results have shown that in rabbits with atherosclerosis induced by high-fat feed, this compound can down-regulate the expression of CD40 and reduce enzymatic activity of MMP-2.

A. Liu · Y. Wang · G. Du (✉) · X. Yang · J. Zhang · M. Tang · J. Chen · Y. Chen · Z. Qu · J. Wang · X. Wang · Y. Sun · P. Chen · C. Li
Institute of Materia Medica, Chinese Academy of Medical Science and Peking Union Medical College, Beijing, China
e-mail: dugh@imm.ac.cn

X. Yan (ed.), *Dan Shen (Salvia miltiorrhiza) in Medicine*,
DOI 10.1007/978-94-017-9463-3_3,

3. Tanshinone II_B: MP 200–204 °C, $C_{19}H_{18}O_3$ (294.4). It is active against *Staphylococcus Aureus* and its drug-tolerant strains.
4. Tanshinone IV: Jacinth needle crystal with MP 185–189 °C (decomposed), $C_{18}H_{16}O_4$ (296.3). It has preventive and protective effects against myocardial ischemia, and can recover myocardial contractility after hypoxia.
5. Cryptotanshinone: MP 191 °C, $C_{19}H_{20}O_3$ (296.4). It is active against *S. aureus* and its drug-tolerant strains, as well as against *Mycobacterium hominis* H_{37} Rv strain. It is cytotoxic to human cancer cell lines HeLa, HepG2, and OVCAR-3 with IC_{50} values of 17.2, 29.2, and 9.12 μM, respectively. It has an inhibitory effect on the drug target protein related to type II diabetes, i.e., tyrosine phosphatase lB (PTPlB), with an IC_{50} of 56.1 mM. It has the activities of preventing blood clotting and restoring blood circulation. It also has anti-diabetic and anti-obesity activities via the activation of protein kinase activated by AMP.
6. Isotanshinone I: MP 219 °C, $C_{18}H_{12}O_3$ (276.3). No studies on its bioactivity have been found.
7. Isotanshinone II_A: MP 208 °C, $C_{19}H_{18}O_3$ (294.4). In vitro studies showed that isotanshinone II_A had a significant inhibitory effect on platelet aggregation induced by ADP or collagen, and on the drug target related to type II diabetes, PTPlB, with an IC_{50} of 11.4 mM.
8. Isocryptotanshinone: MP 121 °C, $C_{19}H_{20}O_3$ (296.4). It inhibits the protein PTPlB with an IC_{50} of 51.2 mM.
9. Neocryptotanshinone II: yellow needle crystal with MP of 129–130 °C, $C_{17}H_{18}O_3$ (270.3). No studies on its bioactivity have been found.
10. 1,2,15,16-tetrahydroxytanshinone: prunosus styloid (methanol) with MP of 140–142 °C, $C_{18}H_{16}O_3$ (280.1). In vitro studies showed that this compound had strong anti-Leukemia P388 cell activity.
11. 3α-hydroxytanshinone II_A: also known as hydroxytanshinone II, one of the active constituents in Danshen, with MP of 187 °C, $C_{19}H_{18}O_4$ (310.4). It has antimicrobial activities against *S. aureus* and its tolerant strains, *M. hominis* H_{37}Rv strain, and *hemolytic Streptococcus*.
12. 3β-hydroxytanshinone II_A: red needle crystal with MP of 202 °C, $C_{19}H_{18}O_4$ (310.4). No reports on its bioactivity have been found.
13. Δ^1-dehydrotanshinone: dark-red needle crystal with MP of 147–148 °C, $C_{19}H_{16}O_3$ (292.3). No reports on its bioactivity have been found.
14. Miltirone: MP 100 °C, $C_{19}H_{22}O_2$ (282.4). It inhibits platelet aggregation induced by collagen with an IC_{50} of 5.76 μM. It has sedative effects on the central nervous system. It is a partial agonist of benzodiazepine receptors.
15. Dihydrotanshinone I: $C_{18}H_{14}O_3$ (278.3). It has anti-bacterial functions, and can inhibit *S. aureus* and *M. hominis* H_{37}Rv strain, with a minimum inhibitory concentration (MIC) of 1.5 μg/ml.
16. Dihydroisotanshinone I: $C_{18}H_{14}O_3$ (278.3). It has an in vitro inhibitory effect on platelet aggregation induced by collagen and anti-bacterial effects, for example strong effects on *S. aureus* and its tolerant strains, and on *M. hominis* H_{37}Rv strain with a MIC of 1.5 μg/ml.
17. Nortanshinone: $C_{17}H_{14}O_5$ (nortanshinone). No reports on its pharmacological activities have been found.
18. Dehydromiltirone: red oily material with MP of 45–46 °C, $C_{19}H_{20}O_2$ (280.4). No reports on its pharmacological activities have been found.
19. 1,2-dihydrotanshinone: red needle crystal with MP of 169 °C, $C_{18}H_{14}O_3$ (278.3). No reports on its pharmacological activities have been found.
20. 3-Hydroxymethylenetanshiquinone: red needle crystal with MP of 185–190 °C, $C_{18}H_{44}O_4$ (294.3). No reports on its pharmacological activities have been found.
21. Methylenetanshinone: $C_{20}H_{24}O_2$ (296.4). No reports on its pharmacological activities have been found.
22. Neotanshinone A: also known as tanshiquinone A, which has antibacterial activities

against *Mycobacterium tuberculosis* H37Rv strain, with a MIC of 0.78 μg/ml.

23. Neotanshinone B: also known as tanshiquinone B. $C_{18}H_{16}O_3$ (280.). It has antibacterial effects against *M. tuberculosis* H37Rv strain, with a MIC of 3.1 μg/ml.
24. Neotanshinone C: also known as tanshiquinone C, $C_{16}H_{12}O_3$ (252.3). It has antibacterial effects against *M. tuberculosis* H37Rv strain, with a MIC of 6.3 μg/ml.
25. Neotanshinone D: dark-red needle crystal with MP of 178–180 °C, $C_{21}H_{20}O_4$ (336.4). No reports on its pharmacological activities have been found.
26. Przewaquinone A: jacinth styloid (methanol) with MP of 173–175 °C (decomposed), $C_{19}H_{18}O_4$ (310.4). It has bioactivity on animal tumors, and has a 35.8–67.8 % inhibitive effect on Lewis lung cancer, melanoma B16 and sarcoma 180 after intraperitoneal administration at 120 or 150 mg/kg. It can prolong survival time by >100 % for mice with Leukemia P388, and is stronger than cryptotanshinone with regard to the inhibition of *S. aureus* 209P.
27. Przewaquinone B: purple-brown tabular crystal with MP of 242–243 °C, $C_{18}H_{12}O_4$ (292.3). No reports on its pharmacological activities have been found.
28. Przewaquinone C: orange-red prism crystal (methanol) with MP of 204–206 °C, $C_{18}H_{16}O_4$ (296.3). No reports on its pharmacological activities have been found.
29. Przewaquinone F: orange-red needle crystal (acetone) with MP of 234–236 °C, $C_{18}H_{16}O_5$ (312.3). No reports on its pharmacological activities have been found.
30. Danshenol A: brown-gray needle crystal with MP of 182 °C, $C_{21}H_{20}O_4$ (336.4). It inhibits aldose reductase with an IC_{50} of 0.1 μM.
31. Danshenol B: yellow needle crystal with MP of 176 °C, $C_{22}H_{26}O_4$ (354.5). It inhibits aldose reductase with an IC_{50} of 1.57 μM.
32. Tanshindiol A: orange-red scale-like material (acetic ether) with MP of 222–223 °C, $C_{18}H_{16}O_5$ (312.3). It inhibits human tumor cells A549, SK-OV-3, SK-MEL-2, XF-498, and HCT-15 with an IC_{50} ranging from 0.2 to 0.8 μg/ml, and *M. tuberculosis* H37Rv strain with a MIC of 5 μg/ml, and recovers posthypoxic myocardial contractility with a recovery rate of 33.3 % at 25 μM in rats.
33. Tanshindiol B: also known as Tanshinol C, orange-red tabular crystal (acetic ether) with MP of 210 °C, $C_{18}H_{16}O_5$ (312.3). It inhibits human tumor cells A549, SK-OV-3, SK-MEL-2, XF-498, and HCT-15 with IC_{50}s ranging from 0.4 to 1.0 μg/ml and *M. tuberculosis* H37Rv strain with a MIC of 5 μg/ml, and recovers posthypoxic myocardial contractility with a recovery rate of 34.3 % at 25 μM in rats.
34. Tanshindiol C: also known as Tanshinol B, jacinth tabular crystal (acetic ether) with MP of 213–215 °C, $C_{18}H_{16}O_5$ (312.3). It inhibits human tumor cells A549, SK-OV-3, SK-MEL-2, XF-498, and HCT-15 with IC_{50}s ranging from 0.3 to 0.9 μg/ml and *M. tuberculosis* H37Rv strain with a MIC of 5 μg/ml, and recovers posthypoxic myocardial contractility with a recovery rate of 27.5 % at 25 μM in rats.
35. Miltionone I: $C_{19}H_{20}O_4$ (312.4). No reports on its pharmacological activities have been found.
36. Miltionone II: $C_{19}H_{20}O_4$ (312.4). It has the effects of preventing blood clotting and restoring blood circulation.
37. Miltipolone: $C_{19}H_{24}O_3$ (300.4). No reports on its pharmacological activities have been found.
38. Danshen spiroketallactone: white needle crystal with MP of 203–205 °C, $C_{17}H_{16}O_3$ (268.3). No reports on its pharmacological activities have been found.
39. Tanshinlactone: $C_{17}H_{16}O_3$ (264.3). No reports on its pharmacological activities have been found.
40. Neotanshinlactone: an isomer of tanshinlactone, $C_{17}H_{12}O_3$ (264.3). It has anti-breast cancer activity.
41. Tanshinaldehyde: dark-red needle crystal (ethanol) with MP of 223–225 °C, $C_{19}H_{16}O_4$ (308.3). No reports on its pharmacological activities have been found.

42. Methyltanshinonate: MP of 175–176 °C, $C_{20}H_{18}O_5$ (338.4). No reports on its pharmacological activities have been found.
43. Lithospermate B: $C_{36}H_{28}$ Mg O_4 (740.9). Its methylated metabolite has strong antioxidant activity.
44. 2-isopropyl-8-methylphenanthrene-3,4-dione: $C_{17}H_{16}O_2$ (252.3). It has inhibitive effects against platelet aggregation.
45. Saloilenone: $C_{20}H_{20}O_2$ (292.4). No reports on its pharmacological activities have been found.
46. Ferruginol: $C_{20}H_{30}O$ (286.5) with MP of 57–59 °C. It has thc function of rcmoving free radicals, antioxidation, and protecting the stomach from injury.
47. Salviol: also known as Tanshinol, $C_{20}H_{30}O_2$ (302.5). No reports on its pharmacological activities have been found.
48. Sugiol: MP 298–299 °C (decomposed), $C_{20}H_{28}O_2$ (300.4). It has antioxidant, antibacterial, and antiinflammatory effects, and inhibits aldose reductase.
49. Salvilenone: a diterpene compound, $C_{20}H_{24}O$ (280.4). No reports on its pharmacological activities have been found.
50. Neosalvianen: one of the nitrogen compounds extracted from Danshen, $C_{21}H_{21}NO_2$ (319.4). It is cytotoxic to human cancer cell lines HeLa, HepG2, and OVCAR-3 with an IC_{50} of 63.9, 59.2, and 74.6 μM, respectively.
51. Salvianen: another nitrogen compound extracted from Danshen as reported recently, an isomer of neosalvianen, $C_{21}H_{21}NO_2$ (319.4). It is cytotoxic to human cancer cell lines HeLa, HepG2, and OVCAR-3 with an IC_{50} of 32.3, 30.4, and 39.5 μM, respectively.
52. Salvianan: a nitrogen compound, $C_{21}H_{23}NO_2$ (321.4). No reports on its pharmacological activities have been found.
53. Salviadione: a nitrogen compound, C_{19} H_{21} NO_2(295.4). It is cytotoxic to human cancer cell lines HeLa, HepG2, and OVCAR-3 with an IC_{50} of 63.9, 59.2, and 74.6 μM, respectively.
54. 5-(methoxymethyl)-1H-pyrrole-2-carbaldehyde: a pyrrole derivative, $C_7H_9NO_2$ (139.2). No reports on its pharmacological activities have been found.

3.1.2 Water-Soluble Constituents

Water-soluble constituents of Danshen include mainly the compounds of phenolic acids, and are introduced below.

1. Salvianolic acid A: yellow amorphous powder, one of the major phenolic acids in Danshen, a derivative of Rosmarinic acid, $C_{26}H_{22}O_{10}$ (494.5). It has antitumor effects, inhibits platelet aggregation, inhibits gastric secretion (rats), antioxidant effects (SalA can intensively inhibit lipid peroxidation induced by vitamin C-nicotinamidc ADP and Fe^{2+}-cysteine in the brain, liver, and kidney in mice), inhibits H^+, K^+-adenosine triphosphatase with an IC_{50} of 0.52 μM, inhibits p-nitrophenylphosphatase (pNPPase), inhibits secretion and ulcers with an IC_{50} of 1.7 μM, inhibits 5-lipoxygenase with an IC_{50} of 0.38 μM, inhibits aldose reductase with an IC_{50} of 9.80 μM in lens, protects heart muscle from injury, can reduce memory disorders in mice induced by perfusion ischemia, and has a two-way role on potassium channels in the myocardial membrane.
2. Salvianolic acid B: also known as salianic acid B or lithospermic acid B, a condensation product of three molecules of danshensu and one molecule of caffeic acid, a polyhydroxy compound of depsides, yellowish amorphous powder, $C_{36}H_{30}O_{16}$ (718.6). It has many pharmacological actions such as improving renal function, preventing cardiovascular diseases, antioxidation, preventing ischemic injury, protecting the liver, anti-fibrosis, preventing nervous toxicity, anticoagulation, and anti-inflammation, and is one of the major phenolic acids in Danshen.
3. Danshensu: also known as salianic acid A, white needle crystal with MP of 84–86 °C. It has the effects of improving heart function and dilating the coronary artery muscle.
4. Salvianolic acid C: a derivative of rosmarinic acid, yellow amorphous powder. $C_{26}H_{20}O_{10}$ (492.4). It has the effects of removing free radicals and resisting platelet aggregation.

5. Salvianolic acid D: a derivative of rosmarinic acid, $C_{20}H_{18}O_{10}$ (418.4). No reports on its pharmacological activities have been found.
6. Salvianolic acid E: a derivative of rosmarinic acid, a condensation product of two rosmarinic acids, $C_{36}H_{30}O_{16}$ (718.6). No reports on its pharmacological activities have been found.
7. Salvianolic acid F: a derivative of caffeic acid, $C_{17}H_{14}O_6$ (314.3). No reports on its pharmacological activities have been found.
8. Salvianolic acid G: $C_{18}H_{12}O_7$ (340.3). No reports on its pharmacological activities have been found.
9. Salvianolic acid H: a derivative of rosmarinic acid, a condensation product of one molecule of danshensu and two molecules of caffeic acid, $C_{27}H_{22}O_{12}$ (538.5). No reports on its pharmacological activities have been found.
10. Salvianolic acid I: a condensation product between one molecule of tanshensu and two molecules of caffeic acid, $C_{27}H_{22}O_{12}$ (538.5). No reports on its pharmacological activities have been found.
11. Salvianolic acid J: a derivative of rosmarinic acid, $C_{27}H_{22}O_{12}$ (538.5). No reports on its pharmacological activities have been found.
12. Salvianolic acid K: a condensation product of danshensu and rosmarinic acid, $C_{27}H_{24}O_{13}$ (556.5). It inhibits aldose reductase.
13. Salianic acid C: white solid, easily absorbs water in air, $C_{18}H_{18}O_9$ (378.3). It is a condensation product of two molecules of danshensu. No reports on its pharmacological activities have been found.
14. Isosalvianolic acid C: $C_{26}H_{20}O_{10}$ (492.4). No reports on its pharmacological activities have been found.
15. Lithospermic acid: $C_{27}H_{22}O_{12}$ (538.5). It inhibits aldose reductase and has contraceptive effects.
16. Butanedioic acid: also known as succinic acid, MP of 150 °C, $C_4H_6O_4$ (118.1). It has antibacterial effects (including *S. aureus, Diplococcus pharyngis communis*, *Pseudomonas aeruginosa*, *Bacillus proteus*, *Bacterium typhosum and Bacillus dysenteriae*), is antiulcerative, relieves pain, tranquilizes the central nervous system, and is antipyretic. It can be used to treat otitis media, paronychia, pustule sore, infected burns, suppurative tonsillitis and enteritis.
17. Caffeic acid: 3,4-dihydroxy-cinnamic acid, $C_9H_8O_4$ (180.2). It has antioxidant, antifungal, anti-hepatotoxicity, anti-inflammatory, anti-pain, antiulcerative, and antiviral effects, and excites the central nervous system.
18. Rosmarinic acid: a condensation product of one molecule of danshensu and one molecule of caffeic acid with MP of 204 °C, $C_{18}H_{16}O_8$ (360.3). It prevents thrombosis (in vivo studies in rats indicated that rosmarinic acid can inhibit venous thrombosis), inhibits platelet aggregation (in vitro studies indicated that rosmarinic acid can inhibit platelet aggregation induced by collagen), has anti-inflammatory activity, can promote fibrinolysis, has antiviral activity (herpes simplex virus) and antioxidative effects (inhibiting lipid peroxidation induced by vitamin C-nicotinamide ADP and Fe^{2+}-cysteine in the brain, liver, and kidney of mice, and removing superoxide anions), can inhibit pathogenic bacteria, and inhibits adenyl cyclase and aldose reductase.
19. Methhyl rosmarinate: the methylesterification product of rosmarinic acid, $C_{19}H_{18}O_8$ (374.4). It inhibits adenyl cyclase.
20. Ursolic acid: white powder (methyl chloride-methanol) with MP of 265–267 °C, $C_{30}H_{48}O_3$ (456.7). It has antitumor and antibacterial effects (inhibiting in vitro multiple *Staphylococci* with MIC of 300 μg/ml, Gram-positive bacteria with MIC of 50–400 μg/ml, Gram-negative bacteria with MIC of 200–800 μg/ml, and yeasts with MIC of 100–700 μg/ml), as well as anticonvulsive (induced by pentetrazole), anti-inflammatory (induced by embedment of tampon in rats, 12.5 mg/kg d, ip, effective within 7 days), and pyretolysis effects, and can reduce serum Transaminase (100 mg/kg in animal, LD_{50} = 680 mg/kg for i.p. in mouse).
21. Monomethyl lithospermate: $C_{28}H_{24}O_{12}$ (552.5). No reports on its pharmacological activities have been found.

22. Isoferulic acid: also known as hesperitinic acid, colorless needle crystal with MP of 225–229 °C, $C_{10}H_{10}O_4$ (194.2). It reduces blood glucose concentrations by regulating the level of β-endorphin in diabetic rats induced by streptozocin.
23. Protocatechualdehyde: para- and meta-dihydroxybenzaldehyde, $C_7H_6O_3$ (138.1). It has the effects of preventing thrombosis, improving blood circulation, preventing oxidative damage, and reinforcing learning memory.
24. Protocatechuic acid: para- and meta-dihydroxybenzoic acid with MP of 199–200 °C, $C_7H_6O_3$ (154.1). It activates prostaglandin synthetase, has anti-asthma (induced by histamine in guinea pigs), antibacterial (*Staphylococcus albus, S. aureus, Diplococcus pneumonia, a-Streptococcus, Escherichia coli, P. aeruginosa, B. proteus and Haemophilus influenzae*), antifungal, antihepatic toxicity, anti-inflammatory, and antitoxin effects, prevents cough (expelling phlegm by oral administration of 4.86 mg in rats), and can reduce oxygen consumption by heart muscles with LD_{50} of 896.4 mg/kg for i.p. in rats.
25. Linolenic acid: an unsaturated fatty acid, $C_{18}H_{30}O_2$ (278.4). It inhibits the decomposing activity of sTF/VIIa (soluble tissue factor/activated factor VII complex) on amides with an IC_{50} of 30 μM.
26. Linoleic acid: an unsaturated fatty acid, $C_{18}H_{32}O_2$ (280.5). It inhibits the decomposing activity of sTF/VIIa (soluble tissue factor/activated factor VII complex) on amides with an IC_{50} of 41 μM.
27. Oleic acid: also known as octadecenoic acid, an unsaturated fatty acid. It inhibits the decomposing activity of sTF/VIIa (soluble tissue factor/activated factor VII complex) on amides with an IC_{50} of 80 μM.
28. Palmitic acid: also known as hexadecoic acid or hexadecanoic acid, a saturated fatty acid, $C_{16}H_{32}O_2$ (256.4). It inhibits the decomposing activity of sTF/VIIa (soluble tissue factor/activated factor VII complex) on amides with an IC_{50} of 120 μM.
29. Salviaflaside: a glycoside of rosmarinic acid, $C_{24}H_{26}O_{13}$ (522.5). It inhibits aldose reductase.
30. Satviaflaside methyl ester: a glycoside of rosmarinic acid methyl ester, $C_{25}H_{28}O_{13}$ (536.5). No reports on its pharmacological activities have been found.

3.1.3 Other Constituents

1. Baicalin: 5,6,7-trihydroxyflavone-7-glucosiduronic acid, $C_{21}H_{18}O_{11}$ (446.4) with MP of 223 °C. It is anti-allergy, antimicrobial, and anti-inflammation, and has antipyretic, choleretic, diuretic, anti-hypertensive, sedative, and anti-toxic effects. It can reduce the death rate of mice from strychnine poisoning.
2. Daucosterol: white powder with MP of 295 °C, $C_{36}H_{62}O_6$ (590.9). It has neuroprotective effects and in vitro antioxidative activity.
3. β-sitosterol: 136–137 °C in MP, $C_{29}H_{50}O$ (414.7). It has antitumor (inhibits Lewis lung cancer and adenocarcinoma 715 in mice), anti-inflammatory, cough-preventing (mouse cough model induced by ammonia water, oral, ED = 250 mg/kg), and antihyperlipemia (reduces cholesterol in mice and chimpanzees) effects.
4. Tigogenin: $C_{27}H_{44}O_3$ (416.6) with MP of 205–206 °C. It is a raw material for corticosteroid synthesis, and it can inhibit tumor cell proliferation.
5. Vitamin E: it has antioxidant effects, preventing the oxidation of unsaturated fatty acids of stable cell membranes.

Besides the 89 constituents in Danshen listed above, there are also 15 free amino acids, e.g., glutamic acid, alanine, aspartate, histidine, and arginine; hydrolyzed amino acids; and mineral elements, e.g., calcium, magnesium, barium, aluminum, nickel, selenium, ferrum, zinc, and tin. Among the 15 amino acids, the content of glutamic acid is the highest, and 7 are essential amino acids for the human body.

In summary, after many years of effort, researchers have discovered more than 100 chemical constituents in Danshen, and new chemical constituents are subject to further isolation and identification. Chemical constituents are the material basis of Danshen's pharmacological actions, and pharmacological studies of these compounds are the basis for the studies of the mechanisms of the drug effects of Danshen, which provides a scientific basis for clinical application and further development of Danshen medicines.

The pharmacological activities of some chemical constituents in Danshen have been reviewed extensively. For example, tanshinone compounds, which have antibacterial activity and prevent and protect against cardiovascular diseases, and salvianolic acids, which have antiplatelet aggregation, antithrombus, and antioxidation effects, are important material bases of Danshen, which play pharmacological actions clinically. However, no study on the synergism between them has been reported, and further studies are still to be conducted.

3.2 Pharmacological Actions of Tanshinones

Guanhua Du and Ailin Liu

Tanshinone is one of the active components of Danshen. Scientific studies and clinical practices in the last several decades demonstrated that tanshinones have good antibacterial activities and a good effect on the cardiovascular system, and good therapeutic effects were achieved clinically. Recent studies have proved that tanshinones have strong antibacterial activities against drug resistant bacteria, and are the most effective, natural, nonantibiotic antibacterial compounds extracted from plants.

3.2.1 Overview of Studies on Tanshinones

It was in the 1930s that modern chemical and medical methods were first used to study the active constituents of Danshen and their pharmacological actions. In the past several decades, chemists, pharmacists, pharmacologists and medical workers have, by close cooperation, succeeded in researching and developing many new Danshen preparations which have been used to treat many diseases.

3.2.1.1 Tanshinone Study Overview

Chemical constituents in Danshen can be divided into water- and lipo-soluble components. Early studies focused mainly on the lipo-soluble constituents of Danshen. In the 1930s, Japanese scientists first studied the lipo-soluble constituents of Danshen. Thereafter, scientists in China made deep-going explorations in chemical study, first demonstrating the chemical structure of tanshinone, then discovering many new compounds. At present, more than 40 liposoluble compounds have been obtained; among them, most are compounds of diterpene quinone. These compounds were divided into tanshinones and royleanones [54–57] based on their skeleton structures.

Previous studies have demonstrated that tanshinones have two main pharmacological effects. One is antibacterial effect, and cryptotanshinone is the most potent in this aspect among tanshinone compounds. Another is the effect on the cardiovascular system, which is used mainly to treat **coronary artery muscle** disease, and tanshinone II_A is the compound, which gives the most obvious results. Based on a large number of studies, the clinical drugs of tanshinone and sodium tanshinone sulfonate were developed successfully, in which the former is used for acne and the latter for coronary heart disease.

3.2.1.2 Study Overview on the Pharmacological Actions of Tanshinones [58]

There had been abundant research on the pharmacological actions of tanshinones prior to the 1990s, which were mainly performed by using total extracts or liposoluble extracts from Danshen and observing various pharmacological activities and providing many clues for further research on active constituents in Danshen. The pharmacological effects of Danshen were shown mainly as follows.

Cardiovascular System [59–61]

Danshen has multiple pharmacological actions on the cardiovascular system, e.g., dilating the coronary artery muscle, increasing coronary flow and its index, protecting ischemic cardiac muscle, promoting the recovery of cardiac muscle with ischemic injury, reinforcing the regeneration capacity of cardiac muscle, reducing infarct range, improving microcirculation disturbance, preventing oxidation injury and apoptosis of cardiac muscle cells, and relieving Atherosclerosis [63].

Nervous System

Experimental results showed that Danshen could reduce spontaneous electric activity in the cerebral cortex of rabbits and autonomic activities of mice. Danshen alone could not make an animal sleep but if combined with a hypnotic drug, it could enhance that drug's efficacy [64]. Tanshinone II_B had a protective effect on the nerves of rats with hemorrhagic stroke [65].

Liver

Danshen had protective effect on livers poisoned by CCl_4; it could promote the regeneration of liver cells, inhibit hepatic fibrosis, and recover liver function. It showed protective effects on both animal models and cultured liver cells [66, 67]. Tanshinone II_A could inhibit the growth of liver cells, and induce apoptosis of liver cells [68].

Antimicrobial and Anti-inflammatory Effect

In vitro studies indicated that tanshinones could inhibit *S. aureus*, hemolytic *Streptococcus, tubercle bacillus*, rusty trichopyton, and red trichopyton. Danshen could also inhibit the formation of experimental inflammation in animals and Chemotaxis of leucocytes, and regulate the immunological function of animals [69].

3.2.1.3 Overview of Clinical Studies [54, 58, 69, 70]

Overview of Clinical Application of Tanshinones

Extensive studies on Danshen provided an experimental basis for its clinical application, which provided further knowledge on the clinical efficacies of Danshen and promoted the development of various preparations of Danshen medicines for clinical use. Commonly used Danshen preparations include Composite Danshen injection (1 ml injection solution contains extracts from 1 g each of Danshen and rosewood), Danshen injection, Danshen 201 injection (sodium tanshinone II_A sulfonate), Danshen Shuxin Tablet, Composite Danshen Tablet, Danshensu injection, Tanshinone Capsule, and so on. In the 1970s, research teams in China were organized to perform large-scale studies, and their results confirmed the clinical efficacy of Danshen.

Danshen is used to treat coronary heart disease, angina pectoris and myocardial infarction. Danshen can improve myocardial contractility, slow the heart rate, and significantly improve clinical symptoms, all without altered oxygen consumption. The preparations of Danshen are superior to low molecular dextran for the therapy of ischemic strokes and also improve other complications. However, its effectiveness in treating hemorrhagic stroke has yet to be reported.

The use of Danshen to treat infectious diseases is based on its antibacterial and anti-inflammatory effects. It has been used clinically to treat whooping cough, epidemic meningitis accompanied by disseminated intravascular coagulation (DIC), pediatric persistent pneumonia, epidemic hemorrhagic fever, and tuberculosis with certain effectiveness. However, Danshen drugs need to be combined with other drugs, especially antibiotics, to cure these diseases.

It is worth to specially note that tanshinones have been used to treat various skin diseases such as acne. Tanshinone Capsule is superior to other existing preparations for acne treatment, and it is one of the most successful Danshen preparations used clinically.

Study on Therapy of Acne with Tanshinones

Long-term clinical observations and studies were conducted on tanshinones' treatment of acne. Studies of a large number of cases demonstrated that tanshinones had good therapeutic effects on common, impetiginous, and cystic acne, and the treatment efficiency was higher than that of tetracycline.

To treat acne with tanshinones, an oral dosage of 2–3 g per day was administered for 8 weeks as a treatment course. A total of 172 patients were treated, and the efficiency was as high as 88.8 %, higher than that of tetracycline treatment (70.7 %).

3.2.2 Antibacterial Activity of Tanshinones

It has been demonstrated long before that tanshinones have strong antibacterial function. Its oral preparations were developed for clinical applications with good effectiveness.

3.2.2.1 Antibacterial Action Against *Staphylococcus Aureus*

The filter paper disk method was used to measure the antibacterial activity of tanshinones, and the results showed that when a disk contained as little as 6.25 μg of tanshinones, the growth of strain 209P was still inhibited.

3.2.2.2 Antibacterial Action Against Drug-Tolerant *Staphylococcus Aureus*

An assay was done using paper disk method on the clinically isolated *S. aureus* strains tolerant to benzylpenicillin, streptomycin, and chlortetracycline. The results showed that tanshinones had strong antibacterial activity, and its strength against the drug-tolerant strains was similar to that against drug sensitive strains. When tanshinones were used to treat *S. aureus* infection, the pathogenic bacterium tolerant to multiple antibiotics (benzylpenicillin, streptomycin, chloramphenicol, chlortetracycline, chlortetracycline, tetracycline, oxytetracycline, kanamycin, and gentamicin) was still sensitive to tanshinones.

3.2.2.3 Comparison of Antibacterial Activities Between Tanshinone and Berberine Against *Staphylococcus Aureus*

It has been demonstrated clinically that berberine has a strong antibacterial effect against *S. aureus*. A comparison of the antibacterial efficacy against *S. aureus* was conducted using different amounts (50, 25, 12.5, 6.25 μg) of tanshinone and berberine. The results showed that the smallest amount of tanshinone had a stronger antibacterial activity than the highest amount of berberine.

3.2.2.4 Antibacterial Action of Tanshinones on Tubercle Bacillus

Bacterial suspension culture was used to determine the antibacterial activity of tanshione against the virulent strain of *M. hominis*, $H_{37}Rv$. The results are shown in Table 3.1. The minimal inhibitory concentration of total tanshinones was lower than 1.5 μg/ml. However, except for neotanshinone A, most members of the tanshinone family are weak in antibacterial activity.

3.2.2.5 Antifungal Activity

Tanshinone at 25 μg/ml has different antifungal activities against rust trichopyton and red trichopyton, with very significant effectiveness at 100 μg/ml.

3.2.2.6 The Antibacterial Activities of Mono-component Tanshinone

Inhibitory Effect on the Growth of Gram-Positive Bacteria

Most components of tanshinones have certain antibacterial activities against Gram-positive bacteria, with a least inhibitory concentration (MIC) of up to 1 μg/ml, which is phenomenal. Among the five Gram-positive bacteria tested, the strongest antibacterial activities were found in

Table 3.1 Antibacterial actions of various liposoluble constituents in Danshen

Sample name	*Staphylococcus aureus* 209P (diameter of inhibition zone, mm)	*Mycobacterium hominis* μg/ml
Tanshinones	–	<1.5
Cryptotanshinone	23	<1.5
Dihydrotanshinone I	23	1.5
Hydroxytanshinone	16	3.1
Tanshinone II_B	17	
Tanshinone II_A	–	<1.5
Neotanshinone A	–	0.78
Neotanshinone B	–	3.1
Neotanshinone C	–	6.3
Tanshinone I	–	100

samples 11 and 13 against *Sarcinae sp.*, with MIC = 1; the second strongest was found in sample 11 against *Bacillus cereus* and *Bacillus subtilis* with MIC = 2. The antibacterial activities of samples 13, 16, and 17 against *B. cereus*, sample 16 against *S. aureus* 209, and total salvianolic acids against *B. subtilis* were also relatively strong with MIC = 4. Samples with relatively strong antibacterial activities against all five bacteria were samples 1, 11, 13, 16, 17, and total tanshinones. The results are listed in Table 3.2.

Inhibitory Effect on the Growth of Gram-Negative Bacteria

Tanshinones have no significant inhibitory activities against Gram-negative bacteria. Some components have weak inhibition against *B. proteus* OX-19 strain.

Inhibitory Effect of Tanshinones on Growth of Fungi

None of the components of tanshinones have significant inhibitory effects on the growth of *Candida albicans* and *Aspergillus niger*, and

Table 3.2 MIC of tanshinones against 5 gram-positive bacteria (μg/ml)

Sample no.	*Sarcinae*	*Staphylococcus* 209	*Staphylococcus* 15	*Bacillus cereus*	*Bacillus subtilis*
1	64	16	16	8	8
2	1,024	1,024	512	512	512
3	256	128	64	64	256
4	512	512	256	128	512
5	512	256	128	64	256
6	–[a]	1,024	512	128	1,024
7	64	64	64	64	64
8	128	128	128	64	128
9	256	256	256	64	256
10	256	64	64	64	128
11	1	89	8	2	2
12	128	16	64	8	16
13	1	16	16	4	8
14	–	512	512	256	512
15	1,024	256	128	128	512
16	64	4	8	4	8
17	8	16	16	4	64
18	64	64	32	32	128
19	64	64	64	32	128
20	–	256	512	64	512
21	–	512	512	256	–
22	–	1,024	1,024	512	–
23	–	–	–	–	–
Total tanshinone	16	16	16	2	4

[a] MIC > 1,024 μg/ml

some components have weak inhibitory effects on wine yeast and *C. albicans*.

Tanshinones were the first major active constituents that were isolated and studied systemically with modern techniques. Early studies demonstrated that tanshinones have significant antibacterial activities, and a clinical drug, Tanshinone, was developed based on such a discovery. Although the antibacterial activities of tanshinones were studied earlier, the knowledge of the antibacterial activities of different components in tanshinones is insufficient, except for scattered reports. Reevaluating the antibacterial activities of different components in tanshinones and studying their differences in antibacterial activities are important for truly understanding the actions of tanshinones.

Experimental results indicated that total tanshinones have strong antibacterial activities, which have exceeded the additive action of all components. The reasons for this may include two aspects: on the one hand, the addition of all components has enhanced activity and produces synergistic actions; on the other hand, there may be components, which are insoluble and thus not isolated yet, which have strong antibacterial activities. Doubtless, further studies are needed to test the hypothesis. The above discussed research results indicate that the perspective of tanshinones as antibacterial drugs is promising.

3.2.3 Anti-inflammatory Activity of Tanshinones

3.2.3.1 Effect of Tanshinones on Infectious Arthrocele of Rats

Tanshinones have significant anti-inflammatory activity against both infectious (infection by *S. aureus*) and noninfectious (albumen arthritis, carrageenan arthritis, and dextran arthritis) inflammations, but have no inhibitory activity on inflammatory granuloma.

3.2.3.2 Inhibitory Effect of Tanshinones on Mouse Ear Swelling Caused by Croton Oil

Tanshinone at 10 mg/ml has an inhibitory effect on mouse ear swelling; the effect is comparable to hydrocortisone at 0.1 mg/ml, indicating that tanshinones have strong anti-inflammatory activity.

3.2.4 Estrogen-like Activity of Tanshinones

3.2.4.1 Effect of Tanshinones on Uterine Development of Young Female Mice

Tanshinones at 400 mg/kg d with oral administration had a slight promoting effect on the uterine development of young female mice; it could increase the uterine weight of young female mouse, which is significantly weaker than subcutaneous injection of estrogen at 25 μg/kg d.

3.2.4.2 Effect of Resisting Male Hormones

Tanshinones and testosterone propionate were administered in male rats following excision of the testis, and the results showed that tanshinones have certain anti-male-hormone-like action.

3.2.5 Effect of Tanshinones on Diseases of Cardiovascular System

The effects of tanshinones on the cardiovascular system are the most important effect of the liposoluble constituents in Danshen. tanshinones in Danshen are also major components used to prevent cardiovascular diseases. Thereby, drugs with wide clinical application and definite effectiveness for cardiovascular diseases have been developed.

Studies have showed that tanshinone II_A plays a good role in preventing cardiovascular diseases. However, due to poor water-solubility, tanshinone II_A is difficult to be absorbed when administered orally. Chemists have modified the structure of tanshinone II_A and prepared the freely soluble injection of sodium tanshinone II_A sulfonate which has been widely used clinically.

However, studies on tanshinones are insufficient so far. Questions like whether every component has activity, whether each component has similar activity, and the relationship among the components of tanshinones are still yet be answered by further studies that are needed.

3.3 Effects of Salvianolic Acid on Myocardial Ischemia-Reperfusion Injury and Cardiac Muscle Cell

Xiuying Yang and Guanhua Du

Myocardial ischemia/reperfusion (I/R)-induced damage is an important cause of sudden death.

The mechanisms causing reperfusion injury mainly include excessive free radicals and calcium overload in cells. Reducing free radicals, inhibiting oxidation and peroxidation reactions, and reducing intracellular calcium are important routes of preventing I/R injury to heart muscle. Salvianolic acid A (SalA) has a very strong antioxidative function, which can capture oxygen radicals and reduce intracellular calcium. Therefore, SalA may have certain protective effects against myocardial I/R injury.

Studies using isolated hearts have demonstrated that SalA does have a protective effect against I/R injury. Animal experiments have also proven that SalA has a significant protective effect against both myocardial I/R arrhythmia and myocardial I/R damage.

3.3.1 Protective Effect of Salvianolic Acid A in Rat Myocardial I/R Injury

Spragne Dawley (SD) male rats of 200–250 g were used for studying acute myocardial I/R to demonstrate the protective effect of SalA against myocardial injury. The rats were randomized into blank, sham operation, model, positive drug (Lidocaine injection, 5 mg/kg), and low-, middle- and high-dose of SalA (0.5, 1.5, 5 mg/kg) groups, 6 animals per group. The rat model of acute myocardial ischemic reperfusion was prepared according to the methods described in literature. The SD rats were first anesthetized using 20 % urethane (0.5 ml/100 g), and a segment of normal electrocardiogram was recorded. Then, their trachea was cut open followed by intubation, and a breathing machine was connected for respiration support with breathing frequency at 74, inspiratory/expiratory ratio of 1:1, and tidal volume of 10–14 ml. Under the anesthetic state, the chest was opened, and the cardiac pericardium cut to expose the heart. Pin insertion was done under the left anterior descending coronary artery muscle, i.e. between the pulmonary conus and left auricle, 1.5 mm in depth. The needle was then withdrawn from the border of the right ventricle. Two thick threads were punctured across the middle of the ligature, and a thick needle was used to make the thick threads and the ligature individually permeated the chest wall of two sides. The thoracic cavity was sutured, breathing machine removed, the thoracic cavity was extruded to discharge accumulated blood and gas, and spontaneous respiration was resumed.

Five minutes after postoperative stabilization, the ligature was drawn tight and stabilized, and changes of the electrocardiogram were recorded, with observable elevation of the ST segment and J point. Ten minutes later, the ligature was loosened and changes of electrocardiogram recorded to observe ventricular fibrillation. Two hours after reperfusion, blood was collected from the femoral artery and centrifuged at 4,000 rpm to separate the serum, and the myocardial zymogram was measured. The hearts were taken out 2 h after reperfusion, and the cardiac atrium and connective tissue were collected, weighted, and cut into pieces and stored at −40 °C for later experiments.

3.3.1.1 ECG Records and Analysis of the Rats with Myocardial Ischemia and Reperfusion

Cardioelectric activities were recorded according to a standard II lead. Two electrodes were connected to the underside of the xiphoid bone and right arm. Then, changes in cardioelectric activities, arrhythmogenesis, and the time and duration of arrhythmia were observed. The arrhythmia was graded according to the standards set by The Lambeth Conventions:

1. no arrhythmia;
2. sporadic premature ventricualr contraction (<3 times/min);
3. frequent premature ventricualr contraction (≥3 times/min);
4. sporadic ventricular tachycardia (<3 times/min);
5. frequent ventricular tachycardia (≥3 times/min) or sporadic ventricular fibrillation (<3 times/min);
6. frequent ventricular fibrillation (≥3 times/min) or death.

The results showed that SalA at a medium dose of 1.5 mg/kg could reduce the tendency of post-ligation arrhythmia. It also could relieve post-reperfusion arrhythmia (Table 3.3).

3.3.1.2 Myocardial Enzyme Activities in Rat with Ischemic Reperfusion

Two hours after reperfusion, blood was collected from the femoral artery and placed at 4 °C for 6 h, then centrifuged at 6,500 rpm for 10 min to separate blood serum. The activities of creatine kinase (CK), creatine kinase isoenzyme MB (CK-MB) and lactate dehydrogenase (LDH) were measured. The results showed that after ischemic reperfusion, SalA at a medium dose of 1.5 m/kg could reduce the levels of LDH, CK and CK-MB in serum, which was statistically significant for CK and CK-MB (Table 3.4).

3.3.1.3 Effect of Salvianolic Acid A on ATPase Activities in Rat Heart Muscle with Myocardial Ischemic Reperfusion

ATPase can degrade ATP into ADP and inorganic phosphorus. Measuring inorganic phosphorus can estimate ATPase activity. The results showed that compared with the model group, the high dose group of SalA (5 mg/kg) had a tendency of increased total ATPase activity, Na–K ATPase and Ca–Mg ATPase. However, the differences were not statistically significant (Table 3.5).

3.3.2 Effect of Salvianolic Acid A on In Vitro Cultured Myocardial Cells of Rats

Myocardial cells of neonatal SD rats were cultured for Pl-P3 generations. Cells were inoculated in a 96-well cell culture plate. Tests were done following full fusion of the cells. 100 μl of hypoxic solution (NaH_2PO_4 0.9, $NaHCO_3$ 6.0, $CaCl_2$ 1.0, $MgSO_4$ 1.2 mmol/L, sodium lactate 40, HEPES 20, NaCl 98.5, KCl 10 mmol/L, pH6.8; pre-saturated with high-concentration nitrogen at 37 °C, 99.9 % 1 L/min × 30 min) was added (SalA at different concentrations was mixed into hypoxic solution). Then, the culture plate was placed into a sealing bag and the bag was filled with nitrogen at 37 °C for 1 h. Serum free DMEM medium

Table 3.3 Protective effect of SalA in rat Myocardial ischemic. Reperfusion Injury ($\bar{X} \pm s$, $n = 6$)

Group	Heart rate (preoperative)	Latency period of postischemia arrhythmia	Duration of postperfusion arrhythmia	Duration of postperfusion fibrillation ventricular
Model	330 ± 56	7.53 ± 1.25	3.04 ± 3.03	2.22 ± 1.65
SalA 0.5 mg/kg	332 ± 54	6.06 ± 3.39	4.68 ± 3.40	4.52 ± 5.07
SalA 1.5 mg/kg	313 ± 40	8.75 ± 1.76	2.12 ± 3.37	2.06 ± 3.38
SalA 5 mg/kg	330 ± 63	6.55 ± 0.99	1.69 ± 0.84	1.51 ± 0.99
Lidocaine 5 mg/kg	324 ± 67	6.92 ± 3.68	1.26 ± 1.07	0.88 ± 1.20

Table 3.4 Effect of SalA on the activities of myocardial relevant enzymes in blood of the rat with ischemic reperfusion ($\bar{X} \pm s$, $n = 6$, U/L)

	Normal	Model	Sham operation	SalA 0.5 mg/kg	SalA 1.5 mg/kg	SalA 5 mg/kg	Lidocaine 5 mg/kg
LDH	9.0 ± 2*	11.8 ± 2	8.9 ± 3*	10.5 ± 2	9.7 ± 1	11.1 ± 3	10.5 ± 2
CK	106 ± 94	201 ± 65	115 ± 92	134 ± 69	114 ± 41*	161 ± 82	181 ± 113
CK-MB	103 ± 83	153 ± 42	93 ± 71	116 ± 54	10531*±	135 ± 62	146 ± 80

Compared with model group, $^{*}P < 0.05$

Table 3.5 Effect of SalA on ATPase activities in the rat heart with ischemic reperfusion ($\bar{X} \pm s$, $n = 6$, μmolPi/mg P/hr)

	Normal	Sham operation	Model	SalA 0.5 mg/kg	SalA 1.5 mg/kg	SalA 5 mg/kg	Lidocaine 5 mg/kg
t-ATPase	7.28 ± 0.94	6.74 ± 1.05	6.82 ± 1.09	6.08 ± 0.59	6.72 ± 1.35	6.92 ± 1.47	6.72 ± 0.58
Na–K ATPase	4.65 ± 2.05	4.95 ± 2.35	5.19 ± 2.14	5.42 ± 2.34	4.83 ± 2.26	5.73 ± 0.56	5.56 ± 0.52
Ca–Mg ATPase	2.86 ± 1.09	2.59 ± 0.68	1.89 ± 0.81	2.72 ± 0.48	2.40 ± 1.17	3.72 ± 1.05	3.07 ± 0.66

oxygenated using 95 % O_2-5 % CO_2 was taken out and SalA was added at the same concentration used in the hypoxic solution. The supernatant in the culture plate was discarded, and the above oxygenated serum-free DMEM medium added. Again the culture plate was placed into a sealing bag and filled continuously with oxygen for 24 h. The cell viability was measured with ATP content and FDA method, and cell injury was detected with LDH method.

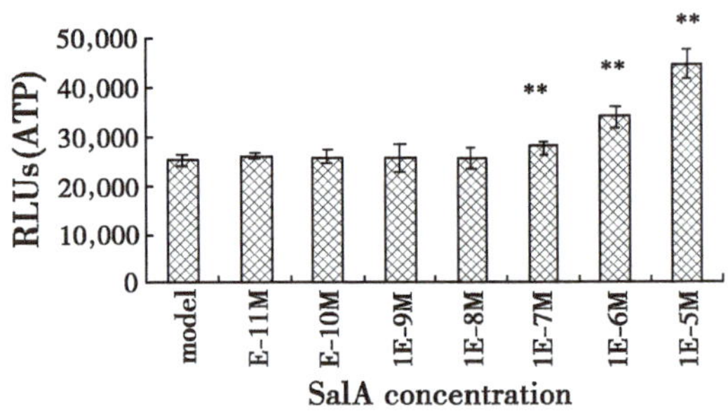

Fig. 3.1 Protective effect of SalA against hypoxia/reoxygenation injury of myocardial cells. SalA concentration

3.3.2.1 Effect of Salvianolic Acid A on ATP Content in Myocardial Cells

Following hypoxia/reoxygenation injury, ATP content in myocardial cells decreases. However, SalA at 10^{-5}–10^{-7} M can increase ATP content in these cells, suggesting that SalA has a protective effect against hypoxia/reoxygenation injury (Fig. 3.1). Similar results were obtained using FDA detection method.

3.3.2.2 LDH Activity in Culture Liquid of Myocardial Cells Treated with Salvianolic Acid A

In the myocardial cells of the hypoxia/reoxygenation injury model, there was increased LDH content in the culture medium supernatant. Compared with the model group, the SalA 10^{-5} group had reduced LDH activity in the cell supernatant, and the difference was statistically significant, suggesting that SalA at 10^{-5} has a protective effect against hypoxia/reoxygenation injury in the myocardial cells of rats (Fig. 3.2).

3.3.2.3 Effect of Salvianolic Acid A on Calcium Content in Myocardial Cells

Dual-wavelength fluorescence method was used to detect calcium in myocardial cells. Results showed that SalA reduced the ratio of excitation wavelength of Fura-2/AM at 340/380 nm, suggesting that SalA could decrease the rise of calcium content in myocardial cells induced by KCl (Fig. 3.3).

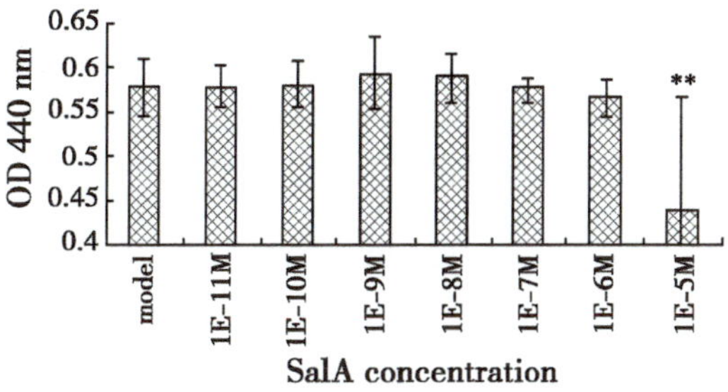

Fig. 3.2 Effect of SalA on LDH activity in culture medium of myocardial cells with hypoxia/reoxygenation injury.SalA concentration

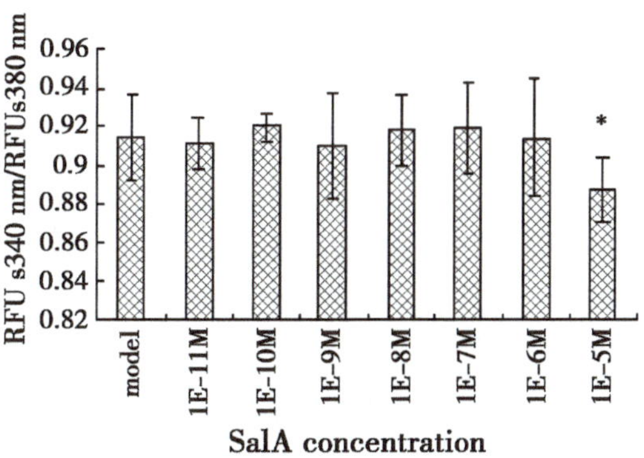

Fig. 3.3 Effect of SalA on free calcium in rat myocardial cells with hypoxia/reoxygenation injury

Danshen is a TCM drug used to activate blood circulation and dissipate blood stasis, which has the effects of nourishing blood, tranquilizing, cooling blood, removing abscesses, expelling pus, and promoting healing, and was used extensively clinically for treating coronary artery muscle disease, cerebral ischemic stroke, and Atherosclerosis. Modern pharmacological studies have found that it has the effects of resisting myocardial ischemia, anti-platelet aggregation, antithrombosis, regulating blood fat, antioxidation, and resisting the formation of atherosclerotic plaques. Studies on the mechanisms of these Functions have shown that Danshen has the effects of removing free radicals, reducing calcium overload, improving capillary function, inhibiting activation of leucocytes, and reducing cell Apoptosis, and has a significant therapeutic effect on myocardial ischemia-reperfusion injury.

Ischemic reperfusion studies were performed on the left anterior descending coronary artery muscle of SD rats with ligation, and the results demonstrated that SalA can reduce the incidence of ischemic reperfusion arrhythmia in ischemic myocardial tissue, and decrease the levels of CK and CK-MB in blood, with the most significant result achieved at a mid-dose of 1.5 mg/kg i.v. Compared with the model group, the high SalA dose (5 mg/kg) group had a tendency of an increase in activities of total ATPase, Na–K ATPase and Ca–Mg ATPase, which is consistent with previous reports.

This study showed that SalA can protect against hypoxia/reoxygenation injury in in vitro cultured myocardial cells of neonatal rats and, meanwhile, has an inhibitory effect against increased calcium in cells induced by KCl, suggesting that SalA inhibits apoptosis caused by injury factors. SalA can increase ATP production in myocardial cells. High levels of SalA ($>10^{-5}$ M) inhibit cell proliferation and beating, and cause injury. However, at low levels ($<10^{-5}$ M), SalA shows no significant effect. SalA at high doses ($>10^{-6}$ M) reduces the mitochondrial membrane potential ($\Delta\psi$) in neonatal rat cardiomyocytes.

3.4 Studies on Treating Nervous Degenerative Diseases with Salvianolic Acid B

Juntian Zhang, Guanhua Du, Mingke Tang and Yonghong Chen

Salvianolic acids A and B are potent antioxidants, with free radical scavenging activities 100–200 times stronger than that of vitamin E. Because of the importance of antioxidants in the treatment of nervous degenerative diseases, we studied the effects of salvianolic acid B (SalB) on preventing cerebral ischemia, inhibiting Aβ aggregation, improving mitochondrial injury and inhibiting cell Apoptosis, and the results are presented below.

3.4.1 Effect of Salvianolic Acid B on Mitochondrial Injury and Nerve Cell Apoptosis Caused by Cerebral Ischemic Reperfusion

3.4.1.1 Improving Effects of Salvianolic Acid B on Focal Cerebral Ischemia and Anti-Platelet Aggregation and Antithrombosis in Rats

Suture method was used to prepare a rat model of focal cerebral ischemic reperfusion injury. A single dose of 810 mg/kg i.v. administered 10 min before cerebral ischemia in rats can significantly improve abnormalities of nerve symptom caused ischemic reperfusion injury and reduce apoplexy index, cerebral infarction area and brain edema.

Both in vivo or in vitro, salvianolic acid has a significant inhibitory effect against platelet aggregation and thrombosis induced by multiple factors, but it does not affect the function of blood coagulation, and does not have risk of hemorrhage.

3.4.1.2 Protective Effect of Salvianolic Acid B on Mitochondrial Damage Caused by Focal Cerebral Ischemic Reperfusion

Intact mitochondria were isolated from ischemic brain mantle tissue. Then, the respiratory function of mitochondria was measured using Clark oxygen electrodes, and biochemical methods were used to detect the activities of complex enzymes in the mitochondrial electron transfer chain. Results indicated that cerebral ischemia reperfusion significantly impaired the respiratory function of mitochondrion, resulting in a P/O ratio reduced from 2.88 to 2.31, state 3 respiratory rate reduced from 99.93 to 63.22 nmol/(min mg), state 4 respiratory rate increased from 28.01 to 37.59 nmol O/(min mg), and respiratory control ratio reduced from 3.65 to 1.70. Salvianolic acid at 3–10 mg/kg i.v. caused dose-dependent P/O ratio and state 3 or 4 respiratory rates to be at approximately normal levels. Cerebral ischemia reperfusion can reduce the activities of complexes I and IV from 1.79 and 0.3 to 0.43 and 0.18 μmol/(min mg protein), respectively. Salvianolic acid at 810 mg/kg i.v. can increase the activities of complexes I and IV to 1.33 and 0.32 μmol/(min mg protein), respectively.

3.4.1.3 Inhibitory Effect of Salvianolic Acid B on the Nerve Cell Apoptosis Caused by Cerebral Ischemia Reperfusion and its Mechanism

Study results indicated that ischemia for 1 h plus reperfusion for 24 h in rat brain can significant induce apoptosis of nerve cells in ischemic brain mantle, and that SalB administered at 3, 6, and 10 mg/kg i.v. before ischemia can inhibit nerve cell apoptosis induced by cerebral ischemia reperfusion, with best effectiveness at 10 mg/kg. Ischemia for 1 h plus reperfusion for 24 h in rat brain can impair the barrier function of mitochondrial membrane, reduce mitochondrial membrane potentials, cause mitochondrial swelling, release cytochrome C, and increase Caspase-3 mRNA and protein in ischemic brain mantle. The above doses of SalB can protect the barrier function of mitochondrial membrane and inhibit production of Caspase-3 mRNA and protein.

It is thus evident that by increasing mitochondrial membrane potentials, increasing antioxidant GSH and antioxidase SOD, and inhibiting the release of cytochrome C, SalB inhibits the activation of Caspase-3, and thus achieves an anti-apoptotic effect.

3.4.1.4 Anti-apoptotic Effect of Salvianolic Acid B on PC12 Cells with High-Level of Caspase-3 Expression

Plasmid *p*EGFP-C containing wild-type Caspase-3 (*p*EGFP-C-Caspase-3) was transfected into PC12 cells via liposomes, and G418 used for screening cell clones. Western blot was used to screen cell strains with a high level of

Caspase-3 expression. Finally, the PC12 cell line with stable and high levels of Caspase-3 expression (PC12-Caspase-3) was constructed successfully. 6-OHDA can significantly increase the cell Apoptosis rate of the cell line and the activity of Caspase. SalB at 10^{-5}, 10^{-6} and 10^{-7} mol/L can significantly decrease the cell Apoptosis rate and Caspase activity, indicating that SalB can inhibit the cell Apoptosis mediated by Caspase-3.

3.4.2 Inhibitory Effect of Salvianolic Acid B on $A\beta_{1-40}$ Fibrogenesis and Its Protective Effect on Mitochondrial Damage and Cell Apoptosis of PC12 Cells Caused by $A\beta_{1-40}$ Self-aggregation

3.4.2.1 Effect on $A\beta_{1-40}$ Self-aggregation and Fibrogenesis

$A\beta_{1-40}$ (100 mg/L) was incubated at 25 °C for 30 h. Aβ aggregation and fibrogenesis were observed with electron microscopy, and from 48 to 100 h, fibrogenesis increased. SalB at 10 nmol/L within 30 h could completely inhibit Aβ aggregation and fibrogenesis, and the inhibitory effect could last until 48–100 h.

3.4.2.2 Inhibitory Effect on PC12 Cytotoxicity Induced by Aging Aβ

Aging (aggregated) $A\beta_{25-35}$ was incubated with PC12 cells for 48 h, and its cytotoxicity to PC12 cells was confirmed by a MTT reduction assay. SalB at 1 μ mol/L can significantly reduce such cytotoxicity.

3.4.2.3 Inhibitory Effect on Aβ Aggregation-Induced Apoptosis of PC12 Cells and the PC12 Cells Expressing Wild-Type or Mutant PS-2

Flow cytometry analysis indicated that SalB at 10^{-8}–10^{-6} mol/L can significantly inhibit cell apoptosis of PC12, PC12PS-2, and PC12PS2N1411 cells caused by $A\beta_{1-40}$ aggregation, with the inhibition rates being 19, 22, and 20.8 %, respectively.

3.4.2.4 Effect on Expression of Apoptotic Factor-4 (Par-4) of Prostatic Cells

Recent studies have found that Par-4 is a new apoptotic factor, which might induce apoptosis by activating Caspase-3. Western blot analysis demonstrated that after exposing PC12 and PC12PS2N1411 cells to $A\beta_{1-40}$ for 24 h, the expression of Par-4 increased. Pretreatment with SalB at 10^{-7} mol/L could prevent the increase.

3.4.2.5 Inhibitory Effect on the Intracellular Increase of Calcium and Reactive Oxygen Species (ROS) Caused by Aβ

After PC12PS2N1411 cells were incubated with $A\beta_{1-40}$ for 24 h, intracellular calcium increased from 188.48 nmol/L at the stationary phase to 326.2 nmol/L. SalB at 10^{-7}–10^{-6} mol/L reduced cellular free Ca^{2+} to 249.22 and 233.14 nmol/L, and could also significantly inhibit the increase of mitochondrial ROS caused by $A\beta_{1-40}$.

3.4.2.6 Protective Effect Against MitochondriaL Membrane Injury

After PC12PS2N1411 cells were exposed to $A\beta_{1-40}$, mitochondrial membrane potentials decreased significantly. SalB at 10^{-7}–10^{-6} mol/L could recover mitochondrial membrane potentials to near the normal level.

3.4.2.7 Effect on Glucose Uptake in Synaptosomes

Fe^{2+} at 50 μmol/L can obviously impair glucose uptake by synaptosomes. As time continues, the uptake rate falls to very low levels. SalB at 10^{-6} mol/L can significantly improve the disorder of glucose uptake.

3.4.2.8 Effect on Release of Cytochrome C

After being exposed to Aβ, the mitochondrial membrane could release cytochrome C. SalB at

10^{-7}–10^{-6} mol/L could significantly inhibit the release.

3.4.2.9 Inhibitory Effect on Caspase-3

It was observed using chemical methods and Western blot that SalB can inhibit the activation of Caspase-3 caused by Aβ.

3.4.3 Effect of Salvianolic Acid B on Neurogenesis in Rat Middle Cerebral Artery with Ischemic Reperfusion

Neurogenesis occurs over the entire life of mammals, including humans. It has been confirmed now that neurogenesis occurs mainly in the sub-granular zone (SGZ) and sub-ventricular zone (SVZ) of the dentate gyrus of Ammon's horn. Factors like ischemia, trauma, learning, exercise, abundant environment, and neurotrophic factors can promote the brain to produce new nerve cells. In contrast, aging, nervous degenerative diseases and β-amyloid proteins impair the brain's capability of producing new nerve cells. In recent years, many studies have found that some small natural molecules from plants also have neurogenesis-promoting effects, e.g., ginsenoside Rgl, bilobalide, and SalB.

Previous studies have found that SalB has the effects of preventing oxidation, cell Apoptosis, amyloid aggregation and amyloid cytotoxicity. In addition, Lay et al. proved that SalB can significantly promote the proliferation of vascular endothelial cells and has a mitogen-like action. Recent studies demonstrated that SalB not only can improve rat nerve cell injury caused by acute cerebral ischemia reperfusion, as well as behavior disorders caused by cerebral ischemia, but also has proliferation- and differentiation-promoting effects on the nerve cells in the SGZ and SVZ of the dentate gyrus of Ammon's horn. This may be an important mechanism of SalB's function of protecting the brain under hypoxic conditions.

In this study, rats with middle cerebral artery occlusion (MCAO) were used. Samples collected 7 days after operation were examined for neurogenesis by BrdU labeling. Samples collected 14 days after operation were observed for nerve cell injury by Nissl body staining.

3.4.3.1 Effect of Salvianolic Acid B on the Number of Newborn Cells in the SGZ of Ammon's Horn in MACO Rats

BrdU positive cells could be seen in Ammon's horn in the groups of sham operation, ischemic reperfusion, and SalB at 1 and 10 mg/kg (Fig. 3.4), indicating that there is neurogenesis in Ammon's horn under physiological conditions. Neonatal cells in the model and salvianolic acid groups were significantly more than those in the sham operation group, indicating that ischemic injury can stimulate neurogenesis in Ammon's horn. Neonatal cells were distributed mainly in the SGZ and showed morphological characteristics of granulocyte, which assumed a cluster distribution and their nuclei displayed irregular, spherical or ellipsoidal morphology. Compared with the model group, positive neonatal cells in SGZ were increased in SalB at 1 mg/kg, but statistically insignificant ($p < 0.05$), and increased significantly in SalB at 10 mg/kg group ($p > 0.01$), indicating that SalB can further promote neurogenesis in Ammon's horn of rats with cerebral ischemia reperfusion (Table 3.6).

3.4.3.2 Effect of Salvianolic Acid B on the Number of Newborn Cells in the SVZ of Ammon's Horn in MACO Rats

BrdU positive cells can be seen in the sham operation, ischemic model, and SalB groups, which are mainly distributed in the dorsolateral angle and lateral wall of the lateral ventricle. Nuclei of BrdU positive cells had two shapes: one was large and spherical and stained deeply with brilliant yellow fluorescence; another was ellipsoidal and stained slightly with red or yellowish fluorescence. Outward extension of BrdU positive cells could be seen in the triangle zone of the SVZ dorsolateral angle. In addition, BrdU positive cells were also observed in the partial capillary endothelium of the SVZ and corpus striatum. BrdU positive cells in SVZ were

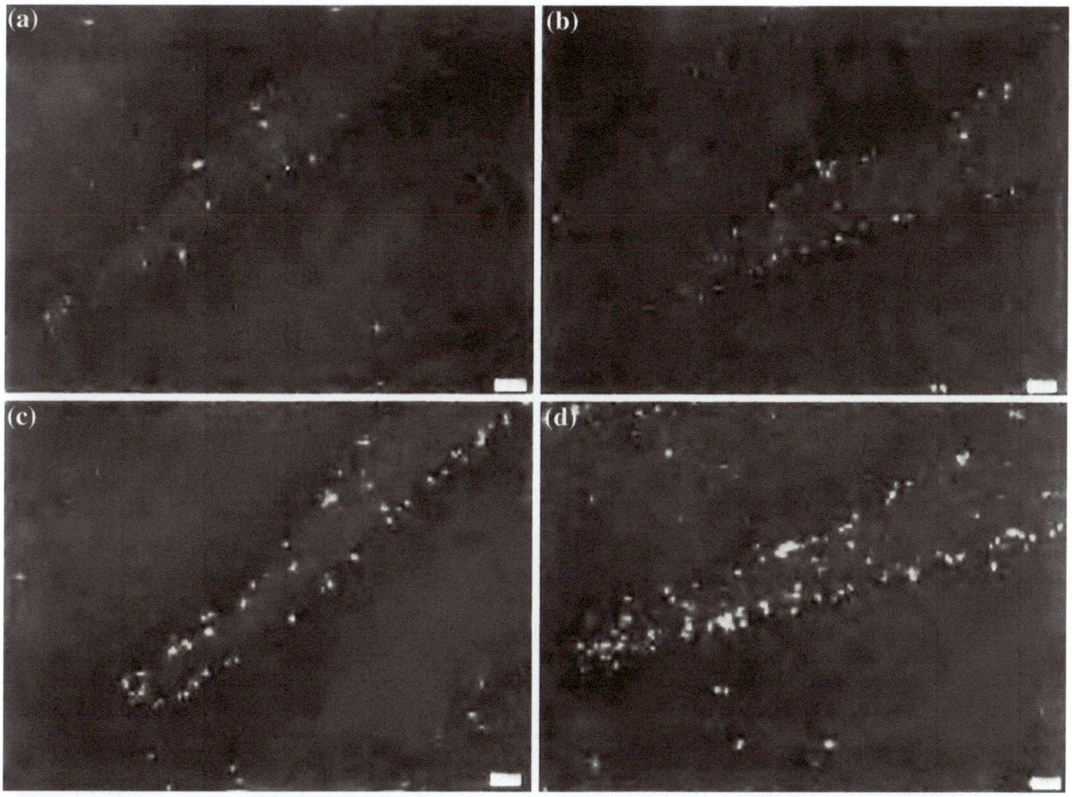

Fig. 3.4 Effect of salvianolic acid B on newborn cell number in the SGZ of Ammon's horn of MACO rat (x200). **a** Sham operation. **b** Ischemia reperfusion. **c** Salvianolic acid B at 1 mg/kg. **d** Salvianolic acid B at 10 mg/kg

Table 3.6 Effect of salvianolic acid B on BrdU positive cells in ischemic brain tissue, $n = 6$, $\bar{X} \pm s$

Group	Dose mg kg^{-1}	SVZ/n mm^{-2}	SGZ (total cells)
Sham operation	–	$75 \pm 20^{*}$	$11 \pm 2^{*}$
Ischemia reperfusion	–	$125 \pm 21^{\triangle}$	$16 \pm 3^{\triangle}$
Salvianolic acid B	1	$210 \pm 11^{**\triangle\triangle}$	$19 \pm 2^{\triangle}$
	10	$300 \pm 27^{**\triangle\triangle}$	$47 \pm 4^{**\triangle\triangle}$

* Compared with ischemia reperfusion: * $P < 0.05$, ** $P < 0.01$
$^{\triangle}$ Compared with sham operation: $^{\triangle}$ $P < 0.05$, $^{\triangle\triangle}$ $P < 0.01$

significantly more in the model group than in the sham operation group ($p < 0.05$, Table 3.6), indicating that ischemic injury can also stimulate neurogenesis in SVZ. Compared with the model group, BrdU positive cells in SVZ increased in the SalB at 1 mg/kg group, and significantly increased in the SalB at 10 mg/kg group ($p < 0.01$, Table 3.6) where nuclei increased significantly and had clearer outlines, demonstrating that SalB can promote neurogenesis in the SVZ of ischemia/reperfusion rats.

3.4.3.3 Effect of Salvianolic Acid B on Nerve Cells in CAl Zone and Cortex of Ammon's Horn in MCAO Rats

In sham operated rats, there were tightly lined pyramidal cells in the CAl zone of Ammon's horn and cortical cells lined up with a large spherical nucleus, clear nucleolus, and abundant and deeply stained Nissl bodies, after Nissl body staining (Fig. 3.5). In the model group, there were uneven, discontinuous, loose cells in the CAl zone of ischemic Ammon's horn, and cell counting showed that neurons decreased significantly ($P < 0.01$). For cortical nerve cells, there was an unstuck and loose alignment, and significantly reduced Nissl bodies in the cytoplasm with decreased coloring; nerve cells decreased significantly ($P < 0.01$) and glial cells increased. In the SalB at 10 mg/kg group, the cell morphology in the CAl zone of Ammon's horn was basically intact; Nissl bodies were numerous and the staining was deep, clear, and abundant; and the number of nerve cells was more than that in the model group ($P < 0.01$) and about the same as in the sham operation (Fig. 3.5g). In SalB groups, there was a great loss of nerve cells in ischemic cortex, accompanied by a great proliferation of glial cells. Compared with the model group, there were more scattered nerve cells in SalB groups, as shown by cell counting (Fig. 3.5h).

3.4.3.4 Effect of Salvianolic Acid B on Recovery of Limb Function in MCAO Rats

After ischemia reperfusion in rats, the motor function of limbs decreased obviously and recovered slowly with time. The recovery was not significant in the first week after ischemia, but accelerated in the second week, when most animals could reach the endpoint through gimbal

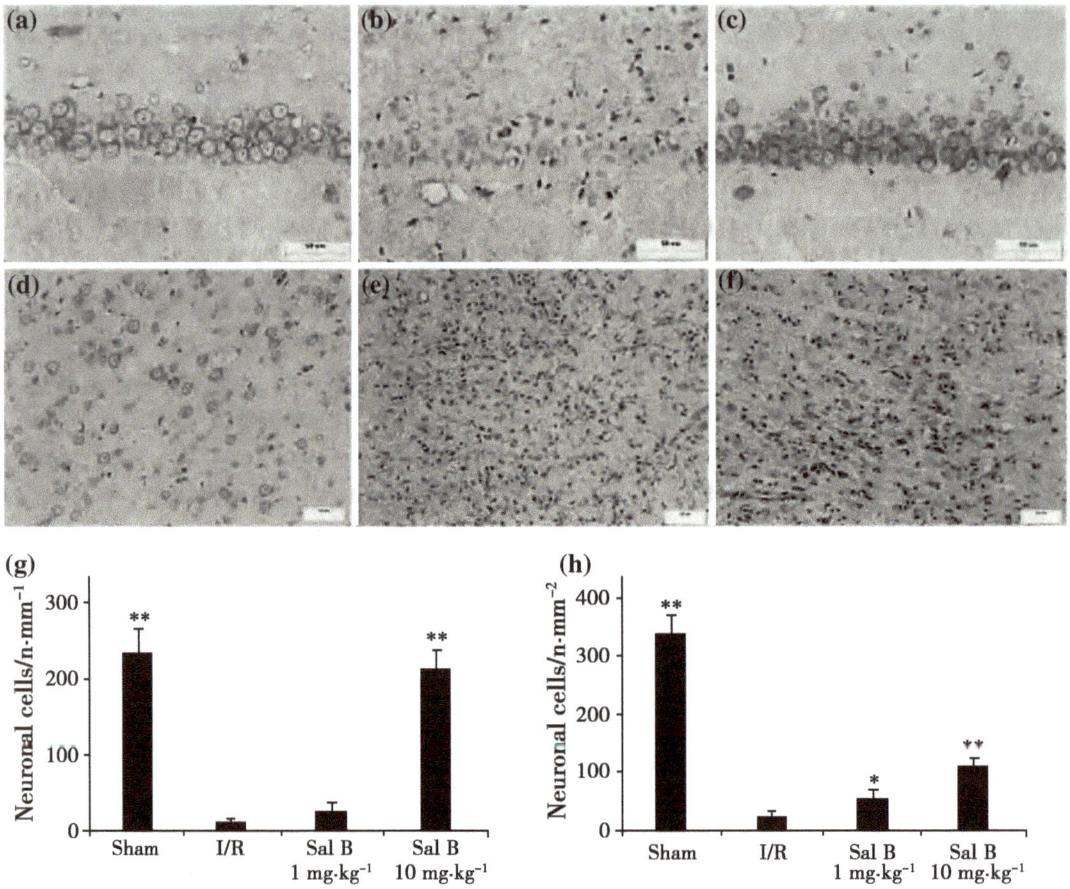

Fig. 3.5 Effect of salvianolic acid B on nerve cells in CAl zone and cortex of Ammon's horn in MCAO rats. Compared with ischemia reperfusion: $^{*}P < 0.05$, $^{**}P < 0.01$; n:6; Scale bar, 50 μm

lever. However, the motor function score was low compared with the sham operation (Fig. 3.6). SalB at 10 mg/kg could significantly promote the recovery of motor function in ischemic animals, the effect of SalB becoming more and more obvious from day 10. On day 14 after ischemia, the motor functions of SalB treated rats were almost the same as the rats in the sham operation group. SalB at 1 mg/kg had no obvious effect.

This study found that salvianolic acid B can improve nerve cell injury in the CAl zone and cortex of Ammon's horn in ischemic rats, the BrdU positive cells in SGZ and SVZ increased compared with the ischemia model, and salvianolic acid B can significantly improve the recovery of motor function. The results suggest that the effect of SalB's neurogenesis promotion in rats with cerebral ischemia is related closely to its protective effect on the brain. For adult mammals, neurogenesis occurs mainly in the SGZ and SVZ. A large number of studies have found that in the rat model of transient focal cerebral ischemia, BrdU positive cells in the SVZ increase significantly on day 4 following ischemia and reach the peak value Es3 on day 7. Also, the rate of newborn cell generation in the SGZ increases significantly on day 7, with peak values on days 7–11. Our study found that on day 7 post-ischemia, all nerve cells in the SVZ and SGZ of ischemic rats increased significantly. However, compared with the ischemia model, SalB (at 1 and 10 mg/kg) can further promote the increase of nerve cells in these zones.

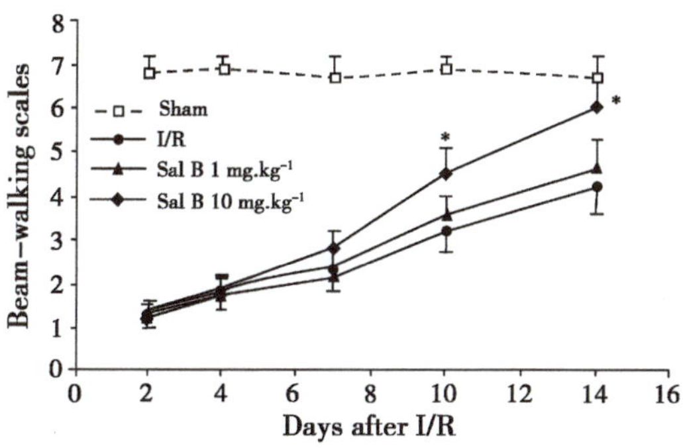

Fig. 3.6 Effect of salvianolic acid B on recovery of limb function in rats. Compared with ischemia reperfusion: $^*P < 0.05$ ($x \pm$ SD $n = 10$)

As a new hot spot in neuroscience research, the study of stem cells or precursor cells is growing rapidly. By now, scientists generally believe that the new nerve cells in Ammon's horn are mainly cells in the SGZ of the dentate gyrus of Ammon's horn, which can continuously regenerate, differentiate, and migrate. After maturation, new cells replace the injured or apoptotic nerve cells. Our study has found that there is a great loss of nerve cells in the CA1 region 2 weeks after ischemia, with an unclear CA1 band. In the groups treated with SalB, there are clearly stained nerved cells in the CA1 zone, with intact morphology. Combined with the fact that SalB can significantly promote the generation of nerve cells in the SGZ, we conjecture that cells in the CA1 region may be differentiated and maturated via the infragranular layer in the dentate gyrus of Ammon's horn. There have been different reports on neurogenesis in the cerebral cortex. Most scientists considered that cortex itself has no neurogenic function and that the nerve cells occurring following ischemia might come from the SVZ. Our study demonstrates that SalB can significantly promote the production of nerve cells in the SVZ zone of ischemic brain. Compared with the model, the Nissl body staining-positive nerve cells in the motor cortex in the SalB group increase significantly, suggesting that the function of SalB in protecting cortical nerve cells may be related to the fact that it promotes neurogenesis in the SVZ. Our study also finds that there is a significant decrease in motor function in rats following cerebral ischemia, and that the motor function recovers obviously after administrating SalB for 14 days. This function recovery may be related to the fact that SalB promotes migration of the new nerve cells in the SVZ toward the neostriatum and motor cortex.

Studies demonstrate that Danshen can promote the regeneration of liver and bone tissues. Some even suggest that the function may be related to its promotion of DNA synthesis, showing mitogen-like action. Our results show that SalB can promote the production of nerve cells, which may be the very reason that Danshen can promote cell proliferation. Lay, et al. discovered that SalB can regulate the levels of

VEGF and VEGF receptors in vitro, promote proliferation of vascular endothelial cells, and accelerate the repair and healing of skin with ischemic injury. These studies provide evidence for the application of salvianolic acids in cerebrovascular disease treatment.

This study demonstrates that SalB can promote neurogenesis in rat brain with ischemic injury, which provides a new target of action for the study on improvement of ischemic injury by SalB, and more possibilities for treating nervous degenerative diseases (e.g., AD), cerebral apoplexy and various memory disorders.

3.4.4 Prospects of the Study on Salvianolic Acid B

Under normal pharmacological conditions, many biochemical reactions may produce oxygen radicals. For example, during metabolism of arachidonic acid, xanthine oxidase acts on the substrate hypoxanthine. The self-oxidative reaction of catechol amine activates granular leukocytes. Especially, mitochondria consume a great amount of oxygen during energy metabolism, where about 1–2 % oxygen is oxidized into superoxide anions during oxidative phosphorylation. However, mitochondria contain a large number of antioxidants (reduced glutathione, NADPH, and vitamins C and E) and antioxidases (SOD, glutathione peroxidase [GSH-Px], catalase [CAT], NADP hydrogen transportase) which can remove the superoxide anions produced in mitochondria. Under aging and pathological conditions, excessive production of superoxide anions or a sharp reduction of antioxidants impairs the balance of oxidation and antioxidation, and thus produces severe harm to tissues and organs. It is realized now that excessive production of free radicals is the cross point of multiple pathogenetic mechanisms of nervous degenerative diseases. It has a reciprocal effect with calcium overload and thus produces a so-called malignant feed-forward cycle, i.e., Ca^{2+} accelerates production of free radicals or, in contrast, free radicals increase the accumulation of Ca^{2+}. SalB and SalA are 100 times more potent in antioxidative activities than vitamin E, mannitol and gingko extract Egb761. Oil-water distribution data suggests that the two compounds can permeate the blood-brain barrier and have a greater inhibitory effect on the lipid peroxidation of brain microparticles than on the lipid peroxidation of liver and kidney microparticles. As a whole, the experimental results demonstrate the application value of SalB in the therapy of nervous degenerative diseases.

Cerebral ischemia is a nervous degenerative disease characterized by the reduction of blood flow in microcirculation and glucose/energy Metabolism disorders. Key early factors causing ischemic cerebral injury include disorders of calcium homeostasis, excitotoxicity, excessive production of free radicals, mitochondrial function damage, and energy depletion. During the middle and late stages of cerebral ischemia, delayed death and delayed inflammation occur in neurons, accompanied by a lack of neurological symptoms, decline in memory and cognitive abilities, cerebral infarction and brain edema. According to the above pathogenesis and mechanism of ischemic cerebral injury, an ideal anti-cerebral-ischemia drug should meet the following basic requirements: (1) it dilates cerebral vessels selectively and does not act on peripheral blood vessels, otherwise, the decrease in systemic blood pressure will reduce cerebral perfusion pressure; (2) it increases cerebral blood flow, but at the same time does not increase cerebral metabolism, or the oxygen and glucose needed for cerebral metabolism will be less than the increase in cerebral blood flow; (3) it has the effect of dilating cerebral vessels but without the cerebrovascular "steal" phenomenon, i.e., it only increases the blood flow in ischemic brain but has no effect on the blood flow in normal regions; (4) it has not only the effect of anti-cerebral ischemia, but also the effects of anti-platelet aggregation and anti-thrombosis; (5) it can inhibit the release of excitatory amino acids, remove cellular calcium overload and obviate excessive production of free radicals; (6) it protects mitochondria, and can increase energy metabolism; (7) it can permeate the blood-brain barrier and achieve higher levels in the brain;

(8) it can inhibit nerve cell Apoptosis and reduce the production and expression of cytokines and adhesion molecules related to inflammation. By a large number of studies, the pharmacological actions of salvianolic acid have the following characteristics: within 0.5 h before or after focal cerebral ischemia, a single intravenous administration has the effects of reducing the cerebral infarction area, relieving brain edema and nervous ischemic symptoms and improving Memory; it increases blood flow in cortex with ischemia caused by MCAO but does not affect the blood flow in normal cortex; although SalA and SalB are water-soluble constituents, they can permeate the blood-brain barrier with an oil-water distribution coefficient of 1.15 and 0.54, respectively; salvianolic acid can also reduce cellular calcium and inhibit the release of calcium-dependent and non-calcium-dependent glutamic acids in addition to very strong effects of antioxidation and removing free radicals; it has a significant protective effect against mitochondrial injury caused by ischemia and anoxemia, increases ATP content and reduces accumulation of lactic acid; salvianolic acid has a significant inhibitory effect on platelet aggregation and thrombosis caused by multiple factors; salvianolic acid can significantly inhibit nerve cell apoptosis caused by cerebral ischemia and Aβ. Based on the above pharmacological characteristics, salvianolic acid meets the prerequisites of an ideal anti-cerebral-ischemia drug.

At present, the drugs used for the prevention and treatment of senile dementia are mainly cholinesterase inhibitors and cerebrovascular vasodilators; both are symptomatic treatments without ideal efficacy. With the continuous extension of human life, the incidence rate of senile dementia is also increasing, and thus it is more urgent to develop a new drug which is superior to existing drugs. Because the studies of senile dementia pathogenesis are deepening, there are already some new therapeutic targets, and the idea of interference with the formation and deposition of Aβ has been accepted by the majority of scholars. The studies include reducing the production of amyloid precursor protein (APP) using antisense technology, inhibiting sedimentation of Aβ, developing β,γ-secretase inhibitors and lysosome inhibitors, and reducing the neurotoxicity of Aβ. The application of $A\beta_{1-42}$ vaccine in APP transgenic animals in the USA has demonstrated that it can prevent the formation of Aβ in early stages and reduce sedimentation of Aβ in late stages, indicating the promising prospect of the therapeutic strategy. Our study proves for the first time that SalB incubated at 10^{-8} mol/L at 25 °C for 30 h can completely inhibit the aggregation of $A\beta_{1-40}$ and fibrogenesis. Currently, the popular opinion on cell apoptosis is: actions of ischemia/anoxemia, Aβ and PS on the mitochondrial membrane lead to decreases in membrane potentials and ATP synthesis; increase of ROS and cellular calcium → increase the release of cytochrome C → activation of apoptosis-related factors, with increased expression of Caspase → DNA fragmentation and cell apoptosis. Sal B not only has the effects of preventing cerebral ischemia, improving glucose transport, and increasing ATP content, but also has the effects of increasing mitochondrial membrane potentials, inhibiting release of cytochrome C, reducing cellular calcium and production of ROS, and inhibiting expression of upstream apoptotic factor Par-4 and apoptosis executing factor Caspase-3.

Neuronal cell death occurs in two ways: cell apoptosis and cell necrosis. The former is caused mainly by increased expression of apoptotic genes, and the latter is due to uncontrollable massive ions, especially internal flow of calcium ions, and cell lysis. It is now confirmed that whether an increase in the expression of apoptotic genes or the flow of a large amount of ions into the cells, the culprit is still the disorder of calcium homeostasis and the excessive formation of free radicals. Cerebral ischemia and glucose metabolism disorder are considered acute neurodegenerative diseases, and senile dementia and Parkinson's are chronic neurodegenerative diseases. Both types of diseases have similar a pathogenesis: cellular calcium overload and excitotoxicity in the early stage; followed by mitochondrial injury and reduced and exhausted energy supply; cell apoptosis, reduced neural plasticity, and nerve degeneration in the late stage. SalB is a strong antioxidant, which can not only remove cellular calcium overload, but also antagonize Aβ toxicity and inhibit cell apoptosis. Undoubtedly, developing it into a drug to treat senile dementia has important theoretical significance and practical value.

3.5 Examples of the Effect of Salvianolic Acid on Focal Cerebral Ischemia

Zhiwei Qu, Jie Wang and Xiaoying Wang

Electric cautery was used to make MCAO rats a focal cerebral ischemia model. The protective effect of salvianolic acid and its powder injection on cerebral ischemia was observed and compared with that of the positive control drug, Composite Danshen injection (CDI). The results showed that salvianolic acid at 20 or 10 mg/kg i.v. and Salvianolic Acid Powder injection (SAPI) (13.5 mg powder contained 10 mg salvianolic acid) at 27 or 13.5 mg/kg i.v. could significantly reduce the area of cerebral infarction caused by MCAO within 24 h, and reduce the water content in the ischemic side of the brain, with comparable effects and no significant difference. The effects of SAPI at 13.5 mg/kg and salvianolic acid at 10 mg/kg (equivalent to 5 times of clinical dosage) are comparable to that of CDI at 2 g/kg (equivalent to 10 times of clinical dosage).

3.5.1 Objective

To observe the effects of salvianolic acid on functional neurological deficit cerebral infarct area and cerebral water content in focal cerebral ischemia rats caused by MCAO.

3.5.2 Experimental Materials

3.5.2.1 Animals

Male Wistar rats, weight 280–330 g, supplied by the Laboratory Animal Center in the Chinese Academy of Medical Sciences.

3.5.2.2 Apparatus

High frequency electric knife, model KW65-3A, manufactured by Beijing Medical Electronic Apparatus Factory; operating microscope, model SXP-18, manufactured by Shanghai Medical Optical Instrument Factory; P/G2003-incubator/carbinet drier, manufactured by Chongqing Experimental Equipment Factory.

3.5.2.3 Chemicals and Reagents

Salvianolic acid: lot 981008.

Salvianolic Acid Powder injection (SAPI): lot 981201 (100 mg salvianolic acid in every 135 mg of powder).

The abovementioned chemicals were provided by Professors Li LN and Tao ZH in our institute. CDI was manufactured by Huangshan Tianmu Pharmaceutical Co. Ltd, lot 990506. Red tetrazoline (TTC) was manufactured by Academy of Military Medical Sciences of PLA, chemically pure, lot 870401.

3.5.2.4 Formulation of Experimental Drug

SAPI: salvianolic acid was prepared with sterile physiological saline to the required concentration.

3.5.2.5 Dose and Administration Route

SAPI: 27, 13.5 or 6.75 mg/kg; salvianolic acid: 20, 10 or 5 mg/kg; CDI: 2 or 1 g/kg. Route: injection at sublingual vein.

3.5.2.6 Setup of Test Groups

Male Wistar rats were randomized into 10 groups, i.e., sham operation, control (normal saline, i.v.), SAPI at 27, 13.5 or 6.75 mg/kg, salvianolic acid (981008) at 20, 10 or 5 mg/kg, and CDI at 2 or 1 g/kg.

3.5.2.7 Storage of the Drugs

Dilute freshly, and store at 4 °C for later use within 3 days.

3.5.3 Experimental Method

3.5.3.1 Test for Focal Cerebral Ischemia in Rats

After anesthetization with chloral hydrate at 350 mg/kg i.p., immobilize the rat in the left-lateral position on the plate. Under the operating microscope, cut open the skin along the middle point of the connecting line between the external auditory meatus and angulus oculi to expose the zygomatic arch, and use a small stretcher to increase the interval between the squamous bone and mandible. Open a 1 cm × 1 cm bone window at the bottom of the skull, and tear the cerebral dura mater to expose the middle cerebral artery. By electric coagulation, use a high frequency electric knife to occlude the middle artery on one side of the brain to reate focal cerebral ischemia, followed by sewing up the incision layer by layer. Intra- and post-operative room temperatures were controlled strictly at 24–25 °C.

Except for the sham operation group, occlusion was done in the middle artery on one side of the brain for other groups. 30 min after the occluding, drug administration was done via the sublingual vein. 24 h later, nerve symptoms were assessed according to the methods described in the reference. Then, perform decapitation to take out the brain and measure water content in brain.

Additional male Wistar rats were grouped and administered as above. Perform decapitation 24 h after occluding, take out the brain, and measure the cerebral infarction area according to the methods described in the reference.

3.5.3.2 Measurement of Cerebral Infarction Area

Place the fully stripped brain into a small cup with physiological saline in a 4 °C refrigerator for 10 min, remove the olfactory bulb, cerebellum and low-set brain stem, then cut the remains into five pieces along the coronal plane. Immediately place them into TTC staining solution, and incubate in a 37 °C water bath for 30 min in the dark. Take out the brain pieces and place into 10 % formaldehyde solution for fixation. Normal tissue assumes a rosy pink color, and ischemic tissue a white color. The infarction area was measured and then calculate the percentage of the infarct area of the area of the whole brain.

3.5.3.3 Neurological Symptoms Score Criteria

1. Left rat tail and observe the two forelimbs. For normal animals, their forelimbs are forward protruding and symmetric. Following operation, the forelimbs contralateral to the ischemic brain hemisphere curl inward and the shoulders rotate inward. Observe the extent and rate from 0 to 4 points.
2. Drag two limbs. The muscle strength is symmetric for normal rats. After operation, the muscle becomes weak in the forelimbs contralateral to the ischemic brain hemisphere. Observe the extent and rate from 0 to 3 points.
3. Push two shoulders. The resistances of bilateral shoulders are symmetric for normal rats. After operation, the resistance is reduced in the shoulder contralateral to the ischemic brain hemisphere. Observe the extent and rate from 0 to 3 points.

According to the above criteria (full score of 10), the higher the scores are, the more severe the brain disorder is, which is used as the index for brain damage.

3.5.3.4 Determination of Brain Water Content

Dry-wet method was used to measure the brain water content. After euthanizing the animal, take out the whole brain and remove the rhinencephalon, low-set brain stem and cerebellum. Immediately, take the wet weight of the brain and bake in a 120 °C baking oven for 18 h until constant weight is reached. Weigh the dry brain. The calculation formula for brain water content is the following:

brain water content (%)

$$= (\text{wet weight of brain} - \text{dry weight of brain}) / \text{wet weight of brain} \times 100\,\%$$

3.5.4 Results

3.5.4.1 Effect of SAPI and Salvianolic Acid on Cerebral Infarction Area of MCAO Rats

As shown in Table 3.7, SAPI at 27 or 13.5 mg/kg and salvianolic acid at 20 or 10 mg/kg could reduce the cerebral infarction area of rats 24 h after MACO. Compared with the control group, the difference was statistically significant. The effects of the two preparations were statistically identical. Similar effects were also observed among the following groups: salvianolic acid at 10 mg/kg, SAPI at 13.5 mg/kg, and CDI at 2 g/kg.

3.5.4.2 Effect of SAPI and Salvianolic Acid on Functional Neurological Deficit in MCAO Rats

After waking from anesthesia, hemiplegia-like symptoms occurred in the control group rats; the symptoms were mainly inward contraction of contralateral forelimbs, intorsion of the shoulder, reduced muscle tone in forelimbs, and reduced resistance of forelimbs when pushed toward the opposite side. SAPI at 27 or 13.5 mg/kg and salvianolic acid at 20 or 10 mg/kg can improve nerve function symptoms. Compared with the control group, the improvement was statistically significant. The results between the two preparations were statistically identical. Similar effects were observed in the following treatment groups: salvianolic acid at 10 mg/kg, SAPI at 13.5 mg/kg and CDI at 2 g/kg (Table 3.8).

Table 3.7 Effect of SAPI (i.v.) and slavianolic acid (i.v.) on cerebral infarction area of MCAO rats ($\bar{X} \pm s$, $n = 10$)

Group	Dose (mg/kg)	Cerebral infarction area (%)
Control		7.25 ± 3.16
Salvianolic acid Powder **Injection**	27	3.18 ± 1.03**
	13.5	4.67 ± 1.29*
	6.75	5.75 ± 1.33
Salvianolic acid	20	3.82 ± 1.07**
	10	4.67 ± 1.44*
	5	5.76 ± 1.37
Composite Danshen **Injection**	2,000	4.59 ± 1.18*

Table 3.8 Effect of salvianolic acid powder injection and salvianolic acid on functional neurological deficit in MCAO rats ($\bar{X} \pm s$, $n = 10$)

Group	Dose (mg/kg)	Score on symptom
Sham operation		3.46 ± 1.44***
Control		7.18 ± 1.75
Salvianolic acid powder **Injection**	27	3.97 ± 1.76**
	13.5	5.10 ± 1.30*
	6.75	5.98 ± 0.98
Salvianolic acid	20	3.98 ± 1.92**
	10	5.05 ± 1.62*
	5	6.01 ± 1.26
Composite Danshen injection	2,000	4.99 ± 1.71*
	1,000	6.08 ± 1.74

Compared with control: * $P < 0.05$, ** $P < 0.01$, *** $P < 0.001$

3.5.4.3 Effect of SAPI and Salvianolic Acid on Brain Water Content of MCAO Rats

As shown in Table 3.9, after occluding the middle artery on one side of rat brain for 24 h, the brain water content was 82.82 ± 1.29 for the control but 80.91 ± 0.79 for the sham operation, with a significant difference between them, suggesting that occlusion and severe edema occurred in the occluded hemisphere of the brain. SAPI at 27 or 13.5 mg/kg and salvianolic acid at 20 or 10 mg/kg can significantly reduce ischemic cerebral edema, with a comparable effect between them and no statistically significant difference. A comparable effect was observed in the salvianolic acid at 10 mg/kg, SAPI at 13.5 mg/kg and CDI at 2 g/kg groups.

Table 3.9 Effect of salvianolic acid powder injection and salvianolic acid on brain water content of MCAO rat ($\bar{X} \pm s$, $n = 10$)

Group	Dose (mg/kg)	Brain water content (%)
Sham operation		80.91 ± 0.79**
Control		82.82 ± 1.29
Salvianolic acid powder injection	27	80.99 ± 0.90**
	13.5	81.52 ± 1.07*
	6.75	82.01 ± 0.85
Salvianolic acid	20	81.00 ± 0.86**
	10	81.50 ± 0.87*
	5	82.07 ± 1.02
Composite Danshen injection	2,000	81.55 ± 0.61*
	1,000	82.42 ± 0.92

Compared with control: * $P < 0.05$, ** $P < 0.01$

3.5.5 Conclusion

1. MCAO by electric cautery results in focal cerebral ischemia and severe edema in the occluded brain hemisphere.
2. SAPI at 27 or 13.5 mg/kg i.v. and salvianolic acid at 20 or 10 mg/kg i.v. can improve functional neurological deficit. Compared with the control, there was a statistically significant effect. The effects of SAPI and salvianolic acid are comparable without a significant difference. A comparable effect was seen in the SAPI at 13.5 mg/kg, salvianolic acid at 10 mg/kg and CDI at 2 g/kg groups.
3. SAPI at 27 or 13.5 mg/kg i.v. and salvianolic acid at 20 or 10 mg/kg i.v. can significantly reduce the cerebral infarction area in rats 24 h after MCAO. Compared with the control, there was a statistically significant difference. The effects between the two preparations are comparable without a statistically significant difference. Comparable effects were seen in the salvianolic acid at 10 mg/kg, SAPI at 13.5 mg/kg and CDI at 2 g/kg groups.
4. SAPI at 27 or 13.5 mg/kg i.v. and salvianolic acid at 20 or 10 mg/kg i.v. can significantly reduce ischemic cerebral edema, and they have no significant effects at low doses. The effects of SAPI and salvianolic acid are comparable. Comparable effects were seen in SAPI at 13.5 mg/kg, salvianolic acid at 10 mg/kg and CDI at 2 g/kg groups.

References

1. Zhen HZ, et al. Current studies and applications of traditional Chinese drug. 2nd ed. Shanghai: Xue Yuan Publishing House; 1997.
2. Ji YB, et al. Pharmacology and application of antitumor active constituents in traditional Chinese drug. Harbin: Heilongjiang Science and Technology Publishing House; 1995.
3. Sun WJ, et al. Concise manual natural active constituents. Beijing: China Medico-Pharmaceutical Sciences and Technology Publishing House; 1998.
4. Don MJ, Shen CC, et al. J Nat. Prod. 2005;68:1066.
5. Adams JD, Wall M, Garcia C. Evid Based Complement Altern Med. 2005;2(1):107.
6. Fang ZY, Lin R, Yuan BX, et al. J Ethnopharmacol. 2008;115(2):217.
7. Takeo S, Tanonaka K, Hirai K, et al. Biochem Pharrmcol. 1990;40(5):1137.
8. Yagi A, Fujimoto K, Tanonaka K, et al. Planta Med. 1989;55(1):51.
9. Kim EJ, Jung SN, Son KH, et al. Mol Pharmacol. 2007;72(1):62.
10. Han YM, Oh H, Na M, et al. Biol Pharm Bull. 2005;28(9):1795.
11. Lee AR, Wu WL, Chang WL, et al. J Nat Prod. 1987;50(2):157.
12. Jiangsu New Medical College. Shanghai: Shanghai Science and Technology Publishing House; 1977.
13. Lin HC, Chang WL. Phytochemistry. 2000;53(8):951.
14. Li ZT, et al. Acta Pharm Sin. 1991;26(3):209.
15. Luo HW, et al. Phytochemistry. 1985;24(4):815.
16. Ryu SY, Lee CO, Choi SU. Planta Med. 1997;63(4):339.
17. Luo HW, Wang N. Acta Pharm Sin. 1988;23(11):830.
18. Wang N, Luo HW, Niwa M, et al. Planta Med. 1989;55(4):390.
19. Lin LZ, Wang XM, et al. Acta Pharm Sin. 1988;23(4):273.
20. Yu SC, Ju XL, Duan GX. J Anhui Health Vocat Tech Coll. 2002;1(2):43.
21. Luo HW, et al. Acta Pharm Sin. 1985;20(7):542.
22. Jiang JH, et al. Manual of active constituents in plant. Beijing: People Health Publishing House; 1986.
23. Yang BJ, et al. Acta Pharm Sin. 1981;16(11):837.
24. Yang BJ, Huang XL, Zhou QR. Acta Pharm Sin. 1984;19(4):274.
25. Yasukawa K, Akihisa T, Tamur T, et al. Biol Pharm Bull. 1994;17(3):460.

26. Chen HF, et al. Dictionary of active constituents in plant, vol. 1. Beijing: China Medico-Pharmaceutical Sciences and Technology Publishing House; 2001.
27. Ikeshiro Y, et al. Phytochemistry. 1989;28(11):3139.
28. Kong DY. Acta Pharm Sin. 1985;20(10):747.
29. Luo HW, et al. Chem Pharm Bull. 1986;34(8):3166.
30. Wang X, Bastow KF, et al. J Med Chem. 2004;47 (23):5826–9.
31. Hase K, Kasimu R, Basnet P, et al. Planta Med. 1997;63(1):22.
32. Zhang Y, Akao T, Nakamura N, et al. Drug Metab Dispos. 2004;32(7):752.
33. Onitsuka M, et al. Chem Pharm Bull. 1983;31:1670.
34. Rodriguez JA, Theoduloz C, Yanez T, et al. Life Sci. 2006;78(21):2503.
35. Chen WS, Jia XM, Zhang WD, et al. Acta Pharm Sin. 2003;38(5):354.
36. Ulubelen A, Oksuz S, Kolak U, et al. Planta Med. 2000;66(5):4,58.
37. Tezuka Y, Kasimu R, Basnet P, et al. Chem Pharm Bull. (Tokyo). 1997;45(8):1306.
38. Chao KP, Hua KF, Hsu HY, et al. Planta Med. 2005;71(4):300.
39. Kusumi T, et al. Phytoehemistry. 1985;24(9):2118.
40. Gu WH. J Chin Herbal 1981;12(2,3):41.
41. Modern study on Chinese herbal compiled by institute of materia medica, chinese academy of medical sciences. 2nd ed. Beijing: Union Publishing House of Beijing University of Medical Science and Peking Union Medical University; 1999. p. 82.
42. Kasimu R, Tanaka K, Tezuka Y, et al. Chem Pharm Bull (Tokyo). 1998;46(3):500.
43. Ma Y, Rena K, Wang XM, et al. J Sinkiang Med Coll. 2004;27(6):577.
44. Wang YS, et al. Beijing: People Health Publishing House, 1998. p. 190.
45. Chinese Materia Medica compilation committee of the State Administration of Traditional Chinese Medicines. Chinese Materia Medica, vol. 1–30. Shanghai: Shanghai Science and Technology Publishing House; 1999.
46. Li LN. J Chin Pharm Sci. 1997;6(2):57.
47. Kohda H, Takeda O, Tanaka S, et al. Chem Pharm Bull. 1989;37(5):1287.
48. Liu IM, Chen WC, Cheng JT. J Pharmacol Exp Ther. 2003;307(3):1196.
49. Wang D, Girard TS, et al. J Nat Prod. 1998;61:1352.
50. Ma C, Wang W, Chen Y-Y, et al. J Nat Prod. 2005;68:1259.
51. Corbiere C, Liagre B, Bianchi A, et al. Int J Oncol. 2003;22(4):899.
52. Peng AZ. J Hunan Med Coll. 1994;19(6):571.
53. Huang YM, Yuan JR, Shi JY, et al. Shandong J Tradit Chin Med. 1992;11(4):36.
54. Zhen TX, Shi ZK. Chin Pharm Sci. 1989;24:6.
55. Lin HQ, Zhang WL. Chinese Pharm Sci. 1993;45:615.
56. Lin HQ, Zhang WL. Chin Pharm Sci. 1993;45:21.
57. Lin HQ, Zhang WL. Chin Pharm Sci. 1993;45:28.
58. Guo JX, et al. Study and clinical application of Salvia miltiorrhiza. Beijing: China Medico-Pharmaceutical Sciences and Technology Publishing House; 1992.
59. Han C, Wang XJ, Zhang GY. Chin J Pathophysiol. 1991;7:337.
60. Zhen RX, Fang SM, Han BM, et al. Chinese J Integr Med. 1990;10:609.
61. Zhu P, Chen SY, Huang L, et al. Chin J Thorac Cardiovasc Surg. 1989;5:103.
62. Fu J, Huang H, Liu J, et al. Eur J Pharmacol. 2007;568(1–3):213.
63. Tang F, Wu X, Wang T, et al. Vasc Pharmacol. 2007;46(6):427.
64. Liu CL, Shi WZ, Sun LH, et al. J Chin Mater Med. 1990;15:112.
65. Yu XY, Lin SG, Zhou ZW, et al. Neurosci Lett. 2007;417(3):261.
66. Deng HJ, Ma XH, Xu RL, et al. J Chin Mater Med. 1992;17:233.
67. Xu JK, Hiroshi K, Zheng JJ, et al. Acta Pharm Sin. 2006;41(7):631.
68. Tang Z, Tang Y, Fu L. J Huazhong Univ Sci Technol Med Sci. 2003;23(2):166.
69. Zhou Z. Chin J Integr Med. 1990;10:242.
70. Xu RS, et al. Biology and application of Salvia miltiorrhiza. Beijing: Scientific Publishing House; 1990.
71. Wang GZ, Yan YH, Zhu CF. J Hebei Tradit Chin Med Pharmacol. 2006;21(4):4.
72. Du GH, Qiu Y. Acta Pharm Sin. 1995;30(10):731.
73. Li ZH, Du GH. J Binzhou Med Univ. 1996;19 (4):315.
74. Xu SY. Experimental method of pharmacology. 3rd ed. Beijing: People's Medical Publishing House; 2002. p. 921.
75. Curtis MJ, Walker MJ. Cardiovasc Res. 1988;22 (9):656.
76. Ji HG, Si L, Si XC, et al. J Tradit Chin Med. 2006;21 (3):52.
77. Aasum E, Hafstad AD, Larsen TS. Mol Ceil Biochem. 2003;249(1–2):97.
78. Yamamoto K, Ito H, 1wakura K, et al. J Am Coll Cardiol. 2002;40(10):1755.
79. Kurisu S, Inotle I, Kawagoe T, et al. Int Med. 2003;42 (7):554.
80. Nosal'ova V, Drabikova K, Zurova-Nedelcevova J, et al. Neuro Endocrinol Lett. 2006;27 Suppl 2:152.
81. Caballero S, Sengupta N, Afzal A, et al. Diabetes. 2007;56(4):960.
82. Zeng J, Min S. Chin J Extracorporeal Circ. 2004;2 (3):180.
83. Ma L, Yang Y, Yao SL, et al. Chongqing Med J. 2007;36(7):50.
84. Wang GZ, Yan YP, Zhu CF. J Hebei Tradit Chin Med Pharmacol. 2006;21(2):4.
85. Li LN, Zhang JT. Water-soluble active constituents in Salvia miltiorrhiza and its sibling plant compiled by Chinese Union Medical University, Chinese Academy

of Medical Sciences. Science annual conference proceedings. 1st ed. Beijing: Chinese Union Medical University Publishing House; 2000. p. 175.
86. Zhang JT. Chin Pharmacol Bull. 2000;16(2):121.
87. Yonghon C, Guanhua D, Juntian Z. Acta Pharmacol Sin. 1999;2l(5):463.
88. Feng Z, Zhang JT. Acta Pharm Sin. 2000;35(12):881.
89. Tan MK, Zhang JT. Acta Pharmacol Sin. 2001;22 (4):380.
90. Tan M, Zhang J. Jpn J Pharmacol. 2002;88(4):422.
91. Tang MK, Ren DC, Zhang JT, Du GH. Phytomedicine. 2002;9(5):405.
92. Du GH, Zhang JT. Basic Clinical Med. 2001;20 (5):10.

Effects of Danshen on the Cardiovascular System

4

Xiaoming Zhu, Lianhua Fang, Guanhua Du, Ran Zhang, Dongxia Wang, Jinglan Xu and Xiaoying Wang

Large amounts of literature and long-term clinical practice have confirmed that Danshen offers protection over myocardial ischemia and reperfusion injury as a kind of traditional Chinese medicine to promote blood circulation and remove blood stasis. Moreover, it can also inhibit myocardial cell hypertrophy, prevent cardiac arrhythmia, and treat viral myocarditis. Danshen is clinically used for the protection of cardiac muscle and as an adjunctive therapy for pneumocardial disease during operations on the heart. Danshen can directly relax the coronary artery muscle muscle, and thus it can be clinically applied to the treatment of coronary heart disease and can improve microcirculation by promoting blood circulation, removing blood stasis, and inhibiting blood clotting, can reduce the viscosity of blood and significantly improve the hyperviscosity of blood, and finally inhibit thrombogenesis. Danshen can inhibit the proliferation and migration of vascular smooth muscle cells, protect the functions of blood vessel endothelium, and indirectly exert its regulatory functions through the inhibitory effects of the "dangerous risks" of hypertension, and its injection can be used for the treatment of acute hypertensive cerebral hemorrhage. The pharmacological action of Danshen on cardiovascular system diseases and its mechanisms will be discussed in full scale in this chapter.

X. Zhu · L. Fang · G. Du (✉) · R. Zhang · D. Wang · J. Xu · X. Wang
Institute of Materia Medica, Chinese Academy of Medical Science and Peking Union Medical College, Beijing, People's Republic of China
e-mail: dugh@imm.ac.cn

4.1 The Protection of Danshen Over Heart

Xiaoming Zhu, Lianhua Fang and Guanhua Du

Danshen is a traditional Chinese medicine to promote blood circulation and remove blood stasis, and after a long period of pharmacological studies, it is generally believed that Danshen's pharmacological actions are mainly on the cardiovascular and cerebrovascular systems; among them, its effect on the heart is one of the important components. Danshen is clinically applied in the treatment of coronary heart disease as a traditional drug with definite therapeutic effects. Coronary heart disease refers to myocardial ischemia, hypoxia, cardiac hypofunction, and a series of diseases that are induced by angiemphraxis due to coronary atherosclerosis. Thrombolytic therapy, percutaneous transluminal coronary angioplasty (PTCA), coronary artery muscle muscle, bypass graft, laser coronary recanalization, and others have been widely used in the treatment of clinical coronary heart disease with the development of interventional heart science since the 1980s, but they cannot reduce the death rate from coronary heart disease. On the contrary, injury that is more serious than

X. Yan (ed.), *Dan Shen (Salvia miltiorrhiza) in Medicine*,
DOI 10.1007/978-94-017-9463-3_4,

myocardial ischemia, namely ischemic reperfusion injury, is induced when the ischemic cardiac muscle resumes blood supply after a certain time. Danshen has shown protective functions against myocardial ischemia and repeat reperfusion injury. This section is an introduction to the protective function of Danshen on the heart.

4.1.1 Improving the Blood Supply of Ischemic Cardiac Muscle

Heart ischemia is induced by blood-supply insufficiency or relative blood-supply insufficiency of the coronary artery muscle muscle, and induction of myocardial ischemia can be caused by myocarditis, cardiac enlargement, hypertrophy, coronary stenosis, and others. Generally, myocardial ischemia changes are induced by blood-supply insufficiency from coronary stenosis due to coronary atherosclerosis or coronary angiospasm, resulting in heart-stroke occurring in the patients. This is the most frequently seen myocardial ischemia clinically. Continuous and serious myocardial ischemia can induce cellular necrosis in some cardiac muscle, which is termed myocardial infarction. Previous investigations have revealed that Danshen has significant functions in preventing myocardial ischemia, and the protection given from its extract is similar to that from the angiotensin converting enzyme inhibitor Ramipril, which can increase the survival rate of rats with myocardial ischemia, reduce the area of myocardial ischemia, and improve the blood supply to ischemic cardiac muscle [1].

The functions of Danshen in myocardial ischemia prevention are related to the dilation of the coronary artery muscle muscle, decrease in peripheral resistance, and increase in coronary flow. Chen et al. [2] established a model for left anterior descending coronary artery muscle muscle stenosis in dogs, and they found that coronary resistance continuously decreased, while coronary flow and myocardial ventilation increased and oxygen consumption decreased when Danshen injection was constantly infused in the coronary artery muscle muscle [2 mg/(kg min)]. Yan et al. [3] confirmed that Composite Danshen Dropping Pill can significantly increase the coronary flow in rats and its effects were better than those of Compound Danshen Tablet. Tanshinone II-A sulfonate and Tanshinol are both active ingredients which dilate the coronary artery muscle muscle; tanshinone II-A sulfonate can activate the ATP-sensitive potassium channels in the coronary artery muscle muscle and calcium-activated potassium channels and inhibit calcium channels, and thus the electrical activity of vascular smooth muscle can be significantly reduced, inducing coronary relaxation [4].

Danshen can also promote the establishment of coronary compensatory circulation to alleviate the injuries to blood capillaries in the ischemic infarction area and renew the blood supply in ischemic cardiac muscle, and it has significant functions in reducing the disturbance in heart circulation induced by myocardial infarction [5]. Liu et al. [6, 7] carried out investigations on the limitation of Danshen in myocardial infarction ranges and its mechanism. They found that the myocardial infarction area in the Danshen group was significantly smaller than that of the control group 21 days after coronary deligation, and the amount of branched blood vessels in the ischemic region of the Danshen group was significantly greater than that in the control group 2 h and 22 days after coronary deligation. At the same time, heart muscle capillaries and the density of blood-supply vessels in the ischemic region of the Danshen group were also significantly increased. In the acute stage of myocardial infarction, platelet aggregation and blood viscosity are significantly increased in comparison to before the myocardial infarction, and Danshen can prevent such changes, with various indexes returning close to the normal range after treatment. The functions of Danshen in the opening and formation of coronary compensatory circulation are related to its dilation of the coronary artery muscle muscle, the upregulation of the expression of vascular endothelial growth factor (VEGF), and the promotion of vascular neogenesis.

4.1.2 Improving the Energy Metabolism of Cardiac Muscle

Energy metabolism disorders are the basis for the morbidity of myocardial ischemic reperfusion. The heart is an organ that requires oxygen and consumes energy, and the supply of oxygen and energy metabolites is not sufficient during myocardial ischemia. Myocardial cells rapidly switch from aerobic metabolism to anaerobic glycolysis, ATP production begins to decrease, and lactic acid accumulates in the tissues, and thus leads to acid-intoxication. With the prolongation of ischemic time and reperfusion, ATP, total amount of adenine nucleotides, and energy charge gradually decrease, while the degradation products of ATP, adenosine diphosphate, and adenosine monophosphate are significantly increased. The decrease in ATP will give rise to a series of metabolic disorders and abnormalities, such as the decrease in ATP-dependent plasma membrane pump activity, changes in membrane potential and the ST segment of electrocardiogram, and others. Previous investigations have suggested that Danshen can improve the energy metabolism of myocardial cells in several ways as described below.

4.1.2.1 It Can Promote the Metabolism of Tricarboxylic Acid Cycle

In the metabolism of the tricarboxylic acid cycle, succinate Dehydrogenase (SDH), malate Dehydrogenase, glucose-6-phosphate Dehydrogenase, phosphorylase and others are the crucial enzymes and will directly affect the total amount of ATP that is finally produced from aerobic metabolism. Previous investigations have revealed that Danshen can protect these enzymes involved in glucose metabolism. Zhang et al. [8] established the myocardial ischemic reperfusion model by ligating and relaxing the anterior descending branch in the left coronary artery muscle muscle of New Zealand white rabbits, and they found that SDH activity was significantly decreased compared t that of the normal group; if Danshen injection was injected intravenously 2 min before the de-ligation and immediately after reperfusion, SDH activity was significantly improved compared to that of the ischemic reperfusion group; no significant difference was found in SDH activity compared to that of the normal group if Danshen injection was combined with basic fibroblast growth factor (bFGF).

4.1.2.2 Protecting Mitochondria and Recovering Their Functions

Danshen can alleviate lipid peroxidation of the mitochondrial membrane and mitochondrial calcium overload, improve the fluidity of the mitochondrial membrane and reduce its permeability, and maintain the integrity of mitochondrial structures. Previous experiments have confirmed that respiratory chain function in the mitochondrial inner membrane was inhibited in the early stage of ischemia, and the pumping rate of H^+ in the inner membrane, electron transfer rate and $H^+/2e^-$ all significantly decreased. Oxidative phosphorylation was partially uncoupled, but no injuries were found in the lipid conformation of the membrane; reperfusion relieved this kind of inhibition and respiratory chain function was partly recovered, but could not be recovered to normal levels. Tanshinol can effectively prevent the decrease in the abovementioned parameters and recover respiratory chain function [9].

4.1.2.3 Eliminating Free Radicals

During the process of ischemic reperfusion, free radicals are one of the important factors that induce the injuries in myocardial cells. Large amount of ATP in the cardiac muscle is consumed and xanthine and hypoxanthine are produced during the myocardial ischemia stage, and the activity of xanthine oxidase increases and then leads to the production of large amount of oxygen free radicals during the reperfusion stage, and these free radicals destroy the membrane structures by inducing the lipid peroxidation of biomembranes, and thus the membrane permeability is changed, the ionic flow and the cell membrane potential are increased, and finally the cell death is induced.

Previous investigations have confirmed that Danshen is a good free radical scavenger; it can increase the content of SOD and eliminate superoxide anions and hydroxyl radicals. In in vitro free radical producing systems such as the xanthine oxidase reaction system and vitamin C–Fe^{2+} reaction system, salvianolic acid A (SalA), salvianolic acid B (SalB) and other water soluble ingredients of Danshen have a stronger capability of eliminating free radicals than do vitamin E, vitamin C and gingko extract EGb761. In experiments on ex vivo rat hearts, SalA can significantly decrease the content of myocardial MDA after ischemic reperfusion, reduce the leakage of LDH in cells, and can decrease the incidence rate of ventricular fibrillation. Tanshinol can increase the activities of SOD and GSH-Px, alleviate injuries to the ultrastructure, significantly alleviate the decrease in mitochondrial membrane fluidity, and reduce lipid peroxidation in cardiac muscle under ischemic reperfusion conditions in rats, and its protective effects are significantly better than those of the well-recognized antioxidant sodium selenite [10].

Fan et al. [11] established a myocardial infarction model by ligating the left anterior descending coronary artery muscle muscle of rats, and they found that after intragastric administration of SalB, the size of the myocardial infarction was significantly reduced compared to that of the natural recovery group. The mechanism may be that SalB is an effective antioxidant which can increase the content of SOD in cells and relieve the toxic effects of free radicals in myocardial cells.

Jiang et al. [12] established a experimental myocardial ischemia model in mice, and they used Danshen injection to investigate the production and elimination of oxygen free radicals, the release of nitric oxide synthase, and other aspects. Their results showed that pretreatment with Danshen injection can increase the activity of SOD in ischemic myocardial cells, indicating that pre-treatment could relieve lipid peroxidation injuries to the cell membrane, stabilize the cell membrane, enhance the capability of eliminating oxygen free radicals, reduce leakage of intracellular creatine kinase, reduce cellular injuries induced by free radicals, and protect ischemic myocardial cells.

4.1.2.4 Relieving Calcium Overload

When myocardial ischemia occurs, both ATP content and Na^+–K^+ pump function decrease, which results in the inflow of large amounts of Na^+ and the outflow of K^+. The decrease in cell membrane potential leads to its depolarization, which in turn leads to the opening of voltage dependent Ca^{2+} channels and the influx of large amounts of Ca^{2+}. At the same time, the receptor-dependent Ca^{2+} channel can be opened due to the release of K^+, protein kinase C (PKC) and other transmitters, and finally gives rise to calcium overload. During the process of reperfusion, functional impairment of the cell membrane is induced due to the injuries to the cell membrane and the abnormalities in ion metabolism, thus the intracellular calcium concentration continuously rises and further aggravates the situation of intracellular calcium overload. The excessive Ca^{2+} in the cells activates phospholipases on the membrane and thus degrades the membrane phospholipids into lysophospholipid, and finally leads to ischemic contracture. In the meantime, calcium overload promotes the production of large amount of free radicals, causing lipid peroxidation of the membrane and damage to the plasma membrane, and thus affects the structure and permeability of myocardial cell membranes.

Danshen can effectively block the L-type calcium channel of ischemic myocardial cells. Recent reports have showed that tanshinone II-A has functions similar to those of L-type calcium channel blockers such as verapamil, and that it can be used for the prevention and treatment of myocardial ischemia reperfusion injury and cardiac arrhythmia. Moreover, Danshen can alleviate injuries to the cell membrane and its permeability as well as Ca^{2+} influx by eliminating free radicals, and it can also recover the mitochondrial function, promote ATP production and provide energy for calcium pumps, Na^{2+}–Ca^{2+} exchanging pump and Ca^{2+}–H^+ exchanging pump, and decrease intracellular Ca^{2+} concentration. At the same time, previous literature has reported that Danshen showed significant inhibitory effects on the phosphatidyl inositol signal transduction pathway after myocardial ischemia reperfusion, and it can significantly decrease 4,5-diphosphate phosphatidyl inositol (PIP_2) and 1,4,5-inositol triphosphate (IP_3) levels. The decrease in IP_3 will inhibit the IP_3 receptor channel, and thus decrease calcium release in the sarcoplasmic reticulum and reduce calcium overload [13].

4.1.2.5 Inhibiting Myocardial Cell Apoptosis

Apoptosis is an autonomous programmed cell death that is regulated by various physiological and pathological factors. Apoptotic cells always undergo morphological changes, such as cell shrinkage, chromatin condensation, and apoptotic body formation.

The mechanism of apoptosis is very complex, and current investigations have found that it is closely correlated with several factors listed below:

(1) Oxidative stress: oxygen free radicals can cause damage to DNA and activate the genes related to apoptosis. Oxygen free radicals can also damage the structure and function of the cell membrane, increase its permeability, and promote the influx of extracellular calcium; they can also activate nuclear transcription factors NF-κB and AP-1, which accelerate the expression of apoptosis related genes and induce apoptosis.
(2) Calcium overload: activates Ca^{2+}/Mg^{2+} dependent endonucleases and degrades DNA; can activate glutamic acid amide transferase and catalyze the formation of covalent bonds between peptides, and then promote the intensive cross-linking of cytoskeletal proteins which facilitates the formation of apoptic bodies; and can also activate nuclear transcription factors and accelerate the expression of Apoptosis related genes.
(3) Damage to mitochondria: both ischemia and anoxemia can inhibit the TCA cycle and respiratory chain function in the mitochondria, and thus accelerate the apoptosis of cells; the permeability of the mitochondrial membrane is increased and thus cytochrome C, Apoptosis protease activating factor (*Apaf*) and apoptosis inducing factor (AIF) are released to the cytosol through the permeability transition pores of mitochondria, and activating the Caspase apoptosis cascade reaction.

Ischemic reperfusion, viral infection, and cytokine stimuli in the cardiac muscle can all induce apoptosis in myocardial cells. In the process of ischemic reperfusion, many factors such as free radicals and intracellular calcium overload can also induce the apoptosis of myocardial cells. The apoptosis of myocardial cells is an important cause of cardiac function abnormalities.

Danshen can protect cardiac muscle by significantly inhibiting the apoptotic effects of myocardial cells. Ke et al. [14] found that Danshen injection could significantly inhibit the apoptosis of myocardial cells in suckling mice, which were induced by sugar and oxygen deficiencies for 6 h. Danshen at either 7.5 or 15 g/L can cause a significant decrease of the peak and its right-shift of the apoptotic diploid in the myocardial cells, and could also inhibit the increase in myocardial enzyme (LDH, α-HBD) activities after sugar and oxygen deficiencies. This function of Danshen may be related to its inhibition of the expression of the p53 gene and *Fas* gene in myocardial cells.

Danshen can also increase the expression of the *Bcl*-2 gene. Previous investigations have found that *Bcl*-2 can improve the capability of resistance to ischemic reperfusion injury in myocardial cells of rats and decrease degraded fragments of DNA; therefore, Danshen can reduce the apoptosis of myocardial cells by upregulating the *Bcl*-2 gene. Further, it can also inhibit the expression of the apoptosis promoting gene *Bax,* and thus protect myocardial cells [15].

Previous investigations revealed that tanshinone II-A is an active ingredient in Danshen that can inhibit the apoptosis of myocardial cells, and it has significant inhibitory effects on the apoptosis of myocardial cells induced by angiotensin II (Ang II) [16] (Table 4.1).

Apart from the direct regulatory effects on the genes related to myocardial cells, Danshen can also inhibit the apoptosis of myocardial cells in indirect ways: (1) Eliminating oxygen free

Table 4.1 The effects of tanshinone II-A on the apoptosi of myocardial cells induced by Ang II ($\bar{X} \pm s$)

Group	Apoptotic rate of myocardial cells (%)
Control group	17.7 ± 1.5
Ang II	35.4 ± 2.6
Ang II + sodium tanshinon II-A silate	23.2 ± 2.4
Sodium tanshinon II-A silate	15.4 ± 1.7
Verapamil	17.2 ± 1.6
Ang II + verapamil	24.3 ± 2.2

radicals, increasing SOD activity, reducing lipid peroxidation, protecting cell membranes, and mitochondrial membranes; (2) Significant protection over mitochondrial respiratory chain function, which can protect mitochondria and improve energy metabolism; (3) Relieving the calcium overload of myocardial cells and regulating calcium homeostasis.

4.1.2.6 Enhance the Protective Effect of Ischemic Preconditioning on Myocardium

After several brief episodes of ischemia-reperfusion, the cardiac muscle can tolerate relatively longer subsequent periods of ischemia. This phenomenon is called ischemic preconditioning of myocardial protection. Ischemic preconditioning is endogenous protective machinery, which can reduce the occurrence rate of arrhythmia induced by ischemic reperfusion and the extent of myocardial infarction, and thus promote the recovery of damaged myocardial function.

Lu et al. [17] established a myocardial ischemia reperfusion model by ligating and relaxing the left anterior descending coronary artery muscle muscle of rats, and observed the protection by the combination of Danshen injection and ischemic preconditioning on the cardiac muscle, and found that the occurrence of reperfusion ventricular fibrillation decreased in the group treated with the combination of Danshen injection and ischemic preconditioning compared to that of the group treated with Danshen injection alone, and both the frequency of premature ventricular contraction and the area of myocardial infarction were reduced. Compared with the single ischemic preconditioning group, the area of myocardial infarction was further reduced in the combination group, indicating that Danshen can reinforce the protection of cardiac muscle from ischemic preconditioning, particularly the protection given by ischemic preconditioning in preventing myocardial infarction.

Wang et al. [18] observed the effect of tanshinone II-A sulfonate on the ischemic preconditioning of cardiac muscle in rabbits, and found that the monophasic action potential durations at 50 % ($MAPD_{50}$) in cardiac muscle were rapidly shortened when ischemic preconditioning was carried out for 5 min, and the combination of tanshinone II-A sulfonate and ischemic preconditioning eliminated the rapid shortening of $MAPD_{50}$ in the early stage of ischemia. Tanshinone II-A sulfonate can also significantly reinforce the reductions of myocardial infarction area and tissue damage, but had no significant effects on the prevention of ischemic reperfusion arrhythmia by ischemic preconditioning.

4.1.3 Inhibition of Myocardial Hypertrophy

Myocardial hypertrophy is one of the most frequently seen symptoms of hypertension, and it is one of the hot spots of hypertension research. It is also an independent risk factor in cardiovascular diseases. Many factors such as growth factors, catecholamine, Ang II, aldosterone, insulin, and various growth hormones are involved in the progression of myocardial hypertrophy, and among them Ang II has the biggest effect. Ang II can cause hypertrophy of myocardial cells and increase mRNA transcription and protein synthesis. Ang II can also stimulate fibroblasts and increase the release of transforming growth factor β1, thus increasing the synthesis of extracellular matrix proteins in the heart (including fibronectin and collagen). Ang II can promote the expression of myc, fos, and other proto-oncogenes, and induce cellular proliferation and hypertrophy; it can also induce the expression of various kinds of growth factors and cellular proliferation. When the heart is overloaded by work, the stretching of the plasma membrane activates Ang II in myocardial cells, which binds to Ang II receptors on the plasma membrane in the form of autocrine, thus activating the phospholipase system to produce IP_3 and diacyl glycerol (DAG). IP_3 can promote the release of Ca^{2+} from the sarcoplasmic reticulum to increase intracellular Ca^{2+}, then mitogen-activated protein kinase (MAPK) and calcineurin (CaN) pathways send hypertrophy signals to the nucleus. Meanwhile DAG activates PKC, which phosphorylates some essential transcription factors leading to the transcription and synthesis of new spectrin. Growth factors and growth factor receptors are triggered to induce the hypertrophy of myocardial cells [19].

After administration of Danshen injection over a long period, the left ventricular mass index, the diameter of myocardial cells, the volume percentage of myocardial collagen, and the ratio of cardiac perivascular collagen area and lumen area in the spontaneous hypertensive rats (SHR) all fell significantly, indicating that Danshen has effects on preventing and reversing left ventricular hypertrophy and myocardial fibrosis, which is related to an improvement in hemodynamics [20]. Previous investigations have reported that tanshinone II-A is an active ingredient in Danshen which can inhibit myocardial hypertrophy. Feng et al. [21] observed the effects of the inhibition of Ang II induced myocardial hypertrophy by tanshinone II-A sulfonate in neonatal rat myocardial cells in primary culture. Tanshinone II-A sulfonate at 10^{-7} g/L can significantly inhibit Ang II induced hypertrophy in myocardial cells, and its effects were equal to those of 10^{-7} mol/L verapamil (Table 4.2).

Table 4.2 The effects of tanshinone II-A sulfonate and verapamil on the size of myocardial cells induced by Ang II ($\bar{X} \pm s, n = 6$)

Group	Cell diameter (μm)
Control group	19.9 ± 1.8
Ang II	29.2 ± 3.1
Ang II + tanshinone II-A sulfonate	21.3 ± 2.7
tanshinone II-A sulfonate	20.5 ± 3.1
Verapamil	20.1 ± 2.7
Ang II + verapamil	20.1 ± 3.2

Tanshinone II-A sulfonate can significantly decrease the increase of intracellular calcium and the increase of protein synthesis rate, both induced by Ang II. Present investigations revealed that the activities of PKC and CaN increased significantly 24 h after Ang II stimulus. Tanshinone II-A sulfonate can significantly inhibit the activities of PKC and CaN in myocardial cells mediated by Ang II, which may be related to the calcium ion antagonistic action of tanshinone II-A sulfonate [21]. Moreover, tanshinone II-A can inhibit the reinforcement of the expression of proto-oncogene c-fos mRNA in myocardial cells that is induced by Ang II, and its effects are similar to those of Valsartan, an Ang II receptor blocking agent [22].

Aldosterone is one of the main causative agents for left ventricular hypertrophy. Zhan et al. [23] reported that 12 weeks after the administration of tanshinone II-A, aldosterone and Ang II levels as well as the expression level of the CYPllB2 gene, which is related to aldosterone synthesis in the myocardial tissues of spontaneously hypertensive rats, all decreased significantly, and the local biosynthesis of aldosterone in the heart was also reduced.

It has been demonstrated that the expression of intercellular adhesion molecule 1 (ICAM-1) in hypertensive rats that undergo myocardial hypertrophy is significantly increased. Zhan et al. [24] found that tanshinone II-A can significantly down-regulate ICAM-1 expression in cardiac muscle; the number of infiltrating macrophages was reduced, and the hypertrophy and fibrosis of myocardial cells were alleviated. NF-κB plays important functions in myocardial hypertrophy, and it produces effects of myocardial cell hypertrophy by inducing atrial natriuretic peptide (ANP) and tumor necrosis factor in myocardial cells, while tanshinone II-A can significantly inhibit the upregulation of NF-κB that is induced by Ang II [25]. The apoptosis insufficiency in left ventricular cells of SHR may play an important function in the morbidity of left ventricular hypertrophy. Liu [26] reported that tanshinone II-A can significantly promote the apoptosis of myocardial cells.

Apart from tanshinone II-A, Tanshinol also has functions in inhibiting myocardial hypertrophy. Tanshinol can significantly inhibit Ang II-induced ANP and β-actin gene expression, and thus it can be used for the prevention and treatment of myocardial hypertrophy [27]. Furthermore, it is found that tanshinone can inhibit myocardial hypertrophy in rats and reduce the rate of protein synthesis induced by endothelin-1 (ET-1), deoxyepinephrine and insulin-like growth factor (IGF), though it cannot reduce the rate of protein synthesis in myocardial cells by itself [28].

4.1.4 Anti-arrhythmia

Previous studies have found that Danshen has potential functions in preventing arrhythmia, and Danshen or Composite Danshen Dropping Pill are used in clinical treatment of this disease.

4.1.4.1 The Effects of Tanshinone II-A Sulfonateon the Effective Refractory Period (ERP) of Myocardial Cells

Using patch clamps, it was found that tanshinone II-A sulfonate could block the L-type calcium channel of myocardial cells. Zhu et al. [29] observed the effects of tanshinone II-A sulfonate on the monophasic action potential (MAP) in the right ventricle and on the effective refractory period (ERP) in rabbits under integral conditions, and they found that sinus rhythms slowed down in the rabbits after they were injected with tanshinone II-A sulfonate, and MAP amplitudes (MAPA) were slightly decreased. No significant difference was found, indicating that tanshinone II-A sulfonate had no significant effect on rapid sodium channels; at the same time, it also had no significant effects on the horizontal distance to the upslope of MAP ($MAPD_{50}$, $MAPD_{90}$) when repolarization reached 50–90 %. However, ERP and the ratio between ERP and MAPD increased with an increase in administered dosage, indicating that tanshinone II-A sulfonate might have potential functions in preventing cardiac dysrhythmia.

4.1.4.2 The Effects of Tanshinol on Sodium Currents in Myocardial Cells

Using the patch clamp technique, Zhao et al. [30] found that Tanshinol at 25–50 μM had significant inhibitory effects on sodium channel currents in guinea pig ventricular muscle cells, and the effect was dose-dependent. There are also data reported that Tanshinol can significantly reduce the myocardial cell action potential overshoot (OS), inhibit of the degree of openness of the fast sodium channel, significantly decrease the amplitude of the action potential in myocardial cells (APA), and decrease the rate of sodium influx.

4.1.4.3 The Effects of Tanshinone II-A on Potassium Current

Using the patch clamp technique, researchers have found that tanshinone II-A can block potassium currents in myocardial cells, blocking both the inward rectification potassium current (I_{K1}) and transient outward current (I_{to}). I_{to} mainly participates in stage 1 of action potential repolarization, while I_{K1} mainly takes part in the maintenance of the resting membrane potential and in stage 3 of action potential repolarization. The inhibition of these two kinds of currents should prolong the action potential duration and the effective refractory period, indicating that it might have functions in preventing cardiac dysrhythmia [31].

4.1.5 Treatment of Viral Myocarditis

Viral myocarditis is the cellular necrosis and interstitial inflammatory changes in myocardial cells induced by the viral invasion of myocardial cells. Danshen can also be used in the treatment of viral myocarditis clinically, and it can gradually restore the myocardial enzymes and the electrocardiogram of patients to normal levels.

Table 4.3 Comparison of the therapeutic effects on the two groups [cases (%)]

Group	N	Excellent	Effective	Ineffective	Total effectiveness
Treatment group	50	45 (90)	5 (10)	0	50 (100)
Control group	50	10 (20)	25 (50)	15 (30)	35 (70)

When inflammation occurs in cardiac muscle, the release of oxygen free radicals is one of the important pathogenetical processes of viral myocarditis. SOD decreases in erythrocytes and the scavenging of free radicals is reduced, which leads to the disruption of nucleic acids and the disintegration of polysaccharides as well as the peroxidation of unsaturated fatty acids in myocardial cells, damaging the plasma membrane of myocardial cells and weakening the oxidative phosphorylation of mitochondria, finally resulting in myocardial damage. The application of Danshen can effectively remove large amounts of oxygen free radicals in the body, thus avoiding damage to cardiac muscle. Previous references have reported that Danshen can significantly improve the active rate of C_3b receptors in the red blood cells (RBC-C_3bRR) of patients with viral myocarditis, reduce circulating immune complex (CIC), reinforce the immunological adhesive functions of red blood cells, and reduce myocardial damage [32].

An [33] studied the therapeutic effects of Danshen on viral myocarditis accompanied with cardiac dysrhythmia (50 cases) and compared the results with those treated with conventional Western drugs (50 cases). The results are listed in Table 4.3.

4.1.6 Summary

In summary, the pharmacological actions of Danshen on the heart involve many aspects, and most studies have focused on Danshen's protective effect on myocardial ischemia-reperfusion injury and its mechanism, such as the improvement of myocardial energy metabolism, the scavenging of oxygen free radicals, the extenuation of calcium overload, the inhibition of myocardial cell Apoptosis, etc. Secondly, the morbidity of hypertension accompanied by left ventricular hypertrophy has been continuously increasing in recent years, and thus Danshen's functions in reversing myocardial hypertrophy have attracted a lot of attention. Several compounds have been discovered in Danshen, which have protective functions on cardiac muscle, but further investigations of the specific pharmacological mechanism are still in great need. It is reasonable to believe that its clinical application values are sure to be further explored with the advancements in the investigations on Danshen.

4.2 The Effects of Danshen on Atherosclerosis

Ran Zhang, Lianhua Fang
and Guanhua Du

Atherosclerosis (AS) refers to lipid deposition under the inner membrane in some regions of the artery, and simultaneous proliferation of smooth muscle cells and fibrous matrix ingredients, which gradually develop into atherosclerotic plaques. The arterial wall in the plague region is thickened and hardened, and the internal tissues in the plaques bind to the deposited lipids after undergoing cellular necrosis, forming a porridge-like substance termed scleratheroma. In recent years, clinical application of Danshen has continuously expanded, supported by large amounts of clinical practices. Its pharmacological actions involve every element of the pathogenesis of AS, so a comprehensive review is given in this section.

4.2.1 Regulation of Lipid Metabolism

AS is a commonly seen and commonly encountered disease that seriously threatens the health and life of hundreds of millions of people. Among the numerous etiological factors including lipoprotein metabolic disorder, genetic variance, improper diet, living habits, and external environment, abnormalities in the metabolism of plasma lipids, particularly the total cholesterol (TC), are most closely related to AS. Almost 100 years ago, Virchow proposed the lipid infiltration theory for atherosclerosis, which assumes that lipids come from the plasma and enter the arterial wall by means of infiltration. Epidemiologic studies have confirmed that the occurrence of atherosclerosis is related to hyperlipemia, particularly the increase of low density lipoprotein (LDL) and very low density lipoprotein (VLDL), while high density lipoprotein (HDL) is negatively related to AS.

LDL is closely related to AS, and the extremely high concentration of LDL in human plasma is one of the high risk factors, which induce coronary heart disease. Some suggest that when LDL levels increase, LDL accumulates on the arterial walls and is taken in by scavenger cells at that site, accelerating the occurrence of AS. LDL is transported by two types of receptors, the LDL receptor and scavenger receptor, on the cell surface. Further investigations revealed that LDL can migrate to the inner membrane of the artery when the LDL level in plasma increases. LDL in the inner membrane of the artery is affected by oxygen free radicals and metabolites produced by arterial intimal cells, and is extremely liable to undergo oxidation and form oxidized LDL (OX-LDL). OX-LDL can inhibit the binding of LDL to its receptor and inhibit the migration of macrophages, so LDL cannot be eliminated by the LDL receptors of macrophages, thus OX-LDL accumulates in large amounts in a local region and the formation of atherosclerosis is accelerated.

Previous investigations have found that the occurrence of AS is likely in animals with relatively low HDL levels in their plasma, such as rabbits, pigs, and quails. Similarly, AS is unlikely to occur in animals with relatively high HDL levels. Many drug research studies and animal experiments reveal that HDL has functions in preventing AS. The specific mechanism is still being studied, but the possible mechanism may be: (1) it can induce the retrograde transportation of cholesterol, and then promote the removal of cholesterol from the arterial wall; (2) it can inhibit the transformation and proliferation of arterial smooth muscle cells; (3) it can protect vascular endothelial cells; (4) it can increase the synthesis of PGI_2 and inhibit platelet aggregation.

Meng et al. studied the effect of Danshen on AS prevention in rabbits. They found that Danshen could not only significantly decrease the size of atheromatous plaques in the main artery, but also reduce total cholesterol (TC), triglyceride (TG) and low density lipoprotein (LDL) in serum, to some extent. Wang et al. [36] induced hyperlipemia in rabbits and then damaged the endodermis of the iliac artery by using a 3F thrombus eliminating balloon catheter, and fed the animals with high-fat feed after the operation and treated them with Danshen. They found that TC in the high-fat model group could be reduced from 3141.6 ± 539.6 to 1260.0 ± 540.0 mg/dl. The length of arterial stenosis was reduced from 28.6 ± 7.7 to 16.8 ± 12.7 mm after 4 weeks of Danshen administration, the diameter at the narrowest site was increased from 1.00 ± 0.54 to 1.77 ± 0.84 mm, and the stenosis degree was reduced from (58.30 ± 18.39) % to (34.90 ± 23.98) %. The differences were statistically significant, indicating that Danshen can inhibit the increase in blood fat of rabbits fed a high-fat diet, postpone the formation of AS induced simultaneously by hyperlipemia and endothelial damage factors, and then alleviate the arterial canal stenosis induced by atheromatous plaques. Composite Danshen preparations can also significantly decrease the serum TC, TG, and LDL in rabbits fed high-fat feeds, and can increase the serum HDL and reduce the thickness of the vascular intima and medial and their ratio. Compared to the model group, the inhibition rate of the simvastatin group and the high and low dosage groups of composite Danshen on the intima-media thickness were 32, 41, and 24 %, respectively. Danshen can regulate blood fat levels by inhibiting the synthesis of endogenous cholesterol, promoting the transportation and scavenging of lipids, and by accelerating its excretion and many other pathways.

4.2.1.1 Inhibition of the Synthesis of Endogenous Lipids

Sun Shi suggested that Tanshinol has functions in inhibiting the synthesis of endogenous cholesterol (Ch). Amphotericin B forms a complex with Ch on the membranes of animal cells, then micropores appear on the cell membrane and cell death is induced. Under the conditions of insufficient exogenous Ch supply, if a Ch synthesis inhibitor is used to inhibit the synthesis of endogenous Ch, the formation of Amphotericin B-Ch complexes can be inhibited and the cells can avoid being killed by amphotericin B. Fibroblasts were first incubated with the culture solution with serum without lipoproteins to create the condition of no exogenous Ch supply, then the experiment was carried out by dividing the cells into three groups; Ch synthesis inhibitor was not added in the first group, lovastatin was added to the second group, and different concentrations of Tanshinol were added to the third group. The results showed that all fibroblasts in the first group died, the fibroblasts in the second and the third groups maintained normal morphology, and the number of surviving cells in the third group was positively related to the concentration of Tanshinol, indicating that Tanshinol can reduce the total cholesterol and triacylglycerol in serum by inhibiting the synthesis of endogenous Ch.

4.2.1.2 Affecting the Transportation and Excrement of Lipids in Vivo

Since lipids are not soluble in water, and they can be circulated and transported only after they bind to apoproteins to form lipoprotein complexes, which have a relatively greater solubility; therefore, apoproteins play crucial functions in lipoprotein metabolism, and its transformation is more important than are lipoproteins and blood fat in the identification of AS. One of the important mechanisms for TCM lipid-regulating anti-AS drugs is possibly related to the significant increase in the ratio of apoprotein AI (ApoAI) to apoprotein B100 (ApoB100). Lipoproteins and apoproteins play important functions in lipid metabolism. Qin Peisen and Sun Ling and others found in their observations that Danshen can not only decrease TC, TG, and ApoB, but also increase HDL, ApoA and ApoA/ApoB, and regulate the disintegration, metabolism and transportation processes of lipids by changing the composition of lipoproteins.

Lipoprotein receptors determine the metabolic outlets of various kinds of plasma lipoproteins; they can regulate plasma lipoprotein and fat levels, and then regulate the intake amount of plasma lipoproteins by adjacent tissues and prevent the formation of a plasma LDL removal barrier, avoiding an abnormal increase in plasma cholesterol and the occurrence of AS. The low density lipoprotein receptor (LDL-R) in the liver plays very important roles in the lipid metabolism of the entire body. Most cholesterol in the plasma exists as LDL, while 65–70 % of LDL is removed by LDL-R in hepatic cells. Liver LDL-R can also affect the rate of LDL production. With the wide applications of molecular biology techniques, it is possible for people to carry out deeper investigations into the expression level of the LDL-R gene and its regulatory factors. Recently, some scholars found that the activity of LDL-R in hypercholesteremia and aged people is lowered, and using drugs to enhance the expression of LDL-R in order to lower blood cholesterol level has been one of the hot spots of current investigations. Experiments have confirmed that compound Danshen preparations can significantly decrease serum TC and LDL cholesterol in hyperlipemic animals. Results from further studies showed that compound Danshen preparations can promote the expression of LDL-R mRNA in rats fed a high-fat diet and rabbit liver. Results from in vitro experiments also revealed that compound Danshen preparations can significantly increase the expression of LDL-R mRNA in human fibroblasts that were cultured for 2 h, indicating that compound Danshen preparations can affect the regulation of LDL-R at the transcriptional level and activate the LDL-R pathway, increasing the number of LDL-R on the surface of the cell membrane to make LDL-R bind to more LDL and other lipoproteins containing apoprotein

ApoB 600 and ApoE. After LDL-R binds to LDL, LDL is disintegrated by the enzymes in the lysosome, cholesterol esters are hydrolyzed into free fatty acids (FAA) by acid esterase. The increased FAA can inhibit HMG-CoA reductase, which reduces the production of cholesterol in cells, and activate acyl coenzyme A-cholesterol acyltransferase (ACAT) to re-esterify FAA into CE and store it in the cytosol, thus accelerating the elimination of LDL in the plasma. Therefore, it is presumed that Danshen can reduce cholesterol in the blood and prevent AS, which may be related to the decrease in positive expression of OX-LDL in the arterial wall and the reinforcement of the expression of liver LDL-R mRNA. Excessive endogenous lipid synthesis and excrement disturbance are major reasons for the formation of hyperlipemia in humans, so research on the effects of TCM on the activities of crucial enzymes in lipid metabolism will become increasingly active.

4.2.2 Prevention of Lipid Peroxidation

Lipid peroxidation induced by **free radicals** is closely related to the occurrence and development of many diseases. With the rapid development of free radical medicine in recent years, damage induced by a decrease in the capability of preventing oxidation and eliminating free radicals and the aggregation of lipid peroxides (LPO) draws more and more attention. It is reported that lipid peroxidation can lead to an increase in negative charge on the surface of the membrane, a decrease in electrical stability, and an increase in permeability. The damages of LPO to the membrane can lead to the abnormal movement of transmembrane ions, particularly Ca^{2+} influx. Clinical data have confirmed that the LPO level is increased in the plasma in hyperlipemia, diabetes, and myocardial infarction.

In the early stage of experimental AS, the LPO plasma level plasma increases significantly, and endothelial cells (EC) are fully liable to injury since they are exposed to high levels of LPO for a long period. The permeability of EC in the early stage of AS increases, the adhesion of monocytes and platelets is closely related to the changes in EC functions, and LPO in the blood can act directly on EC. Furthermore, LPO produced by EC themselves, **free radicals** released by the foam cells in lipid plaques, and LDL that undergoes oxidation modification can all cause damage to EC and promote the progress of AS. LPO has inhibitory effects on the prostacyclin (PGI_2) synthase system. The synthesis of PGI_2 inevitably decreases when EC is significantly damaged. PGI_2 has functions in preventing platelet aggregation and relaxing blood vessels, which is closely related to the occurrence of pathological changes in AS. Oxygen free radicals not only can act directly on the cell membrane and induce lipid peroxidation injuries to the cell membrane, but also can change LDL into OX-LDL, and thus lead to damage in target cells; these are the reasons why free radical lipid oxidation causes changes in EC function at the early stage. The changes in the functions of endothelial cells help the monocytes to penetrate and the intake of LDL, which leads to TC aggregation and the formation of foam cells. Thus the endothelial cells undergo cellular necrosis and defluxion, and then it may develop into AS and other complex pathological changes. Therefore, the reinforcement of the antioxidative capability of the organism, the prevention of lipid peroxidation, and the protection of vascular endothelial cells are an important way to prevent and treat AS. Free radical scavengers can inhibit the oxidation modification of lipoproteins, prevent lipid peroxidation in the membrane, inhibit the occurrence of AS, and change the progress of lesions.

4.2.2.1 Preventing Lipid Peroxidation and Protecting Endothelial Cells

Recent studies have revealed that Danshen is an effective antioxidant. Meng et al. found that Danshen can increase the fluidity of cell membranes by eliminating **free radicals**, decreasing the content of LPO, and thus the cell structures and functions are kept normal, which is in accordance with other reports. Scanning electron microscopy revealed that Danshen can protect the endothelial cells in arteries and inhibit the deposition of cellulose, platelet and lipid particles on the inner membranes of arteries. Danshen can significantly antagonize damage to human vascular endothelial cells induced by bacterial endotoxin and hydrogen peroxide, significantly increase the activity of cells after injury, inhibit the production of lipid peroxides, and regulate the production of nitric oxide (NO) and the expression of nitric oxide synthase (NOS) in dual ways. The antioxidant activity may be an important mechanism for Danshen preventing the occurrence and development of AS. The study reported by Liu Hong showed that Danshen could reduce the extent of lipid peroxidation in blood vessel endothelia in rabbits, improve NO levels in serum, decrease the concentration of plasma endothelin (ET) to regulate the balance between NO and ET, and improve vasomotor functions. Lin Qi et al. reported that Danshen can regulate the metabolism of blood fat, particularly by inhibiting lipid peroxidation modification and protecting EC, and it alleviates AS plaques already formed.

4.2.2.2 Preventing LDL Lipid Peroxidation and Reducing OX-LDL Formation

The aggregation of large amount of lipids in macrophages in the subendothelial layer of arteries and the formation of foam cells are important features of the early stage of AS. Smooth **muscle** cells and monocytes in blood circulation are all involved in the formation of foam cells. The formation of foam cells is closely related to OX-LDL, which is formed by LDL after oxidation modification by many kinds of cells and lipoxygenase. Its chemical composition and physico-chemical properties are significantly changed. The oxidation modification of LDL in vivo is a complex process, and it cannot be summarized by a kind of definite mode now, but it is generally assumed that it has relatively a strong relationship to oxygen fee radicals in cells, and that the lipid peroxides produced by LDL oxidation modification process can inhibit the synthesis of HDL. OX-LDL is cytotoxic and it can cause denaturation, cellular necrosis and defluxion in endothelial cells, the permeability damage to endothelial cells is increased and thus large amount of lipoproteins in the blood can enter the arterial wall. Previous investigations also revealed that damaged endothelial cells can release Platelet-derived growth factor (PDGF) and endothelin to stimulate the proliferation, migration and penetration into the inner membrane of medial smooth **muscle** cells, thickening the inner membrane, and monocyte chemoattractant protein-l (MCP-1) can be released to promote the aggregation of monocytes under the endodermis. OX-LDL can also decrease the activity of antioxidases in endothelial cells of human umbilical veins. While the content of lipid peroxides is increased, it can also directly induce platelet aggregation, promote thrombogenesis, and the adhesion of monocytes to endothelial cells, all of which can promote the occurrence and development of AS.

Danshen can reinforce the organism's capability of preventing oxidation and eliminating **free radicals** by increasing the activities of Glutathione peroxidase (GSH-Px) and superoxide dismutase (SOD) in endothelial cells, thus effectively inhibiting lipid peroxidation, and then preventing and treating oxidative damage from OX- LDL to endothelial cells. SalB can eliminate 1,1-diphenol -2-m-dinitrohydrazyl free radicals, and it can effectively inhibit the oxidation of LDL when compared with Probucol. In order to evaluate the degree of its anti-artherosclerosis effects, New Zealand white rabbits were used as an animal model. One group was fed a standard diet, one group was fed a high cholesterol diet, one group was fed a high cholesterol diet containing 1 % probucol, and one group was fed a high cholesterol diet containing 5 % SalB. The animals were raised for 12 weeks. The results revealed that plasma cholesterol in the SalB group and probucol group both decreased, and LDL in the rabbits from the SalB group can well resist Cu^{2+} induced oxidation compared with the high cholesterol diet groups, and it contained more VitE. The degree of endothelial damage was determined at the sixth week, and it was found that the SalB group decreased by 53 %, the scleratheroma degree in the abdominal aorta decreased by 56 %, and the cholesterol deposition in the thoracic aorta decreased by 50 %. This study showed that the decrease in the degree of artherosclerosis in hypercholesteremic animals by the use of SalB was related to not only the function of lowering cholesterol, but also to its antioxidative functions, which prevented endothelial injuries and the oxidative modifications to LDL.

Wu Lan et al. investigated the inhibitory effects of SalB on Cu^{2+} induced LDL oxidation modification, and the results showed that the addition of SalB before incubation with Cu^{2+} and LDL can decrease the production of thiobarbituric acid reacting substance in LDL, the fluorescence intensity was inhibited, and the elimination of VitE content was significantly slowed. This showed that SalB showed relatively strong inhibitory effects on Cu^{2+} induced LDL oxidation modification, indicating that SalB had important significance in the prevention and treatment of artherosclerosis as a kind of antioxidant agent. Wu Weiping et al. dialyzed the LDL in human plasma together with Cu^{2+} and then thin-layer chromatography was used to analyze the OX-LDl band and the contents of free fatty acids, phospholipids, cholesterol, and TG. The results showed that no obvious OX-LDL band formed after Danshen was added, and no significant change was found in the lipid ingredients compared with the control group before dialysis, indicating that Danshen had functions in preventing LDL lipid peroxidation and reducing the formation of OX-LDL. Wang Nan et al. found in their investigations that the abovementioned changes in LDL can be prevented if a certain amount of Tanshinol was added during the process of the preparation of OX-LDL by using H_2O_2 and horseradish peroxidase in vitro; the content of lipid peroxides in various kinds of LDL was significantly decreased after Danshen was added, indicating that Tanshinol had functions in preventing LDL oxidation. Tanshinol can inhibit the oxidation modification of LDL in the model of smooth muscle cells in the main artery of cattle, significantly decrease malondialdehyde (MDA) in LDL, and the antioxidant ability is positively related to the amount of Tanshinol, indicating that Tanshinol can be used in the prevention and treatment of atherosclerosis.

4.2.2.3 Scavenging Free Radicals

It has been considered in recent years that oxygen **free radicals** have very important functions in the oxidation modification of LDL. Some scholars reported that Danshen had very significant eliminating functions for the superoxide anions that were produced by the xanthine-xanthine oxidase system and the superoxide anions that were produced by stimulating white blood cells with phorbol myristate acetate (PMA). One of its active ingredients, Tanshinol, showed stronger eliminating effects on superoxide anions than superoxide dismutase, but it could not effectively eliminate hydroxyl radicals, while another active ingredient, tanshinone, could effectively eliminate lipid **free radicals** produced by mitochondria, and thus it was considered that the antioxidant functions of Danshen may be realized by eliminating different free radicals at different sites through its different active ingredients. The effects of quercimelin on the production of superoxide anions by xanthine-xanthine oxidase and alkaline sodium dithionite system was observed by chemiluminescence, which showed that quercimelin had significant inhibitory effects on the luminous intensity of both of these luminous systems, indicating that quercimelin can reduce the production of superoxide anions by inhibiting that activity of xanthine oxidase separately from its direct eliminating effects on superoxide anions.

Investigations in recent years have found that peroxynitrite is one of the important mediators of arteriosclerosis and other diseases. The experiments carried out by Liao et al. confirmed that Tanshinol can effectively protect the organism from being injured by nitrite and prevent the formation of artherosclerosis. In general, all of the experimental results confirmed that Danshen had significant protection over vascular endothelial cells, and its mechanism was related to the antioxidant functions and the regulations of the production of endogenous active substances.

4.2.3 Improvement in Functional Disturbance of Blood Vessel Endothelium

The endodermis used to be considered an inactive cell layer that existed in the blood vascular system and imported or exported various kinds of substances from the blood. Now, endothelial cells are considered multifunctional cells with very complex functions, and they can produce different reactions in response to the ever changing physiological and pathological environment. The endodermis not only is the selective and permeable barrier for material exchange in the blood and tissues, but also can synthesize and secrete various kinds of biologically active substances and take part in many kinds of pathological and physiological processes. The endodermis synthesizes and secretes endothelin, prostacyclin (PGI_2) and endothelium-derived relaxing factor (EDRF), and they maintain the relaxation and contraction of blood vessels and change the activity of platelets. The endodermis can regulate blood clotting and thrombolysis by producing various kinds of regulatory mediators, such as thrombomodulin, heparin sulfate, tissue plasminogen activator, and plasminogen activator inhibitor-1. Furthermore, the endodermis is involved in the metabolism of platelet aggregates adenosine diphosphate, 5-hydroxytryptamine, prostaglandin and angiotensin, and then inhibits the activity and adhesion of platelets. The endodermis produces intercellular adhesion molecules (ICAM) and vascular cell adhesion molecules (VCAM), and these adhesive molecules can regulate the adhesion of endothelial cells to monocytes and lymphocytes. The endodermis also produces growth stimulating factor (basic fibroblast growth factor) and tissue growth factor B-1, and they function in the growth of plaques by repeatedly stimulating the cells.

A large number of investigations have showed that endothelial injuries and endothelial hypofunction are initiating factors for the morbidity of artherosclerosis, and thus the injuries or functional changes in endothelial cells play an important role in the development of AS. Ross et al. formally proposed in the early 1970s that endothelial injuries took part in the process of atherosclerosis, termed the "reaction to injury theory". He suggested that endothelial injuries were caused by various factors, particularly by endothelial stripping causing the lipid ingredients of the plasma to enter the endodermis. Platelets adhere to the exposed collagen under the endodermis, and release many factors inducing smooth **muscle** to proliferate, leading to the development of AS, and thus endothelial injuries are the initiating element for the development of AS. Current investigations have found that many factors can induce endothelial injuries in arteries, including mechanical factors (such as hemodynamic effects), chemical factors (such as hypercholesteremia, OX-LDL, smoking, and homocysteic acidmia), immunity, and viral infection. When the blood vessel endothelium is damaged, the collagen under the endodermis is exposed and platelets adhere to the collagen, accumulating and releasing various kinds of biological active substances, such as 5-hydroxytryptamine, ADP, and histamine, which can all increase the permeability of vessel walls. Thus, LDL and other macromolecular substances in the blood can penetrate the endodermis and accumulate under the endodermis, which aids the further development of AS.

With scientific development, the role of endothelial dysfunction in AS formation has received more and more attention. Ross published "the revised reaction to injury theory" in 1986, and pointed out that endothelial injuries not only referred to endothelial denudation, but also included endothelial functional changes without morphological changes, that is, functional disturbance, and endothelial functional disturbance may occur in the early stage of artherosclerosis. With further advancements in the understanding of endothelial function in the 1990s, people paid more attention to the significance of non-denudating functional injuries in the development of AS. Therefore, functional disturbances in blood vessel endothelia act as an initiating factor in the occurrence and development of artherosclerosis.

4.2.3.1 Affecting the Release of ET and NO

Previous investigations have showed that blood vessel endothelium-dependent relaxation weakened or even disappeared in the development AS, which may be related to the decrease in NO synthesis and/or the increase in endothelin (ET) secretion in vascular endothelial cells (VEC) during the development of AS. ET is a kind of circular polypeptide composed of 21 amino acids, and it was isolated from endothelial cells in the main artery of pigs by Japanese scholar Yanagisawa and his colleagues in 1988. It has been known that ET not only has strong and continuous blood vessel contraction functions in vitro and in vivo, but also is a strong growth stimulating factor with prolonged activity. It can promote the proliferation of vascular smooth **muscle** cells, and previous experiments have confirmed that the extent and range of artherosclerosis are positively related to the level of ET in plasma. NO is produced by NOS catalysis and it can increase the content of cyclic guanosine monophosphate (cGMP) and expand blood vessels. It inhibits the biological synthesis of ET and has antagonistic functions to ET. Therefore, it is considered that the disequilibrium of ET and NO is closely related to the occurrence and development of artherosclerosis.

Jiang and Liu found that under hypoxic conditions, the content of ET in the supernatant of pulmonary artery endothelial cell culture was 2.8 times that under normal conditions. The metabolites of NO significantly decreased, indicating that hypoxia could increase the expression of ET in and the release from the endothelial cells of pulmonary artery, and the synthesis and release of NO were inhibited.

Danshen can correct hypoxia-induced abnormalities of ET and NO secretion by promoting NO synthesis and inhibiting ET release in vascular endothelial cells, and thus alleviate hypoxic pulmonary vasoconstriction. The investigations carried out by Liu Hong revealed that Danshen could decrease ET-1 levels in the plasma and vessel wall tissues of artherosclerotic rabbits, and it could increase the content of NO in the serum and vessel wall tissues. Shi Yifan found that composite Danshen preparation showed similar functions, indicating that Danshen could protect blood vessel endothelium function in AS animals, and thus inhibit the formation of experimental AS. In a further investigation into the effects of various kinds of ingredients from Danshen on endothelial cell function in artherosclerosis model rabbits, it was found that the water soluble but not liposoluble ingredient SalB in Danshen could decrease the ET concentration in the plasma of artherosclerotic rabbits. Li Weiqing et al. suggested that the mechanism for Composite Danshen Dropping Pill (CDDP) in treating ischemic heart disease may be that it regulates NO and ET levels in the plasma, promotes the repair of blood vessel endothelia, reduces microthrombus, and improves blood supply in the coronary artery muscle muscle.

With the advancement of molecular biology in recent decades, the molecular pathological studies of Danshen's functions against AS have also made considerable progress. Using RT-PCR, Feng Peifang et al. determined the expression level of the ET-1 gene in the peripheral blood circulation endothelial cells of coronary heart disease patients, and the results showed that compared with healthy people, the expression of ET-l mRNA in coronary heart disease and heart-stroke patients increased significantly. The symptoms of heart-stroke in 29 cases were alleviated after treatment with CDDP. Electrocardiograms revealed that myocardial ischemia improved and the expression of ET-l mRNA in the vascular endothelial cells also significantly decreased at the same time, indicating that CDDP can alleviate the onset of heart-stroke by regulating the expression of ET-l mRNA in vascular endothelial cells. Xie Meilin et al. also demonstrated that the Stasis Eliminating Tablet composed of Danshen and hawthorn extracts could inhibit the expression of ET-l mRNA in the vessel wall tissues in artherosclerotic rabbits. Luo and Wang reported that hypoxia can induce the release of NO in endothelial cells and the expression of endothelial nitric oxide synthase (eNOS) mRNA and induced nitric oxide synthase (iNOS) mRNA were upregulated ($P < 0.01$). SalB magnesium salt could improve hypoxia-induced injuries in endothelial cells, inhibit the release of NO in endothelial cells, and increase mRNA expression of eNOS and iNOS under hypoxic conditions; it could also inhibit the expression of iNOS mRNA in vessel wall tissues, and thus gain the functions of AS prevention from many aspects.

4.2.3.2 Affecting the Balance of TXA2/PGI_2

Vascular endothelial cells are not only a physiological barrier between the blood and vascular smooth muscle, but also are a highly active metabolic pool. They have certain anti-thrombus functions themselves and can synthesize many vessel active substances and inhibit the activation of the blood clotting system and thrombogenesis in order to maintain the flowing status of blood. VEC can be injured by many causes, such as endotoxin, mediators of inflammation, free radicals and others, and injuries in VEC can lead to their functional imbalance. The activated endothelial cells change from anticoagulation status to coagulation promoting status, tissue factor (TF) is synthesized and expressed, the balance of TXA_2/PGI_2 is interfered with, and finally microthrombi are formed due to the thrombus promoting functions, and thus cardiovascular and cerebrovascular diseases are induced and aggravated. Therefore, VEC injuries occupy a very important place in the formation of thrombus disease.

Ang II is an active substance for blood vessel contraction. It not only regulates blood pressure, but also directly induces the functional disturbance of endothelial cells and promotes the expression of TF, and reduces the antithrombotic function of endothelial cells. Zu Shuyu et al. used restriction enzymes to construct the luciferase reporter plasmid that contains different upstream regulatory sequences of the tissue factor gene in vascular endothelial cells. After liposome transfection, they found that the active component of Danshen could inhibit the reinforcement of TF gene expression in human endothelial cells induced by Ang II. Zhang Yanjun et al. showed that Danshen could reduce TXB_2 concentrations in artherosclerotic rabbits and increase the content of G-ket0-PGF, tanshinone II-A could antagonize LDL-induced injuries in blood vessel endothelia, and the content of MDA was reduced and the level of 6-ket0-PGF_2 was increased.

4.2.4 The Inhibition of the Expression of Adhesive Molecules and the Antagonism of the Adhesion Between Cells

The development of AS is a very complex pathological process. Monocytes adhere to the blood vessel endothelium and migrate into the endodermis, then intake lipids and transform into foam cells. This is one of the key steps in the development of AS. In animal models fed with high cholesterol, the adhesion of monocytes to the blood vessel endothelium can be observed within several weeks, and then monocytes infiltrate the inner membrane of blood vessels and intake cholesterol, change into lipid-rich macrophages, and form lipid strips, which are the earliest atherosclerotic plaques. The adhesion molecules play very important functions in mediating the adhesion and migration of monocytes to the endodermis. Due to high cholesterol, hemodynamics and other factors, blood vessel endothelium function is weakened, and the lipids activate the complement chemotactic monocytes, and the endothelial cells are induced to express adhesion molecules. Monocytes adhere to vascular endothelial cells and migrate into the endodermis, and many adhesion molecules including the integrin family, immunoglobulin superfamily, and selectin family are all involved in and mediate the abovementioned processes. These adhesion molecules can promote the white blood cells and monocytes in the blood to migrate to the endodermis at early stages of plaque formation, and they promote the enlargement of plaques in preexisting plaque tissues. The adhesion molecules in the plaques can also affect the replication, division, and proliferation of smooth **muscle** cells in the middle layer of the blood vessels.

4.2.4.1 Inhibiting the Expression of Intercellular Adhesion Molecule

Intercellular adhesion molecule (ICAM-1), a member of the immunoglobulin superfamily, is one of the adhesion molecules that are closely related to cardiovascular functions. It exists on the surface of white blood cells, endothelial cells, smooth **muscle** cells, epithelial cells, and many other cells. ICAM-1 can mediate the adhesion of white blood cells in the circulation to endothelial cells, promote the adhesion of white blood cells to vascular smooth muscle cells and their retention in the AS region, which contributes to AS during the formation of chronic inflammation. Moreover, ICAM-1 is also involved in cell proliferation and migration. So, it can be concluded that ICAM-1 is involved in numerous aspects of AS formation. The results from the investigations of Zhang Mei et al. showed that in the vessel walls of normal rats, almost no ICAM-1 protein or its mRNA was detected, but in the vessel walls of AS model rats, ICAM-1 protein and its mRNA expression were significantly increased. Danshen injection could inhibit the blood fat, ICAM-1 protein, and the expression of its mRNA in artherosclerotic rats, indicating that the effects of Danshen injection in preventing artherosclerosis were related to the decrease in ICAM-l. Jiang Kaiyu et al. found that the incubation of human umbilical vein endothelial cells with tanshinone II-A sulfonate could antagonize the function of tumor necrosis factor-induced HUVEC adhesion molecules ICAM-1 and E-selectin, and the antagonizing function was tanshinone II-A sulfonatedose-dependent. Further investigation revealed that Tanshinol showed inhibitory effects on the adhesion of white blood cells by inhibiting the expression of CD11 in activated white blood cells. The abovementioned experiments indicated that Danshen can protect blood vessel endothelia and prevent AS by reducing the adhesion of white blood cells. Wu Jinghai et al. found that Tanshinol could inhibit the adhesion of peripheral blood monocytes to human umbilical vein endothelial cells in patients with psoriasis, and it could also decrease the expression of ICAM-l on the surface of monocytes in peripheral blood. Therefore, it can be concluded that the inhibition of ICAM-1 expression by Danshen has experimental evidence.

4.2.4.2 Inhibiting the Expression of Vascular Cell Adhesion Molecule

Vascular cell adhesion molecule-1 (VCAM-1) plays a very important role in mediating the adhesion and migration of monocytes to the endodermis. VCAM-1 is highly expressed in the AS plaques of patients and lipid plaques of rats and rabbits. Investigations by Chen Jianzong et al. demonstrated that the expression of VCAM-1 in the carotid wall of AS rabbits was significantly increased. Composite Danshen Dropping Pill can significantly inhibit or downregulate the expression of VCAM-1 in the carotid wall, which may be one of the mechanisms for the prevention and treatment of carotid artherosclerosis.

4.2.4.3 Inhibiting the Expression of Monocyte Chemoattractant Protein-1

Monocyte chemoattractant protein-1 (MCP-1) is a monocyte chemotactic activity factor, and its high affinity specific receptor exists on monocytes, which can cause the monocytes in the blood to migrate into the inner membrane and then activate into macrophages to swallow lipids and change to foam cells. The expression levels of MCP-1 mRNA and protein were significantly increased in AS genesis, indicating that MCP-1 is involved in the occurrence and development of AS. Xu Tao et al. found that Danshen can downregulate the expression of MCP-1 in vascular endothelial cells and smooth **muscle** cells. It can thus reduce the aggregation of inflammatory cells in vascular endothelial cells and smooth muscle cells, and further alleviate the damage from high TG to blood vessel endothelium, indicating that the effects of Danshen in the prevention and treatment of AS are related to the inhibition of MCP-1 expression.

Tanshinone II-A sulfonate can inhibit the expression of P-selectin on the surface of human platelets that is induced by thrombosin in a dose-dependent manner after incubation with platelets. Investigations by Wang Ling et al. demonstrated that Danshen reduced both the number and intensity of adhesions between red blood cells and endothelial cells by simultaneously acting upon these two types of cells, and blood circulation was thus improved.

4.2.5 Inhibition of Vascular Smooth Muscle Cell Proliferation

Vascular smooth muscle cells (VSMC) are located in the middle membrane of blood vessels and are the only cell component in the middle layer of blood vessels. Under physiological conditions, vascular smooth muscle cells regulate the tension of blood vessels by contracting and relaxing, and they maintain the normal functions of blood vessels by secreting blood vessel regulatory factors. Since the 1960s, with advancements in immunohistochemistry and ultrastructural investigations, researchers have found that most of the cells in AS plaques are derived from the proliferation of smooth Mmd, then the signals are transmitted into the nucleus, and gene transcription is initiated during the process of AS development. Smooth muscle cells in the tunica media of the artery migrate to the endodermis, and both their phenotype and gene expression undergo significant changes, which are involved in intercellular substance synthesis, lipid deposition, and plaque calcification, and then lead to the thickening of blood vessel walls. Therefore, smooth muscle cells play an important role in the formation and development of AS.

4.2.5.1 Regulating Cell Cycles

Morphologically, smooth muscle cells show two phenotypes: contractile phenotype and synthetic phenotype. VSMC of normal adult mammals are of the contractile phenotype, and their main function is to regulate the tension of blood vessels. When the inner membrane is injured or artherosclerosis occurs, VSMC undergo changes in their characteristics and phenotypic flexibility under the effects of various kinds of stimulating factors and growth factors, such as Ang II, Platelet-derived growth factor (PDGF), and interleukin (IL). The cells proliferate, undergo hypertrophy and generate functions of synthesis and secretion. At this point, the Golgi apparatus, rough endoplasmic reticulum, ribosomes, and others in the cytosol of VSMC increase, cytoskeletal proteins are expressed in high levels, and protein kinase C increases, while the expressions of myosin heavy chain and actin decrease, transforming the cells from the contractile phenotype to synthetic phenotype. The smooth muscle cells of the synthetic phenotype are dedifferentiating cells, which have very strong proliferative and secretive capabilities. They can proliferate in large amounts under the effects of various kinds of factors, produce large amounts of intercellular substances such as collagen fibers, elastic fibers, proteoglycan, glycoprotein, and they swallow cholesterol and cause lipids to accumulate inside the cells and form muscle derived foam cells. The proliferative smooth muscle cells and foam cells are liable to undergo cellular necrosis, and after cell death, they, together with intercellular substances and lipids, create typical athero-pathological changes. Danshen can inhibit the proliferation of VSMC and exert its functions in decreasing the thickness of the arterial wall and plaque formation by regulating the cell cycle of smooth muscle cells.

Cell proliferation is a well-organized progress of the cell cycle under the influence of extracellular signals, and the cell cycle is the process of the growth of single eukaryotic cells and the division into daughter cells, which is composed of the dividing phase (M phase) and intermediate stage (G_1, S, G_2 stages). The cell grows and becomes large in G_1 stage, and it enters S stage and undergoes the synthesis and replication of DNA when it reaches a certain volume; stage V is the post-synthetic phase of DNA, which prepares for mitotic division. The chromosome is divided and the cells are divided into two daughter cells in M stage, the cells return to G_1 stage after the division, and thus the cell cycle is finished. Under some circumstances, the cells can temporarily exit the cell cycle and enter the resting stage (G_0 stage). Interference in the cell cycle is a key element in the control of cell proliferation, and some experiments show that tanshinone has significant inhibitory effects on VSMC proliferation stimulated by 5 % fetal bovine serum. It significantly decreases the metabolic rate of MTT, decreasing the synthesis of total protein, and the proportion of cells in G_0/G_1 phase increases, indicating that tanshinone can stop the proliferative cycle of VSMC in G_0/G_1 phase, and then inhibit VSMC proliferation.

Danshen can significantly inhibit the H_2O_2 induced expression of NF-κB and PCNA proteins, and thus DNA cannot replicate and the cells stop in G_0 phase. It could also be found from the results of flow cytometry that the inhibition of H_2O_2 induced VSMC proliferation by Danshen results from the inhibition of the transition from G_0/G_1 phase to S phase in cells, and thus the number of cells in G_2/M phase decreases, This kind of effect may be an important mechanism of the inhibition of VSMC proliferation.

4.2.5.2 Inhibiting Mitogen-Induced Smooth Muscle Cell Proliferation

Among the factors that can stimulate VSMC proliferation, PDGF is one of the hot spots in current investigations, since it is the major mitogen in serum, which can promote the mitotic division and growth of smooth muscle cells and fibroblasts. For example, proliferative smooth muscle cells can migrate from the middle layer of blood vessels to the inner membrane under the stimulation of PDGF, promote the infiltration of LDL and TC into cells and induce the accumulation of lipids in cells, promote the synthesis of collagen, elastic fibers and proteoglycan, inhibit myosin synthesis, and finally induce cellular damage or necrosis, and altogether constitute intra-arterial atherosclerotic plaques. PDGF is the product of *sis* gene expression in endothelial cells, but monocytes, fibroblasts and smooth muscle cells also have the *sis* gene, which can produce and secrete PDGF-like growth factors. LDL, tumor necrosis factor and many factors can all promote *sis* gene expression in these cells and produce PDGF, and experimental results have demonstrated that PDGF can significantly promote SMC proliferation and significantly increase the amount of SMC and H^3-TdR incorporation. Danshen can inhibit basal state SMC proliferation and the proliferation stimulated by PDGF, which may be one of the mechanisms for the inhibition of AS by Danshen.

4.2.5.3 Inhibiting Proto-Oncogene Expression

It is found in recent investigations that oncogenes c-*sis*, c-*jun*, c-*myc,* and c-*fos* are major proto-oncogenes that are related to the regulation of cell proliferation, and their activation can lead to excessive PDGF production and the deactivation of anti-oncogene p53 and Rb, and then the inhibition in SMC proliferation is lost and the development of AS is promoted. Therefore, VSMC proliferation can be inhibited if the expression of some genes related to the regulation of VSMC proliferation is inhibited. Danshen can inhibit the expression of genes that are related to cell proliferation, and thus inhibit DNA synthesis in the cells, which provides an experimental basis for the treatment of cardiovascular disease induced by VSMC proliferation.

An experiment was conducted by utilizing *in situ* hybridization to investigate the effects of tanshinone II-A sulfonate on the expression of c-*myc*, which is related to the proliferation of VSMC. It was found that macrophage derived growth factor can significantly promote the high expression of c-*myc* in SMC, suggesting that the infiltration and accumulation of macrophages (MP) around the lesion in AS pathogenesis may be through the promotion of abnormal c-*myc* expression, which stimulates the proliferation of SMC and promotes the occurrence and development of AS. Tanshinone II-A sulfonate can inhibit the stimulation of macrophage-derived growth factors to SMC and decrease c-*myc* expression, indicating that it may inhibit the proliferation of smooth muscle cells in artherosclerosis.

The results from the investigation by Wang Wei et al. revealed that the expression of proto-oncogene c-*jun* in in vitro cultured VSMC decreased under the effects of compound

Danshen preparation, indicating that Danshen could inhibit mitogens (such as growth factors, endothelin and others), and thus inhibit the expression of proto-oncogene c-*jun* and regulate cell proliferation. DNA synthesis also decreased in VSMC that were cultured in serum containing Danshen, indicating that the proliferation of cells decreased. endothelin can activate phospholipase D, phospholipase C, protein kinase C, protein tyrosine kinase and others by the receptor coupled to the G-protein in VSMC as a kind of potent substance to promote mitotic division, and the activation of these enzymes can lead to an increase in intracellular inositol triphosphate and Ca^{2+} concentrations. Inositol triphosphate and Ca^{2+} are both secondary messengers in the cells, which can induce the expression of c-*fos* and c-*jun* expression by activating some transcription factors. c-*fos* and c-*jun* synthesized in the cytosol are transmitted into the nucleus and bind to the AP-1 site of some genes in the form of heterodimers, and thus the stimulating signal of endothelin is transformed into the signal for gene expression and cell proliferation in VSMC. The expression products of some proto-oncogenes such as c-*fos* and c-*jun* are located in the nucleus, and these products are DNA-binding proteins, namely transcription factors. These oncogenes do not express, or express at low levels, in nondividing resting cells, and their expression increases significantly under the effects of cell proliferation signals. The expression product of c-*jun* can form homodimers and bind to DNA, and c-*jun* can also connect with the expression product of c-*fos* by a leucine zipper and form heterodimers. This complex can bind to DNA through their corresponding basic amino acid regions and promote the expression of cell division genes, and thus lead to cell proliferation. Danshen can inhibit endogenous c-*jun* expression and decrease the production of c-*fos*/c-*jun* heterodimers, and thus the expression activity of endothelin gene is decreased. It is presumed that the inhibition of Danshen on DNA synthesis of VSMC results from the inhibition of endogenous c-*jun* expression.

4.2.6 Regulation of Antithrombotic System, Inhibition of Thrombosis

AS is closely related to blood clotting and fibrin degradation, and thrombogenesis is an important factor in AS development. Clinical observations have found that Danshen can change the hemorrheology, decrease the "sticky, accumulating and stagnant" tendency of the blood in patients that suffer from coronary heart disease, and that it has excellent therapeutic activity. Animal experiments have demonstrated that Danshen has good effects on in vitro thrombogenesis, platelet functions, blood clotting functions and fibrinolytic enzyme activity in rabbits.

4.2.6.1 Increasing the Anticoagulation and Fibrin Degradation Activities

The fibrin degradation system is one of the important anti-thrombus components in organisms. Plasminogen is activated into fibrinolysin under the effects of its activator, and the latter can degrade fibrin, eliminating excessive fibrin deposits in blood vessels, tissues, and body fluids, and maintaining the smooth flow of blood circulation in tissue fluid canals and secretory ducts. Investigations into the changes in blood clotting characteristics and the occurrence and development of AS have gradually intensified, and many data have confirmed that fibrin deposition and the weakening of local anticoagulation and thrombogenesis can promote the occurrence and development of AS.

Using a self-control approach, blood was collected from rat hearts at different intervals before and after the rats were subjected to intravenous Tanshinol injections, and the corresponding parameters were measured. The results showed that after the injection, the in vitro thrombus formation time was prolonged, the length of thrombus was shortened, and the weight was reduced; the number of platelets and the aggregation decreased, the clotting time was prolonged, and serum fibrinogen breakdown products increased 30 min after the injection of

Tanshinol, indicating that Danshen had significant anti-thrombosis functions. The anti-thrombosis functions of Danshen showed obviously inhibitory effects on the occurrence and development of AS. 6-Keto-PGF_2 was determined by radioimmunity assays, chromogenic substrate S2390 was used to determine plasminogen activity (PA) and its inhibitor (PAI), and the activity of thrombus regulating protein was indirectly determined by using chromogenic substrates. The results of these investigations showed that Tanshinol can promote the secretion of PA and increase the activity of thrombus regulating protein in endothelial cells of cattle. Wang et al. observed that Danshen could inhibit not only the increase in serum TC in a high cholesterol diet and alleviate lumen stenosis induced by scleratheroma of iliac artery, but also the increase in plasma plasminogen concentration and fibrinolysis fibrinogen concentration. Shen et al. found that Danshen could inhibit the activation of endotoxin of tissue factors in endothelial cells, and thus increase the activities of anticlotting and antithrombosis, which may also be one of the mechanisms for Danshen to promote blood flow and eliminate stasis.

4.2.6.2 Preventing Blood Clotting and Inhibiting in Vitro Thrombogenesis

Li Chengzhu et al. observed the effects of intravenous injection of Danshen on thrombogenesis, platelet functions and blood clotting functions in rats and mice. The results showed that Danshen could prolong the in vitro thrombus formation time, reduce the length and weight of thrombus (wet weight and dry weight), inhibit platelet adhesion and aggregation, and prolong recalcification time, prothrombin time and kaolin partial thromboplastin time. These results indicated that Danshen had significant anticoagulative functions. Jiang Kaiyu et al. investigated the anti-thrombogenesis functions of Danshen extracts, and the results showed that Danshen had significant inhibitory effects on thrombogenesis induced by injuries in the blood vessel endothelia in arteries and veins. Intravenous injection of Danshen in rats and mice significantly reduced the length and weight of thrombus (wet weight and dry weight), and platelet adhesion and aggregation functions were decreased.

4.2.6.3 Improving Hemorrheological and Microcirculation Disturbance

In ischemic cardiovascular and cerebrovascular diseases, microcirculation disturbance, accompanied by increased plasma viscosity and reduced red blood cell deformability, frequently occurs. It has been found that Danshen can increase blood flow and decrease the viscosity of blood, and it has achieved certain therapeutic effects in the treatment of ischemic cardiovascular and cerebrovascular diseases. Both Qin Peisen et al. and Zhou Yixuan have found in their studies that Composite Danshen Dropping Pill (CDDP) could improve hemorrheological parameters in patients with angina pectoris, significantly decrease the specific viscosity of whole blood and plasma fibrinogen in the patients, decrease the hematocrit, erythrocyte aggregation index, and shorten the erythrocyte electrophoretic time, thus accelerating blood flow and improving microcirculation, and inhibit platelet functions and regulate the balance between prostacyclin (PGI_2)/Thromboxane A2 (TXA2).

Li Yanping utilized fluorescence polarization to determine the membrane fluidity of platelets in rabbits, and investigated the mechanism for the treatment of cardiovascular disease with CDDP. The results showed that CDDP could increase the membrane fluidity of platelets and decrease the microviscosity of platelets. Li Jie et al. carried out investigations to examine the effects of CDDP on the hemorheology of experimental hyperlipemia rats. The results showed that CDDP could significantly decrease the viscosity of whole blood, the adhesive rate of platelets, and the thrombus index in high-fat model rats, indicating that CDDP could prevent and treat artherosclerosis and coronary heart disease. Guo Jun et al. carried out investigations on the effects of CDDP on the hemorheology and microcosmic rheology of red blood cells in hyperlipemic dogs as well as on its mechanisms. The dog hyperlipemia model was duplicated, and 6 months later, CDDP was administered for 4 months. The blood was then collected to determine the contents of triglyceride and cholesterol, hemorheology, erythrocyte electrophoretic rate, osmotic fragility, membrane fluidity and the molar ratio of cholesterol and phospholipids in red blood cell membranes (Ch/Pl). The results showed that after treatment with CDDP, the blood fat level in hyperlipemic dogs was significantly decreased, the microcosmic rheological characteristics were significantly improved, the cell electrophoretic rate was increased, the membrane fluidity of the cells and the deforming and orientation in the flowing field were enhanced, and the osmotic fragility was decreased. Thus the high shear and low shear viscosity of whole blood was significantly decreased and the high-fat stagnant status was improved, the microcirculation was improved, and lipid deposition and thrombogenesis were decreased. The in vitro effects of Danshen injection on the adhesion of human umbilical vein endothelial cells were quantitatively investigated with flow cytometry, and it was found that Danshen can decrease the number and the intensity of adhesions between red blood cells and umbilical vein endothelial cells. Another investigation found that Danshen can improve the deformability of red blood cells and improve the viscosity and elasticity of blood, and thus improve microcirculation.

In summary, Danshen, through multiple targets, can improve the hemorheology under pathological conditions, change the sticky, thick, aggregate, and stagnant status of hyperlipemic animals, and protect the functions of vascular endothelial cells, and thus improve blood microcirculation and facilitate the treatment of various kinds of cardiovascular and cerebrovascular diseases.

4.2.6.4 Regulating the Balance of PGI_2/TXA_2 and Improving Platelet Functions

The developments of coronary heart disease, angina pectoris, artherosclerosis and thrombus are closely related to platelets, and platelets play very important functions in the blood clotting system, smooth muscle proliferation, and the formation of atheromatous plaques. Under normal conditions, platelets are circulated in the blood vessels in a dispersive status, but they undergo a series of interconnected changes, such as morphological changes, adhesion, aggregation, and release, when the blood vessel endothelium is damaged or the blood flow is changed or stimulated by chemicals, and thus the effects of thrombus promotion are brought into play. The factors that can regulate platelet functions mainly include the arachidonic acid system, the cyclic nucleotide system, including cyclic adenosine monophosphate and cyclic guanosine monophosphate, the inositide system, and the Ca^{2+} system.

Danshen can reduce the level of cAMP in platelets in in vitro experiments on rabbits and rats. Ye et al. found that Danshen could significantly inhibit the platelet aggregation induced by arachidonic acid (AA), adenosine diphosphate (ADP), collagen and other inducers, in rats and rabbits, indicating that the effects of Danshen on platelet functions was through improving the functions of cAMP, but may be through inhibiting Thromboxane A synthase.

TXA_2 is a kind of biological active substance that is produced by platelets, which can promote platelet aggregation, blood clotting and thrombogenesis. PGI_2, which is released by vascular endothelial cells, has a very strong function of relaxing blood vessel, anti-platelet aggregation, anti- clotting, and antithrombogenesis, which antagonize the effects of TXA_2. Therefore, it is very important to keep the ratio of PGI_2/TXA_2 relatively stable to inhibit the development of AS. Studies have found that one of the mechanisms for the inhibition of platelet aggregation by Danshen is achieved by intervening in the AA metabolism pathway to specifically block the production of the aggregation-inducing Thromboxane TXA_2. In addition, some have suggested that Danshen can decrease TXA_2 synthesis by inhibiting the activity of thrombus ethylene oxide synthase, and it can also weaken platelet aggregation and thrombogenesis by regulating the synthesis of active substance PGI_2 in the vascular endothelial cells and the release of ET. Another ingredient of Danshen, rotocatechualdehyde, can inhibit the dense granules of platelets to release

5-hydroxytryptamine. It is parallel to the inhibitory effects on platelet aggregation, and the detailed mechanism may be related to arachidonic acid metabolism.

It can be found that Danshen may achieve its function of preventing thrombogenesis by reinforcing the functions of the thrombolysis system of the organism, improving hemorheology, and regulating PGI_2/TXA_2 balance. It can not only prevent the occurrence of AS, but also rapidly eliminate pre-existing AS.

4.2.7 Calcium Antagonism

Almost all of the known calcium antagonists can affect the important elements of AS development. Calcium antagonists can affect the formation of AS by changing intracellular or extracellular metabolic or physiological processes. The concentrations of calcium in the plasma and in the cell differ by tens of thousands folds, and thus Ca^{2+} can naturally enter damaged cells along the concentration gradient under any circumstance that the cell membrane is damaged, as is the case for the formation of AS plaques. Some scholars suggest that a high-fat status can increase cholesterol in the endarterium and then increase the amount of calcium in the cells via calcium channels, which is liable to induce calcium overload in the cells and damage the cells. Calcium can activate some enzymes and then block the functions of the cells, and finally lead to the loss of cell structure and functions. Cell death, particularly the disruption and death of lipid carrying cells, can release membrane active substances, promote the proliferation and phenotypic alternations of smooth muscle cells (SMC), and induce synthetic type SMC to secrete various kinds of matrix, increasing the amount of elastic fibers and collagen fibers. Therefore, it can be assumed that the decrease in calcium influx into the cells can alleviate cell injuries and alleviate the degree of pathological changes of AS.

Many investigations have confirmed that Danshen is a kind of calcium antagonist. For example, a proper concentration of tanshinone II-A sulfonate (DS-201) can decrease the protein release of and calcium intake into cardiac muscle tissue in a dose-dependent manner; therefore it is thought that the protective effect of DS-201 against calcium abnormality is achieved through calcium antagonism. DS-201 can increase coronary flow, expand capillary vessels, and slow the heart rate. It also has negative inotropic and anti-arrhythmia functions, improves hemorheological characteristics, and alleviates damage in the intercalated disk of cardiac muscle, muscular fibrils and mitochondria, which are the common characteristics of calcium antagonists. Chen Sihong et al. pointed out that the effects of DS-20 1 on the active intake of calcium by mitochondria showed effects that were similar to those of calcium channel inhibitors. It can decrease or inhibit the mitochondrial calcium overload induced by adnephrin in cardiac muscle, and thus protect the mitochondria and myocardial cells from being injured by calcium abnormalities. Zhu Ping et al. [118] found in his investigations on the reperfusion injuries of cardiac muscle that large amounts of calcium accumulated in myocardial cells in the early stage of reperfusion. After Danshen was applied, the transfer of calcium ions inward was blocked and the plasma calcium level increased, thus it was considered that Danshen may block slow calcium channels in the plasma membrane by a certain pathway.

Many animal experiments have confirmed that calcium antagonists can significantly affect the aggregation of calcium, cholesterol and matrix components in atherosclerotic arterial walls, and they can significantly decrease AS size. Henry and Bentley orally administered nifedipine at 40 mg/d for 8 weeks to rabbits which had been fed on cholesterol. The cholesterol levels in the arterial wall decreased by about 40 %, and the sizes of the plaques that underwent pathological changes in the arterial wall were decreased. Mu Yongfang et al. carried out preliminary observations on the prevention of experimental AS by using Danshen and small doses of aspirin, and the results showed that AS lesions were reduced and that cholesterol content in the main artery was lowered. Stein et al. carried out investigations on the effects of verapamil on arterial endothelial cells and SMC in 1985, and the results demonstrated that this drug can increase LDL intake and degradation, and significantly increase the number of LDL receptors on the cell surface, thus increasing the elimination of LDL and decreasing the aggregation of cholesterol on the arterial wall. Qin et al. [124] carried out investigations on the prevention of experimental AS in Chinese small pigs by using nifedipine, and the results showed that nifedipine could inhibit the occurrence and development of AS, induced by high cholesterol foods, in the main artery and coronary artery muscle muscle of small pigs. This may be due to nifedipine inhibiting the transformation of SMC from the contractile phenotype to synthetic phenotype, and thus inhibiting SMC proliferation and LDL deposition in the positions of AS pathological changes. The anti-AS functions of calcium antagonists are the results of multifaceted interactions between lipoproteins and arterial wall metabolism. They can change the aggregation of cholesterol and lipoproteins in the arterial wall and inhibit the excessive production of matrix ingredients in the arteries.

4.2.8 Treatment of Coronary Heart Disease and Heart-Stroke

Coronary heart disease and heart-stroke are commonly seen diseases in cardiovascular medicine. They are induced when the blood supply from the coronary artery muscle muscle cannot satisfy the needs of the cardiac muscle, which results in rapid and temporary ischemia and anoxemia in the cardiac muscle, thus causing heart-stroke. The clinical drugs used to treat these diseases at present are mainly dilating coronary agents and anticoagulants. Wang Minjia utilized CDDP to treat 38 coronary heart disease and heart-stroke patients, and the results showed that the clinical symptoms were effectively controlled and that the electrocardiograms were also improved. The total effective rate was 91.7 % with significant therapeutic effects and mild adverse effects, so it deserves to be widely used clinically. Chen Shiliang et al. made micrometer constrictors by themselves to prepare the left anterior descending coronary artery muscle muscle model in dogs. Compound Danshen injection (CDI), which is composed of Danshen and rosewood, was infused at a constant flow, and it was found that the coronary resistance decreased continuously and the coronary blood flow increased while oxygen consumption decreased, the myocardial contractility increased, and the mean coronary pressure increased after the infusion was carried out for 30 min. The results indicated that CDI had functions in directly expanding the coronary artery muscle muscle and improving blood supply in the ischemic region.

C-reactive protein is a sensitive and reliable parameter to reflect the level of inflammation in the organism; the higher its concentration is, the larger the inflammatory scale is, and the more serious the coronary damages are. CDDP can significantly decrease the level of C-reactive protein in patients that suffer from acute coronary syndrome and stabilize the angiosclerotic plaques. Thus CDDP reduces the occurrence of acute coronary syndrome by inhibiting the inflammatory reactions and alleviating coronary injuries in the patients. Nie Guili et al. investigated the material basis and targets of the major components of Danshen SalB and tanshinone II-A which have effects in the treatment of artherosclerosis development, and the results showed that both SalB and tanshinone II-A can achieve their effects of preventing and treating artherosclerosis by inhibiting the expression of inflammatory mediators in artherosclerosis.

CDDP is composed of Notoginseng, Danshen and rosewood oil. It can promote blood flow and eliminate stasis, regulate qi to alleviate pain, and can alleviate chest distress, chest pain, palpitation and other symptoms. Investigations showed that twice sublingual administration of CDDP can significantly eliminate or alleviate acute chest distress and chest pain in uremic patients when they undergo dialysis. The effective rate was 86 % and the effective rate for eliminating or alleviating palpitation was 55 %. Therefore, CDDP is more suitable for the patients who undergo acute chest distress and chest pain during dialysis, and it is also effective in the patients that undergo palpitation in dialysis.

4.2.9 Summary

In summary, Danshen, its effective components, and compound formulas all have definite functions in AS prevention and treatment. The mechanism involves many aspects, such as the regulation of lipid metabolism, protection of endothelial cells, interference in platelet functions, improvement in hemorheology, inhibition of SMC proliferation and expression of cell adhesion molecules, calcium antagonism, etc. Researchers have studied the mechanisms of Danshen's functions in preventing and treating AS from the perspectives of cellular biology and molecular biology, which lead the research to the molecular level. The investigations of the pharmacological machinery should be further deepened, and its effectiveness should be examined in clinical applications. This has important significance for both basic and clinical investigations.

4.3 The Effects of Danshen on Hypertension and Its Risk Factors

Lianhua Fang and Guanhua Du

Li et al. [132] found that Danshen had no significant inhibitory effects on the dose-response curve of noradrenaline (NE) and KCl in isolated thoracic aortas of rabbits, indicating that Danshen did not directly relax in vitro vascular smooth muscle, but rather than Danshen's ability to relax blood vessels may be realized by indirectly affecting neural and body fluid factors. The water soluble components of Danshen Tanshinol can significantly expand the coronary artery muscle muscle in dogs and cats that undergo experimental acute myocardial infarction. They can also dilate the isolated coronary artery muscle muscle of cats and pigs, increase coronary blood flow, and antagonize the effects of morphine and propranolole hydrochloride in contracting the coronary artery muscle muscle [133]. Thus Tanshinol is considered one of the active ingredients in dilating the blood vessels of coronary arteries.

Danshen and its water soluble ingredient SalB can decrease sodium-induced high blood pressure in hypertensive rats, significantly increase renal blood flow, glomerular filtration rate, sodium and sarcosine anhydride discharge in rats, and it can also increase total renal kallikrein, the activity of kallikrein and prostaglandin E_2 (PGE_2) in rats. Thus the functions of SalB in reducing pressure are considered to be mainly related to kallikrein and the prostaglandin system [134, 135]. Kamata et al. [136] reported that SalB can significantly lower blood pressure in hypertensive rats, and these effects were thought to be related to endodermis dependent vasodilation. The present investigation showed that Danshen and its active ingredients had antagonistic effects on angiotensin II (Ang II), and the significant effects of Danshen on reducing blood pressure also indicate that it may be related to the rennin-angiotensin system. Gao et al. [137] used angiotensin-conversion enzyme (ACE) activity determination methods to screen ACE inhibitor (ACEI) from Danshen's water soluble extracts and liposoluble extracts. Their results revealed that the active ingredients of Danshen which inhibit ACE exist in the water soluble extracts. Its major water soluble ingredients, total phenolic acids, were further isolated and their ACEI functions were determined. SalB is an ingredient with a relatively high concentration amongst total phenolic acids. It was prepared and isolated by HPLC and its ACEI activity was tested; the results showed that SalB was one of the active components with ACEI activity.

Among the multiple risk factors that cause hypertension, the inflammatory infection hypothesis of arteriosclerosis has received more attention in recent years. Many reports point out that the inflammatory markers of microbial pathogens and C-reactive protein (CRP) are related to the occurrence of arteriosclerosis and hypertension and other cardiovascular diseases, but the definite mechanism is still unknown. However, the increase in CRP reflects that slight reactive inflammation exists in the blood vessels, the cytokines that are related to the inflammation are activated, and the inflammation is directly or indirectly involved in the damage to endothelial function or in other pathological and physiological changes in the blood vessels. Xu et al. [138] found that Danshen and Tanshinol showed significant inhibitory effects on many cytokines such as interleukin-1(IL-1), interleukin-6 (IL-6), interleukin-8 (IL-8) and tumor necrosis factor (TNF) which are secreted by the Kupffer cells (KC) of rats after endotoxin stimulation. The inhibitory mechanism includes the promotion of immunological suppressor PGE_2 secretion by activated KC, and the inhibition of granulocyte macrophage colony stimulating factor (GM-CSF) secretion. Recent investigations showed that Danshen has anti-inflammatory effects, the mechanism of which may involve the inhibition of the release of cytokine and lysosomal enzymes, phagocytosis and adhesion, and alleviation of endothelial cell damage. Thus blood vessel endothelium function in hypertensive patients is improved, the content of plasma ET is decreased and the content of NO is increased, the normal morphology of blood cells is maintained, the movement of blood cells which flowed slowly or underwent stasis before the treatment is accelerated, the aggregation of blood cells is disintegrated to a certain extent, and normal hemodynamics and stable blood pressure are achieved.

Among the numerous new achievements in the study of hypertension, epidemiologists have proposed the concept of "risk factors". They explicitly stated the circadian rhythm law of blood pressure fluctuations and the idea of multiple causes of hypertension, and its progressively increasing effects are emphasized. Fang et al. [139] found that the effects of Danshen on total cholesterol (TC), triacylglycerol (TG), blood glucose (Glu), hematocrit (HCT), CRP, fibrinogen (Fg), left ventricular ejection fraction (LVEF), cardiac index (ci), cardiac output (CO), heart rate and other risk factors of hypertension were significant. Though it did not show obvious effects in blood pressure reduction, it showed significant effects in the stabilization of ambulatory blood pressure and the maintenance of stable rhythmic changes. Previous investigations have found that besides expanding the intra-acinar pulmonary artery, Danshen can more importantly alleviate hypoxic damage to endothelial cells, inhibit the phenotypic change of medial smooth muscle cells from contractile to synthetic, and inhibit the proliferation of vessel wall cells (endothelial cells, smooth muscle cells, adventitial cells), which may be the key point in preventing an increase in pulmonary arterial pressure [140]. After intravenous infusion of Danshen injection, the erythrocytes of these hypertensive patients electrophorese at a faster rate or return to a normal rate, and the hematocrit and whole blood viscosity decrease or return to normal. All these factors play important roles in the stabilization of blood pressure [141]. Recent investigations have found that Danshen cannot significantly change systolic blood pressure, but can inhibit the development of left ventricle hypertrophy in spontaneous hypertensive rats (SHR) and decrease the expression of c-*fos* in cardiac muscle tissues [142] (Table 4.4). Thus it is assumed that Danshen can prevent left ventricle hypertrophy by reducing c-*fos* expression in the cardiac muscle of SHR and inhibiting myocardial cell hypertrophy and collagen

Table 4.4 The effects of Danshen on the systolic pressure, myocardial hypertrophy and myocardial fibrosis in spontaneous hypertensive rats ($\bar{X} \pm s$)

Group	Systolic pressure (mm Hg)	Ponderal index of left ventricle (mg/g)	Diameter of myocardial cells (μm)	Area of myocardial cells (μm^2)	Collagen volume index (mg/g)	Area of collagen/area of lumen (mg/g)
Control	146 ± 9	2.85 ± 0.23	16.3 ± 2.1	218 ± 52	3.8 ± 0.6	2.4 ± 0.4
Hypertension	179 ± 3$^{##}$	4.31 ± 0.66$^{##}$	25.2 ± 4.4$^{##}$	490 ± 119$^{##}$	6.8 ± 0.7$^{##}$	4.8 ± 0.7$^{##}$
Danshen	171 ± 6	3.23 ± 0.25**	16.3 ± 1.8**	231 ± 69**	4.6 ± 0.9**	3.0 ± 0.6**

$^{##}$ The comparison between the hypertension group and the control group ($P < 0.01$)

** The comparison between the Danshen group and the hypertension group ($P < 0.01$)

Table 4.5 The effects of tanshinone on myocardial aldosterone content, CYP11Bl and CYP11B2 gene expression in hypertension rats ($\bar{X} \pm s$)

Group	Aldosterone (ng/g)	Angiotensin (ng/g)	CYP11B1 (ng/g)	CYP11B2 (ng/g)
Control	0.018 ± 0.009	0.043 ± 0.012	1.973 ± 0.102	0.786 ± 0.121
Hypertension	0.056 ± 0.014##	0.093 ± 0.016##	2.774 ± 0.138#	1.573 ± 0.106##
Danshen	0.031 ± 0.010#*	0.088 ± 0.024##	2.533 ± 0.127#	1.024 ± 0.113#*
F	5.46	3.99	3.52	6.55

synthesis. Some investigations suggested that the inhibition of left ventricle hypertrophy in hypertensive rats by tanshinone II-A may be related to its down regulation of the expression of CYPllB2, the gene associated with myocardial aldosterone synthesis, and to the decrease in aldosterone biosynthesis in local regions of the heart [23] (Table 4.5).

Plasma fibrinogen (Fg) is the most positive forecasting factor for coronary heart disease. Hypertension is always accompanied by an increase in fibrinogen, and the mechanism for the development of hypertension may be related to the involvement in the inflammatory mechanism of artherosclerosis, increase in blood viscosity and the stimulation of vascular smooth **muscle** cell proliferation. Compound Danshen injection (CDI) may regulate the metabolic imbalance of ET and calcitonin gene-related peptide (CGRP) by improving the status of blood circulation, thus regulating vasomotor functions and facilitating the reduction of blood pressure in primary hypertension patients [23]. The results of clinical investigations have confirmed that the treatment of hypertensive cerebral hemorrhage patients with CDI at the early stage can promote the recovery of neural function in the ischemic regions near hematomas, reduce disability rates, and improve the living standards of patients [143–145].

Hypertensive cerebral hemorrhage has the characteristics of acute onset, fast progress, and serious sequelae. Tang and Chen [146] observed the therapeutic effects of CDI in the treatment of acute hypertensive cerebral hemorrhage. All patients received an integrative treatment, including intracranial pressure reduction, blood pressure regulation, transfusion, anti-infection, and others. 45 patients were administered with 10 ml/day Danshen injection 6 h after morbidity (crude drug 15 g/10 ml). The dosage was adjusted into 20 ml/day on the third day and the total course of treatment lasted for 21 days. The results showed that the effective rate of the Danshen treatment group was 78 % (35 cases), compared with the control group's 57 % (24 cases); absorption of hematoma, treatment group 78 % (35 cases), control group 52 % (22 cases). The plasma fibrinogen and hematocrit indexes were also significantly improved in the treatment group.

Danshen can improve hemodynamics, promote hematoma absorption, eliminate free radicals, and reduce the release of excitatory amino acids in brain tissue, and reduce hyperexcitable blood sugar. In this manner it protects the tissues by expanding cerebral vessels, increasing cerebral blood flow, increasing anti-anticoagulation and fibrinolytic activity, decreasing blood viscosity, and reducing platelet aggregation. Liu et al. [147] observed the therapeutic efficacy of CDI on 71 acute hypertensive cerebral hemorrhage patients. The patients were randomized into two groups; those in the treatment group were subjected to intravenous drop infusion of CDI within 6–72 h on the basis of conventional treatment (16 ml CDI was added to 500 mL 5 % glucose solution, once a day for continuous 14 days, and then the patients were observed for 21 days), and those in the control group were subjected to conventional treatment with Western medicines. The results showed that the size of cephalophyma in the treatment group decreased by 54.5 %, while in the control group it decreased by 39.7 %; the score for neurologic impairment in the treatment group was significantly lower than that of the control group ($P < 0.01$); and both significant effective rate and effective rate in the treatment group were higher than those in the control group ($P < 0.05$). Li et al. [148] observed the therapeutic effects of Danshen injection at early stage on 124 acute hypertensive cerebral hemorrhage patients, and the results showed that Danshen treatment of patients with a small or medium amount of hypertensive intracerebral hemorrhage at the early stage achieved good therapeutical effects, and could improve daily locomotive activity and promote the absorption of intracranial hematoma. Yu et al. [149] investigated the therapeutic efficacy of combined Astragalus Root injection and CDI on acute hypertensive cerebral hemorrhage, and the results showed that the healing rate and improvement rate in the treatment group were significantly higher than those in the control group ($P < 0.05$); the absorption of cephalophyma in the treatment group was better than that of the control group ($P < 0.01$); the incidence rate of major complications in the treatment group was also lower than that of the control group ($P < 0.01$), confirming that the effects of combined Astragalus Root **injection** and CDI in the treatment of acute hypertensive cerebral hemorrhage was better than that of conventional treatment alone.

Kim et al. [150] found that tanshinone II-A in Danshen had significant functions in reducing blood pressure and expanding blood vessels when administered to hypertensive hamsters; the mechanism was related to the stimulation of nitric oxide synthase and the promotion of nitric oxide synthesis in vivo. The authors suggested that further investigations should be focused on the continuous pressure reducing effects of Danshen, because although Western medicines can quickly lower blood pressure, the duration of the effect is very short, while Danshen and other traditional Chinese medicines have the opposite effect. Their research results provided a basis for the formal application of Danshen as the treatment medicine for hypertension. Lu and Zhou [151] made observations on the therapeutic effects of CDDP in the treatment of refractory hypertension and found that the levels of cholesterol, triacylglycerol, low density lipoprotein cholesterol in the CDDP group decreased while high density lipoprotein cholesterol increased, which was statistically significant compared to those of the placebo group ($P < 0.05$); systolic and diastolic blood pressures significantly decreased after treatment, which was also statistically significant compared to before the treatment ($P < 0.05$) and compared to those of the placebo group ($P < 0.05$), suggesting that CDDP has effects in reducing blood pressure in refractory hypertension patients by lowering the blood fat level. Zhan et al. [152] investigated the effects of tanshinone on the activation of peripheral blood monouclear cells (PBMc) at the early stage of essential hypertension (EH), and their results showed that PBMCs at the early stage of the disease course in EH patients were in pre-activated status, and tanshinone can inhibit the further activation of pre-activated PMBC.

Zhou et al. [153] studied the inhibitory effects of SalB salt on the increase in portal venous pressure in rats induced by endothelin-1 (ET-1), and investigated the mechanism. The results showed that the **injection** of ET-1 led to an increase in portal venous pressure in rats, and among the various treatments, ET-1 injection alone gave the biggest increase. In rats pre-administered with SalB solution or pre-injected with ET receptor blocker, ET-1 injection also increased the portal venous pressure, but the increase was significantly smaller than that of the ET-1 group, indicating that SalB can significantly inhibit the portal venous pressure increase induced by ET-1, with an effect similar to that of ET receptor blocker. Hu [154] investigated the effects of tanshinone II-A sulfonate on the prothrombotic state and pulmonary artery hypertension in the acute exacerbation stage of chronic cardiopulmonary disease, and results showed that the whole blood low shear viscosity, fibrinogen, D dimmer, and pulmonary arterial systolic pressure in patients suffering from pneumocardial disease were statistically different compared with those of the control group ($P < 0.05$). The whole blood viscosity, fibrinogen, D dimmer, and pulmonary arterial systolic pressure were significantly lowered in patients after receiving tanshinone II-A sulfonate treatment ($P < 0.05$), with no significant difference found in these parameters before and after conventional treatment ($P > 0.05$), indicating that tanshinone II-A sulfonate can significantly inhibit the prethrombotic status and reduce pulmonary arterial pressure in the acute exacerbation stage of pneumocardial disease. Zou and Wang [155] studied the relationship between peroxidation injuries and hypertension as well as the effect of Compound Danshen injection intervention. The results showed that peroxidation injuries existed in hypertension patients and that Danshen could intervene with this relation. The mechanism may be related to its protection of endothelial functions by preventing oxidation. The therapeutic effects of Danshen are excellent and stable, and it is safe and convenient, so it is a type of effective adjunctive therapy for hypertension patients and deserves to be promoted and applied clinically.

Hypertensive retinopathy is characterized by angiospasm, angiostenosis, and thickening of vessel walls, and exudation, bleeding and cytoid bodies or even retinal venous embolism may occur in serious cases. It is a commonly seen retinal vessel disease that can seriously threaten eyesight. The treatment for retinal arterial and venous embolism should place emphasis on drug administration at the early stage. The possibility of re-obstruction still exists after the thrombus is dissolved; decoagulants and platelet agglutination inhibitors can be used in combination in order to prevent re-obstruction, but the range of the application is not intensive enough and it is not suitable for patients who also have hemorrhagic diseases, serious liver and renal damage, cerebral vessel disturbance, alimentary tract hemorrhage, or serious hypertension. CDDP is mainly composed of Danshen, Notoginseng and bornyl alcohol. Of these, the active ingredient is Danshen's water soluble Tanshinol, which can improve anticoagulative and fibrinolytic activities as well as inhibit thrombogenesis. It can also release prostacyclin substances, inhibit platelet aggregation, reduce blood viscosity, improve microcirculation disturbance, block hydroxyl radical production and prevent lipid peroxidation, and it is an effective oxygen free radical scavenger. Yang et al. [156] made clinical observations and confirmed that CDDP had certain improving effects on hypertension, arteriosclerosis, and retinal venous embolism at the early stage of diabetes.

In summary, as a TCM drug used to promote blood flow and eliminate stasis, Danshen and its active ingredients exert their regulatory effects on hypertension by inhibiting "risk factors", and Danshen **injection** has definite therapeutic effects on acute hypertensive cerebral hemorrhage.

4.4 The Effects on Blood Vessel Endothelium and Smooth Muscle

Lianhua Fang and Guanhua Du

4.4.1 Danshen's Protective Effect on Vascular Endothelial Cells

Vascular endothelial cells are located between flowing blood and blood vessel tissue, and constitute an intact cell monolayer barrier. However, the functions of these cells are not limited to protection, anticoagulation, and anti-cell adhesion; they also secrete many kinds of active substances and involved in physiological and pathological regulatory processes. The injuries and functional disturbances of endothelial cells are related to the development of many diseases, particularly to the development of artherosclerosis (AS), acute cardiovascular and cerebral incidents, ischemia-reperfusion damage, hypoxic injuries, and others. Injuries and functional disturbances in endothelial cells are initial factors for the development of AS, while oxidized low density lipoprotein (OX-LDL) plays important roles in endothelial cell injury. OX-LDL has effects on endothelial cells and leads to their functional disturbance, such as reduction in NO production, increase in the expression of inflammatory factors, promotion of the adhesion of lymphocytes and monocytes to endothelial cells and migration to subendothelial tissues, endocytosis of lipid products, and transformation into foam cells, thus promoting the occurrence and development of arteriosclerosis. OX-LDL can also lower antioxidase activity in cultured human umbilical vein endothelial cells and increase lipid peroxide levels, all of which can promote the occurrence and development of AS. Xu et al. [157] confirmed that Compound Danshen **injection** can alleviate the extent of oxidative injuries on macrophages, and that its protection of superoxide dismutase (SOD) was even better than that of sodium selenite. Ge et al. [158] found that the endothelium-dependent vasodilation functions of the brachial artery in 35 cases of defective endothelial function were significantly improved after oral administration of CDDP for 3 months, and the percentage of flow-mediated brachial artery blood flow was significantly increased (Table 4.6). However, some reports showed that Danshen **injection** had no inhibitory effects on the growth of the endothelial cell line of human umbilical vein ECV304 cells [159].

Adhesion molecules play important roles during the process of endothelial cell injury, and their high expression can induce the release of a series of cytokines that are related to the development of cardiovascular diseases. Ren et al. [160] studied human umbilical vein endothelial cells, and they found that Danshen and its extracts can inhibit the expression of vascular cell adhesion molecule-l (VCAM-1), intercellular adhesion molecule-l (ICAM-1) and endothelial cell selectin (E-selectin), all induced by tumor necrosis factor (TNF-α) (Table 4.7). When arterial inflammation is active, matrix metalloproteinases (MMPs) can be further activated, and the activated MMPs can degrade extracellular matrix, weaken the fibril cap of arteriosclerotic plaques, and cause stable plaques to become vulnerable, thus increasing the risk of malignant angiocardiopathy incidents. Recent investigations have found that Danshen injectable powder can inhibit the degradation of extracellular matrix by down-regulating MMP9 expression in endothelial cells induced by OX-LDL. The inhibition of extracellular matrix degradation can further protect endothelial cell function [161]; the extracts of Danshen in ethanol (50 μg/ml) and SalB (50 μg/ml) can significantly inhibit the binding rate of human monocyte U937 to TNF-α activated human thoracic aorta endothelial cells, and it can significantly inhibit NF-κB activity induced by TNF-α. Experiments confirmed that SalB and ethanol extracts of Danshen had anti-inflammatory functions, and all of its anti-inflammatory activities and inhibitory effects on LDL can be used to explain its effectiveness in AS prevention. Jiang et al. [90] showed that Tanshinol can inhibit vascular endothelial cells and granular cells to express cellular adhesion molecules, which may be one of the mechanisms for Tanshinol to exert its effects in preventing thrombogenesis. Experimental studies on the protection functions of Danshen and its active ingredients on endothelial cells showed that salvianolic acids had relatively potent antagonistic effects on vascular endothelial cell damage induced by oxidative stress, which can protect endothelial cells, and that they also inhibited the adhesion of neutrophils to endothelial cells [160]

Table 4.6 The comparison of the brachial artery indexes after blood flow-mediation and sublingual administration of glyceryl trinitrate in the patients that suffered from endothelial functional defect ($\bar{X} \pm s$)

Index	Before drug administration	After drug administration	*t*	*P*
The inner diameter of brachial artery at rest (mm)	4.31 ± 0.62	4.58 ± 0.65	0.672	>0.05
The inner diameter of brachial artery that was mediated by blood flow (mm)	4.46 ± 0.65	5.06 ± 0.71	0.687	>0.05
The inner diameter of brachial artery after sublingual administration of glyceryl trinitrate (mm)	5.17 ± 0.60	5.54 ± 0.78	0.703	>0.05
Endothelium-dependent relaxation functions (%)	3.50 ± 1.28	10.38 ± 3.04	6.826	<0.01
Endothelium-independent relaxation functions (%)	20.13 ± 9.39	20.15 ± 3.79	0.721	>0.05
The increasing percentage of blood flow that was mediated by blood flow (%)	203.2 ± 134.0	369.2 ± 247.1	15.3	<0.01

Table 4.7 The effects of water soluble components of Danshen on the expression of adhesion molecules in human umbilical vein endothelial cells induced by TNF-α ($\bar{X} \pm s$)

Group	Dosage	E-selectin	ICAM-1	VCAM-1
Control group		0.06 ± 0.04	0.15 ± 0.03	0.07 ± 0.01
TNF-α	400 U/ml	0.62 ± 0.09###	0.59 ± 0.05###	0.69 ± 0.01###
Water soluble ingredients	0.001 mg/ml	0.54 ± 0.09	0.45 ± 0.03**	0.68 ± 0.02
Water soluble ingredients	0.01 mg/ml	0.50 ± 0.04**	0.37 ± 0.02***	0.62 ± 0.02*
Water soluble ingredients	0.1 mg/ml	0.47 ± 0.05***	0.32 ± 0.04***	0.56 ± 0.02**

Compared with the control group, ### $P < 0.001$
* Compared with the TNF-α control group, * $P < 0.05$, ** $P < 0.01$, *** $P < 0.001$

(Table 4.8). Salvianolic acids also had excellent protection over endothelial cells induced by OX-LDL, confirming that they can protect cells through their antioxidation function, but the existence of other functions cannot be excluded [160, 162, 163]. Lin et al. [164] studied endothelial cells damage in coronary heart disease with blood stasis syndrome after intervention with drugs which can promote blood flow and eliminate stasis. They found that endothelial functional disturbances existed in patients suffering from coronary heart disease with blood stasis syndrome and that Danshen **injection** and Xiangdan injection intervention down-regulated the expression of endothelial cell adhesion factors, indicating that these two drugs can promote the proliferation of endothelial cells and improve cell arrangement and intercellular adhesion as well as the adhesion between cells and extracellular matrix. This may be the basis for how they improve microcirculation, promote the formation of compensatory circulation, and exert therapeutic actions in promoting blood flow and eliminating stasis.

In the early stage of AS development, minimally modified low density lipoprotein (mm-LDL) has important functions. It has been found in recent years that mm-LDL has effects on endothelial cells which can on one hand increase the adhesion of monocytes to endothelial cells, and on the other promote the secretion of monocyte chemoattractant protein-1 and colony stimulating factor by endothelial cells, thus accelerating the migration of monocytes, causing them to deposit in endothelial cells and transform into macrophages, then intake LDL and transform into foam cells. mm-LDL activates large conductance calcium-activated potassium channels (BK_{ca}), and then increases the electrochemical driving force of resting calcium influx and intracellular calcium concentration, and induces functional disturbances in endothelial cells. Water soluble extracts of Danshen protect endothelial cells by weakening the excessive activation of BK_{ca} at the early stage of mm-LDL-induced functional disturbance of endothelial cells [165].

Table 4.8 The effects of water soluble ingredients of Danshen on the adhesion of neutrophils to endothelial cells induced by fMLP (N-formyl-metleucyl-phenylalanine) ($\bar{X} \pm s$)

Group	Dosage	OD value
Control group		0.10 ± 0.01
fMLP	10^{-5} mol/L	0.38 ± 0.03###
Water soluble ingredients	0.001 mg/ml	0.36 ± 0.04
Water soluble ingredients	0.01 mg/ml	0.28 ± 0.02**
Water soluble ingredients	0.1 mg/ml	0.27 ± 0.03***

Compared with the control group, ### $P < 0.001$
* Compared with the fMCP group, ** $P < 0.01$, *** $P < 0.001$

Table 4.9 The effects of tanshinone on restenosis after endothelial cell deligation of carotid artery in mice ($\bar{X} \pm s$)

Group	Area of tunica intima/μm^2	Area of tunica media/μm^2	Relative area of lumen/%	The ratio between the areas of tunica intima and tunica media/%
Normal group	410 ± 141	56637 ± 33224	329.73 ± 1.73	1.237 ± 1.15
Model group	37778 ± 22.534$^{\#\#}$	125311 ± 62403$^{\#\#}$	32.25 ± 14.97$^{\#\#}$	31.85 ± 11.236$^{\#\#}$
Low tanshinone dosage	5334 ± 2706**	58746 ± 30753**	43.42 ± 6.94*	10.07 ± 4.95**
High tanshinone dosage	4787 ± 2679**	44984 ± 23420**	46.82 ± 9.08*	10.90 ± 5.09**

$^{\#}$ Compared with the control group, $^{\#\#}$ $P < 0.001$
* Compared with the model group, * $P < 0.05$, ** $P < 0.01$

Endothelin (ET), which is angiotensin and has strong blood vessel contraction functions, and nitric oxide (NO), which is a relaxing factor of blood vessels, are important factors which can affect cardiovascular and cerebrovascular diseases, particularly coronary heart disease and cerebrovascular diseases. Significant increases in ET activity can be observed in these diseases, while the NO level decreases. Treatment with Compound Danshen **injection** and CDDP can decrease the activity of plasma ET and increase NO activity, and these effects are related to the expression of endothelin in vascular endothelial cells [166].

The major pathological change in restenosis after percutaneous tranluminal coronary angioplasty is abnormal hyperplasy of tunica intima. Hung et al. [167] utilized Danshen rich in SalB to treat angioplasty model rabbits, and found that SalB can induce the apoptosis of neointima cells in blood vessels and then inhibit the thickening of the tunica intima, thus taking effects in preventing and treating AS. Li et al. [168] found that the areas of endarterium and tunica media as well as their ratio were significantly decreased after restenotic mice were subjected to tanshinone treatment for 4 weeks. The relative lumen area increased, and the differences were statistically significant, indicating that tanshinone can inhibit the proliferation of tunica intima and has active functions in the prevention and treatment of blood vessel restenosis (Table 4.9).

Sun et al. [169] studied the intervention effect of Danshen's mono-component (764-3) on the tumor necrosis factor–α (TNFα) induced tissue factor expression in human vascular endothelial cells, as well as the effects on free intracellular calcium ($[Ca^{2+}]_i$) levels in individual endothelial cells, and discussed the potential mechanism of the inhibitory effect of 764-3 for the prevention and treatment of thromboembolic diseases of the cardiovascular system. Their results showed that 764-3 could inhibit the increase in tissue factor gene expression induced by TNFα in endothelial cells, and they speculated that intracellular free calcium ions may be involved in this inhibitory effect. Li et al. [170, 171] observed the effects of angiotensin II (AngII) on endothelial cells of the main artery of pigs and investigated the protection of tanshinone II-A on vascular endothelial cells, and the results showed that tanshinone II-A affected the protection of vascular endothelial cells by increasing eNOS expression and NO secretion in vascular endothelial cells as well as by decreasing $[Ca^{2+}]_i$ levels in endothelial cells.

Zhang et al. [172] utilized gene microarrays to investigate the effects of Compound Danshen **injection** (CDI) on the gene expression profile in vascular endothelial cells. The results showed that CDI can regulate the functions of endothelial cells in many ways at the genetic level. In the range of cardiovascular disease and cytokines, CDI up-regulates NOS, MTHFR and GST genes, and down-regulates six genes, including TNFα, ET-1,

ICAM, VCAM, Pal-1 and LipoA. The expressions of these genes are related to vasomotion, blood clotting, inflammatory and immunological reactions, free radical scavenging and substance metabolism, etc. Therefore, compound Danshen may regulate gene expression in vascular endothelial cells in the following pathways to achieve its protective functions: (1) it can regulate the expression of oxidation reduction functional genes in the cells, and reduce the production of oxygen **free radicals**, lipid peroxides, OX-LDL, etc. It is assumed that Danshen can protect blood vessel endothelia by direct or indirect antioxidative effects; (2) it can regulate the expression of the blood clotting and fibrinolytic system, decrease the expression of thrombinogen and endothelin-l mRNA, and increase the expression of endothelial nitric oxide synthase mRNA, thus improving the hypercoagulable state; (3) it can regulate the expression of cytokine genes. Injuries in blood vessel endothelia are the result of the interactions of many kinds of cytokine. TNF-α, INF-γ, MCP-l, MIPl-α and others can promote the expression of endothelin-1 and endothelin convertase mRNA in the endothelial cells of human umbilical veins, and then mediate and take part in inflammatory and other reactions. Its functional disturbance is one of the important factors that induce vascular endothelial cell injuries. Compound Danshen regulates cytokine endocrine and especially has significant effects in protecting vascular endothelial cells by down-regulating TNF-α expression.

Wei [173] carried out investigations on the effects of Danshen on changes in the adhesion of white blood cells to endothelial cells in infection lesions after cerebral ischemia-reperfusion injury, and the results showed that cerebral ischemia-reperfusion injury can induce an increase in the mural index of arteriole leucocytes in infection lesions and a decrease in the fracture stress between leucocytes and endothelial cells, as well as a significant decrease in adhesion. Danshen can significantly alleviate the adhesion of endothelial cells after cerebral ischemia-reperfusion injury. By using advanced microcirculation observation methods, they confirmed that mural leucocytes significantly increased in the arteriole after cerebral ischemia-reperfusion injury, and that the fracture force decreased significantly. While the number of mural leucocytes decreased and fracture force increased, adhesion decreased. Danshen is a commonly used TCM drug to promote blood flow and eliminate stasis, and has gained excellent therapeutic effects in clinical practice in treating ischemic cerebrovascular diseases. This investigation showed that Danshen can significantly alleviate the local infiltration of leucocytes, and this may be related to Danshen's function in expanding contracted arterioles, improving microcirculation, and inhibiting the adhesion of vascular endothelial cells to leucocytes in the ischemic region.

4.4.2 The Effects of Danshen on the Proliferation and Migration of VSMC

The abnormal proliferation of vascular smooth **muscle** cells (VSMC) is the common pathological basis for many vascular proliferating diseases such as hypertension, coronary arteriosclerotic heart disease, and restenosis after angioplasty. Thus, the inhibition of abnormal VSMC proliferation is one of the main approaches for the treatment of vascular proliferating diseases. Danshen is a frequently used drug to promote blood flow and eliminate stasi, which is used in clinical investigations and has achieved excellent therapeutic effects. Previous investigations have revealed that Danshen has certain protection over capillary endothelial cells and myocardial cells of the heart, but its effects on VSMC proliferation are not very clear. Li et al. [174] found that total salvianolic acids can significantly increase the survival rate of vascular smooth **muscle** cells in rats and significantly decrease the content of lactate dehydrogenase (LDH) in the supernatant of H_2O_2 damaged cell culture. The significant decrease in the release of NO induced by H_2O_2 damage indicated that total salvianolic acid had protective effects on vascular smooth **muscle** damaged by H_2O_2. Wu et al. [110] utilized immunohistochemistry, fow cytometry and in situ nucleic acid hybridization to observe the effects of tanshinone II-A sulfonate on the proliferation of pulmonary arterial smooth muscle cells, and results showed that it could significantly inhibit the transition of pulmonary arterial smooth muscle cells from G_0/G_1 stage to S stage and significantly decrease the expression of proliferating cell nuclear antigens, PDGF-A and B chain mRNA. This indicates that tanshinone II-A sulfonate may effectively inhibit the proliferation of pulmonary arterial smooth muscle cells induced by a conditioned medium for hypoxic endothelial cells by down-regulating the expression of PDGF mRNA. Wang et al. [175] showed that tanshinone had significant inhibitory effects on the proliferation of arterial vascular smooth muscle cells in rats stimulated by 5 % fetal bovine serum, which significantly decreased the metabolic rate of MTT and the synthesis of total proteins in cells, and increased the proportion of cells in G_0/G_1 stage and decreased the proportion of cells in S stage. This indicated that tanshinone can stop the proliferative cycle of VSMC in G_0/G_1 stage, and thus inhibit the proliferation of VSMC. In the same experiment, however, salvianolic acid did not show the above-mentioned effects on VSMCs (Table 4.10). Recent investigation [103] revealed that the liposoluble active ingredient of Danshen, tanshinone, inhibited the proliferation of VSMC induced by fetal bovine serum. It could decrease the expression of extracellular signal regulatory protein kinase pERKl/2 in a dose-dependent manner, indicating that tanshinone inhibits VSMC proliferation by inhibiting the ERK pathway (Table 4.11). Chen and Xu [176] carried out investigations on the inhibitory effects of CDDP on the proliferation of VSMC in rabbit thoracic aortas induced by high sugar/high insulin. They also discussed the possible mechanism of the effects. Their results showed that CDDP extract can inhibit VSMC proliferation through the mediation of NO. Tanshinone II-A had significant inhibitory effects on the proliferation of VSMC induced by high sugar, and the mechanism may be that it partially blocks the signal transduction pathway of mitogen activated protein kinase [177].

Apoptosis is a dynamic process regulated by a series of apoptosis related genes. *Bcl*-2 and *Bax* are a pair of positive and negative apoptosis regulating genes; *Bcl*-2 has a relatively weak ability to promote cell proliferation, while overexpression of *Bcl*-2 can specifically inhibit apoptosis and the inhibition of *Bcl*-2 expression can promote cell apoptosis. Li et al. [178] found that H_2O_2 down-regulated the expression of the anti-apoptosis gene *Bcl*-2 by its hydroxyl radical metabolites and the antioxidation ability of VSMC was decreased, and thus apoptosis was induced. *Bax* is also a member of the *Bcl*-2 family, and its functions are contrary to those of *Bcl*-2; overexpression of *Bax* can antagonize the effects of Apoptosis inhibition by *Bcl*-2, and external reactive oxygen (RO) increases and up-regulates the expression of *Bax* under the effects of stimulating signals. *Bax* causes **mitochondria** to release cytochrome and apoptosis inducing factors by affecting mitochondrial permeability transition pores and decreasing the transmembrane potential of mitochondria. Therefore, the ratio of *Bax*/*Bcl*-2 has a crucial function in determining whether cells can survive after receiving stimulating signals; if the relative amount of *Bcl*-2 is higher than that of *Bax*, then apoptosis is inhibited, otherwise apoptosis is promoted. Du et al. [179] selected primarily cultured VSMC from the 3rd to 5th generations of rats, using H_2O_2 as ROS to induce VSMC apoptosis, MTT colorimetric method to determine the survival rate of cells, flow

Table 4.10 The effects of tanshinone and salvianolic acid on VSMC cycle ($\bar{X} \pm s$)

Group	Concentration (μm/ml)	*G0/G1*	*S*	*G2/M*
Control	0	59.9 ± 2.4	11.6 ± 2.5	28.6 ± 0.2
Tanshinone	50	73.8 ± 1.3*	4.2 ± 1.9*	21.8 ± 1.2
Salvianolic acid	50	61.6 ± 1.8*	10.0 ± 1.8	28.4 ± 1.2

* Compared with the control group, $P < 0.05$

Table 4.11 The effects of tanshinone on the proliferation of VSMC and the expression of ERK ($\bar{X} \pm s$)

Group	Concentration (μm/ml)	A_{570nm}	Increasing rate (%)	Protein amount in each hole (/g)	ERK inhibition rate (%)
Control	0	0.510 ± 0.045	100	49.6 ± 2.5	
Tanshinone	0.4	0.459 ± 0.048*	89.8*	44.5 ± 3.3**	
	2	0.434 ± 0.029**	85.1**	43.6 ± 1.6**	17.7 ± 3.6*
	10	0.410 ± 0.031**	80.4**	42.5 ± 2.6**	26.2 ± 4.1*
	50	0.348 ± 0.025**	68.1**	31.5 ± 3.2**	72.7 ± 3.8*

* Compared with the control group, * $P < 0.05$, ** $P < 0.01$

cytometry to determine the apoptotic rate, and immunocytochemical methods to determine the expression of *Bcl*-2 and *Bax* proteins in VSMC simultaneously. They studied the intervention of Danshen on the Apoptosis of VSMC, and results showed that the survival rate in the Danshen group was significantly higher than that of the H_2O_2 group after intervention, while the apoptotic rate and the positive index (PI) for *Bax*/*Bcl*-2 protein expression were significantly decreased, indicating that Danshen can antagonize H_2O_2 induced VSMC Apoptosis by down-regulating the expression of *Bax*/*Bcl*-2 protein (Table 4.12).

The migration and proliferation of VSMC from the tunica media to the tunica intima of blood vessels is the pathological basis for vascular remodeling, and MMP-2 and MMP-9 are the important enzymes involved in this process. It has been confirmed that the process not only includes the degradation of extracellular matrix (ECM), but also the synthesis of adhesion proteins and chemotatic factors in large amounts, and osteopontin (OPN) is one of the major ingredients [180]. One of the mechanisms for the promotion of VSMC proliferation and migration by OX-LDL is the stimulation of MMP-2, MMP-9 and, OPN expression [181, 182]. The regulations on ECM metabolism can also inhibit VSMC proliferation and migration, and postpone the excessive proliferation of new intima. Liu et al. [183] confirmed that Danshen or CDI had significant down-regulating effects on MMP-2 and MMP-9 expression and on the synthesis and secretion of OPN induced by OX-LDL. The induction of OX-LDL on the above-mentioned genes almost disappeared after OX-LDL was incubated with Danshen or CDI for 1 h and then used to treat VSMC, indicating that Danshen or CDI have antioxidant functions and protect the cells from being injured by oxidative stress by weakening the oxidative toxicity of OX-LDL. Thus, Danshen or CDI can be considered scavengers of OX-LDL. Shi and Wen [184] found that Danshen had significant inhibitory effects on VSMC migration induced by newborn calf serum; the results of Northern blotting indicated that Danshen can inhibit the expression of genes related to cell migration; the analysis for creatine kinase activity showed that Danshen can significantly decrease the activity of this enzyme. The above-mentioned results indicate that Danshen's inhibition of VSMC migration is related to the inhibition of OPN gene expression and the slowing of energy conversion.

Table 4.12 The effects of Danshen on the apoptosi of VSMC and the expression of apoptosi related proteins induced by H_2O_2 ($\bar{X} \pm s$)

Group	Absorbance	Apoptotic rate (%)	*Bcl*-2 (PI)	*Bax* (PI)	*Bax*/*Bcl*-2
Normal group	0.296 ± 0.04	2.84 ± 0.25	4.45 ± 1.59	4.26 ± 1.62	0.96 ± 0.28
H_2O_2	0.143 ± 0.02##	15.69 ± 1.83##	2.14 ± 1.87##	8.63 ± 2.72##	4.01 ± 3.12##
Danshen + H_2O_2	0.215 ± 0.05*	5.62 ± 0.79**	3.85 ± 1.24**	6.32 ± 1.74**	1.91 ± 0.25**

Compared with the normal group, ## $P < 0.01$
* Compared with the H_2O_2 group, * $P < 0.05$, ** $P < 0.01$

4.4.3 Summary

Danshen and its active ingredients can prevent restenosis after coronary intervention by inhibiting various kinds of growth factors and the expression of some genes, inducing the differentiation, maturation and apoptosis of smooth muscle cells of tunica intima, preventing the adhesion, aggregation and release of platelets in the position of endothelial injuries, and inhibiting the migration and proliferation of vascular smooth muscle cells. Danshen has achieved excellent therapeutic effects in clinical applications in preventing and treating cardiovascular and cerebrovascular diseases, while the inhibition of VSMC proliferation is an effective pathway for the prevention and treatment of hypertension, artherosclerosis and other angiostenosis diseases.

In summary, as a TCM drug used to promote blood flow and eliminate stasis, Danshen and its active ingredients can prevent and treat vascular proliferating diseases such as coronary arteriosclerotic heart disease, restenosis after angioplasty and others, and its effects are achieved through protecting vascular endothelial cells and inhibiting the proliferation and migration of VSMC and many other pathways.

4.5 The Molecular Mechanism of Danshen's Protection on Myocardial Ischemia-Reperfusion Injuries

Dongxia Wang, Jinglan Xu and Xiaoying Wang

4.5.1 The Scavenging of Free Radicals and Prevention of Lipid Peroxidation

Free radical is the general term for molecules with atoms and atomic groups which contain unpaired electrons in the outer electron orbit. Among them, the most important are oxygen free radicals (including O_2^-, OH). The chemical properties of free radicals are very active, and are liable to lose electrons (to be reduced), so they have strong oxidizing abilities and intense effects on inducing lipid peroxidation. Under physiological conditions, free radicals can be removed in a timely manner by the anti-oxidative substances in cells and the production and degradation of free radicals is kept in balance, therefore it has no hazardous effects on the organism. However, under pathological conditions, free radicals are produced excessively and the anti-oxidative capabilities of the organism are insufficient, so lipid peroxidation chain reactions can be induced and the cell membrane damaged, so cell death occurs.

Using electron spin resonance to directly detect oxygen free radicals, it is found that the content of oxygen free radicals in the blood and myocardial tissues can be increased by several fold within several seconds or several minutes after myocardial reperfusion. In ischemic tissue, oxygen is in short supply and there is not enough oxygen as electron receptors. When reperfusion occurs, the oxygen supply is recovered, which provides large amounts of electron receptor and causes oxygen free radicals to increase by a large amount within a short period. Furthermore, free radicals can continuously expand into new free radicals through its intermediate metabolites once they are produced, and thus forming chain reactions. free radicals can react with various kinds of cell components, such as membrane phospholipids, proteins, and nucleic acids, which leads to cell structure damage and causes functional metabolic disturbances. Danshen plays important roles in the elimination of oxygen free radicals and the prevention of lipid peroxidation.

4.5.1.1 The Prevention of Lipid Peroxidation

Zhao et al. [185] prepared in vivo acute myocardial ischemia models by ligating the left ventricle branch of the coronary artery muscle muscle, and thiobarbituric acid fluorometry was used to determine the content of lipid peroxides in cardiac muscle. They found that the increase in the cardiac muscle lipid peroxide content was related to the extent of the elevation of the ST segment in the electrocardiogram. If intravenous administration of Danshen injection was carried out before deligation of the left ventricle branch of the coronary artery muscle muscle as the protective factor, the content of lipid peroxides in the ischemic region would significantly decrease, as would the extent of the elevation in the ST segment at 2–5 min after deligation. These results indicate that Danshen inhibits the process of lipid peroxidation by a certain mechanism and alleviates the damage caused by lipid peroxidation to the cardiac muscle. It showed that Danshen has effects on oxidation prevention in the ischemic models of the organisms, alleviates damage to the membrane, and blocks calcium influx, which is an important mechanism for protecting cardiac muscle and alleviating abnormal electrical activity.

Tang et al. [186] utilized the Langendorff perfusion model to observe changes in the oxygen free radical scavenger enzyme system, superoxide dismutase (SOD), Glutathione peroxidase (GSH-Px), the ultrastructures, as well as the anti-oxidative effects of Tanshinol on myocardial ischemia-reperfusion injury. The results showed that the activities of SOD and GSH-Px gradually decreased as the ischemic time prolonged. 40 min after myocardial ischemia, oxygen-enriched reperfusion was carried out, but SOD activity continued to fall 20 min after reperfusion. At this point, the activity of GSH-Px was significantly lower than that of the group with 60 min of ischemia, and ultrastructural damage at this time was the most serious. In the group administered with Tanshinol and sodium selenite before ischemia-reperfusion, the activities of myocardial SOD and GSH-Px were significantly higher than those of the ischemia-reperfusion alone group, and the ultrastructural damage was also relatively less severe. It is well recognized that the protective function of Tanshinol is better than that of the anti-oxidative sodium selenite.

Sun et al. [187] induced ischemia-reperfusion injuries in isolated rat hearts using the oxygen radical induced production of lipid peroxides as the index (thiobarbituric acid was used for detection), and observed the protective effects of sodium tanshinone II-A silate. The results showed that tanshinone can inhibit the production of reactive oxygen species, and thus decrease the production of lipid peroxides, indicating that sodium tanshinone II-A silate may be an excellent antioxidant.

Han et al. [188] prepared acute myocardial ischemia models by ligating the branch of left ventricle of the coronary artery muscle muscle under the conditions of general anesthesia, chest opening, spontaneous breathing and spontaneous heart rate, and then produced the reperfusion injury model by loosening the ligation. The content of lipid peroxides in the tissues that underwent myocardial ischemia and reperfusion injury was determined, and Danshen injection was used as the protective agent to observe its effects. The results showed that lipid peroxides levels in tissues from the myocardial ischemia region of the ischemia group gradually increased and lipid peroxidation of the cell membrane was gradually reinforced; membrane damage was gradually aggravated and reperfusion further promoted the increase in lipid peroxide levels in the ischemic region. The occurrence of the excessive activation of reperfusion-induced lipid peroxidation is dependent on the duration of ischemia, i.e., the extent of ischemic injuries, indicating that reperfusion injuries were induced on the basis of ischemic injuries. In the protection group administered with Danshen injection in advance, lipid peroxide levels in the tissues in the ischemic region decreased by 56 %, indicating that the protection of Danshen over myocardial ischemia and reperfusion injury was closely related to the inhibition of the excessive activation in lipid peroxidation. Therefore, it is considered that the protective mechanism of Danshen is by breaking the vicious cycle of cell damage mediated by oxygen free radicals and by inhibiting the production of lipid peroxides and the large amount of Ca^{2+} influx, thus inhibiting oxygen free radical mediated cell damage, protecting the structural and functional integrity of the cell membrane, decreasing the elevation of the ST segment in electrocardiograms at the early stage of ischemia, and delaying the occurrence of myocardial ischemia and reperfusion injuries.

4.5.1.2 Free Radical Scavenging

It has been found by many investigations in recent years that after myocardial ischemia and reperfusion injury, oxygen free radicals increase abnormally and lipid peroxidation is aggravated in cardiac muscle. The oxygen free radical theory is believed to be an important pathogenesis for myocardial ischemia and reperfusion injury. The functions of free radicals in myocardial ischemia and reperfusion injury include the reinforcement of lipid peroxidation in the membrane, inhibition of protein function, and damage to nucleic acids and chromosomes.

Chang et al. [189] utilized the $FeSO_4$/ascorbic acid system (·OH system) to observe damage by oxygen free radicals to mitochondrial H^+-ATPase of rat cardiac muscle, and the protective effects of Tanshinol (DS-182). Their results showed that $FeSO_4$/ascorbic acid could decrease the hydrolyzing activity of mitochondrial H^+-ATPase, and that Tanshinol showed an obvious protective effect which was concentration dependent and had a "saturation effect". The mechanism for Tanshinol's protection on H^+-ATPase may be achieved by eliminating oxygen free radicals, and thus protecting normal mitochondrial structure and function. Based on the above study, Su et al. [190] utilized the ·OH production system to directly damage myocardial mitochondria in rats, and then observed the protective effects of Tanshinol. The results showed that Tanshinol had protective effects on mitochondrial ADP oxygen ratio (P/O), respiratory control ratio (PCR), and the activities of cytochrome oxidase, indicating that Tanshinol was an excellent ·OH scavenger.

Zhang Li et al. carried out investigations on the effects of Tanshinol (DS-182) on mitochondrial changes in myocardial ischemia and reperfusion injury in rats, and the results showed that Tanshinol may be used as a kind of potent $\cdot O_2^-$ scavenger to protect myocardial mitochondria from being injured by lipid peroxidation induced by oxygen free radicals during myocardial ischemia and reperfusion. Tong et al. [191] further confirmed at the cellular and organ levels that Danshen showed inhibitory effects on the production of free radicals and the induction of lipid peroxide toxicity by catechol amine, which protected the integrity of myocardial cells.

4.5.2 Protection of the Cell Membrane

4.5.2.1 Danshen and Its Active Ingredients Can Act as Membrane Stabilizing Agents

The Effects on the Fluidity of Cell Membrane System

The lipid bilayer and mosaic proteins have fluidity in the plane, which is very important for material exchange, ion transportation, cell interactions and other biological characteristics. Sun et al. [187] utilized the in vitro heart model of rats to determine the changes in mitochondrial membrane fluidity during ischemia and reperfusion. They labeled the membrane lipid phase using 1,6-dibenzyl- 1,3,5-hexatriene (DPH) as the probe, and determined the changes in membrane fluidity by marking the membrane fluorescence polarization, fluorescence anisotropy, and calculating lipid microviscosity. The results showed that under the ischemia-reperfusion condition, the membrane fluidity decreased, the microviscosity increase in the lipid region appeared later than the change in membrane permeability, and the ultrastructure of myocardial cells underwent significant changes (the mitochondria were swollen, the arrangement of cristae was disordered, floss with low electron density and thick electron-dense particles appeared, nuclear chromatin became peripheral). However, when tanshinone was administered, lipid peroxides in the myocardial tissues would not increase and the fluidity of mitochondrial membrane would not decrease, and the data were statistically the same as those from the control group. The ultrastructural changes in myocardial tissues paralleled the results. These results confirmed that tanshinone can protect the liquid status of the mitochondrial membrane and avoid decrease of mitochondrial membrane fluidity during reperfusion.

The Effects on the Changes in Membrane Receptors

Receptors are membrane proteins on the cell surface which bind to specific ligands, and thus lead to special reactions in the cells. When the membrane structure or microenvironment changes, the receptors, as membrane proteins, will inevitably undergo corresponding changes.

Fu et al. [192] utilized the left ventricle branch in the coronary artery muscle muscle of rabbit heart as a myocardial ischemia model, and used radio-ligand binding to observe the changes in the α_1 receptors on cardiac muscle cell membranes under the conditions of myocardial ischemia and reperfusion. The results showed that the receptor density increased while the affinity did not change under ischemic conditions, and α_1 receptor maintained previously high levels at the early stage of reperfusion. Danshen could stabilize the α_1 receptor on the myocardium and avoid an increase in the number of receptors during ischemia-reperfusion. Su et al. [190] found that Tanshinol had protective effects on cytochrome oxidase, indicating that it could protect both membrane lipids and membrane proteins, i.e., it had protective effects on the membrane structures.

In summary, it can be concluded that Danshen and its active ingredients exhibit excellent effects on stabilizing cell membranes and organelle membranes.

4.5.2.2 The Effects on the Electric Activity of Cell Membranes

As we all know, the resting potential (RP) is depends mainly on the distribution of resting intracellular and extracellular K^+. The action potential amplitude (APA), over shooting (OS), and the maximum depolarization rate of phase O (V) in rapid response myocardial cells can represent the extent and rate of change in the rapid channels, which are mainly related to the rapid influx of Na^+. The phase 2 platform of the action potential is mainly related to the opening of slow channels, Ca^{2+} influx, and slow K^+ efflux. The duration of the action potential of myocardial cells, APD_{90}–APD_{50}, can be used in the computation of phase 3 duration, since the proportion of the duration in the platform stage is relatively large in the total duration of the action potential, thus the changes in phase 2 duration are inevitably reflected in APD_{90}–APD_{50}.

Xu et al. [193] recorded the action potential of in vitro ventricle papillary muscle in guinea pigs by using intracellular microelectrodes and computer real-time analysis, observing and the effects of different dosages of Danshen injection on related indexes. The results showed that Danshen had no significant effects on the phase 3 duration of the action potential in guinea pig in vitro ventricle papillary muscle, and the decrease in ADP_{90} was mainly caused by the decrease in ADP_{50}. That is to say, Danshen had no significant effects on the depolarization induced by rapid influx of Na^+ and the repolarization induced by K^+ efflux in the guinea pig ventricle papillary muscle, but it can inhibit the slow influx of Ca^{2+} and thus significantly decrease the duration of platform stage. The effect was dose-dependent.

Xu et al. [194] also carried out investigations on the effects of Danshen on the electric activity of myocardial cells in the prevention of myocardial damage induced by isopropylarterenol in the right ventricle papillary muscle in rats. The results demonstrated that isopropylarterenol could lead to significant changes in the process of depolarization in rat right ventricle papillary muscle, and the action potential amplitude, over shooting, resting potential, and the maximum depolarization rate of phase O significantly decreased. Its effects on the process of repolarization in myocardial cells could prolong the duration of the action potential, which was mainly induced by the prolongation of ADP_{50}. The electrical changes induced by isopropylarterenol administration after Danshen treatment were significantly improved in comparison to those of isopropylarterenol treatment alone, and the improvements were mainly in the repolarization of myocardial cells and the effects on the action potential duration, which could recover to the control group level. The decrease in action potential duration of myocardial cells induced by Danshen was mainly induced by the decrease in

ADP_{50}. Danshen can improve the depolarization rate of myocardial cells; computer real-time analysis was used to analyze dV/dT at each point in the process of depolarization in myocardial cells, and the continuous and dynamic changes in the depolarization rate of each point in the action potential upstroke were observed. It was found that the depolarization rate of every point in the Danshen group was higher than that in the isopropylarterenol group, and the depolarization of myocardial cells also showed certain improvements at the same time. The liposoluble ingredient of Danshen sodium tanshinone II-A silate can block the slow channel of the cell membrane and prevent large amounts of calcium from entering the myocardial cells and shortening the platform stage. Danshen injection also showed the same effects, which was to inhibit the slow-phase calcium ion influx in the platform stage, which decreased ADP_{50} and ADP_{90}, preventing the calcium overload of myocardial cells, and thus showed protective effects on the cardiac muscle. During the process of depolarization, Danshen could antagonize isopropylarterenol and increase Vmax and OS, indicating that Danshen can affect other pathways apart from the slow channels.

Su et al. [195] studied the effects of ischemia-reperfusion on the transmembrane potential of mitochondria and mitoplasts. The protective effects of Tanshinol in the Langendorff non-acting non-cyclic isolated heart perfusion model, using succinic acid as the substrate of the respiratory chain and flowing dialysis with Rh123 fluorescent dye, were observed. The results showed that the pre-energized transmembrane potential of myocardial mitochondria increased and the energized transmembrane potential did not change 20 min after ischemia; the pre-energized transmembrane potential of myocardial mitochondria did not change and the energized transmembrane potential decreased 20 min after reperfusion. The increase in pre-energized transmembrane potential during ischemia disappeared in mitoplasts without the mitochondrial outer membrane, and the tendency for changes in other parameters was the same as that of mitochondria, indicating that the changes in the mitochondrial transmembrane potential in the early stage of ischemia-reperfusion were related to the outer membrane. Tanshinol had significant protective and recovering effects on the transmembrane potentials of mitochondria and mitoplasts.

Xu et al. [196] utilized the whole cell patch clamp technique to study the effects of tanshinone II-A on type I calcium currents and transmembrane potentials in isolated single ventricular myocardial cells of guinea pigs, and Langendorff devices and protease circulatory digestion were used to obtain 70–80 % calcium-tolerant myocardial cells with clear transverse striations and rod-like shape. pCLAMP5.51 software and EIP-7 patch clamp amplifiers were used to record the changes in the membrane current of myocardial cells and the transmembrane potential; tanshinone additive administration was used and all of the parameters for observation were accomplished in the same cell. The results showed that tanshinone II-A had effects similar to those of verapamil and other L-type calcium channel blockers and that the blocking effects were not voltage-dependent, which can be used in the prevention and treatment of myocardial ischemia-reperfusion injury and heart rate aberrations.

4.5.3 The Prevention of Calcium Overload in the Cells

Calcium ions are involved in the normal physiological activities of many cells in the human body, including cell proliferation, information transduction, myocardial excitation-contraction coupling, transmission of nerve impulses, secretion, morphogenesis, cell senescence and many other aspects. The execution of all of these functions is dependent on the energy-dependent distribution of calcium ions which differs by 10,000 fold in and out of the cells. Myocardial reperfusion injury is a kind of important pathophysiological phenomenon, and its development is closely related to damage to the normal distribution of calcium ions, i.e., intracellular calcium overload.

Since Robert B. Jennings discovered the possible relationship between “myocardial ischemia-reperfusion injury” and a significant increase in intracellular calcium concentration in the early 1970s, a large number of studies have supported this view. It is assumed that “calcium overload” is an important factor for reperfusion injuries, and intracellular calcium overload is one of the important pathological changes in myocardial ischemia-reperfusion injury, and is considered the final common path to cell death.

Currently, two factors are considered the major causes for the occurrence of calcium overload: (1) increase in Na^+/Ca^{2+} exchange; (2) damage to membrane structures by large amounts of oxygen free radicals produced during reperfusion, and the action of catecholamine.

4.5.3.1 The Transportation of Ca^{2+} in and Out of the Cells

At present, most people believe that Na^+/Ca^{2+} exchangers perform intracellular and extracellular dual directional transportation of these ions at a ratio of three Na^+ to one Ca^{2+}. Under physiological conditions, the main transportation direction of the Na^+/Ca^{2+} exchanger is to export intracellular Ca^{2+}, and the low calcium concentration that maintains the resting state of myocardial cells with calcium pumps on the cell membrane is regulated by Ca^{2+}, ATP, Mg^{2+}, and H^+ concentrations. The activity of Na^+/Ca^{2+} exchangers is mainly regulated by the transmembrane Na^+ concentration gradient.

Observations of transmembrane calcium transportation during myocardial ischemia-reperfusion injuries have revealed that the calcium content in cardiac muscle began to increase several minutes after ischemia. Large amounts of calcium entered the myocardial cells within several minutes after reperfusion began, and the calcium influx was maintained for a relatively long period while calcium efflux only showed related increase, indicating that calcium overload mainly occurs in the reperfusion stage, mainly due to the increase in calcium influx rather than the decrease in calcium efflux. Many experiments have confirmed that the Na^+/Ca^{2+} exchanger is the main pathway for Ca^{2+} to enter the cells in ischemia-reperfusion injuries and calcium abnormalities.

4.5.3.2 Calcium Distribution in Ischemia-Reperfusion Myocardial Cells and the Effects of Tanshinol

Ma et al. [197] made a myocardial ischemia-reperfusion injury model by ligating and reopening the coronary artery muscle muscle of rabbits, and monitored total myocardial calcium by atomic absorption spectrography. They measured calcium contents in the ischemic and non-ischemic regions in the experimental animals in three groups: ischemia group, ischemia-reperfusion group and ischemia-reperfusion plus Danshen protection group. They studied the mechanism for the protection of ischemia-reperfusion cardiac muscle from Danshen preparations by combining with the pathological and biochemical processes of ischemia-reperfusion injuries. The results showed that the myocardial calcium contents in non-ischemic regions among the 3 experimental groups were statistically the same as those of the control group, and that the calcium content in ischemic regions was different among the 3 groups: the contents in the ischemia group and ischemia-reperfusion plus Danshen group were similar, and both were lower than that of the ischemia-reperfusion group.The experiments demonstrated that Danshen acted as a Ca^{2+} antagonist, which can prevent extracellular calcium from entering the cells and block the transformation of Xanthine dehydrogenase (D type) to xanthine oxidase (O type). Therefore, cytoplasmic Ca^{2+} can be lowered while at the same time the production of free radicals is decreased, and calcium overload can be alleviated. This can facilitate the recovery of mitochondrial function to some extent and promote an increase in energy reserves, and thus alleviate ischemia-reperfusion injuries and exert protective effects on myocardial cells.

Fu et al. [192] determined the total calcium amount in myocardial cell membranes and myocardial tissues by atomic absorption spectrography, using the myocardial ischemia-reperfusion model of rabbits. They found that the calcium content in the cell membrane did not change in the first fifteen minutes after ischemia, but it significantly increased as ischemia went on (30–40 min). Though the calcium level decreased a little 4 min after reperfusion, which was given after 40 min of ischemia, it was maintained at a high level. It recovered to the control level 20 min after reperfusion, and the total calcium amount in the cardiac muscle increased. The change in the calcium content in the myocardial cell membrane was completely consistent with the trend of the change in $\alpha 1$ receptor density. This confirmed that membrane-bound calcium increased during ischemia and fluxed into the cells during reperfusion, membrane calcium recovered to a normal value, and intracellular calcium increased (it can be proven by the increase in total calcium).

Yue et al. [198] used the Langendorff infusion method for in vitro hearts in guinea pigs to prepare the model for calcium abnormalities in cardiac muscle. Different concentrations of DS-201 (tanshinone) were used as the protective agent for cardiac muscle, and the release of proteins and calcium intake in myocardial tissues was determined. The protection of DS-201 against calcium abnormalities in cardiac muscle was observed and then compared with that of the known Ca^{2+} antagonist verapamil hydrochloride to investigate the calcium antagonism of DS-201. The results showed that DS-201 had significant protection over injuries induced by myocardial calcium abnormalities, and that it could inhibit calcium influx and alleviate calcium deposition during the process of calcium abnormality and protein (protease) release induced by myocardial damage. This effect was dose-dependent in a certain range more effective than verapamil hydrochloride.

4.5.3.3 The Effects of Tanshinol on the Intracellular Free Calcium Concentration in Myocardial Cells that Undergo Oxygen Abnormalities

Zhu et al. [199] built an oxygen abnormality model to mimic reperfusion injuries by using the myocardial cells of rats that were isolated in vitro. The fluorescent probe (Fluo-3) was used to determine the changes in the intracellular free calcium concentration, and the effects of Tanshinol on the intracellular free calcium concentration in myocardial cells that underwent oxygen abnormalities were measured. The results showed that Tanshinol can significantly decrease the intracellular free calcium concentration in myocardial cells during oxygen abnormalities and improve the rate of rod shaped cells. It was confirmed again that Danshen can inhibit the influx of large amounts of Ca^{2+} and achieve the effects of protecting cardiac muscle and avoiding reperfusion injuries by breaking the morbidity chain of vicious cycles like damage induced by oxygen free radicals.

It is considered that the mechanism for calcium overload mainly includes: (1) increase in sodium/calcium exchange; (2) damage to membrane structures induced by large amounts of oxygen free radicals during reperfusion; (3) the effects of catechol amine; (4) the involvement of local renin angiotensin systems in the cardiac muscle.

Danshen can block the formation of lipid peroxides, which is one of the links in the malignant, cell damaging cycle mediated by oxygen free radicals. It has significant functions of recovering and protecting the membrane potential of mitochondria and ATPase synthesis reactions, and it also inhibits the metabolism of phospholipid system in myocardial tissues, and thus inhibits the release of calcium ions from sarcoplasmic reticulum and alleviates the extent of calcium overload in myocardial cells.

4.5.4 The Effects of Danshen on Post-ischemia-Reperfusion "No-Reflow"

Some people believe that a long period of ischemia can lead to damage to the structure and function of the myocardial capillary system, and thus cause the blood in the ischemic region to become incapable of full recovery when the blood is reperfused, which is equivalent to continued ischemia or superimposed ischemia. This phenomenon is called "no-reflow". The formation of "no-reflow" is generally considered to be related to the following factors: (1) the edema of myocardial cells oppresses the microvascular vessels; (2) the cytosol protrusion of vascular endothelial cells stretches to the lumen, which hampers blood flow and promotes the adhesion of leucocytes to blood vessel endothelium, platelet aggregation or the formation of microthrombus; (3) the loading capacity of vascular bed functions decreases and the capacity for reactive hyperemia decreases when the blood vessels are opened after vascular occlusion.

Previous studies have suggested that the occurrence of "no-reflow" is closely related to lipid peroxidation reactions, and the mechanism is (1) lipid peroxides may damage microvascular endothelium; (2) lipid peroxides or lipid peroxidation of platelet membrane can induce platelets to release thromboxan (TX) and promote platelet aggregation and angiospasm. In addition, the functional disturbance of microvascular vessels may also be involved.

Han et al. [188] observed the effects of Danshen preparations on local capillary flow in the myocardial ischemia-reperfusion model of rabbits, and the results showed that the administration of Danshen preparations before ischemia and before reperfusion can almost eliminate "no-reflow regions" and recover capillary flow in 85 % of the control group.

4.5.5 The Effects of Danshen on Myocardial Energy Metabolism

4.5.5.1 The Effects of Danshen on ATP Degradation and Re-Synthesis in Myocardial Tissues

When severe ischemic myocardial disease occurs, the **mitochondrial** energy metabolism in myocardial cells malFunctions, which results in serious deficiencies or depletion of high-energy phosphate, leading to the inhibition of phosphorylation of membrane proteins, activation of sarcolemma phOspholipase, damage to sarcolemma, the dysfunction of intracellular and membrane energy-dependent ion transport, overload of intracellular Ca^{2+}, and eventually cell death. Therefore, investigations into the changing status of adenine nucleotide in ischemic myocardial cells are of great importance for the elucidation of the mechanism for, and directed protection from, ischemic damage in myocardial cells.

Ma et al. [197] made a myocardial ischemia-reperfusion model by ligating and reopening the coronary artery muscle muscle of rabbits, and observed the changes in adenylates and lipid peroxides in the myocardial tissues of ischemic and non-ischemic regions in three groups of experimental animals: ischemia group, ischemia-reperfusion group, and ischemia-reperfusion plus Danshen protection group. The results showed that there was a common characteristic in ATP content and energy charge (E) in the ischemic regions of myocardial tissues among the three experiment groups: ATP and E levels in the ischemia group were close to those in the Danshen protection group, and levels in both groups were higher than those in the ischemia-reperfusion group. The values of the non-ischemic regions among the three groups were Danshen protection group > ischemia group > ischemia-reperfusion group. This result showed that Danshen preparations can improve myocardial energy reserve capacity, and thus prevent myocardial ischemia and reperfusion injuries and protect cardiac muscle. From the integrative analysis on the contents of ADP, AMP and AN (adenylate total) in the three experiment groups, the contents of ADP and AMP in the ischemic region of the Danshen protection group were both higher than those of the ischemia group and the ischemia-reperfusion group, and the differences were significant. Thus it can be concluded that Danshen preparations can slow the gradual degradation of ATP, and thus improve myocardial energy reserve capacity. As far as AN content was concerned, it was higher in the Danshen protection group in comparison to in the ischemia-reperfusion group; particularly, it was higher in the Danshen protection group in comparison to in the ischemia group, indicating that it was possible for Danshen preparations to re-synthesize adenine nucleotide using degraded adenine nucleotide. In summary, Danshen preparations can reduce the hydrolysis of ATP and use the hydrolyzed products of ATP to re-synthesize ATP, so that the content of myocardial ATP is increased relatively and the ion pump disturbance is alleviated.

4.5.5.2 The Effects of Ischemia-Reperfusion on the Coupling of MitochondriaL Electron Transfer and Proton Pumping in Cardiac Muscle of Rats

In 1961, Peter D. Mitchell proposed the chemiosmotic hypothesis to explain the mechanism of the coupling of oxidation and phosphorylation in mitochondria. According to this hypothesis, the electron transfer chain is coupled with transmembrane energy conversion for electrochemical potential and ATP synthesis. The chemical energy of the oxidized material is converted into the electrochemical proton gradient spans the mitochondrial inner membrane by the respiratory chain, and this electrochemical proton gradient not only can drive ATP synthesis, but also affect the transportation of calcium ions, substrates, and others, and it can also regulate mitochondrial metabolism. Adolfas Toleikis carried out investigations on the changes in the electrochemical proton gradient in ischemic myocardial mitochondria. $H^+/2e$ can directly reflect the status of energy conversion in mitochondria.

Zhang et al. [200] utilized the Langendorff infusion device to produce a myocardial ischemia-reperfusion model in rats. mitochondria were isolated by differential centrifugation, Clark oxygen electrodes were used to determine the RCR (the oxygen consumption rate of mitochondria) and the P/O ratio of purified mitochondria (ADP/O), and potassium ferricyanide impulse method was used to determine the initial rates of electron transfer and proton pumping in the respiratory chain. The results showed that $H^+/2e$ in ischemia did not show a statistically significant difference with the control group. While it was significantly lower in the reperfusion group in comparison to the control group, the decrease in $H^+/2e$ was earlier than the decrease in ADP/O. RCR and ADP/O in the early stage of ischemia and reperfusion were significantly higher than those of the control group, indicating that the coupling degree of energy conversion of myocardial mitochondria in reperfused rats decreased.

4.5.5.3 The Effects of Tanshinol on Proton ATPase Hydrolysis and Synthetic Activity of Myocardial Mitochondria

Chang et al. [189] utilized $FeSO_4$/ascorbate as the oxygen free radical producing system, and measured the changes in the hydrolyzing activity of mitochondrial H^+-ATPase after incubation with myocardial mitochondria by the phosphorus method, and observed damage to H^+-ATPase of myocardial mitochondria induced by oxygen free radicals from $FeSO_4$/ascorbate and the protective effects of Danshen. The results showed that $FeSO_4$/ascorbate can significantly decrease the hydrolyzing activity of mitochondrial H^+-ATPase, while Tanshinol had significant protective effects, preventing the decrease in the hydrolyzing activity of H^+-ATPase in a certain dose-dependent manner with a "saturation effect".

Zhu et al. [201] utilized the Langendorff infusion device to prepare a myocardial ischemia-reperfusion model in rats, isolated mitochondria by differential centrifugation, determined the protein concentration in the mitochondrial suspension by the Biuret method, the oxygen consumption, RCR and P/O by Clark oxygen electrodes, the activity of ATP hydrolysis by H^+-ATPase by inorganic phosphorus, and the activity for ATP synthesis by H^+-ATPase by fluorescein-luciferase chemiluminescence. The results showed that the decrease in ATP synthesis in H^+-ATPase appeared in the early stage of ischemia (10 min) and reperfusion (20 min), and continued to decrease with the prolongation of ischemic time. However, the capacity of ATP hydrolysis and mitochondrial respiratory function in the early ischemic stage showed a compensatory rise, and decreased with the prolongation of ischemic duration, indicating that Tanshinol (DS-182) had significant protective effects on H-ATPase activity. At the same time, Zhu Shijun et al. also observed the damage caused by oxygen free radicals in the hydrolyzing and synthesizing activities of H^+-ATPase in rat myocardial mitochondria, and the protective effects of DS-182 by using the $FeSO_4$/ascorbate system. The results showed that oxygen free radicals produced by the $FeSO_4$/ascorbate system could significantly increase the hydrolyzing activity of mitochondrial H^+-ATPase and significantly decrease the synthesizing activity. Tanshinol could prevent the increase in hydrolyzing activity and the decrease in synthesizing activity of H^+-ATPase, and exhibited good protective effects.

Su et al. [195] utilized the Langendorff non-acting non-cyclic isolated heart perfusion model to determine the effects of ischemia-reperfusion on the synthesizing Activity of atpase, with succinate as the respiratory chain substrate and fluorescein-luciferase chemiluminescence to detect ATPase synthesis activity during ischemia-reperfusion. They observed the protective effects of Tanshinol. The results showed that the synthesis activity of mitoplast ATPase did not change during ischemia, but decreased during reperfusion, and Tanshinol showed significant protective and recovering effects on ATP synthesis activity.

4.5.5.4 The Protective Effects of Danshen on Energy Substances in Cardiac Surgery

Ma and Wang [203] carried out investigations on the effects of continuous infusion with Danshen and warm blood on the changes in energy substances in **cardiac muscle** in 7 patients suffering from rheumatic heart disease that were subjected to a valve replacement operation. They used HPLC to determine the energy substances at different time points during the operation (arrest, arrest 40 min, resuscitation, resuscitation 20 min). The results showed that creatine phosphate (PCr) and ATP contents did not show significant changes except for a significant decrease 20 min after the re-beat. The total amount of adenine nucleotide also did not show changes, which was better than with the continuous infusion of warm blood alone.

4.5.6 The Effects of Danshen on the Apoptosis of Myocardial Cells and the Expression of Apoptosis Related Genes During Myocardial Ischemia-Reperfusion of Rats

Zhao Ming et al. utilized braid ligation in the left coronary artery muscle muscle to prepare a myocardial ischemia-reperfusion model, and TdT-mediated dUTP nick end labeling (TUNEL) and S–P immunohistochemical methods were used to determine changes in the expressions of *Fas* and *Bcl*-2 in myocardial apoptotic cells and myocardial cells. The results showed that the apoptotic index (AI) of myocardial cells and the positive expression indexes for *Fas* and *Bcl*-2 protein were significantly increased in the ischemia reperfusion group in comparison to those in the sham operation group; the PE1 of Al and Fas protein decreased after intervention with Composite Danshen Dropping Pill (CDDP) in comparison to that of the ischemia reperfusion group, and it decreased further with an increase in the dosage of CDDP. CDDP can protect myocardial cells from damage caused by ischemia-reperfusion by inhibiting myocardial Apoptosis, down-regulating *Fas* gene expression, and up-regulating *Fas* gene expression after ischemia-reperfusion.

Zhang Daishou et al. prepared a myocardial ischemia-reperfusion model by ligating the left coronary artery muscle muscle of rats, and the results showed that the Apoptosis of myocardial cells was aggravated during acute myocardial ischemia-reperfusion, while *Bax* and *Bcl*-2 were involved in the Apoptosis of myocardial cells during the regulation of ischemia-reperfusion. Danshen can inhibit Apoptosis and accelerate the expression of *Bax*.

References

1. Sun J, Huang SH, Tan BK, et al. Life Sci. 2005;76(24):2849.
2. Chen SL, Wu XH, Wang XD. Chin Med Res Clin. 2005;3(5):13.
3. Yan XJ, Wu TF. Tianjin Pharm. 1992;4(4):22.

4. Zhang J, Zeng XR, Yang Y, et al. J Luzhou Med Coll. 2000;23(3):177.
5. Liu QG, Wang L, Lu ZY, et al. Chin J Hosp Pharm. 1999;19(11):653.
6. Liu QG, Sun YM, Cheng YZ, et al. China J Mod Med. 2005;5(8):2257.
7. Liu QG, Sun YM, Cheng YZ, et al. Chin Pharmacol Bull. 2005;21(9):1078.
8. Zhang ZX, Ren CL, Chen SX, et al. J Jiangsu Univ (Med Ed). 2003;13(5):391.
9. Su XH, Wang XM. Chin J Pathophysiol. 1996;12(2):186.
10. Tang LH, Wang XM, Liang DQ. Chin J Pathophysiol. 1989;5(2):65.
11. Fan YC, Zhan GF, Zhang WZ. J Tianjin Univ Tradit Chin Med. 2004;23(1):12.
12. Jiang HP, Wu YT, Chen JG. Chin J Microcirc. 2006;16(1):40.
13. Ke SS, Zhao ZD. Tianjin Med. 1996;24(12):715.
14. Ke YH, Qu SB. J Beijing Coll Acupuncture-Moxibustion. 2001;8(1):17.
15. Zhang DF, Ruan CW, Liu ZM, et al. Shanghai Med J. 2003;26(9):669.
16. Xie H, Zheng Z. Chin J Hypertens. 2004;12(4):359.
17. Lu WF, Xia Q, Zhang X. Chin Tradit Herbal Drugs. 1998;(29):460.
18. Wang R, Chen LB, Zeng XR. J Shandong Univ. 2005;43(4):280.
19. Wang GJ. J Jinzhou Med Coll. 2003;24(1):47.
20. Tang ZZ, Tang Y. J Fourth Mil Med Univ. 2004;24(2):100.
21. Feng J, Li SS. Chin J Hypertens. 2005;13(8):488.
22. Chen BH, Gong LY, Xu W, et al. Chin J Prim Med Pharm. 2005;12(7):775.
23. Zhan CY, Tao XL, Han SJ, et al. Chin J Clin Rehabil. 2005;9(23):243.
24. Zhan CY, Tao XL, Tian C, et al. Chin Clin Rehabil. 2005;9(19):254.
25. Sun LP, Zheng Z. Pract Geriatr. 2004;18(1):25.
26. Liu F, Zheng Z, Zhong Q. Chin J Pract Chin W Med. 2004;4(17):1096.
27. Guo ZQ, Wang SR, Zhu LQ, et al. Chin J Integr Med. 2005;25(4):342.
28. Maki T, Kawahara Y, Tanonaka K, et al. Planta Med. 2002;68(12):1103.
29. Zhu LM, Zeng XT. Shandong Med J. 2005;15(20):22.
30. Zhao YW, Xie WL, Sun WJ, et al. Acta Acad Med CPAF. 2005;14(4):256.
31. Yu HB, Xu CQ, Shan HL, et al. J Harbin Med Univ. 2002;36(2):112.
32. Cheng GZ, Hu XH, Li Y. Jilin J Tradit Chin Med. 2000;20(4):17.
33. An XF. Tianjin Med J. 1997;25(8):502.
34. van der Wal AC, Das PK, Tigges AJ, et al. Am J Pathol. 1992;141(1):161.
35. Meng ZX, Zhu FX, Han CL, et al. J Cardiovasc Pulm Dis. 1995;14(2):101.
36. Wang P, Zhang Y, Fan YM, et al. J Cardiovasc Pulm Dis. 1996;15(1):52.
37. Tian JY, Chen JZ, Gu Y, et al. Chin J Clin Rehabil. 2000;9(8):1708.
38. Sun XM. Chin Tradit Herbal Drugs. 1991;22(1):22.
39. Han F, Xie ML, Zhu LJ, et al. Chin J New Drugs Clin Remedies. 2002;21(7):389.
40. Jiang CC, Cai HH, Fan LM, et al. Acta Univ Med Nanjing. 1994;14(4):522.
41. Qin PS, Fan LJ, Liu KY, et al. Chin Tradit Pat Med. 2000;22(9):630.
42. Sun L, Dai PS, Shao J, et al. Tianjin J Tradit Chin Med. 2000;17(3):61.
43. Li L. Chin J Geriatr. 2003;22(11):706.
44. Liu J, Zhang HJ. Tianjin Pharm. 2003;15(3):38.
45. Whorton AR, Montgomery ME, Kent RS. J Clin Invest. 1985;76(1):295.
46. Lin RX, Liu GP, Zhao LC. J Guangzhou Univ Tradit Chin Med. 1997;14(4):288.
47. Lin Q, et al. Chin J Basic Med Tradit Chin Med. 2001;7(5):54.
48. Liu H. Hubei J Tradit Chin Med. 2001;23(8):11.
49. Lin Q, Lu JG. Shanxi J Tradit Chin Med. 2002;18(4):56.
50. Yu GY, Deng ZD, Qu ZL. Chin J Pathol. 1998;27:174.
51. Chen TZ, Lai JK. Chin J Pathol. 1989;18:15.
52. Xu DB, Zhang JP, Zhang WZ, et al. J Tianjin Med Univ. 2001;7(1):21.
53. Wang XX. Chin J Pharmacol Toxicol. 1993;7(2):157.
54. Stringer MD, Görög PG, Freeman A, et al. BMJ. 1989;298(6669):281.
55. Wu YJ, Hong CY, Lin SJ, et al. Arterioscler Thromb Vasc Biol. 1998;18(3):481.
56. Wu L, et al. Chin J Gerontol. 1996;16(4):239.
57. Wu WP. Beijing Med J. 1995;17(3):133.
58. Wang N, Cai HJ, Zhu Y, et al. Acta Uni Med Nanjing. 1994;14(4):529.
59. Zhang L, Wang XM, Zhao BL, et al. J Harbin Med Univ. 1992;26(4):255.
60. Zhao Y, Jiang W, Hou JW, et al. Acta Biochimica et Biophysica Sinica. 1995;27(6):610.
61. Xie HL, Gu ZL, Qian ZN, et al. Chin J Mod Appl Pharm. 1992;9(1):1.
62. Pryor WA, Squadrito GL. Am J Physiol. 1995;268(5Pt l):L699.
63. Liao LF, Zhou X. Chin J Mod Appl Pharm. 2002;4(2):132.
64. Raines EW, Ross R. BrHeart J. 1993;69(1 Suppl):S30.
65. Kuzuya M, Kuzuya F. Free Radic Biol Med. 1993;14(1):67.
66. Anderson TJ, Gerhard MD, Meredith IT, et al. Am J Cardiol. 1995;75(6):71B.
67. Cai HJ, editor. Basis and clinical investigations on artherosclerosis. JiangSu Science and Technology Press; 1996. p. 102.
68. Ross R. N Engl J Med. 1986;314(8):488.
69. Collins P, Fox K. Eul Heart J. 1992;13(4):560.
70. Xie HL, Gu ZL, Zhu LJ, et al. Chin Tradit Herbal Drugs. 2001;32(7):1007.

71. Shi YF. Chin J Cardiovasc Rehabil Med. 2002;11(6):226.
72. Xu JL, Wang XM, Wang DX. Chin J Pathophysiol. 2005;21(8):1659.
73. Lao J, Gu YD. Chin J Surg. 1993;9(1):39.
74. Jiang Z, Liu HX. Chin J Intern Med. 1997;36(11):778.
75. Zhang YY. Chin J Pathophysiol. 1998;14(6):626.
76. Li WQ, Xie MZ, Wei P, et al. Shanxi J Tradit Chin Med. 2000;16(4):25.
77. Feng PF, Qin NP, Qin YG, et al. Chin J Integr Med. 1999;19(5):286.
78. Luo WB, Wang YP. Acta Pharmacol Sin. 2001;22(12):1134.
79. Xie ML, Lu Q, Zhu LJ, et al. J Chin Med Mater. 2000;23(8):474.
80. Zhang YJ, Zhang FY, Fan YC. Tianjin J Tradit Chin Med. 2005;22(4):328.
81. Zu SY, Gao YM, Zhu GJ. Chin J Geriatr Heart Brain Vessel Dis. 2004;6(4):257.
82. Finkelstein JD. Eur J Pediatr. 1998;157(Suppl 2):S40.
83. Boushey CJ, Beresford SA, Omenn GS, et al. Probable benefits of increasing folic acid intakes. JAMA. 1995;274(13):1049.
84. Huang RX, Su ZP. Cerebral apoplexy. Beijing: The People's Health Press; 2001. p. 48.
85. Bo QL. Chin Pharmacol Bull. 2003;7(24):3286.
86. Ross R. Transplant Proc. 1993;25(2):2041.
87. Plow EF, D'Souza SE. Circulation. 1997;95(6):1355.
88. Entwistle J, Hail CL, Turley EA. J Cell Biochem. 1996;61(4):569.
89. Zhang M, Wen JK, Sun HC, et al. Chin J Pathophysiol. 2004;20(10):1871.
90. Jiang KY, Gu ZL, Ruan CG. Chin Pharmacol Bull. 2000;16(6):682.
91. Stampfer MJ, Malinow MR, Willett WC, et al. JAMA. 1992;268(7):877.
92. Yuan XL, Hai CX, Liang X, et al. Chin J Clin Rehabil. 2003;7(24):3281.
93. Zhang XC. Chin J Arterioscler. 1999;7(4):179.
94. Chen JZ, Tian JY, Gu Y. Chin J Clin Rehabil. 2004;8(24):5048.
95. Xu T, Yu L, Zheng Z. J Clin Cardiol. 2005;21(1):54.
96. Jiang KY, Ruan CG, Gu ZL, et al. Acta Pharmacol Sin. 1998;19(1):47.
97. Wang L, Huang X, Ding ZH, et al. West China J Pharm Sci. 1995;10(4):193.
98. Cheng SW. Hunan Guiding J Tradit Chin Med Pharmacol. 2003;9(7):66.
99. Zhou HD, Ju YF. Sichuan Med J. 1997;18(3):177.
100. Wang H, Gao XM, Zhang BL. Tianjin J Tradit Chin Med. 2004;21(3):231.
101. Ma XQ, Shi D. Chin J Gen Surg. 2002;11(8):488.
102. Zhang WJ, Bao XF, Wang XF. Chin J Arterioscler. 1999;4(1):45.
103. Wang H, Gao XM, Zhang BL, et al. Chin Pharm J. 2005;40(2):106.
104. Du XH, Yang FJ, Wang WM. J Chin Med Mater. 2003;26(9):647.
105. Wang W, Xu ZP, Lui HY. J Tianjin Coll Tradit Chin Med. 2003;22(2):23.
106. Yang HP. J Hengyang Med Coll. 1991;19:123.
107. Sun BH, Zhang WR, Che DY. Acta Univ Med Tongji. 1995;24(1):51.
108. Du JR. West China J Pharm Sci. 1999;14(1):11.
109. Zhou XM, Lu ZY, Wang DW. Chin J Integr Med. 1996;16(8):480.
110. Wu HM, Yin WH. Acta Univ Med Tongji. 2001;30(2):106.
111. Chen ZW, Ma CG, Fang M, et al. Acta Pharm Sin. 1994;29(1):15.
112. Ge SR. J Chin Med Mater. 2002;25(9):683.
113. Shi Y, Wen JK. J Cardiovasc Pulm Dis. 1999;18(3):216.
114. Zu ZW, Xu LN, Tian JY, et al. Acta Pharm Sin. 1993;28(4):241.
115. Wu ZZ, Ni Z, Xu HQ, et al. J Nanjing Univ Tradit Chin Med. 1995;11(6):35.
116. Gu YH, Zhang CY, Huang GQ, et al. J Shanghai Second Med Univ. 1990;10(3):208.
117. Gu YH, Zhang CY, Huang GQ, et al. Chin J Hematol. 1993;14(10):540.
118. Zhu P, et al. Chin J Clin Thorac Cardiovasc Surg. 1989;5(2):103.
119. Chen SH, et al. J Biol Chem. 1988;4(1):88.
120. Parmley WW, Blurnlein S, Sievers R. Am J Cardiol. 1985;55(3):165B.
121. Henry PD, Bentley KI. J Clin Invest. 1981;68(5):1366.
122. Mu YF, et al. Chin J Gerontol. 1987;4:256.
123. Stein O, Leitersdorf E, Stein Y. Arteriosclerosis. 1985;5(1):35.
124. Qin SC, et al. Chin J Gerontol. 1989;9(supplement):19.
125. Sun XM, et al. Chin J Integr Med. 1989;9(10):804.
126. Wang MJ. Chin J Synthetical Med. 2006;7(3):234.
127. Chen SL, Wu XH, Wang XD. Chin Med Invest Clin Invest. 2005;3(5):13.
128. Zhuge LM, Wu Q, Lou ZJ. J Zhejiang Tradit Chin Med Coll. 2005;29(4):13.
129. Nie GL, Guo MJ, Li YP, et al. Chin Geraeology J. 2007;27(9):823.
130. Li FJ. Mod J Integr Tradit Chin W Med. 2006;15(20):2803.
131. Wu JH, Yang SY, Jin LD, et al. J Shanghai Med Univ. 1998;25(1):47.
132. Li DM, Li HF, Li W, et al. J Gansu Coll Tradit Chin Med. 2000;17(12):15.
133. Yin J, editor. Modern investigations and clinical applications of traditional Chinese medicine. Beijing: Xueyuan Publishing House; 1993. p. 174.
134. Yokozawa T, Lee Tw, Chung HY, et al. J Pharm Pharmacol. 1990;42(10):712.
135. Yokozawa T, Lee Tw, Oura H, et al. Nephron. 1992;60(4):460.
136. Kamata K, Iizuka T, Nagai M, et al. GeFt Pharmacol. 1993;24(4):977.
137. Gao XP, Xu DY, Deng YL, et al. Chin J Tradit Chin Med. 2004;29(4):359.

138. Xu F, Lu BG, Yao Z, et al. Chin Crit Care Med. 1996;8(5):262.
139. Fang KF, Yang YX, Gong NS, et al. Chin J Integr Med Cardio-/Cerebrovasc Dis. 2004;2(3):133.
140. Che DY, Zhang WR. Acta Univ Med Tongji. 1994;23(2):81.
141. Yang Y, Deng CS. Hubei J Tradit Chin Med. 1999;21(6):284.
142. Cheng JY, Lu ZQ, Li ZP, et al. Chin J Clin Pharmacol Ther. 2004;9(7):807.
143. Wu YS, Gao XF, Yang JH, et al. Chin J Integr Tradit W Med Intensive Crit Care. 1998;5(2):69.
144. Huang W. Mod J Integr Tradit Chin W Med. 2004;13(20):2702.
145. Yu JG, Yu ZR. J Nantong Med. 2000;20(2):172.
146. Tang XX, Chen XL. Guangxi Med J. 2004;26(2):251.
147. Liu H, Zheng ZE, Huang X. J Chin Mod Med. 2005;2(6):511.
148. Li JX, Mi JP, Zhao SQ, et al. J Tradit Chin Med. 2007;22(9):57.
149. Yu CL, Yao XQ. Chin J Integr Med Cardio-/Cerebrovasc Dis. 2007;5(6):478.
150. Kim DD, Sanchez FA, Duran RG, et al. Am J Physiol Heart Circ Physiol. 2007;292(5):H2131.
151. Lu YJ, Zhou M. Chin J Integr Med Cardio-/Cerebrovasc Dis. 2007;5(4):291.
152. Zhan CY, Tao XL, Zhou SX, et al. Chin Pharm J. 2006;41(17):1309.
153. Zhou Y, Gu J, Xu LM. J Chin Integr Med. 2007;5(1):61.
154. Hu SX. Pract J Card Cereb Pneumal Vasc Dis. 2006;14(9):697.
155. Zou P, Wang XY. J Guiyang Coll Tradit Chin Med. 2007;29(5):25.
156. Yang Y, Wang SX, Ma SY, et al. Jilin J Tradit Chin Med. 2006;26(5):37.
157. Xu DB, Zhang JP, Zhang WZ, et al. J Tianjin Med Univ. 2001;7(1):21.
158. Ge YX, Duan YY, Ruan LT, et al. Chin J Clin Rehabililation. 2004;8(6):1096.
159. Yang LR, Xu XY, Chen G. Pharmacol Clin Chin Mater Med. 2004;20(2):7.
160. Ren DC, Du GH, Zhang JT. Jpn J Pharmacol. 2002;90(4):276.
161. Ju G, Chen WL, Dong X. J China-Japan Friendship Hosp. 2005;19(4):230.
162. Ren DC, Du GH, Zhang JT. Chin J Pharmacol Toxical. 2003;17(5):333.
163. Ren DC, Du GH, Zhang JT. Chin Med J. 2003;116(4):630.
164. Lin GY, Wang ZJ. Shaanxi J Tradit Chin Med. 2007;28(6):647.
165. Lin L, Zheng YF, Qu JH, et al. Acta Acad Med Sin. 2001;23(4):386.
166. Hua SM, Guan GY, Jin H, et al. Chin J Clin Pharmacol Ther. 2000;5(4):363.
167. Hung HH, Chen YL, Lin SJ, et al. Histol Histopathol. 2001;16(1):175.
168. Li X, Du JR, Zhang R, et al. China J Chin Mater Med. 2004;29(3):255.
169. Sun Y, Gao YM, Zhu GJ. Basic Clin Med. 2003;23(6):634.
170. Li YS, Wang ZH, et al. Chin J Crit Care Med. 2007;27(8):720.
171. Li YS, Gong ZH, Liang QS, et al. Chin J Hypertens. 2006;14(11):882.
172. Zhang XG, Yu ZH, Wen SY. Chin Pharm J. 2006;41(16):1276.
173. Wei LS. Med J Nat Defending Forces Southwest China. 2006;16(6):596.
174. Li YJ, Du GH. Chin Pharm J. 2003;88(11):844.
175. Wang H, Gao XM, Zhang BL, et al. Tianjin J Tradit Chin Med. 2004;21(3):281.
176. Chen P, Xu XJ. Chin J Integr Tradit W Med Intensive Crit Care. 2007;14(5):813.
177. Li XR, Li YP, Deng XL, et al. Chin J Clin Rehabil. 2006;10(31):55.
178. Li PF, Dietz R, von Harsdorf R. FEBS Lett. 1997;404(2–3):249.
179. Du XH, Xiang H, Yu LS, et al. J Chin Med Mater. 2004;27(9):659.
180. Zhou XX, Wen JK, Han M. Chin J Biochem Mol Biol. 1999;15(6):997.
181. Liu HB, Wen JK, Han M. Chin J Arterioscler. 2001;9(1):10.
182. Liu HB, Wen JK, Han M. Chin J Gerontol. 2001;3(21):22.
183. Liu HB, Wen JK, Han M. Chin J Integr Med. 2002;22(10):764.
184. Shi Y, Wen JK. J Cardiovasc Pulm Dis. 1999;18(3):216.
185. Zhao GC, et al. Chin J Pathophysiol. 1987;3(4).
186. Tang LH, et al. Chin J Pathophysiol. 1989;5(2).
187. Sun DD, et al. J Harbin Med Univ. 1986;(2):6.
188. Han C, et al. Chin J Pathophysiol. 1991;7(4):337.
189. Chang YZ, et al. Chin J Pathophysiol. 1991;7(5).
190. Su XH, et al. Chin J Pathophysiol. 1992;8(2):122.
191. Tong LJ, et al. Chin J Pathophysiol. 1988;22(3):1.
192. Fu GH, et al. J Harbin Med Univ. 1990;24(5):334.
193. Xu CQ, et al. J Harbin Med Univ. 1991;25(6):409.
194. Xu CQ, et al. Chin J Pathophysiol. 1993;9(1):17.
195. Su XH, Wang XM, et al. Chin J Pathophysiol. 1996;12(2):190.
196. Xu CQ, et al. Chin J Pathophysiol. 1997;13(1):43.
197. Ma LY, Gao WD, Wang XM, et al. Chin J Pathophysiol. 1989;5(4):212.
198. Yue P, et al. Chin J Pathophysiol. 1987;3(3).
199. Zhu FX, Lui GM, Yang ZH, et al. Chin J Pathophysiol. 1999;15(10):877.
200. Zhang H, et al. Chin J Pathophysiol. 1993;9(5):561.
201. Zhu SJ, Wang XM, et al. Chin J Pathophysiol. 1994;10(7 supplement).
202. Zhu SJ, Wang XM. Chin J Pathophysiol. 1995;11(1).
203. Ma LY, Wang XM et al. Chin J Pathophysiol. 1993;9(4):485.

Protective Effects of Danshen on Cerebral Vessels and the Nervous System

5

Guangliang Han, Yuehua Wang, Guanhua Du, Hongmei Guang and Xinrui Cheng

Brain injury is a complex process involving multiple factors. Traditional Chinese medicine features holistic views and dialectical treatments. Meanwhile, the compositions of TCM drugs are complicated, and treat cerebral ischemia injury through multiple pathways and with multiple targets. Therefore, TCM has some advantages in the prevention and treatment of brain injury over Western medicine.

5.1 Pharmacological Action of Danshen to Treat Cerebral Hemorrhage

Guangliang Han, Yuehua Wang and Guanhua Du

5.1.1 Overview of Cerebral Hemorrhage

Cerebral hemorrhage is also termed cerebrovascular hemorrhage or hemorrhagic cerebral stroke. Cerebrovascular diseases are the major threats to the health of Chinese people, among which cerebral hemorrhage is the most severe.

According to TCM theories, cerebral hemorrhage is caused by yin-yang disharmony in zang-fu organs and disorder of qi and blood, leading to local cerebral blood vessel damage and belongs to rence and onset in recently YYblood overflowing outside of the blood vessel. The etiology and pathogenesis of cerebral hemorrhage are complicated. The internal factors in them are organ dysfunction, and the external factors affecting it, such as excess of emotion in five minds, excess of strain and damage, climate change, etc. The treatment of acute cerebral hemorrhage has been mainly to stop bleeding hemostasis.

The mortality and disability rates of cerebral hemorrhage are high and have led the board among various Cerebrovascular diseases. There is still a lack of effective treatment of the disease at present. Clinically, the pathogenesis of the disease can be divided into three stages: acute stage, recovery stage, and sequela stage. The acute stage is mainly caused by pathogenic wind, the symptoms are qi and blood deficiency, and the treatment is mainly by Western medicine. The treatments in the latter two stages mainly rely on TCM. Traditionally, medical treatment mainly included joint dehydration (mannitol, furosemide, and hormone) with hemostasis, in order to drain the blood and lower the blood pressure. However, the curative effect of these treatments was not satisfactory. Although surgical treatment has been used recently, the curative effect has still not increased significantly. Could Danshen be used to treat cerebral hemorrhage?

G. Han · Y. Wang · G. Du (✉) · H. Guang · X. Cheng
Institute of Materia Medica, Chinese Academy of Medical Science and Peking Union Medical College, Beijing, People's Republic of China
e-mail: dugh@imm.ac.cn

X. Yan (ed.), *Dan Shen (Salvia miltiorrhiza) in Medicine*,
DOI 10.1007/978-94-017-9463-3_5,

The following paragraphs attempt to answer this question.

5.1.2 Pharmacological Action of Danshen

Modern pharmacological investigations have already confirmed that Danshen has effects of improving microcirculation and decreasing blood viscosity, as well as of central sedation and hypnosis [1]. The water-soluble components of Danshen can inhibit the activity of cyclic adenosine phosphodiesterase in noncompetitive and irreversible ways to increase the level of cyclic adenosine monophosphate (cAMP), which might be related to the inhibitory effect of Danshen on the cerebral cortex [2]. Sedation has some effect on the stabilization and recovery of the patient's condition. Danshen's effects on promoting blood flow and removing blood stasis are achieved through multiple-layer and multiple-target regulation and treatment of the body's tissues and organs under physiological and pathological conditions. Clinically, Danshen is generally used in compound prescriptions such as Compound Danshen **injection**, Compound Danshen Tablet, etc. The Composite Danshen Dropping Pill has advantages of convenient administration, high solubility, rapid dissolution, and high bioavailability, which overcomes the disadvantages of traditional Chinese drugs, i.e., slow onset. Also, the sublingual administration of CDDP could avoid the first-pass effect in the liver through oral administration and achieve maximum pharmacological efficacy, so that it can also be used to treat comatose patients with cerebral hemorrhage, which may be another ideal form of medication besides **injection**.

5.1.3 The TCM Basis for Using Danshen to Treat Cerebral Hemorrhage

It is recorded in the *Treatise on Hemorrhagic* (《血证论》) that blood outside of vessels is in blood stasis no matter if it is clean or fresh. In TCM, there is a theory which says that stroke treatment should start with treating the blood; if blood flows, the stroke will be gone. When cerebral vessels are broken, the blood flows outside of the vessel, and accumulates in that location, becoming extravascular blood, which is also called blood stasis. During the acute stage of cerebral hemorrhage, the patient shows the following symptoms: hemiplegia, limb numbness, facial distortion, cyanosis of tongue and lips, acerbity and stagnation of arterial and venous pulse, etc. All of these are symptoms of blood stasis, stasis obstruction in the channels and collaterals, and inhibited qi movement.

According to TCM theories, fresh blood will not be generated until the stagnated blood is removed; stagnated blood in the body is not beneficial to normal blood, and meanwhile will hamper the generation of fresh blood. Therefore, the treatment of hemorrhagic disease always focuses on the clearing of blood stasis, repairing the brain collateral so that the movements of qi and blood can be smooth, yin and yang can be balanced, and zang-fu organs can be in harmony. So, one of the important principles in treatment of acute cerebral hemorrhage is to invigorate blood movement and remove blood stasis. Moreover, as is known to everyone, the brain cavity volume is fixed with limited extension. Therefore, cerebral hemorrhage will cause increased intracranial pressure and the cephalophyma will also compress the surrounding brain tissues, and prolonged compression will result in irreversible damages to the tissues. Research has showed the direct correlation between the duration of compression and the rates of lethality and disability. Thus, it is of great significance for the sake of patient recovery to remove the compression of brain tissues as early as possible.

5.1.4 The Theoretical and Experimental Basis of Danshen's Treatment of Cerebral Hemorrhage

Promoting blood flow and removing blood stasis are usually applied in treatment during the recovery and sequela stages of stroke and the acute stage of ischemic stroke. However, these treatments are prohibited during the acute stage of hemorrhagic stroke. During the acute stage of cerebral hemorrhage, the clinical application of Danshen is greatly limited due to the consideration of hemorrhage aggravation. Nevertheless, with the further investigation of blood stasis and microcirculation, people gradually recognized that there is a close relationship between the degree of blood stasis and microcirculation [3]. It is generally considered that rehemorrhagia in pure hypertensive cerebral hemorrhage after the acute stage is rare. Studies indicated that there is no significant relationship between rehemorrhagia and age of onset, the site of first hemorrhage, the amount of hemorrhage, severity of the disease, presentations of perforation of cerebral ventricle, and complications, etc. However, recurrence is related to the poor control of hypertension. The presence of amyloidosis also increases the risk of rehemorrhagia. Also, in the hemorrhage site, the existences of reduced blood flow, decreased intravascular pressure, organization of hematoma and contracture, as well as the protective effect of sludge blood to small vessels, causes rehemorrhagia to usually not occur at the first bleeding site, but at the other part of the cerebral hemisphere of the opposite or same side [1].

Glutamic acid is a kind of main excitatory transmitter in brain tissue. It is released by the presynaptic membrane, and is deactivated mainly through the uptake of neurocytes. In brains injured by ischemia and anoxemia, a great deal of glutamic acid is released, which activates *N*-methyl-*D*-aspartic acid receptors to open calcium channels and results in a massive inflow of calcium ions to induce cytotoxicity. Through the patch clamp technique, people found that tanshinone II_A can block the L-type calcium channels of ventricular muscular cells in guinea pigs and antagonize the calcium channels of ureteric smooth **muscle**s to dilate small arteries [4, 5]. Tanshinol can dilate mesenteric arterioles under contraction states, speed up blood flow, prolong prothrombin time, and have a significant anticoagulative effect [6]. In patients with cerebral hemorrhage treated by Compound Danshen **injection**, the XG-8C multi-site and multifunctional color microcirculation detector was used to observe the microcirculation of the nail fold of the left ring finger at 18–25 °C to find that the morphous scores (diameter ratio of artery-vein, distinctness of vessel), fluid state scores, and peri-loop state scores were all significantly improved, which indicated that Danshen can improve the angiospasm state in patients [7]. Furthermore, Danshen also has bidirectional regulation effects on the fibrinolytic system so that it can regulate both the hypocoagulable and hypercoagulable states [8].

The possible mechanisms of compound Danshen preparations to treat cerebral hemorrhage include: (1) improving microcirculation surrounding the hemorrhage site to promote hematoma absorption; (2) opening compensatory circulation to increase the capillary network and decrease vascular pressure at the hemorrhage site in order to prevent rehemorrhagia; (3) relieving angiospasm and cerebral edema and increasing the tolerance of nervous tissue to hypoxia to avoid neurocyte injury; (4) increasing the activity of erythrocuprein to inhibit the production of peroxidized lipids in order to relieve peroxidation injury; (5) relieving inflammatory reactions of the surrounding tissue, decreasing intracranial hypertension, improving local blood circulation, and promoting the functional recovery of neurocytes [9].

5.1.5 Inseparable Relationship Between Cerebral Hemorrhage and Cerebral Ischemia

In cerebral hemorrhage, ischemic damage to brain tissue can be induced by the positioning effect of the hematoma on local microcirculation, angiospasm caused by release of vasoactive substances (e.g. adnephrin, serotonin, Thromboxane A2, etc.), vascular blockage during reperfusion stage, and lasting increased intracranial pressure. Therefore, cerebral hemorrhage and cerebral ischemia are inseparable.

Dysfunction of blood circulation plays an important role in the damaging process of strokes, which exists even in hemorrhagic stroke.

Thus, improving cerebral blood circulation is the focal point of the treatment. Removing blood stasis and promoting blood flow have ideal therapeutic effects in both cerebral infarction and cerebral hemorrhage, and usually, the earlier the treatment is, the better the outcome will be. So, early treatment to improve cerebral blood circulation can decrease the risk of vascular dementia. In clinical treatment, compound Danshen preparations have the functions of promoting blood flow, removing blood stasis, improving circulation, opening compensatory circulation, increasing capillary networks, and decreasing vascular pressure at the bleeding location or infarction site to promote the absorption of the hematoma. Meanwhile, the applications are safe. Therefore, in grass-roots hospitals without CT instruments and have difficulties in the discrimination between ischemic and hemorrhagic cerebral accidents, Danshen can be used as a treatment with fewer side effects and important practical value [10–12].

5.1.6 Decreasing Intracranial Pressure and Promoting the Absorption of Cephalophyma

The functional recovery after cerebral hemorrhage is not only related to the bleeding site but is also more dependent on the rate of absorption of the hematoma. Treating cerebral hemorrhage with compound Danshen preparations can significantly accelerate the rates of reduction in edematous areas surrounding hematomas and the absorption of hematomas [13]. In treatments with Danshen (24 g/day, intravenous injection) and amidocaproic acid (12 g/day, intravenous injection), it was found that Danshen could significantly promote the absorption of hematomas and the effect was better than that of amidocaproic acid [14].

Danshen is an oxygen free radical scavenger. Its sodium sulfonate and Tanshinol can decrease the content of lipid peroxides, increase the activities of erythrocuprein and glutathione peroxidase, and stabilize cellular membranes [15]. In an experiment to detect free radicals with magnetic resonance, Yang Chen [16] confirmed that compound Danshen preparations can efficiently clear the oxygen free radicals produced during reperfusion, and the effect was similar to that of erythrocuprein. Danshen can significantly inhibit the production of Lipid peroxides (LPO) to efficiently prevent the increase of cerebral vasopermeability and generation of cerebral edema, relieve angiospasm, improve cerebral circulation and promote the absorption of hematomas. Tanshinone II_A inhibits the interaction between intracellular lipid peroxidase and DNA, and clears lipid-derived free radicals to block the chain reaction of lipid peroxidation and inhibit the production of DNA affixtures to reduce their toxicity [17]. It was also found that the active component 764-3 in Danshen can reduce the production of malonaldehyde, decrease the permeability of cerebral vessels to macromolecules, and inhibit the formation of cerebral edema after cerebral hemorrhage in rats. The mechanism of these actions might be related to its antioxidative effect [18]. Pharmacological studies indicated that the main liposoluble component in Danshen, tanshinone, can inhibit the chemotaxis of leukocytes and transmigration of neutrophilic granulocytes, and weaken the activities of granulocytes and the functions of phagocytes in the ischemic site in rats to reduce leukocytic infiltration and damage due to lipid peroxidation in the ischemic site and decrease the scope of cerebral infarction [19].

5.1.7 Improving Hemorheological Characteristics

The bleeding in cerebral hemorrhage is usually not caused by coagulation disorders, so the application of hemostasis methods to treat this disease could cause the extravascular blood to clot and retain in the brain, prolong the compression of the hematoma on brain tissue. Therefore, the treatment should aim to promote blood flow and remove blood stasis so that the hematoma can be absorbed, microcirculation improved, cerebral blood flow increased, vascular endothelial cells protected, the permeability of blood vessels reduced, and cerebral edema alleviated. Then, increased therapeutic effect, shortened disease course, decreased disability rate, and improved prognosis can finally be achieved. Modern hemorheological inspection indicated that in the hemorrhagic strokes of both low hematocrit type and normal hematocrit type, the plasma viscosity, erythrocyte electrophoretic time, and fibrinogen and K value in erythrosedimentation formulas all increased significantly, and they are powerful evidence of the presence of blood stasis in hemorrhagic stroke. In patients with cerebrovascular diseases and increased blood viscosity, Danshen can shorten erythrocyte electrophoretic time, significantly decrease blood viscosity and high and low shear rates, improve hemorheological and hemodynamical states, regulate vascular pressure and resistance, and promote microcirculation, which is beneficial to the oxygen supply to tissues and functional recovery.

Clinical studies also suggest that in the acute stage of cerebral hemorrhage, the blood is in a hypercoagulable state. Danshen, as one of the Chinese crude drugs with properties of promoting blood flow and removing blood stasis, has the functions of anticoagulation, anti-thrombogenesis, anti-free radicals, etc. Therefore, it is considered that the application of Danshen in the acute stage for patients with cerebral hemorrhage is rational and also consistent with the following standpoints: *the blood circulating out of vessels is stagnated blood* and the treatment should firstly consider *removing blood stasis*; meanwhile, clinical applications have obtained satisfactory effects [20–22].

Ninety-four cases of patients with intracerebral hemorrhage diagnosed by CT or MRI scans were treated with conventional therapy (dehydration, decreasing intracranial pressure, anti-infection, operation on patients with indications for surgery, and prevention of complications) or compound Danshen preparations, and the results showed that hemorheological indexes were all improved after 15 days of treatment. Compared with those in the conventional treatment group, each index in the Danshen group was significantly better. Also, by the detection of intracranial pressure after cerebral hemorrhage, it was found that Danshen can shorten the duration of intracranial hypertension [23].

5.1.8 The Timing of Danshen Treatment for Cerebral Hemorrhage

There have been many clinical reports about the timing of Danshen application in patients with cerebral hemorrhage. For example, in treatment during the stable stage of cerebral hemorrhage, Zhu Zuofeng [24] used deer velvet as the chief drug and Danshen as the deputy drug in 30 patients when their conditions were stabilized, or 2 weeks after onset. His clinically significant efficiency was 83 %. In treatment of patients with cerebral hemorrhage, there is a report showing that the effect of early application of compound Danshen preparation through intravenous injection was better than that at 10 days after onset [15]. In 46 cases of patients with cerebral hemorrhage complicated by perforation of the cerebral ventricle, which was detected by CT scans, intravenous injection of 20 ml of compound Danshen preparation was started on the day of onset, lasting for 14–42 day. CT scans were conducted on days 14, 28, and 48, and no re-hemorrhagia was found [26]. The therapeutic effect of *ultra-early* application of compound Danshen injection to treat cerebral hemorrhage was better than that in the *subacute stage* [27]; the application of compound Danshen injection at 6 h after onset also obtained satisfactory therapeutic effects [28]. The case of re-hemorrhagia and hematoma expansion was not observed in treatment with the drug promoting blood flow and removing blood stasis applied 3 h after onset [29]. However, studies have proved that there is continuous bleeding in early cerebral hemorrhage, especially within the first several hours after onset. Therefore, it is considered that application after 24 h from the onset of cerebral hemorrhage might be safer [30].

5.2 The Effect of Danshen on Learning and Memory Abilities

Yuehua Wang and Guanhua Du

Senile dementia is a syndrome characterized by cognitive disorder. Its major pathological changes include senile plaques, neurofibrillary tangles, and neuron loss [31]. In TCM, this syndrome is also called *dementia, short memory, silent aphrenia*, etc. It is generally divided into three categories: (1) Senile primary degenerative dementia, which is named Alzheimer's disease (AD); (2) Vascular dementia (VD); (3) Mixed dementia. Among them, AD is the most common and its morbidity increases with age; therefore, this disease is becoming an important health issue concerning everyone with the aging of the society. Foreign data indicated that AD had become the fourth leading cause of death following heart disease, tumor, and stroke in Western countries. AD has become a severe health problem and serious social problem, and the whole world is increasing research efforts [32]. VD is a kind of chronic brain syndrome caused by brain tissue injury due to a series of cerebrovascular factors, and patients have significant disorders in the abilities of memory, calculation, thinking, orientation, judgment, etc. The course of the disease shows stepped progress and morbidity is high [33]. For many years, the treatment of VD using compound Danshen preparations as the main drug has acquired satisfactory results.

Existing drugs for the treatment of AD have some therapeutic effects on the early and middle stages of AD patients, but hardly any drugs have therapeutic effects on advanced AD. The currently used drugs in Western medicine are mainly cholinesterase inhibitors which inhibit the degradation of acetylcholine, increasing the content of acetylcholine in the brain in order to improve the cognitive function of the patients [34]. The drug treatment of VD emphasizes the dilation of cerebral vessels and regulation of cerebral metabolism. Through clinical observation, the determination of treatment based on pathogenesis, obtained through differentiation of symptoms and signs in TCM, has already made great breakthroughs in the diagnosis and treatment of AD. The theory of TCM considers that its pathological mechanisms include weakness of the brain and spinal cord, disorder of vital activity, insufficiency of vital energy and blood, disorder of mind, stasis of heart and blood, dementia, stagnation of phlegm in the upper orifices, and clouded and indistinct mind [34].

5.2.1 The Effect of Danshen on Promoting the Ability of Learning and Memory

The shuttle-box test was applied to mice to observe the effect of Danshen on their learning abilities and memory, as well as the improving effect of Danshen on learning and memory disorders in mice after receiving electroconvulsive shocks, in order to further study its mechanism of action and prospects of application. The experimental results show that Danshen can significantly increase learning ability and memory in normal mice, and it also antagonizes learning and memory disorders caused by electroconvulsive shock [35].

Learning and memory are higher nervous activities of the central Nervous system and are also important indexes for measuring intellectual development. Hypoxic-ischemic brain damage (HIBD) in neonates usually results in intellectual development disorders. In neonatal rat models of HIBD, Compound Danshen injection was injected intraperitoneally at a dose of 15 g/kg, and the maze experiment was used to observe changes in long-term memory and learning ability in order to study the effect of Danshen on long-term intelligence. The experimental results showed (see Table 5.1) that the times to meet the standard in the Danshen treatment group were significantly shorter than those in the model group; meanwhile, accuracy in the treatment group was also significantly higher than in the model group, which suggests that the rats in the HIBD group presented learning and memory disorders and that treatment with Danshen significantly improved the disorders. Morphological and histopathological

Table 5.1 Changes in learning and memory after HIBD in rats ($\bar{X} \pm s$)

Group	Times to meet standard		t	Accuracy (%)		t
	$d1$	$d2$		$d1$	$d2$	
Control	46.00 ± 5.16	30.00 ± 6.66	9.81	75.30 ± 3.31	81.50 ± 2.79	4.18
HIBD	94.00 ± 8.43	85.00 ± 8.90	4.74	62.40 ± 2.22	62.70 ± 2.45	0.52
Danshen	50.00 ± 8.16	34.00 ± 5.16	7.23	76.10 ± 4.72	80.76 ± 2.78	3.87
F	78.05	81.92		55.84	74.14	
P	<0.05	<0.05		<0.05	<0.05	

studies showed that the rats in the Danshen treatment group had only slight cortical atrophy on the ligated side but no significant pathological changes. However, the rats in the HIBD group had cortical atrophy on the left side, liquefaction, and cavitation, as well as degeneration and necrosis of neurocytes, softening foci, colloid nodules and calcified foci in histopathological observations. There is currently no specific treatment for severe HIBD, and finding an effective drug treatment is always the hot spot of scientific research. Investigations showed that Danshen has the functions of antagonizing calcium ions, clearing free radicals, decreasing blood viscosity, reducing cerebral edema after ischemia and hypoxia, increasing ATP content in the brain, and regulating the levels of monoamine neurotransmitter and excitatory amino acids after cerebral hemorrhage. These functions are the basis of Danshen's protective effect against HIBD.

Infarction in the left temporal lobe cortex of rats was selectively induced by stereotaxic photochemical induction to observe the effect of Danshen on the improvement of spatial cognition disorder [37]. The Morris water maze test was applied in the ethological experiment to detect behavioral changes in rats. Results showed that the response time and the time it took to find the target for rats in the Danshen group were shortened significantly, and they used normal cognitive strategies to a greater extent and transferred from a random style to a straight-line style. These results suggest that Danshen can significantly improve the spatial cognition disorder process after temporal infarction in rats.

5.2.2 The Effect of Tanshinone on Alzheimer's-Like Disease in Rats

It has been shown that tanshinone II_A can improve disorders in learning and memory caused by scopolamine in mice, and antagonize aging [38]. The micro-injection of amyloid β-peptide ($A\beta_{1-40}$) into the dorsal side of the dentate gyrus in rats was conducted to establish an AD-like animal model with memory disorder. The changes in cholinergic fibers in each region of the hippocampus in the rats were observed through histochemical methods, and the expressions of inducible nitric oxide synthase (iNOS) and neuronal nitric oxide synthase were measured through immunohistochemical method. The results showed that the administration of tanshinone at 50 mg/kg for 10 days can decrease the area of acetylcholinesterase (AChE) inhibitor-positive fibers caused by $A\beta_{1-40}$ and improve the expression of NOS in the hippocampus, which suggests that the mechanism of action of tanshinone to treat AD rats might be related to the protective effect on the intracephalic cholinergic system and regulation of NOS expression (see Table 5.2).

The micro-injection of amyloid β-peptide ($A\beta_{1-40}$) into the dorsal cell band of the dentate gyrus in rats was conducted to establish an intended human AD-like animal model. The drugs (50, 100 mg/kg) were administered by intragastric administration in the AD-like rats, and the passive avoidance test in light-dark box and shuttle-box test were applied to study the improving effect of tanshinone on learning and memory disorders in AD-like rats [39]. The results showed that 14 days after injection of $A\beta_{1-40}$ (20 μg) in the hippocampus, the rats in the model group displayed significant learning and memory disorders. However, the administration of tanshinone at 50 or 100 mg/kg could prolong the latent period of passive avoidance to different extents; compared with the model group, the times of active avoidances increased significantly for the rats in the 50 mg/kg group and the times of escape failure also decreased, which suggests that tanshinone has significant improving effects on learning and memory disorders induced by $A\beta_{1-40}$ in rats, and 50 mg/kg of tanshinone has the best efficacy (see Table 5.3).

In some other studies, condensed amyloid β-peptide was micro-injected into the hippocampus of rats to establish the animal model. The passive avoidance test in light-dark box and shuttle-box test were applied to test learning and memory, and HE staining, gentian violet staining, and Nissel staining were used to observe the pathological tissue [40] in order to study the effect of tanshinone on learning and memory disorders in rats. Through the examination of the passive

Table 5.2 Effects of tanshinone on AChE and NOS expression in hippocampus of the rats treated with $A\beta_{1-40}$ ($\bar{X} \pm s$)

Group (*n* = 10)	Area of AChE-positive fibers		nNOS-positive neurons		iNOS-positive neurons	
	Optical density of fibers	Area of fibers (%)	Number of cells	Optical density	Number of cells	Optical density
Control group						
CA1	0.45 ± 0.08	19.28 ± 5.62	35.26 ± 9.24	0.38 ± 0.07	10.98 ± 5.59	0.20 ± 0.03
CA2~3	0.46 ± 0.07	23.82 ± 6.41	67.76 ± 12.18	0.41 ± 0.08	8.58 ± 5.05	0.18 ± 0.03
CA4	0.48 ± 0.08	27.06 ± 6.71	23.84 ± 7.06	0.37 ± 0.06	15.22 ± 6.20	0.23 ± 0.04
$A\beta_{1-40}$ Model group						
CA1	0.36 ± 0.05***	13.04 ± 4.49***	21.44 ± 6.88***	0.28 ± 0.04***	139.58 ± 17.67***	0.30 ± 0.05***
CA2~3	0.37 ± 0.05***	16.52 ± 4.95***	46.48 ± 9.72***	0.32 ± 0.05***	127.32 ± 17.03***	0.29 ± 0.06***
CA4	0.38 ± 0.06***	10.68 ± 4.18***	13.68 ± 4.27***	0.25 ± 0.04***	181.68 ± 25.52***	0.33 ± 0.07***
Tanshinone treatment group						
CA1	0.40 ± 0.06**	16.44 ± 4.94**	27.54 ± 8.29**	0.32 ± 0.06**	98.92 ± 13.08**	0.26 ± 0.04**
CA2~3	0.41 ± 0.06**	20.48 ± 5.98**	54.32 ± 11.59**	0.37 ± 0.06**	90.10 ± 11.80**	0.25 ± 0.05**
CA4	0.42 ± 0.07*	13.76 ± 4.84*	17.22 ± 6.92*	0.28 ± 0.05*	133.14 ± 17.34**	0.29 ± 0.06**

Compared with the Model group: * $P < 0.05$, ** $P < 0.01$
Compared with the Control group: *** $P < 0.01$

Table 5.3 Effects of Tanshinone on light-dark box passive avoidance response and shuttle-box test in AD-like rats ($\bar{X} \pm s$)

Group	Latent period of passive avoidance (s)	24 h After the shuttle-box training		48 h After the shuttle-box training	
		Times of active avoidances	Times of escape failure	Times of active avoidances	Times of escape failure
Model group	112 ± 39	1.36 ± 0.81	9.43 ± 4.13	7.66 ± 3.18	2.62 ± 1.01
Vector Control group	21 ± 11*	4.83 ± 2.25*	2.76 ± 1.02*	1.41 ± 0.94*	9.45 ± 4.09*
Danshen (50 mg/kg) group	85 ± 31**	3.65 ± 2.06	4.23 ± 1.95	5.43 ± 2.61**	3.22 ± 1.29**
Danshen (100 mg/kg) group	65 ± 27*	3.32 ± 1.83	5.39 ± 2.17	4.52 ± 2.20	4.88 ± 2.02

Compared with the Model group: * $P < 0.01$, ** $P < 0.05$

avoidance test in light-dark box, it was found that tanshinone at 50 and 100 mg/kg could prolong the latent period of entering the hole to different extents (see Fig. 5.1); the results of shuttle-box test showed that tanshinone at 50 mg/kg can significantly increase the times of active avoidances and reduce the times of escape failure. However, there was no statistical significance in the other treatment groups (see Tables 5.4 and 5.5), which suggests that tanshinone at 50 mg/kg has the best efficacy in improving learning and memory in rats. Morphological studies discovered that the extent of injured hippocampal neurons near the pin hole in the model group was significantly larger than that in the control group. Compared with the treatment group and model group, different dosages of tanshinone can reduce the amount of injured neurons to different extents (see Figs. 5.2, 5.3, 5.4 and 5.5). These results indicated that, for the protective effect of tanshinone on the improvement of learning and memory in rats, it should have other pathways besides protecting the hippocampal neurons. Further studies showed that tanshinone also protects hippocampal cholinergic fibers, indicated by the improvement in learning and memory, reduction of the hyperplasy of hippocampal colloid cells, protection of hippocampal neurons, and decreasing Aβ accumulation. These results suggest that tanshinone antagonizes the neurotoxicity of $A\beta_{1-40}$. However, the phenomenon of the effect of 100 mg/kg dosage being lower than that of 50 mg/kg dosage might be due to the adverse effect of the drug.

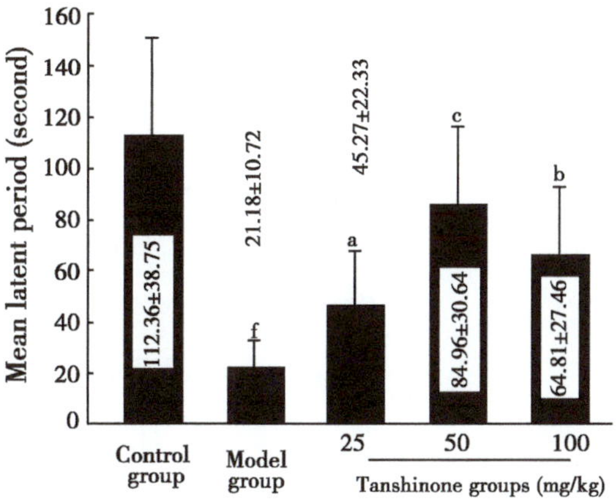

Fig. 5.1 Effect of tanshinone on the light-dark box passive avoidance response in rats. Compared with the Control group: *f* $P < 0.01$, *a* $P < 0.05$; compared with the Model group: *b* $P < 0.01$, *c* $P < 0.05$

Table 5.4 Effect of tanshinone on shuttle-box test response in rats after 24 h of training ($\bar{X} \pm s$)

Group	*n*	Dosage	Times of active avoidances	Times of escape failure
Control group	15		4.83 ± 2.25	2.76 ± 1.02
Model group	15	20 μg	1.36 ± 0.81*	9.43 ± 4.13*
Tanshinone group	15	25 mg/kg	2.47 ± 1.42	6.55 ± 2.58
	15	50 mg/kg	3.65 ± 2.06	4.23 ± 1.92
	15	100 mg/kg	3.32 ± 1.83	5.39 ± 2.17

Compared with the Control group: * $P < 0.05$

Table 5.5 Effect of tanshinone on shuttle-box test response in rats after 48 h of training ($\bar{X} \pm s$)

Grouping	*n*	Dosage	Times of active avoidances	Times of escape failure
Control group	15		7.66 ± 3.18	2.76 ± 1.02
Model group	15	20 μg	1.41 ± 0.94*	9.45 ± 4.09*
Tanshinone group	15	25 mg/kg	3.25 ± 1.78	6.09 ± 2.47
	15	50 mg/kg	5.43 ± 2.62**	3.22 ± 1.29**
	15	100 mg/kg	4.52 ± 2.20	4.88 ± 2.02

Compared with the control group: * $P < 0.05$
Compared with the model group: ** $P < 0.05$

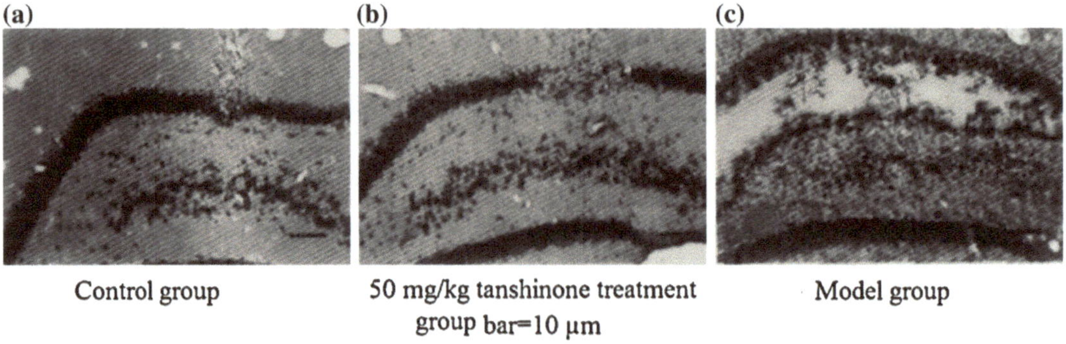

Fig. 5.2 Effect of tanshinone on hippocampal colloid cells in rats treated with $A\beta_{1-40}$ (HE staining)

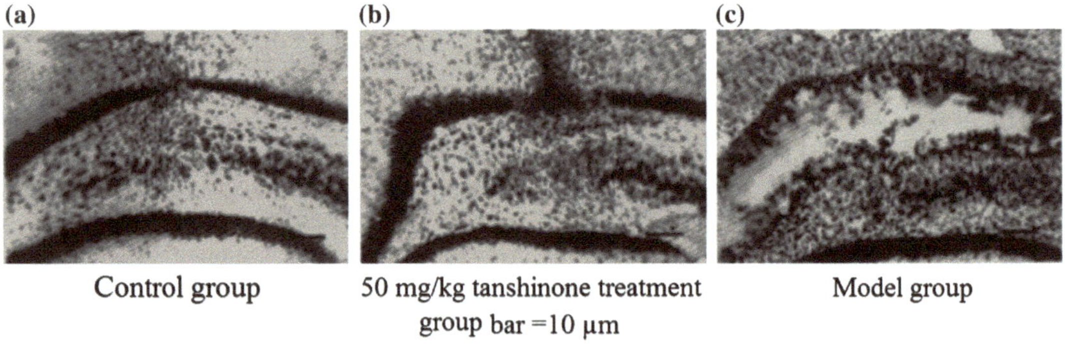

Fig. 5.3 Effect of tanshinone on hippocampal colloid cells in rats treated with $A\beta_{1-40}$ (nissel staining)

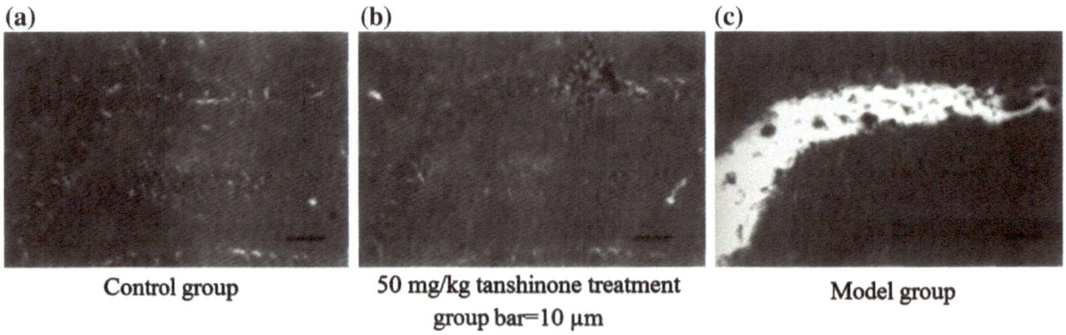

Fig. 5.4 Effect of tanshinone on Aβ accumulation in the hippocampus in rats treated by $A\beta_{1-40}$ (crystal violet staining)

The effect of tanshinone on the expressions of interleukin 1β (IL-1β) and interleukin-6 (IL-6) mRNAs in AD rats was further observed [41] to investigate the molecular mechanism of tanshinone's antagonizing effect on the neurotoxicity of $A\beta_{1-40}$. The results showed that learning abilities and memory decreased significantly in ethological examinations 14 days after the **injection** of $A\beta_{1-40}$ into the hippocampus (see Tables 5.6 and 5.7). Meanwhile, the expressions of both IL-1β and IL-6 mRNAs in the hippocampus were significantly higher than those in the control group (see Figs. 5.6, 5.7 and 5.8). These results suggest that tanshinone can significantly inhibit the decrease in

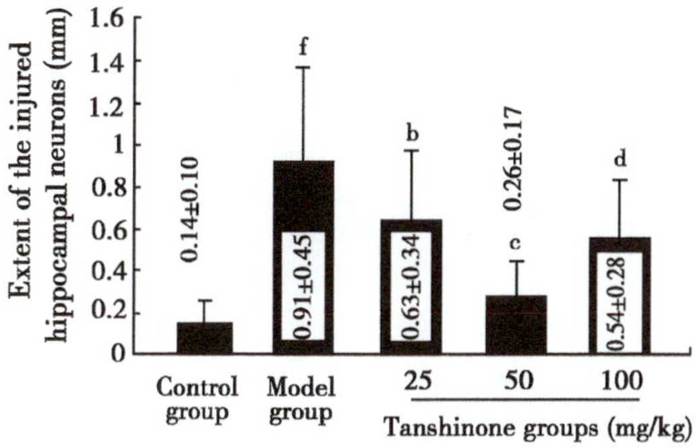

Fig. 5.5 Effect of tanshinone on the extent of injured hippocampal neurons in rats treated with $A\beta_{1-40}$. Compared with the Control group: *f* $P < 0.01$; compared with the model group: *b* $P < 0.05$, *c* $P < 0.01$

Table 5.6 Effect of tanshinone on the shuttle-box test response in rats after 24 h of training ($\bar{X} \pm s$)

Group	Times of active avoidances	Times of escape failure
Control	4.83 ± 2.25	2.76 ± 1.02
Model	1.36 ± 0.81*	9.43 ± 4.13*
Treatment	3.65 ± 2.06	4.23 ± 1.95

Compared with the Control group: * $P < 0.05$, $n = 15$

Table 5.7 Effect of tanshinone on shuttle-box test response in rats after 48 h of training ($\bar{X} \pm s$)

Grouping	Times of active avoidances	Times of escape failure
Control	7.66 ± 3.18	2.62 ± 1.01
Model	1.41 ± 0.94*	9.45 ± 4.09*
Treatment	5.43 ± 2.61**	3.22 ± 1.29**

Compared with the Control group: * $P < 0.05$
Compared with the Model group: ** $P < 0.05$, $n = 15$

learning ability and memory and the above-mentioned pathological changes in rats.

Recently, it has been reported that anti-inflammatory agents or immune depressants can decrease AD morbidity or delay the appearance of AD symptoms, which implies that there may be inflammatory injuries in the brains of AD patients. Studies have indicated that the generation of proinflammatory cytokines and complement proteins, as well as the proliferation of microglia cells, is the cause of degenerative changes in neurons, rather than just the response to the process of degenerative changes. For example, brains of transgenic mice overexpressing IL-6 showed significant degenerative changes in neurons. Studies conducted in the transgenic mouse model of β amyloid protein precursor (βAPP) indicated that Aβ might continuously activate the reparation process of inflammation and change the acute response under normal conditions to chronic inflammatory injuries. The results of in vitro culture suggest that Aβ can activate microglia cells to produce and release some cytokines, including IL-1β, IL-6, etc. However, IL-1β has an autocrine effect on microglia cells, which can not only promote the proliferation of microglia cells in neuron cell culture, but also can increase the expressions of IL-6, tumor necrosis factor (TNF-α) and IL-1β in microglia cells. Moreover, IL-1β also has paracrine effects which can activate astrocytes to up-regulate the expressions of cytokines [42].

In this experiment, it was discovered by using RT-PCR that the expressions of proinflammatory cytokines IL-1β and IL-6 mRNA in the

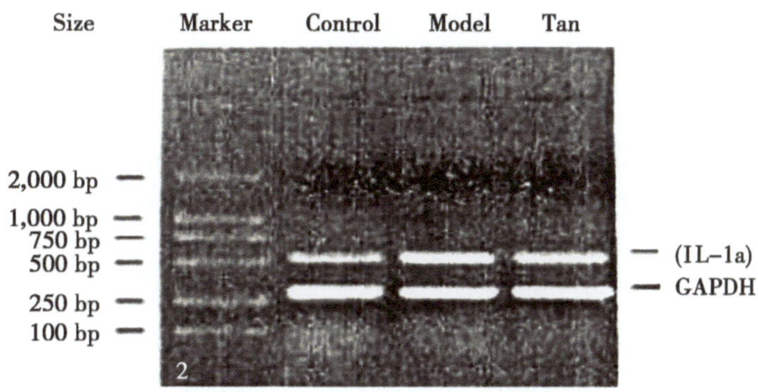

Fig. 5.6 Electrophoresis of IL-1β RT-PCR products

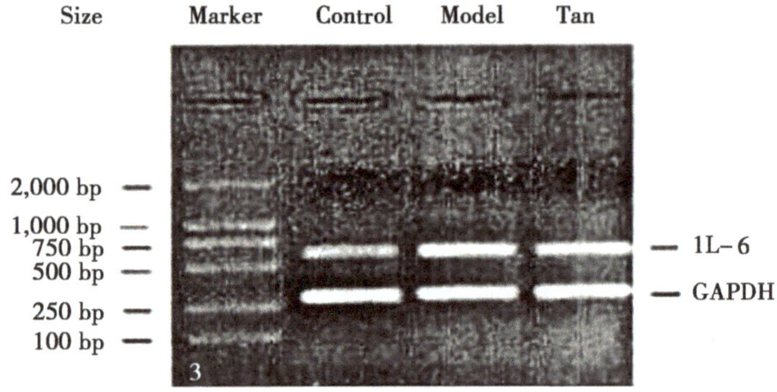

Fig. 5.7 Electrophoresis of IL-6 RT-PCR products

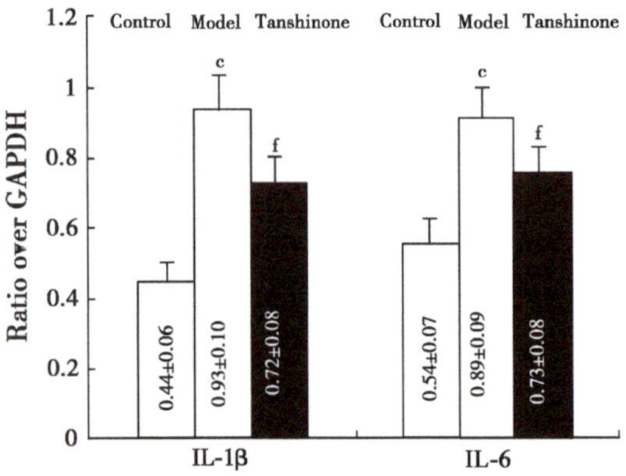

Fig. 5.8 Effect of tanshinone on the expressions of IL-1β and IL-6 mRNA in the hippocampus in rats

hippocampus of rats injected with Aβ in the hippocampus were significantly higher than those in the control group, which indicated that the injection of Aβ caused inflammatory reactions in the hippocampus and that the strong reaction might damage neurons, causing rats to show significant decreases in learning and memory. However, treatment with tanshinone significantly inhibited the expressions of the proinflammatory cytokines in the brains of rats. Meanwhile, the learning and memory of the rats in the treatment group improved significantly compared to those in the model group. Therefore, the effect of tanshinone on improving learning and memory in rats might be related to its function of alleviating inflammatory reactions in the brain.

5.2.3 Effect of Compound Danshen Preparations on Learning and Memory in Dementia Rats

Compound Danshen injection contains mainly Danshen and rosewood. Danshen can relieve cerebral edema, inhibit thrombogenesis, increase blood flow in coronary arteries, and improve heart function, so it is extensively used in the clinical treatment of cerebrovascular diseases. Danshen also has the functions of significantly relieving the disorder of monoamine neurotransmitters after cerebral hemorrhage, regulating the contents of neuropeptides and excitatory amino acids, eliminating or decreasing oxygen free radicals, improving cerebral microcirculation, etc.

Compound Danshen Tablet (CDT) is composed of three Chinese crude drugs: Danshen, Notoginseng and borneol. The formula is unique and its indications include coronary heart disease and angina. In the recent years, the effect of Compound Danshen Tablet on learning and memory disorders in experimental AD rats has been further studied. The following is the method of establishing animal models: Rats receive one intracerebroventricular injection of D-galactose 6 days after intraperitoneal injection of D-galactose. The effect of CDT was observed. Huperzine A was used as the positive control. The results showed that the latent period to enter the dark chamber was prolonged significantly for rats in the groups treated with

Table 5.8 Effect of compound Danshen tablet on memory disorder in senile dementia rats ($\bar{X} \pm s$)

Group	*n*	Dosage (g/kg)	First test			Test after 1 week		
			Latent period (s)	Times of entrances within 5 min	Rats entering dark chamber	Latent period (s)	Times of entrances within 5 min	Rats entering dark chamber
Operation control	9	–	7.80 ± 1.75	1.50 ± 0.53**	9	219.6 ± 130.9*	0.50 ± 0.84*	3
Model	9	–	9.78 ± 3.92	3.78 ± 2.16	9	59.9 ± 92.6	3.33 ± 3.04	8
Huperzine A	8	2.3×10^{-5}	10.30 ± 5.09	2.86 ± 1.07	8	184.4 ± 148.1	0.71 ± 1.11	3
Compound Danshen tablet	8	0.315	9.11 ± 3.30	2.10 ± 0.99*	8	232.3 ± 125.5**	0.88 ± 1.35*	3
	8	0.630	11.25 ± 5.18	2.25 ± 0.89	8	259.7 ± 76.8**	0.42 ± 0.79*	2

Compared with the Model group: * $P < 0.05$, ** $P < 0.01$

large or small doses of CDT, compared with the model group; the time to enter the dark chamber within 5 min was also significantly less than in the model group. The results suggest that CDT has significant beneficial effects on mid-term memory in senile dementia rats [43] (see Table 5.8).

Animals with learning and memory acquirement disorder and memory reappearance disorder, induced by hyoscine and ethanol, respectively, were tested with the maze method and diving platform method to compare the effects of Compound Danshen injection (CDI) and piracetam on learning and memory in rats and mice [44]. The results showed that both CDI and piracetam improved the disorders. Compared to the model group, the effect of CDI was slightly better than that of piracetam, which suggests that CDI can ameliorate learning and memory disorders in animals (see Tables 5.9, 5.10 and 5.11).

Vascular dementia (VD) is a chronic brain syndrome caused by brain tissue damage from a series of cerebrovascular factors. The patients show disorders of memory, calculation, thinking, orientation, judgment, etc. Compound Danshen preparation is composed of Danshen, desert cistanche, mulberry, acorus, etc., which is boiled to make a water decoction. Both carotid arteries were occluded with noninvasive bulldog clamps and nitroprusside sodium (2.5 mg/kg) was injected intraperitoneally to establish an animal model of vascular dementia, in order to investigate the effect of compound Danshen preparation on learning and memory in vascular dementia

Table 5.9 Effect of compound Danshen injection on learning and memory in rats with memory acquirement disorder ($\bar{X} \pm s$)

Group	Number of cases	Times of correct responses during training	Times of correct responses during test
Control	9	6.2 ± 2.8**	6.8 ± 2.1**
Model	9	4.0 ± 1.9*	4.6 ± 1.9*
Piracetam	9	5.4 ± 1.4	6.9 ± 1.5**
Compound Danshen injection	9	5.9 ± 1.7**	7.9 ± 1.1**

Compared with the control group, respectively: * $P < 0.05$
Compared with the model group, respectively: ** $P < 0.05$

Table 5.10 Effect of compound Danshen injection on memory disorder induced by ethanol in rats ($\bar{X} \pm s$)

Group	Number of cases	Times of correct responses during training	Times of correct responses during test
Control	10	5.8 ± 1.7**	6.3 ± 1.8**
Model	10	4.0 ± 1.9*	4.6 ± 1.5*
Piracetam	10	5.6 ± 1.7	6.4 ± 2.1
Compound Danshen **injection**	10	6.1 ± 2.0**	6.7 ± 1.8**

Compared with the control group, respectively: * $P < 0.05$
Compared with the model group, respectively: ** $P < 0.05$

Table 5.11 Effect of compound Danshen injection on learning and memory in mice with memory acquirement disorder ($\bar{X} \pm s$)

Group	Number of cases	Times of correct responses during training			Times of correct responses during test	
		Times of errors within 5 min	Electrical injury latency (s)	Latent period (s)	Times of errors within 5 min	Electrical injury latency (s)
Control	13	3.4 ± 2.0****	91.6 ± 55.5**	117.7 ± 94.5****	2.4 ± 1.35	48.5 ± 34.6**
Model	14	8.1 ± 5.1**	157.1 ± 88.9*	12.8 ± 10.8**	7.4 ± 4.0**	91.1 ± 64.2*
Piracetam	14	3.6 ± 2.5****	105.5 ± 77.2	108.3 ± 85.9****	2.6 ± 1.8****	76.0 ± 35.9
Compound Danshen injection	14	3.9 ± 2.9**	96.4 ± 67.2	86.9 ± 91.2****	1.8 ± 1.1****	43.0 ± 29.5**

Compared with the control group, respectively: * $P < 0.05$, ** $P < 0.01$
Compared with the model group, respectively: *** $P < 0.05$, **** $P < 0.05$

Table 5.12 Comparison of the mean times of electric injury and mean electric injury latency of the rats in each group during training stage ($\bar{X} \pm s$)

Group	Number of cases	Mean times of electric injury	Mean electric injury latency (s)
Normal	10	5.1 ± 0.9**	18.55 ± 3.25**
Model	9	7.9 ± 1.2	25.54 ± 5.21
Western medicine	9	7.2 ± 1.1	21.54 ± 4.30*
Compound Danshen large dosage	10	6.8 ± 1.5**	20.37 ± 2.57**
Compound Danshen small dosage	9	7.5 ± 1.7	21.60 ± 4.64*

Compared with the model group: * $P < 0.05$, ** $P < 0.01$

Table 5.13 Comparison of the mean times of electric injury of the rats in each group ($\bar{X} \pm s$)

Group	Number of cases	Before medication	Day 7	Day 34	Day 35
Normal	10	6.8 ± 1.1*	5.7 ± 1.0**	5.6 ± 1.1**	4.6 ± 0.9
Model	9	9.5 ± 1.2	8.5 ± 1.3	9.1 ± 1.4	7.8 ± 1.3
Western medicine	9	8.9 ± 0.5	8.7 ± 1.4	7.9 ± 1.3	6.5 ± 1.4*
Compound Danshen large dosage	10	9.6 ± 1.3	7.0 ± 1.3*	8.3 ± 1.3**	5.4 ± 1.3**
Compound Danshen small dosage	9	8.7 ± 1.4	7.9 ± 1.5	8.4 ± 1.2*	6.9 ± 1.4*

Compared with the model group: * $P < 0.05$, ** $P < 0.01$

rats [45]. The results showed that compound Danshen preparation can significantly ameliorate learning and memory in vascular dementia rats (see Tables 5.12, 5.13 and 5.14).

Apoptosis is an important physiological pattern for the selective death of excessive cells in the development of the nervous system. Apoptosis of neurocytes is related to the pathological process of nervous lesions including cerebral

Table 5.14 Comparison of the mean electric injury latency of the rats in each group ($\bar{X} \pm s$)

Group	Number of cases	Before medication	Day 7	Day 34	Day 35
Normal	10	$17.81 \pm 8.24^{**}$	$15.37 \pm 5.48^{**}$	$13.02 \pm 6.13^{**}$	$11.93 \pm 5.21^{**}$
Model	9	28.75 ± 8.31	28.89 ± 6.24	24.25 ± 5.22	23.32 ± 5.63
Western medicine	9	28.93 ± 7.37	23.32 ± 6.52	$19.04 \pm 4.15^{*}$	$16.43 \pm 4.32^{*}$
Compound Danshen large dosage	10	28.97 ± 8.24	$21.73 \pm 4.56^{*}$	$15.53 \pm 5.24^{**}$	$12.42 \pm 5.53^{**}$
Compound Danshen small dosage	9	27.98 ± 6.37	23.87 ± 5.61	$18.34 \pm 5.51^{*}$	$16.92 \pm 5.27^{*}$

Compared with the model group: $^{*}P < 0.05$, $^{**}P < 0.01$

Table 5.15 The Effect of Compound Danshen on the Apoptosis of rat brain neurocytes ($\bar{X} \pm s$)

Group	Section field of vision/cells	Counting of apoptotic neurocytes/cells
Normal	20	9.2 ± 2.0
Model	20	37.1 ± 7.9
Hydergine	20	$32.5 \pm 6.5^{*}$
Compound Danshen preparation	20	$27.5 \pm 4.6^{**}$

Compared with the model group: $^{*}P < 0.05$, $^{**}P < 0.01$

hemorrhage, brain injury, etc. Therefore, the role of neurocyte apoptosis in vascular dementia has attracted increasing attraction. The apoptosis of cerebral neurocytes was detected by the TUNNEL technique to observe the effect of compound Danshen preparations. The results indicated that compound Danshen preparations can decrease the apoptosis of neurocytes (see Table 5.15).

The physiological process of learning and memory is complicated and needs sophisticated integration. Acetylcholine (Ach) is an important transmitter which exerts and maintains advanced nervous function and has a close relationship with memory, thinking, and intelligence. Some data indicate that neurotransmitters decreased significantly in the brain tissue during the process of vascular dementia. Particularly, the decrease in acetylcholine and monoamine neurotransmitters in the cerebral cortex was closely related to the decrease in intelligence in vascular dementia. Large doses of compound Danshen preparations can significantly increase the content of Ach in brain tissue and inhibit the activity of AchE, which suggests that compound Danshen preparations can improve the function of the cholinergic system and regulate the physiological metabolism of Ach in brain tissue to improve memory. The contents of monoamine neurotransmitter 5-HT in the hippocampus and hypothalamus in dementia rats decreased with increase in cholinesterase activity, which means that the severity of dementia is consistent with the decrease of monoamine neurotransmitter 5-HT in the hippocampus and hypothalamus. Some researchers observed the effect of compound Danshen preparation on intracephalic neurotransmitters in vascular dementia rats [47], and the experimental results suggest that compound Danshen preparations increased the content of 5-HT in the hippocampus of rats. This phenomenon was especially significant in the large dosage group, which suggest that compound Danshen preparations can significantly ameliorate learning and memory disorders in rats, which might be related to its function of increasing 5-HT levels in the hippocampus in rats (see Table 5.16).

Compound Danshen preparations have functions of nourishing and promoting blood, improving microcirculation, decreasing the oxygen consumption of myocardial cells and brain cells, increasing the activities of enzyme systems

Table 5.16 Changes in the contents of Ach and AchE in brain tissue and 5-HT in hippocampus in dementia rats of each group ($\bar{X} \pm s$)

Group	*n*	Ach (mg/g)	AchE (μmol h^{-1} g^{-1})	5-HT (μg/g)
Normal	10	214.45 ± 21.65	354.54 ± 54.62	5.47 ± 0.87
Model	9	134.74 ± 24.31**	530.12 ± 85.17**	3.54 ± 0.72**
Western medicine	9	163.30 ± 28.54*	397.27 ± 74.72**	4.55 ± 0.47**
Compound Danshen large dosage	10	161.55 ± 20.51*	456.27 ± 72.57*	5.47 ± 0.71**
Compound Danshen small dosage	9	157.30 ± 27.64	486.57 ± 91.27	4.54 ± 0.64*

Compared with the model group: * $P < 0.05$, ** $P < 0.01$

to clear free radicals, increasing the RNA activities of nerve cells in the brain, and antagonizing thrombogenesis.

5.3 Pharmacological Effects of Danshen for Treatment of Acute Ischemic Cerebrovascular Disease

Hongmei Guang, Yuehua Wang and Guanhua Du

In recent years, traditional Chinese medicines have obtained remarkable achievements in the treatment of ischemic cerebrovascular diseases, among which Danshen preparations have been used for over 30 years in the treatment of cerebrovascular diseases. As a common drug used to promote blood flow and remove blood stasis, its property is mild, and particularly, it has satisfactory therapeutic effects on a series of symptoms with long courses and slow effects caused by ischemic cerebrovascular diseases.

5.3.1 Mechanism of Action of Danshen in Ischemic Cerebrovascular Disease

Danshen and its lipid-soluble component, tanshinone, and water-soluble components, phenolic acids, have extensive pharmacological actions, the most important of which are the satisfactory effects on antagonizing thrombi and protecting the brain from injury. Ischemic brain injury is a complicated pathological process. Consistent with the diversity of neurocyte damage, as a neuro-protective agent, Danshen has extensive and diverse mechanisms such as improving the microcirculation of brain tissue, reducing damage caused by oxygen free radicals, decreasing calcium overload, antagonizing Apoptosis, inhibiting inflammatory reactions, inhibiting the release of excitatory amino acids, regulating neuroendocrine function, etc.

5.3.1.1 Improving Cerebral Microcirculation

Disorder of cerebral microcirculation is an important physiopathological process of acute ischemic cerebrovascular disease. This process mainly includes platelet and leucocyte activation, erythrocyte aggregation, thrombin production and activation, fibrin thrombogenesis, hemorheological abnormality, functional disorders of vascular endothelial cells and smooth muscle cells, etc. In ischemic brain tissue, Danshen improves cerebral microcirculation mainly through the following mechanisms:

Improving Hemorheology

The main pathological characteristics of hemorheological abnormality include an increase in blood viscosity caused by decrease of the surface charge in erythrocytes, increase of aggregation, acceleration of sedimentation rate, decrease of deformability, increase of fibrinogen, and increase of plasma viscosity. Abnormal hemorheology is not only the most definite risk factor for cerebrovascular disease, but also one of the important factors aggravating cerebral hemorrhage and hypoxia. To prevent and treat cerebrovascular diseases, it is essential to promote blood flow, remove blood stasis, and improve blood viscosity. Danshen is a traditional drug used to promote blood flow and remove blood stasis, and has versatile biological effects. Many basic and clinical experiments have been conducted to study its pharmacodynamic actions and related mechanisms. It is found that compound Danshen injection can significantly increase blood flow in the cerebral capillary vessels of mice with disordered cerebral microcirculation, through its functions of increasing erythrocyte deformability and improving blood viscosity. In acute cerebral hemorrhage rat models, compound Danshen injection can significantly decrease the cerebral index and water content of brain tissue. Meanwhile, it can also decrease the permeability of intracephalic blood capillaries and maintain the normal appearance and structures of neurocytes and brain tissue. In the cerebral hemorrhage rabbit model, the effect of Danshen on the microcirculation of the cerebral pia mater was observed, and the results indicated that Danshen can accelerate cerebral blood flow, improve the fluid state, and induce erythrocyte disaggregation to some extent to improve cerebral circulation.

In the Department of Neurology of Shanghai Huashan Hospital, Danshen has been used in the treatment of ischemic stroke since 1970. It has been discovered that blood viscosity was reduced when patients were treated with compound Danshen injection, and the erythrocyte electrophoretic time was decreased to almost a normal level compared with data from before treatment, and the results were statistically significant. Moreover, comparison between the total therapeutic effects at the end of treatment in the Danshen group and Western medicine group suggest that the effect of Danshen was even better than that of Western medicine. Hou Tingting et al. used compound Danshen injection to treat 28 patients with stroke and measured the hemorheological indexes. They found that all indexes, including apparent blood viscosity, plasma viscosity, erythrosedimentation, hematocrit, platelet adhesion rate, and fibrinogen decreased significantly after treatment, and the differences were statistically significant compared to those before treatment. It has also been discovered that early application of Danshen in the treatment of cerebral infarction can significantly improve the kinetic indices of cerebral circulation, accelerate cerebral blood flow, improve elasticity of cerebral vessels, and decrease blood viscidity and it is also confirmed that early application can increase the therapeutic efficacy and decrease disability rate. Those above-mentioned basic and clinical experimental results confirmed the positive therapeutic effect of Danshen in the treatment of cerebrovascular diseases, and its functions include increasing the surface charge of erythrocytes, promoting erythrocyte plasticity, improving erythrocyte deformability, decreasing erythrocyte aggregation, inhibiting platelet aggregation, reducing blood viscosity, and improving hemorheology, so that it can antagonize coagulation, prevent thrombogenesis, increase blood flow in the cerebral hemorrhage tissue, improve the blood supply to brain tissue, and promote cerebral metabolism.

Improving Body's Fibrinolytic and Anticoagulant Activity

The fibrinolytic system is composed of mainly plasminogen, plasminogen activator, activator-specific inhibitor, fibrinolysin, and plasmin inhibitor. For thrombogenesis in cerebral blood vessels, there is a great amount of plasminogen and other plasma proteins in the blood clot, in which the plasminogen is converted into fibrinolysin by plasminogen activator, which has the fibrinolytic activity to digest fibrin, fibrinogen, factor V, factor VIII, and factor XII. Fibrinolysis digests fibrin in the thrombus and removes sludge blood slowly to recover the blood supply of the blocked vessel. Clinical studies also discovered that there were anticoagulation disorders and damage to the fibrinolytic system in patients with acute cerebral infarction, which suggests that abnormal coagulation status is also a risk factor of cerebral infarction. Correction of abnormal coagulation status can reduce the risk factor of stroke and decrease damage caused by cerebral hemorrhage to protect the brain. Anticoagulation refers to the prevention of blood clot formation. Anticoagulation and fibrinolysis are prominent at the normal physiological state, which is important for the prevention of stroke. In the recent years, thrombolysis has been taken as the method with the most potential to treat ischemic stroke; however, the results of foreign and domestic studies showed that only intraarterial thrombolysis within 3 h, intraarterial catheter-directed thrombolysis, and supplement of percutaneous transluminal angioplasty (PTA) have certain effect, and the applied drugs are mainly tissue-type plasminogen activator (t-PA) and recombinant t-PA. It is still an effective therapeutic method to conduct thrombolysis and anticoagulation with effective drugs at the proper time, although thrombolysis is limited by multiple factors such as the therapeutic time window, medical apparatus, expense, etc., which limits its clinical application. Long-term studies of Danshen preparations have discovered that they can increase thrombolysis and anticoagulation activities of the body to effectively improve clinical symptoms and decrease the mortality and disability rate. Platelets contain multiple enzymes and bioactive substances which can be released selectively by stimulation of various factors to participate in many in vivo physiological and physiopathological processes. The increase in platelet activity plays an important role in the process of cerebral arterial thrombosis, in which damaged endothelial cells release adenosine diphosphate (ADP) to induce platelet adhesion and aggregation, and platelets release ADP and thromboxan (TXA_2) to induce platelet coagulation in the injured site and result in irreversible damage if the aggregating factors exist and increase continuously. The main pathogenesis of ischemic cerebrovascular disease is Atherosclerosis, and its pathological changes are all closely related to the functional abnormality of platelets. Studies indicated that in patients with cerebrovascular disease, the activity of platelet aggregation showed increased aggregation and adhesion, decreased threshold of aggregation, and high spontaneous aggregation. The increase of platelet aggregation products in circulation is the reason for the occurrence of ischemic cerebrovascular disease. However, the functional changes in platelets initiate ischemia. Clinical studies showed that all of the metabolites of Arachidonic acid in the plasma of patients with cerebrovascular disease increased, and platelet release reactions also increased. It has been reported that the active components in Danshen can significantly increase cAMP levels in platelets, inhibit the production of prostaglandin vasoconstrictor class materials, such as thromboxan A_2, to increase the membrane fluidity of platelets and inhibit platelet releasing, adhesion and aggregation, and antagonize thrombogenesis. In vitro experiments further demonstrated that Danshen has inhibitory effects on blood clotting at all three stages, and can rapidly transform fibrinogen into fibrin monomers and split fibrin into fibrin degradation products (FDP), which in turn inhibit platelet aggregation to antagonize coagulation. It can also inhibit the platelet aggregation induced by ADP and collagen to shorten the duration of thrombus and significantly decrease the dry weight and wet weight. The mechanism of action is related to the inhibition of actin activity in platelets to activate Mg^{2+} -ATPase and the inhibition of production of TXA_2 in platelets. The effect of compound Danshen preparations to antagonize platelet aggregation is achieved through the effects on the shifting of glycoprotein CD62P in the platelet granule membrane and thrombin-sensitive protein CD63 in lysosomal granules.

The capillary endothelial cells in cerebral vessels can produce anticoagulant activated protein and thrombomodulin (TM). TM is an important transmembrane glycoprotein with anticoagulative functions, and is produced by endothelial cells and macrophages and covers the surfaces of endothelial cells and platelets. TM is a high-affinity receptor for thrombin, which is located on the inner surface of endothelial cells, and the thrombin-TM complex has strong activating effect on protein C (PC) in circulating blood. TM can accelerate the activating effect of thrombin on PC, and affects the activation of prothrombin. Studies have found that Tanshinol can promote bovine endothelial cells to secrete plasminogen activator (PA), increase the production of Thromboxane and PGI_2, and decrease the activity of plasminogen activator inhibitor (PAI) in bovine endothelial cells. Meanwhile, it can also increase the activity of thrombomodulin in the membrane of bovine endothelial cells.

Regulating Vasoconstriction

The metabolism of arachidonic acid (AA) produces prostaglandins (PGs) with multiple biological functions, among which thromboxan A_2 (TXA_2) and prostacyclin I_2 (PGI_2) are two of the most important products. TXA_2 has the effect of contracting blood vessels, but on the contrary, PGI_2 has the effect of dilating blood vessels. In cerebral hemorrhage rat models, there are apparent TXA_2-PGI_2 balance disorders. Danshen can decrease the production of TXA_2 and increase the content of PGl_2 to regulate the TXA_2-PGl_2 equilibrium in order to regulate vasomotor function and improve blood and oxygen supplies to the brain tissue.

Endothelin (ET) is a vasoactive peptide composed of 21 amino acid residues and is produced and released by vascular endothelial cells and mainly distributed in vascular endothelial cells and neurocytes. Basic studies and clinical data indicated that ET participates in the physiopathological process of the damage caused by cerebral hemorrhage, in which the effect of ET to contract blood vessels aggravates the ischemia and hypoxia of local brain tissue; meanwhile, ET can also promote the production of vasoconstrictor substances such as prostaglandin endoperoxide H_2 (PGH_2), TXA_2, etc., accompanied with the generation of free radicals, which all aggravate brain injury. The studies found that Danshen can decrease the level of plasma Endothelin in patients with cerebral hemorrhage and increase the activity of nitric oxide to dilate blood vessels and improve blood circulation.

Calcitonin gene-related peptide (CGRP) is widely present in brain tissue with the same distribution as ET. They have biologically antagonistic effects on each other, and can antagonize sympathetic nervous excitation, dilate blood vessels, and improve cerebral hemorrhage status after cerebral infarction. Sheng Wenhua et al. observed 60 hospitalized patients with cerebral infarction. CGRP levels in the compound Danshen preparation (CDP) treatment group were significantly higher than in the non-CDP treatment group ($P < 0.01$). However, ET levels in the CDP treatment group were lower than in the non-CDP treatment group ($P < 0.05$), and there was no significant dependability between CGRP and ET in both groups. Therefore, it is considered that one of the mechanisms for CDP to treat cerebral infarction is to decrease the production and secretion of ET and promote the production and secretion of CGRP in vascular endothelial cells.

5.3.1.2 Antioxidation

The active components tanshinone, Tanshinol and Salvianolic acids in Danshen are all effective antioxidants. In cerebral hemorrhage, the production of oxygen free radicals and hydroxyl radicals increases, the activity of the endogenous antioxidant system superoxide dismutase (SOD) decreases, and the levels of reduced glutathione hormone peroxydase (GSH-PX) and catalase (CAT) decrease. Oxygen free radicals have a strong effect on lipid peroxidation, which causes damage to the cellular membrane and production of a great amount of malondialdehyde (MDA), leading to neurocyte death and aggravation of brain tissue injury. The scavenging effect of Danshen on oxygen free radicals has been confirmed by many experiments. Tanshinol can clear superoxide anions produced by xanthine-xanthinoxidase system. Tanshinone is an effective inhibitor of the interaction between intracellular lipid peroxide and DNA, and its protective effect on DNA might be through the clearance of lipid free radicals to block the chain reaction of lipid peroxidation, inhibit the formation of DNA complexes, and finally decrease the cytotoxicity of lipid free radicals. Meanwhile, tanshinone can also clear lipid free radicals produced during lipid peroxidation in the mitochondrial membrane so that the respiratory function of mitochondria will not be affected. In vitro experiments have shown that salvianolic acid A can inhibit the generation of MDA in synaptosome and mitochondria induced by $Fe2^{+}$-Cys, decrease the genesis of lipid peroxidation, reduce the damage to brain tissue caused by oxygen free radicals, effectively decrease the area of cerebral infarction, ameliorate neurologic impairment, and improve impaired memory. In the cerebral hemorrhage model established through middle cerebral artery occlusion (MCAO), the in vitro effects of total salvianolic acids on lipid peroxidation, superoxide anion, and hydroxyl radicals were detected. The experimental results indicated that total salvianolic acids can decrease the cerebral infarction area and reduce cerebral edema, and the effects were better than those of vitamin E in the same dose. Compound Danshen injection can significantly decrease the cerebral infarction area and content of MDA in serum, and increase the content of SOD in serum in acute cerebral hemorrhage rats. It was found in clinical experiments that with the basic therapies of dehydration, anti-infection, and maintenance of water-electrolyte and acid–base equilibriums, the application of compound Danshen injection in the treatment of patients with acute cerebral infarction can significantly improve the neurological function scores of patients, with an essential cure rate and obvious improvement rate of 75 %. So, the clinical symptoms of patients with cerebral infarction were ameliorated.

5.3.1.3 Protecting Mitochondria and Improving Energy Metabolism

The role of mitochondria in the damage process of cerebral hemorrhage cannot be ignored. mitochondria are the major site of energy metabolism and free radical production. The pumping of protons out of the mitochondria and electron transport in the respiratory chain are not only processes to generate energy, but also the main step in producing oxygen free radicals. In cerebral hemorrhage, intracellular ATP production is severely reduced, which affects the function of energy-dependent ion pumps. Since the transmembrane ion gradient cannot be maintained due to the functional decrease of Na^+–K^+ pumps, extracellular K^+ concentration is increased and Na^+ is retained inside cells. This results in the swelling of subcellular structural organelles, severely damaging mitochondrial function and leads to a vicious cycle. Protecting mitochondrial structures and functions from damage is an important precondition of reducing the severity of damage to ischemic brain tissue and improving its functional recovery. Studies have shown that Tanshinol has certain recovery effects on the change in mitochondrial transmembrane potential in cerebral hemorrhage and also has apparent protective effects on respiratory chain function, which is of significance to the protection of ischemic and hypoxic brain tissue. This researcher proposed for the first time that the mitochondria outer membrane has regulatory effects on the inner membrane, and he considered that the disorder of energy metabolism is the basis of ischemia and reperfusion injury and interacts with free radicals, and that these two factors are the cause and effect of each other and promote the progression of damage.

Compound Danshen injection and tanshinone can significantly prolong the survival time under atmospheric hypoxic conditions in mice. Some components in Danshen can increase not only the time of hypoxia tolerance to ordinary oxygen deficit, but also the time of hypoxia tolerance to acute and repeated oxygen deficit. Hu Hebin et al. found that, compared with mice in the control group, the time of hypoxia tolerance of mice in the Danshen treatment group prolonged significantly ($P < 0.05$), which might be due to the inhibition of the transformation of adenosine monophosphate (AMP) to inosine monophosphate (IMP), improvement of energy metabolism and adenylate metabolism under hypoxic conditions, etc. Cui Shufen et al. observed the protective effect of compound Danshen preparations on ischemic brain injury in rats and they found that Danshen could significantly decrease the water content in the brain, and the protective effect on ischemic brain injury is related to the increased activity of Na^+–K^+–ATPase in brain tissue.

5.3.1.4 Decreasing Calcium Overload and Regulating Calcium Homeostasis

As a secondary messenger, metabolic regulatory factor, and membrane stabilizer, intracellular Ca^{2+} plays an important role in normal cellular functions and also mediates cell death under multiple pathological conditions. In ischemic brain tissue, an excessive accumulation of intracellular Ca^{2+} is an important inductor of tissue damage in cerebral hemorrhage, and there is a direct correlation between the concentration of intracellular Ca^{2+} and the degree of injury to the cells.

The relationships between Ca^{2+} overload, ATP depletion which leads to ischemic injury, and oxygen free radicals which lead to lipid peroxidation are interdependent on each other, and each one reinforces the others. This kind of relationship creates a vicious cycle which eventually leads to delayed neuronal death. Studies confirmed that as a receptor-operated Ca^{2+} blocking agent, Danshen can reduce neural injury in the hippocampal CA1 region after brain injury through its functions of improving ATPase activity and inhibiting Ca^{2+} aggregation. Tanshinol can inhibit the metabolism of the phospholipids-inositol system in brain tissue, relatively decrease the production of inositol triphosphate (IP_3) in tissue with acute cerebral hemorrhage, and inhibit the release of intracellular Ca^{2+}, so that the extent of overload in neurocytes is decreased, the amplitude of intracellular Ca^{2+} and lipid peroxyl radicals is reduced, and the degree of neural injury in cerebral hemorrhage is limited. Tanshinone can significantly increase the neurocyte survival rate, decrease the level of LDH released by cells, and reduce intracellular Ca^{2+} concentration in the Ca^{2+} overload damage model. These results suggest that Danshen regulates the homeostasis of Ca^{2+} concentration through the reduction of Ca^{2+} overload, which is also one of its mechanisms to antagonize cerebral hemorrhage.

5.3.1.5 Anti-apoptosis

Apoptosis is a cell death process regulated by gene programming. The literature in recent years has reported that delayed neurocyte death occurs after cerebral hemorrhage, especially in the ischemic penumbra. However, delayed selective cell death is mainly completed through the apoptosis pathway. The occurrence of apoptosis is related to many factors including mitochondria, cytochrome c (Cyt-c), transcription factor, *Bcl*-2 protein family, Caspase, etc. Among them, *Bcl*-2 and *Bcl*-X, with anti-Apoptosis effects, affect proteins in the mitochondrial outer membrane and prevent the release of Cyt-c from mitochondria, and inhibit Apoptosis. However, *Bax*, *Bak*, *Bcl*-Xs, etc., which have Apoptosis-promoting effects, promote the opening of voltage-dependent anion channels (VDAC) to induce the release of Cyt-c from mitochondria to activate Caspase-3, the final protein to induce apoptosis, and it can also prevent the anti-apoptosis effect of *Bcl*-2. In the focal cerebral hemorrhage rat model, Qu Youzhi et al. found that x, the number of *Bcl*-2 positive cells increased in the Danshen treatment group with statistical significance, which suggests that Danshen can up-regulate the expression of *Bcl*-2 in cerebral hemorrhage to inhibit apoptosis. Wang Xiaoxia et al. experimentally proved that Danshen can up-regulate the expression of *Bcl*-2 and decrease the expression of *Bax* to inhibit neurocyte apoptosis in cerebral hemorrhage.

Transcription factors are also critical in the process of neurocyte apoptosis. Experiments have confirmed that acute and transient cerebral hemorrhage can activate the cAMP response element binding protein to form phosphorylated CREB (pCREB) and further exert extensive biological functions through the expression of downstream genes. Shen Xia et al. found that Danshen can up-regulate the expression of pCREB in the hippocampus and increase the ratio of pCREB/CREB to inhibit apoptosis in cerebral hemorrhage rats. In the focal cerebral hemorrhage rat model, Danshen can decrease the expression of cyclin D1 in neurocytes to inhibit the occurrence of neurocyte apoptosis after cerebral hemorrhage. Danshen can also inhibit apoptosis to protect neurocytes through its functions of partially inhibiting the abnormal expression of Endothelin-1 (ET-1) caused by ischemia, and down-regulating the expression of proteinase interleukin-1β converting enzyme (ICE) after cerebral hemorrhage.

5.3.1.6 Inhibiting the Release of Excitatory Amino Acids

Excitatory amino acids (EAA) include glutamic acid (Glu), aspartic acid (Asp), glycine (Gly), cysteine (Cys), etc. Under physiological conditions, EAA play important roles in the information transfer between neurons as normal neurotransmitters. When brain tissue is in an ischemic and hypoxic state, a large amount of glutamic acid is released and its reuptake mechanism is blocked. Glutamic acid accumulates in the synapse to activate postsynaptic *N*-methyl-*D*-aspartate (NMDA) receptors and α-amino-3-hydroxy-5-methyl-4-isoxazole-propionic acid (AMPA) receptors, which induces the outflow of a large amount of K^+ and inflow of Na^+ and Ca^{2+}, resulting in intracellular Ca^{2+} overload and initiating a series of pathological reactions of neural damage which leads to cell necrosis and apoptosis. Some researchers considered that the excitatory toxic effect induced by neurotransmitters is the final pathway of ischemic brain injury, and studies found that the level of EAA increased significantly in the hippocampal extracellular fluid in cerebral hemorrhage rats and treatment with Danshen could significantly decrease the level of EAA in extracellular fluid. Li Ju et al. measured the contents of Glu and Asp in the hippocampal extracellular fluid of rats and their results showed that the contents of these EAA in the Danshen group were significantly lower than those in the normal saline group ($P < 0.01$). The dynamic contents of Cys in the striatal extracellular fluid in the striate body area in the ischemia-reperfusion model of gerbils were measured and the effect of Danshen was observed. The results showed that compared with the ischemia control group, the content of Cys in the Danshen group decreased significantly, which suggests that Danshen can inhibit the release of EAA in cerebral hemorrhage. In animal experiments with hypobaric hypoxia and normobaric hypoxia, the survival time was significantly prolonged in both Danshen treatment groups, the recovery of the orthodromic population spike (OPS) of brain slices in these two treatment groups was significantly better than in the control groups, and the frequency of hypoxia injury potential (HIP) was low due to the fact that the transmission of excitation in the synapses between the collateral out of the CA3 region and hippocampal pyramidal cells in the CA1 region was accomplished by glutamic acid. Therefore, the inhibition of glutamic acid will block synaptic transmission, which also inhibits the OPS of the hippocampal pyramidal cell layer in the CA1 region induced by stimulation of the collateral. These results suggest that Danshen can inhibit the release of EAA to antagonize brain tissue damage induced by hypoxia.

5.3.1.7 Inhibiting Toxic Effect of Nitric Oxide

In ischemic brain tissue, the content of intracellular nitric oxide (NO) increases and the molecules are transferred out of the cells to participate in the early cytotoxic reaction after hypoxia. For hypoxia caused by cerebral hemorrhage, the increase in NO has both beneficial and harmful effects: (1) NO can dilate cerebral vessels, maintain cerebral blood flow, inhibit aggregation and adhesion of platelets and leucocytes, and block *N*-methyl-*D*-aspartate (NMDA) receptors to reduce damage to cells; (2) NO can bind to the iron-sulfide centers of some enzymes to deactivate them; (3) NO activates guanylate cyclase to increase the level of cGMP. cGMP has at least three functions: directly affecting ion channels and cGMP-dependent protein kinase; activating or inhibiting phosphodiesterase to change the content of cAMP; NO combines with superoxide anions to form $ONOO^-$ which breakdowns to generate OH^- with strong toxicity to produce cytotoxicity.

During hypoxia caused by cerebral hemorrhage, a great amount of excitatory neurotransmitter glutamic acid is released, which activates NMDA receptors to produce superoxide anions. Meanwhile, the inflow of Ca^{2+} activates NOS to increase the amount of NO. The superoxide anion reacts with NO to produce $ONOO^-$, which might be a new pathway of oxidative stress response when the brain is under hypoxia stress. They can be divided into two forms based on its functions: constitutive NOS (cNOS) and inducible NOS (iNOS). cNOS is located in the brain and endothelial cells, and can be activated to generate NO and exert its physiological functions under normal conditions. iNOS is located in macrophages, vascular smooth muscle cells, activated Neutrophilic granulocytes and endothelial cells, and the expression of iNOS in the cells is very low under normal conditions. However, during hypoxia-ischemic injury of brain tissue, the expression of iNOS increases and the NO produced reacts with peroxides to produce peroxide nitro products with neurotoxicity, mediates the excitation induced by NMDA receptors and glutamic acid, and results in damage to neurons. Animal experiments also indicated that the expression of NOS increased significantly and the production of NO also increased in brain tissue after cerebral hemorrhage and Danshen could partially down-regulate the abnormal expression of NOS induced by hypoxia to reduce the production of NO and decrease ischemic brain injury. Compared with the control group, the expression of iNOS in the ischemic cerebral cortex was significantly lowered in the compound Danshen preparation treatment group ($P < 0.05$, $P < 0.01$). Meanwhile, compound Danshen preparation could also reduce the symptoms of nerve injury ($P < 0.05$). In clinical experiments, when hypoxia-ischemic cerebral diseases were treated with compound Danshen injection, the level of NO in the cerebrospinal fluid decreased with the recovery of patients' conditions. However, the extent of decrease in the Danshen treatment group was larger, which suggests that inhibiting the production of NO is one the mechanisms of Danshen to treat ischemic cerebrovascular diseases.

5.3.1.8 Inhibiting Inflammatory Reaction

In the recent years, most researchers considered that inflammatory reactions were an important factor aggravating neurocyte damage caused by ischemia. Cerebral hemorrhage may cause a series of inflammatory reactions such as leucocyte activation, expression of adhesion molecules, and the actions of multiple cytokines and the mechanisms are complicated. After cerebral hemorrhage, proinflammatory cytokines produced in the ischemia area induce the expression of many adhesion molecules, which further promote leukocyte infiltration to generate inflammatory reactions and aggravate brain injuries. Interventions in certain steps of the inflammatory reactions in ischemic stroke might become the new method for treatment of cerebral infarction.

Inflammation after cerebral hemorrhage is a process of cascade reactions, in which the expression of adhesion molecules is the precondition for immunocytes to assemble, block capillaries, and migrate to the surrounding brain tissue to exert cytotoxic effects. Liu Jun et al. found that Danshen can significantly increase blood flow in the distributing area of the middle cerebral artery in the ischemic side of the brain, decrease the number of CD18/CD11b positive immunocytes in peripheral blood, and inhibit leukocyte infiltration and neurocyte necrosis in the ischemic area of the middle cerebral artery. This suggests that its mechanism of inhibiting the adhesion of leucocytes to vascular endothelial cells is related to the blocking of CD11b/CD18 in the ligand-integrin family of leukocyte adhesion molecules in peripheral blood. In cerebral hemorrhage, the expression of intercellular adhesion molecule-1 (ICAM-1) is closely related to neutrophilic granulocyte infiltration, and studies found that Danshen could inhibit the expression of ICAM-1, prevent the adhesion between vascular endothelial cells and leucocytes, and inhibit neutrophilic granulocyte infiltration. It has been also reported that Danshen can inhibit the adhesion between polymorphonuclear leucocytes (PMN) and cerebral capillary endothelial cells to reduce inflammatory reactions in the tissue with cerebral hemorrhage. Therefore, the protective effect of Danshen on neurons after cerebral hemorrhage is achieved through its functions of dual blockade of the adhesion between vascular endothelial cells and leucocytes, decreasing neutrophilic granulocyte infiltration in ischemic brain tissue, and reducing damage to neurocytes. The participation of inflammatory reactions in cerebral ischemia-reperfusion injury is a result of complex actions of multiple mechanisms, which interact with each other and form a vicious cycle. Therefore, treatment aimed directly at a certain mechanism usually cannot obtain satisfactory therapeutic results. Danshen can protect cerebral hemorrhage brain tissue by inhibiting the inflammatory reactions from numerous aspects.

5.3.1.9 Regulating Neuroendocrine Function

Increasing the Expression of Neurotrophic Factors

There are many neurotrophic factors (NTF) related to neuron regeneration in the dentate band, such as epidermal growth factor (EGF), basic fibroblast growth factor (bFGF), insulin-like growth factor (IGF), ciliary neurotrophic factor (CNTF), etc. Of these, bFGF is known for its effect on the regeneration and differentiation of precursor cells of intracerebral neurons in adult rats, and its function in the nervous system has been attracting increasing attention. In in vitro cultures of the cerebral cortex, hippocampus, hypothalamus, cerebellum, and striatum neurons, bFGF maintains survival and promotes axonal growth. In cultured hippocampal cells of rats, bFGF can increase the neurotoxicity threshold of glutamic acid and reduce Glutamate-induced intracellular calcium increase. In incubated hippocampal brain slices, bFGF can enhance the activity of MK-801 to antagonize ischemic damage to neurocytes and inhibit the enhancement of presynaptic microcurrents induced by oxygen-glucose deprivation, which indicates that bFGF can decrease the release of glutamic acid. In vivo experiments also showed that bFGF can decrease the area of infarction after cerebral hemorrhage and antagonize the neurotoxicity of NMDA. After cerebral ischemia, as the autoprotection mechanism of the brain, the expression of bFGF increases, which promotes neuronal survival and axon growth through high-affinity tyrosine kinase receptors. Kuang Peigen et al. found that in the ischemic brain region in rats, the bFGF-like immune reaction increased and nerve cells degenerated. When treated with Danshen, the bFGF-like immune reaction was significantly stronger than that in the model group and the extent of neurocyte degeneration was milder. Meanwhile, the non-ischemic side also showed a mild increase in bFGF-like immune reaction, suggesting that Danshen might up-regulate the expression of bFGF to enhance the reparative effect of bFGF in cerebral hemorrhage. Wu Weiping et al. applied the stereotaxic photochemical method to selectively induce cortical ischemia in the left temporal lobe of rats. The disorder of spatial memory was significantly improved with the treatment of Danshen, and its mechanism is related to its functions of reducing ischemic damage to the temporal lobe and up-regulating the expression of bFGF.

Decreasing the Expression of Platelet-Derived Growth Factor

Platelet-derived growth factor (PDGF) can promote the mitosis of fibroblasts, smooth muscle cells, and glial cells. It also can act on vascular endothelial cells and nerve cells, causing cellular responses such as proliferation, chemotaxis, actin protein reorganization, and calcium flow. In cerebral ischemia, the level of PDGF expression is increased, but its significance is not fully understood. It may be related to angiogenesis and gliosis in the infarcted region. Studies have discovered that Danshen can significantly decrease the intracerebral expression of PDFG-A, improve the disorder of spatial cognition, and reduce the infarction volume in rats with ischemic damage of the temporal lobe.

Increasing the Expression of Heat Shock Proteins

Heat shock proteins (HSP) are a group of proteins which are newly synthesized by cells in response to stress. HSPs are secreted into the plasma or interstitial fluid by cells to exert their effects. Inducible HSPs have the functions of protein reparation and removal. In normal brains, HSP70 mRNA or HSP70 are almost undetectable. However, during hypoxia-ischemic brain injury, the expression of HSP70 (including the gene, mRNA, and its protein) in the hippocampus, cerebral cortex, and hypothalamus can protect nerve cells, maintain cell homeostasis, enhance the stress endurance capability of the cells, and generate resistance to various damages. Studies suggest that the expressions of HSP70 and Hsc70 can inhibit the activation process between Caspase-9 and Caspase-3, so they are involved in the inhibition of Apoptosis. In the recent years, more studies have focused on the pretreatment of cerebral hemorrhage with drugs, and the results suggest that the initiation of the HSP gene and the increase of HSP expression could have protective effects. Some investigators have observed the morphology of cultured rat hippocampal neurons under hypoxic conditions, the changes of HSP70 expression, and the effect of Danshen, and the results show that the survival rate and rate of HSP positive hippocampal neurocytes in the Danshen group were significantly higher than those in the injury group, suggesting that Danshen might directly protect neurocytes by increasing the expression of HSP70.

Decreasing the Level of β-Endorphin

β-endorphin (β-EP) is a neuropeptide which belongs to the endogenous opioid system and is extensively distributed in the central nervous system with multiple physiological functions and pathological effects. Recent studies have shown that β-EP is closely related to the pathological process of acute ischemic brain injury. Opioid peptide antagonists, either the receptor antagonist naloxone or functional antagonist thyroid stimulating hormone, can reduce ischemic injury and relieve cerebral edema. In the cerebral hemorrhage gerbil model, the β-ET content changes in ischemic brain tissue and the effects of Danshen on the β-ET content changes and symptoms of stroke after ischemia were observed, and the results show that Danshen can improve the symptoms of stroke, reduce cerebral edema, and inhibit the increase of β-EP after ischemia, suggesting that the increase of β-EP in the central nervous system after ischemia might aggravate ischemic injury. The inhibitory effect of Danshen on the endogenous opioid peptide system might be one of its mechanisms to antagonize ischemic brain injury.

5.3.2 Summary and Prospect

The pathogenesis of ischemic cerebrovascular disease is complicated and there is no satisfactory therapeutic drug yet, although some drugs such as rt-PA, streptokinase, urokinase, acenterine, etc., have certain therapeutic effects. Some researchers propose that the combination of thrombolytics and neural protectants might be an effective therapeutic way to treat stroke, and the search for a thrombolytic drug with a prolonged therapeutic time window is also very important. In China, many traditional Chinese medicines with properties of promoting blood flow and removing blood stasis are the focus of research, and Danshen preparations are commonly used drugs, which are extensively applied in the treatment of acute ischemic stroke. Many years of basic studies and evaluations of clinical therapeutic effects indicate that the pharmacological actions of Danshen are extensive and complicated, and can not only prolong the thrombolytic therapy time window, but also protect neurocytes. So, Danshen is important in the treatment of acute ischemic cerebrovascular disease. Moreover, Danshen also has the following features: adequate resources, low price, almost no toxicity or adverse effect, wide safety range of clinical application, etc., so it has very promising prospects.

However, most reports on the pharmacological actions of Danshen presented with the "possible mechanism of its action" due to the uncertainty of the related mechanisms, which limited the development of new active components with definite effects. Most studies on its active components focus on total salvianolic acids, salvianolic acid A and salvianolic acid B. However, only a few studies are on the other active components. Therefore, the scientific and systematic study of Danshen should be further enhanced to promote the effective development of new drugs and sufficient utilization of the resource.

In the clinical treatment of ischemic cerebrovascular disease, applications of Compound Danshen injection and Composite Danshen Dropping Pill are relatively common. With the progress of studies, people will further understand the pharmacological actions of Danshen and develop more effective preparations of Danshen to make the applications more extensive in the prevention and treatment of ischemic cerebrovascular disease.

5.4 Protective Effect of Danshen on Cerebral Hemorrhage-Reperfusion Injury

Xinrui Cheng, Yuehua Wang and Guanhua Du

Ischemia reperfusion injury (IRI) is related to the genesis and development of multiple diseases and this injury is a complicated physiopathological process involving many factors, including primary injuries during the ischemic stage and reperfusion injuries after the recovery of blood circulation. The mechanisms of the injuries include mainly energy exhaustion, excitatory toxic reactions, free radical injuries, Ca^{2+} overload, inflammatory reactions, etc. The injuries in the early stage of ischemia include energy exhaustion, toxic effect of excitatory amino acids, Ca^{2+} overload, free radical injuries, etc.; the injuries in the later stage include inflammatory reactions, which aggravate the free radical injuries, and the long-term deficiency of energy will enlarge the injury extent; in the reperfusion stage, effects of free radical injuries and Ca^{2+} overload are further enhanced and the injuries enter into a vicious cycle [135]. Alleviation of the tissue and organ injuries caused by IRI and the promotion of recovery are important issues.

Experimental studies have shown that traditional Chinese medicines can antagonize IRI and have a favorable application prospect. Among traditional Chinese medicines, Danshen can affect those aspects of IRI mentioned above, and has extensive regulatory functions. Therefore, Danshen can essentially block each link in the pathogenesis of cerebral ischemia-reperfusion, which includes some channel proteins on the cell membrane, adhesion molecules, intracellular information transferring molecules, transcriptional factors of gene expression, and some cytokines in body fluid. Danshen has satisfactory effects on the prevention and treatment of cerebral ischemia-reperfusion injury and has also attracted extensive interest. This section will review the main mechanisms of Danshen's action against cerebral ischemia-reperfusion injury.

5.4.1 Improving Energy Metabolism

Currently, the majority of researchers believe that the disorder of energy metabolism is the basis of cerebral IRI. mitochondria are the main site of energy metabolism and free radical production, and pumping of protons out of the mitochondria and electron transport in the respiratory chain are both processes of generating energy and the main step of producing oxygen free radicals. In the early stage of ischemia, the mitochondria show stress response reactions such as increased oxygen consumption, increased respiratory depression rate, and increased hydrolysis Activity of atpase. As ischemia prolongs and reperfusion occurs, the total contents of ATP and adenine nucleotide and energy charges gradually decrease. Salvianolic acid A has significant protective effects on mitochondria by its increasing their respiratory function, increasing the activities of respiratory chain enzymes, increasing ATP levels, maintaining the mitochondrial membrane potential, reducing mitochondrial membrane swelling, etc. [136].

Danshen has significant protective effects on respiratory chain function after cerebral IRI in rats [137], and it has been proposed that the mitochondrial outer membrane has regulatory effects on the inner membrane, and that the disorder of energy metabolism is the basis of IRI, and interacts with free radicals. These two factors, the disorder of energy metabolism and free radicals, are the cause and effect of each other, and they together promote the progression of damage. The studies also confirmed that Danshen can increase the content of ATP in brain tissue and ameliorate cerebral edema after ischemia. The water-soluble components of Danshen show typical noncompetitive irreversible inhibition of cyclic adenine nucleotide phosphodiesterase in the brain tissue of rabbits. It is postulated that the inhibitory effect on the cerebral cortex might be achieved through the inhibition of cyclic adenine nucleotide phosphodiesterase activity and the increase in adenosine cyclophosphate levels. Moreover, studies have proved that sodium tanshinone II_A silate has inhibitory effects on Na^+–K^+–ATPase in the microsomes of rat hearts and brains [138, 139].

Based on a great amount of research data, it can be postulated [141] that the detailed mechanisms for Danshen to maintain energy and increase the energy reserve might be: (1) reducing mitochondrial lipid peroxidation and intramitochondrial calcium overload, improving mitochondrial membrane fluidity, decreasing membrane permeability, and maintaining integrity of mitochondrial structure; (2) preserving the activities of enzymes related to glycometabolism, such as succinate Dehydrogenase (SDH), malate Dehydrogenase, glucose-6-phosphate Dehydrogenase, phosphorylase, etc. to promote the tricarboxylic acid cycle in the early stage of ischemia and Hypoxia; (3) recovering the activities of important components in two intramitochondrial respiratory chains—coenzyme Q and SDH and the respiratory function of the complex II + III + IV in the respiratory chain in the mitochondrial inner membrane; (4) enhancing electron transfer and proton transfer coupling, increasing the synthesis activity of atpase and decreasing the hydrolysis activity of atpase to further promote oxidative phosphorylation and increase the energy supply in hypoxia-ischemic tissue.

5.4.2 Clearing Free Radicals

The increase of free radicals is an important causative factor of IRI. The targets of oxygen free radical attack include unsaturated fatty acids of the lipids in the cell membrane, membrane proteins, and DNA. This makes them lose normal biological structures and functions, induce disorders of cellular functions and energy metabolism, and cause a series of pathological changes. Therefore, antagonizing free radical injuries is also an important way to prevent and treat cerebral ischemia-reperfusion injury. In vitro experiments have shown that Danshen is a strong antioxidant and its protective effect on cerebral IRI is related to its strong antioxidant functions and increasing the overall capability of antioxidation in the body.

A large amount of free radicals are generated after the occurrence of acute cerebral infarction, and the cell damage mediated by free radicals plays an important role in the occurrence and development of cerebral edema, which leads to further deterioration of cerebral perfusion. Malondialdehyde (MDA) is the metabolite of lipid peroxidation between oxygen free radicals and unsaturated fatty acids in biological membranes; therefore, measuring the superoxide dismutase (SOD) and MDA levels in the blood can indirectly reflect the capability of clearing free radicals and the contents change of free radicals in the brain tissue after acute cerebral infarction. Clinical trials have proven that the ability of Compound Danshen injection to remove MDA is significantly better than that of citicoline, that Tanshinol and tanshinone are ideal free radical scavenging agents, and that ability of Tanshinol to clear free radicals is better than that of SOD.

Cerebral ischemia-reperfusion produces a great amount of oxygen free radicals which destroy cellular membranes and the blood–brain barrier. Meanwhile, Ca^{2+} influx further damages the cellular membrane, depletes energy from cells, and destroys the structures of brain cells, aggravating brain damage. Danshen can block Ca^{2+} influx and clear oxygen free radicals. It can also regulate microcirculation and improve blood supply in the ischemic area [142]. The findings of Mo Zhixian et al. show that pretreatment with Compound Danshen injection for 3 days through intraperitoneal injection could decrease water content in the brain, decrease MDA levels in the cerebral cortex and hippocampal tissue, and significantly increase the activities of Glutathione peroxidase (GSH-Px) and SOD in ischemia-reperfusion rats. This suggests that it has the functions of inhibiting oxygen free radical reactions and reducing lipid peroxidation, and that it can also maintain comparatively high levels of ATP and reduce the accumulation of lactic acid, thus decreasing damage to brain tissue. Xu Luoling et al. further confirm that tanshinone could directly act on neutrophilic granulocytes to inhibit respiratory bursts and decrease the production of oxygen free radicals [142]. The recent findings indicate that the protective effects of tanshinone II_A on focal cerebral IRI might be

Table 5.17 Effects of salvianolic acids on the activities of SOD, GSH-Px, CAT, and GSH-ST 2 h after reperfusion in rats with 24 h of ischemia ($\bar{X} \pm s$)

Group	Dose (mg/kg)	SOD (kU/g protein)	GSH-Px (kU/g protein)	CAT (U/g protein)	GSH-ST (kU/g protein)
Sham operation		5.02 ± 0.64	12.5 ± 0.9	155 ± 18	29.7 ± 1.5
Control		7.84 ± 0.65*	15.9 ± 1.6*	204 ± 19*	23.5 ± 1.2*
Salvianolic acids	2.5	8.81 ± 0.87**	16.3 ± 1.7	229 ± 16**	23.1 ± 1.6
	5.0	8.95 ± 0.82***	17.8 ± 1.8**	246 ± 22***	24.2 ± 2.0
	10.0	9.45 ± 1.21***	19.4 ± 2.1***	273 ± 45***	24.8 ± 2.3

Compared with the sham operation group: * $P < 0.01$
Compared with the control group: ** $P < 0.05$, *** $P < 0.0 1$

related to its function of antagonizing the neurotoxic effect of nitric oxide [54]. Total salvianolic acids also have significant protective effects on cerebral IRI in mice, and the mechanism is related to their antioxidant activities. For example, total salvianolic acids can increase the activity of SOD, decrease the content of MDA, and increase the content of GSH-Px in the brain tissue of ischemia-reperfusion mice; salvianolic acid B has significant protective effects on cerebral IRI in rats, and this effect is related to its functions of reducing lipid peroxidation, clearing free radicals and increasing energy metabolism [137].

It is presently believed that the increase in antioxidant enzyme activity in the early stage of ischemia is the body's self-defense response to the cell damage caused by oxygen free radicals. However, in the late stage of ischemia-reperfusion, the production of a great amount of oxygen free radicals greatly consumes the antioxidant enzymes in the body, leading to the significant decrease in SOD, GSH-Px, and catalase (CAT) activity. Salvianolic acids can significantly inhibit antioxidant enzyme activity in the later stage of reperfusion by increasing enzyme activity in the early stage of reperfusion to protect the brain tissue from IRI, which is consistent with their strong in vitro effect of antioxidation. It is postulated that the antioxidation mechanism of salvianolic acids might be due to their abundant phenolic hydroxyl groups and large conjugated π bonds, which makes them reducing substances that can accept the attack of oxidative substances such as oxygen free radicals, or act as the medium of electron transport to change the oxidizing reaction pathway of free radicals and protect the body from oxidative damage. Glutathione S-transferase (GSH-ST) has the functions of antagonizing lipid peroxidation and clearing exogenous toxins. Different stages of cerebral IRI have different effects on the activities of SOD, GSH-Px, and CAT. The activities of SOD, GSH-Px, and CAT increase significantly in the early stage of reperfusion; however, the activities of these three enzymes decrease significantly in the later stage, and the level of GST-ST activity is also low in both the early and later stages of reperfusion. The neuroprotective effect of salvianolic acids are achieved through their function of increasing SOD, GSH-Px, and CAT activities.

The activity changes of the above-mentioned enzymes and protective effects of salvianolic acids [74] and salvianolic acid B at 2 and 4 h after reperfusion in rats with 24 h of cerebral hemorrhage are listed in Tables 5.17 and 5.18.

5.4.3 Reducing Calcium Overload

Under normal conditions, calcium homeostasis is maintained by Ca^{2+} trafficking through calcium channels, Na^{+}–Ca^{2+} exchanging, Ca^{2+} release from intracellular calcium stores, etc. In cerebral IRI, this homeostasis is destroyed, which causes calcium overload inside the cells and cell organelles. In cerebral IRI, intracellular Ca^{2+} concentration is directly correlated with the degree of cell injury, and intracellular calcium overload is an important cause of tissue and organ injuries. As a receptor-operated Ca^{2+}

Table 5.18 Effects of salvianolic acid B on the activities of SOD, GSH, and MDA 4 h after reperfusion in rats with 24 h of ischemia ($\bar{X} \pm s$)

Group	*n*	SOD/kU g^{-1} protein	GSH/mg g^{-1} protein	MDA/mol g^{-1} protein
Sham operation	6	$0.92 \pm 0.08^{**}$	$4.31 \pm 0.82^{**}$	$0.270 \pm 0.037^{**}$
Model	7	$0.55 \pm 0.11^{*****}$	$1.78 \pm 0.62^{*****}$	$0.432 \pm 0.041^{*****}$
Nimodipine 4 mg·kg^{-1}	6	$0.82 \pm 0.06^{**,****}$	$3.82 \pm 0.96^{**,***}$	$0.333 \pm 0.037^{**,****}$
Salvianolic acid B 3 mg·kg^{-1}	5	$0.76 \pm 0.10^{*,****}$	$3.00 \pm 0.94^{*,****}$	$0.368 \pm 0.058^{**,****}$
Salvianolic acid B 10 mg·kg^{-1}	6	$0.84 \pm 0.05^{**,***}$	$3.34 \pm 0.67^{**,***}$	$0.327 \pm 0.062^{**,***}$

Compared with the control group: $^{*}P < 0.05$, $^{**}P < 0.01$
Compared with the sham operation group: $^{***}P > 0.05$, $^{****}P < 0.05$, $^{*****}P < 0.01$

blocking agent, Danshen can reduce cell injury in ischemia-reperfusion through its function of decreasing the intracellular concentration of Ca^{2+} and the amount of lipid peroxidation free radicals. Danshen can reduce calcium overload through the following ways:

5.4.3.1 Decreasing Membrane Permeability

The increase in cell membrane permeability after damage is mainly due to intracellular calcium overload. By clearing free radicals, Danshen can not only reduce cell membrane damage, thus decreasing calcium influx through the cell membrane, but also increase the oxygen utilizing capability of mitochondria so that the production of ATP increases due to oxidative phosphorylation coupling. This provides energy for calcium pumps, Na^{+}–Ca^{2+} pumps, Ca^{2+}–H^{+} pumps, etc., so that Ca^{2+} is pumped out of the cells or into the endoplasmic reticulum, mitochondria and microsomes, which further decreases the intracellular calcium concentration.

5.4.3.2 Blocking Calcium Channels in the Cell Membrane

Calcium channels are the major gateway for controlling the transmembrane influx of calcium. Danshen and its active components, such as tanshinone II_A and component 764-3, can block the voltage-dependent calcium channels in single guinea pig cardiac cells and isolated rabbit Purkinje voltage-dependent calcium channels in the wild fibers and cortical synaptosomes [145]. Danshen also has the effects of certain hormones or mediators and directly acts on receptor-operated Ca^{2+} channels to affect calcium transmembrane influx [146].

5.4.3.3 Blocking the Calcium Release Channels of Intracellular Calcium Stores

The calcium release channels of intracellular calcium stores belong to receptor-operated calcium channels which include inositol triphosphate (IP_3)-operated calcium channels and Ryanodine-sensitive calcium channels. It has been shown that Danshen has significant inhibitory effects on the hyperfunctional state of the myocardial phosphatidate-inositol information-transmission system after acute myocardial IRI in rats, and can decrease the contents of phosphatidylinosital biphosphate (PIP_2) and inositol triphosphate significantly. This suggests that Danshen might reduce calcium overload by decreasing the synthesis of inositol triphosphate to inhibit inositol triphosphate-operated calcium channels and decrease the release of calcium from the sarcoplasmic reticulum [147].

5.4.3.4 Increasing the Content of Cytochrome P450 or Promoting the Synthesis of Cytochrome P450-Dependent Mixed-Function Oxidase

By increasing the content of cytochrome P450 or promoting the synthesis of cytochrome P450-dependent mixed-function oxidase, Danshen can recover the function of microsomes to actively transfer calcium and protect the activities of Na^{+}/K^{+}-ATPase and Ca^{2+}-ATPase, thus blocking

Table 5.19 Protective effect of tanshinone on cortical neurocyte damage in rats induced by caffeine ($\bar{X} \pm s$, $n = 8$)

Group	Drug concentration (μg ml^{-1})	Crystal violet staining of living cells (OD)	LDH measurement (U ml^{-1})
Normal		0.39 ± 0.01	43.17 ± 0.92
Model		0.24 ± 0.02*	56.61 ± 1.73*
Tanshinone	200	0.40 ± 0.14**	46.27 ± 0.26**
	50	0.31 ± 0.01**	54.19 ± 0.76**
	12.5	0.28 ± 0.02**	56.17 ± 0.90
	3.125	0.25 ± 0.02	59.47 ± 0.71

Compared with the normal group: * $P < 0.01$
Compared with the model group: ** $P < 0.01$

Table 5.20 Protective effect of tanshinone on cortical neurocyte damage in rats induced by KCl ($\bar{X} \pm s$, $n = 8$)

Group	Drug concentration (μg ml^{-1})	Crystal violet staining of living cells (OD)	LDH measurement (U ml^{-1})
Normal		0.38 ± 0.01	15.16 ± 0.53
Model		0.15 ± 0.03*	43.21 ± 0.56*
Tanshinone	200	0.25 ± 0.01***	23.87 ± 0.86***
	50	0.24 ± 0.02***	32.58 ± 0.83***
	12.5	0.23 ± 0.02***	38.85 ± 1.64***
	3.125	0.18 ± 0.01***	41.46 ± 2.57**

Compared with the normal group: * $P < 0.01$
Compared with the model group: ** $P < 0.05$, *** $P < 0.01$

calcium influx and promoting the pumping of calcium out of the cell.

Many basic studies and clinical trials demonstrate that Danshen can reduce calcium overload in many ways. In the calcium overload model induced by caffeine, KCl and NMDA, it is found that tanshinone can significantly increase the number of surviving neurocytes, decrease the amount of lactate Dehydrogenase released by cells, and reduce intracellular Ca^{2+} concentration in those above-mentioned three types of injury models, showing significant protective effects in these three different types of intracellular calcium overload injury models. Another report shows that the major active component of tanshinone, sodium tanshinone II_A silate can inhibit the influx of large amounts of Ca^{2+} during calcium overload, suggesting that Danshen is a kind of Ca^{2+} antagonist and realizes its protective effect by blocking the migration of Ca^{2+} into cells, which is one of the mechanisms of Danshen to antagonize cerebral hypoxia-ischemic injury [148]. The experimental results of the protective effects of tanshinone on the above-mentioned three injury models are listed in Tables 5.19, 5.20 and 5.21.

5.4.4 Inhibiting the Release of Excitatory Amino Acids (EAA)

EAA include glutamic acid (Glu), aspartate (Asp), etc. After cerebral IRI, a large amount of extracellular Glu accumulates, which stimulates *N*-methyl-*D*-aspartate receptors (NMDAR) and induces overexcitation and the influx of a large amount of Ca^{2+}, initiating a series of pathological reactions of neurocyte damage, which causes cell necrosis and apoptosis. Li Ju et al. measured the contents of Glu and Asp in hippocampal extracellular fluid in rats and the results indicate that their contents in the Danshen group were significantly lower than in the normal saline group. The content of cysteine (Cys) in striatal extracellular fluid in the striate body area and the effect of Danshen were dynamically

Table 5.21 Protective effect of Tanshinone on cortical neurocyte damage in rats induced by NMDA ($\bar{X} \pm s$, $n = 8$)

Group	Drug concentration ($\mu g\ ml^{-1}$)	Crystal violet staining of living cells (OD)	LDH measurement ($U\ ml^{-1}$)
Normal		0.36 ± 0.01	37.15 ± 0.55
Model		$0.14 \pm 0.01^{*}$	$50.00 \pm 1.06^{*}$
Tanshinone	200	$0.23 \pm 0.01^{**}$	$35.73 \pm 0.81^{**}$
	50	$0.20 \pm 0.01^{**}$	$38.58 \pm 0.86^{**}$
	12.5	$0.18 \pm 0.02^{**}$	$40.00 \pm 1.78^{**}$
	3.125	$0.16 \pm 0.02^{**}$	$48.58 \pm 1.03^{**}$

Compared with the normal group: * $P < 0.01$
Compared with the model group: ** $P < 0.01$

detected in the cerebral ischemia-reperfusion gerbil model, and the results indicated that Danshen can inhibit the release of EAA by brain cells during ischemia [149].

5.4.5 Regulating Immunity and Effects on Cytokines

Danshen can regulate immune response after cerebral ischemia-reperfusion through its effects on multiple immune cells and cytokines [150].

5.4.5.1 Regulatory Effect on Immune Cells

In the cerebral ischemia-reperfusion rat model, Liu Jun et al. found that Danshen can inhibit the infiltration of multinuclear granular leukocytes and decrease the release of cytokines by microglia cells activated by cerebral hemorrhage tissue. On the other hand, some components in Danshen can directly act on the surface of microglia cells to inhibit the expression of cell surface antigens or block the signal transduction through external factors in order to reduce immunological injury to the body after ischemia-reperfusion [105, 151].

5.4.5.2 Effect on the Expressions of Intercellular Adhesion Molecules

It has been discovered that in the pathological processes of cerebral hemorrhage and cerebral ischemia-reperfusion, intercellular adhesion molecule-1 (ICAM-1) is critical to the injuries caused by cerebral hemorrhage and cerebral ischemia-reperfusion. Animal experiments confirmed that after cerebral ischemia-reperfusion, the number of ICAM-1 positive capillary endothelial cells and neutrophilic infiltration of the cerebral hemorrhage region in the Danshen treatment group were significantly lower than those in the control group. The results of another experiment showed that the number of surface CD18/CD11b positive peripheral blood leucocytes in the Danshen treatment group was significantly lower than in the control group, suggesting that Danshen blocks CD18/CD11b antigens on the leucocyte surface [152]. Therefore, Danshen can block the adhesion between vascular endothelial cells and leucocytes in two ways, and decrease neutrophilic infiltration in ischemic brain tissue, thus reducing neurocyte damage.

5.4.5.3 Effect on Interleukin-1β (IL-1β)

The application of Danshen can significantly decrease the number of IL-1β converting enzyme (ICE) positive cells and improve the morphology of neurocytes, indicating that Danshen has inhibitory effects on the expression of ICE, thereby reducing IL-1β-mediated brain injury [153].

5.4.5.4 Effect on Nuclear Factor NF-κB

In recent years, data have indicated that after cerebral IRI, the activation of NF-κB leads to increase in expression of multiple adhesion molecules to induce inflammatory cascade reactions, which plays an important role in cerebral ischemia-reperfusion neurocyte injury. Therefore, it is believed that NF-κB is the core regulatory factor for inflammation and one of the initiating factors resulting in vascular endothelial cell damage after cerebral ischemia-reperfusion. Research results show that Danshen can significantly inhibit the expression of NF-κB in primary neurocytes of rats with ischemia-reperfusion, thus inhibiting the adhesion molecules and inflammatory reactions induced by ischemia-reperfusion, indicating that it might have a therapeutic effect on cerebral ischemia-reperfusion [154].

5.4.5.5 Basic Fibroblast Growth Factor (BFGF)

The expression of bFGF is up-regulated after cerebral hemorrhage, and bFGF promotes neurocyte survival and the growth of the neural axis through high-affinity tyrosine kinase receptors. Danshen can enhance bFGF-like immune reactions and alleviate the degree of neurocyte degeneration in rats after cerebral ischemia-reperfusion, suggesting that Danshen can up-regulate the expression of bFGF and enhance the reparative effect of bFGF after cerebral ischemia-reperfusion [155].

5.4.5.6 Platelet-Derived Growth Factor (PDGF)

PDGF can promote the mitosis of fibroblasts, smooth muscle cells, and glial cells. It also can act on vascular endothelial cells and nerve cells, causing cellular responses like proliferation, chemotaxis, actin protein reorganization and calcium flow, etc. In cerebral ischemia, the level of PDGF expression is increased but its significance is not fully understood; it may be related to angiogenesis and gliosis in the infarcted region. Studies have discovered that Danshen can significantly decrease the intracerebral expression of PDFG-A, improve the disorder of spatial cognition, and reduce infarction volume in rats with ischemic damage to the temporal lobe [156].

5.4.6 Affecting the Expression of Heat Shock Protein

Heat shock protein (HSP) is a kind of important molecular chaperone in cells and is of important significance in the maintenance of homeostasis in tissues and cells, environmental adaptation, and resistance to external injury. In vivo and in vitro studies indicate that the protective effects of Danshen on neurocytes after ischemia-reperfusion are always accompanied by increase in immune positive reaction to HSP70. Not only did the number of HSP70 positive cells increase, but also the staining intensity was significantly higher than in the control group [157]. By contrast, studies have found that the improving effect of Danshen on the disorder of spatial cognition is accompanied with a decrease in the expression of HSP32, and it is postulated that HSP32 might degrade hemoglobin to produce an excess of bilirubin during ischemia-reperfusion which mainly damages the neurocytes. So, Danshen reduces the expression of HSP32 by clearing free radicals and decreasing the levels of oxide products, thus reducing stress injury to the temporal lobe. Therefore, Danshen might regulate the synthesis of HSP through different pathways. The protective effect of Danshen is achieved by preventing the cytotoxic effect induced by overexpression of HSP32 on one hand, and increasing the expression of HSP70 in local tissue on the other [156].

It was found in ischemia and/or ischemia-reperfusion models that Danshen could ameliorate the metabolic disorders of mediators of monoamines, peptides, and amino acids, decrease the concentrations of thromboxan B_2 (TXB_2) and arginine vasopressin (AVP), reduce cerebral edema, partially inhibit the expressions of c-*fos* gene and Endothelin-1 (ET-1) gene, and decrease the disability rate and mortality of animals, which has been confirmed by pathohistological results. However, because the results of in vivo experiments are influenced by many factors, it is difficult to clarify whether the protective effect of Danshen is caused indirectly by its already known functions of promoting blood flow and removing blood stasis, or if it has direct protective effects on the damaged neurocytes. Therefore, rat hippocampal neurocyte culture was used to monitor the changes of HSP70 expression and cell morphology. The results show that Danshen has definite direct neuroprotective effects under hypoxic conditions. Therefore, we propose that during ischemia-reperfusion, Danshen not only has an indirect protective effect on neurocytes through many functions such as "promoting blood flow and removing blood stasis", etc., but also a direct protective effect. The latest findings indicated that Danshen can partially inhibit the expression of ET-1 in the cerebral cortex between ischemia and reperfusion, which might be a protective mechanism of Danshen on cerebral regional ischemia and reperfusion.

5.4.7 Improving Hemorheology and Microcirculation

Danshen has improving effects on the hemorheological parameters and microcirculation in tissues and organs during ischemia-reperfusion. The mechanisms include the following several aspects [158, 159]:

1. Regulating the ratio of thromboxan A2 (TXA2)/ prostacyclin I2 (PGI2). In the cerebral ischemia-reperfusion rabbit model, it was discovered that the significant disorder of TXA_2/PGI_2 balance presents in cerebral IRI, and Danshen can increase the content of PGI_2 and decrease the production of TXA_2 and TXA_2/PGI_2 ratio, reducing pathological damage to the brain tissue. It is thus believed that the regulation of TXA_2/PGI_2 balance might be one of the mechanisms for Danshen's effective treatment of cerebral IRI.
2. Regulating the synthesis and release of ET-1 at the cellular and gene levels to eliminate its harmful effects. The level of plasma ET-1 increases significantly after neonatal asphyxia. However, Danshen has partial inhibitory effects on the expression of ET-1 induced by ischemia.
3. Promoting the expression of nitric oxide synthase (NOS), especially the expression of constitutive NOS mRNA, and increasing the activity of NOS and secretion of NO by vascular endothelial cells (VEC) at the cellular and molecular levels to correct the abnormalities of ET-1 and NO synthesis and secretion by vascular endothelial cells induced by IRI to a certain extent.
4. Improving rheological properties of blood cells. For example, by the effects of clearing free radicals, decreasing cAMP and Ca^{2+} concentrations in red blood cells (RBC) and ameliorating the ion disorders of Na^+, K^+ and Ca^{2+}, Danshen maintains the biconcave disk-shape of RBC, decreases the rigidity index, recovers deformability, inhibits leucocyte chemotaxis, activation, adhesion and aggregation, inhibits platelet activation, aggregation and releasing reactions, and blocks the P-selectin mediated adhesions between platelets and monocytes, neutrophilic granulocytes, killer cells, etc., and the following cascade reactions of blood clotting by inhibiting the expression of P-selectin in platelets.
5. Dilating blood vessels. The components in Danshen include lithospermic acid B and its derivate α-hydroxyl-3,4-dihydroxy lithospermic acid. They dilate blood vessels to improve circulation by relaxing endothelium-dependent or non-endothelial-dependent vascular smooth muscles.

5.4.8 Inhibiting Apoptosis

Experimental studies have confirmed that Apoptosis is the main mechanism for the death of neurocytes in mild cerebral hemorrhage and reperfusion. The necrosis of cells and tissues is an irreversible event; however, Apoptosis is controlled by a series of programs and involves the activation and expression of a series of genes. Therefore, the event can be rescued by intervention. More and more experiments have demonstrated that Danshen can protect damaged neurocytes after cerebral ischemia-reperfusion by the regulation of some gene expressions. Experimental results indicate that Danshen can down-regulate the expression of ICE after cerebral hemorrhage and up-regulate the expression of anti-Apoptosis gene *Bcl*-2 to protect the neurocytes [160]. However, the mechanism for the regulatory effects of Danshen on the expressions of ICE and *Bcl*-2 in brain cells after ischemia is not clear yet and needs further study. Recently, it has been found that Danshen can affect the expressions of immediate early genes c-*fos* and c-*jun* to protect neurocytes.

With progress in studies of the mechanisms for Danshen's function, some signal pathways and targets of gene regulation can be found. Danshen can significantly increase the DNA binding activity of activator protein-1 (AP-1) in the hippocampal CA1 region after ischemia, and its protective effect might be related to its regulation of the DNA binding activity of AP-1 [161].

5.4.9 Improving Learning and Memory Disorders

The step down test and decapitation method were used to observe the effects of total salvianolic acids on learning and memory disorders and resistance to hypoxia in ischemia-reperfusion mice [54], and the chemical method was used to observe the effects of total salvianolic acids on the activity of SOD and contents of MDA and GSH in their brain tissue. The results show that total salvianolic acids can improve learning and memory disorders induced by ischemia-reperfusion shorten the time of passive avoidance response, reduce the times of errors, prolong the latent period, and prolong the duration of breath after decapitation. Total salvianolic acids can also increase the activity of SOD, decrease the content of MDA, and increase the content of GSH in brain tissue in ischemia-reperfusion mice.

The step down test and light-dark box test were applied to study the protective effect of salvianolic acid A (SalA) on the memory impairment caused by cerebral ischemia-reperfusion in mice, and the relationship between its antioxidant activity and its protective effect on the brain injury induced by reperfusion. The results showed that intravenous injection of SalA can improve impaired memory in mice. In the step down test, the mean times of errors in the treatment group were significantly lower than those in the control group, and the latent period in the treatment group was longer than that in the control group. Meanwhile, intravenous injection of SalA can decrease the content of MDA in the cerebral cortex, hippocampus and striate body in cerebral ischemia-reperfusion mice. The in vitro test indicated that intravenous injection of 10 mg/kg SalA can block cerebral lipid peroxidation and clear free hydroxyl radicals. These results indicate that the improving effect of SalA on the memory impairment induced by cerebral ischemia-reperfusion in mice might be related to its antioxidative function. Moreover, studies suggest that SalB can also improve learning and memory disorders induced by transient cerebral hemorrhage in mice [162, 165]. The ischemia-reperfusion mice showed learning and memory disorders, and the hypoxia caused by cerebral hemorrhage shortened the duration of breath after decapitation. Total Salvianolic acids can improve learning and memory disorders and prolong the duration of breath after decapitation, suggesting that total salvianolic acids have protective effects on ischemia-reperfusion injury. The protective effects of salvianolic acids on learning and memory disorders in mice with cerebral ischemia-reperfusion are shown in Table 5.22.

The escaping reflection of mice in the shuttle box was used to measure changes in the active avoidance response rate, and experimental results show that Danshen could significantly increase the times of active avoidance responses in mice in the shuttle box. Danshen can also significantly improve learning and memory disorders induced by electroconvulsive shock. In the experiment on spatial memory disorder due to single temporal ischemic injury in rats, it was found that treatment with Danshen can significantly improve spatial memory disorder in rats with temporal ischemic injury, significantly reduce the temporal ischemic injury, and significantly decrease the expression of 5-hydroxytryptamine (5-HT) in the ischemic lesions. Therefore, Danshen can significantly improve spatial memory disorders in rats with single temporal ischemic injury and the mechanism might be related to its functions of reducing temporal injury and down-regulating

Table 5.22 Effects of total salvianolic acid on the latency and error number of step down test in ischemia-reperfusion mice ($\bar{X} \pm s$, $n = 10$)

Group	Dose/ mg kg^{-1}	Learning		Memory	
		Reactive time/s	Error number/ counts	Latency/s	Error number/ time
Sham operation	–	$7.2 \pm 5.5^{**}$	$2.9 \pm 1.5^{*}$	$249.8 \pm 99.7^{**}$	$0.4 \pm 1.0^{**}$
Model	–	24.1 ± 15.3	6.2 ± 3.5	91.4 ± 114.7	2.5 ± 1.7
Total salvianolic acids	3	$11.4 \pm 8.7^{*}$	3.7 ± 2.5	$195.5 \pm 78.3^{*}$	$0.9 \pm 0.5^{**}$
	6	17.6 ± 15.1	3.5 ± 3.3	$243.9 \pm 86.2^{**}$	$0.5 \pm 0.7^{**}$
	10	$4.9 \pm 2.8^{**}$	$3.4 \pm 2.6^{*}$	$245.3 \pm 81.1^{**}$	$0.7 \pm 1.0^{**}$

Compared with the model group: * $P < 0.05$, ** $P < 0.01$

the expressions of c-*jun*, bFGF and HSP70. It has three significances: (1) down-regulation of c-*jun* gene can decrease the damages of ischemia, hypoxia, EAA, free radicals, depolarization and ion influx to the treated temporal lobe; (2) down-regulation of the bFGF gene further confirms the significant decrease of temporal ischemic injury; (3) the significant decrease in HSP70 expression suggests that the hypoxia-ischemic attacks on temporal neurons are reduced significantly. In the temporal infarction animal model, spatial cognition disorders were significantly improved in mice in the Danshen treatment group, the infarction volume of the temporal lobe in the Danshen treatment group was significantly less than in the model group, and the expression of HSP32 in the infarction site and surrounding tissue induced by ischemia and free radicals decreased significantly.

The combination of many experimental data indicates that Danshen might achieve its protective effect on the brain through the decrease of neurocyte apoptosis. Its main anti-apoptosis pathways mainly include: (1) reducing the expression of ICE; (2) blocking apoptosis-inducing signal transduction, such as the signal transduction mediated by death factor and its receptor, apoptosis induced by the second messenger calcium, and apoptosis initiated after mitochondrial injury; (3) increasing the expression of *Bcl*-2, which is a gene inhibiting apoptosis, and reducing the expression *Fas*, which is a gene promoting apoptosis, to regulate the balance among each apoptosis gene.

In summary, the functions of Danshen are very extensive and have regulatory effects on each step of IRI and systematically block a series of “vicious cycles” in the processes of the occurrence and development of IRI, so that it has significant protective effects against IRI. Therefore, Danshen is an ideal drug for the prevention and treatment of IRI. The clinical application has more extensive prospects. These results provide an experimental and theoretical basis for the development of potential and characteristics of traditional Chinese medicines, and for the treatment of hypoxia-ischemic diseases with Danshen and its preparations.

In ischemia-reperfusion injury, multiple factors interact with each other, such as free radicals, functional disorders of mitochondria, energy metabolism disorders, intracellular calcium overload, etc., and Danshen has effects on all of these aspects. The efficacy of drug treatment in the acute stage of ischemic stroke is dissatisfactory, and only acenterine, t-PA, etc., have some therapeutic effect. In China, many traditional Chinese medicines with the effect of promoting blood flow and removing blood stasis are the focus of studies. The preparations of Danshen are common drugs extensively used in the treatment of acute ischemic stroke; the literature has reported the application of Danshen in the treatment of cerebrovascular diseases for over 30 years. The pharmacological studies, clinical trials, economic evaluations, and the evaluations of the clinical therapeutic effect of Danshen preparations indicate that it has extensive and complicated pharmacological actions and potential application prospects. The preparations of Danshen are considered to have few adverse effects and relatively high safety. Presently, systematic evaluations should be conducted to confirm whether the preparations of Danshen have definite therapeutic effects on acute ischemic stroke.

On the whole, with the progress of research, the molecular mechanism for Danshen to antagonize IRI and to interfere with intracellular signal transduction to protect tissues and organs still needs further elucidation. Further studies on unknown fields (e.g. whether Danshen has effects on mitochondrial permeability channels, DNA fragment deletion, and abnormal expressions of respiratory chain related genes; the definite intracellular signal pathway through which Danshen realizes its protective effects; the mechanisms of Danshen to regulate proto-oncogene and anti-apoptosis genes, etc.) of the mechanisms of actions of Danshen will provide novel ways for medical workers to broaden its application fields and bring new hopes to the treatment of IRI related diseases. Meanwhile, the application of Danshen as a medicine will be extensively appreciated.

References

1. Fu XQ, Cai ZW, Li XC, et al. Study J Tradit Chin Med. 2003;21(3):373.
2. Yin J (ed) Modern study and application of traditional chinese medicine. Academy Press, Beijing, 1997. p. 1162.
3. Gong MF, Zou J. Heilongjiang J Tradit Chin Med. 1996;1:3.
4. Xu CQ, Wang XM, Fan JS, et al. Chin J Pathophysiol. 1997;13(1):443.
5. Chang FB, Hou SW, Wu TH. Chin J Zool. 1997;32 (5):126.
6. Li CZ, Yang SC, Zhao FD, et al (1983) J Integr Tradit West Med 3(5):297.
7. Zhao XH, Gu CZ, Huang HY. Chin J Microcirc. 2003;13(1):27.
8. Fan MZ, Zhao RT. Chin Tradit Patent Med 1988; (1):39.
9. Shen WL, Lu SZ. Chin J Integr Tradit West Med Intensive Crit Care. 1999;6(10):473.
10. Peng YZ. Jiangsu J Tradit Chin Med. 1997;18(1):43.
11. Promoting Blood Flow And Removing Blood Stasis Are Important Roles to Treat Acute Cerebral Hemorrhage. Clinical Focus, 1991;6(5):230.
12. Yang SD, Xue F, Yang JJ, et al. Chin J Integr Tradit West Med. 1992;12(2):111.
13. Qiu QX, Liu Z, Zeng NG. Youjiang Medical Journal, 1999;27(1):6.
14. Ye YR, Liang DH. Clin. Med. China. 2002;18 (6):521.
15. Li ZC. Chin J Med. 2001;28(11):43.
16. Yang WD, Song HL, Zhao BL. Chin Pharmacol Bull. 1990;8(2):77.
17. Yang C. Chin J Cardiovasc Med. 1980;17(3):179.
18. Cao EH, Liu XQ, Li JF, et al. Acta Biophys. Sinica. 1996;12(2):3392.
19. Wei G, Chen FS, Yan WJ. Chin Pharmacol Bull. 1997;13(2):129
20. Liu JD, Tang RR. Chin J Pathophysiol. 1993;9 (3):369.
21. Chen RQ. Relationship between Syndromes and Treatment of Hemorrhagic Disease and Modern Study. Wuhan: Hubei Science and Technology Press; 1990. p. 291.
22. Li WH. Chin J Integr Tradit West Med. 1984;9:372.
23. Tang KR. Shandong J Tradit Chin Med. 1987;5:32.
24. Jin AG, Han YM, Chang BL. Clinical Focus. 2000;15(22):1035.
25. Zhu ZF. Shaanxi J Tradit Chin Med. 2002;23 (8):687.
26. Xu WR, Chen JH, Cheng Q. Shanxi Clinical Medicine. 2000;9(10):740.
27. Shen WP, Wang ZX, Li JY, et al. J Nanjing Univ Tradit Chin Med. 1997;13(1):15.
28. Li RK, Zhao H, Zhang YY et al. Chin J Integr Tradit West Med Intensive Crit Care. 1999;6(10):462.
29. Huang Y. Chin J Int Tradit West Med Intensive Crit Care. 2000;7(5):279.
30. Kazui S, Narifomi H, Yammnoto H, et al. Stroke. 1996;27:1783.
31. Zhang JT. Pharm. J. 2000;35(8):635.
32. Xing XJ. J Fujian College Tradit Chin Med. 2002;12 (1):58.
33. Yu SX, Guo MX. Chin J Pract Intern Med. 1993;13 (6):328.
34. Zong GF. New J Tradit Chin Med. 2005;37(6):84.
35. Wang ZX, Ning J, Sun XD. J Shanxi Med Univ. 1997;28:5.
36. Cheng YX, Wang LS, Zhu CL. J Appl Clin Pediatr. 2002;17(6):635.
37. Yang JJ, Sui N. Chin J Neuroimmunol Neurol 1998; 5(4):199.
38. Li LB, Lin CY, Ru LQ, et al. J Huazhong Univ Sci Technol. 2003;3(2):19.
39. Liu TY, Li L. J Xianning Coll. 2004;18(1):24.
40. Li LX, Ru LQ. Chin J Gerontol. 2003;23(5):294.
41. Li LX, Xu ZE, Zhao B. Chin J Histochem Cytochem. 2004;13(2):168.

42. Sheng SL, (ed.) Senile dementia: from molecular biology to clinical diagnosis and treatment. Beijing: Science & Technology Literature Press, 1999. p. 263.
43. Qin RA, Luo GB, Huang ZY, et al. China J Tradit Chin Med Pharm. 2005;20(6):377.
44. Wang LJ, Wang Y, et al. China Pharmaceuticals. 2002;11(9):29.
45. Huang Q. Hubei J Tradit Chin Med. 2003;25(8):50.
46. Huang Q. J Hubei Inst Nationalities. 2002;19(4):29.
47. Yuan DP, Huang Q, Zhang ML. Shanxi J Tradit Chin Med. 2002;18(6):47.
48. Yan XC, chief editor. Practical traditional Chinese medicine brain diseases. Beijing: Academy Press; 1993. p. 784.
49. Du GH, Zhang JT. Basic Med. Sci. Clin. 2000;20 (5):394.
50. Caplan LR. JAMA. 2004;292(15):1883.
51. Du GH, Zhang JT. Pharmaceutical J. 1995;30:184.
52. Du GH, Qiu Y, Zhang JT. Pharm J. 1995;30 (10):731.
53. Bendixen BH, Ocava L. Curr. Cardiol. Rep. 2002;4 (2):149.
54. Red DC, Du GH, Zhang JT. Chin Pharmacol Bull. 2002;18(2):219.
55. Hickenbottom SL, Vn Barsan G. Neurol Clin. 2000;18(2):379.
56. Lain BY, I.o AC, Sun X, et al. Phytomedicine,2003;10(4):286.
57. Wang Q, Wu JF, Zhang JT. Chin Pharmacol Bull. 1999;15(2):164.
58. Wang J, Zhang JT. Chin Pharmacol. Bull. 1999;15 (3):237.
59. Hong XR, Wu AQ, You ZD. Chin J Integr Tradit West Med. 2002;22(8):607.
60. Wu B, Liu M. Chin J Clin Rehabil. 2004;8 (10):1925.
61. Chai RX. Chin J Tradit Med. Science and Technology. 2003;10(6):390.
62. Li XH, Guan YQ, Bing SL. J Tradit Chin Med. 1999,40(9):121.
63. Gong SJ, Li DG, Shi RH, et al. J Chin Microcirc. 2000;4(4):213.
64. Ziv H, Flemingger G, Dialdetti R. Stroke. 1992;23:74.
65. Li CZ, Lin JB, Yang SC, et al. Chin J Tradit Med. Science and Technology. 1984;4(9):565.
66. Dong JC. Chin Pharm J. 1996;31(1):6.
67. Fan ZD. Chin J Coal Ind Med. 2003;6(4):380.
68. Zhang YH. Jiangsu J Tradit Chin Med. 1998;6:19.
69. Xin Qin, Li XF, et al. Chin Tradit Pat Med. 2004;26 (3):222.
70. Li XH, Guan YQ, Bing SL. West China J Pharm Sci. 2001;16(1):70.
71. Qiu Y, Rui YC, Zhang L, et al. Pharm J Chin People's Liberation Army. 2002;18(1):17.
72. Sheng WH, Li XH, Yao HY, et al. Chin J Neuroimmunol Neurol. 2004;11(2):121.
73. Yu WG, Xu LN. Pharm. J. 1998;33(1):62.
74. Wu JF, Wang J, Zhang JT. Chin J New Drugs. 2000;9(70):452.
75. Cheng TJ, Yang ZY. Chin J Clin Pharmacol Ther. 2003;8(1):23.
76. Li JY, Lu CN. Jiangsu J Tradit Chin Med. 2002;22 (4):14.
77. Dai XH. J Bethune Military Medical Coll. 2003;1 (4):216.
78. Yang WD, Zhu HL. Chin Pharmacol Bull. 1990;6 (2):118.
79. Zhang JT, Du GH. Progression in neuropharmacological study. People's Medical Publishing House, 2002. p. 49.
80. Ovbiagele B,Kidwell C.S,Starkman S, et al. Curr Neurol Neurosci Rep. 2003;3(1):9.
81. Li L, Liu GT. Chin J Pharmacol Toxicol, I998;12 (4):275.
82. Su XH, Wang XM, Jiao XM, et al. Chin J Pathophysiol. 1996;12(2):186.
83. Shen LH. J Fudan Univ (Med Sci), 2001;(4):297.
84. Kobayashi T, Kuroda S, Tada M, et al. Brain Res. 2003;960(1–2):62.
85. He LN, He SB, Yang J. Chin J Mod Appl Pharm. 2001;18(3):179.
86. Li ZZ, Wu WP, Kuang PG. Chin J Geriatr Heart Brain and Vessel Dis. 2001;3(3):190.
87. Mo ZX, Zheng YS, Fan ZQ, et al. Pharmacol Clin Chin Mater Med. 1998;14(4):24.
88. Plesnila N. Acta Neurochir Suppl. 2004;89:15.
89. Gao YZ, Hu PZ, et al. Med J Nat Defending Forces Northwest China. 2001;22(3):268.
90. Zhu DY, La LL, Liu SH, et al. Proc Natl Acad Sci. 2004;101(25):9453.
91. Shen X, Ge W, Liu YH, et al. Chin J Pract Chin Mod Med. 2003;12(4):45.
92. Zhang JT, Lang SY, Kuang PG, et al. Chin J Geriatr Heart Brain Vessel Dis. 2001;3(4):272.
93. Zhang JN, Yi SY, Shangguan XF. J Fourth Military Medical Univ. 2002;3:189.
94. Zhang JT, Li YS, Zhao SP, et al. J Apoplexy Nerv Dis. 2001;18(1):26.
95. Zhang JT, Wu WP, Liu JX, et al. Chin J Geriatr Cardiovasc Dis. 2000;2(2):124.
96. FeLIstel PJ, Jin Y, Kimelberg HK. Stroke, 2004;35 (5):1164.
97. Wang J, Zhang JT. Chin J Pharmacol Toxicol. 1999;13(3):197.
98. Kimelberg HK, Nestor NB. Neurochem Res. 2004;29(1):267.
99. Li J, Yang CD. J Apoplexy Nerv Dis, 1995;12 (5):258.
100. Lerouet D, Beray-Berthat V, Palmier B, et al. Brain Res. 2002;958(1):166.
101. Cao S, Guo ZK, Zhang ZL, et al. Heilongjiang Med J. 2003;27(9):653.
102. Keynes RG, Garthwaite J. Curr Mol Med. 2004;4 (2):179.
103. Li ZZ, Kuang PG, Wu WP. Stroke Ner Dis. 1997;4 (1):1.

104. Wang ZQ, Chen XC, Yang GY, et al. Chin Med J (Engl). 2004;117(4):586.
105. Liu J, Kuang PG, Wu WP, et al. Chin J Geriatr Heart Brain Vessel Dis. 2002;4(2):127.
106. Alvaro-Gonzalez LC, Freijo-Guerrero MM. Rev Neurol. 2002;35(5):4522.
107. Wang WJ, Wu CZ, Yao Z, et al. Chin J Immunol. 1995;11(6):370.
108. Gao CY, Zhou HD, Chen ME, et al. J Apoplexy Nerv Dis. 2001;18(2):113.
109. Wada K, Sugimori H, Bbide P.G., et al. Stroke, 2003;34(11):2722.
110. Barth A, Barth L, Morrison R.S, et al. Neuroreport, 1996;7:1461.
111. Bethel A, Kitsch J.R, Koehler R.C, et al. Stroke, 1997;28(3):609.
112. Kazubiko N, Finkiestein SP, Beal MFJ. J Cereb Blood Flow Metab. 1993;13(2):221.
113. Maiese K, Boniece l, DeMeo 13, Wagner J.A. J Neurosei, 1993;13:3034.
114. Nakata N, Kato H, Kogure K. Brain Res. 1993;605:354.
115. Liu J, Wu WP, Kuang PG. Chin J Neuroimmunol Neurol. 1996;3(4):225.
116. Jiang SJ, Wu WP, Zhang XS, et al. Chin J Neurol. 1999;5(32):290.
117. Olsson T, Hanssono, Nylandsted J, et al. Exp Brain Res. 2004;154(4):442.
118. Kuang PG, Liu J, Wang FY, et al. Journal of Brain and Nervous Diseases. 1999;5(1):1.
119. Jin X, Xiao C, Tanguay R.M., et al. Am J Med, 2004;117(6):406.
120. Xu J, Zhou Y. Modern Medicine & Health. 2004;20 (5):310.
121. Hong X, Zheng L, Chen X. Chin J Med. 2002;82 (23):1637.
122. Li CJ, Xia CQ. Jiangxi J Tradit Chin Med. 2003;34 (6):27.
123. Geng ZB, Ying JY. J Tradit Chin Med Res. 2000;16 (4):30.
124. Hou TT. J Tradit Chin Med Study. 1998;11(1):13.
125. Hu L, Zhang Y. Chin J Clin Med. 1998;18(3):36.
126. Min LQ, Ma WY, et al. Chin J Integr Tradit West Med. 2002;22(5):353.
127. Shen J, Jiang WH. Chin J Integr Med Cardio. 2004;2(2):85.
128. Wang QJ, Hu SL, He LY, et al. Basic Med Sci Clin. 1997;17(6):74.
129. Xu BZ, Gao M. Chin J Pract Chin West Med. 2004;17(4):2406.
130. Zhang HS. Shaanxi J Tradit Chin Med. 1991;12 (4):153.
131. Zhang YY, Chen RX. Chin J New Drugs Clin Remedies. 2000;19(1):30.
132. Chen S.D, Lee J.M,Y ang D.I, et al. Am J Cardiovasc Drugs, 2002;2(5):303.
133. Atkinson RP, DeLemos C. Thromb Res. 2000;98 (3):97.
134. Ma LY, Chen W. Chin Tradit Herbal Drugs, 1999;30(5):391.
135. Meng XJ, Zhao XQ, Yan SX, et al. Chin J Pract Chin West Med. 2003;3(16):2015.
136. Zhang RF, Li X. Journal of Shanxi Medical University. 2003;34(2):182.
137. Wang SX, Xie SH. Chin Tradit Herbal Drugs. 1986;17(10):19.
138. Dong LP, Wang TY. Chin J Appl Physiol. 1996;12 (2):107.
139. Wang YL, Chen NH, Ding JH. Chin J Pharmacol Toxicol. 1994;8:19.
140. Li MN, Dong WB. Chin J Crit Care Med. 2005;25 (5):351.
141. Zhang RX, Zhang XX, Wu WP, et al. Med J Chin People's Liberation Army. 1999;24(6):447.
142. Xu LL, Wu Q, Wang BY, et al. Chin J Pathophysiology. 1994;10(6):635.
143. Ye LB, Xi T, Chen F, et al. J China Pharm Univ. 2004;35(3):267.
144. Chen YH, Du GH, Zhang JT. Chin Pharmacol Bull. 2000;21(5):463.
145. Yao RL, Tian L, Liu LB, et al. Chin J Publ Health. 1999;15(12):1059.
146. Wang T, Li GS, Tang QZ, et al. J Hubei Med Uni. 1996;17(2):134.
147. Lu QP, Shi CZ, Zhang ZL, et al. Chin J Exp. 1996;130:59.
148. He SB, He LN, Yang J. Chin Tradit Pat Med. 2002;24(5):371.
149. Chang JJ, Kuang PG, Liu J, et al. Stroke and Nerv Dis. 1997;4(1):8.
150. Liang H. Liaoning Pharm Clin Remedies. 2004;7 (2):103.
151. Liu J, Wu WP, Kuang PG, et al. Chin J Neuroimmunol Neurol. 1998;3(4):225.
152. Liu J, Kuang PG, Wu WP, et al. Chin J Nerv Mental Dis. 1999;25(4):198.
153. Zhang JT, Wu WP, Liu JX, et al. Chin J Geriatr Cardiovasc Dis, 2000;2(2):124.
154. Tian Y, Di ZL, Liang HD, et al. Chin J Pharmacoepidemiol. 2004;13(4):169.
155. Liu J, Wu WP, Kuang PG. Chin J Neuroimmunol Neurol, 1996;3(4):225.
156. Jiang SJ, Wu WP, Zhang XS, et al. Chin J Neurol. 1999;32(5):290.
157. Kuang PG, Liu J, Wang FY, et al. J Brain Nerv Dis. 1999;5(1):1.
158. Zhong MY. J Guangdong Coll Pharm. 2004;20 (2):203.
159. Cheung F, Sung Fl, et al. Mol Cel Biochem. 2000;207(1–2):35.
160. Zhang JT, Li YS, Zhao SP, et al. Stroke Nerv Dis. 2001;8(1):26.

161. Shen X, Cui GY, Liu YH, et al. Chin J Neurol. 2004;37(2):174.
162. Du GH, Zhang JT. Pharm J. 1995;30:184.
163. Du GH, Wiu Y, Zhang JT. Chin Med J. 1997;110:65.
164. Du GH, Zhang JT. Facilitation of and induced by salvianolic acid A and B in mice. The Third China-Japan Joint Meeting on Pharmacology, Beijin 9, 1993, 111.
165. Du GH, Zhang JT. Protective effects of salvianolic acid A against impairment of induced by cerebral ischemia reperfusion in mice. The 6th Symposium of Neuropharmacology, Beijing, 1994, 28.

Effect of Danshen on the Blood System and Microcirculatory Function

6

Jingyan Han, Zhixin Guo, Jinhua Wang, Li Zhang and Guanhua Du

Danshen, its main active components, and Danshen preparations all have the function of improving microcirculation, and these functions are the pharmacological basis for their clinical application in the prevention and treatment of cardio-cerebral-vascular diseases.

Microcirculation covers the areas of small arteries, capillaries, and small veins in the circulatory system, and it also includes the blood, lymph and tissue fluid circulating in these areas. Microcirculation is the location, inside and outside of blood vessels, for the exchange of materials, energy, and information. The microcirculatory system accounts for 90 % of the vascular system in vivo, and distributes to all organs except for cartilage.

Microcirculatory dysfunctions can be caused by any factor which can influence blood vessels, blood components, blood volume, and blood flow velocity. Ischemia-reperfusion (I/R) causes the production of excessive peroxides by vascular endothelial cells and leukocytes, resulting in microcirculatory dysfunction [1–4]. Endotoxin (1ipopolysaccharide, LPS) damages vascular endothelial cells directly and activates leukocytes, leading to microcirculatory dysfunction [5–8]. The overexcitement of the autonomic nervous system causes repeated arteriole contraction, inducing microcirculatory dysfunction [9]. Surgery, interventional therapy, trauma, and organ transplants impair vascular endothelial cells directly, or induce I/R, causing microcirculatory dysfunction [10–14].

Microcirculatory dysfunction is a complex process of pathological change, including damage to vascular endothelial cells [5], changes in blood vessel diameters and blood flow velocity [5, 15, 16], and activation and adhesion of red blood cells [17], leukocytes [4, 6, 8], monocytes [18], lymphocytes [19], platelet [20–22], mast cells degranulation [4, 23], and involving peroxides [5, 16], selectin [15, 24], adhesion molecules [5, 8], vasoactive substances [25], and inflammatory factors [26, 27].

To improve microcirculatory dysfunction, research efforts in the past were focused on the roles of certain links or a single component, such as antiperoxide superoxide dismutase (SOD), catalase (CAT), glutathione precursor (*N*-acetyl-l-cysteine, NAC) [16–31], antiselectin antibody [32, 33], antiadhesion molecule antibodies [32], protease inhibitors [34, 35], anti-inflammatory factor antibodies [36], mast cell stabilizers [37], etc. However, since inhibitors aimed at a single target cannot improve the overall dysfunction of microcirculation, this approach is yet to achieve satisfactory results. The characteristics of microcirculatory dysfunction are the complex disease process of multitarget injuries, which need multitarget interventions. Danshen, its major water-soluble components, and Danshen preparations

J. Han · Z. Guo · J. Wang · L. Zhang · G. Du (✉)
Institute of Materia Medica, Chinese Academy of Medical Science and Peking Union Medical College, Beijing, China
e-mail: dugh@imm.ac.cn

X. Yan (ed.), *Dan Shen (Salvia miltiorrhiza) in Medicine*,
DOI 10.1007/978-94-017-9463-3_6,

have the functions of improving microcirculatory dysfunction at multiple targets [5, 8], and have good effects on the prevention and treatment of microcirculatory dysfunction caused by (I/R) [1, 38–40], LPS [5, 8], and other factors [22].

6.1 The Effect of Danshen on Microcirculatory Dysfunction Caused by I/R

I/R refers to recanalization after blood flow occlusion. Contracture of the small arteries caused by sympathetic nervous overexcitement, vasoconstriction induced by cold temperature, vascular obstruction by thrombosis, blood vessel ligation by surgery and organ transplantation; all of these can lead to the blockage of blood flow, which is also called ischemia. When ischemia occurs, ATP is degraded to AMP, which is then broken down into hypoxanthine. On the other hand, xanthine dehydrogenase is transformed into xanthine oxidase [3].

Antispasmodic drugs remove vasospasm, warm and heat herbs expand constricted blood vessels, and thrombolytic therapies or interventional therapies recanalize obstructed blood flow. When the blood vessels are recanalized, the blood provides the ischemic area with a large amount of O_2 and water, which help xanthine oxidase to oxidize hypoxanthine to produce a large number of negative oxygen anions ($\cdot O_2^-$), which is converted by SOD into hydrogen peroxide (H_2O_2), and then into H_2O by CAT. However, portions of H_2O_2 are changed into hydroxyl radicals ($\cdot$OH) [3] by the Haber-Weiss reaction. Nitric oxide (NO) produced in vascular endothelial cells combines with $\cdot O_2^-$ to form peroxynitrite ($ONOO^-$) [41]. H_2O_2, $\cdot OH^-$, and $ONOO^-$ are highly toxic peroxides which can cause lipid peroxidation, DNA breakage, and damage to vascular endothelial cells and perivascular cells [42–44].

The peroxide products generated during I/R can degrade the transcription inhibitor kappa B (I-κB) [45] and activate the nuclear factor kappa B, (NF-κB) [46], which causes the translocation of its subunits P65 and P50 to the nucleus [45]. These changes in turn activate adhesion factors, inflammatory factors, apoptosis protein genes, and promote the synthesis, expression, and release of adhesion molecules [47] and inflammatory factors [48, 49].

The expression of vascular endothelial selectin (E-selectin) and leukocyte selectin (L-selectin) leads to leukocytes rolling along the vessel wall [32, 50, 51]. The expression of vascular endothelial adhesion molecule-1 (ICMA-1) and leukocyte adhesion molecules CD11b/CD18 causes the adhesion of leukocytes to the vessel wall [51, 52]. Leukocytes attached to vascular endothelia can, on the one hand, produce peroxides explosively through coenzyme II (nicotinamide adenine dinucleotide phosphate, NADPH) oxidase [53–56], and on the other, release a wide range of proteases [3, 57, 58] which dissolve and damage vascular endothelial cells and the vascular basement membrane [57, 59]. Furthermore, leukocytes attached to vascular endothelia pass the vascular endothelial cells and vascular basement membrane and emigrate out of the vessel through injured vascular endothelial cells, or through the cell junction (see Fig. 6.1).

Leukocytes that have emigrated from blood vessels release peroxides and proteases, which further damage cells or tissues in the surrounding regions. Platelet adhesion molecule CD31 activated by I/R can adhere to the vessel wall, but also to red blood cells (see Fig. 6.2a). Leukocytes adhered to vascular endothelial cells can also adhere to platelets and red blood cells (see Fig. 6.2b). Adhesion among platelets, leukocytes, red blood cells, and vessel walls are the early form of thrombosis.

The injuries of vascular endothelial cells, the adhesion between platelets and vascular endothelial cells, and the release of active substances activate thrombin, which turns fibrinogen into fibrin. The latter forms a network structure in which blood cell components forms a hard thrombosis (see Fig. 6.2c and d). The mast cells outside of blood vessels degranulate after being stimulated by peroxides produced by I/R [16], releasing vasoactive substances such as histamine, serotonin, platelet-activating factors, and inflammatory factors such as TNF-α and

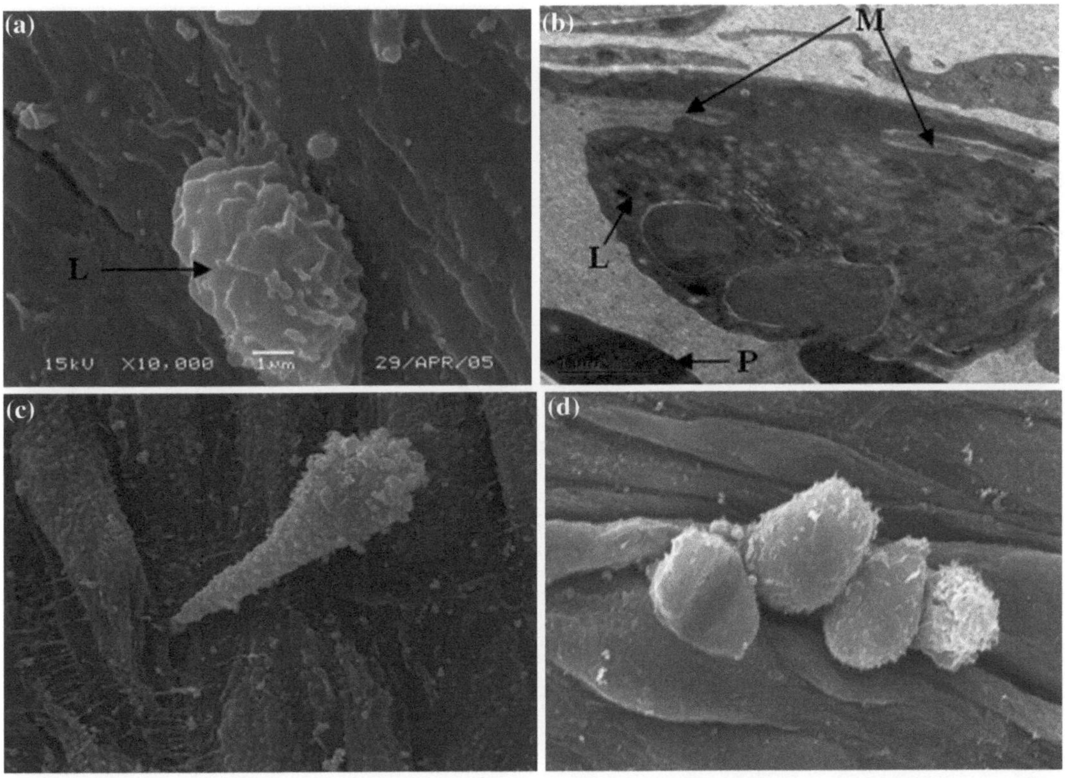

Fig. 6.1 Leukocyte adhesion to vascular walls and migration. **a** The scanning electron microscope image of leukocyte adhered to the blood vessel wall *L* leukocyte. **b** The transmission electron microscope image of leukocyte adhered to the blood vessel wall emigrating out of the blood vessels. *M* The broken vascular endothelium. *L* Leukocyte adhered to the blood vessel wall emigrating through the blood vessels. Portion of the leukocyte has been emigrating through the disconnected region of vascular endothelium. *P* Intravascular platelet. **c** and **d** Leukocyte emigrating out of the blood vessels from a gap along the endothelium. (Images provided by Zhao Bai-Huu, Zhao Xin-Rong, and Chang Xin, Tasly Microcirculation Research Center of Health Science Department, Peking University)

interleukins [60, 61]. These vasoactive substances and inflammatory factors attack blood vessels from the outside, increasing cell adhesion inside and worsening vascular permeability [37]. Inflammatory factors released by vascular endothelial cells and leukocytes are also involved in the process of damage to blood vessels and surrounding tissue cells. Monocytes and T lymphocytes are involved in the process of damage to blood vessels and surrounding tissue cells in the late reperfusion stage [62].

The peroxidase, proteases, and inflammatory factors produced in the process of I/R attack vascular endothelia and the vascular basement membrane, and vasoactive substances and inflammatory factors released by mast cells attack blood vessels from the outside, causing the leakage of plasma albumin, perivascular edema, blood cell emigration, hemorrhage, and thrombosis. Thrombosis, peroxides, inflammatory factors, edema, and vascular inflammatory cell emigration may lead to apoptosis and cell death in the surrounding tissues (see Fig. 6.3). I/R-induced microcirculatory dysfunction is a pathological process consisting of multiple links and targets, so its improvement needs interference on multiple levels.

Since 1993, the authors have devoted themselves to microcirculation research in a medical digestion microcirculation lab in the medical department of Keio University, under the guidance of the late professor Tsuchiya Masaharu, the

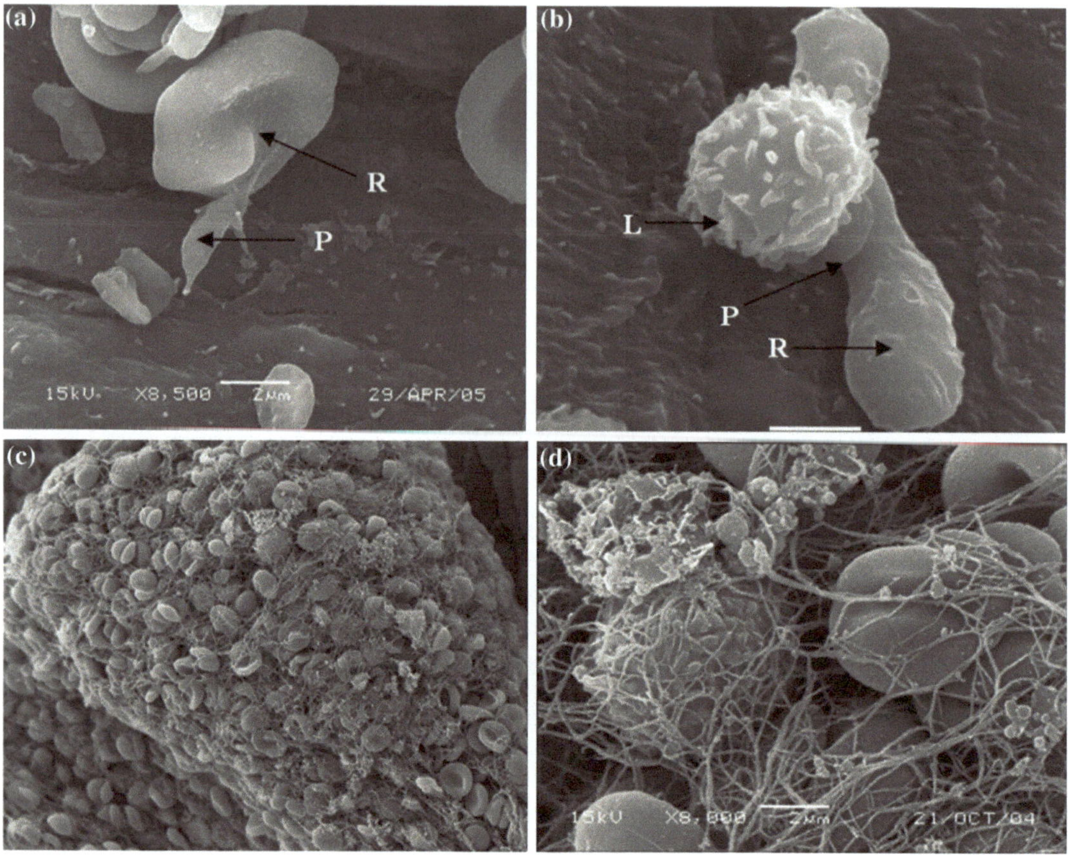

Fig. 6.2 Thrombosis. **a** The scanning electron microscope image of a platelet (*P*), it adheres to the vessel wall on one end and to a red blood cell (*R*) on the other. **b** The scanning electron microscopy images of a leukocyte (*L*) adhered to a blood vessel wall, platelet, and red blood cell. **c** and **d** Thrombosis formation by a network of blood cells connected by fibrin. (Images provided by Zhao Baihuu, Zhao Xin-Rong, and Chang Xin, Tasly Microcirculation Research Center of Health Science Department, Peking University)

chairman of the Japan Microcirculation Association, professor Hiromasa Ishii, the incumbent chairman of the Japan Microcirculation Association, and professor Souitirou Miura. Using microcirculation visualization research systems, the authors have made systematic and dynamic studies on the effects of Danshen, Tanshinol, total salvianolic acids, compound Danshen injection (CDI), and Compound Danshen Dropping Pill on the improvement of I/R-induced microcirculatory dysfunction. Tasly Microcirculation Research Center of Peking University Health Science Center was founded in 2004, and study has been continued since then, with results proving that Danshen, Danshen active components, and Danshen preparations can improve the microcirculatory dysfunction induced by I/R.

6.1.1 Danshen Improved the Microcirculatory Dysfunction Caused by I/R

An I/R model was established by ligating the mouse mesenteric artery and vein for 10 min, then releasing the ligation to restore blood flow. High-sensitivity CCD cameras and fluorescence cameras were connected to an inversion biomicroscope to observe and record the dynamic processes in the rat mesenteric microcirculation system.

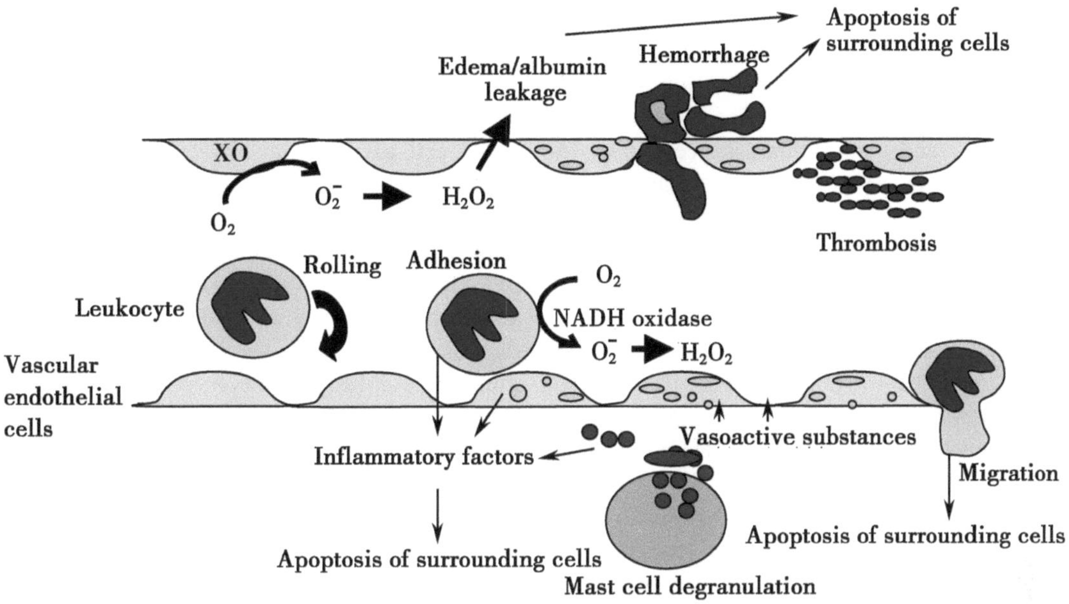

Fig. 6.3 Multitarget damage of microcirculatory dysfunction. (Figure is drawn by the authors)

6.1.1.1 Effect on the Small Artery and Vein Vessel Diameter

During the first 30 min of I/R, the diameters of the mesenteric small artery and vein in rats had no significant changes. Intragastric administration of Danshen extract (0.4 g/kg), Dantonic™ (Composite Danshen Dropping Pill) (0.4 g/kg), CDI [fivefold dilution, 6 ml/(kg h)], Tanshinol [5 mg/(kg h)], total salvianolic acids [5 mg/(kg h)] before I/R and 10 min after reperfusion (continuous intravenous drip) did not significantly change the diameters of the rat mesenteric small artery and vein. The results suggest that Danshen's improving effects on I/R-induced microcirculatory dysfunction may be achieved through ways other than changing blood vessels.

However, Danshen did have significant improving effects on the contraction of small arteries induced by local dropping of norepinephrine [63]. The results indicate that Danshen does not affect the vascular diameter in normal circumstances or in pathological circumstances where changes in the vascular diameter were absent. However, it can inhibit and improve vasoconstriction when blood vessels do constrict.

6.1.1.2 Effect on Peroxides

In vitro studies have shown that the Danshen and its active components can inhibit the production of a variety of peroxides. Tanshinol can reduce I/R-induced mitochondrial membrane damage and lipid peroxidation, and clear $\cdot O_2^-$ generated from xanthine and xanthine oxidase [64]. Salvianolic acid A can inhibit I/R-induced brain lipid peroxide (LPO) increase and remove $\cdot O_2^-$ [65]. Salvianolic acid B can inhibit peroxide production in PC12 cells induced by amyloid peptide A [66] and in human aortic smooth muscle cells (SMC) induced by TNF-α, and inhibit NADPH oxidase activity [67]. Magnesium lithospermate B can clear $ONOO^-$ [68].

In order to study Danshen's effects on rat mesenteric small vein peroxides after I/R, 10 μmol of peroxide fluorescent probe Rhodamine Dihydroartemisinin 123 (dihydrorhodamine 123, DHR) was added by continuous dropping onto the surface of the mesentery. DHR reacts with peroxide to generate rhodamine, which penetrates into cells and combines with the mitochondrial membrane. The dynamic process of peroxide generation and location can be observed under fluorescent microscopy with a wavelength of 420–490 nm and emitting light at 520 nm to stimulate luciferase luminescence. Starting from 10 min after reperfusion, the fluorescence intensity of small veins increased significantly, and the fluorescence intensity increased with reperfusion time. The results suggest that I/R causes an increase in peroxide production in small vein endothelia. Intragastric administration of Danshen extract, Dantonic™ [69], intravenous administration of CDI [70], Tanshinol, and total salvianolic acids before I/R showed significant inhibitory effects 10 min after reperfusion on peroxide production in rat mesenteric small veins induced by I/R. Intravenous administration of CDI [70], Tanshinol, and total salvianolic acids 10 min after reperfusion can also significantly inhibit peroxide production in venule endothelia, and the effects were the same as the result of pretreatment and posttreatment with SOD [12,000 u/(kg h)] and CAT [20 mg/(kg h)].

These results suggest that the improving effects of Danshen injection, Tanshinol, total salvianolic acids, Dantonic™, CDI on I/R-induced microcirculatory dysfunction are related to their functions of inhibiting peroxides.

6.1.1.3 Effect on the Expression of Adhesion Molecules in Leukocytes and Vascular Endothelia and Cell Adhesion

In vitro studies have shown that Danshen's water extract [71, 72], protocatechuic aldehyde [73], salvianolic acid B [72], and Dantonic™ [74] can all inhibit the TNF-α-induced expression of ICAM-1 and VCAM-1. The role of protocatechuic aldehyde was related to its inhibitory effect on the activation of NF-κB, AP-1 [73]. We used high-sensitivity CCD camera systems connected to an inverted biological microscope to continuously count the number of leukocytes adhered to the rat mesenteric small artery and vein wall after I/R (leukocyte adhesion in the same part of the vessel wall for more than 10 s). Sham groups had almost no leukocyte adhesion. In the I/R group, after the start of reperfusion, there was a significant increase in the adhesion of white blood cells in thin vein walls. Pretreatment with intragastric administration of Danshen extract, Dantonic™ [69], and CDI [70], and intravenous administration of Tanshinol and total salvianolic acids significantly reduced the amount of white blood cell adhesion in thin veins 10 min after reperfusion. Leukocytes were found adhered to the rat mesenteric artery and vein wall 10 min after reperfusion, and intravenous administration of SOD and CAT could only inhibit the new adhesion of leukocytes, with no effect on the dissociation of leukocytes already adhered to the thin vein wall [70]. However, posttreatment with Tanshinol and CDI could dissociate leukocytes from the thin vein wall [7, 75].

Leukocytes adhered to the blood vessel wall produced peroxides through NADPH oxidase. At the same time, they secrete proteases, which increases the damage to the nearby endothelial cells and basement membrane. Leukocytes adhered to the blood vessel wall emigrate from the vessel through the gaps or injuries in the wall, which can cause perivascular cell damage. Inhibition and dissociation of the adhesion between leukocytes and endothelial cells is the key to improving I/R-induced microcirculatory dysfunction and blood vessel damage. Danshen injections, Tanshinol, total salvianolic acids, Dantonic™, and CDI all have such effects, which are related to Tanshinol's function of inhibiting the expression of leukocyte adhesion molecule CD11b/CD18 and ICAM-1 [8]. In other words, Danshen and its major components inhibit the adhesion between leukocytes and endothelial cells by inhibiting the expression of adhesion molecules in these cells.

6.1.1.4 Effect on Mast Cell Degranulation

After I/R, mast cells degranulate, releasing vasoactive substances and inflammatory factors which attack blood vessels extravascularly, increasing the rolling and adhesion of leukocytes as well as vascular permeability. 30 min after I/R, 0.1 % toluidine blue dye solution was added dropwise onto the mesenteric surface. 30 s later, the dye was slowly washed off with normal saline, and the mast cell granules were stained blue-violet (see Fig. 6.4). The high-sensitivity camera system connected to the microscope was used to

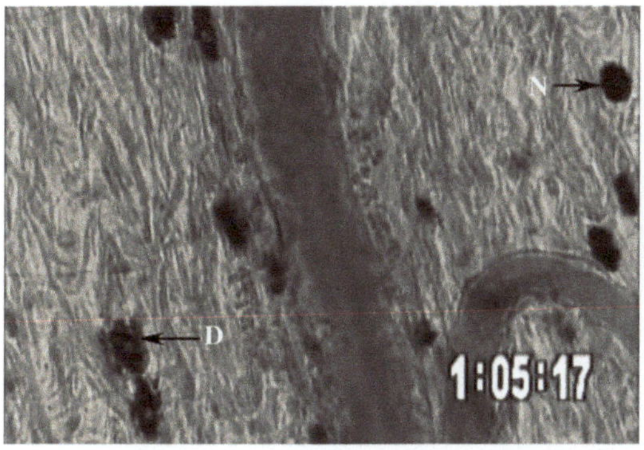

Fig. 6.4 Degranulation image of mast cell outside venule vessel and inside the mesenteric interstitial. *N* Nondegranulation mast cells, *D* degranulation mast cells. (Image provided by the authors)

observe five fields along the small veins, and the degranulated and nondegranulated mast cells were counted to calculate the rate of degranulation [6]. The results showed that the degranulation rate of rat mesentery interstitial mast cells was significantly increased after I/R, and intragastric administration of Danshen extract could not inhibit the process [69]. Intragastric administration of Dantonic™ [69] and pretreatment with CDI, Tanshinol, total salvianolic acids, and SOD could inhibit mast cell degranulation, but showed no inhibitory effect posttreatment. Neither pretreatment nor posttreatment with CAT could inhibit the degranulation. Both pretreatment and posttreatment with total saponins, Rbl, and Rl of Notoginseng in Dantonic™ could inhibit I/R-induced mast cell degranulation. The results suggest that the inhibitory effect of Dantonic™ on mast cell degranulation was achieved by Notoginseng and its active ingredients. 15,16-hydroxytanshinone and cryptotanshinone could inhibit mast cell degranulation in vitro [76], but whether these lipid-soluble components of Danshen have any improving effects on the mast cell degranulation induced by I/R is unknown.

6.1.1.5 Effect on the Leakage of Venule Albumin

After I/R, the peroxides produced by vascular endothelial cells and adhered leukocytes, along with the inflammatory factors and proteases released from the latter, can attack blood vessels from the inside and damage vascular endothelial cells and the vascular basement membrane. Vasoactive substances and inflammatory factors released by the degranulation of extravascular mast cells not only can increase the adhesion between vascular endothelia and leukocytes, but also can expand the gap between vascular endothelial cells and increase vascular permeability, resulting in the leakage of serum albumin, and causing perivascular edema. FITC (fluorescein) labeled albumin, 50 mg/kg, was injected slowly through the jugular vein, and the changes within and outside the same small vein were observed and recorded using a fluorescence microscope camera system (Fig. 6.5). The plasma albumin leakage situation through the small veins was calculated according to FITC fluorescein change in leakage between inside and outside vessels [5].

I/R can cause a significant increase in FITC-labeled albumin leakage in rat mesenteric small venules. Intragastric administration of Danshen extract and Dantonic™ [69] and pretreatment with Tanshinol, total salvianolic acids, and CDI [70] showed significant inhibitory effects on leakage 20 min after reperfusion. Albumin leakage had already occurred at 10 min after reperfusion, and administration of CDI [70], Tanshinol, and total salvianolic acids at this time still can inhibit leakage. SOD and CAT treatments showed similar results.

Danshen, Tanshinol, total salvianolic acids, and CDI have inhibitory and improving effects on I/R-induced peroxide production and adhesion between leukocytes and endothelial cells. The Danshen and water-soluble components in Dantonic™ can inhibit peroxide production and adhesion between leukocytes and endothelial cells, and the Notoginseng and total saponins components can inhibit adhesion between leukocytes and endothelial cells and degranulation of mast cells. So, Dantonic™ can protect vascular endothelial cells and the basement membrane from albumin leakage and aggressive factors either intravascularly or extravascularly by targeting multiple links.

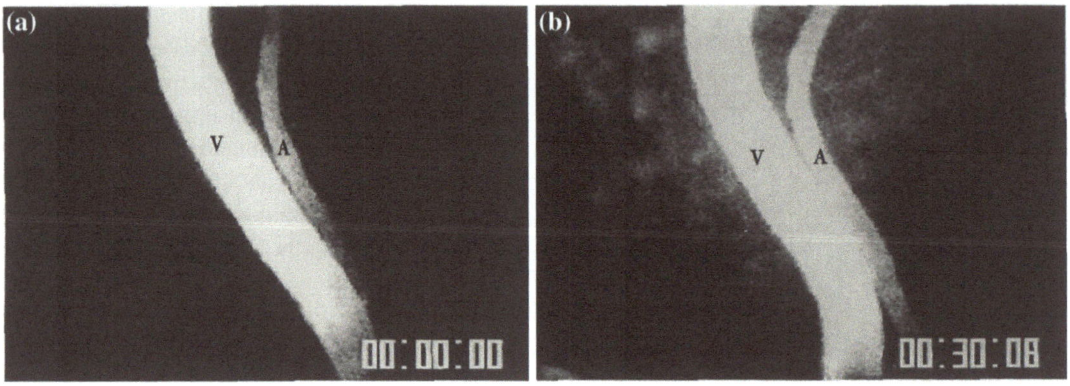

Fig. 6.5 Image of FITC-labeled plasma albumin leakage. **a** The image before I/R, no FITC-labeled plasma albumin leakage. **b** The image 30 min after I/R, the FITC-albumin leakage can be observed around the blood vessels. *V* venule, *A* arteriole. (Image provided by the authors)

6.1.2 The Improving Effect of Danshen on I/R-Related Injury

After the recanalization of blood blockage, dysfunction in the microcirculatory system and damage to related organs could occur. Danshen and Danshen formulations can improve the dysfunction and damage [77].

6.1.2.1 Protective Effect on Myocardial Cells After I/R

The water-soluble extract of Danshen can reduce coronary left ventricular myocardial infarction size and increase survival rate after I/R in rats [78, 79]. Lithospermate B can reduce myocardial injury after I/R in rabbits.

Clinical treatment with Dantonic™ for I/R-related diseases has achieved good therapeutic efficacy. Wang et al. [80] reported that when using Dantonic™ to treat coronary heart disease patients with left ventricular hypertrophy, the update rate of microcirculation and update time, and retention time were significantly improved compared to before the treatment. By analyzing the microcirculation and thromboelastography of coronary heart disease in patients with bulbar conjunctiva, Chen et al. [81] found that the medium-dose (10 pills each time, three times per day) and high-dose (15 pills each time, three times per day) of Dantonic™ had dose-dependent improving effects on the bulbar conjunctiva arteriole vessel diameter, microvein diameter, microartery blood flow velocity, and microartery flow in patients with coronary artery muscle muscle disease. Fu Xiao-Liang et al. [82] observed the nail fold microcirculation of 92 patients with coronary heart disease, and found that Dantonic™ can significantly improve the number of nail fold capillary loops, reduce tube-loop cross deformities and pipe loop diameter, increase the linear and granular flow, reduce erythrocyte aggregation, etc.

6.1.2.2 Protective Effect on I/R Brain

Danshen and Danshen preparations were used in clinical treatments of cerebrovascular disease [83]. Danshen injected at dosages of 15 and 30 mg/kg could reduce the area of cerebral infarction and peripheral blood peroxide production after I/R [84]. Tanshinone II_A and tanshinone II_B administered by intraperitoneal injection could pass the blood-brain barrier and reduce the area of cerebral infarction induced by middle cerebral artery I/R in mice [85]. Salvianolic acid A or B (3 and 10 mg/kg) by intravenous administration could improve the content change of MDA induced by I/R in mouse brain cortex, hippocampus and striatum, and improve memory [64, 86]. Salvianolic acid B by intravenous administration (10 mg/kg) also could inhibit the decrease of SOD and GSH and the increase of MDA and lactic acid in I/R-induced brain tissue [87].

6.1.2.3 Protective Effect on Liver After I/R

Chronic alcohol intake (CAI) can cause hypertriglyceridemia and fatty liver, and consume glutathione, which could decrease the body's capacity of removing peroxides. During the oxidative stress response, the capacity of removing peroxides decreases, which leads to the accumulation of oxides and results in serious microcirculatory dysfunction. Supplements of glutathione precursor NAC can reduce mesenteric microcirculatory dysfunction induced by reperfusion in rats [16]. In order to study the effects of Dantonic™ on the mesenteric and hepatic microcirculatory dysfunction induced by I/R in rats with CAI, the authors used the method invented by Professor Liber to feed rats with hypertriglyceridemia and fatty liver induced by CAI. A rat I/R model was established by ligation of the mesenteric artery and vein. The results showed that the peroxide fluorescence intensity and FITC-labeled albumin leakage in CAI rats were more severe than those in non-CAI rats. SOD and NAC could significantly inhibit these changes [16], and one pretreatment with Dantonic™ (0.4 g/kg) could also significantly inhibit the changes [88]. We further observed the microcirculation situation in rat liver one hour after ligation/reperfusion of mesenteric artery. Compared with the sham operation group, the number of leukocytes adhered to the liver after I/R increased significantly. Compared with the non-CAI group, the number of leukocytes in the CAI group increased markedly. Also, the triglyceride level in CAI rats was significantly higher than that in non-CAI rats. Administration of Dantonic™ (0.4 g/kg) 12 h and 90 min before reperfusion could significantly inhibit I/R-induced leukocyte adhesion to the liver, suppress the increase of serum ALT and TNF-α, and lower serum triglyceride [38]. In addition, intramuscular injection of Danshen aqueous extract could improve morphological damage in rat liver caused by I/R, and suppress the increase of serum ALT and MDA [89].

6.1.2.4 Protective Renal Effects After I/R

Intravenous administration of lithospermate B, one of the water-soluble components of Danshen, at dosages of 10, 30, and 60 mg/kg could improve renal cortex micro flow after I/R in SD rats [39]. A dosage of 40 mg/kg could improve renal function and reduce renal tissue injury after I/R in rats [90].

6.1.2.5 Improving Effects on Microcirculatory Dysfunction After Small Intestine Transplantation

Li et al. [91] observed the protective effect of Danshen on microcirculation in hybrid pigs after small intestine transplantation. The supplied intestine was preserved in either Eurocollina or Eurocollina plus Danshen preservation solution (Danshen injection 20 ml/L), and stored at 4 °C for 24 h before autologous transplantation. Intestines preserved in Danshen preservation solution showed a lower score of intestinal mucosa injury 30 min after transplantation than those in Eurocollina solution, and microcirculation permeability in the former was also significantly decreased compared with the latter. These results suggest that Danshen can effectively inhibit the increase of small intestine microcirculation permeability when stored at low temperature, and reduce the damage to vascular endothelial cells.

6.1.2.6 Improving Effects on Lower Limb Injuries After I/R

Fan et al. [92] built a left lower limb I/R injury model in New Zealand white rabbits using pneumatic tourniquets, and observed the protective effect of Danshen injection on limb injuries. After I/R, blood flow in the left lower limb skeletal muscle of New Zealand white rabbits was slow, erythrocytes aggregated severely, a large number of leukocytes attached to the blood vessel wall, white microthrombi were formed, and the no-reflow phenomenon was serious. Pretreatment with Danshen injection by intravenous administration before ischemia and reperfusion could alleviate the microcirculatory dysfunction and reduce skeletal muscle tissue injury. The results suggest that Danshen has a protective effect on I/R-induced microcirculation dysfunction in lower limbs.

6.1.2.7 Effect on Flaps

Liang et al. [93] observed the effect of Danshen injection on the vitality and microcirculation of random pattern skin flaps. The results showed that remote flap necrosis was induced by microcirculatory dysfunction, which was caused by the gradual decrease in vascular perfusion pressure, and Danshen could significantly improve the microcirculatory dysfunction and improve flap survival length. Yan et al. [94] also observed the improving effect of Danshen injection on the microcirculation of rabbit skin tissue. After lifting the flap, blood flow in the microcirculatory system of the flap tissue decreased significantly. 1 h later, plasma ET/NO increased markedly, and returned to preoperational levels in 5 days. In the Danshen intravenous injection group, after lifting the flap, the blood flow in the flap microcirculation also decreased significantly, but after that, the blood flow was significantly higher than that in control group at each time point. Plasma ET/NO increased markedly at first, then decreased 3 h later, and returned to preoperational levels in 1 day. The flap survival length in the Danshen group was markedly increased l4 days after operation. The results indicate that Danshen improves the microcirculation of flap tissue and increases flap survival length by optimizing the ratio of plasma ET/NO.

Danshen, its major active components, and compound Danshen preparations improve microcirculatory dysfunction induced by I/R, and protect organs by acting on multiple links and targets [1]. Besides being used for the treatment of cardiovascular and **cerebrovascular diseases**, it is expected to become a mainstream medicine for the improvement of microcirculatory dysfunction caused by surgery, trauma, thrombolytic therapy, and interventional therapy.

6.2 Effect of Danshen on Microcirculatory Dysfunction Caused by Endotoxins

Lipopolysaccharide (LPS) is the major component of the outer membrane of Gram-negative bacteria. It can bind plasma CD14, form complexes with MD2, and act on TLR-4 receptors in the membranes of white blood cells and vascular endothelial cells. Through the IKK system, LPS can also degrade I-κB-α, activate NF-κB, cause PS0, P65 nuclear translocation, and initiate the gene expressions of adhesion molecules and Inflammatory factors [95–102]. LPS can cause the degranulation of mast cells by binding to their TLR-4 receptors [103]. LPS can induce the expressions of selectin and adhesion molecules, resulting in the migration of WBC along the vessel wall and adhesion between leukocytes and vascular endothelial cells, respectively [7, 8, 102]. The peroxides and proteases released by adhered leukocytes and the granules released by peripheral vascular mast cells damage vascular endothelial cells and the vascular basement membrane, leading to the leakage of albumin [5] and bleeding. Vascular endothelial injuries and platelet activation lead to thrombosis. LPS can cause complex microcirculatory dysfunction, and even cause disseminated intravascular coagulation (DIC), so it is a major cause of infectious disease fatality. Despite the fact that the treatments of dilatancy, blood vessel diameter regulation, antibiotics, TNF-α antibodies have saved many patients, drugs which can comprehensively (including antiperoxides, anti-cell adhesion, antialbumin leakage, antiplatelet adhesion, antimast cell degranulation) improve the microcirculatory dysfunction caused by LPS are yet to be developed. So, we have been studying the improving effects of Danshen, Tanshinol, lithospermate B, and Danshen preparations on endotoxin-induced microcirculatory dysfunction in rats, using visualization methods [5, 104].

6.2.1 Effect on Diameter of Small Artery and Vein

The endotoxemia-induced microcirculatory dysfunction model was built by continuous infusion of endotoxin at a dosage of 2 mg/(kg h) in Wistar rats through the femoral vein. Before the infusion, Tanshinol [5 mg/(kg h)], lithospermate B [5 mg/(kg h)], and CDI [5-fold dilution, 6 ml/(kg h)] were administered through the jugular vein. Rat mesenteric small artery and vein diameters showed no significant change after endotoxin infusion. Pretreatments with Tanshinol, lithospermate B, and CDI had no significant impact on the diameters of the mesenteric small artery and vein in rats [5, 8].

6.2.2 Effect on Erythrocyte Velocity of the Small Veins

Red blood cell flow velocity in the rat mesenteric small artery and vein was recorded with high-speed (2,000/s) cameras. The flow velocity in the rat mesenteric small vein was reduced significantly after endotoxin infusion for 40 min. Tanshinol, lithospermate B, and CDI markedly inhibited the velocity [5, 8].

6.2.3 Effect on the Adhesion of Leukocytes to the Small Veins

After endotoxin infusion, the amount of leukocytes attached to the mesenteric venule of rats increased significantly 10 min after the infusion was initiated, and the increase continued. Pretreatments with Tanshinol, lithospermate B, and CDI significantly inhibited the endotoxin-induced adhesion and leukocyte migration out of blood vessels [5, 8]. The results from in vitro studies also showed that Tanshinol, lithospermate B, and CDI could inhibit endotoxin-induced expression of adhesion molecule CD1 lb/CDl8 in neutrophils [5, 8]. CDI could inhibit endotoxin-induced expression of ICAM-1 in vascular endothelial cells [5]. Zhao et al. [105] have demonstrated that Danshen could inhibit endotoxin-induced expression of CD11a/CD18 in rabbit peripheral blood neutrophils, thus inhibiting the adhesion between neutrophils and endothelial cells.

6.2.4 Inhibitive Effect on Peroxides

We observed the effects of Tanshinol, lithospermate B, and CDI on the endotoxin-induced production of peroxides in peripheral blood neutrophils, and the results showed that they could inhibit the production [5, 8]. Furthermore, we observed the dynamic process of peroxide production in small veins of rats after intravenous infusion of endotoxin by dropping peroxide dihydrorhodamine fluorescence probes on the surface. 20 min after endotoxin infusion, peroxide production in the wall of rat mesenteric small veins increased significantly. Pretreatment with CDI could significantly inhibit production [5]. These results suggest that the major water-soluble components of Danshen and its preparations can inhibit endotoxin-induced peroxide production in vascular endothelial cells and leukocytes.

6.2.5 Protective Effect on Endovascular Cell Injury

Vascular endothelial cell injury can be observed in vivo by intravenous injection of the macromolecular pigment Monastral Blue B (MBB) (0.1 ml/100 g). CDI could significantly inhibit the increase in MBB stained regions in endotoxin-induced rat mesenteric venules [5]. Under transmission electron microscopy, the capillary endothelium of rats was uneven with a number of vesicles in endothelial cells after endotoxin administration. CDI can reduce the unevenness in the vascular endothelium and the number of vesicles in endothelial cells [5]. The above study proves that CDI has a protective effect on vascular endothelia.

6.2.6 Effect on Mast Cell Degranulation

After continuous endotoxin infusion, the mast cell degranulation rate outside blood vessels increased significantly. Pretreatment with Tanshinol, lithospermate B, and CDI could inhibit endotoxin-induced mast cell degranulation [5, 8]. However, it is unknown if these compounds still can inhibit mast cell degranulation after endotoxin-induced microcirculatory dysfunction has already occurred.

6.2.7 Inhibitory Effect on Serum Albumin Leakage

20 min after continuous infusion of endotoxin, FITC-albumin leakage along the small vein started to increase significantly. 60 min after CDI infusion, the leakage was reduced significantly [5].

Liu et al. [106, 107] have proved that CDI at the dosage of 0.5 g/2 ml with gavage administration had preventive effects on microcirculatry disorders, such as endothelial edema, gap increasing, wall thickening and injuries, cell adhesion, leukocyte emigration, red blood cell migration, blood flow slowing, etc.

Although Danshen is not the drug used for heat clearing and blood cooling, due to its anti-peroxides, anti-cell adhesion, and protection of vascular endothelium, it can improve endotoxin-induced microcirculatory dysfunction. The study results provided a pharmacological basis for the application of Danshen and its preparations in the treatment of infectious diseases.

6.3 Improving Effects of Danshen on Microcirculatory Dysfunction Caused by Other Factors

Danshen and Danshen preparations can improve microcirculatory dysfunction caused by photochemical reactions, norepinephrine, heat injury, noise, etc.

6.3.1 Improving Effect on Microcirculatory Dysfunction Caused by Photochemical Reactions

We used photochemical methods established by Mr. Oshima Norio, i.e., intravenous injecting light-sensitive agent hematoporphyrin monomethyl ether (HMME) (0.25 mg/kg), and irradiation with 100 W mercury lamps of fluorescence microscopes (facula diameter of 250 μm). The time from irradiation to thrombosis was recorded continuously with a CCD camera connected to an inverted biological microscope. Thrombosis in thin veins started to form 9.8 ± 1.02 s after irradiation in the photochemical reaction group, and in rats pretreated with Dantonic™ (0.4 g/kg), the corresponding time was 20.8 ± 2.56 s. The thrombosis area in photochemical reaction rats increased significantly, while in rats pretreated with Dantonic™, the increase was significantly inhibited [22]. The results suggest that Dantonic™ which contains Danshen can inhibit the formation of thrombosis.

Jin et al. [108] proved that Danshen injection can inhibit the reduction of the compound action potential amplitude after the photochemical reaction in guinea pigs.

6.3.2 Improving Effect on Microcirculatory Dysfunction Caused by Noradrenaline

Chen et al. [63] observed the anticontraction effect of synthetic or isolated Tanshinol on mouse mesenteric small arteries and veins induced by the local application of norepinephrine. Dropping norepinephrine (4 μg/0.1 ml) on the mesentery could cause strong arteriole contraction, while vein contraction was not obvious. Three minutes later, isolated and synthetic Tanshinol (1 mg/0.1 ml), respectively, were dripped. Compared with the normal saline (0.1 ml) group, isolated and synthetic Tanshinol can significantly restore vascular diameter. It suggests that Danshen has inhibitory effects on vasoconstriction induced by norepinephrine.

6.3.3 Improving Effect on Microcirculatory Dysfunction Caused by Scalding

Yan et al. [109] observed the improving effect of Danshen injection on microcirculatory dysfunction in deep second-degree burn wounds in rats. After deep second-degree burns, the blood in the microcirculation system in the wounds decreased sharply, and reached a minimum after 2 h. On the contrary, the contents of endothelin (ET) and NO increased, reaching peak values 4 and 8 h later, respectively. 8 h after injury, the ET/NO ratio of the Danshen group was 2.87 times the preinjury value. In the saline group, the ratio was still significantly higher than preinjury 72 h after injury. The SOD activity of the saline group dropped first, then recovered slowly, while in the Danshen group, the activity came back much sooner. In the Danshen group, the wounds healed in a shorter period of time, there was no scar formation, and the hair grew normally. By regulating the ratio of plasma ET/NO and reducing the synthesis of ET and NO, Danshen can improve microcirculation in wound tissues, thereby reducing the progressive damage of early burns, which is beneficial to good healing.

6.3.4 Improving Effect on Microcirculatory Dysfunction Caused by Noise

Jiang et al. [110] established acoustic organ microcirculation capillary dysfunction models by noise exposure (120 db SPL) for 2 h, and observed the improving effect of Danshen on noise-induced acoustic organ microcirculatory dysfunction. The results showed that Danshen could improve hemorheological properties and the vascular microcapillaries' congestion state in noise-damaged acoustic organs. It suggests that Danshen has protective effects on noise-induced microcirculatory dysfunction and cochlear noise damage in acoustic organs. Zhang et al. [111] also proved that Danshen injection has improving effects on the diameters of auricle microcirculatory fine arteries and veins, and on the velocity of blood flow in mice [111].

6.3.5 Improving Effect on Microcirculatory Dysfunction Caused by Dextran Polymer

Using dextran polymer, Wang et al. [112] established a mice microcirculation dysfunction model. They used a dynamic analyzer of Doppler laser microcirculation to observe the effect of Danshen injection. The results showed that Danshen injection could improve the microblood flow in mouse brains treated with dextran polymer.

In addition, Li et al. [113] studied the effect of Dantonic™ on the hemorheology of hyperlipemic New Zealand white rabbits, and their results showed that Dantonic™ could reduce the whole blood low-shear viscosity, whole blood middle-shear viscosity and hematocrit. Xu et al. [114] also proved that the Dantonic™ could improve erythrocyte deformability in acute blood hyperviscosity syndrome model rats.

6.3.6 Summary

Danshen, its major water-soluble components, and its compound preparations have an inhibitive effect on peroxides and the expression of adhesion molecules, and protect vascular endothelial cells. They can inhibit a variety of microcirculatory dysfunctions associated with the peroxide generation and adhesion molecule expression induced by I/R, LPS, transplantation, burns, photochemical reactions, and protect the target organs (heart, brain, liver, kidney, intestines, etc.).

6.4 Research Progress in Danshen's Effects on the Blood System, Microcirculation, and Hemorheology

Jinhua Wang, Li Zhang and Guanhua Du

Many in-depth studies have been carried out to observe the effects of Danshen on the blood system, microcirculation, and hemorheology. In this section, the research progress of recent years is reviewed.

6.4.1 Anticoagulation, Promoting Fibrinolysis, and Antithrombosis

6.4.1.1 Increase Anticoagulant and Fibrinolytic Activity

Shen Chuan-Lu et al. found that Danshen could inhibit the activation of endothelial cell tissue factors induced by endotoxin, thereby enhancing anticoagulation and fibrinolysis activity, which may also be one of the mechanisms for Danshen's function of promoting blood circulation. Gu Yang-Hong et al. found that Tanshinol could promote the secretion of plasminogen activator (PA) in bovine endothelial cells, reduce the activity of plasminogen activator inhibitor (PAI), and increase the activity of thrombomodulin. Tang Zhi-Zhong et al. found that Danshen injection can increase serum SOD levels and decrease LPO markedly, so that the tissue-type PA (t-PA) increased and PAI activity decreased, achieving the following goals: expanding the coronary artery muscle, improving the cardiac blood supply, reducing myocardial oxygen consumption, enhancing myocardial contractility, improving left ventricular function and the fibrinolytic system, and preventing thrombosis in patients with coronary heart disease. Wang Pen et al. observed that Danshen inhibited the increase of serum total cholesterol (TC) in rabbits on a high-cholesterol diet, reduced cavity narrowness of the skeleton artery induced by atherosclerosis, while inhibiting the increase of plasminogen and fibrinogen concentration.

6.4.1.2 Anticoagulation and Antithrombosis

Chen Zhi-Yong et al. found that Danshen component 764-3 inhibited platelet aggregation in vitro and inhibited the release of platelet Thromboxane A2 (TXA2), but also promoted the production of prostacyclin (PGI_2). Another report showed that 764-3 increased the release of t-PA from aortic endothelial cells, and it could affect both the expressions of platelet GMP-140 and platelet glycoprotein II_b/III_a complex as well as the anticoagulant function of human umbilical vein endothelial cells (HUVEC). The study found that 764-3 inhibited the expression of GMP-140 and glycoprotein complexes on the platelet membrane surface, and increased the number and activity of thrombomodulin on the surface of endothelial cell membrane, which provided new experimental evidence to illustrate the antithrombosis mechanism of 764-3.

Experiments have shown that Tanshinol could inhibit thrombosis in vitro. After 30 min of Danshen administration, the platelet base was reduced, platelet aggregation function was decreased, and prothrombin time was prolonged, suggesting that Danshen has a definite anticlotting function. Ye Gui-Wen et al. found that acetyl salvianolic acid A (ASAA) could inhibit platelet aggregation in rats and rabbits induced by Arachidonic acid (AA), adenosine diphosphate (ADP), or collagen. Their further study found that one of the mechanisms for ASAA to inhibit platelet aggregation was to block induced aggregation Thromboxane, TXA_2, via the AA pathway.

In addition, some people speculated that Danshen reduces the synthesis of TXA_2 by inhibiting the activity of thrombosis propane oxygen synthase. Also, Danshen could weaken the effects of platelet aggregation and thrombosis by regulating the synthesis and release of the active substances PGI_2 and ET in vascular endothelial cells. Jiang Kai-Yu et al. found that Danshen extract WH505 could inhibit the formation of venous thrombosis, and its mechanism may be related to suppressing ET-1 expression in cells and protecting vascular endothelial cells. They also found that Danshen had inhibitory effects on platelet thrombosis, which may be related to its suppressing the expression of platelet P-selectin.

Jin Xi-Lu et al. found that eight Tanshinol derivatives have antiplatelet aggregation activity in vitro. Luo Xiao-Qin et al. found that CDI has antithrombosis effects, increasing platelet and plasma cAMP levels. The increased levels of cAMP could inhibit phosphoesterase and cyclooxygenase, reduce the generation of prostaglandins peroxide, and activate protein kinase, so that membrane proteins were phosphorylated, changing the conformation of these proteins. Thus, the adaptability of the membrane proteins to platelet aggregation molecules also changed, thereby inhibiting platelet aggregation and achieving the effect of antithrombosis. Therefore, one of the mechanisms for CDI's antithrombosis can be considered the increase of cAMP levels in platelets and plasma. Zhang XJ et al. also found that compound Danshen extract significantly inhibited the formation of thrombosis and reduced platelet aggregation. Intravenous injection of CDI in rats at dosages of 50 and 100 mg/kg resulted in thrombosis inhibition rates of 41.9 and 54.8 %, respectively. It proved that Danshen has a moderate function of antivenous thrombosis, which may be related to its functions of inhibiting platelet aggregation and increasing plasma fibrinolytic activities.

6.4.2 Effect on Lipids

6.4.2.1 Effect of Reducing Blood Lipids

Meng Zhen-Xing et al. found that Danshen could decrease TC, triglycerides (TG), and low-density lipoprotein (LDL) to a certain extent. Wang Pen et al. found that Danshen can inhibit the increase of blood lipids in rabbits on a high-fat diet, postpone atherosclerosis formation caused by hyperlipidemia and endothelial injury, and alleviate lumen narrowness caused by atherosclerotic plaque.

6.4.2.2 Effect on Antilipid Peroxidation

Experiments have shown that Danshen has the function of anti-LDL peroxidation and reducing the formation of complete oxidized LDL (OX-LDL). Wang Nan et al. reported that Danshen could inhibit the oxidative modification of LDL by aortic SMC, and its antioxidant potency was equivalent to that of vitamin E. Another active ingredient of Danshen, tanshinone, could eliminate lipid free radicals produced by mitochondria.

6.4.2.3 Effect on Cell Adhesion Molecules

Jiang Kai-Yu et al. found that HUVEC cultured with tanshinone II_A sulfonate had an antagonistic effect on surface adhesion molecule ICAM-1 induced by TNF-α. The antagonism was dose-dependent on tanshinone. Tanshinone incubated with platelets could inhibit the expression of thrombin-induced platelet surface P-selectin in a dose-dependent fashion. Wu Jing-Hai et al. found that the expression level of ICAM-l on the surfaces of peripheral blood mononuclear cells (PBMC) in patients with psoriasis was significantly higher than in the normal group, and the level was significantly reduced after tanshinone treatment. This showed that Danshen inhibited expression of ICAM-1.

In conclusion, Danshen, a drug used to invigorate blood and dissolve stasis in TCM, has functions of anticoagulation, fibrinolysis promotion, and antithrombosis. It can reduce blood lipid levels, antagonize lipid oxidation, and inhibit the expression of blood adhesion molecules. In recent years, Danshen and its agents have been widely used in the treatment of cardiovascular and thrombotic diseases, and clinical practice has proven that Danshen does have curative effects on thrombotic diseases.

6.4.3 The Function of Danshen on Improving Microcirculation and Promoting Hemorheology

6.4.3.1 Improving Microcirculation

Microcirculatory disturbance is characterized by the reduced velocity of capillary blood flow and the increased blood viscosity and aggregation of red blood cells. The characteristic hemorheology of microcirculatory disturbance is dense, i.e., increased blood concentrations or changed composition; sticky, i.e., increased blood viscosity; condensate, i.e., increased blood coagulation; and aggregating, i.e., increased blood aggregation. Blood viscosity is an important indicator of hemorheology.

Danshen injection was administered to peripheral microcirculatory dysfunction rabbit models, and the blood flow in the microcirculatory system was significantly faster and the number of open network of capillaries was increased. 60 % of the animals had an

Table 6.1 The changes in hemorheology in coronary heart disease patients before and after Danshen treatment ($\overline{X} \pm S$)

Indicators	Before treatment	After treatment
η_b, 20 s^{-1} (whole blood viscosity)	9.52 ± 1.41	9.08 ± 1.85*
mPa · s^{-1} (whole blood viscosity shear rate)	6.71 ± 1.45	6.05 ± 1.19*
ESR (erythrocyte sedimentation rate)	17.45 ± 2.56	20.01 ± 8.09**
HCT (mm/h) hematocrit	47.92 ± 5.84	40.71 ± 3.63**
Fib (g/L) plasma fibrinogen	4.05 ± 0.21	2.98 ± 0.25*
Pag (T) (%) platelet adhesion rate	53.03 ± 7.46	36.68 ± 8.78*
η_b (plasma viscosity)	1.97 ± 0.53	1.73 ± 0.24*

Compared with before treatment: * $P < 0.01$, ** $P < 0.05$

improvement of blood flow state, blood cells had certain degree of deaggregation, and the blood flow changed from granular or broken thread-like to normal. Both Danshen injection and tanshinone injection (Tween-free) at different dosages could increase the counts of eye conjunctiva bulbi and mesenteric vascular intersections in rabbits, which could help the increase in local tissue microcirculatory blood perfusion and the establishment of collateral circulation. Such changes were not found in the rabbits treated with normal saline. Experiments on microcirculation in the hamster cheek pouch showed that Danshen had improving effects on microcirculatory diameter, flow rate, fluid and capillaries after local infusion of norepinephrine, and submicrocirculation veins had similar improvement. After treatment with Danshen or CDI, the red blood cell flow rate and liquid state, hematocrit, whole blood viscosity, viscosity of plasma, and red blood cell electrophoresis velocity in capillary were all improved or back to normal in patients with coronary heart disease, compared to before treatment (see Table 6.1).

Children with bronchial asthma were treated with Danshen or Ketotifen, and the results showed that the nail fold microcirculation in 15 patients in the Danshen group had varying degrees of improvement after treatment. Danshen also could improve microcirculatory dysfunction caused by some other drugs.

The most prominent and wide range of pathological changes in patients with diabetes was systemic diffuse vascular disease, which has symptoms of microcirculatory dysfunction, microvessel tumor formation, and microvascular basement membrane thickening. Dantonic™ can antagonize angiotensin, improve blood flow and lower the levels of blood viscosity, density, and aggregation, inhibit platelet aggregation, increase the solubility of fibrin ferment, antagonize calcium and inhibit its accumulation in the cells, prevent calcium overload, remove oxygen free radicals, inhibit the endogenous synthesis of cholesterol, and lower blood cholesterol and triglyceride levels. After oral administration of Dantonic™ for 3 months, the nail fold microcirculation and other indicators in 38 patients had all improved to different extents, and the integrated scores declined. 18 patients of the original 26 cases with severe abnormalities improved to moderate abnormalities, 10 of 12 patients with originally moderate abnormalities improved to mild abnormalities, and the difference between pre- and posttreatment as statistically significant, which showed that Dantonic™ can be used to treat microcirculatory dysfunction in old diabetes. Danshen has significant effects in every treatment group, and its mechanism are:

1. Expanding the microarteries and increasing microcirculatory perfusion by regulating the production and release of vasoactive substances. Danshen can increase the concentrations of blood vessel expanding substances prostacyclin and NO in plasma, and decrease the concentration of blood vessel constricting substances thrombosis-TXA_2 and ET, thus inhibiting platelet aggregation.
2. Reducing whole blood viscosity, improving the deformability of red blood cells, thereby

reducing microcirculation resistance and increasing the perfusion flow. It was reported that Danshen injection could increase the content of 2,3-biphosphoglycerate in red blood cells and enhance the stability and deformability of red blood cells, leading to reduced blood viscosity.

Some studies have shown that Danshen could inhibit vascular endothelial cell injury and reduce the permeability of microcirculation in small intestine transplants. These functions were related with Danshen's inhibitory effect on the adhesion, aggregation, and oxygen free radical production of neutrophils and platelets. A study also confirmed that Danshen could preserve energy substances (such as ATP) in the small intestine at low temperatures, and ATP plays an important role in maintaining the integrity of cells. Preservation solutions with Danshen injection can effectively reduce the tissue damage of small intestine preserved at low temperatures and inhibit the increase in microcirculation permeability.

Recent studies showed that patients with chronic measles had adverse changes in microcirculation, which provided the integrative medicine treatment of chronic measles with a scientific foundation. CDI was used in the treatment of chronic measles; hemorheology indices closely related to chronic measles were selected to evaluate relapses within 6 months. The results showed that after CDI treatment, whole blood viscosity, plasma viscosity, erythrocyte rigidity aggregation index, and the index of red blood cells were reduced in patients with chronic measles, while the hemorheology indicators in the control group were unimpressive. The difference between the two groups was statistically significant, and the efficacy of the treatment group was better than the control group.

Liu Cheng-Huang et al. reported that among 52 cases of measles, 90.4 % had abnormal microcirculation, and among the patients with abnormal microcirculation, 91.7 % of them were chronic measles patients. The patients had abnormal microcirculation not only in skin lesions, but their nail folds in the lesions also had the same abnormal microcirculation potential. The abnormal hemorheology of chronic measles indicates the existence of blood stasis. Blood stasis is a microcirculatory dysfunction, and the degree of blood stasis is closely related to the degree of microcirculatory dysfunction. When microcirculatory dysfunction occurs, mast cell degranulation releases vasoactive substances such as histamine, which would speed up the leakage of serum substances such as albumin and thereby aggravate the microcirculatory dysfunction, leading to the ineffective treatment of chronic measles. The mechanism of CDI's treatment is to invigorate blood and dissolve stasis to influence the hemorheology, thus improving microcirculation. It can also inhibit the release of vasoactive substances by mast cell degranulation, thus inhibiting the leakage of serum substances such as albumin, improving microcirculation, and finally inhibiting the recurrence of measles.

Wang Ling et al. studied the effect of Danshen injection on in vitro adhesion between cultured HUVEC and health human RBC with a streaming quantitative system. It was found that Danshen could decrease the number of RBC adhered to endothelial cells, and weaken the intensity of adhesion by acting on both RBC and endothelial cells simultaneously, which may be another important mechanism for Danshen's improving effect on blood circulation. It was also found that Danshen used in vitro had a better effect than when used in clinical treatment. This discovery has important clinical significance.

In conclusion, the mechanisms of Danshen injection's improving effect on microcirculation are: (1) lower plasma and whole blood viscosity, improve the deformability of red blood cells, thereby reducing the resistance of microcirculation; antagonize hypercoagulable red blood cell aggregation, increase microcirculation perfusion flow; (2) expand microarteries, increase microcirculation perfusion flow by regulating the production and release of vasoactive substances.

6.4.3.2 Promoting Hemorheology

Chen Hui-Zong et al. observed the changes in hemorheology in 43 patients with chronic pulmonary heart disease and hyperviscosity before and after treatment with CDI, and the results indicated that the drug had a certain degree of effect on the blood flow state, velocity, red blood cell deaggregation, etc. Wang Yi et al. discussed the effect of the components in Guanxin II (Chuanxiong, Danshen, safflower, Paeonia, Dalbergia) on the hemorheology parameters, and they found that Chuanxiong, Danshen, safflower, and Dalbergia have varying degrees of improvement on red blood cells rheology, and were able to significantly reduce the mutual attraction among erythrocytes, increase their surface charges, and thereby reduce their aggregation. Chuanxiong, Danshen, and Paeonia could significantly change platelet adhesion, and Chuanxiong, Paeonia, and Dalbergia could reduce fibrinogen, all showing the effect of reducing platelet aggregation. It is thus evident that the improving effects of Guanxin II on hemorheology are achieved by different targeting. Some studies have shown that Danshen plays a significant role in reducing viscosity, and has a significant impact on red blood cell deformability and aggregation. Besides the effect on reducing red blood cell aggregation, Danshen has a significant role in lowering plasma viscosity, suggesting that Danshen not only can influence red blood cells, white blood cells, and platelets in blood, but can also influence plasma components such as plasma proteins, fiber proteins, cells with larger viscosity coefficients, and microparticles.

In patients with chronic renal failure, blood coagulation and aggregation of red blood cells and platelets are increased, due mainly to a variety of uremic toxin accumulation, increased blood viscosity, declined red blood cell deformability, and increased aggregation. Lei Chang-Kun et al. treated 24 patients with chronic renal failure with Tetramethylpyrazine, Danshen, and heparin, and found that a significant improvement in microcirculation was found in the kidney, that the aggregation of red blood cells and platelets, the adhesion rates of fibrinogen and platelets, plasma viscosity, and the whole blood reduction viscosity, viscosity and length of thrombosis in vitro all had decreased, red blood cell electrophoresis time was increased, and whole blood viscosity and hematocrit had increased or remained at normal levels.

Tang Zhi-Zhong et al. observed the effect of long-term Danshen consumption on cardiomyopathy and hemorheology in spontaneously hypertensive rats (SHR). Seven hemorheology indicators, such as low-shear and high-shear whole blood viscosity, plasma viscosity, plasma Fibronectin protein content, hematocrit, red blood cell deformation index, the aggregation index, were determined. It was found that these indicators in the Danshen group were significantly improved compared to in the control group. The results showed that Danshen had the effects of lowering blood viscosity, reducing red blood cell aggregation, enhancing red blood cell deformability, improving microcirculation, increasing myocardial organ blood perfusion flow, improving microcirculatory dysfunction in local tissue, promoting the exchange and discharge of metabolites in local tissue, improving hypoxia in local tissue cells, reducing the generation of free radicals, and preventing and inhibiting myocardial fiber hypertrophy and myocardial fibrosis. Yan Chang-Kai et al. found that Danshen capsule could inhibit the activity of platelet aggregation induced by ADP in rats, delay thrombosis after the electrical stimulation of rat carotid artery, reduce whole blood viscosity, plasma viscosity, hematocrit, and red blood cell electrophoresis time, as well as the red blood cell aggregation index in "Blood-stasis" rats, thus improving the hemorheology.

Tan Jian-San used an acute necrotizing pancreatitis (ANP) model built by retrograde perfusion. They observed that there was a series of abnormal hemorheology in the early stage of ANP, such as the increased viscosity of whole blood under low-shear rate, enhanced red cell aggregation, reduced red blood cell deformability, etc. Danshen treatment could significantly improve abnormal hemorheology, improve the survival rate of ANP, and reduce the extent of pancreatic necrosis. It suggests that Danshen capsule can significantly decrease the viscosity, aggregation, and condensation of the blood, having a good function of promoting blood circulation.

Xu Hong-Yan et al. used an orthogonal design to study the effect of different combinations of polysaccharide sulfate and compound Danshen on the hemorheology of rats. They found that polysaccharide sulfate and compound Danshen had synergistic effects on reducing the whole blood viscosity, and the better combinations were those of a middle dosage of polysaccharide sulfate with a middle or low dosage of Danshen.

The effect of intravenous administration of Danshen on the hemorheology of acute spinal cord injury patients was studied. The hemorheology indicators (whole blood viscosity, low-, medium- and high-shear values, fibrinogen, red blood cell aggregation index, red blood cell deformation index) in the patients before and after treatment were collected. It was found that after Danshen treatment, the hemorheology of the patients was improved, and the difference was statistically significant. Some studies used only compound Danshen to treat the disease, and the results showed not only a drop in high-shear value, but also a significant fall in the low-shear value and red blood cell aggregation index. The low-shear value indicates the aggregative force of red blood cells.

Danshen is a plant drug with the function of invigorating blood and dissolving stasis. Compound Danshen can improve hemorheology by dilating the coronary artery muscle, increasing blood flow, ad increasing cardiac contractility. Besides these, the main function of compound Danshen is to suppress the synthesis of prostaglandin in platelets, inhibit platelet adhesion and aggregation, and inhibit the activity of platelet factor 3. It can also extend the serum prothrombin time.

6.4.4 Summary

The research advancement in Danshen's effects on the blood system and microcirculation has shown that all of the existing Danshen preparations, including injections and dropping pills, can reduce whole blood and plasma viscosity, improve the deformability of red blood cells, antagonize the aggregation of hypercoagulable red blood cells, and increase the perfusion flow of microcirculation. Thereby, they have a significant effect in reducing the resistance of microcirculation. In addition, Danshen has marked effects of promoting fibrinolysis, anticoagulation, and antithrombosis, and all of these form the pharmacological foundation for Danshen's clinical application in cardio-cerebral-vascular diseases.

References

1. Han JY, Fan JY, Horie Y, et al. Pharmacol Ther. 2008;117(2):280.
2. Granger DN. Microcirculation. 1999;6(3):167.
3. Granger DN. Am J Physiol. 1988;255(6 Pt 2):H1269.
4. Han JY, Horie Y, Li D, et al. Clin Hemorheol Mi4. crocirc. 2006;34(1-2):145.
5. Han JY, Horie Y, Miura S, et al. World J Gastroenterol. 2007;13(26):3581.
6. Sun K, Wang CS, Guo J, et al. Life Scl. 2007;81 (6):509.
7. Sun K, Wang CS, Guo J, et al. Clin Hemorheol Microcirc. 2006;34(1–2):103.
8. Guo J, Sun K, Wang CS, et al. Shock. 2008;19 (2):205.
9. Lomax AE, O'Reilly M, Neshat S, et al. J Physiol. 2007;583(Pt 2):719.
10. Kanadasi M, Kudaiberdieva G, Birand A. Ann Noninvasive Electrocardiol. 2002;7(2):106.
11. Lindal S, Vaage J, Oisen R, et al. Scand Cardiovasc J. 1999;33(3):143.
12. Menger MD, Vollmar B. Microcirculation. 2000;7 (5):291.
13. Kacila M, Bevilaqua S, Matteucci S, et al. Bosn J Basic Med Sci. 2006;6(2):59.
14. Mathur A, Baz M, Staples ED, et al. Ann Thorac Surg. 2006;81(5):1844, discussion 9.
15. Jeppesen P, Knudsen ST, Poulsen PL, et al. Acta Ophthalmol Scand. 2007;85(3):280.
16. Han JY, Miura S, Akiba Y, et al. Am J Physiol Gastrointest Liver Physiol. 2001;280(5):G939.
17. Ademoglu E, Tamer S, Albeniz I, et al. Acta Haematol. 2004;112(4):184.
18. Nishimura S, Manabe I, Nagasaki M, et al. J Clin Invest. 2008;118(2):710.
19. Fujiyama Y, Hokari R, Miura S, et al. J Gastroenterol Hepatol. 2007;22(11):1838.
20. Stokes KY, Calahan L, Russell JM, et al. Microcirculation. 2006;13(5):377.
21. Tailor A, Cooper D, Granger DN. Microcirculation. 2005;12(3):275.
22. Wang F, Liu YY, Liu LY, et al. Clin Hemorheol Microcirc. 2006;34(1–2):131.
23. Singh M, Saini HK. J Cardiovasc Pharmacol Ther. 2003;8(2):135.
24. Vollmar B, Slotta JE, Nickels RM, et al. Microcirculation. 2003;10(2):143.

25. Montrucchio G, Alloatti G, Camussi G. Physiol Rev. 2000;80(4):1669.
26. Meldrum DR, Donnahoo KK. J Surg Res. 1999;85 (2):185.
27. Huang J, Upadhyay UM, Tamargo RJ. Surg Neurol. 2006;66(3):232.
28. Ishikawa M, Cooper D, Russell J, et al. Stroke. 2003;34(7):1777.
29. Lee HT, Kim M, Jan M, et al. Am J Physiol Renal Physiol. 2006;291(1):F67.
30. Rakhit RD, Seller C, Wustmann K, et al. Coron Artery Dis. 2005;16(3):147.
31. Dwivedi VK, Chandra M, Misra PC, et al. J Enzyme Inhib Med Chem. 2006;21(1):43.
32. Kurose I, Anderson DC, Miyasaka M, et al. Circ Res. 1994;74(2):336.
33. Kurose I, Yamada T, Wolf R, et al. J Leukoc Biol. 1994;55(6):771.
34. Kim YI, Chung HJ, Song KE, et al. Am J Surg. 2006;191(1):72.
35. Young LH, Ikeda Y, Lefer AM. Methods Find Exp Clin Pharmacol. 2001;23(3):107.
36. Sack M. Pharmacol Ther. 2002;94(1–2):123.
37. Kurose I, Argenbright LW, Wolf R, et al. Am J Physiol. 1997;272(6 Pt 2):H2976.
38. Horie Y, Han JY, Mori S, et al. World J Gastroenterol. 2005;11(4):511.
39. Chen CG, Wang YP. Acta Pharmacol Sin. 2006;27 (2):217.
40. Zhang WD, Chen H, Zhang C, et al. Planta Med. 2006;72(1):4.
41. Miles AM, Bohle DS, Glassbrenner PA, et al. J Biol Chem. 1996;271(1):40.
42. Witting PK, Liao WQ, Matthew JH, et al. Biochem Biophys Res Commun. 2006;348(2):485.
43. Xue JH, Nonoguchi K, Fukumoto M, et al. Free Radic Biol Med. 1999;27(11–12):1238.
44. Hemnani T, Parihar MS. Indian J Physiol Pharmacol. 1998;42(4):440.
45. Yeh KY, Yeh M, Glass J, et al. Gastroenterology. 2000;118(3):525.
46. Nurmi A, Lindsberg PJ, Koistinaho M, et al. Stroke. 2004;35(4):987.
47. Howard EF, Chen Q, Cheng C, et al. Neurosci Lett. 1998;248(3):199.
48. Iho S, Tanaka Y, Takauji R, et al. J Leukoc Biol. 2003;74(5):942.
49. Williams AJ, Dave JR, Tortella FC. Neurochem Int. 2006;49(2):106.
50. Arndt H, Palitzsch KD, Anderson DC, et al. Gut. 1995;37(3):374.
51. Zirrmerman BJ, Holt JW, Paulmn JC, et al. Am J Physiol. 1994;266(3 Pt 2):H847.
52. Van der Vieren M, Crowe DT, Hoekstra D, et al. J Immunol. 1999;163(4):1984.
53. Salas A, Panes J, Elizalde JI, et al. J Leukoc Biol. 1999;66(1):59.
54. Panes J, Granger DIN. Gastroenterology. 1996;111 (4):981.
55. Hayashi Y, Sawa Y, Fukuyama N, et al. J Thorac Cardiovasc Surg. 2003;126(6):1813.
56. Zimmerman BJ, Granger DN. Am J Physiol. 1990;259(2 Pt 2):H390.
57. Carden D, Xiao F, Moak C, et al. Am J Physiol. 1998;275(2 Pt 2):H385.
58. Yamaguchi Y, Matsumura F, Liang J, et al. Transplantation. 1999;68(10):1459.
59. Parks DA, Granger DN, Bulkley GB, et al. Gastroenterology. 1985;89(1):6.
60. Frangogiannis NG, Lindsey ML, Michael LH, et al. Circulation. 1998;98(7):699.
61. Rocha MF, Aguiar JE, Sidrim JJ, et al. Toxicon. 2003;42(2):183.
62. Hoffmann SC, Kampen RL, Amur S, et al. Transplantation. 2002;74(7):916.
63. Zhanghua Cheng, et al. J Shanghai Med Univ. 1987;14(1):25.
64. Zhao BL, Jiang W, Zhao Y, et al. Biochem Mol Biol Int. 1996;38(6):1171.
65. Du GH, Zhang J. Chin Med J (Engl). 1997;110 (1):65.
66. Lin YH, Liu AH, Wu HL, et al. Biochem Biophys Res Commun. 2006;348(2):593.
67. Zhang HS, Wang SQ. J Mol Cell Cardiol. 2006;41 (1):138.
68. Soung DY, Rhee SH, Kim JS, et al. J Pharm Pharmacol. 2003;55(10):1427.
69. Han JY, Akiba Y, Suzuki H, et al. Microcirc Annu. 2003;9:39.
70. Han J-Y, Akiba Y, Suzuki H, et al. Microcirc Annu. 2001;7:33.
71. Ren DC, Du GH, Zhang JT. Jpn J Pharmacol. 2002;90(3):276.
72. Chen YH, Lin SJ, Ku HH, et al. J Cell Biochem. 2001;82(3):512.
73. Zhou Z, Liu Y, Miao AD, et al. Eur J Pharmacol. 2005;513(1–2):1.
74. Ling S, Dai A, Guo Z, et al. Clin Exp Pharmacol Physiol. 2005;32(7):571.
75. Han JY, Kemegaya Y, Wang L, et al. Microcirculatory aspect of asian traditional medicine. 1998;109.
76. Ryu SY, Oak MH, Kim KM. Planta Med. 1999;65 (7):654.
77. Fung KP, Zeng LH, Wu J, et al. Life Sci. 1993;52 (22):PL239.
78. Sun J, Huang SH, Tan BK, et al. Life Sci. 2005;76 (24):2849.
79. Ji X, Tan BK, Zhu YC, et al. Life Sci. 2003;73 (11):14131.
80. Wang XH, et al. Reversion effects of composite salviae dropping pill on the thickening of left ventricle in coronary heart disease patients. In: National academic symposia of the modernization of traditional Chinese medicine (compound Salvia Miltiorrhiza preparations) of 1999, 1999. p 52.
81. Chen XY, et al. Preliminary study on the effects of composite Salviae dropping pill on the

microcirculation of bulbar conjunctiva and thrombelastogram in coronary heart disease patients. In: National academic symposia of compound Salvia Miltiorrhiza preparations of 1996, 1996. p 3.
82. Fu XL, et al. Observation on the therapeutic effects of composite Salviae dropping pill on coronary heart disease based on the change in the microcirculation of nail folds. In: National academic symposia of the modernization of traditional Chinese medicine (compound Salvia Miltiorrhiza preparations) of 1999, 1999. p 89.
83. Wu B, Liu M, Zhang S. Cochrane Database Syst Rev. 2004;(4):CD004295.
84. Lo CJ, Lin JG, Kuo JS, et al. Am J Chin Med. 2003;31(2):191.
85. Lain BY, Lo Ac, Sun X, et al. Phytomedicine. 2003;10(4):286.
86. Du GH, Qiu Y, Zhang JT. J Asian Nat Prod Res. 2000;2(2):145.
87. Chen YH, Du GH, Zhang JT. Acta Pharmacol Sin. 2000;21(5):463.
88. Han JY, Akiba Y, Miura S, et al. Microcirc Ann. 2004;20:23.
89. Zhang WH, Wang JS, Zhou Y, et al. World J Castroenterol. 2003;9(9):2040.
90. Kang DG, Oh H, Sohn EJ, et al. Life Sci. 2004;75 (15):1801.
91. Li YS, et al. Chin Natl J New Gastroenterol. 1997;5 (9):554.
92. Fan L, et al. Chin J Tradit Med Traumatol Orthop. 2003;11(5):9.
93. Liang J, et al. Chin J MicroSurg. 1994;17(3):207.
94. Yan Z, et al. Mod J Integr Chin Tradit West Med. 1999;8(8):l211.
95. Dejkhamron P, Thimmarayappa J, Kotlyarevska K, et al. Mol Cell Endocrinol. 2007;274(1–2):35.
96. Yang Y, Zhou H, Yang Y, et al. Mol Immunol. 2007;44(5):984.
97. Honjo T, Inoue N, Shiraki R, et al. J Vasc Res. 2006;43(2):131.
98. Dauphinee SM, Karsan A. Lab Invest. 2006;86(1):9.
99. Kokubo Y, Matson GB, Derugin N, et al. Brain Res. 2002;947(1):1.
100. Hassan F, Islam S, Tumurkhuu G, et al. BMC Cancer. 2006;6:81.
101. Simonyi A, Wang Q, Miller RL, et al. Mol Neurobiol. 2005;31(1–3):135.
102. Harris NR, Russell JM, Granger DN. Circ Shock. 1994;43(4):155.
103. Kubo Y, Fukuishi N, Yoshioka M, et al. Inflamm Res. 2007;56(2):70.
104. Guo YX, Zhang DJ, Wang H, et al. J Pharm Biomed Anal. 2007;43(2):435.
105. Zhao Y, et al. J Chengdu Univ Tradit Chin Med. 2003;26(40):33.
106. Liu YY, et al. J Chin Microcirc. 2002;6(5):302.
107. Liu YY, et al. J Chin Microcirc. 2002;48(2):83.
108. Jin YM, et al. Chin J Otorhinolaryngol Integr Tradit West Med. 1997;5(4):156.
109. Yan Z, et al. Pract J Med Pharm. 1999;12(2):1.
110. Jiang LX, et al. J Clin Otolaryngol. 1992;6(1):14.
111. Zhang ZF, et al. Liaoning J Tradit Chin Med. 1999;26(6):281.
112. Wang SJ, et al. J Chin Microcirc. 2002;4(4):213.
113. Li T. et al. Heilongjiang Anim Sci Vet Med. 2000;3:22.
114. Xu ZP, et al. Chin Herbal Med. 2000;31(4):283.
115. Han JY, Akiba Y. Suzuki H, et al. Symp Life Sci Med Tokyo. 2005;13:247.
116. Gu YH, et al. J Shanghai Second Med Univ. 10 (3):208.
117. Wang P, Zhang Y, Fan Y, et al. J Cardiovasc Pulm Dis. 1996;5(1):52.
118. Shen C, et al. Chin J Hematol. 1993;14(10):540.
119. Ye Wengu W, et al. Acta Pharm Sin. 1994;29:412.
120. Ye W, et al. Acta Pharm Sin. 1998;33(1):62.
121. Ye W, et al. Acta Pharm Sin. 1998;25(6):467.
122. Wang N, Cai H, et al. J Nanjing Med Univ. 1994;14 (4):529.
123. Jiang K, et al. Acta Pharmacol Sin. 1998;19(1):47.
124. Wu J, Yang S, et al. J Shanghai Med Univ. 1998;25 (1):47.
125. Jiang K, et al. Chin Tradit Patent Med. 2000;22: 782.
126. Zhang X, Wang S, Guo J, et al. Chin J Mod Appl Pharm. 2007;24(1):15.
127. Liu H. Heilongjiang Med J. 2007;20(6):624.
128. Yang H, et al. J West Chin Univ Med Sci. 2000;31 (2):30.
129. Xiao Z, et al. J Tongji Med Univ. 1997;26:450.
130. Pan J, et al. Strait Med. 1998;10:79.
131. Wang C, et al. Chin J Clin Pharmacol Ther. 2002;7 (1):30.
132. Hua S, et al. Chin J Clin Pharmacol Ther. 2000;5 (4):363.
133. Chen H, et al. New J Tradit Chin Med. 1997;8 (29):34.
134. Wang Y, et al. Pract J Integr Tradit Chin West Med. 1997;10(15):1427.
135. Xu H, et al. Chin J Hemorheol. 2001;11(1):27.
136. Yan L, et al. J West Chin Univ Med Sci. 1991;6 (8):333.
137. Luo X, et al. Chin Tradit Patent Med. 1996;16 (10):36.
138. Yin J, et al. Modern study and application of traditional Chinese medicine. Beijing: Xueyuan Press; 1997. p. 1162.
139. Yang W, Song H, Zhao B. Chin Pharmacol Bull. 1990;8(2):77.
140. Li C, Yang S, Zhao F, et al. J Integr Tradit Chin West Med. 1983;3(5):297.
141. Wu Z. Tradit Chin Med Res. 1994;2:57.
142. Chen R, et al. Treatment of blood diseased and relevant modern studies. Wuhan: Hubei Science and Technology Press; 1990. p. 291.

143. Gong M, Zou J. Heilongjiang J Tradit Chin Med. 1996;1:3.
144. Wang Y, Liu B. Hunan Guiding J Tradit Chin Med Pharmacol. 1998;4(1):9.
145. Yang C. Chin J Cardiol. 1980;17(3):179.
146. Huang Y. Chin J Integr Tradit West Med Intensive Crit Care. 2000;7(5):279.
147. Fan M, Zhao R. Chin Tradit Patent Med. 1988;1:39.
148. Xiaohui Z, Gu C, Huang H. J Chin Microcirc. 2003;13(1):27.

7 Preventive and Therapeutic Effects of Danshen on Digestive System Diseases

Yanqiao Zang, Ying Dai, Guanhua Du and Mei Gao

In recent years, modern medical methods combined with chemical methods, molecular and cellular biological techniques have been used in the systemic study of Danshen, obtaining a great deal of new knowledge and experimental results, which indicate that Danshen can be widely used to treat digestive system diseases.

7.1 Pharmacological Effects of Danshen on Diseases of Stomach and Intestine

Yanqiao Zang, Ying Dai and Guanhua Du

7.1.1 Pharmacological Effects on Peptic Ulcers

Peptic ulcers are the inflammatory ulcers (round or oval) resulting from the erosion of the stomach mucous membrane by gastric acid and pepsin due to an imbalance between the secretion of gastric acid and pepsin and the protective effects of the mucous membrane. It is also related to the producing factors of prostaglandin, and there are Helicobacter pylori in the mucous membrane of the gastric antrum in most peptic ulcer patients.

The occurrence of peptic ulcers requires at least the following two factors: various nonsteroidal anti-inflammatory drugs which inhibit the production of prostaglandin and damage the stomach mucous membrane, and excessive secretion of gastric acid which erodes the stomach and intestine mucous membrane.

It has been shown that Danshen has good therapeutic effects on peptic ulcers, and the antiulcerative effects of its water-soluble components have been proven by a great number of basic and clinical studies.

The mechanisms of Danshen's therapeutic effects on peptic ulcers are as follows: (1) Scavenging oxygen **free radicals** and inhibiting lipid peroxidation. Danshen can protect the stomach and duodenum mucous membrane from injury and facilitate the healing of ulcers by scavenging oxygen **free radicals** and inhibiting lipid peroxidation. (2) Lowering the level of endothelin (ET) in plasma, improving microcirculation, and increasing the blood flow in the mucous membrane of the stomach and duodenum. It has been found in recent years that the epithelial cells of the mucous membrane of stomach and intestine can produce ET. The mechanisms by which ET induces ulcers are as follows: ET can lead to the constriction of the vessels in the stomach and duodenum so as to result in ischemia, hypoxia, and acidosis of the mucous membrane, and ischemia, hypoxia, and endothelial cell injury can stimulate the release of

Y. Zang · Y. Dai · G. Du (✉) · M. Gao
Institute of Materia Medica, Chinese Academy of Medical Science and Peking Union Medical College, Beijing, China
e-mail: dugh@imm.ac.cn

X. Yan (ed.), *Dan Shen (Salvia miltiorrhiza) in Medicine*,
DOI 10.1007/978-94-017-9463-3_7,

ET, forming a vicious circle. As an antagonist of ET, Danshen can lower the level of ET in plasma and improve the blood supply to the mucous membrane of the stomach and duodenum so as to prevent ulcer formation. (3) Facilitating the reparation of the ulcerous area. Salvianolic acid can improve microcirculation, increase blood flow in the stomach mucous membrane, stimulate the secretion of mucus, facilitate the synthesis of DNA in the stomach mucous membrane cells on the ulcer borders, and induce the synthesis and release of endogenous prostaglandin, so it is involved in the overall process of mucous membrane repair. (4) Promoting the expression and release of growth factors. It has been proven that Danshen can facilitate the expression of basic fibroblast growth factor (bFGF) and its receptor bFGFR during the healing process of gastric ulcers induced by acetic acid in rats, which is one of the mechanisms of its therapeutic effects. (5) Increasing the levels of serum NO, plasma PG and mucus adhered to the stomach wall, significantly inhibiting apoptosis of the mucous membrane cells on the gastric ulcer border, and facilitating the expression of EGFR and the apoptosis-inhibiting gene *Bcl*-2. Therefore, Danshen can accelerate ulcer healing by activating the defense factors of the stomach mucous membrane and inhibiting apoptosis pathways.

7.1.2 Pharmacological Effects of Danshen on Intestinal Tract

7.1.2.1 Improving Intestinal Tract Microcirculation by Dilating Blood Vessels

Danshen can increase the openings of capillaries, dilate arterioles, increase microcirculatory blood flow, regulate the balance between Thromboxane B2 (TXB2) and prostaglandin E_2 (PGE_2), inhibit the production of ET, increase the concentration of NO in plasma, lower levels of ET, and optimize the ratio of ET/NO so as to influence the contraction and relaxation of capillaries. It has been demonstrated that after injection of Danshen, both the blood flow rates of periphery microcirculation in rabbits and mesenteric microcirculation in dogs and the number of open capillaries increased significantly, and aggregative red blood cells (RBC) showed various degrees of disaggregation. Besides the direct effects on the effector organs, tanshinone II_A can also decrease the tension of sympathetic ganglia by inhibiting sympathetic ganglionic cells so as to relax the vascular smooth muscles innervated by these nerves and dilate blood vessels.

Danshen can significantly inhibit the aggregation of platelets induced by ADP and collagens both in vivo and in vitro, prolong the prothrombin time, kaolin-activated partial thromboplastin time and the time of thrombosis, shorten the length of thrombosis, and significantly lower the dry weight and wet weight of thrombosis. Danshen and Tanshinol can stimulate the secretion of tissue plasminogen activator (tPA) by bovine endothelial cells, increase the content of Prostaglandin I2 (PGI2), and decrease the activity of plasmiongen inhibitor. Danshen can increase the activity of thrombosis-conditioning proteins in the surfaces of bovine endothelial cell membranes, inhibit the activation of endotoxin on the surface tissue factors of endothelial cells, and enhance the anticoagulation and fibrinolytic activity of the body.

Compound Danshen preparations can significantly lower the level of TXA2 in plasma and increase the production of PGI_2 so as to improve microcirculation.

7.1.2.2 Protecting the Intestinal Tract by Calcium Antagonism and Antioxidation

Danshen can directly act on the calcium channel controlled by cell membrane receptors, increase the active transport of Ca^{2+} pumps, and facilitate the excretion of Ca^{2+}, so as to lower the concentration of free calcium ions inside the cell, reduce the transformation from xanthine dehydrogenase to xanthine oxidase, and decrease the production of free radicals in the cells via lipid peroxidation.

Danshen injection can significantly lower the content of lipid peroxides (LPO) in serum, and increase the activity of superoxide dismutase

(SOD) in serum. Danshen can resist the peroxidation of low density lipoprotein (LDL) and reduce the formation of oxidized low density lipoprotein (OX-LDL). Danshen has significant scavenging effects on superoxide anions produced by the xanthine-xanthine oxidase system and on superoxide anions produced by white blood cells under the stimulation of phorbol myristic acetic acid, and Tanshinol's scavenging of superoxide anions is stronger than that of SOD. Salvianolic acid A (SalA), salvianolic acid B (salB), protocatechualdehyde, protocatechuic acid, caffeic acid, and Rosmarinic acid isolated from Danshen have strong protective effects on the biomembrane against peroxidation injury. Among them, the antioxidant activity of salA is the highest. Tanshinone (one of the active components of Danshen) can effectively scavenge the lipoid free radicals produced by mitochondria.

7.1.2.3 Protecting the Digestive Tract Mucous Membrane Cells, and Facilitating the Reparation of Injured Cells

Compound Danshen preparations can facilitate the synthesis of proteins and DNA in gastrointestinal tract mucous membrane cells and accelerate the division and proliferation of undifferentiated epithelia in the membrane, which leads to the formation of new mucous membrane cells in the gastrointestinal tract and accelerates the recovery of the injured membrane.

Platelet-derived growth factor (PDGF) is a strong growth stimulation factor for interstitial cells (such as fibroblast cells) and tissue cells (such as smooth muscle cells) in serum, which can stimulate macrophages to synthesize bFGF and transforming growth factor β (TGF-β). bFGF has extensive biological activities and is involved in almost the entire process of tissue reparation. It is the attraxin and strong growth stimulation factor of fibroblast cells (FBs), and has significant promoting effects on the proliferation of FBs and the synthesis of collagens. bFGF can inhibit the division of FBs, which is dependent on local concentrations of bFGF. When the concentration is too high, proliferation is inhibited, and the deposit of collagen is postponed. PDGF has chemotaxis and proliferation effects on epithelial cells, and can directly or indirectly increase the synthesis and secretion of extracellular matrix (ECM) proteins, facilitate the adhesion of collagen matrix, and stimulate the formation of connective tissues. Therefore, PDGF is acknowledged as an important injury reparation factor.

Danshen's effect on the injury healing of the granulation barrier and tissue cell regeneration of injured digestive organs has a clear temporal relation. In the early and middle stages of the healing process, the expressions of PDGF, bFGF, and TGF-β are positive or strongly positive, and in the late stage, their expressions are weak and statistically the same as those in the control group. This shows that the wound repair mechanisms of Danshen, in particular in the regulation of FBs proliferation and suppression, are in line with the inherent process of the body's repair.

7.1.2.4 Inhibiting the Expression of ICAM-1 and Blocking the Activation of PMN

Acute inflammatory reactions can lead to the release of cell factors by Polymorphonuclear Neutrophils (PMNs) and endothelial cells (ECs), such as Interleukin-1 (IL-1) and tumor necrosis factor-α (TNF-α). In turn, these factors can stimulate ECs and PMNs and increase the expression of intercellular adhesion molecule-1 (ICAM-1). The upregulation of ICAM 1 expression can facilitate the adhesion and activation of PMNs and ECs and produce large amounts of reactive oxygen species (ROS) and proteolytic enzymes, causing a series of pathological changes such as vascular obstruction, microcirculatory disorders, and damage in tissues surrounded by ECs. ICAM-1 and its ligand in PMNs, CD11b/CD18, are the adhesion molecules mediating PMNs and ECs.

After early application of Danshen for treatment, the production and content of TNF-α in tissues and plasma are reduced, the number of CD18 and CD11b positive cells in the periphery blood is reduced, the expression of ICAM-1 in blood vessel endothelium is inhibited, the adhesion of PMNs and ECs is suppressed, tissues and organs are protected, and the absorption of endotoxin is reduced. After treatment with Danshen, the aggregation of PMNs is not changed, indicating Danshen can inhibit ICAM-1 expression so as to reduce PMNs adhesion and block the pathway of PMNs activation. Danshen inhibits the expression of ICAM-1 protein, but does not completely block the aggregation of PMNs in intestinal tissue and does not block the normal participation of PMNs in inflammatory reactions. It only decreases the adverse reactions caused by excessive activation of PMNs.

7.1.2.5 Improving the Respiratory Functions of Mitochondria and Increasing the Energy of Mitochondria

The important functions of mitochondria include biological oxidation and supplying energy required for the activity of cells. Cytochrome aa3 and cytochrome C, the intermediate link of electron transfer chain and the terminal link in the respiratory chain, can be used to evaluate the functions of the respiratory chain in the mitochondria of intestinal mucous membrane epithelial cells. The change in energy charge is consistent with the change in respiratory chain and similar to the change of cytochrome C.

In burning or serious acute pancreatitis, the levels of cytochrome C and cytochrome aa_3 in the mitochondria of intestinal mucous membrane epithelial cells decrease significantly, the production of oxygen free radicals and the consumption of SOD increase, and SOD activity decreases. Danshen can significantly increase the levels of cytochrome aa_3 and cytochrome C in intestinal mucous membrane epithelial cells and the activity of SOD after injury, significantly improve respiratory chain function in the mitochondria of intestinal mucous membrane epithelial cells, increase the energy of mitochondria, and significantly reduce the production of oxygen free radicals. All of these are favorable for the protection of the intestinal mucous membrane.

7.1.2.6 Antibacterial Effects

Total tanshinone has relatively strong antibacterial effects on sensitive bacteria and on the penicillin, aureomycin, and erythromycin resistant bacterium *Staphylococcus aureus*. Cryptotanshinone, dihydrotanshinone, hydroxytanshinone II_A, salvianolic acid ester, and tanshinone II_B have significant inhibitory effects on *S. aureus* and its drug-resistant strains. Therefore, it has significant effects on infections caused by drug-resistant *S. aureus*. Danshen has a strong inhibitory effect against *Escherichia coli*, *Bacillus proteus,* and dermatophytes, and has inhibitory effects against *Salmonella typhosa* and *Bacillus dysenteriae*.

7.1.3 Clinical Application of Danshen in Intestinal Tract Diseases

7.1.3.1 Intestinal Adhesion

Intestinal adhesion is an independent surgical clinical illness, and is the most common cause of intestinal obstruction; it was reported that 40–80 % of intestinal obstruction cases are caused by intestinal adhesion. There are many causes for intestinal adhesion, and most acquired intestinal adhesions occur after abdominal surgery. Inflammation, injury, hemorrhage, tumor, and foreign matters also can cause intestinal adhesion, but its mechanism has not been completely understood yet. Most people believe that after injury of the visceral organs in the abdominal cavity, new vessels grow to the ischemic area and form adhesive cords, then forming fibrous adhesions. They are gradually organized to form an adhesive strap, producing adhesions. Therefore, in the developing process of intestinal adhesions, local application of compound Danshen injection can block the pathway of occurrence, and the intestinal adhesion is prevented.

The mechanisms of Danshen injection's preventive effect on adhesions are as follows: (1) Dilating blood vessels, improving postoperative microcirculatory disturbance, relieving or preventing the increase of blood viscosity after operation, preventing ischemia, and quickening the blood and transforming stasis. (2) Activating the fibrinolytic system, facilitating fibrinolysis, reducing fibrinogen, inhibiting the proliferation and secretion of fibrocytes, decreasing the formation of collagen fibers and proliferation of fibrous tissues, antifibrosis, preventing adhesion between exudative fibrins in the surface of the abdominal cavity and the organization of fibrins, and preventing the formation of intestinal adhesions. (3) Decreasing capillary permeability, reducing inflammatory exudation, improving local blood circulation, and accelerating the absorption of inflammatory exudates. (4) Activating the monocyte/phagocyte system, secreting a great variety of growth factors, increasing the proliferative activity of reparative cells, facilitating cell regeneration, and injury recovery in the early stage. (5) Inhibiting *Staphylococci* and *E. coli*, and relieving adhesions caused by inflammatory reactions.

Aside from the above mechanisms, Danshen can prevent intestinal adhesion by alleviating reperfusion injury to the stenotic intestinal tract. The calcium antagonist effect and significant antioxidant activity of Danshen can prevent lipid peroxidation in the mucous membrane and exert protective effects against reperfusion injury to the mucous membrane after recovery of blood circulation in the stenotic intestinal tract. Danshen also has the functions of maintaining low temperatures and preserving energy substances such as ATP in the small intestine. If Danshen injection is added to the preservative fluid, the injury to blood vessel endothelium in the small intestine preserved at low temperature can be effectively controlled, and the permeability of transplanted small intestine microcirculation can be decreased. Compound Danshen injection was directly injected into the abdominal cavity during the closing of the abdomen after enterolysis, and the recurrence rate was significantly lower than in the control group over 1–9 years of follow-up.

Danshen can significantly inhibit the synthesis of tumor cell DNA, directly kill tumor cells, and prevent and treat the metastasis and recurrence of tumors, so it can be used as an adjunctive therapy for abdominal cavity chemotherapy in the early stage after operation.

After abdominal operation, the recovery of gastrointestinal tract function is slow, the patient's discomfort and pain is obvious, and the quality of life is poor. The combination of conventional nonsurgical methods, such as fasting, gastrointestinal decompression, nutritional support, maintenance of water, electrolyte, and acid-base balances, with traditional Chinese medicines that can promote blood flow and remove blood stasis, such as Danshen, has significant therapeutic effects.

7.1.3.2 Intestinal Obstruction

Intestinal obstruction is a common surgical and emergency illness. If not treated in a timely manner, it will result in bacterial peritonitis and toxic shock. Clinically, Danshen is usually used as an adjunctive medicine for this disease, and the pharmacological effects of Danshen are as follows: (1) Improving the hypoxia and blood circulation of the intestinal wall, accelerating the absorption of gas, facilitating the disappearance of edema, speeding up the recovery of peristalsis in the intestinal tract, reducing the hemorrhage and edema of intestinal mucosa caused by embolism, and facilitating the formation of new vessels and injury recovery. (2) Promoting blood flow and removing stasis, inhibiting the production of Thromboxane, reducing the aggregation of blood cells, increasing the opening of capillaries, improving microcirculation, mitigating congestion of the intestinal tract, and preventing the formation of thrombosis. (3) Antibacterial and antioxidant effects, enhancing immunity, protecting tissues and organs, scavenging free radicals, and preventing the translocation of bacteria caused by ischemia and hypoxia of the intestinal tract.

7.1.3.3 Burns and Enterogenic Infection

After a burn, vascular ECs release more vasoconstrictor substances than diastolic materials, increasing the ratio of ET/NO. The microvessels, particularly venules, in tissues and organs continuously contract, blood stagnates, the slow blood flow causes microcirculatory disorder, there is insufficient blood supply in the intestinal tract, the intestinal mucosa is short of blood, oxygen, and energy supply, the mucosal barrier is damaged, and many bacteria in the intestinal tract penetrate the intestinal wall and invade other tissues and organs from the intestine.

Although Danshen treatment in the early stage of burn injuries cannot change the trend of intestinal tract ischemia and bacterial translocation, it can alleviate the decrease in blood flow in the posterior intestine mucous membrane, reduce the ratio of ET/NO, and significantly reduce the translocation rate of enterobacteria. In addition, the mechanisms for Danshen's functions might include the following:

Anti-Inflammatory Effects

The antibacterial effects of Danshen reduce bacteria in the intestinal tract and thus reduce the translocation of bacteria from the intestine. It also can directly kill the dislocated bacteria and decrease the translocation rate of bacteria.

Regulating Immunity

Danshen can regulate the immune system and increase the antibacterial capability of the body.

Improving Ischemia-Reperfusion Injury

Danshen can scavenge oxygen free radicals, protect mitochondria, improve energy metabolism, and regulate the balance of TXA_2/PGI_2 so as to ameliorate ischemia-reperfusion injury (IRI) to the intestinal tract, maintain the intestinal mucosa barrier, and prevent the translocation of enterobacteria.

7.1.3.4 Acute Radiation Enteritis

Acute radiation enteritis is a serious complication caused by radiation therapy of abdominal and pelvic tumors. The regeneration of gastrointestinal tract mucous membrane cells is inhibited, the intestinal villi are damaged, mucous membrane ulcers are caused, and the digestion and absorption of food are affected. The epithelial cells of the small intestine are regenerative tissues which are highly sensitive to radiation. Within a certain range of exposure, the epithelia can undergo the process of injury and reparation, and the epithelial cells of the intestinal villi are supplemented by the proliferation of the undifferentiated cells in the intestinal gland epithelia so as to maintain the integrity of the structure and functions of intestinal villi. The injury features tissue atrophy in the early stage and proliferation of tissue cells in the reparation stage.

Intraperitoneal injection of Compound Danshen injection has good therapeutic effects on acute radiation enteritis. The possible mechanisms are as follows: Danshen can dilate the small blood vessels in the gastrointestinal tract mucous membrane, increase the blood flow to the mucous membrane, provide enough oxygen and nutritional substances to mucous membrane tissues, facilitate the energy metabolism of mucous membrane tissues, promote the regeneration of gastrointestinal tract mucous epithelia and gland cells, exert indirect effects by increasing the levels of cell growth factors in microenvironment, promote the synthesis of proteins and DNA in gastrointestinal tract mucous membrane cells, form new gastrointestinal tract mucous membrane cells, and repair the injured mucous membrane. Intraperitoneal injection of Compound Danshen injection can accelerate the division and proliferation of undifferentiated epithelial cells of the gastrointestinal tract mucous membrane in the area injured by radiation, and speed up reparation of the injured gastrointestinal tract mucous membrane. The results showed that in the Danshen treatment group, the mucous epithelium of small intestine and colon are more integral, have more villi, deeper crypts, and thicker mucous membranes, indicating that Compound Danshen injection can accelerate the repair of the small intestine and colon mucous membrane damaged by radiation.

7.1.3.5 Ischemia-Reperfusion Injury of Intestine

IRI can delay the functional recovery of transplanted organs and is the main reason for organ dysfunction after transplantation, having an adverse effect on the long-term survival of transplanted organs. PMN-EC adhesion is the major factor for the tissue damage caused by ischemia-reperfusion.

During the process of ischemia-reperfusion, Danshen injection can inhibit ICAM-1 expression so as to decrease PMNs adhesion, block the pathway of activation of PMNs, effectively relieve the injury to small intestine ECs, maintain low temperatures and preserve energy substances such as ATP in the small intestine, effectively inhibit injury to the blood vessel endothelium of the small intestine preserved at low temperature, and decrease the permeability of transplanted small intestine microcirculation. Danshen can scavenge oxygen free radicals, protect mitochondria, improve energy metabolism, and regulate the balance of TXA_2/PGI_2 so as to ameliorate the IRI of intestinal tract, maintain the intestinal mucosa barrier, and prevent the translocation of enterobacteria, and has a protective effect against intestinal injury caused by ischemia-reperfusion.

7.1.3.6 Rotavirus Enteritis

Rotavirus enteritis accounts for 12–71 % of infantile diarrhea. In China, of the stool specimens of acute infectious diarrhea of children in the autumn and winter, the positive rates of rotavirus were 50 and 44 %, respectively. The occurrence of rotavirus enteritis and diarrhea is due to the impairment of enteric epithelial cells directly caused by rotavirus. Intestinal villi cells with absorption functions fall off and the secretory immature cells in the crypts increase, leading to hypersecretion. The secretory functions of the cells in the crypts are regulated by cAMP, cGMP, and Ca^{2+}. Platelet activating factors (PAF) can facilitate the occurrence of endoenteritis. Phospholipase A_2 (PLA_2) is the key enzyme for the synthesis of PAF, which is regulated by Ca^{2+}, and calcium channel blocker can inhibit the synthesis of PAF.

Danshen can block the inflow of Ca^{2+}, and is a calcium channel blocker. It can decrease the secretion by the cells in crypts, inhibit the injury of PAF to intestinal mucosa, relieve vasospasm, accelerate the blood perfusion rate of intestinal microcirculation, increase the number of capillary networks, recover the perfusion rate of red blood cells in blood capillaries to a normal level, significantly ameliorate microcirculation, and improve the status of cellular metabolism. It can prevent edema from endoenteritis, facilitate the reparation of intestinal villi, recover their functions, and relieve diarrhea. In addition, Danshen has nonspecific anti-inflammation, antiexudation, and antiendotoxemia effects. It can maintain the integrity of mitochondria 1 structure, increase microvilli, inhibit excessive allergic reactions and the actions of oxygen free radicals so as to prevent rotaviral injury to intestinal mucosa, and facilitate the recovery of intestinal function. The adjunctive therapy of Compound Danshen injection is effective for rotavirus enteritis.

7.1.3.7 Acute Diarrhea, Acute Bacterial Dysentery, and Acute Necrotic Enteritis

Acute Diarrhea

Danshen injection is used to treat infectious diarrhea. It can reduce fever, alleviate abdominal pain, and cure diarrhea in a short time. It has strong antibacterial, anti-inflammatory, spasmolytic, and analgesic effects. Danshen can improve microcirculation and effectively prevent and cure shock caused by diarrhea. It has antiviral effects and can relieve diarrhea by regulating immunologic function, improving the microcirculation of the intestinal tract, and facilitating the recovery of the transport and absorption functions of intestinal mucosa.

Acute Bacterial Dysentery

Dysentery is an acute infectious disease of the intestinal tract caused by *B. dysenteriae*. Danshen can promote blood flow and remove stasis, inhibit *B. dysenteriae*, and relieve the symptoms of diarrhea.

Acute Necrotic Enteritis

Acute necrotic enteritis is an acute necrotic hemorrhagic disease of the intestinal tract caused by bacterial infection and the allergic reactions of the body. The pathogen is the β toxin of *Bacillus perfringens* type O, which can lead to necrotic changes in the intestinal tract mucous membrane. Bacterial endotoxin is not only is the antigen causing allergic reactions, but also can simulate the ends of sympathetic nerves, result in the constriction of periphery vessels, increase heart load, cause the heart to release myocardial depressant factors, injure myocardial structure and function, and lead to heart failure.

Danshen can activate blood circulation to dissipate blood stasis, and can especially improve the blood circulation of the intestinal tract, enhance metabolism, facilitate the recovery of intestinal tract functions, significantly dilate coronary arteries, reduce blood flow resistance, increase coronary flow, improve the blood circulation of heart, protect cardiac function, and has bacteriostatic effects against many bacteria. Danshen has significant therapeutic effects on acute necrotic enteritis in clinical practice.

7.1.3.8 Ulcerative Colitis

It has been found in recent years that patients with ulcerative colitis are hypercoagulative and have obvious microcirculatory disturbances. There is submucosal thrombosis, and even if the intestinal segment is not affected, there are microthrombi in capillaries. In the active stage of ulcerative colitis, there is significant change in blood composition, the platelets are in a significantly activated state and they participate in the pathological process by releasing inflammatory mediators. The more severe the disease is, the more obvious the platelet activation is, and the more serious vascular endothelial injury is, aggravating the inflammatory reactions of the intestinal wall and causing a disturbance of blood circulation. The disturbance of blood circulation facilitates and aggravates inflammatory reactions. Thus, a vicious circle is formed. These intestinal tract diseases occur repeatedly and are difficult to treat.

The potential mechanisms of the therapeutic effects of Compound Danshen injection on ulcerative colitis are as follows: (1) Compound Danshen injection is a strong scavenging agent for free radicals, and it can inhibit the generation of oxygen free radicals by neutrophils in the inflammatory response. It has antibacterial and anti-inflammatory effects, and can significantly inhibit the chemotaxis and random migration of leukocytes, stabilize cell membranes, and reduce the rupture of lysosomes so as to protect the intestinal mucosa. (2) Danshen can promote blood flow and remove stasis, improve microcirculation, inhibit the aggregation of platelets, decrease the viscosity of blood, rectify the hypercoagulative state and prethrombosis state, increase local blood supply and nutrition, and facilitate the reparation and regeneration of tissues. (3) Danshen has immunoregulatory functions, and has a strong inhibitory effect on cellular and humoral immune function. Compound Danshen injection can improve the clinical symptoms of ulcerative colitis, facilitate the absorption of inflammation, and speed up the healing process.

7.1.3.9 Digestive Tract Tumors

Danshen has been used to treat various tumors in clinical practice, and significant therapeutic effects have been reported. However, since it has been reported that Danshen has promoting effects on the metastasis of tumor cells, which has attracted the attention of clinical workers, the study on the antitumor function of Danshen has been negatively affected. It has been shown by subsequent studies, however, that sodium tanshinone II_A sulfonate and the aqueous extracts of Danshen cannot facilitate the growth and metastasis of cancers. There has been no definite conclusion about the therapeutic effects on tumors, but it has been used in combined treatments for a great variety of tumors, and it has synergistic effects with anticancer drugs. The studies by Chen Xiaodong et al. have shown that Danshen has significant inhibitory and cytotoxic effects on the synthesis of tumor cell DNA, and has preventive and curative effects on the metastasis and recurrence of tumors. Danshen has inhibitory effects on the voltage-dependent calcium channels and receptor-operated calcium channels of arterial smooth muscle cells, and can dilate blood vessels, decrease the viscosity of blood, improve the local blood circulation and blood oxygen status of tumors, and increase the sensitivity of tumor blood vessels to chemotherapeutic drugs. At present, there have been no consistent reports about the antistomach cancer effects of tanshinone II_A. It has been proven that tanshinone II_A can inhibit the proliferation of stomach cancer SGC7901 cells by halting the cell cycle and inducing apoptosis, and the induction of the upregulation of p53 gene expression in cells is also one of the molecular mechanisms. It has been found that tanshinone II_A can stop cells at G_0/G_1 phase, and prevent cells from entering S phase. In addition, it has been proved that Danshen has protective effects against injuries caused by chemotherapy and radiotherapy, which may be one of the mechanisms of the enhancing effect of Danshen on the therapeutic effects of antineoplastic drugs.

7.2 Pharmacological Effects of Danshen on Acute Pancreatitis and Its Mechanisms

Mei Gao and Guanhua Du

In recent years, it has been demonstrated by clinical and pharmacological studies that Danshen has significant therapeutic effects on acute pancreatitis (AP). AP is a common acute disease in clinical practice. Severe acute pancreatitis (SAP), in particular, is characterized by acute onset, serious clinical symptoms, and high fatality rates. According to the clinical analysis of Zhang Jianxin et al. using Danshen in a combined treatment for SAP can reduce the incidence rate of complications and the death rate, shorten the period of treatment, and increase the cure rate. We review the mechanisms of the therapeutic effects of Danshen on AP in this section.

7.2.1 Improving Hemorrheology

The role of microcirculatory disturbance of the pancreas in the occurrence of AP has received more and more attention. Danshen improves microcirculation, promotes blood flow and removes stasis, and it has certain therapeutic effects on pancreatitis. Experiments by Fan Linjun et al. showed that there was serious damage to blood capillary beds at 6 h after the onset of SAP in dogs, and with the increased damage, the pancreatic lesions was progressively aggravated. However, after application of Danshen, the surface density of pancreatic blood vessels significantly increased, and the pathological changes of pancreatic tissues were significantly improved. The study by Zhao Xiaoyan et al. has shown that after administration of Danshen injection to SAP rats, the hemorrheological indices (such as blood viscosity, the indices of aggregation of red blood cells and fibrinogen factors) were significantly lower than those of the model group, and the microthrombi in the pancreas decreased, indicating that Danshen can improve the rheological properties of blood, regulate the viscosity of blood, and ameliorate the local microcirculation and hemorrheology in the necrotic area of the pancreas. Danshen can ameliorate microcirculatory disturbances, which is concerned with its enhancement of the deformability of red blood cells and the improvement of blood viscoelasticity. Studies have shown that the active components of Danshen, sodium tanshinone II_A sulfonate (DS201) and Tanshinol (DS182), can inhibit the inflow of calcium to red blood cells, decrease the concentration of free calcium in red blood cells, and enhance the deformability of red blood cells.

7.2.2 Antiplatelet Aggregation

The factors influencing the formation of microthrombi in the pancreas during AP include Thromboxane A2 (TXA2/Prostaglandin I2 (PGI2), an abnormal increase of PAF, and injury to blood vessel endothelium. Danshen can protect blood vessel endothelia, facilitate the balance of TXA_2/PGI_2, and decrease PAF level, so it has an antithrombotic effect. Xu Wengui et al. suggested that acetyl salA (a derivative of an active component of Danshen) could act on the metabolism of arachidonic acid and specifically block the production of aggregation-inducing TXA_2. Moreover, it has a mild promoting effect on the production of prostacyclin in vessel walls, so as to exert antiplatelet functions. The study by Zhao Xiaoyan has shown that the level of PAF in plasma increased significantly 6 h after the establishment of SAP models in dogs and the value of TXA_2/PGI_2 gradually increased, and after application of Danshen injection, the levels of PAF and TXA_2 decreased, and the number of microthrombi in the pancreas was reduced. Liu Yuying et al. observed the in vivo injury of rat mesentery blood capillary endothelia caused by endotoxin and the adhesive behaviors of platelets, and found that Compound Danshen injection could protect blood vessel endothelia from injury and prevent platelet adhesiveness and thrombosis.

7.2.3 Against Angiotensin II (Ang II)

Renin-angiotensin system (RAS) is also a stress response system. It has been shown that Ang II can participate in not only the development of SAP by regulating periphery blood circulation and the microcirculation of the pancreas, but also in the inflammatory reactions of SAP by upregulating intercellular adhesion molecule-1 (ICAM-1) before inflammatory reactions. It has been shown by animal experiments that Danshen can antagonize the effects of Ang II, resisting the excessive activation of RAS at multiple levels and targets, and mitigate or block the systemic inflammatory reaction syndrome after resuscitation of SAP due to ischemia of visceral organs, such as the liver, intestine, etc.

7.2.4 Regulating Endothelin Level

ET is a biologically active peptide which can cause intense constriction of blood vessels and contraction of vascular smooth muscles. Recently, it has been shown that the increase of ET plays an important role in AP and in injury to extrapancreatic organs. The excessive production of ET can cause blood circulation disorders, a decrease in blood flow in the pancreas, ischemia and necrosis of the pancreas, intense constriction of pulmonary vessels, and pulmonary hypertension and pulmonary edema. Danshen can lower ET levels and improve the microcirculation of the pancreas. The animal experiment study by Foitzik et al. showed that the increase in ET levels could significantly reduce the blood flow in the pancreas and accelerate the necrosis of acinar cells, and the preventive use of an ET antagonist could alleviate the injury to the pancreas and extrapancreatic organs. The clinical study by Liu Jiansheng et al. also showed that the ET and NO levels in AP patients increased significantly, and the degree of increase was correlated with the severity of AP. There was a significant difference in the decrease of ET and NO between the Danshen treatment group and control group, indicating that Danshen could improve the microcirculation of the pancreas and resist the actions of ET and NO. Jiang Zhi et al. used in vitro cell culture to investigate the effects of Danshen on the release of ET and PGI_2 by ECs in pulmonary arteries injured by free radicals, and found that Danshen could promote the recovery of the functions of the synthesis and release of ET and PGI_2 by ECs of pulmonary arteries injured by free radicals, which might inhibit the occurrence and development of pulmonary hypertension.

7.2.5 Scavenging Free Radicals

During AP, neutrophils are activated, a great deal of oxygen free radicals are produced, and a series of oxidative stress responses are induced, causing injuries to pancreatic tissues and extrapancreatic organs. Many studies have shown that Danshen has the functions of scavenging oxygen free radicals and antioxidation. The study by Zhang Tie et al. has shown that Danshen can decrease the plasma level of malondialdehyde (MDA) in AP rats, increase the level of SOD, mitigate pathological damage to the pancreas, and has a synergistic effect on ulinastatin. The plasma level of MDA reflects the degree of peroxidation in the body, i.e., the level of oxygen free radicals, while SOD can scavenge oxygen free radicals, indicating that Danshen can enhance the scavenging effect on oxygen free radicals. The study by Yang Hongjie et al. has shown that both Danshen and astragalus root have the capability of scavenging lipid free radicals and superoxide anion free radicals, but have different capabilities. As for the elimination of lipid free radicals, astragalus root is stronger than Danshen, while as for the elimination of superoxide anion free radicals, both of them are strong scavenging agents, but Danshen is much stronger than astragalus root. The study by Lei et al. showed that Danshen could increase the activity of SOD in multiple organs (pancreas, liver, kidney) of rats with SAP so as to mitigate lipid peroxidation in tissues. It has been shown by other studies that all of the active components of Danshen (tanshinone, Tanshinol, Salvianolic acids) are effective antioxidants, salvianolic acids being the strongest among them.

7.2.6 Inhibitory Effects on Ca^{2+}

Calcium homeostasis plays an important role in the AP pancreatic acinar cell injury. Calcium ions can inhibit the secretions of the pancreas, which leads to the accumulation of zymogen in cells, the impairment of acinar cell polarization, and caerulin-like cell injury. The increased Ca^{2+} level can activate PLA_2 which hydrolyzes phospholipids. The increased Ca^{2+} level can also facilitate the transformation of xanthine dehydrogenase to xanthine oxidase, and increase oxygen free radicals. Danshen can block calcium channels and prevent the inflow of calcium in cells, and mitigate cell injury caused by the increase in calcium concentration in cells. Qin Renyi applied Danshen to rats with AP, which was induced by ligation of the bile duct and pancreatic duct and infusion of high doses of cholecystokinin octopeptides (CCK-OP), and found that Ca^{2+} levels in the acinar cells of the pancreas decreased significantly. The pathological changes of the pancreas were significantly mitigated, but there was no significant change in the concentration of serum calcium, indicating that Danshen could prevent the inflow of calcium ions in cells, alleviating hypercalcemia in cells and injury to the pancreas. It has been suggested by other studies that Danshen inhibits the secretion and release of the postsynaptic neurotransmitter SP of neurons by lowering the concentration of Ca^{2+} in the postsynaptic membrane, and mitigates the degree of "autodigestion" of pancreatic tissues by reducing the secretion of pancreatin through the neuroendocrine regulation pathway, so the inflammatory exudation is reduced, tissue edema is alleviated, and the injury to cellular structure and functions and the degree of tissue ischemia and necrosis are reduced.

7.2.7 AntiBacterial and Anti-Inflammatory Effects

Infection is a common complication of SAP, and its incidence rate is around 40–70 %. The infections are mainly caused by enterogenous Gram-negative bacteria, but it has been found that the incidence rate of infections by Gram-positive bacteria (mainly including *S. aureus* and *S. epidermidis*) is increasing. Danshen has strong inhibitory effects against Gram-positive bacteria (especially *S. aureus*); the active antibacterial components are total tanshinones, (cryptotanshinone, tanshinone B, methyltanshinonate, hydroxytanshinone A, and dihydrotanshinone), and among them cryptotanshinone is the most potent. It has been shown that the active components of Danshen have strong bacteriostatic effects on *S. aureus* and its drug-resistant strains. Luo Houwei et al. found that total tanshinone had good therapeutic effects on the first and second stages of inflammation and acute and subacute inflammation, while it had no therapeutic effect on the third stage (chronic stage) of inflammation. It has been shown that Tanshinol can eliminate necrotic tissues of the pancreas in a short time, facilitate the regeneration of pancreatic cells, and dilate blood vessels in the pancreas so as to promote the recovery of pancreatic function.

7.2.8 Inhibiting the Aggregation of White Blood Cells

The occurrence and development of AP are closely related to the excessive aggregation and activation of leukocytes (PMNs), the release of various cell factors, and the systemic inflammatory response syndrome (SIRS) mediated by the cascade reactions of these cell factors. The aggregation of PMNs requires the participation of adhesion molecules, and Danshen can decrease the expression of adhesion molecules and inhibit the adherence of PMNs so as to mitigate the injury of each affected organ in AP. Sha Jianping et al. performed an experimental study on the inhibitory effects of Danshen on the adhesion between PMNs and ECs in the early stages of AP in rabbits, and found that Danshen could decrease the expression of adhesion molecules CD11a/CD18 and CD11b/CD18, inhibit the adhesion between PMNs and ECs, ameliorate microcirculation, and mitigate tissue damage caused by the adhesion between PMNs and ECs. All of these effects were helpful for the early treatment of AP. Chen Haihua applied Danshen to rats with lung injuries caused by endotoxin, and found that the levels of P-selectin and soluble intercellular adhesion molecule-1 (sICAM-1) significantly decreased, and that lung injury was mitigated, indicating that Danshen could suppress the expression of P-selectin and ICAM-1 so as to reduce the recruitment and aggregation of white blood cells in lungs and alleviate lung injury.

7.2.9 Protective Effects on Extrapancreatic Organs

Many studies have shown that Danshen has a protective effect on multiple extrapancreatic organs in AP cases. It has been shown by animal experiments that Danshen can inhibit the excessive release of NO in serum and liver tissues so as to mitigate hepatic injury and inhibit systemic inflammatory reactions, and has a protective effect on liver against injury in experimental AP. It has been shown by clinical studies that after application of Danshen as an adjunctive therapy in patients with AP complicated by liver impairment, the recovery time of liver function and hospitalization time were significantly shortened compared to the control group, indicating that Compound Danshen injection has preventive and curative effects on liver function impairment complicated by AP. Yang Xiaojun used sodium taurocholate to establish a SAP rat model, and found that blood flow in the intestinal tract mucous membrane of the rats in the Danshen treatment group was significantly better than that of the model group, while the enteric bacteria translocation rate and death rate in the Danshen treatment group were significantly lower than those of the AP model group, indicating Danshen could mitigate intestinal injury by improving the blood flow in the intestinal tract mucous membrane and lowering the enteric bacteria translocation rate and death rate of AP rats. Zhang Jianxin et al. found that after Danshen was administered to rats with SAP complicated by lung injury, the injury to lung tissues was mitigated, the blood flow was significantly improved, and the activities of serum PLA_2 and TXA_2/PGI_2 were significantly decreased, indicating that Danshen has good protective effects against lung injury from SAP through the above mechanisms. The study by Zhang Ying et al. showed that Danshen injection could inhibit the expression of ET-1 mRNA in the kidneys of rats with SAP, decrease the concentration of ET-1 in blood, lower the permeability of blood capillaries, increase blood flow, and improve the ischemic state of the kidney so as to mitigate the injury to kidneys.

To sum up, Danshen has affirmative therapeutic effects on AP, but it must be combined with other drugs to enhance the therapeutic effects and reach the goal of shortening the period of treatment, reducing complications, and lowering the fatality rate.

7.3 Progress in the Study on Pharmacological Effects and Relevant Mechanisms of Danshen in the Liver

In recent years, Danshen has been widely used to treat different kinds of acute and chronic hepatitis, liver fibrosis, liver cirrhosis with portal hypertension and hepatic IRI. With the development of biomedicine and pharmacology, studies on the mechanisms of Danshen's effects are deepening, and important progress has been made.

7.3.1 Pharmacological Effects of Danshen on the Liver

7.3.1.1 Effects Against Peroxidation Injury by Free Radicals

Oxidative stress plays an important role in viral hepatitis, drug hepatitis, the occurrence and development of liver fibrosis, and liver cirrhosis. Inflammatory cells, including injured hepatic cells, activated Kupffer cells, and neutrophils, can produce ROS such as H_2O_2, and these ROS and the degraded products of lipid peroxidation can stimulate the activation and proliferation of hepatic stellate cells (HSCs) through paracrine secretion, resulting in liver fibrosis. Both in vivo and in vitro experiments have demonstrated that Danshen can significantly inhibit peroxide reactions of the lipids in hepatic cells, and can protect the integrity of cell membrane structure. It has been shown that Danshen can decrease the content of MDA in the hepatic tissues of rats with liver fibrosis induced by carbon tetrachloride (CCl_4) and dimethyl nitrosamine (DMN), significantly increase the activity of SOD, and prevent injury caused by free radicals and lipid peroxidation. It was found in primary cultures of human fetal liver cells that when subjected to CCl_4 injury, the production of MDA in hepatic cells increased, the membrane fluidity decreased, and the release of alanine aminotransferase (ALT) increased. In the Danshen treatment group, the injury was relieved and membrane fluidity was recovered. Danshen's antioxidant effect has been demonstrated repeatedly, and it is one of the important common effects of its various water-soluble components.

7.3.1.2 Inhibiting the Increase of Ca^{2+} Caused by Injury and Protecting Hepatic Cells

Fura-2 single cell microscopic fluorescence detection was used to observe the effects of Danshen injection on the concentration of free calcium (Ca^{2+}) in hepatic cells during hepatic ischemia-reperfusion, electron spin resonance was used to observe the maximal amplitude value (Y_{max}) of the lipid peroxidation free radical Roo· in hepatic tissues, and transmission electron microscopy was used to observe the pathological changes of the ultrastructure of hepatic cells. The results showed that 45 min after hepatic ischemia and 10 min after reperfusion, Ca^{2+} levels began to increase; 60 min after reperfusion, Ca^{2+} and Roo·Y_{max} levels increased significantly; under electron microscopy, there was mitochondrial swelling, partial vacuolization, texture changes in the matrix pattern, a decrease in granules in the endoplasmic reticulum, and excessive aggregation of nuclear chromatin. In the Danshen injection pretreatment group (equivalent to 6 g/kg body weight of crude drug), 60 min after reperfusion, Ca^{2+} and Roo·Y_{max} values were statistically the same as those of the normal control group, but were significantly different from those of the injury group and normal saline group, and a little lower than those of the Verapamil treatment group. Under electron microscopy, there was mild swelling of the mitochondria of hepatic cells and there was no vacuolization, the cytoplasm was abundant, the nuclear membrane was intact, and the cellular structure and morphology were generally normal. The results of the clinical experiments agreed with the results of the animal experiments. In vitro experiments showed that high concentrations of Ca^{2+} added to rat hepatic cell suspensions could increase the inflow of Ca^{2+} across the membrane, causing a sudden increase in the level of Ca^{2+} in hepatic cells, and Verapamil and Danshen injection could prevent the increase of Ca^{2+}. Therefore, Danshen injection can prevent the Ca^{2+} overload of hepatic cells with ischemia-reperfusion and protect hepatic cells.

7.3.1.3 Immunoregulatory Effects

Plasma Fibronectin (FN) is the main opsonin for the monocyte-macrophage system, and one of its major functions is as a mediator in the phagocytosis of the reticuloendothelial system, and it has opsonic effects on endotoxin and many factors that impair the liver. *D*-galactosamine was used to establish an acute hepatic injury rat model, and the pathological changes and changes in the level of plasma FN were observed with light microscopy and electron microscopy. It was found that Danshen could stimulate the increase of plasma FN levels to enhance the phagocytosis and opsonic activity of the reticuloendothelial system, prevent immunologic injury of the liver, and protect hepatic cells. In addition, Danshen injection could significantly inhibit the activation of the complement system of patients with serious viral hepatitis, reduce the immune complex and antibody, enhance lymphocyte transformation, and prevent immunologic injury in rats with immunological liver fibrosis. Danshen could also inhibit the production of tumor necrosis factor-α (TNF-α) and relieve inflammatory reactions.

7.3.1.4 Promoting Liver Cell Regeneration

Rats with partial excision of the liver after partial hepatic ischemia were used to investigate the effects of Danshen on the regeneration of remnant liver, and it was found that Danshen could improve the liver function of the animals after operation and significantly increase the survival rate, the regenerative rate of remnant liver, the mitotic index of hepatic cells, and the positive rate of serum AFP.

Rat hepatic injury models were established with CCl_4 and D-galactose, and then treated with Danshen. The results showed that Danshen could significantly lower the level of MDA in hepatic tissues, induce the synthesis of cytochrome P450 in hepatic cells, improve the biotransformation functions of the liver, and prevent hepatic injury. Danshen could also significantly increase DNA, RNA, and protein contents in hepatic cells, facilitate the synthesis of ceruloplasmin, ameliorate hypoxia, and facilitate the reparation and regeneration of liver cells.

7.3.1.5 Improving Microcirculation of the Liver

Research data has shown that there are various degrees of hepatic microcirculatory disorders in the developing process of various hepatitis diseases, so the improvement of microcirculation is the important link in the treatment of hepatitis. Danshen can increase the concentrations of NO and prostacyclin in plasma, decrease the concentrations of vasoconstriction substances, such as Thromboxane and Endothelin, reduce the aggregation of platelets, inhibit phosphodiesterase, lower the viscosity of blood, and increase the content of 2,3-diphosphoglyceric acid in red blood cells. Therefore, it could ameliorate the stability and deformability of red blood cells and decrease blood viscosity. With the above effects, Danshen ameliorates the ischemia and hypoxia of liver cells and improves the hemodynamics of the portal vein and the rheological properties of blood. These effects are the basis for its application in the treatment of liver cirrhosis with portal hypertension, and it has better effects than β receptor blockers and there were no adverse reactions.

7.3.1.6 Antiliver Fibrosis

Liver fibrosis is a common pathologic process of a great variety of chronic liver diseases, which is the excessive deposit of hepatic extracellular matrix (ECM) due to the imbalance between the production and degradation of ECM. The pathological mechanisms involve many factors. The activation of HSCs by oxidative stress can significantly increase the production of ECM (in particular collagens) and matrix metal protease (MMP) inhibitor; as a result, the activity of MMP is decreased, and the degradation of ECM is reduced. Since the synthesis of ECM is greater than its degradation, liver fibrosis occurs. It has been proven by animal experiments and clinical practice in recent years that Danshen has an antiliver fibrosis effect. The mechanisms of the effect mainly include the following:

Protecting Hepatic Cells, Relieving Hepatic Injury, Indirectly Inhibiting the Formation of Fibrosis

Danshen injection can significantly decrease the level of MDA in hepatic tissues and serum in the hepatic injury rat model induced by CCl_4, prevent the degeneration and necrosis of hepatic cells, inhibit the proliferation of collagen fibers and formation of pseudolobuli, and effectively avoid the occurrence of liver cirrhosis. Human serum albumin (HSA) was used to prepare a rat immunological liver fibrosis model. The preventive and therapeutic administration of Danshen could significantly lower the positive rate of animal serum immune complexes (IC), anaphylactic shock mortality, serum enzymes, the level of hyaluronic acid (HA) and the content of hydroxyproline (Hyp) in hepatic tissue homogenate, and inhibit the proliferation of collagen fibers, especially collagen Type I. It is suspected that Danshen exerts these functions by relieving cell-mediated immunological injury, terminating the induction of the starting factors for liver fibrosis, and inhibiting the growth and proliferation of fibroblasts.

Inhibiting the Formation of Collagens, and Facilitating the Degradation of the Pathological Collagenous Deposit

CCl_4 was used to injure cultured human embryo hepatic cells, and the results showed that the content of Hyp in hepatic cells and the level of type III precollagen (PC III) in the cell culture medium increased significantly. Pretreatment with Danshen could alleviate these changes, indicating that Danshen could inhibit the synthesis of collagens in hepatic cells. Rats with chronic hepatic injury in different stages were treated with Danshen injection, and at the end of the 6th week, the degree of liver fibroplasia of rats in the early stage treatment group was significantly milder than that of the control group, and there was no liver cirrhosis, so Danshen could inhibit the proliferation of hepatic collagens. Rats in the intermediate stage treatment group showed no liver cirrhosis at the end of the 9th week after treatment, and the rats in the corresponding control group all had nodular cirrhosis of the liver, so Danshen could facilitate the degradation and absorption of sedimentary fibers. After liver cirrhosis was established, Danshen injection was used for 3 weeks, and the excretion amount of hydroxyproline (the degradation products of collagens) in the urine of animals was significantly higher than that of the control group, and the content of collagens in the liver was significantly lower than that of the control group, so Danshen could facilitate the degradation of collagens. The degradation of collagens is mainly regulated and controlled by the activity of MMPs, and Wasser et al. found that Danshen could increase the expression of MMP mRNA in hepatic tissues with liver fibrosis, and inhibit the expression of metal protease tissue inhibitor-1 (TIMP-1) mRNA.

Inhibiting the Proliferation and Activation of Hepatic Stellate Cells (HSCs), and Facilitating Their Apoptosis

The activation and transformation of HSCs is the necessary pathway for liver fibrosis. Electron microscopy, TUNEL staining, and flow cytometry were used to investigate the effects of Danshen on the proliferation and apoptosis of HSCs cultured in vitro, and it was found that Danshen could inhibit the proliferation of HSCs and facilitate their apoptosis. Danshen could inhibit the expression of α-smooth muscle actin (α-SMA), the marker of activation of subcultured HSCs in vitro, and inhibit the production of type I collagen. Other experiments have proven that compound Danshen preparations can inhibit the proliferation of HSCs, significantly decrease the mRNA levels of type I, III, and IV collagens in HSC-T6 cells, and Danshen could upregulate the gene expression of interstitial substance collagenase and inhibit the gene expression of TIMP-1 so as to increase the degradation of collagens. Danshen can significantly inhibit the activity of NF-κB in the HSCs of rats and downregulate the expression of TNFα mRNA and proteins; this is one of the important mechanisms of its inhibiting effect on the proliferation of HSCs. Transforming growth factor β1 (TGFβ1) is an important fibroplasia stimulating factor, and TGβ1 can facilitate the activation of HSCs through autocrine and paracrine secretion. It has been shown that Danshen can inhibit the expression of TGFβ1 mRNA and type I collagen mRNA, indicating that Danshen can inhibit the activation of the autocrines of HSCs. Danshen can inhibit the proliferation of rat hepatic stellate cell strain IG12 cells and rat fibroblast cell strain WEB cells and their collagen synthesis. Recent studies have shown that Danshen can enhance the expression of the apoptosis-related proteins *Fas*L/

Fas and *Bax* in HSCs cultured in vitro, reduce the expression of *Bcl*-2, and induce their apoptosis.

7.3.2 Mechanisms of the Effects of Danshen's Chemical Components on the Liver

The active components of Danshen include liposoluble compounds and water-soluble compounds. Both components have extensive and different pharmacological activities. According to TCM theory, the effects of Danshen alone are equivalent to the effects of four different drugs. With the in-depth investigation of the mechanisms of the effects of Danshen's active components, its clinical application should be based on pattern differentiation, so that Danshen's functions can be better utilized.

7.3.2.1 Mechanisms of the Effects of the Liposoluble Components of Danshen

Since tanshinone was found in the 1930s, more than 40 liposoluble compounds have been identified, most of them belonging to tanshinone type compounds, and among them, tanshinone II_A and tanshinone I were studied more thoroughly.

Antioxidation

Tanshinone II_A is a natural antioxidant which has protective effects against acute hepatic injury in mice induced by CCl_4 and D-galactosamine, and against liver fibrosis in rats induced by CCl_4. It has been proven by in vitro experiments that tanshinone II_A has a significant protective effect on rat primary hepatic cells injured by CCl_4, and can significantly inhibit the increase of ALT, LDH, MDA and NO levels and the decrease of SOD levels in injured hepatic cells, and significantly increase the survival rate of hepatic cells. In addition, the coculture of FeNTA and HSCs could produce oxygen stress, and the effect of tanshinone II_A on this system was investigated. It was found that FeNTA could facilitate the proliferation of HSCs, increase the level of collagens in the culture medium supernatant, and that tanshinone II_A can reverse this effect, which may be related to its antioxidant effects.

Inhibiting the Proliferation of Liver Cancer Line Bel-7402 Cells and Inducing Their Apoptosis

An experiment was conducted to compare the effects of tanshinone II_A, Danshen extract, and Danshen injection on the growth of liver cancer cell line Bel-7402 and periphery blood stem cells, and the results showed that tanshinone II_A and Danshen extract could significantly inhibit the proliferation of Bel-7402 cells, but their inhibitory effects on periphery blood stem cells were weak. Danshen injection had no inhibitory effects. On the contrary, it could promote the growth of tumor cells and periphery blood stem cells. The drug might postpone cell senescence and death by supporting and protecting the injured and senescent cells and facilitating their metabolism. Danshen extract contains liposoluble components, while Danshen injection contains water-soluble components, so it is suspected that the liposoluble components of Danshen (especially tanshinone) can inhibit the growth of tumor cells. These results agree with the previous report that Danshen injection can facilitate the growth of tumor cells. Therefore, Danshen injection and the various active components of Danshen should be properly differentiated and used accordingly. For the present, the antitumor effects of tanshinone have been confirmed, and the mechanisms may be related to the cytotoxic effects, the induction of the differentiation of tumor cells by inhibiting the expression of proto-oncogenes, and the facilitation of the apoptosis of tumor cells by regulating the expression of apoptosis-related genes and facilitating the release of cytochrome C and activation of Caspase-3.

Induction of the Apoptosis of HSCs and Antiliver Fibrosis

It has been confirmed by recent studies that tanshinone I can induce the apoptosis of HSC T-HSC/Cl-6 by facilitating the release of cytochrome C from mitochondria.

Effects on the Drug-Metabolizing Enzymes of the Liver

Tanshinones can inhibit the activity of the drug-metabolizing enzymes of the liver. Among them, tanshinone II_A is the major active component for the inhibition of the drug-metabolizing enzymes of the liver. Experimental results showed that tanshinone II_A could selectively inhibit human cytochrome P4501A2 with an inhibition constant K(i) of (7.2 ± 0.7) nmol/L, indicating a strong competitive inhibitory effect. The inhibitory effects of tanshinone I and cryptotanshinone on human cytochrome P4501A2 are of low selectivity, which may influence the effects of combined medication and are important for clinical application.

7.3.2.2 Mechanisms of Effects of Danshen's Water-Soluble Components

All of the water-soluble components of Danshen have a phenolic acid structure. Tanshinol was first discovered among them, with the chemical name 3,4-dihydroxy phenyl-lactic acid. Thereafter, a series of phenolic acid compounds were identified and named as salvianolic acids. They were sorted in alphabetic order and mainly include salvianolic acids A, B, C, D, E, G, H, I, J, tetramethyl salvianolic acid F, isosalvianolic acid C, rosmarinic acid, lithospermic acid, etc. Most salvianolic acids are formed through the combination of Tanshinol and other organic acids. The total phenolic acids of Danshen has strong antioxidative effects. Among them, the two components with the highest contents, SalA and SalB, have the strongest activities. The water-soluble extract of Danshen (SM) can scavenge superoxide anion and hydroxyl free radicals, inhibit lipid peroxide reactions, prevent hepatic injury, significantly inhibit the proliferation of human liver cancer cell strain HepG2, change cellular morphology, and eventually lead to the apoptosis of cancer cells. These effects are time-dependent and dose-dependent. Further studies have found that the inhibitory effects of SM on the growth of cells are the leading cause for its induction of the apoptosis of liver cancer cells. This process may decrease the GSH in cells through the direct combination between SM and GSH so as to open the mitochondrial permeability transition (MPT) pathway, cause the mitochondrial membrane to lose function, lead to the production of active oxygen, and result in the eventual apoptosis of cancer cells.

Protective Effects of Danshen's Monocomponent IH 764-3 on the Liver

IH 764-3 is a water-soluble component of Danshen with a molecular weight of 158. It was discovered by the Institute of Hematology, Chinese Academy of Medical Sciences. Experiments have shown that IH764-3 can significantly prevent liver fibrosis in rats induced by CCl_4, lower the contents of hepatic hydroxyproline and the mRNA of type I and type III precollagens, decrease the levels of serum HA and laminin (LN), and inhibit the proliferation of HSCs and the synthesis of collagens in a time-dependent, dose-dependent manner. It has been shown by in vitro experiments that IH 764-3 can increase the level of cAMP in cells, decrease the level of calmodulin, and inhibit the proliferation of cultured HSCs. Pretreatment with IH 764-3 could significantly alleviate ischemia-reperfusion (I/R) injury in I/R rat models. The mechanism might be based on the excessive consumption of ATP. Therefore, it could reduce the formation of ROS, inhibit the inflow of Ca^{2+}, prevent inflammation, maintain the stability of the cell membrane and improve microcirculation. Proline hydroxylase is an important enzyme for the hydroxyl modification post collagen translation. The inhibition of this enzyme's activity can reduce the stabilization and synthesis of collagens. It has been found by in vitro studies that IH764-3 can effectively inhibit the activity of proline hydroxylase and inhibit the synthesis and secretion of collagens.

ROS and their induced lipid peroxidation (LPO) are an important cause for the activation and proliferation of HSCs and the synthesis of collagens. Stimulating cultured HSCs with H_2O_2 in vitro can increase cell proliferation and collagen synthesis, and IH764-3 can inhibit these two events by downregulating the expression of focal adhesion kinase (FAK). FAK is a type of non-receptor tyrosine kinase existing in various types of cells. H_2O_2 can lead to the phosphorylation of

FAK so as to facilitate the adherence, proliferation and migration of the cells, and inhibit apoptosis. Experiments have shown that IH764-3 can induce the expression of MMP-13 in HSCs stimulated with H_2O_2 and inhibit the expression of TIMP-1, thus effectively promoting the degradation of collagens and reversing liver fibrosis. It has been shown that IH764-3 can significantly facilitate the expression of caspase-3 proteins and induce the apoptosis of activated stellate cells.

Mechanisms of the Liver Protecting and Antiliver Fibrosis Effects of Salvianolic Acid A

SalA has strong antioxidant activity, and it has been proven that the seven phenolic acids isolated from Danshen have protective effects against peroxidation injury of the microsomes in the liver, liver parenchyma cells, and red blood cells of mice. SalA has the strongest effect among them, and its antilipid peroxidation effect is more than 1000 times higher than that of vitamin E. It has been found by electron microscopy that Fe^{2+}-cysteine can cause serious injury to the microvilli on the surface of hepatic cell membranes, resulting in ulceration and exfoliation. SalA can prevent such damage and inhibit the lipid peroxidation of serous membrane and the consumption of reduced coenzyme II. When CCl_4 is used to induce rat hepatic injury and liver fibrosis, intragastric administration of SalA can significantly lower the activities of ALT and aspartate aminotransferase (AST) in the sera of these rats, and lower the content of MDA in liver tissues. Therefore, SalA has significant antilipid peroxidation and antiliver injury effects. On the other hand, SalA can significantly decrease the content of Hyp (the unique amino acid of collagen) in the liver, prevent liver fibrosis, and inhibit the deposit of types I and III collagen in the matrix. All of this indicates that it has significant antiliver fibrosis effects. SalA can inhibit the proliferation of the HSCs of rats cultured in vitro, the secretion and deposit of collagen type I, and suppress the expression of procollagen α2(I) mRNA, indicating that it can directly inhibit the activating function of HSCs.

Mechanisms of the Antiliver Fibrosis Effects of Salvianolic Acid B

The molecular formula and molecular weight of SalB are $C_{36}H_{30}O_{16}$ and 718, respectively. The Institute of Hepatic Disease of Shanghai University of Traditional Chinese Medicine proved that SalB could effectively reverse the liver fibrosis of chronic hepatitis B and that its therapeutic effects were better than those of γ-interferon by a double-blind clinical trial and the pathological biopsy of hepatic tissues before and after treatment. It has been proven that the mechanisms of antiliver fibrosis effects are as follows:

Anti-lipid Peroxidation Damage

Lipid peroxide reactions are the bridge linking tissue damage and fibrosis. It has been found in recent years that lipid peroxides can cause fibrosis by influencing the metabolism of collagens. In liver fibrosis caused by CCl_4 and alcohol, the peroxidized products of lipids are the principal mechanism of hepatic injury. SalB can alleviate the necrosis of liver cells of rats caused by D-galactosamine and CCl_4, significantly decrease the activities of ALT and AST, and increase the production of prostaglandin in nonparenchymal cells. SalB can significantly decrease the content of MDA and hydroxyproline in hepatic tissues with fibrosis induced by DMN, inhibit the synthesis of collagens by liver with fibrosis, and facilitate the degradation of types I and III collagens. Its antiliver fibrosis effects are related to its antilipid peroxidation effects. SalB has a chemical structure similar to catechol which is shared by other 5-lipoxygenase inhibitors, such as flavone and caffeic acid, and SalB's inhibitory effect on 5-lipoxygenase is related to its antioxidative effects. Its mechanism may be to inhibit the generation of ROS in the initial stage of arachidonic acid oxidation, thus inhibiting the activity center of 5-lipoxygenase.

MDA is the main product of lipid peroxidation and can severely injure cell membrane structure and result in the swelling and necrosis of hepatic cells. It has been shown that MDA or the inducers of lipid peroxide reactions can significantly activate the HSCs in the rest state and stimulate the HSCs to enter into S phase. 2′,7′-dichlorodihydrofluorescein (DCFH) incorporation was used to investigate the protective effects of SalB on the HSC of primary culture injured by MDA. The results showed that SalB could reduce fluorescence after oxidation in cells, and decrease the level of intracellular oxidative stress when compared with the MDA injury group. There were significant changes under microscopy observation.

Inhibiting the Proliferation and Activation of Stellate Cells, and Intervening Intracellular Signal Transduction

It has been shown that after the stimulation of HSCs by MDA, the expression of proliferating cell nuclear antigen (PCNA) increases. SalB can inhibit the proliferation of HSCs after MDA stimulation and the expression of PCNA, suppress the activation of HSCs subculture in vitro, inhibit the synthesis of collagens, and reduce the deposition of collagen fibers. After activation, HSCs can synthesize and secrete LN, which, with hyperplastic type I collagen, forms the intact basement membrane, leading to the capillarization of dieldrin gap and further activating HSCs. It has been proven that SalB can inhibit the synthesis of LN A in HSCs and interfere with the formation of capillaries by reducing the synthesis of type I collagen and LN, so it not only can mitigate liver fibrosis, but also can inhibit the further activation of HSCs.

TGF-β1 is an important cell factor facilitating fibrosis, which can significantly promote the activation of HSCs and the production of collagens in these cells. The signal transduction pathway in HSCs is mainly mediated by Smad proteins. When TGF-β1 was used to stimulate the activation of rat primary HSCs culture and SalB was applied, SalB could inhibit the secretion of collagens and the expression of the proteins of Q-SMA and plasminogen activator inhibitor, downregulate the gene expression of type I precollagen, suppress the expression of Smad2 and Smad3 proteins in cytoplasm and nuclei, and inhibit the phosphorylation and nuclear translocation of Smad2 proteins in cells. Therefore, its effects of antiliver fibrosis and inhibition on the activation of HSCs are based on the signal transduction of TGF-β1 in HSCs. AP1 protein is composed of the gene products of c-*fos* and c-*jun* and it is involved in the TGF-β1 effects on collagen gene initiation. It has been proven by experiments that TGF-β1 can increase the expression of c-*fos* in HSCs and that SalB has a significant inhibitory effect on it, indicating SalB can intervene in the expression of transcription factors in the TGF-β1 signal transduction in HSCs and their biological effects. It has been found that SalB could inhibit the phosphorylation of ERK1/2 in normal primary HSCs cultured for 9 days, or in HSCs stimulated by TGF-β1, but it had no influence on the expression of TβRI and TβRII. SalB can inhibit the synthesis of type I collagen in normal primary HSC culture and the synthesis and secretion of type I collagen in the TGF-β1-stimulated HSCs. SalB has no significant effects on the activities of MMP-2 and MMP-13 in HSC culture fluid, but has significant promoting effects on the activity of MMP-9. Therefore, SalB can reduce the synthesis and secretion of type I collagen by HSCs by inhibiting the ERK signal transduction pathway in HSCs.

To sum up, with the development of research on Danshen, the pharmacological effects of the active components of Danshen and corresponding mechanisms become more and more clear. The study on the pharmacological effects of Danshen on the liver and their relevant mechanisms will further facilitate the development of Danshen preparations and their application in liver diseases.

References

1. Ruizhen C. Chin J Tradit Med Sci Technol. 2003;10(6):390.
2. Shizheng T, Jianguo W, Jiye Z. J Chin Microcirc. 2004;8(2):87, 102.
3. Hongyu W, Yajuan L, Ying W, et al. Heilongjiang Med J. 2000;13(3):172.
4. Zhiwei W, Yinai W, Xiantang L, et al. China J Mod Med. 2003;13(15):88, 90.
5. Xiangrong C, Jingbo L, Hanping S. Chin J Hosp Pharm. 2001;21(1):44.
6. Yuanzhen Z, Chunfang G. Chin J Surg Integr Tradit W Med. 2004;10(1):8.

7. Liqun P, Yi Z, Haoru Z, et al. Chin Arch Tradit Chin Med. 2003;21(1):62.
8. Bairong L, Zhiwei C, Huiyu X. Bull Med Res. 2002;31(5):49.
9. Jianping S, Bide Z, Yan Z, et al. Chin J Integr Tradit W Med Dig. 2004;12(1):3.
10. Han W, Guanglin M, Weidong H. Acta Acad Med Nantong. 2004;24(1):50.
11. Musheng L, Huilai L, Shiting B, et al. China J Exp Surg. 2003;20(2):130.
12. Zhenzhong Q, Rui L, Zhenxue W. Chin J Integr Tradit W Med Intensive Crit Care. 2003;10(2):90.
13. Jianping S, Bide Z, Ruhong G, et al. J Chengdu Univ Tradit Chin Med. 2003;26(3):1.
14. Maoming G, Hongbo G. Chongqing Med J. 1998;27 (5):308.
15. Yicong X, Ruirong G. Chin J Rural Doct. 1998;9:16.
16. Jiyun S. Chin J Integr Tradit W Med Gastro-spleen. 2000;8(6):380.
17. Guangcheng Z. Chin J Clin Pediatr. 2004;12(3):959.
18. Xijun D, Xinjia Z, Huiqing W. Chin J Misdiagn. 2003;3(3):355.
19. Zhihong Z, Guojie L, Changqing G. Mod J Integr Chin Tradit W Med. 2004;13(7):860, 863.
20. Dingyu C, Lixia G, Minghui X. Pract Clin J Integr Tradit Chin W Med. 2003;3(3):14.
21. Yande W, Benqing H. Jiangsu J Tradit Chin Med. 2004;25(4):27.
22. Shizheng T, Jianguo W, Fang H. Chin J Crit Care Med. 2004;24(3):185.
23. Xiuqi D. Zhejiang J Prev Med. 2003;15(7):60.
24. Lei L. Mod J Integr Chin Tradit W Med. 2000;9 (23):2346.
25. Xianhua Z. Herald Med. 1996;15(6):337.
26. Songmei W, Guiping X. J Zhangjiakou Med Coll. 2002;19(3):66.
27. Guihua L, Keping J, Xinya C. Mod J Integr Chin Tradit W Med. 2003;12(19):2062.
28. Qiming C. Acta Acad Med Zunyi. 1998;21 (Supplement):59.
29. Fenglian G. Chin J Hosp Pharm. 1999;19(6):363.
30. Yanping Z, Xueren C. Tradit Chin Drug Res Clin Pharm. 1999;10(4):253.
31. Guozhong W, Li J, et al. Mod Med J. 2004;32(1):14.
32. Chao Z, Jun J, et al. J Emerg Tradit Chin Med. 2003;12(3):2545.
33. Yanying L. Harbin Med J. 2003;23(1):28.
34. Liqun R, Aiqun C. Zhejiang Med J. 2003;15(5):78.
35. Makino O, wakushima H, et al. Boil Pharm Bull. 2002;25(4):523–5.
36. Yun B, Kang S, Chung W, et al. Immunopharmacology. 2000;49:355.
37. Xijun D. Shandong Med J. 2004;44(10):24.
38. Lei L. Mod J Integr Chin Tradit W Med. 2000;9 (23):2346.
39. Wudong T. Acta Acad Med Suzhou. 1997;17 (3):469.
40. Dingyu C, Lixia G, et al. Pract Clin J Integr Tradit Chin W Med. 2003;3(3):14.
41. Xiaodong C, Qilian L. Chin J Integr Trad Med Dig. 2004;12(1):42.
42. Guanhua D, Juntian Z. Herald Med. 2004;23(7):435.
43. Yanquan L, et al. Shanghai J Tradit Chin Med. 2005;39(12):51.
44. Hongquan Z, et al. Chin J Clin Pharm Ther. 2005;10 (7):812.
45. Jian C, et al. Fudan Univ J. 2007;34(1):57.
46. Xiaoli Z, et al. J Xi'an Jiaotong Univ (Med Sci), 28 (4):511.
47. Li L, Hongquan Z. Chin Tradit Pat Med. 2006;28 (3):394.
48. Chao Z, et al. World Chin J Digestol. 2004;12 (3):694.
49. Xiaoli H, Shunying L, Guopin W. J Southeast Univ (Med Sci Ed). 2006;25(6):464.
50. Jianxin Z, Guozha C, Long L, et al. Chin J Emerg Med. 2002;11(2):101.
51. Linjun F, Hongguang G, Aochuan W. Chin J Bases Clin Gen Surg. 1997;4(1):7.
52. Xiaoyan Z, Shihai X, Ping G, et al. J Chin Microcirc. 2001;11(3):31.
53. Liping F, Jinxi L, Daosheng Z, et al. J Chin Microcirc. 1997;7(4):17.
54. Wengui X, Lina X. Acta Pharm Sinica. 1998;33 (1):62.
55. Yuying L, Xiumei Z, Shirong Z, et al. J Chin Microcirc. 2002;6(5):302.
56. Hunley TE. KON V Pediatr Nephrol. 2001;16(9):752.
57. Foitzik T, Eibl G, Hotz B, et al. Dig Dis Sci. 2002;47:130.
58. Foitzik T, Faulhaber J, Hubert G, et al. Pancreas. 2001;22(3):248.
59. Jiansheng L, Xiaohong Z, Ji F, et al. J Postgrad Med. 2003;26(1):21.
60. Zhi J, Huanxing L, Jiusheng Q, et al. Chin J Pathophysiol. 1999;15(2):121, 192.
61. Tie Z, Tieliang C. Chin J Surg Integr Tradit W Med. 2004;10(1):34.
62. Hongjie Y, Chengchuan C, Min Z. Fudan Univ J. 2003;42(6):935.
63. Lei L, Guimin W, Junming J, et al. Chin J Integr Tradit W Med. Digestion. 2005;13(1):1.
64. Renyi Q, Shengquan Z, Zaide W, et al. China J Exp Surg. 2000;17(4):333.
65. Xiaobing Y, Yushi C. Chin J Nosocomiol. 2004;14 (2):174.
66. Yong L, Yiming Y, Shulan Y. Chin Herbal Med. 2000;31(4):304.
67. Houwei L, Baoyang W. J China Pharm Coll. 2002;33(1):6.
68. Jianping S, Bide Z, Yan X, et al. Chin J Integr Tradit W Med Intensive Crit Care. 2003;10(5):279.
69. Haihua C, Xiyong X, Zhui Y, et al. J New Med. 2003;13(4):220.
70. Gang L, Kaicheng W, Haiping C, et al. Chin J Surg Integr Tradit W Med. 2001;7(6):394.
71. Donghai W, Longgen F, Jimin L. Jiangxi J Tradit Chin Med. 2005;36(268):21.

72. Xiaojun Y, Jing L, Shizheng T, et al. Chin J Integr Tradit W Med Intensive Crit Care. 2005;12(4):245.
73. Ying Z, Chengxian S, Ping B, et al. J Dig Surg. 2005;4(3):215.
74. Xiaona Z, Jing H. World Chin J Dig. 2007;15(1):72.
75. Yan L, Bi T, Xiaopeng J, et al. China J Mod Med. 2006;6(9):1294.
76. Guanhua D, Juntian Z. Herald Med. 2004;23(7):435.
77. Jang Sl, Jeong Sl, Kim KJ, et al. Planta Med. 2003;69(11):1057.
78. Lee TY, Wang GJ, Chiu JH, et al. J Pharm Pharmacol. 2003;55(11):1561.
79. Mosaddik MA. Phytomedicine. 2003;10(8):682.
80. Ueng YF, Kuo YH, Wang SY, et al. Life Sci. 2004;74(7):885.
81. Jinping Q, Peiling H, Yawei L, et al. J Pharm Pharmacol. 2003;55(8):1163.
82. Ueng YF, Kuo YH, Peng HC, et al. Xenobiotica. 2003;33(6):603.
83. Kim JY, Kim KM, Nan JX, et al. Pharmacol Toxicol. 2003;92(4):195.
84. Liu C, Liu P, Hu Y, et al. Chin Med J. 2002;82 (18):1267.
85. Oh SH, Nan JX, Sohn DW, et al. Toxicol Appl Pharmacol. 2002;182(1):27.
86. Zhang XL, Liu L, Jiang HQ. World J Gastroenterol. 2002;8(3):515.
87. Liu P, Hu Y, I。 iu C, et al. Liver, 2001, 21(6):384.
88. Zhiwei Z, Zaide W. J Tongji Med Univ. 1995;24 (4):272.
89. Jingjing S, Xinda Z, Yinkun L. Chin J Integr Tradit W Med. 1999;19(5):292.
90. Youwen T, Yumei Y, Xuejun Y, et al. J Chin Microcirc. 2001;5(4):281.
91. Hejun D, Xuehui M, et al. China J Chin Mater Med. 1992;17(4):233.
92. Yiyang H, Ping L, et al. Shanghai J Tradit Chin Med. 1999;10:7.
93. Eryu L, Baoen W, et al. Chin J Hepatol. 1993;1 (2):93.
94. Hongjun Y, Liqun S, Xiuxin Y, et al. Chin J Exp Clin Immunol. 1997;9(3):25.
95. Xinguang Q, et al. J Integr Tradit Chin W Med. 1991;11(2):102.
96. Youwen T, et al. J Integr Tradit Chin W Med. 1999;9(2):41.
97. Xuemei M, Baoen W, et al. Chin J Inf Tradit Chin Med. 2000;7(7):25.
98. Zhihua X, et al. J Tongji Med Univ. 1997;26(6):450.
99. Xuehui M, et al. Chin J Hepatol. 1994;2(2):79.
100. Shuixiang H, et al. J Chin Acad Med Sci. 1996;18 (1):70.
101. Xuehui M, et al. J Integr Tradit Chin W Med. 1988;8(3):161.
102. Fanggang C, et al. China J Mod Med. 2002;12 (13):17.
103. Qingyan L, et al. Pract J Cancer. 2001;16(5):467.
104. Yuexiang C, et al. Chin Med J. 1998;78(8):636.
105. Yuexiang C, et al. Chin J Hepatol. 2002;6(10):3.
106. Shen M, Chen Y, De-Kai, et al. World Chin J gestol. 2001;9(10):1121.
107. Liu J, Shen HM, Ong CN. Life Sci. 2001;69 (16):1833.
108. Liu CH, Liu P, Hu YY, et al. Aeta Pharmacol Sin. 2000;21(8):721.
109. Huang YT, Lee TY, Lin HC, et al. Can J Physiol Pharmacol. 2001;79(7):566.
110. Liu J, Yang CF, Wasser S, et al. Life Sci. 2001;69 (3):309.
111. Nan JX, Park EJ, Kang HC, et al. J Pharm Pharmacol. 2001;53(2):197.
112. Liu J, Shen HM, Ong CN. Cancer Lett. 2000;153 (1–2):85.
113. Liu J,·Yang CF, Lee BL, et al. Free Radic Res, 1999; 3l(6):559.
114. Wasser S, Ho JM, Ang HK, et al. J Hepatol. 1998;29 (5):760.
115. Shulan Y, et al. Chin J Cancer. 1996;25(6):408.
116. Shulan Y, et al. Oncology. l997;17(5):268.
117. Xiujie W, et al. Chin J Oncol. 1996;18(6):412.
118. Renmin H, et al. Chin J Cancer. 1998;17(3):164.
119. Zhongzhi T, et al. Acta Acad Med Mil Tertiae. 2003;25(9):774.
120. Lianjian C, et al. J Chin Med Mater. 2003;26(7):504.
121. Cao EH, Lu XQ, Wang JJ, et al. Free Radic Biol Med. 1996;20(6):801.
122. Guanhua D, Juntian Z. Herald Med. 2004;23(6):355.
123. Yiyang H, et al. Acta Pharmacol Sin. 1997;18 (5):478.
124. Cheng L, Yiyang H, Runping W, et al. Chin J Integr Tradit W Med Dig. 2001;9(1):l5.
125. Yiyang H, Runping W, Xuejun Z, et al. Chin J Hepatol. 2000;8(5):299.
126. Gengtao L, Tiemei Z, Baoen W, et al. Chin J Pharm Toxicol. 1992;6(1):77.
127. Zhishen H, Juntian Z. Acta Pharm Sin. 1992;27 (2):96.
128. Chenghai L, Ping L, Yiyang H, et al. Acta Pharmacol Sin. 2000;21(8):721.
129. Yunhua C, Xiaoling W, Zhenguo Y, et al. Chin J Integr Tradit W Med Liver Dis. 2003;13(4):210.
130. Zhang XL, Liu L, Jiang HQ. World J Gastroenterol. 2002;8(3):515.
131. Yun L, Lili D, Dingming S, et al. Chongqing Med J. 2003;32(12):170l.
132. Li L, et al. Basic Clin Med. 2006;26(7):724.
133. Fuchun H, et al. Herald Med. 2005;24(2):103.
134. Dongying X, et al. Chin J Hepatol. 2004;12(8):471.

The Effects of Danshen on Respiratory Diseases and Their Mechanisms

8

Guorong He, Guanhua Du and Danshen Zhang

The respiratory system is composed of the respiratory tract and lungs. The respiratory tract includes the nose, pharynx, larynx, trachea, and all levels of bronchi, and the lung consists of the pulmonary parenchyma (bronchial tree and alveoli of lung) and pulmonary interstitial substance (connective tissues, blood vessels, lymphatic vessels, lymph nodes, and nerves). The respiratory system is the interface between ambient air and the intracorporeal organs and tissues. During respiration, various harmful substances contained in the inhaled air enter the lungs, and the pulmonary circulation receives venous blood from the entire body, so injuring factors entering the human body from other areas can affect the lungs. Therefore, the respiratory system has an intact composite defense system, including mechanical defense (elimination via mucous, cilia, and cough), phagocytic defense, and acquired specific immune defense [1]. At present, the conditions of respiratory diseases are very serious due to air pollution, indoor contamination, smoking, inhalation of various harmful substances due to industrial development, misuse of antibacterial drugs, aging of population, and the influence of some behavioral factors.

G. He · G. Du (✉) · D. Zhang
Institute of Materia Medica, Chinese Academy of Medical Science and Peking Union Medical College, Beijing, China
e-mail: dugh@imm.ac.cn

Danshen has a great variety of pharmacological actions in the treatment of respiratory diseases: (1) Decreasing the production and release of free radicals, inhibiting free radical reactions, and lowering the permeability of histiocytes and blood vessels. (2) Improving blood rheology, invigorating blood and dissolving stasis, decreasing blood viscosity, and improving microcirculation. (3) Antagonism against calcium. (4) Increasing tissue hypoxia tolerance, enhancing oxygen utilization by tissues, and reducing the consumption of ATP. (5) Stabilizing and reinforcing the cell membrane. Therefore, a comprehensive review of Danshen's pharmacological actions is of great significance in its application to respiratory diseases.

8.1 The Main Pharmacological Effects of Danshen's Treatment of Respiratory Diseases

Guorong He, Guanhua Du and Danshen Zhang

8.1.1 Scavenging Free Radicals

Danshen can significantly improve ischemia-reperfusion injury in the lungs by scavenging oxygen free radicals [2]. Oxygen free radicals have powerful lipid peroxidation actions, and can cause lipid peroxide (LPO) reactions and injure cell membranes, leading to cell death and the

X. Yan (ed.), *Dan Shen (Salvia miltiorrhiza) in Medicine*,
DOI 10.1007/978-94-017-9463-3_8,

injury of organs or tissues under ischemia-reperfusion conditions. The mechanisms for Danshen's scavenging oxygen free radicals are as follows:

(1) Danshen can significantly lower the content of LPO and increase the activity of superoxide dismutase (SOD) in serum, so Danshen is a good extraneous oxygen free radical scavenger [3].
(2) Danshen can inhibit the aggregation of white blood cells in the lungs, significantly reducing the production of oxygen free radicals in the lungs and reducing lung injury during cardiopulmonary bypass [4]. After ischemia-reperfusion injury of limbs, lung tissue shows similar changes with the ischemia-reperfusion injury of limbs, i.e., the injury caused during the reperfusion stage is more serious than that during the ischemic stage. The pathologic changes are similar to those of acute respiratory distress syndrome (ARDS): there is congestion of pulmonary capillaries, leukocyte aggregation, interstitial and alveolar edema, and inflammatory cell infiltration. However, during ischemia-reperfusion injury of limbs, there is no platelet microthrombus in pulmonary capillaries. The main manifestations are as follows: there is aggregation of Polymorphonuclear Neutrophils (PMN), the PMNs adhere to the wall and cause impaction, leading to microvascular expansion and deposition of red blood cells, a few leukocyte micro thrombi, and alveolar membrane cell degeneration, and necrosis. With the prolongation of ischemic time, the tissues consume a great deal of ATP, which is broken down into xanthine and hypoxanthine, promoting the transformation of Xanthine dehydrogenase into xanthine oxidase. After blood flow resumes in the limbs, the "respiratory burst" occurs. Under the actions of xanthine oxidase, oxygen generates a great deal of oxygen free radicals, the ischemic limbs are injured by reperfusion, and free radicals can be transported to distant tissues and organs via blood flow and produce oxygen free radicals and particulate matters harmful to tissues directly or through the activation of Neutrophils. The free radicals cause LPO reactions. Among the particulate matters, Elastase can cause the structural impairment of lung tissues, which is confirmed by the fact that the number of Neutrophils in the alveolar septum increases significantly [5–9].

In the early stage of serious burns, the body shows noninfectious systemic inflammatory reactions. The lung is one of the organs to be injured first, and shows mainly an increase in lung microvascular permeability and pulmonary interstitial edema, and the formation of acute lung injury (ALI), and in severe cases it develops into ARDS, which is one of the substantial causes for death resulting from burns. Danshen can be used as the adjunctive therapy for ALI and ARDS because it can eliminate oxygen free radicals, mitigate the oxidative stress to lungs, increase the capability of enduring oxidation stress, and it can inhibit exudation and pulmonary edema.

Leukocytes participate in the inflammatory reactions after burning. After being activated, leukocytes' adhesion to endothelial cells increase, and by releasing oxygen free radicals, Elastase, platelet activating factors (PAF), and Thromboxane A2 (TXA2), they injure alveolar epithelial cells and interstitium, increase the permeability of pulmonary blood capillaries, and induce pulmonary Edema. Danshen can increase SOD, significantly prevent injury by oxygen free radicals, inhibit peroxide reactions, and has protective actions on cells.

8.1.2 Improving Hemorrheological Indices

Danshen can be effective in the treatment of pulmonary hypertension (PHT). Its antihypertensive effect has something to do with Danshen's actions of selectively dilating pulmonary arterioles, opening the capillary network, and accelerating microvascular flow. The configurational reorganization of the intra-acinar pulmonary arteriolar (IAPA) is the pathomorphological basis for chronic PHT, so inhibition of the reorganization is important for the prevention and treatment of PHT. Danshen not only can dilate the blood vessels of IAPA, but also can mitigate hypoxic injury to Endothelial cells, inhibit the phenotypical change of medial smooth Muscle cells (from contraction type to synthetic type) and the proliferation of vessel wall cells (Endothelial cells, smooth Muscle cells, and pericytes), which may be the key to the inhibition of the increase of pulmonary arterial pressure [10, 11].

Various doses of Danshen were injected into the pulmonary arteries of rats with significantly increased mean pulmonary artery pressure (mPAP) and pulmonary blood vessel pressure (PVR), and it could significantly increase CO and dose-dependently lower the increased pulmonary arterial pressure. When a dose of 0.5 g/kg was used, the decrease in pulmonary arterial pressure was significantly more than that from a 0.1 g/kg dosage, and similar to that from a 1.0 g/kg dosage. Danshen at a dose of 1.0 g/kg showed no side effects, indicating that Danshen is very safe. It has been reported that Danshen could significantly inhibit the proliferation of porcine pulmonary artery smooth Muscle cells (PASMC) and the synthesis of collagens, so as to inhibit the remodeling of pulmonary blood vessels and lower PHT [12, 13].

According to TCM, blood stasis is one of the major causes of chronic pulmonary heart diseases. This concept agrees with the following knowledge of modern medicine: there is long-term hyoxemia in patients with pulmonary heart disease, which can increase the number of red blood cells and the viscosity of whole blood, resulting in the stasis of pulmonary microcirculation, aggravation of heart failure and respiratory failure, and even leading to disseminated intravascular coagulation (DIC) and multiple organ failure [14]. Dantonic™ used as an adjunctive therapy can significantly improve the clinical symptoms and hemorrheological indices in patients with pulmonary heart disease.

In a test of the adhesion of high-metastasis human pulmonary giant cell carcinoma ($PGCL_3$) cells to cell culture plate (coated with Fibronectin), tanshinone II_A could significantly inhibit this process. It has been shown by clinical observation that sclerEdema neonatorum is associated with high blood viscosity and an increased incidence of pulmonary hemorrhage. High molecular weight dextran can cause high blood viscosity and microcirculatory disturbance in rabbits, and the pathological changes in lungs are most significant and are similar to those of pulmonary hemorrhage caused by sclerEdema neonatorum, so it can be used as the animal model for pulmonary hemorrhage. It has been proven that Danshen can prevent pulmonary hemorrhage in rabbits. Compared to those in the control group, rabbits in the Danshen treatment group showed milder hemorrhage and significantly less thrombosis and myocardial necrosis lesions. Therefore, Danshen can prevent and treat the pulmonary hemorrhage caused by high blood viscosity [15].

8.1.3 Calcium Antagonist

In recent years, the relation between Ca^{2+} concentrations and asthma has attracted a lot of attention, and it is generally believed that the onset of asthma is related to the change in Ca^{2+} concentration in the smooth muscle cells of the airway, which can lead to the contraction of bronchial smooth muscles, degranulation of mast cells, release or secretion of inflammatory mediators, and the opening of calcium channels in the cell membrane. The concentration of free Ca^{2+} in cells could be rapidly increased to $(1–50) \times 10^{-6}$ mol/L, which could trigger calcium-dependent excitation–contraction coupling and stimulus–secretion coupling. The pathophysiological changes of asthma (including bronchospasm, release of a series of mediators induced by the activation of inflammatory cells, mucus secretion hyperthyroidism, and the chemotaxis of inflammatory cells) are realized by increasing the concentration of free Ca^{2+} in the corresponding cells. This is the common pathway for the pathogenesis of asthma, so it is believed that any drugs which could prevent the inflow of Ca^{2+} can prevent the onset of asthma [16, 17]. Danshen can relax the trachea strip in rest state in vitro and inhibit the spasmogenic actions of histamine and acetylcholine. It can also prolong the latent period of drug-induced asthma in guinea pigs, and has antagonistic actions against calcium ions.

8.1.4 Regulating the Secretion of Cell Factors [18, 19]

8.1.4.1 Antagonistic Action Against Endothelin

Endothelin (ET-1) is mainly produced in the blood vessel endothelium, and is the effective stimulator for the production and release of tumor necrosis factor-α (TNF-α) and interleukin-8 (IL-8) by monocytes. In the respiratory system, besides pulmonary blood vessel endothelia, both airway epithelial cells and alveolar phagocytes can synthesize and release ET-1. The increase in ET-1 levels is mainly caused by its own increased expression and decreased degradation, because there is no storage of ET-1 in the body. After chlorine treatment, the level of ET-1 continuously increases, reaches a peak in 3 h, and begins to decrease after 6 h. Therefore, ET-1 is the initiating factor for pulmonary edema caused by chlorine. Danshen can significantly inhibit the level of ET-1, so it can mitigate pulmonary edema caused by chlorine injury by lowering the ET-1 level in plasma.

The therapeutic actions of Danshen on hepatopulmonary syndrome (HPS) are related with its inhibition of endothelin secretion by pulmonary artery endothelial cells. Experiments have proven that hypoxia could enhance the gene expression of ET in pulmonary artery endothelial cells, which leads to the increase of ET secretion. After the activation of endothelin receptor type B (ETB) in blood vessel endothelial cells, NO is released by endothelial cells, which causes vasodilation and triggers the physiopathologic process of HPS. There are two types of NO synthase (NOS) in lungs: inducible (iNOS) and endothelial (eNOS). During HPS, the activity of eNOS in lung tissues and pulmonary blood vessels increases, but there is no significant change in the activity of iNOS, indicating the increased production of NO in the lungs is the result of the increased activity of eNOS in lungs. The hypoxemia caused by the increased activity of eNOS can be reversed by NOS inhibitors. It has been found that the level of ET-1 in lungs and plasma increases in HPS, which agrees with the increase of ET-1 mRNA in liver and ET-1 proteins in the epithelial cells of the biliary tract, and these changes are in positive correlation with the increase of the level of eNOS in lungs and the increase of alveolar-arterial oxygen tension difference ($AaDO_2$), indicating that ET-1 is related to the increased production of eNOS in lungs, and NO can be regarded as the result of HPS. Experiments have shown that Danshen can inhibit the secretion of ET by pulmonary artery Endothelial cells in SD rats, influence the process of liver fibrosis and mitigate the extent of fibrosis, and reduce the level of ET in peripheral blood without an increase in the contents of NO and NOS in the lungs, alleviating the symptoms of HPS [20].

8.1.4.2 Antagonistic Actions Against E-Selectin

E-selectin, also known as Endothelial-leukocyte Adhesion molecule-1 (ELAM-1), is derived from endothelial cells, and is a highly glycosylated single strand transmembrane glycoprotein. E-selectin is involved in the adherence of monocytes, neutrophils, and eosinophils. The rolling of leukocytes along the endothelium is realized through the close binding of the intracellular cytoplasmic tail domain of selectin to its ligands. E-selectin is involved in the occurrence of pulmonary diseases, and it is one of the main factors of pulmonary edema. The level of E-selectin is increased in the serum of patients with chronic obstructive pulmonary disease and patients with asthma, so E-selectin can be used as an indicator for the investigation of the protective action of Danshen on lung tissue [21].

In an experiment on pneumonia in mice, the lymphocytic infiltration of pulmonary parenchyma mediated by E-selectin is the key to the establishment of an experimental model. E-selectin is only expressed in activated endothelial cells. After endothelial cells cultured in vivo and in vitro are stimulated by Interleukin-1 (IL-1), TNF-α, and endotoxin, the level of E-selectin reaches a peak in 2–4 h and falls back to baseline in 24 h, while the endothelial cells in rest state do not express E-selectin. So, E-selectin in plasma can be used as a specific mark for the activation of blood vessel endothelial cells and used to indicate the degree of blood vessel endothelium injury.

In pulmonary edema, the level of E-selectin is significantly increased, and the peak value may be 6 times higher than that in the normal group. Danshen was used to treat rats with pulmonary edema. The level of E-selectin in plasma is significantly lowered in the treatment group, about one fold lower than that of the model group. It suggests that Danshen can decrease the expression of E-selectin in blood vessel endothelial cells, mitigating injury to pulmonary vessels and exerting its protective function on lungs, and this is an important mechanism for Danshen's treatment of pulmonary diseases.

8.1.4.3 Antagonistic Action Against Interleukin-8

IL-8 comes from many sources. The principal targets of IL-8 are neutrophils, and it can widely influence the biological activities of granulocytes, so it is involved in lung injury [22]. It is strongly chemotactic toward granulocytes, and can cause the degranulation of granulocytes, produce oxygen free radicals and proteases, increase the penetration of the endothelial layer by granulocytes and the permeability of blood vessel endoderm, and directly or indirectly cause granulocytes to enter the interstitial space or inflammatory area via chemotaxis. The level of IL-8 increases in many pulmonary diseases. Its high concentration in a limited area can cause lung injury, and the levels of IL-8 and ET-1 increase significantly after chlorine treatment. Danshen can reduce the production and content of IL-8, mitigate lung injury, and decrease the occurrence of organ failure.

8.1.5 Antiendotoxin Actions [23–25]

Infection by Gram-negative bacilli and infection-induced systemic inflammatory reaction syndrome are a common cause for ALI in clinical practice, and are mainly caused by bacterial endotoxin. The physiopathological basis for lung injury caused by endotoxin is the sequestration and migration of a great deal of PMNs in the lungs. After activation, a great number of oxygen free radicals are produced and a great variety of proteolytic enzymes are released, causing the degeneration and necrosis of pulmonary blood vessel endothelial cells and alveolar epithelial cells, an increase in capillary permeability, and infiltration and edema in the alveoli and interstitium. The clinical symptom is ALI. Indirect lung injury from endotoxin is induced through a great variety of inflammatory cells and inflammatory mediators. It is currently believed that PMNs play an important role in this process. endotoxin and the other inflammatory mediators (such as IL-8) induced by endotoxin can cause the aggregation and activation of PMNs in lungs. At the time of the adhesion of blood vessel endothelial cells, the concentration of single electron transferring reduced coenzyme II oxidase increases, causing a "respiratory burst", generating large amount of oxygen free radicals which, by their strong oxidation actions, cause the injury and dysfunction of the cell membrane and membranous structures. Danshen can significantly reduce the aggregation of PMNs in lungs and decrease the production of MDA, mitigating lung injury. The manifestations include the decrease in the pulmonary coefficient and the exudation of proteins in bronchial alveolar lavage fluid (BALF). In addition, it has been proven that Danshen can prevent lung injury caused by endotoxin, and the mechanism is related to the inhibition of the aggregation of neutrophils in the lungs and antioxygen free radicals. Danshen can significantly decrease the levels of blood lactate and malondialdehyde (MDA) (the metabolite of arachidonic acid), significantly lower the lung water content, mitigate pulmonary blood stasis, and reduce inflammatory cell infiltration.

8.1.6 Others

The pathogenesis of pulmonary interstitial fibrosis is not yet known, and it is believed that it may be concerned with free radical injury, the unbalanced regulation of collagens, and immunologic abnormality. According to TCM theory on pulmonary interstitial fibrosis, the stagnancy of *qi* and blood stasis is present in the entire process of pulmonary interstitial fibrosis, and blood stasis mainly includes microcirculatory disturbance and the abnormal proliferation of connective tissues. Danshen can invigorate blood and dissolve stasis, resolve hard lumps, improve pulmonary microcirculation, decrease the contents of collagens types I, III, and IV, lead to the degradation and reabsorption of collagen fibers already formed, and it has been demonstrated that Danshen has antioxidative function [26]. In addition, Danshen has inhibitory effects on calcium channels, and can suppress the release of inflammatory mediators and block the stimulating action of cell inflammatory factors on pulmonary fibroblasts by inhibiting the inflow of calcium into inflammatory cells so as to prevent pulmonary fibrosis. Therefore, the anti-liver-fibrosis actions of Danshen include eliminating free radicals, improving blood rheological indices, and antagonistic actions against calcium ions [27].

28 days after perfusion of bleomycin into the trachea of mice, the pulmonary coefficient significantly increased, and there was significant proliferation of fibrous connective tissues, which is in line with the characteristics of early stage pulmonary fibrosis. During the formation of pulmonary fibrosis, the proliferation of fibrous connective tissues of the alveolar wall and the infiltration of Lymphocytes and monocytes result in an increase in lung weight. After intratracheal perfusion of bleomycin in mice, the animals showed various degrees of hair erection, listlessness, and lags in response. In the early stage, the body weight significantly decreased, and symptoms were gradually mitigated about 10−15 days later. After using Danshen, the body weight of mice recovered faster, similar to that of the control group, so it had good adjunctive therapeutic effects on pulmonary fibrosis [28].

8.2 Protective Effects of Danshen on Acute Lung Injury Caused by Various Factors

Guorong He, Guanhua Du and Danshen Zhang

Acute lung injury (ALI) refers to the syndrome with a pathological basis of disseminated capillary membrane and pulmonary alveolus damage, and clinical symptoms of pulmonary inflammatory cell infiltration, edema, gas interchange disturbance, respiratory distress, and refractory hypoxemia after the body is exposed to serious infection, injury, shock, and toxication. ARDS is serious ALI or the terminal phase of ALI. The principal pathological changes include diffusive lung injury and pulmonary shunt, which cause serious disproportions of ventilation and blood flow, hypoxemia, and result in respiratory failure. Trauma, infection and shock can also cause severe and multiple organ function impairment. ALI is the leading cause for the death of patients with serious injuries, extensive burns, and severe infection. A majority of literature reports the death rate as high as 50–60 %. It has been confirmed by recent research that when ALI develops to ARDS, Multiple organ dysfunction syndrome (MODS) occurs, that is, ALI/ARDS is the impairment/failure of multiple organs (including the heart, kidney, intestinal tract, liver, muscles, and brain). The typical pathological changes in lungs are also present in the other organs, and the pulmonary symptoms and signs are only part of the manifestation of MODS during ALI/ARDS.

According to TCM theory, patients with ARDS have various degrees of stagnation of *qi* and stasis of blood. The pharmacological actions of Danshen are very extensive, and it has protective effects on multiple organ injury, especially lung injury.

8.2.1 Protective Effects of Danshen on Traumatic Lung Injury

Different types of trauma can lead to various degrees of injury in lung tissues, the synthesis and release of inflammatory mediators, such as TNF-α, the accumulation of PMN in lungs, the activation of complements, and the production of oxygen free radicals. All of these pathological changes play important roles in the pathogenesis.

8.2.1.1 Lung Injury Caused by Various Types of Ischemia Reperfusion

The injury of pulmonary capillary endothelial cells caused by ischemia reperfusion can induce the adherence and activation of PMNs during the reperfusion stage and lead to a "respiratory burst". A great number of free radicals and proteases are released during the "respiratory burst", which aggravate the injury, increase the permeability of pulmonary microcirculation, reduce the rate of blood flow in microcirculation, and produce stasis. It has been proven that the injury of lung tissues after ischemia reperfusion has 2 peaks: One is 30 min after reperfusion, and the other is 4 h after reperfusion [31]. The absences of PMNs can mitigate the latter lung injury, but cannot alleviate the former. Lung injury in the early stage is caused by the obstruction of pulmonary blood capillaries by blood cells or the actions of macrophages, while the peak of tissue damage 4 h after reperfusion mainly results from PMNs.

Lung Injury Caused by Hemorrhagic Shock—Ischemia Reperfusion [32]

New Zealand white rabbits with body weights of 1.9–2.3 kg had blood let from the femoral artery for about 10 min, and the average arterial pressure was reduced to (40 ± 5) mmHg. Danshen injection (5 mg/kg) was administered via intravenous infusion at the same time. 90 min after shock, shed whole blood and Ringer lactate solution were retransfused for resuscitation (the mean arterial pressure recovered to 80 mmHg). The pulmonary coefficient (pulmonary wet weight/body weight × 100 %), the levels of myeloperoxidase (MPO) and MDA in lung tissues, and the pathological changes in lung tissues were observed. The results showed that in hemorrhagic shock in rabbits, Danshen injection could inhibit the production of MPO and MDA in lung tissues, decrease the pulmonary coefficient, and mitigate the lung injury caused by ischemia reperfusion.

Danshen can accelerate the flow rate of blood cells in blood capillaries, decrease the viscosity of blood, increase the number of opened capillary networks, facilitate the disaggregation of blood cells, and has the actions of anticoagulation and regulating fibrinolysis. It can improve microcirculation and has preventive and therapeutic effects on tissue ischemia-reperfusion injury. Wang Zhiwei performed a study on the optimal administration time for hemorrhagic shock, and found that if Danshen was infused into the artery after blood loss and equilibrium liquid treatment was applied, cell injury could be mitigated [31]. Therefore, Danshen could ameliorate the perfusion of lung tissues, inhibit the activities of MPO and oxygen free radicals, block the cascade reactions of inflammatory reactions, alleviate lung injury after ischemia reperfusion, and had good protective actions on visceral organs.

Lung Injury Caused by Limb Ischemia Reperfusion [32]

Wistar rats with body weights of 350–450 g were used for limb ischemia-reperfusion experiments. Rubber tourniquets were used to ligate the middle parts of the bilateral posterior limbs for 3 h. Compound Danshen injection (CDI) (4 mg/kg) was administered via intraperitoneal injection 30 min before the recovery of blood flow in the limbs. The morphological changes in lung tissues were observed and the PMNs in alveolar septum were counted 3 h after ischemia and 1, 2, 3, and 24 h after reperfusion. The results showed that after limb ischemia reperfusion, the lung tissues showed changes similar to those in the limb injured by ischemia reperfusion, i.e., the injury in the reperfusion stage was more serious than that in the ischemic stage. Danshen could effectively mitigate the lung injury caused by limb ischemia reperfusion.

The mechanisms for the injury of distant tissues and organs caused by limb ischemia reperfusion are not known yet, but it is currently believed that reactive oxygen species are involved. With the prolongation of the ischemic time of limbs, the tissues consume a great deal of ATP, ATP is broken down into xanthine and hypoxanthine, the Xanthine dehydrogenase is transformed into xanthine oxidase, and a "respiratory burst" occurs after recovery of the blood flow in limbs. Under the actions of xanthine oxidase, oxygen produces a great deal of free radicals, reperfusion injury is caused in the ischemic limbs, and free radicals can be transported to distant tissues and organs via blood flow and injure sensitive tissues directly or through the activation of neutrophils. The activation of neutrophils may be another cause for the injury of distant organs. The ischemia of limbs stimulates the activation of PMNs, and a "respiratory burst" is caused after the recovery of blood flow in the limbs. The activated neutrophils can be transported to distant tissues and organs to activate the neutrophils in distant organs, and produce and release a great deal of free radicals and particulate matters harmful to tissues. Free radicals can cause LPO reactions, and the elastase in particulate matters can result in the structural impairment of lung tissues. In addition, OH and H_2O_2 can decompose collagens and hyaluronic acids, lead to the swelling of cells, damage the basement membrane of epithelial tissues, and increase vascular permeability.

Lung Injury Caused by Cerebral Ischemia Reperfusion [33]

Cerebral ischemia-reperfusion injury is a commonly encountered disease after decompression in patients with serious head injury. The injury to the functions of multiple organs can complicate the patient's condition, and the lung is one of the vulnerable organs.

Electrocautery was used to coagulate the bilateral vertebral arteries of Wistar rats with body weights of 200–250 g, and the bilateral common carotid arteries were clamped in order to lead to whole cerebral ischemia. Thirty min later, the pinch was loosened, and the bilateral common carotid arteries were perfused for 60 min. Danshen (2 mg/kg, 1.5 g/ml) was administered via the vena caudalis 30 min before operation. The blood gas indices, the pulmonary coefficient, and the activities of SOD, MDA, and Lactate dehydrogenase (LDH) in lung tissues were measured 90 min after cerebral ischemia. The results showed that Danshen could decrease the content of MDA in the lung tissues and plasma in rats, increase the activity of SOD, significantly mitigate the leakage of LDH and pulmonary edema, and improve pulmonary respiratory functions and increase arterial partial pressure of oxygen (pO_2). Li et al. [34] suggested the lung injury from cerebral ischemia reperfusion was mainly caused by the membrane lipid peroxidation resulting from free radicals and the decrease in ATPase activity.

Danshen is a traditional Chinese medicine that can promote blood flow and dissolve stasis, and experiments have shown that it has good protective effects on lung injury caused by ischemia reperfusion. The mechanisms of the effects are as follows: (1) Invigorating blood and dissolving stasis, decreasing blood viscosity, and improving microcirculation. (2) Calcium antagonist actions. (3) Improving the capability of tissues to endure hypoxia, enhancing oxygen use in tissues, and reducing the consumption of ATP. (4) Directly stabilizing and reinforcing the cell membrane. (5) Decreasing the production and release of free radicals, inhibiting free radical reactions, and lowering the permeability of tissues, cells, and vessels.

8.2.1.2 Lung Injury by Impact [35]

New Zealand white rabbits (body weight: about 2 kg) were fixed onto an impact table. The impact velocity was 5.7 m/s, the force was 75.8 N/cm, and the impact energy was 204 J. The thoracic wall was compressed by 40 % (about 3 cm). The nonimpact side of the rabbit was supported by a hard board, and the body of the rabbit could only compress the table but could not displace. Then, 50 μg/kg of *Escherichia coli* Endotoxin was administered via intravenous injection. 30 min before impact injury, a bolus dose of 10 g/kg (diluted with 10 ml of 5 % glucose) Danshen was administered, and then intravenous infusion of 3 g/(kg h) of Danshen was used as the maintenance dose. The content of TNF-α in the arterial blood of rabbits was measured at 0, 0.5, 1.5, 3, and 6 h after injury, the ratio of pulmonary wet weight/dry weight, arterial blood gas indices, the content of TNF-α, the number of white blood cells, the contents of complement C3a, C5a, and albumin in BALF were measured, and the expression of TNF-α protein and its relative content in lung tissues were determined.

The results showed that Danshen could significantly improve the blood gas indices of animals with traumatic ALI, significantly increase the PaO_2 and decrease $PaCO_2$ in peripheral blood, inhibit the exudation of albumin and aggregation of neutrophils in the alveolar space, reduce the activities of C3a and C5a, and alleviate pulmonary Edema in animals with ALI. Danshen could effectively lower the content of TNF-α in peripheral blood and the contents of TNF-α, C3a, C5a, and albumin in BALF, inhibit the expression of TNF-α proteins in lung tissues, reduce the injury to multiple organs and tissues mediated by cell factors such as TNF-α, mitigate the injury to visceral organs and tissues, and alleviate the occurrence and development of traumatic ALI and MODS. Treatment with Danshen in the early stage of trauma could mitigate the pulmonary acute inflammatory reactions mediated by TNF-α.

8.2.1.3 Acute Lung Injury After Burning [36]

In the early stage of serious burning, the body shows noninfectious systemic inflammatory reactions. The lung is the first organ injured, and the manifestations include increased permeability of pulmonary blood capillaries and pulmonary interstitial edema, formation of ALI and ARDS in serious cases, and this is one of the substantial causes for death caused by burning.

The hair of SD rats with body weight of 200–250 g was removed, then the back was submerged in 100 °C water for 12 s, and immediately washed with cold water in order to produce 30 % Total body surface area (TBSA) third-degree burns. Ringer's solution (40 ml/kg) was administered via intraperitoneal injection for antishock, and iodophor was applied to the wound for anti-infection. Danshen injection (84 mg) and/or sodium β-escin (0.33 mg) was administered via sublingual vein injection at 0, 4, and 12 h after burning. After 24 h, the effects of Danshen alone, sodium β-escin alone, and of the combined use of the two drugs on the adhesion and aggregation of white blood cells in the periphery blood, and the contents of MPO, MDA, and SOD in the lung tissue of the rats were determined. The results showed that both Danshen and sodium β-escin could decrease the adhesion and aggregation of white blood cells in peripheral blood, reduce the aggregation and adhesion of PMNs in lung tissues, mitigate the injury to lung tissues caused by free radicals and metabolites, enhance the protective actions of SOD, and alleviate lung injury. The effects were even better when the two drugs were used together.

It is believed at present that ALI is related to the activation of PMNs and injury caused by free radicals. After serious burning, PMNs are activated and accumulated in the microcirculation of various organs throughout the body, adhere to capillary walls, release inflammatory mediators, cell factors, free radicals and proteases, and result in the disorder of cell membrane function and energy metabolism. Among them, free radicals can increase the systemic and local vascular permeability, and protein denaturation and decomposition. Therefore, reducing the number of PMNs and their adhesion and aggregation and decreasing the production of free radicals are important for the treatment of ALI. Research has shown that both Danshen and sodium β-escin can inhibit the aggregation and adhesion of PMNs in the lungs, decrease the injury to lung tissue caused by free radicals and their metabolites, and enhance the protective actions of SOD on the body. Among them, the increase in SOD content in lung tissues caused by Danshen is greater than that caused by sodium β-escin. Combined use of the two drugs can decrease the injury by PMNs and free radicals to the body, and enhance the protective actions of SOD on the body. Therefore, Danshen and sodium β-escin may have synergistic effects in reducing the adhesion and aggregation of PMNs, inhibiting oxidation injury, and eliminating free radicals. The mechanisms of the actions are as follows: (1) Both drugs could decrease the adhesion and aggregation of PMNs on blood vessel endothelial cells, reduce the quantity and activity of PMNs so as to mitigate the injury of pulmonary blood vessel endothelial cells, and decrease vascular permeability and exudation. (2) Both drugs can reduce the formation of LPOs in lung tissue, increase the concentration of SOD (oxygen free radical scavenger), weaken the injuring actions and strengthen the anti-injury actions. Combined use of the two drugs has synergistic effects, and the therapeutic effects are more significant. Experimental results have shown that after burning, the adhesion and aggregation of white blood cells are enhanced, a great number of PMNs accumulate in the lungs, and many oxygen free radicals are released. These factors exceed the defensive capacity of the protective factors, such as SOD, and ALI is caused. Combined treatment with Danshen and sodium β-escin in the early stage of burning can alleviate the inflammatory reactions of lungs, and prevent the occurrence and development of ALI and ARDS.

8.2.1.4 Lung Injury Induced by Radiation [37–39]

Radiotherapy is an important treatment measure for thoracic tumors, and the injury to lung tissues caused by radiation is a common and severe complication. Radiation-induced lung injury usually occurs 2–8 weeks after radiotherapy, including radiation pneumonitis and chronic pulmonary fibrosis. In particular, acute radiation pneumonitis usually occurs 1 month after radiotherapy, and the manifestations include irritable cough, short breath, abundant phlegm, fever, chest distress, dyspnea, and cyanosis. When it is complicated by infection, there is yellow purulent sputum. In serious cases, radiotherapy is interrupted, the life of the patients is threatened, and the control of tumors and the patients' quality of life are significantly impaired.

Kunming mice with body weights of 18–24 g were anesthetized and fixed. The right thoracic region was exposed, and the other area of the body was covered with a 3 mm-thick lead plate. X-ray (180 kV, 12 mA s, distance 30 cm, total dose 2024 rad) was used to irradiate the chest. Intraperitoneal administration of Danshen injection was started 20 days before irradiation. 0.2 ml (containing 0.16 g of crude drug) was administered to each animal, once per day, and 10 doses were administered to each animal. The changes in lung tissues and thymus were observed for 10 days and 1, 3, and 6 months after irradiation. The results showed that Danshen injection could improve the general conditions of mice, mitigate lung tissue injury caused by radiation, and facilitate reparation, so it had preventive effects on radiation-induced lung injury.

The pathological changes of radiation-induced lung injury include the following in the early stage: the stasis of blood, thrombosis in blood vessels, and edema in blood vessel endothelial cells and the peripheral tissues of blood vessels. In the late stage, the changes include the thickening of the alveolar septum and pulmonary fibrosis. The mechanisms for Danshen's preventive effect on radiation-induced lung injury may be related to its functions of enhancing the resistance of blood vessel endothelial cells to radiation, decreasing capillary permeability, and improving microcirculation. Therefore, Danshen can prevent radiation-induced lung injury, and the application of Danshen preparations during radiotherapy can mitigate or avoid the clinical symptoms and imaging changes of radiation-induced lung injury.

The formula of Fu Fang Huo Xue Tang (Composite Decoction Promoting blood circulation, CDPBC) contains traditional Chinese medicine that can quicken the blood and transform stasis, such as Danshen, Angelica, chuanxiong rhizome, and red peony root. CDPBC was administered to mice exposed to radiation via intraperitoneal injection, and the results showed that this decoction could facilitate the reparation of bone marrow sinuses in mice with acute radiation injury, significantly increase the number of blood capillaries in bone marrow, and dilate the area of blood capillaries. According to the measurement of partial pressure of oxygen in the bone marrow of ulna in situ in vivo, CDPBC not only could dilate blood capillaries, but also significantly increase the partial pressure of oxygen in bone marrow, facilitate the oxygen supply to bone marrow, and mitigate and treat the radiation injury of bone marrow matrix cells. Moreover, it could enhance the adhesion functions of the bone marrow matrix cells in mice with acute radiation damage.

8.2.2 Protective Actions of Danshen Against Infectious Toxic Lung Injury

8.2.2.1 Acute Lung Injury Caused by Endotoxin [40–44]

Lung injury induced by endotoxin is often secondary to serious infections and infectious systemic inflammatory reaction syndrome. ARDS may develop in serious cases. The pathological changes are the injuries to the parenchyma cells of the lung.

Male SD rats with body weights of 180–270 g were used to prepare the model by intravenous injection of *E. coli* Endotoxin (*E. coli* O_{111}:B_4) at 4–5 mg/kg. Danshen was administered via intravenous injection at 8–10 mg/kg 15–30 min before the establishment of the model. The pulmonary coefficients of the animals at 1, 2, and 4 h after the injection of endotoxin, the levels of P-selectin (the transmembrane glycoprotein stored in the α-granules of platelets and the Weibel–Palade bodies of endothelial cells, which plays a key role in the rolling of PMNs along Endothelial cells and the start of adhesion) and soluble intercellular adhesion molecule-1 (ICAM-1, the ligand in blood vessel endothelial cells) in plasma, hemorrheologcial changes, changes in the permeability of pulmonary capillaries, and changes in the MDA and protein contents of BALF were determined.

The results showed that Danshen can mitigate lung injury caused by endotoxin, and the effects are most significant at 2 h after administration of endotoxin. From 2 h after endotoxin administration, the levels of P-selectin and ICAM-1 in the Danshen group were significantly lower than those in the lung injury group. At 4 h, there was no statistically significant difference in the level of ICAM-1 in plasma between the Danshen group and the control group. Other research [45] showed that in lung injury caused by the endotoxin of salmonella, rats with double defects in the expression of P-selectin and ICAM-1 had milder lung injuries and longer survival times when compared with wild type rats. It is suspected that Danshen can inhibit the expression of P-selectin and ICAM-1, and reduce the aggregation of white blood cells so as to mitigate lung injury. Moreover, Danshen can significantly reduce the aggregation of neutrophils in lungs, decrease the production of MDA so as to mitigate lung injury, and manifestations include the decrease of the pulmonary coefficient and the exudation of proteins in BALF.

E. coli endotoxin (*E. coli* O_{111}:B_4) was injected into the femoral vein at 5 mg/kg, and tanshinone II_A was administered via intraperitoneal injection at 5 mg/kg 30 min before and 1 h after the administration of endotoxin in two groups of animals. The animals were observed for 6 h, the expression of PMN-CD18 was detected, the plasma prothrombin time, activated partial thromboplastin time, levels of fibrinogen and protein C, and the content of MDA were measured, the ratio of the pulmonary wet weight/dry weight was calculated, the expression of PMN-CD_{18} of rats was observed, the coagulation and anticoagulation indices were measured, the pathomorphological changes were determined, and blood gas analysis was conducted. The results showed that the basic features of ALI were the formation of microthrombi in lungs and the infiltration of a great number of PMNs. The increased expression of PMN-CD_{18} could aggravate ALI, and tanshinone II_A not only could inhibit lipid peroxidation, improve microcirculation and hypercoagulative state, but also could block the antigenicity of PMN-CD_{18}.

Recently, some have proposed the idea that although the pathological changes induced by endotoxin cause direct injury, the effects of endotoxin are mainly achieved by activating monocytes and macrophages which release TNF-α, IL-1, and a series of other cell factors and inflammatory mediators [48]. Many cells involved in inflammation can produce IL-1. In the acute stage, macrophages synthesize and release IL-1, which is the key substance for local and systemic reactions (fever, increase of granulocytes, synthesis of acute proteins by lungs, etc.) during inflammation, and IL-1 is an important factor for shock, ARDS, and MODS. TNF-α is the key factor for ARDS, and TNF-α can stimulate monocytes and endothelial cells to release IL-1, cause the release of leukocyte antibody-dependent cytotoxin, and impair blood vessel endothelial cells and alveolar epithelial cells.

8.2.2.2 Acute Lung Injury of Sepsis [47]

The manifestations of sepsis are systemic inflammatory reactions. The first organ injured is usually the lung, so induced ALI and ARDS are one of the main causes for patient death. It is believed at present that sepsis can induce ALI due to the in vivo inflammation and relevant chain reactions caused by endotoxin. The inflammatory cells gather in the lungs and release inflammatory mediators, injuring the pulmonary blood capillary endothelial cells and alveolar epithelial cells. Among them, the release of tumor necrosis factors plays an important role in the occurrence and development of ALI and ARDS.

Male Wistar rats with body weights of 220–250 g were used, and cecal ligation and puncture (CLP) was conducted. Acute intestinal perforation was produced, which naturally developed into systemic sepsis. Thus, the rat sepsis model with ALI was prepared. 10 g/kg of Danshen injection was administered via intraperitoneal injection after operation, and 40 ml/kg of balanced salt solution was administered via hypodermic injection in order to supplement the lost liquid during operation. The changes in TNF-α at 3, 6, and 12 h after establishment of the model were observed, and the indices of pulmonary artery partial pressure of oxygen, pulmonary coefficient, pulmonary permeability coefficient, and the cytomorphological changes of visceral organs were used to observe the effects of Danshen on ALI in sepsis and its protective actions on the lungs. The results showed that Danshen could significantly decrease the expression of TNF-α in the peripheral blood and lung tissues of animals with sepsis and ALI, increase PaO_2, lower the pulmonary coefficient and pulmonary permeability coefficient, and mitigate lung injury.

During sepsis, under the actions of causative agents, TNF-α is expressed and produced in the tissues of visceral organs such as the lungs. The aggregation of a large number of leukocytes in the lungs, as well as the activation and uncontrolled release of inflammatory mediators leading to the excessive inflammatory response in the body, are the key factors for the onset of ALI. As a type of acute inflammatory disease, the pathological changes in the ALI of sepsis have the characteristics of inflammatory reactions, including hemangiectasis, increased vessel wall permeability, and edema. Danshen can protect the lung by inhibiting or reducing the expression of cell factors (such as TNF-α) in blood and lung tissues and mitigating acute inflammatory reactions in the lungs so as to prevent the occurrence and development of ALI.

8.2.3 Protective Effects of Danshen Against Chemical Lung Injury

8.2.3.1 Oleic Acid Induced Acute Lung Injury [45, 48–50, 52–55]

The principal pathological changes of ARDS include diffusive lung injury and bypassing in the lungs, which may lead to serious disproportions of ventilation and blood flow, hypoxemia, and result in respiratory failure.

Oleic acid is a fatty acid with relatively strong toxicity. It can activate complements, resulting in the aggregation of polynuclear white blood cells in the lungs, the release of a great deal of free radicals and proteases which can cause lung injury, and the release of inflammatory cell factors such as TNF-α, leading to chemical inflammation and producing lung injury. Although oleic acid can only cause the fat embolism type of ARDS, which cannot completely reflect the etiological model of ARDS, in clinical practice, in ARDS in the various acute high-risk serious diseases, the lung injury is caused by the injury of pulmonary capillary endothelia, pulmonary interstitial Edema, alveolar collapse, and/or the filling of the alveoli with protein-rich liquid, which leads to refractory hypoxemia due to pulmonary bypass and decrease in lung compliance. The ALI animal model prepared with oleic acid is the most used chemical lung injury model at present.

Wistar rats (body weight: 220–260 g) were injected with 0.04 ml/kg of oleic acid, and Danshen extract (prepared by water decoction and alcohol purification method, 1 g/ml) was used for intragastric administration, twice per day, 0.1 ml/kg each time. The changes in blood gas indices (PaO_2 and $PaCO_2$ in the blood of left heart) were observed 6 days later, the pulmonary coefficient was calculated, and the activities of TNF-α and IL-1 secreted by alveolar macrophages were measured. The results showed that Danshen could inhibit the overactivation and secretion of TNF-α and IL-1 by alveolar macrophages, and reduce inflammatory cell infiltration.

New Zealand white rabbits (body weight: 2–2.5 kg) were administered via the pulmonary artery (the catheter was inserted into external jugular vein) or ear vein with 0.1 ml/kg of oleic acid, and 3.0 ml/kg of Compound Danshen injection (CDI) was administered. The pulmonary vascular resistance within 60 min after injection and CO were measured, dynamic observation (30, 60, 120, and 240 min) was performed on the activities of blood gas and plasma TNF and histopathological changes, and the survival rate of rabbits at 6 h was calculated.

The results showed that CDI could effectively inhibit the increase of pulmonary arterial pressure and decrease of CO, so as to improve the hemodynamics of pulmonary circulation. CDI significantly lowered the activity of TNF and increased the survival rate of rabbits. The mechanisms of the improvement of pulmonary Hemodynamics in rabbits with oleic acid induced lung injury are not yet completely known, but they may be related to the following factors: (1) Danshen can significantly enhance the tolerability of the body to oxygen under normal pressure and low pressure conditions, and protect blood vessel endothelial cells so as to decrease capillary permeability. (2) The tanshinone in CDI is a negative inotropic drug, and can inhibit the inflow of calcium and smooth muscle spasms.

Wistar rats (body weight: 220–240 g) were intragastrically administered with Danshen aqueous extract [12 g/(kg d), 3 days]. At 1 h after the last administration, 0.15 ml/kg of oleic acid was injected through the vena caudalis. PaO_2, pulmonary coefficient, the ratio of pulmonary wet weight/dry weight, level of ICAM-1, content of proteins in plasma and BALF, and lung permeability index (LPI = BALF proteins/plasma proteins) were determined, the changes of lung tissues were observed, and the expression of ICAM-1 mRNA was measured.

ALI is a local inflammatory response involving a great variety of inflammatory cells and injury to the pulmonary capillary membrane caused by uncontrolled inflammatory reactions. It has been found in recent years that the adhesion of ICAM-1 is the important pathophysiological basis for inflammatory diseases. This experiment proved that Danshen had protective effects on oleic acid induced ALI, and the mechanism for the effects might be as follows: Danshen could inhibit the expression of ICAM-1, and mitigate inflammatory reactions and LPO reactions so as to protect the morphological structures and functions of lung tissues. It is suspected that preventive treatment with Danshen in the early stage may have preventive and curative effects on ALI/ARDS.

New Zealand white rabbits (body weight: 2–2.5 kG) were slowly injected via the jugular vein with oleic acid at 0.15 ml/kg to prepare a chemical ALI/ARDS model. 10 mg/kg of total phenolic acids of Danshen was administered via intravenous injection 30 min before oleic acid administration. The changes in pulmonary arterial pressure and cardiac functions were observed. The results showed that PHT occurred at 10 min after injection of oleic acid in the ALI/ARDS group, indicating that pulmonary heart functions were impaired. There was right ventricle dilation and the wall of the right ventricle turned thin 2 h after oleic acid injection. Pulmonary capillary endothelial cell injury could lead to the disturbance of pulmonary circulation, which is the pathological basis for the occurrence and development of ARDS. The heart is the primary organ in blood circulation, and pulmonary circulation disturbance will certainly influence cardiac functions, and various lung injury factors can also impair cardiac functions.

The total phenolic acids of Danshen could protect the cardiac functions of rabbits with ALI/ARDS, and maintain the cardiac function indices at normal levels in the critical phase (60 and 120 min after injection of oleic acid). This was important for the improvement of the oxygen deficient conditions of tissue cells during ALI/ARDS and to slow down the development of ALI/ARDS. The total phenolic acids of Danshen could significantly lower the plasma levels of MDA in rabbits in the ALI/ARDS group. This suggests that the total phenolic acids of Danshen could mitigate the LPO reactions in the rabbits, and that it might be related to the improvement of cardiac functions. Some research has found that the total phenolic acids of Danshen can inhibit the aggregation of platelets and thrombosis. Therefore, when the total phenolic acids of Danshen are used to treat ALI/ARDS, they can improve abnormal hemodynamics and pulmonary and systemic blood circulation, and decrease the preload and afterload of the left ventricle. Oxygen free radicals may be involved in the formation of PHT by damaging pulmonary blood vessel endothelial cells. Danshen can recover the Endothelin (ET) synthesis and release functions of the endothelial cells of the pulmonary arteries injured by free radicals so as to inhibit the occurrence and development of PHT. This is another mechanism of the improving effects of Danshen on ALI/ARDS.

Wistar rats (body weight: 194–210 g) were injected intravenously with oleic acid at 0.1 ml/kg to prepare an ALI model. 1.0 g/kg of leech Danshen mixture was administered intragastrically at 1, 12, and 24 h before oleic acid injection. The leech was identified as *Whitmania pigra*, and was directly used after drying and comminution. Danshen was the dry root of Danshen, which was condensed with 90 % alcohol to prepare the extract. A 1:1 ratio of leech and Danshen was used to prepare the leech Danshen mixture. 4 h after the oleic acid injection, the abdomen was opened and aortic blood was taken. The plasma was isolated in order to measure SOD, LPO, TXB2, and 6-ketone prostaglandin Fla (6-keto-pgfla). The pathomorphological changes of lung tissues were observed, and the pulmonary coefficient was calculated. The results showed that leech and Danshen could improve the levels of free radicals and arachidonic acid, and the pathomorphological changes of lungs in oleic acid induced ALI.

It is generally believed that the pathogenesis of ALI includes the aggregation of PMNs in pulmonary blood capillaries, the adherence of PMNs to blood vessel endothelial cells, the release of inflammatory mediators, the injury of alveolar-capillary membrane, and pulmonary edema due to increased permeability. After adherence of PMNs to blood vessel endothelial cells, the blood vessel endothelial cells, alveolar epithelial cells, and interstitial substances are impaired by the release of free radicals, PAF, arachidonic acid metabolites, and proteases. All of these are the basic causes for the pathological changes in the lungs during ALI. Both leech and Danshen are traditional Chinese medicines that can quicken the blood and transform stasis. The preventive mechanisms of ALI have not been completely determined yet, but may be concerned with the following factors: (1) Leech could inhibit platelet aggregation and improve microcirculation. (2) Leech could influence the metabolism of arachidonic acid, inhibit the production of TXA_2, and correct the imbalance of PGI_2/TXA_2. (3) Danshen could inhibit the production of free radicals by PMNs during inflammatory reactions.

8.2.3.2 Amycin-Induced Lung Injury [56]

Amycin is a type of anthracene ring antitumor antibiotic containing a quinoid structure, and can be inserted into DNA and inhibit the synthesis of RNA, so it has intense cytotoxic effects. Free amycin is reduced into a semiquinone free radical under the actions of hepatic drug metabolic enzymes. Semiquinone free radicals react with oxygen and produce a great deal of oxygen free radicals, which directly damage DNA and the structures and functions of proteins and polysaccharides. Oxygen free radicals and the abundant multivalent unsaturated fatty acids in the biomembrane undergo LPO reactions so as to cause membrane dysfunction. The metabolic products of lipid peroxidation (such as MDA and hydrogen peroxide groups) are free radicals and can further amplify the injury of LPO reactions. After the initiating actions of amycin disappear, these secondary injuries still exist, which are the important factors for the continuous development of the injuries to the lungs and bronchi.

Male Wistar rats (body weight: 160–200 g) were injected with doxorubicin hydrochloride in the dorsal vein at 7.5 mg/kg under nonanesthetic conditions to prepare an amycin-induced rat pulmonary and bronchial disease injury model. From the first day of the establishment of the model, 1.6 g/kg CDI (equivalent to 0.8 g/kg of Danshen and 0.8 g/kg of rosewood) was administered via intraperitoneal injection for 14 days. The effects of CDI on the activity of serum SOD, MDA content and the histopathological changes of the lungs and bronchi were observed. The results showed that CDI had protective actions on the injury to the lungs and bronchi, and the mechanism of action was concerned with enhancing the antioxidant ability of the body, inhibiting lipid peroxidation, and facilitating the proliferation and reparation of alveolar epithelial cells and alveolar wall capillary endothelial cells.

The cytotoxic actions of amycin and the induced free radicals and lipid peroxidation can damage organelles containing DNA and RNA and membranous structures (such as cell membrane, nuclei, mitochondria, and lysosome) in the mucous epithelium cells of bronchi and bronchioles, alveolar epithelial cells, and alveolar wall capillary endothelial cell. This leads to disorders in the structures and functions of mitochondrial enzymes and decreases ATP production, resulting in the dysfunction of Na^+–K^+–ATPase and causing edema, degeneration, and necrosis of the mucous epithelia of bronchi and bronchioles, alveolar epithelial cells, and alveolar wall capillary endothelial cells. There is local inflammatory cell infiltration in the mucous membrane of bronchi and bronchioles and alveolar walls, production of a great deal of active oxygen species, and aggravation of the oxidation injury to bronchopulmonary histiocytes. After activation, inflammatory cells can synthesize and secrete a great number of cell factors, such as TNF-α, IL-1, and IL-8, which can lead to inflammatory reactions and aggravate the injury to the bronchopulmonary histiocytes.

One week after CDI was administered, the food intake of rats increased, the body weight gradually recovered, the activity of serum SOD increased, the serum MDA level was significantly lower than that of the model group, and the pathological injury of lungs and bronchi was milder than that of the model group. Danshen can facilitate the absorption of inflammation, inhibit the activation of polymorphonuclear leucocyte and neutrophils, reduce the production of free radicals, increase the activity of endogenous antioxidases such as SOD in order to eliminate oxygen free radicals, suppress the semiquinone free radicals of free amycin and the free radicals formed through the lipid peroxidation and relevant inflammatory chain reactions mediated by amycin, and mitigate the pathological injury of lungs and bronchi. The protective effects of CDI on amycin-induced toxic injury to lungs and bronchi are also concerned with its facilitating actions on the proliferation and reparation of alveolar epithelial cells and alveolar wall capillary endothelial cells.

8.2.3.3 Bleomycin Induced Lung Injury [57, 58]

Free radical injury is the initial step of bleomycin (BLM) induced pathologic processes, such as lung injury, inflammatory reactions, and fibrosis. After entering the body, BLM can produce free radicals, causing the peroxide reactions of cell membrane lipids, increasing the production of LPO in lung tissues, and damaging the functions of cells, especially endothelial cells, which is one of the key factors for BLM induced lung injury.

SD rats (body weight: 200–250 g) were intratracheally injected with BLM (5 mg/kg). In the first 3 days after establishment of the model, 12.5 g/kg of Danshen (equivalent to the amount of crude drugs) was administered via intraperitoneal injection. The effects of Danshen on the pathological changes of BLM induced lung injury, the LPO content in serum, and the changes in the pulmonary wet weight/body weight ratio were observed. The results showed that

Danshen could significantly mitigate pulmonary edema and alveolitis and lower the pulmonary coefficient (pulmonary wet weight/body weight), so it had protective actions on BLM induced ALI.

The pathological developing process of the lung injury caused by single intratracheal injection of BLM is the reparative reaction of lung tissues to injury, and it appears as alveolitis in the early stage and pulmonary fibrosis in the late stage. The alveolitis is the predecessor of pulmonary fibrosis, and the severity of the alveolitis directly influences its prognosis, so inhibiting alveolitis can prevent the formation of pulmonary fibrosis. Danshen can significantly lower the BLM induced increase of MDA, and inhibit free radical reactions and the production of LPO. Therefore, eliminating free radicals and blocking free radical reactions may be an important defense mechanism of Danshen against the lung injury caused by BLM.

8.2.3.4 Lung Injury Caused by Ventilation of Pure Oxygen [59]

Oxygen supports all vital activities, but exposure to pure oxygen can cause excessive responses or tissue damage in the human body. Under normal conditions, the process of aerobic catabolism in tissue cells in the human body is also associated with the production of active oxygen species, so when the production amount is small, tissue damage will not be caused. The amount of this highly active substance determines the degree of injury to tissue cells.

SOD, choline acetyltransferase (CAT), and the oxidation–reduction compounds of glutathione are the most important anti-free-radical substances in cells, and the generation of oxygen free radicals and their toxic effects are the biochemical basis for lung injury caused by pure oxygen. Pure oxygen can significantly increase the production of O_2 in cells, which may be one of the causes for the inclination of patients toward ARDS under mechanical ventilation of pure oxygen.

General anesthesia, tracheal intubation, and mechanical ventilation with pure oxygen were used for 40 patients receiving operation due to esophagus cancer. 10 ml of CDI was administered. Arterial blood was taken before anesthetization and at 1, 2, 4, and 6 h after anesthetization for the measurement of SOD, glutathione peroxidase (GSH-Px), and MDA levels, and the changes in lung tissue ultrastructure were observed. The results showed that CDI could eliminate oxygen free radicals, increase SOD and GSH-Px activities, and alleviate the lung injury caused by pure oxygen ventilation.

CDI is the water-soluble extract of Danshen and rosewood, the principal components of which are water-soluble phenolic acids. The structures of these phenolic acids are similar to the free radical scavengers alpha-tocopherol (Vitamin E) and butyl hydroxyl toluene. All of them have phenolic structures. This structure could interfere with the hydrogen-seizing process of lipid peroxidation, and further react with lipids to form a stable, inactive product so as to stop LPO reactions. Pharmacological research has shown that CDI can inhibit the inflow of Ca^{2+} and the activity of chymotrypsin to reduce the transformation of Xanthine dehydrogenase into xanthine oxidase and the production of free radicals, increase the content of 2, 3-DPG in red blood cells, and suppress platelet aggregation, and the release of 5-HT and platelet TX so as to inhibit pulmonary vasoconstriction reactions caused by H_2O_2. This may be the mechanism of CDI's protection of lungs from injury by pure oxygen.

8.2.4 Protective Effect of Danshen on Mixed Lung Injury

8.2.4.1 Two-Hit (Hemorrhagic Shock + Endotoxin)-Induced Acute Lung Injury [60–62]

New Zealand white rabbits (body weight 2.0–2.8 kg) were bled from the femoral artery to produce acute hemorrhagic shock (mean arterial pressure ≤45 mmHg) for 60 min. The autologous blood and/or Ringer lactate solution were infused for resuscitation (mean arterial pressure ≥75 mmHg). 60 min after stabilization, 10 μg/kg of Endotoxin was administered via intravenous injection, and 5 g/kg of CDI was administered via intravenous injection. The animals were then observed for 4 h. Arterial blood gas analysis was conducted, the level of C reactive protein (CRP) was measured, and pathological changes were observed. The effects of CDI on the CRP level and pathological changes in the "two hit" induced ALI models were determined. The results showed that CDI could significantly increase the level of anti-inflammatory factor CRP in ALI models and mitigate ALI.

New Zealand white rabbits (body weight: 2.0–2.8 kg) were used, and the experimental process included 5 stages: (1) basic stage: 30–60 min after operation. (2) hypovolemic shock stage: exsanguination from the femoral artery was conducted, the blood was mixed with 12 kU/L heparin at room temperature for anti-coagulation and preservation, the arterial blood pressure was continuously monitored, and exsanguination or back infusion of autologous blood or Ringer's solution was performed to maintain the mean arterial pressure at 35−45 mmHg for 1 h. (3) shock recovery stage: back infusion of autologous blood or Ringer's solution was conducted to increase the mean arterial pressure to 75 mmHg, which was stabilized for 1 h. (4) 2-h intravenous injection of LPS stage. (5) 4-h intravenous injection of LPS stage. The activities of TXB_2 and 6-keto-PGFla in circulation during each stage were detected, and arterial blood gas analysis and pathological observation were performed.

The results showed that CDI could significantly decrease the concentration of TXA_2 in the circulation of rabbits with ALI, increase the level of 6-keto-PGFla in serum, and had protective actions on "two hit" ALI.

During ALI, there is an aggregation of neutrophils and platelets in the lungs. neutrophils degranulate and release cytotoxic substances, which activate platelets and blood vessel endothelial cells to synthesize and release a great deal of arachidonic acid metabolic products. Among them, the most important products are TXA_2 and PGI_2. There is a relative balance between TXA_2 and PGI_2 in the plasma or tissues under physiological states. The imbalance of TXA_2 and PGI_2 can result in platelet aggregation, thrombosis, and angiospasm, which is one of the causes for the formation of PHT and microthrombus during ARDS. After injection of endotoxin, a great number of neutrophils are activated, and their degranulation and release of cytotoxic substances stimulate platelet microsomes and blood vessel endothelial cells to synthesize and release TXA_2 and PGI_2.

CDI can eliminate oxygen free radicals, protect the cell membrane, facilitate the degradation of fibrinogens, disaggregate the aggregated red blood cells, improve microcirculation, protect the endothelial cells of normal blood vessels, repair damaged blood vessel endothelial cells, improve the functional state of blood vessel endothelial cells, increase the secretion of thrombin III (AT-III) by reducing the synthesis and release of PAF, NO and ET, so as to protect platelets and exert antithrombosis effects, and improve hemorrheology. It also can decrease the concentration of TXA_2 and increase the release of PGI_2 so as to inhibit the adherence of PMNs to the endothelial cells of pulmonary capillaries and arteries, inhibit the activation of leukotrienes and PMNs, suppress platelet aggregation and microthrombus formation, lower PHT, and mitigate lung injury.

8.2.4.2 Pancreatitis Complicated by Lung Injury [62–64]

Serious acute pancreatitis can cause systemic inflammatory response syndrome (SIRS) and injuries to the lungs, heart, liver, kidney, and circulation system, and even result in organ failure and patient death. Among these, ARDS is an important complication and cause of death. The lung is a vulnerable visceral organ, and about 70 % of the patients dying from serious acute pancreatitis have serious ARDS. Among serious acute pancreatitis patients dying within 2 weeks after onset of the disease, 80 % of the patients have ARDS. Therefore, the effective treatment of serious acute pancreatitis in the early stage is still the key to decreasing death rate.

Five percentage sodium taurocholate (1 ml/kg) was administered to dogs (body weight: 7.5–11 kg) through retrograde pancreatic duct injection (a catheter was inserted into the opening of main pancreatic duct in intestinal lumen for injection), and the injection pressure was more than 30 mmHg. Pancreatic edema was observed after injection. 15 min after injection, hemorrhage was observed in the pancreas. Finally, the pancreas showed extensive hemorrhage and necrosis. After establishment of the model, Danshen injection (5 g/kg) or anisodamine (5 mg/kg) was added to 5 % glucose saline for an intravenous drip. The blood PaO_2, $PaCO_2$ and pH before and 0, 1, 3, 6, 10, and 20 h after the establishment of the model were measured, and the levels of amylase and pancreatic lipase in serum and LDH, acid phosphatase (ACP), albumin and LPO in BALF were determined. The results showed that Danshen had protective effects on lung injury caused by acute hemorrhagic necrotizing pancreatitis.

Balls et al. administered monkeys with *E. coli* endotoxin, and "pulmonary leukocytosis syndrome" and "pulmonary blood capillary injury" were caused. Demling et al. injected sheep with Escherichia coli endotoxin and caused shock in the animals and an increase in pulmonary blood capillary permeability. The results showed that the lysosomal enzyme in pulmonary lymph significantly increased, which was believed to be released by the pulmonary white blood cells gathered in the lungs. In acute hemorrhagic necrotizing pancreatitis, a great deal of digestive enzymes and Endotoxin were released so that the number of stimulated white blood cells increased significantly and white blood cells accumulated in pulmonary capillaries, which was another important source for free radical peroxides.

8.2.4.3 Preventing Lung Injury Caused by Extracorporeal Circulation [65, 66]

Before extracorporeal circulation, radial artery catheterization and internal jugular vein or subclavian vein–right atrium catheterization were conducted. The blood from the radial artery was used to represent left atrium blood. Sarns 7400 blood pump and Xijing 87 bubbling-type oxygenator were used for systemic extracorporeal circulation. Under the conditions of moderate dilution (the concentration of the diluted hemoglobin was 60–70 g/L) and moderate low temperature (the lowest nose temperature during operation was 26–28 °C), cardiac Surgery was performed. The intraoperative perfusion pressure was 56.3–90.0 mmHg, the arterial PaO_2 was 135.82–256.5 mmHg, and Danshen injection was administered before extracorporeal circulation via intravenous infusion after general anesthesia. As for the 30 patients receiving heart valve replacement operations, the number of white blood cells and the concentration of peroxidized lipids in the blood of the left and right atrium were determined before extracorporeal circulation and 10, 45 min, 2 h, and 6 h after the heart beat again. Lung tissues before and after extracorporeal circulation were taken for pathological examination.

It has been shown by experimental and clinical research that lung injury induced by extracorporeal circulation is closely related to the aggregation of white blood cells in lungs. There is aggregation of white blood cells in the lungs in the early stage of pulmonary reperfusion of extracorporeal circulation. The mechanism is as follows: the systemic inflammatory reactions caused by mechanical injury during extracorporeal circulation lead to the activation of complement systems and the production of allergic toxins, and these harmful factors increase the chemotaxis and aggregation of white blood cells. In the early stage of reperfusion in lungs, when the activated white blood cells go through the pulmonary capillary bed, they gather in the lungs due to adhesion, infiltration, and exudation. After activation, the white blood cells gathered in the lungs consume a great deal of oxygen. A "respiratory burst" occurs, and a great deal of oxygen free radicals is produced. Research has shown that there is white blood cell aggregation in the early stage of pulmonary reperfusion, and the production of oxygen free radicals in lungs increases.

The effects of Danshen on mitigating the lung injury of extracorporeal circulation are concerned with the inhibition of white blood cell aggregation in the lungs. It is suspected that the mechanisms may be as follows: Danshen can inhibit leukocyte chemotaxis and aggregation reactions, protect blood vessel endothelial cells, unobstruct the microcirculation, improve the rheological properties of blood, and prevent the aggregation of white blood cells in pulmonary blood capillaries so as to reduce the production of oxygen free radicals and other harmful products in lungs and mitigate lung injury.

It can be seen above that Danshen has anti-inflammatory and antiendotoxin effects, and can inhibit thrombosis, ameliorate the microcirculation of organs, increase blood flow, reduce platelet aggregation, increase the oxygen intake of tissues, eliminate free radicals and superoxide anions, and has protective actions on tissue cells.

References

1. He S, Ma Z. Continuing Med Edu. 2003;17(3):39.
2. Yang W, Zhu H, Zhao B. Chin Pharm Bull. 1990;6(2):118.
3. Yang H, Chen X, et al. Fudan Univ J. 2003;42(16):935.
4. Huang Z, Liao C, Chen D. Chin J Integr Tradit West Med. 1996;16(8):4–51.
5. Chen J, Wang L, et al. China J Mod Med. 2001;11(11):12.
6. Chen J, Wang L, et al. Chin J Surg Integr Tradit West Med. 2002;6(6):412.
7. Liu B, Tian L, et al. China Occup Med. 1992;19(5):267.
8. Lu Q, Sun Z. Chin J Crit Care Med. 1999;19(7):389.
9. Tang D, Hou J, et al. Chin J Burns. 2003;19(5):271.
10. Mo X, Chen W, et al. J Clin Pulm Med. 2002;7(2):1.
11. Minfang Xu, Hongping Sheng, Yuedong Chen. Shanxi J Tradit Chin Med. 2003;19(5):16.
12. Xi X, Che D, et al. J Tongji Med Univ. 1994;23(2):81.
13. Yu W, Pan X, Li Y, et al. Med J Qilu. 2003;18(3):260.
14. Wang Y, Wu A, et al. Chin J Mod Appl Pharm. 2003;20(5):343.
15. Shi Y, Cai D, et al. J Shanghai Med Univ. 1991;18(4):312.
16. Wang L, Xing L, et al. Acta Chin Med Pharm. 1995;5:13.
17. Cao CM, Xia Q, et al. Lire Sci. 2003;72:2451.
18. Dong J, Miao W. Chin Pharm. 2003;20(3):49.
19. Pei S, Cui J, et al. Chin J Tuberc Respir Dis. 2003;26(10):639.
20. Shaowei Pei, Cui J, et al. Chin J Clin Rehabil. 2006;10(3):72.
21. Wu S, Sheng Y, et al. Liaoning J Tradit Chin Med. 2003;30(3):172.
22. Wu s, Sheng y, et al. Liaoning J Tradit Chin Med. 2003;30(10):796.
23. Li m, Li F, et al. Chin J Inf Tradit Chin Med. 2007;14(3):30.
24. Liu S, Yin K, et al. J Emerg Med. 1998;7(3):153.
25. Chen H, Xu X, et al. J New Med. 2003;13(4):220.
26. Miao W, Dong J, et al. Pharm Clin Chin Mater Med. 2003;19(5):24.
27. Chen J, Ruan Y, Li L, et al. Chin Med J. 2003;116(5):757.
28. Zhao N, Guo Z, et al. Foreign Med Sci (Plant Drugs). 2007;22(4):155.
29. Chen J, Wang L, Shi Z et al. Chin J Crit Care Med. 2001;2l(1):8.
30. Tan C, Zheng H, Jiang J et al. Hunan J Tradit Chin Med. 2005;2l(4):87.
31. Wang Z, Gao S, Cheng B, et al. China J Exp Surg. 1994;11(3):170.
32. Eppinger MJ, Lemmon G, Miller S, et al. Am J Surg. 1999;178:147.
33. Duan G, Peng J, Men X, et al. Mod J Integr Chin Tradit West Med. 2004;13(14):1843.
34. Li J, Zhang W, Zhao J et al. Chin J Integr Tradit West Med Intensive Crit Care. 2001;8(4):520.
35. Hu X, Yin W, Liang J, et al. Chin Crit Care Med. 2000;12(9):515.
36. Li Z, Tang D, Jiang M, et al. Chin J Surg Integ Tradit West Med. 2002;8(4):290.
37. Du H, Qian Z, Wang Z. J Integr Tradit Chin West Med. 1990;l0(4):230.
38. Quan X, Zhang F, Kong L. Acta Univ Med Anhui. 2002;37(6):456.
39. Zhang Y, Li N, Wu W. Pharm J Chin People's Liberation Army. 2001;17(6):324.
40. Chen H, Xu X, Yuzui, et al. J New Med. 2003;13(4):220.
41. Xu X, Huang W, Zhang J. J Clin Anesthesiol. 2001;17(11):616.
42. Liu S, Yin K, Wang X, et al. J Emerg Med. 1998;7(3):153.
43. Li M, Li F, Hu B, et al. J Chin Med Mater. 2007;30(4):442.
44. Shi X, Huang L, Cao C, et al. Chin J Tuber Respir Dis. 2005;28(5):350.
45. Kamochi M, Kamochi F, Kim YB, et al. Am J Physiol. 1999;277(2):310.
46. Lu Q, Sun Z. Chin J Crit Care Med. 1999;19(7):389.
47. Xia Y, Liu X. Pharm Clin Chin Mater Med. 2004;20(6):14.
48. Shen B, Xu Y, Qi H. Tianjin Med J. 2000;28(5):293.
49. Tang J, Jin L. J Wenzhou Med College. 1992;22(1):4.
50. Xu M, Zhang G, Wen X, et al. Chin J Tuber Respir Des. 2004;27(2):124.
51. Xing J, Zhao J, Tu P, et al. Chin J Ind Med. 2005;18(3):132.
52. Huo W, Huang L, Zhao R, et al. Chin Crit Care Med. 1998;10(2):89.
53. Jiang Z, Liu H, Qi J et al. Chin J Pathophysiol. 1999;15(2):121.

54. Ye L. Chin J Clin Healthc. 2007;10(3):272.
55. Sun Y, Zhao H. Chin J Pract Chin West Med. 2004;4 (17):2292.
56. Wang X, Zhang R. J Tongji Med Univ. 1993;22 (5):319.
57. Luo W. Chin J Pract Chin West Medicine. 2004;4 (17):1457.
58. Yu W, Pan X, Li Y, et al. Med J Qilu. 2003;18(3):261.
59. Cao C, Huang L. Jiangxi Med J. 2007;42(3):208.
60. Cao C, Huang L, Xiao J. Lingnan J Emer Med. 2004;9(1):1.
61. Wu L, Huang X, He Y, et al. Chin J Emer Med. 2007;16(9):949.
62. Zheng S, Wei J, Wu H. J Integr Tradit Chin West Med. 1989;9(3):158.
63. Liu X, Pan C, Liu Q. Chin J Crit Care Med. 2002;22 (2):109.
64. Zhang J, Cheng G. Chin J Emer Med. 2002;11(2):101.
65. Huang Z, Liao C, Chen D. J Fujian Med Univ. 1996;30(4):316.
66. Qian Y, Wang Z, Zhou J, et al. China J Exp Surg. 2003;20(8):753.

9 The Effects of Danshen on Kidney Diseases

Bin Zhang, Tiantai Zhang and Guanhua Du

The kidney is an important organ with multiple functions, primarily to excrete metabolic wastes through urination and to maintain the equilibrium of water, electrolytes, osmotic pressure, and acid–base balance in order to keep the internal environment steady. In addition, the kidney can secrete a variety of bioactive substances, such as renin, prostaglandin, erythropoietin, and the metabolic active product of Vitamin D3 [1,25-$(OH)_2$-D_3] and can inactivate gastrin and parathyroid hormone so as to regulate the important physiological functions of the body. The kidney is composed of nephrons, the basic physiological unit, and the nephron is composed of glomeruli and renal tubules. Therefore, the pathological changes of glomeruli or renal tubules, such as inflammation, ischemia, or ischemia-reperfusion and toxic injury can result in the abnormality of, renal function change the volume and quality of urine, lead to a series of changes in circulatory, respiratory, nervous, digestive, endocrine, and metabolic functions and may cause uremia in serious cases. Danshen has protective effects on glomeruli and renal tubules against various types of injuries, can increase the glomerular filtration rate and renal blood flow, and can maintain the dilution and concentration effects of the renal tubules on urine so as to improve renal functions.

9.1 Preventive and Curative Effects of Danshen on Renal Inflammatory Diseases

9.1.1 Definition of Renal Inflammatory Diseases and Pathogenesis

Renal inflammatory diseases mainly refer to glomerulonephritis. Glomerulonephritis is an allergic inflammation, with impairment of glomeruli as the main manifestation. It can be divided into primary glomerulonephritis and secondary glomerulonephritis. The former is an independent disease of the kidney, while the latter refers to renal lesions caused by other diseases and is a part of a systemic disease, such as acute pancreatitis complicating nephritis, lupus nephritis, or allergic purpuric nephritis. In addition, vascular diseases (such as hypertension) and metabolic diseases (such as diabetes) can also result in pathological changes in glomeruli.

The clinical manifestations of glomerulonephritis mainly include albuminuria, hematuria, Edema and hypertension. The early symptoms are not obvious, but in the late stage it may lead to renal failure, which seriously threatens the health and life of patients. Glomerulonephritis is the most common cause for renal failure. Most types

B. Zhang · T. Zhang · G. Du (✉)
Institute of Materia Medica, Chinese Academy of Medical Science and Peking Union Medical College, Beijing, China
e-mail: dugh@imm.ac.cn

X. Yan (ed.), *Dan Shen (Salvia miltiorrhiza) in Medicine*,
DOI 10.1007/978-94-017-9463-3_9,

of glomerulonephritis are immunological diseases caused by antigen-antibody responses. An immune complex causes glomerulonephritis in two ways: (1) Antibodies form immune complexes with the inherent insoluble glomerular antigen or non-glomerular antigen in glomeruli; (2) The soluble antigen-antibody complexes formed during blood circulation deposit in glomeruli. After the deposit of immune complexes in glomeruli, glomerular injury may be produced through antibodies, complements, neutrophils, mononuclear macrophages, platelets, mesenteric cells and the blood coagulation system [3].

9.1.2 Preventive and Curative Effects of Danshen on Renal Inflammatory Diseases

Research has shown that Danshen has therapeutic effects on primary glomerulonephritis and secondary glomerulonephritis complicated by other diseases, such as acute necrotic pancreatitis, lupus erythematosus nephritis and purpuric nephritis. Danshen can decrease urine protein and alleviate pathological glomerular changes [4–6]. The mechanisms for these effects are as follows:

9.1.2.1 Anti-inflammatory

In vitro experiments have shown that both tanshinone (the lipo-soluble component of Danshen) and the magnesium salts of salvianolic acid B (the water-soluble component of Danshen) can inhibit inflammation. This action may be related to tanshinone's inhibitory effect on the production of cytokines from inflammatory cells.

9.1.2.2 Inhibition of Arachidonic Acid Metabolism [7–9]

Studies have shown that there is an imbalance of Thromboxane A2 (TXA2) and Prostaglandin I2 (PGI2) in patients with chronic glomerulonephritis, which can lead to renal vascular and mesangial contraction, reduce the area of glomerular filtration, and facilitate the adhesion and aggregation of platelets and white blood cells, which are involved in thrombosis, inflammatory reactions, and histiocyte injury, which become one of the causes of sustained development of chronic glomerulonephritis, glomerular lesions, and eventually renal failure. Experiments have shown that Danshen can improve microcirculation, increase the content of adenosine monophosphate in capillary vessels, decrease the synthesis of PGI_2, inhibit the synthesis of TXA_2 so as to improve the balance of TXA_2 and PGI_2, increase renal blood flow, and decrease urine protein levels.

9.1.2.3 Anti-kidney-fibrosis

Danshen can decrease the expression of Transforming growth factor β1 (TGF-β1) the indicator for fibrosis, and Plasminogen, activator inhibitor type 1 (PAI-1) which improves microcirculation in the kidney, mitigates the fibrosis of renal tissues, and reduces the excretion of proteins in urine [10].

9.1.2.4 Hemorrheology

The mechanism for Danshen's effect on hemorrheologyis as follows: (1) Improving glomerular microcirculation; (2) Anticoagulation, antiplatelet aggregation, anti-thrombosis, so as to prevent and reduce the formation of microthrombi in glomerular capillary vessels and decrease the viscosity of blood; (3) Inhibiting or weakening the allergic impairment of the kidney; (4) Regulating immunologic functions, facilitating the repair and regeneration of injured tissues, and mitigating the metabolism of the body [10, 11].

9.2 Preventive and Curative Effects of Danshen on Renal Failure

9.2.1 Classification and Clinical Manifestation of Renal Failure

Renal failure refers to a series of clinical manifestations as a result of disorder in renal functions, which can be due to different causes. There is an imbalance of water, electrolytes, osmotic pressure, and acid–base, accumulation of metabolic waste, as well as abnormal endocrine function of the kidney, which not only changes the volume and quality of urine, but also causes the dysfunction of multiple systems and organs.

9.2.1.1 Classification Based on Causes and Pathogenesis

Renal failure can be divided into acute renal failure and chronic renal failure according to the cause and pathogenesis.

Acute Renal Failure

Acute renal failure (ARF) refers to the simultaneous, bilateral and rapid decrease of urinary function in the kidneys in a short time so that homeostasis cannot be maintained. There is also a disturbance in water, electrolyte, and acid–base equilibria, and an accumulation of wastes in the body. The clinical manifestations include oliguria or anuria, hyperpotassemia, metabolic acidosis, and azotemia. The urinary dysfunction of ARF is serious, the patient's condition is severe, and the development is very rapid.

Chronic Renal Failure

Chronic renal failure (CRF) refers to the progressive destruction of the renal parenchyma by a variety of chronic kidney diseases. The number of functional nephrons gradually decreases, leading to serious urinary dysfunction and internal environment disturbance in several months or several years. Its clinical manifestations include the retention of metabolic waste and toxic materials in vivo; water, electrolyte, and acid—base disequilibrium; and kidney dysfunction in secretion of bioactive substances. The development of CRF is slow with obvious stages of development, and the manifestations of endocrine function disturbance are apparent.

9.2.1.2 Classification Based on Injury Location

A variety of lesions in and out of the kidney can cause renal failure, and renal failure can be classified based on the injured area.

Renal Tubule Lesions

The renal tubules are the main affected area, and the lesions are mainly caused by ischemia or ischemia-reperfusion injury and the impairment of renal toxic substances.

Glomerulonephritis

The main affected area is the glomeruli.

Interstitial Nephritis

The main affected area is the renal interstitial space, such as in pyelonephritis, etc. [3].

9.2.2 Protective Effects of Danshen Against Renal Failure

Long-term studies have shown that the various active components contained in Danshen have effects in antioxidation, anti-platelet aggregation and anti-inflammation. Oxygen free radicals and inflammation play important roles in kidney injury caused by the above diseases, so Danshen and its active components, including monomeric compounds, have been used to prevent and treat renal failure [12, 13].

9.3 Protective Effects of Danshen on the Ischemia and Ischemia-Reperfusion Injury in Kidneys

9.3.1 Antioxidation

Ischemia and ischemia-reperfusion injury are important factors leading to renal failure. Kidneys are high perfusion organs and are very sensitive to ischemia and ischemia-reperfusion injury. After renal ischemia, ATP is degraded to produce a large amount of hypoxanthine. After reperfusion, through catalysis by xanthine oxidase, hypoxanthine is transformed into xanthine due to restoration of the oxygen supply, producing a large number of oxygen free radicals and hydroxyl free radicals. These free radicals cause damage to kidney tissue mainly through membrane lipid peroxidation, i.e., degradation of the multivalent unsaturated fatty acids in the biomembrane so that the structure and function of the cell membrane and mitochondria l membrane are impaired. The resulting dysfunction and disorder of energy metabolism in various cells leads to renal function impairment. Since the antioxidant ability of the tissues decreases during the ischemic stage, the tissue cells are more sensitive to active oxygen species. Cytochrome C oxidase is the terminal enzyme in the respiratory chain of mitochondria and plays a key role in the aerobic catabolism and oxidative phosphorylation of cells, and is easily damaged during perfusion. Its dysfunction not only can reduce ATP production but also result in the collapse of the mitochondrial membrane potential, leading to the formation of Permeability transition pores (PTP), leakage of cytochrome C into the cytoplasm, activation of the caspase system, and Apoptosis.

It has been shown that a variety of Danshen's components have powerful antioxidant effects and can scavenge free radicals from various sources. Danshen has a protective effect against ischemia and the decrease of Na^+–K^+–ATPase activity in kidneys, so it can protect kidneys against ischemia-reperfusion injury and significantly improve renal function [14].

9.3.2 Affecting Inflammatory Factors

The increased expression of endothelial cell adhesion molecules during ischemia-reperfusion in kidneys plays an important role in the mediation of leukocyte infiltration in kidney tissues and the genesis of renal failure. P-selection is the key component mediating the rolling and initial adhesion of neutrophils on blood vessel endothelia in the early stage of inflammation, while ICAM-1 plays an important role in the further adhesion, activation and infiltration of white blood cells. After injury, the increased expression of P-selectin and E-selectin and ICAM-1 results in the local adhesion of circulating white blood cells, the formation of obstructions in the medulla externa, the increased permeability of endothelia mediated by white blood cells, and the aggregation of red blood cells, resulting in a lack of perfusion in the medulla externa after reperfusion, and the infiltrated, activated white blood cells may cause direct tissue damage.

Some investigation showed that Danshen can inhibit the expression of P-selectin and ICAM-1 in the kidney after hemorrhagic shock, lower the activity of peroxidase in kidney tissues, reduce local leukocyte infiltration, and improve renal function [15].

9.3.3 Effects on Nitric Oxide Synthase

After ischemia-reperfusion in kidneys, there is an increase in the expression of all types of nitric oxide synthases (NOS), including endothelial type (enos), neural type (nnos) and inducible type (inos) . The NO from eNOS plays an important role in the recovery of damaged renal circulation after reperfusion, while the excessive expression of iNOS can directly cause cell injury and result in further impairment of organ function.

Studies showed that Danshen can up-regulate the expression of eNOS and nNOS in the cortices, and inhibit the expression of iNOS in order to maintain the blood flow in the cortices at a normal level. The decrease of NO from iNOS prevents the tubules in the cortices from injury and leads to a gradual recovery in renal function. Therefore, the promoting effect of Danshen on eNOS and its inhibitory effect on the expression of iNOS are important molecular mechanisms for Danshen's protective effects on the kidney [16].

In addition, Danshen has effects in anti-thrombosis, anti-platelet agglutination and anti-coagulation. It has been found that Danshen acts as a calcium antagonist, and calcium antagonists have been proved to effectively alleviate the impairment caused by ischemic acute renal failure. The effect is considered to operate through alleviation of the accumulation of calcium ions in affected cells, as the increase in concentration of calcium in cells is a substantial cause of cell death. Therefore, renal function is protected due to the pharmacological effects of multiple active components in Danshen [17–19].

9.4 Preventive and Curative Effects of Danshen Against the Impairment of Renal Functions Caused by Toxic Substances

Danshen protects against renal function impairment caused by a variety of substances, such as gentamicin, cyclosporine A, adenine, and grass carp gallbladder. Danshen can improve the function of kidneys with renal failure, and increase the filterability and renal blood flow of kidneys. Research has shown that Danshen has a significant antagonistic effect on gentamicin's inhibition of Na^{+}–K^{+}–ATPase activity [20, 22]. Cyclosporine A can cause acute functional constriction of the small vessels of the kidney, leading to hemodynamic change which can result in ischemia in medullary tissue in the early stage. The histological manifestations include the proliferation of interstitial cells of renal tubules and infiltration of macrophages. Cyclosporine A can directly cause the increase of the level of renin-angiotensin in kidneys, and angiotensin can lead to the chronic mild ischemia in the renal cortex and the low perfusion area between cortex and medulla. As a traditional drug of increasing blood flow and reducing stasis, Danshen can protect the kidney against the chronic toxicity of cyclosporine A by decreasing the viscosity of whole blood or plasma, dilating small blood vessels, improving the hemodynamics of capillary vessels in the renal cortex, significantly changing the morphology of fibroblast cells, inhibiting the nuclear fission and proliferation of cells, and promoting the apoptosis of fibroblasts, reducing the cyclosporine A-induced chronic renal interstitial fibrosis and improving renal function [23].

9.5 Preventive and Curative Effects of Danshen Against the Impairment of Renal Function Caused by Renal Interstitial Substance Injury

Renal interstitial fibrosis is characterized by the improper accumulation of connective tissues, including the aggregation of interstitial collagens type I, II, and IV, fibronectin and various proteoglycans, which damages normal structure and leads to a loss of kidney function. Kidney fibrosis will appear in all kinds of kidney diseases when the final phase of renal failure is reached. Therefore, the prevention and treatment of renal fibrosis is one of the main measures in the postponement of renal function aggravation. There is significant renal interstitial fibrosis in lupus nephritis, and kidney scarring (the result of fibrosis) is an important factor in determining the prognosis of lupus nephritis. Fibroblast cells play an important role in renal interstitial fibrosis. In clinical practice, Danshen has been used to treat lupus nephritis and obstructive renal interstitial fibrosis with good therapeutic effects [24, 25].

It is well known that excessive deposits of extracellular matrix (ECM) in the renal interstitial space are the leading cause for renal interstitial fibrosis, and that the deposit of ECM is caused by the increased production and decreased decomposition of ECM in the renal interstitial space. The ECM in the renal interstitial space is produced mainly by fibroblast cells, and the capacity for fibroblast cell multiplication is consistent with the degree of renal interstitial fibrosis, so they play an important role in renal interstitial fibrosis. In areas with inflammation or damage in the renal interstitial space, inflammatory cells such as lymphocytes monocytes and macrophages appear rapidly in a short time, and a great number of cell factors are produced, such as, Platelet-derived growth factor (PDGF). Epidermal growth factor (EGF), Interleukin 4 (IL-4), Transforming growth factor (TGF), and tumor necrosis factor (TNF), which have direct regulatory effects on the differentiation, proliferation, and secretion of fibroblast cells. Among them, TGF-β is the most important regulatory factor of the currently known cytokines. Some research showed that Danshen can (1) inhibit the synthesis and accelerate the degradation of ECM; (2) inhibit the expression of TGF-β, the fibrosis-promoting factor; (3) lead to the high expression of proto oncogene (c-myc) and induce the apoptosis of human kidney fibroblast cells, so as to inhibit the proliferation of human kidney fibroblast cells. Therefore, it can be used to treat renal interstitial fibrosis [24, 25].

Danshen has protective and therapeutic effects on various kinds of kidney diseases, and can prevent and treat kidney diseases (including glomerulonephritis, renal failure, etc.) through its pharmacological functions of antioxidation, circulation improvement, and tissue protection, so as to decrease proteins in urine, increase glomerular filtration rates, and ameliorate various nephrotic syndromes. However, since Danshen has a great variety of active components and the pharmacological effects of many active components are complicated, the mechanisms for Danshen's effects must be studied further.

References

1. Du G, Zhang J. Herald Med. 2004;23(6):355.
2. Du G, Zhang J. Herald Med. 2004;23(7):435.
3. Wu Q, et al. New pathophysiology. Beijing: Chinese Academy Of Medical Sciences & Peking Union Medical College Press; 1999. p. 9.
4. Wang G, Wang J, Tian J. Shanxi J Tradit Chin Med. 1995;11(3):14.
5. Zhang J, Qu J, Li L, et al. Chin J Emerg Med. 2003;12(2):97.
6. Lin S, Ye R, Yang N, et al. Acad J Sun Yat-sen Univ Med Sci. 1999;20(3):161.
7. Kang BY, Chung SW, Kim SH, et al. Immunopharmacology. 2000;49(3):355.
8. Kim SY, Moon TC, Chang HW, et al. Phytother Res. 2002;16(7):616.
9. Ryu SY, Oak MH, Kim KM. Pianta Med. 1999;65 (7):654.
10. He X, Liu X, Zhao X, et al. Chin J Microcirc. 2007;17 (4):29.
11. Huifang Yu, Ding L, Cheng X, et al. Mod J Integr Chin Tradit West Med. 2007;16(27):3985.
12. Du G, Juntian Z. Basic Clin Med. 2000;20(5):10.
13. Zhang W. Med Recapitulate. 1998;4(12):643.
14. Shi H, Zhang H, Chen B. Heilongjiang Med Pharm. 2003;26(2):11.
15. Yanhong L, Keke J. Zhejiang J Integr Tradit Chin West Med. 2003;13(5):289.
16. Liang G, Shen Y, Chen D, et al. Chin J Integr Tradit West Med. 2002;8(2):107.
17. Zhang H, Shi H, Chen B, et al. Chin J Tradit Med. Science and Technology. 2002;9(3):148.
18. Lu F, Lin S, Qiu C. Shanghai Med. 1989;12(4):187.
19. Zhang B, L Huang, Ruan L. China J Tradit Chin Med Pharm. 1991;6(2):22.
20. Li R, Chuang X, Jiang H, et al. J Guiyang Med Coll. 1990;15(3):224.
21. Wang J, Deng W, Yaling Q, et al. Chin J Reg Anat Op. 2002;11(3):247.
22. Qiao B, Zhang Y, Li D, et al. J Henan Med Univ. 2000;35(4):307.
23. Lu F, Zhu C. Shanxi J Tradit Chin Med. 2002;18 (3):53.
24. Zhang G, Ye R, Kong Q, et al. China J Tradit Chin Med Pharm. 1997;12(1):19.
25. Lin Q, Yu J, Deng Y, et al. Nephrology. Chin J Integr Tradit West. 2003;4(2):71.

The Anti-tumor Effects of Danshen 10

Xiuping Chen, Tiantai Zhang and Guanhua Du

In the 1960s, some studies found that Danshen could prolong the survival of mice suffering from Ehrich ascites carcinoma. Over the last 40 years, numerous experimental studies and clinical practices have been undertaken which demonstrated that Danshen (mainly tanshinone and Tanshinol) has the ability to inhibit the growth of a great variety of tumor cells. As an antitumor auxiliary drug, it has good prospects for clinical application and has been used as the adjunctive therapy for tumors in clinical practice.

10.1 Application of Danshen in Clinical Tumor Therapy

Currently, Danshen preparations have been widely used in the treatment for tumors in China. In most conditions, Danshen is used in combination with other drugs, seldom used alone. As one of the adjunctive therapies for malignant tumors, Danshen is used for the following aspects in antitumor therapy:

10.1.1 Alleviating the Pain of Malignant Tumors

The pain caused by malignant tumors is one of the principal symptoms in patients with advanced carcinoma. It has been found that high dosage Danshen injections can significantly alleviate the cancer pain. Twentythree patients with primary and secondary liver carcinoma received an intravenous drip of Danshen, and the total effective rate was 78.3 %. Daily intravenous drips of 20–40 ml of high-dose Danshen injections in 500 ml glucose normal saline was used to treat the pain of late cancer, and it was found that Danshen could significantly alleviate the pain of cancer; the remission rate was as high as 93.3 %, with no adverse reactions.

10.1.2 Enhancing the Therapeutic Effects of Radiotherapy

When chemotherapy combined with Dantonic™ was used to treat middle and late stage breast cancer, the combined therapy could increase the effective rate, alleviate the adverse reactions of chemotherapy, enhance the immunologic functions of the patients, and improve quality of life. When radiotherapy combined with Danshen was used to treat esophagus cancer, Danshen could enhance short-term therapeutic effects, increase immunologic function, stimulate hematopoiesis in bone marrow, and increase the number of white blood cells so as to allow radiotherapy to

X. Chen · T. Zhang · G. Du (✉)
Institute of Materia Medica, Chinese Academy of Medical Science and Peking Union Medical College, Beijing, China
e-mail: dugh@imm.ac.cn

X. Yan (ed.), *Dan Shen (Salvia miltiorrhiza) in Medicine*,
DOI 10.1007/978-94-017-9463-3_10,

be performed. When radiotherapy plus Danshen was used to treat pulmonary carcinosis, Danshen effectively protected cardiac function.

10.1.3 Prevention and Treatment of Radiotherapy and Chemotherapy-Induced Myelosuppression

Ninetyfour patients with malignant tumors were treated with Carboplatin chemotherapy and radiotherapy; Danshen and blood cell increasing drugs were used to prevent and treat bone marrow suppression. The effective rates were as follows: Danshen group, 87.9 %; blood cell increasing drug group, 29 %; control group, 0 %. Therefore, Danshen can effectively prevent and treat bone marrow suppression caused by Carboplatin.

In addition, good therapeutic effects have been obtained when Danshen is used to treat stomach cancer with 5-FU, and liver cancer, and digestive tract cancers with anticancerogen.

Compound Danshen preparations have also been widely used as the adjunctive therapy for tumors in clinical practice.

In clinical practice, Danshen has not been used alone in oncotherapy, only used as an auxiliary drug. However, it can alleviate the pain of malignant tumors, enhance the effects of other therapies, mitigate the side effects of operation, radiotherapy, and chemotherapy, and significantly improve patients' quality of life.

10.2 Actions of Danshen on Malignant Tumors and Possible Mechanisms

The chemical components of Danshen are complex, and the research on its antitumor effects has been mainly focused on tanshinone (the principal fat-soluble component) and Tanshinol (the principal water-soluble component). Research data has proved that the principal antitumor component is tanshinone, and Tanshinol has some effect. It has been found that tanshinone and Tanshinol have direct cytotoxic actions, which can inhibit the proliferation and induce the differentiation and apoptosis of tumor cells. In addition, Danshen also acts on the metastasis and diffusion of tumors.

10.2.1 Effects of Danshen on Tumorigenesis and Relevant Mechanisms

Tumorigenesis is a complicated and long-term biological process, and is the result of the interaction between a great variety of carcinogenic factors in the environment and human body. It has been shown that Danshen has some influence on the genesis of tumors.

10.2.1.1 Inhibition on the Proliferation of Tumor Cells and DNA Synthesis

Flow cytometry results have shown that tanshinone can stop cells in G_0/G_1 phase, and prevent cells from entering S phase so as to inhibit DNA synthesis and cell proliferation. Wang Xiujie et al. detected the in vitro anticancer activity of tanshinone II_A against tumor cells by using 5-bromodeoxyuridine (Brdu) incorporation and antiproliferating cell nuclear antigen (PCNA) immunohistochemical staining. The Brdu labeling rates of the human liver cancer SMMC-7721 cells and human leukemia HL-60 cells were 19.5 and 19.0 %, respectively, which were significantly lower than that of the control group; the PCNA positive rates were 57.0 and 30.3 %, respectively, which were also significantly lower than that of the control group. It has been shown by in vivo and in vitro experimental studies that tanshinone can influence the activity of DNA polymerase δ and inhibit DNA synthesis by suppressing the expression of PCNA and the genes related to cell proliferation so as to inhibit the proliferation of tumor cells.

10.2.1.2 Inducing the Differentiation and Apoptosis of Tumor Cells

Tanshinone can regulate a variety of genes related to the differentiation and apoptosis of tumor cells, so it can induce tumor cells' differentiation in morphology and function and their apoptosis.

Yuan Shulan et al. observed the reversion actions of tanshinone on some phenotypes of human liver cancer cell lines. After human liver cancer cells (SMMC-7721) were treated with 0.5 mg/L tanshinone for 4 days, the structure and morphology of cells turned benign, the growth of cells were significantly inhibited, and the Brdu labeling rates and PCNA positive rates were significantly lower than those of the control group. FCM detection showed that the cells in the tanshinone treatment group were stopped in G_0/G_1 phase, the number of cells in S phase was significantly decreased, and the expression of c-*myc* protein was significantly decreased, while the expression of c-*fos* protein was significantly increased. Clinical experiments have shown that tanshinone II_A could induce the differentiation of a variety of leukemia cells. Liang Yong et al. investigated the effect of tanshinone on the differentiation of the NB4 strains of acute promyelocytic leukemia (APL) cells and relevant molecular mechanisms. After the cells were treated with 0.5 mg/L tanshinone II_A for 5 days, 91.3 % of NB4 cells become myelocytes, metamyelocytes, and neutrophils with better differentiation, and its effect on differentiation was similar to that of ATRA. FCM showed that NB4 cells were stopped in G_0/G_1 phase, and the synthesis of DNA was inhibited (the proportion of cells in S phase decreased). After treatment with tanshinone II_A, the expression of c-*myc* and p53 in NtM cells was downregulated, while the expression of c-*fos* and *Bcl*-2 was upregulated. Research by Wu Yu et al. showed that tanshinone II_A could inhibit the growth of erythroleukemia cell strain K562 of chronic myelocytic leukemia, and these results were statistically significant when compared with the control group.

Yoon et al. reported that induction of apoptosis was a new treatment method in the research on tumors. Tanshinone II_A can induce the apoptosis of HL-60 by inducing the DNA of cells to form multiple fragments of 180 bp. The flow cytometry by PI staining showed that the percentage of diploid cells increased, and that the apoptosis induced by tanshinone II_A was associated with the expression of special ADP-ribose polymerase and caspase-3, the latter playing an important role in the induction of apoptosis by tanshinone II_A.

10.2.1.3 Effects on the Activity of Telomerase in Tumor Cells

It has been reported that when HL-60, K562, and APL cells were treated with 0.5 μg/ml of tanshinone II_A for 5, 9, and 15 days, PCR-TRAP detection showed the inhibition rates on the telomerase activity were 30.8, 50.8 and 37.7 %, respectively, indicating that inhibition of the activity of telomerase was related to the antitumor action mechanism of tanshinone.

10.2.1.4 Effects on the Surface Antigens on Tumor Cells

Tanshinone II_A can change cell surface antigens, induce the differentiation of tumor cells, increase the expression of CD9, decrease the expression of CD63 and CD42a, and lower the number of CD42a and TSP cells so as to inhibit the invasion and metastasis of tumor cells.

10.2.2 Effects on the Development of Tumors

10.2.2.1 Inhibiting Actions on the Proliferation of Tumor Cells

Danshen can kill Ehrich ascites carcinoma cells in mice, prolong the survival of mice suffering from Ehrich ascites carcinoma, and has a synergistic effect on the antineoplastic activity of camptothecin, hydroxyl camptothecin, and cyclophosphamide. It has been reported that the tanshinone isolated from the extracts of Danshen root and tanshinone derivatives can inhibit the proliferation of a great variety of tumor cell strains in vitro, exert cytotoxic actions, and show relatively strong antineoplastic activity. The main active component of Danshen is tanshinone. Yuan Shulan, Wang Xiujie et al. reported that in vitro experimental studies showed that 0.25–100 μg/ml of tanshinone II_A had inhibitory and cytotoxic actions on a variety of tumor cells (HL-60, NB4, K562, SMMC-7721, etc.). The research on the effects of tanshinone II_A on mouse liver cancer H22 showed that tanshinone II_A had significant inhibitory actions on this cell strain, and the average inhibition rates of the 3 tests were 50.0, 38.5 and 40.6 %, respectively. Some components of tanshinone had various degrees of killing effects on cancer cells even at 1 μg/ml. Clinical research has found that tanshinone has various degrees of therapeutic effects on all kinds of tumors (including leukemia, lung cancer, brain glioma, sarcoenchondroma, ovarian cancer, colorectal cancer, stomach cancer, pancreatic cancer, laryngeal carcinoma, and nasopharyngeal carcinoma) (Table 10.1).

In addition, research has proven that Tanshinol (the water-soluble component of Danshen) has antineoplastic activity in vitro. Experiments using a stomach adenocarcinoma NGCC cell strain showed that Tanshinol had significant inhibitory actions in serum-free medium, and when the concentration reached 80 μg/ml, there was no statistical difference between its inhibitory action and that of Alacinomycin. The mechanism of Tanshinol's inhibition on the stomach adenocarcinoma cell strain is related to the blockage of the synthesis of cholesterol in tumor cells. When Tanshinol is present, the activity of HMG-CoA (the key enzyme for the synthesis of cholesterol) decreases so that the growth of tumor cells is inhibited due to the lack of cholesterol. Similar results were obtained for breast cancer cell strain MCF-7.

The inhibition of Danshen on the proliferation of tumor cells has a close relation with its structure. The analysis of the structure–activity relationship has shown that the phenanthrene-quinone structure of tanshinone is the basis for its cytotoxic actions. The phenanthrene ring can bind with DNA molecules, and the free radicals produced by the furan ring and quinones can damage DNA and inhibit the synthesis of tumor cell DNA. If the ring A of tanshinone is substituted by hydroxyl or alkene structures, the derivatives have higher biological activities. Research has shown that tanshinone can inhibit DNA synthesis of sarcoma S180 cells, and the inhibition by tanshinone on the synthesis of tumor cell DNA is related to PCNA. PCNA is the coenzyme of DNA polymerase δ, which begins to increase in the late stage of G_1 phase of the cell cycle and reaches peak in S phase. It is directly concerned with DNA reproduction

Table 10.1 The tumor cell strains that can be inhibited by the extracts of Danshen

Colo-205 cell strain (human colon cancer)	Caco-2 cell strain (human colon cancer)	HC-F-15 cell strain (human colon cancer)
HepG2 cell strain (human liver cancer)	SMMC-7721 cell strain (human liver cancer)	H22 cell strain (mouse liver cancer)
BEL-7402 cell strain (human liver cancer)	A549 cell strain (human lung adenocarcinoma)	W138 cell strain (human lung squamous cell carcinoma)
Hep-2 cell strain (human laryngeal carcinoma)	$_K$B cell strain (human oral epithelial carcinoma)	Hela cell strain (human cervical carcinoma)
SKOV3 cell strain (ovarian cancer)	SK-MEL-2 cell strain (human melanoma)	HL60 cell strain (human leukemia)
NB4 cell strain (promyelocytic leukemia)		PaCa cell strain (human pancreatic cancer)
BcaCD885 cell strain (human buccal carcinoma)	Tca8113 cell strain (tongue cancer)	NGCC cell strain (stomach adenocarcinoma)
MCF cell strain (human breast cancer)	XF498 cell strain (central Nervous system tumor)	K562 cell strain (human chronic erythroleukemia)
CNEl cell strain (human nasopharyngeal carcinoma)	SPC-A-1 cell strain (human lung adenocarcinoma)	PLA-801D cell strain (human giant cell carcinoma of lung)

during the cell proliferation process, and the expression and content of it reflects the proliferative activity of cells. The research on SMMC-7721 cell strain, HL-60 cell strain, and H22 mouse liver cancer cell strain indicates that tanshinone can inhibit the activity of DNA polymerase δ, and that the expression of PCNA stops the cells in G_0/G_1 phase and makes cells enter S phase so as to inhibit the synthesis of DNA, suppressing the growth and proliferation of cells and exerting cytotoxic actions.

10.2.2.2 Effects on the Apoptosis of Tumor Cells

Apoptosis is also termed programmed cell death (PCD) when the specific "suicide" program in the cell is activated under physiological or pathological conditions. There is a balance between proliferation, differentiation, and apoptosis in normal cells. However, in tumor cells, proliferation is infinite, differentiation is blocked, and apoptosis is suppressed. Therefore, inhibition of proliferation, induction of differentiation, and/or apoptosis of tumor cells are the mechanisms of the anticancer effects of many anticancer drugs.

It has been found that hydroxyl camptothecin (HCPT) and 5-FU can cause a similar apoptotic rate of large intestine tumor cells, but the combined use of HPCT and Danshen can significantly increase the apoptotic rate of large intestine tumor cells caused by HCPT.

Present research has shown that nontoxic doses (0.5 μg/ml) of tanshinone II_A can induce the differentiation and maturation of HL-60 cells and result in apoptosis. Yuan Shulan et al. reported that 0.5 μg/ml of tanshinone II_A could significantly inhibit the proliferation of human nasopharyngeal carcinoma cells (CNEl), human lung cancer cells (SPC-A-1), and human leukemia cells (NB_4, K562), resulting in changes in the morphology of cells and leading to apoptosis. Researchers from other countries reported that tanshinone II_A could induce apoptosis of the APL HL-60 cell strain and the K562 cell strain in a dosage and time dependent manner, and that the activity of caspase-3 increased during this process. The caspase family is the most important family of proteases in the apoptosis pathway, indicating that the induction of apoptosis by tanshinone II_A is related to the activation of the caspase family. The apoptosis-inducing actions of tanshinone II_A increases with the expression of the p53 gene and significantly decreases with the expression of the *bcl*-2 gene. It is well known that p53 protein (the expression product of p53 gene) is a DNA-binding protein which can induce apoptosis and prevent the generation of cells with cancer tendencies. *Bcl*-2 gene is the inhibiting gene of apoptosis, the primary function of which is to prolong the life of cells by inhibiting apoptosis. Therefore, tanshinone II_A can prevent the generation of cells with cancer tendencies by upregulating p53 and facilitate apoptosis by downregulating *bcl*-2.

10.2.2.3 Effects on the Infiltration and Metastasis of Tumors

Infiltration and metastasis are both characteristics of malignant tumors and the major difficulties for clinical cancer treatment. There have been no consistent conclusions on whether or not traditional Chinese medicine with the effect of promoting blood flow by removing blood stasis can facilitate the generation and metastasis of tumors. In modern medicine, this problem is investigated based on the relations between tumor metastasis, blood coagulation, and the promotion of blood flow by removing blood stasis. Some researchers suggested that the action of promoting blood flow by removing blood stasis led to fibrinogen's dissolution and disaggregation of platelets, reduction of the viscosity of blood, improvement of microcirculation around cancerous lesions, avoidance of the adhesion of cancer cells, and prevention of the metastasis of tumors. These viewpoints have been supported by many researchers. However, there are some contrary research results which indicate that the traditional Chinese medicine of promoting blood flow by removing blood stasis facilitated the metastasis of cancer cells. At present, there is still a dispute over whether Danshen has actions on the infiltration and metastasis of malignant tumors. Some researchers think Danshen can facilitate the metastasis of tumors, some think Danshen can inhibit the metastasis of tumors, and others think

Danshen has no significant facilitative or inhibitory actions on tumors.

Fu Naiwu et al. reported the effects of Danshen on the growth and metastasis of experimental tumors, and preliminarily investigated the mechanisms. The research showed that application of Danshen alone had no significant antitumor effects on mouse S-180 cells, but that it enhanced the inhibitory effects of cyclophosphamide. The effects of Danshen on the metastasis of tumors were observed through injection of Walker-256 cancer cells into the vena caudalis of rats, and the results showed that Danshen increased the incidence of pulmonary metastasis and lesions of metastasis. In research on the $C_{57}BL$ mice receiving intramuscular inoculation of Lewis lung cancer cells, Danshen had no significant effects on the growth of primary tumors, but significantly facilitated its pulmonary metastasis. ^{51}Cr oncocyte in vivo distribution and detention experiments showed that after Danshen was applied, the detention of oncocytes in the lungs decreased, the detention of oncocytes in the liver increased, and the oncocytes in the blood were maintained at a high level. The author suggested that mechanisms for the facilitating actions of Danshen on pulmonary metastasis include the following: (1) effects of Danshen on the surface of cancer cells: it could inhibit the agglutination of cancer cells induced by phytohemagglutinin (PHA) and maintain cells in a dispersed state; (2) effects of Danshen on Hemodynamics and increasing the permeability of vessels: the increase in vascular permeability could facilitate the metastasis of tumors.

Li Chengzhu compared the Walker-256 group and control group, and found that Danshen injections could significantly decrease the body weight of rats, cause serious anemia, increase the weight of the lungs and the ratio of lung weight to body weight, increase the number of tumor tubercles in the surface of the lungs, and produce metastatic tubercles in other visceral organs. Taken together, Danshen had promoting effects on the hematogenous spread of tumor cells.

Ding Gang et al. investigated the mechanisms of the effects of Danshen and red peony root on the hepatic metastasis of rat Walker cancer, and found that Danshen and red peony root could facilitate carcinomatous metastasis. The rate of hepatic metastasis, the incidence rate of hemorrhagic ascites, and the rate of abdominal cavity dissemination in the Danshen group were significantly higher than those in the control group. In addition, after rats were inoculated with oncocytes, the concentration of vascular endothelial growth factor (VEGF) in serum significantly increased, and the serum concentration was maintained at a high level over the following 2 weeks. The density of blood capillaries in tumor tissues and the expression of VEGF were significantly higher than those of the normal saline group. Therefore, Danshen and red peony root can facilitate the hepatic metastasis of rat transplantation tumor model, and the mechanism may be related to the fact that Danshen and red peony root can facilitate the higher expression of VEGF in rat transplantation tumors and promote the formation of tumor vessels.

Some researchers investigated the actions of Danshen on the expression of PCNA in SD rats with obstructive jaundice caused by liver cancer, and used 5-FU as control. They found that after treatment with Danshen, the liver cancer cell proliferation was limited, the heteromorphism was lowered, the differentiation was good, the angiopoiesis of tumors decreased, and the expression of PCNA in liver cancer and the tissues around cancer significantly decreased. The inhibitory effects of Danshen on liver cancer cells were similar to those of 5-FU. Danshen could significantly inhibit the inside and outside metastasis of liver cancer. In particular, its inhibitory action on the pulmonary metastasis of liver cancer was significantly better than that of 5-FU.

Chen et al. isolated tumor infiltrated lymphocytes (TIL) and autologous tumor cells from the malignant pleural fluid of cancerous patients, and observed the effects of Danshen injection on the long term survival in vitro, amplification and antitumor actions. The results showed that Danshen injection could significantly prolong the survival time of TIL in vitro, and make TIL continually amplify over 10 days and maintain a high level during 11–15 days. The amplified TIL had a stronger damaging effect on K562 cancer cells in vitro, and could prolong the survival time of mice with Hepa cancer cells. Its actions were similar to those of IL-2. Since TIL cells are a new generation of antitumor effector cells, after lymphokine-activated killer (LAK) cells, it can significantly enhance the activity of human TIL and strengthen the damaging effect of TIL on cancer cells.

Zheng Landong implanted Walker-256 liver cancer cells into liver parenchyma to make transplanted liver cancer, thus establishing a transplanted liver cancer model of SD rats, and observed the effects of Danshen on the development of liver cancer through intraperitoneal injection of Danshen into model rats. The results showed that Danshen could significantly reduce the volume of tumors, and the effects were similar to those of 5-FU. Pathomorphological observation showed that the tumor cells in the Danshen group grew poorly, with higher maturation of differentiation, and cell morphology showed benign differentiation tendencies. It was also shown that there were few vessels in the tissues around tumors, and that the diameters of the blood vessels were similar and sporadically distributed. The expression of VEGF in liver cancer and in the tissues around cancer and lung tissues was detected, and the results showed that the expression rates of VEGF in liver cancer and in the tissues around liver cancer in the Danshen treatment group were significantly lower than those in the normal saline group and Inosine + VitC group, and similar to those of the 5-FU group. Therefore, inhibition of liver cancer cell proliferation and the expression of VEGF in liver cancer and the hepatic tissues around cancer may be one of the mechanisms of liver cancer growth inhibition.

As for the effects of Danshen on the metastasis of tumors, not only is there controversy between the schools of facilitation and inhibition, there are also contradicting research results.

Chen Gang et al. compared the inhibiting actions of Ligustrazine injection and Danshen injection on the growth of mouse Lewis lung cancer and the relations with angiogenesis inhibition. The results showed that when compared with the model group, the volume and mass of the Lewis lung cancer transplantation tumor, the number of pulmonary metastasis lesions, the tumor blood capillary density, and the expression of VEGF of the mice in the Ligustrazine injection group were significantly lowered, while there was no statistically significant difference between the Danshen group and the model group. Thus, Danshen had no significant effects on the growth and metastasis of Lewis lung cancer transplantation tumor of mice.

Liu Mingzhang et al. investigated the actions of sodium tanshinone II_A sulfonate (DS-201), which was formed from the sulfonation of tanshinone II_A, on the growth and metastasis of tumors, and found that DS-201 had no effects on the pulmonary metastasis of Lewis cancer. However, after intravenous injection of Walker 256 cancer cells, the number of pulmonary metastasis lesions increased in the DS-201 group and Danshen injection group, indicating that both DS-201 and Danshen injection could increase Walker 256 lung cancer lesions. In this way, the results of sodium tanshinone II_A sulfonate on Lewis cancer myoma and on lung cancer lesions of Walker 256 resulting from intravenous inoculation were inconsistent. The authors gave the following explanations: Lewis cancer generally formed myomas in situ, then invaded blood vessels and went into the lungs via blood, forming secondary tumors, which is typical tumor metastasis. The intramuscular inoculation of Lewis cancer was a good model for observing whether the drugs could facilitate the pulmonary metastasis of cancer. However, injecting Walker 256 cancer cells into the vein led the cancer cells to enter the lungs directly through blood flow and form cancer lesions; since there were no in situ tumors formed before metastasis, this does not meet the definition of cancer metastasis. Therefore, it was not metastatic carcinoma. The author used only the results of DS-201 on pulmonary metastasis through intramuscular inoculation of Lewis cancer to evaluate whether Danshen could facilitate carcinomatous metastasis, and suggested that DS-201 had no promoting effects on either the growth or metastasis of Lewis cancer.

In a word, the mechanisms of the actions of Danshen on the metastasis of tumors are as follows: (1) Enhancing the immunity of the human body to inhibit tumor metastasis; (2) Inhibiting the proliferation of tumor cells and inducing the differentiation and apoptosis of tumor cells; (3) Inhibiting the separation and exfoliation of tumor cells and the formation of a tumor embolus; (4) Preventing the entrance of tumor cells into secondary organs from the circulation system.

10.2.3 Other Potential Effects on Tumors

10.2.3.1 Inhibition on the Activity of Telomerase

Telomeres are an important part of the eukaryotic chromosome. It has been proven that the length of the telomere is positively correlated with the life cycle of cells. Telomerase is the reverse transcriptase which maintains the length of telomeres, and its active expression is important for the proliferation, senescence, immortalization, and canceration of cells. The activity of telomerase increases in most tumor cells, so one of the critical steps of canceration is the expression of telomerase and the maintenance of telomeric DNA.

It has been reported that 0.5 μg/ml tanshinone II_A can significantly inhibit the activity of telomerase in HL-60 cells, K562 cells, and APL cells. The inhibition of telomerase activity in tumor cells and the recovery from the mortality of tumor cells may be one of the mechanisms for the antitumor actions of Danshen.

10.2.3.2 Antioxidation Actions

Research has shown that the processes of cell transformation and tumor proliferation are accompanied by the production of reactive oxygen species (ROS), which can facilitate the growth of tumor cells. The peroxidation of lipids and free radicals may be related to the generation and development of malignant tumors.

The water-soluble components of Danshen have potent effects on the inhibition of lipid peroxidation and scavenging of free radicals. The 7 water-soluble components of Danshen have significant scavenging actions on the superoxide anions produced by the xanthine-xanthine oxidase system. Among them, the eliminating actions of Tanshinol on superoxide anions are better than those of SOD, and salvianolic acids A and B have strong inhibitory effects on lipid peroxidation caused by various factors. ROS can facilitate the proliferation of liver cancer cell strain SMMC-7721, while Tanshinol can inhibit the growth of the tumor cells and lower the level of ROS in the cells. Further research has found that ROS can facilitate the growth of liver cancer cells by activating the protein kinase B (PKB) pathway and regulating and controlling the expression of transcription factor AP-1. Therefore, Tanshinol can inhibit the growth of liver cancer cells by inhibiting the PKB pathway.

10.2.3.3 Reversing the Multidrug Resistance of Tumors

Multidrug resistance (MDR) is one of the major reasons for chemotherapy failure. MDR refers to tumor cells which are not only resistant to one type of antineoplastic drugs, but also resistant to antineoplastic drugs with different structures and mechanisms of action. Its generation is mainly concerned with the excessive expression of P-glycoprotein (P-gp). P-gp is an energy-dependent drug pump, which can remove from cells natural drugs with various structures and action mechanisms, thus producing MDR.

Research has shown that Danshen can reverse the MDR of tumors mainly by decreasing the expression of P-glycoprotein and increasing the concentration of chemotherapeutics in tumor cells.

10.2.4 Discussion and Prospects

Cancer is a serious disease endangering people's health. Little is known about it due to its complicated pathogenesis, and we have few good methods to treat cancer. Operation, radiotherapy, and chemotherapy are the main therapeutic approaches at present. The antitumor actions of Danshen have been gradually discovered and applied in clinical practice. Danshen and its components can effectively mitigate the pain of patients with malignant tumors, enhance immunity, reduce the side effects of radiotherapy and chemotherapy, and improve the quality of life for patients. However, as for basic research, there is disagreement regarding the actions of Danshen on the occurrence and metastasis of tumors. The actions of the mono-components of Danshen (such as tanshinone) on tumors are relatively consistent. Therefore, no matter for clinical practice or basic research, Danshen alone cannot be used as antineoplastic drug due to the following reasons: first, the in vitro inhibiting actions of the simple drug on tumor cells are weak, and the in vivo inhibiting actions on tumors are not yet definite and have low therapeutic effects; second, its effects on the metastasis of tumors have not been determined yet. Its components are complicated and the contents of many components are low. Different experiments have diverse research methods, so the actions of each component on tumors are different, and even the compositions of Danshen produced in different areas show different effects. Its actions on the metastasis of tumors are probably varied, which may depend on the relative strengths of the actions promoting metastasis, the actions inhibiting metastasis, and the state of the body.

At present, some components extracted from Danshen may become new antineoplastic drugs. Although the pharmacological effects of the components extracted from plants are usually weaker than those of synthetic chemical drugs, structural optimization may help to enhance the pharmacological effects of the active components isolated from plants. The composition of Danshen is complex, but it is not impossible to isolate active components from Danshen and develop new antineoplastic drugs. For the present, 2 new compounds have been isolated from Danshen. One is named neo-tanshinlactone, the other salvinal. They have significant antitumor activities in vitro. It should be mentioned that the latter is a new microtubule inhibitor and can induce Apoptosis. Furthermore, it has no cross-resistance with vincristine because it is not a substrate of P-glycoprotein.

References

1. Erutti PA. Science. 1985;227:375.
2. Yangchang J, Yenchang C, Chuankuo C, et al. Mol Pharmacol. 2004;65:77.
3. Liu J, Yang CF, Wasser S. Life Sci. 2001;69:309.
4. Ryu SY, Lee CO, Choi SU. Planta Med. 1997;63(4):339.
5. Yuan SL, Wei YQ, Wang XJ, et al. World J Gastroenterol. 2004;10(14):2024.
6. Sung HJ, Choi SM, Yoon Y, et al. Exp Mol Med. 1999;31(4):174.
7. Wu WL, Chang WL, Chen CF. Am J Chin Med. 1991;19(3):207.
8. Wang XH, Kenneth FB, Chang M, et al. Miltiorrhiza. J Med Chem. 2004;47:5816.
9. Yoon Y, Kim YO, Jeon WK. J Ethnopharmacol. 1999;68(1–3):121.
10. Jihai B, Zhongyu F, Zeling W. Chin J Clin Oncol Rehabil. 1995;2(2):39.
11. Chenghua D, Shaomin Q. Trends Tradit Chin Med. 1994;4:5.
12. Zhangshen D. J Hubei Coll Tradit Chin Med. 2000;2(4):56.
13. Shuangsheng H, Yongjie W. Chin J Inf Tradit Chin Med. 2002;9(1):86.
14. Lianliang L, Juntian Z. Bull Med Res. 2001;30(7):2.
15. Li Yuehong W, Hui LJ, et al. J Med Theory Pract. 2000;1:47.
16. Yong L, Yiming Y, Shulan Y. Chin Tradit Herb Med. 2000;31(4):304.
17. Mingzhang L, Yisui H, Weiqi X. Acta Pharm Sin. 1991;12(6):534.
18. Yicong X, Ruirong W. Chin J Rural Doct. 1998;9:16.
19. Qun Y. Shandong J Tradit Chin Med. 1991;19(4):45.
20. Weidong Y, Hongliang Z, Baolu Z. Chin Pharm Bull. 1990;6(2):118.
21. Yi Y. Hubei J Tradit Chin Med. 1999;21(6):284.
22. Jianning Z, Yunxiang D, Xiaoping H. Mod J Integr Chin Tradit West Med. 1999;8(11):1763.
23. Sheng Z, Hongliang Z, Changhe S, et al. Henan J Oncol. 1994;7(2):134.
24. Lijun X, Qingsheng Y. Clin J Anhui Tradit Chin Med. 1998;10(4):197.
25. Jinqi Y, Qiaofeng Y, Yan L. Zhejiang J Integr Tradit Chin West Med. 2002;12(5):299.
26. Qingsheng Y, Ziyi C, Wei W, et al. Chin J Integr Tradit West Med. 2000;20(3):210.
27. Weixing Z, Yadong W. Jiangsu J Tradit Chin Med. 1997;18(1):23.
28. Youfeng D, Jianan W, Jianping S. Shaanxi J Tradit Chin Med. 1997;18(10):433.
29. Cuicai R. J Guangxi Tradit Chin Med Univ. 1990;7(2):21.
30. Zhenquan Z, Qifu L. Guangxi J Prev Med. 1996;2(4):240.
31. Zhenquan Z, Qifu L, Shicong Z, et al. Guangxi Med J. 1995;17(3):1791.

32. Guangqi H, Shulan Y, Hongyuan Z, et al. Chin J Pharm Toxicol. 1996;10(4):285.
33. Shulan Y, Guangqi H, Renmin H, et al. Chin J Cancer. 1996;15(6):408.
34. Shulan Y, Renmin H, Xiujie W, et al. Oncology. 1997;17(5):268.
35. Dongping H, Miaoqian L, Leida P. Chin Med Fact Mine. 2001;14(4):265.
36. Zhang Haobo Y, Baoming Y, Hao W, et al. Henan J Oncol. 1998;11(3):1679.
37. Shulan Y, Yanping W, Xiaohe C, et al. J West China Univ Med Sci. 2002;33(1):84.
38. Shulan Y, Renmin H, Yi S, et al. J Pract Oncol. 1997;12(6):253.
39. Yong L, Yiming Y, Shulan Y, et al. Chin J Hematol. 2000;21(1):23.
40. Wu Y, Yiming Y, Wentong M, et al. J West China Univ Med Sci. 2002;33(1):77.
41. Renmin H, Shulan Y, Yi S, et al. Chin J Clin Oncol. 1997;24(7):500.
42. Renmin H, Shulan Y, Yi S, et al. Chin J Cancer. 1998;17(3):164.
43. Shulan Y, Yong L, Yiming Y, et al. Exploration on the regulation and control of cells, 1999; 179.
44. Shulan Y, Yiming Y, Guangqi H. The application of tanshinone of traditional chinese medicine extract for curing Tumor. Application for a Patent for Invention CN1264580A. 2002 Aug 30, p. 15.
45. Naiwu F, Xianjun F, Yongquan W, et al. Chin J Oncol. 1981;3(3):165.
46. Shulan Y, Yi S, Xiujie W, et al. West China J Pharm Sci. 2003;18(5):327.
47. Xiujie W, Shulan Y, Renmin H, et al. J West China Univ Med Sci. 1996;27(4):390.
48. Yuwu Z. J Xi'an Med Univ. 1986;7:403.
49. Xiujie W, Shulan Y, Chaojun W, et al. Chin J Oncol. 1996;18(6):412.
50. Zhongzhi T, Ying T, Libo F. J First Mil Med Univ. 2003;23(6):595.
51. Zhongzhi T, Ying T, Libo F. Acta Academiae Medicinae Militaris Tertiae. 2003;25(9):744.
52. Jintao H, Qinghua Z, Shulan Y, et al. Chin J Lung Cancer. 2002;5(2):1235.
53. Jian S, Feng L. J Zhenjiang Med Coll. 1997;7(4):384.
54. Jie F. Chin J Gerontol. 2003;3(23):168.
55. Jingjing S, Xinda Z, Yinkun L. Chin J Integr Tradit West Med. 1999;19(5):292.
56. Enhua C, Xiaoqi L, Jingfu L, et al. Acta Biophysica Sinica. 1996;12(2):339.
57. Shanlin L, Guanzhong L, Huili C, et al. Acta Biochmica Et Biophysica Sinica An. Int J. 2002;34 (1):66.
58. Shulan Y, Xiujie W, Yuquan W. Chin J Cancer. 2003;22(12):1363.
59. Jian C, Gengjin L. Fudan Univ J (Med Sci). 2003;30 (6):626.
60. Shicheng L, Yan Z, Rujian W, et al. Heilongjiang. J Tradit Chin Med. 2005;4:62.
61. Yong L, Zhaogui S. Chin J Pract Tradit Chin West Med. 2005;18(4):576.
62. Yi S, Rao T, Xiaoou P. Chin J Hosp Pharm. 2004;24 (10):637.
63. Gang D, Mingzhi S. China Oncol. 2001;11(4):364.
64. Xiaodong C, Qilian L, Xiaoying L, et al. J Oncol. 2005;11(1):46.
65. Qizhi H, Wanqiang F, Ruijing W, et al. Sichuan J Cancer Control. 2006;19(3):190.
66. Peifeng C, Luming L, Zhen C, et al. Zhejiang J Tradit Chin Med. 2007;31(5):532.
67. Junli X, Shuiliang H. Northwest Pharm J. 2006;21 (4):188.

Studies on the Antibacterial and Anti-inflammatory Actions of Danshen and Its Effects on the Immune System

11

Li Zhang and Guanhua Du

Danshen-based Chinese herbal compound formulas or Danshen active component-based preparations are mainly used for the treatment of coronary heart disease and ischemic cerebrovascular disease in clinical practice. They are also used for noncardiovascular and cerebrovascular diseases such as vasculitis, hard skin disease, thromboangiitis obliterans, burn scars, etc., with good efficacy. Danshen medicines have been widely used to treat infectious diseases such as bronchiolitis, asthmatic bronchitis, viral pneumonia, tuberculosis, pleurisy, viral myocarditis, newborn, psoriasis, acne, and others, and the results are satisfactory Zhang (Chin J Clin Pediatr 12(3):959, 2004, [1]). The pharmacological studies on Danshen and its active components are very broad, and the following review focuses on their anti-inflammatory, antibacterial, and immunological effects.

11.1 Pharmacological Effects of Danshen in Anti-inflammation and Immunity

There have been reports about the anti-inflammatory effects of Danshen in the late 1970s and early 1980s [2, 3]. It has been found that local application of total tanshinone can inhibit the swelling of the ears caused by croton oil in mice. In the first-stage inflammation model, total tanshinone can significantly inhibit the increase in the capillary permeability of rats caused by histamine, the acute joint swelling of rats caused by egg whites, carrageenin and dextran, and exudative formaldehyde peritonitis of rats. In the second-stage inflammation model, it showed that total tashinone can significantly inhibit the migration of white blood cells of mice caused by gelatin and the subacute formaldehyde joint swelling of rats, but has no inhibitory action on the cotton ball granuloma model (third-stage inflammation model). It has been shown by other studies that Danshen injection can inhibit the swelling of the foot in rats caused by carrageenin, and the inhibition rate was as high as 51 % [4]. The studies on the mechanisms of Danshen's anti-inflammatory effects are very active; some studies are about the immunoregulatory effects of Danshen on the normal body, and some are about the anti-inflammatory effects of Danshen during the process of disease, including the influence on immunocyte function and cytokine levels.

11.1.1 Effects on Immunocyte Function

11.1.1.1 Effects on T Lymphocyte Function

T Lymphocytes mainly represent cellular immune function, and lymphoblast experiments [5, 6] have shown that both Danshen injection

L. Zhang · G. Du (✉)
Institute of Materia Medica, Chinese Academy of Medical Science and Peking Union Medical College, Beijing, China
e-mail: dugh@imm.ac.cn

X. Yan (ed.), *Dan Shen (Salvia miltiorrhiza) in Medicine*,
DOI 10.1007/978-94-017-9463-3_11,

Table 11.1 Effects of Danshen on the transformation of Lymphocytes in rats ($\bar{X} \pm s$) [6]

Group	Dosage (g/kg × days)	Number of animals	Counts per minute
Control	0 × 8	10	1310 ± 450
Danshen	1.2 × 8	10	3283 ± 1354**
Compound Danshen	1.6 × 8	10	2514 ± 744*
Ligustrazine	0.04 × 8	10	4479 ± 2052***

Compared with the control group: $^{*}P < 0.05$, $^{**}P < 0.01$, $^{***}P < 0.001$

Table 11.2 Comparison of the PFC values between the administration groups and the control group ($\bar{X} \pm s$) [10]

Group	PFC/10^6 splenic cells/OD_{540} value	*P* value
Saline control	0.33 ± 0.03	
Danshen	0.60 ± 0.08	<0.01
Siwutang (decoction of four drugs)	0.25 ± 0.03	<0.01
Mulberry	0.06 ± 0.01	<0.01

and Compound Danshen injection (CDI) can significantly increase the transformation of Lymphocytes in the blood of normal rats, compared to in the control group, $P < 0.01$ % (Table 11.1). When Danshen injection + cyclophosphamide were used to treat S180 mice, both the weight of the thymus and the number of T Lymphocytes were significantly higher than in those treated with cyclophosphamide alone, indicating that Danshen could alleviate the decrease in Immunologic function caused by cyclophosphamide in mice with tumors [7]. When Danshen injection was applied to the liver fibrosis model prepared with human serum albumin (HSA), it could make the lymphocyte transformation significantly higher than in the model control group and the colchicine (the traditional anti-liver-fibrosis drug) group, and the results were similar to those of normal animals. The author suggested that the effects of Danshen against immunologic liver fibrosis might be better than those of colchicine [8]. Helicobacter pylori infection is closely related to stomach tumors. It has been reported that after the administration of Danshen's water-soluble extract F to the helicobacter pylori sensitized mouse model, splenic Lymphocytes, and stomach lamina propria T Lymphocytes (LPL) significantly increased and the level of interleukin-2 (IL-2) significantly increased, indicating that Danshen water-soluble extract F could be used as the adjuvant for the helicobacter pylori vaccine and could be used to prevent and treat tumors related to helicobacter pylori [9].

11.1.1.2 Effects on B Lymphocyte Function

B Lymphocytes mainly represent humoral immune functions, and they can secret antibodies. It has been shown by hemolytic plaque assays that the water decoction of Danshen significantly increased immunoglobulin secretion by mouse B Lymphocytes and the value of plaque-forming cells (PFC), which was statistically significant compared to those of the control group [10] (Table 11.2). Danshen injection and CDI can significantly increase the semi-hemolytic titer of sensitized red blood cells (SRBC) in normal mice, indicating that they could significantly increase the production of antibodies [6]. The effects of the decoction of Danshen, astragalus root, and Chinese date (Shen Qi Tang) on the immunologic functions of normal mice prove that it can increase the humoral immunity of mice [11].

11.1.1.3 Effects on the Functions of Macrophages

Macrophages are the key cells of nonspecific immunity and are the most important antigen-presenting cells. External stimuli are processed by macrophages and then presented to T Lymphocytes, which is the prerequisite condition for immune response. Macrophages have bidirectional functions in immunoregulation. On the one hand, they can phagocytize and eliminate pathogenic microorganisms and initiate the immune response. On the other hand, their excessive activation may result in the secretion of a great variety of inflammatory factors and inflammatory mediators, such as prostaglandin.

After incubation with 0.03 mg of Tanshinol for 48 h, human mononuclear macrophages and mouse abdominal cavity macrophages showed inhibition of their endotoxin (LPS) induced secretion of 4 inflammatory cytokines, indicating that Tanshinol might have bidirectional regulatory effects: on the one hand, it may have anti-inflammatory effects; on the other, it may enhance immunologic functions [12] (Tables 11.3 and 11.4). Tanshinol can significantly inhibit the synthesis of prostaglandin E_2 (PGE_2) and Thromboxane B2 (TXB2) induced by zymosan, and resist the increased synthesis of PGE_2 induced by $A_{23,187}$, because all of the inflammation stimulants (carrageenin, silica gel, zymosan, etc.) and $A_{23,187}$ can stimulate the synthesis of PGE_2. The synthesis of PGE_2 by macrophages requires calcium ions, so part of the anti-inflammatory mechanism of Tanshinol is related to the inhibition of calcium ion inflow [13]. Chemiluminescence was used to observe the effects of various concentrations of Danshen aqueous extracts on the functions of mouse J774 macrophages, and the results showed that low dose extracts had significant inhibitory effects and that high doses had significant enhancing effects, so Danshen might have bidirectional regulatory effects [14] (Table 11.5). The extracts of Tanshinol can dose-dependently increase the secretion of inflammatory factors, such as IL-1, IL-6, and TNF-α, by rat alveolar macrophages in vitro [15]. The results of chemotaxis experiments and superoxide measurement experiments showed that Danshen could increase the production of neutrophil and macrcophage superoxides in human peripheral

Table 11.3 Effects of Danshen on cytokine secretion by human monocytes ($\bar{X} \pm s$) [12]

	n	TNF-α (Cytotoxic effect, %)	IL-1 (U/ml)	IL-6 (U/ml)	IL-8 (U/ml)
Endotoxin	20	46.0 ± 9.4	21.0 ± 4.4	1.75 ± 4.8	8.4 ± 3.8
Ibuprofen	20	13.5 ± 5.4*	9.4 ± 3.8*	5.4 ± 2.1*	3.2 ± 0.4*
Endotoxin + ibuprofen	20	37.0 ± 7.3**,***	16.3 ± 3.7**,***	12.8 ± 3.2*,***	3.7 ± 0.9*,***
Tanshinol	20	18.5 ± 3.2*	5.9 ± 3.4	5.8 ± 2.1*	3.1 ± 1.2*
Endotoxin + Tanshinol	20	27.3 ± 6.9*,***	8.9 ± 2.7*,***	12.8 ± 1.6*,***	4.2 ± 1.0*,***
Endotoxin + hydrocortisone	20	19.9 ± 6.6*	6.6 ± 2.5*	7.0 ± 3.1*	2.4 ± 0.7*

Compared with the LPS group: * $P < 0.01$, ** $P < 0.05$
Compared with the LPS + HCS group: *** $P < 0.01$

Table 11.4 Effects of Danshen on cytokine secretion by mouse abdominal cavity macrophages ($\bar{X} \pm s$) [12]

Treatment	*n*	TNF-α (Cytotoxic effect, %)	IL-1 (U/ml)	IL-6 (U/ml)	IL-8 (U/ml)
Endotoxin	9	60.5 ± 10.6	67.5 ± 11.9	90.2 ± 12.2	47.8 ± 5.6
Ibuprofen	9	14.3 ± 3.9*	26.4 ± 5.1*	18.6 ± 4.5*	11.2 ± 2.5*
Endotoxin + ibuprofen	9	31.9 ± 5.2*,***	34.8 ± 5.2*,***	49.7 ± 13.9**,***	35.2 ± 4.6*,***
Tanshinol	9	21.5 ± 5.9*	19.4 ± 3.2*	9.6 ± 0.3*	8.6 ± 0.8*
Endotoxin + Tanshinol	9	29.9 ± 7.5*,***	58.1 ± 7.2**	37.3 ± 9.8*,***	29.5 ± 5.9*,***
Endotoxin + hydrocortisone	9	17.7 ± 5.6*	28.8 ± 7.5*	21.5 ± 5.7*	18.9 ± 3.4*

Compared with the LPS group: * $P < 0.01$, ** $P = 0.06$
Compared with the LPS + HCS group: *** $P < 0.01$

Table 11.5 Regulatory effects of Danshen on J774 macrophages function ($\bar{X} \pm s$) [14]

Dosage (mg/L)								
0	10	50	100	500	1,000	2,000	3,000	4,000
152.6 ± 5.2	89.4 ± 12.7* (↓41.4 %)	90.2 ± 2.9* (↓40.9 %)	109.8 ± 5.4* (↓28.0 %)	189.2 ± 3.4* (↓24.0 %)	213.0 ± 9.1* (↓39.6 %)	208.8 ± 10.4* (↓36.8 %)	203.4 ± 5.1* (↓33.3 %)	201.3 ± 9.1* (↓31.9 %)

*Note** The difference is significant when compared with the nondrug control group ($P < 0.05$). The bracketed figures are inhibiting (↓) or enhancing (↑) rates

Table 11.6 Effects of Danshen on the superoxidation of two types of cells (nmoles/1.25 × 10^4 cells, $\bar{X} \pm s$) [16]

Drug concentration (PB/mg ml^{-1})	Neutrophils		Monocytes	
	Zymosan	No zymosan	Zymosan	No zymosan
0 (control group)	23.2 ± 1.5	3.5 ± 0.8	14.2 ± 1.5	3.1 ± 0.6
0.5	22.5 ± 1.8**	8.4 ± 1.08	14.8 ± 1.3**	10.3 ± 1.4*
1	22.4 ± 1.6**	12.5 ± 1.3	14.6 ± 1.6**	8.8 ± 1.3*

Compared with the control group: $^*P < 0.01$, $^{**}P < 0.001$

Table 11.7 Phagocytic percentage and phagocytic index of 4 groups of macrophages ($\bar{X} \pm s$) [10]

Group	Phagocytic percentage	*P* value	Phagocytic index	*P* value
Normal saline control (*n* = 10)	13.58 ± 1.59		0.15 ± 0.02	
Danshen (*n* = 10)	25.45 ± 1.89	<0.01	0.42 ± 0.06	<0.01
Siwutang (*n* = 10)	12.42 ± 1.04	>0.05	0.17 ± 0.03	>0.05
Mulberry (*n* = 10)	26.89 ± 5.49	<0.01	0.25 ± 0.11	<0.01

Table 11.8 Effects of Danshen on the functions of mouse macrophages ($\bar{X} \pm s$) [17]

Group	Number of animals	Phagocytic percentage	Phagocytic index
Control	10	46.63 ± 6.32	1.57 ± 0.47
Danshen	12	37.68 ± 7.23	0.47 ± 0.38*
Cyclophosphamide	12	33.86 ± 7.01	0.68 ± 0.30**
Danshen + cyclophosphamide	12	21.37 ± 5.63	0.53 ± 0.31***

Compared with the control group: $^*P < 0.05$, $^{**}P < 0.01$, $^{***}P < 0.001$

blood, and had inhibitory effects on the chemotaxis of the two types of cells [16] (Table 11.6).

It was shown that administration of Danshen **injection** for 10 days at 1.2 g/kg could significantly increase the phagocytic indices of the macrophages of normal mice [6]. Danshen **injection** could also significantly increase the phagocytic percentage and phagocytic index of the macrophages of mice with low immunity, and the difference from the control group was significant [10] (Table 11.7). Intraperitoneal **injection** of Danshen for five consecutive days at 7.5 g/(kg d) in mice could significantly lower the phagocytic percentage and phagocytic index of mouse macrophages, and it had coordinated repression effects when used with cyclophosphamide, so Danshen might be an immunosuppressant [17] (Table 11.8).

It has been found that Danshen can inhibit the infiltration of polynuclear granulocytes and reduce the release of cytokines that can activate microglia in ischemic cerebral tissues [18, 19]; some components can directly act on the surface of microglia, or suppress the expression of cell surface antigens, or block the signal transmission to external factors so as to mitigate Immunologic injury after ischemia reperfusion. At present, microglia is acknowledged to originate from

Table 11.9 Effects of Danshen on the Chemotaxis of two types of cells (cell count/field, $\bar{X} \pm s$) [16]

Drug concentration (PB/mg ml^{-1})	Neutrophils		Monocytes	
	Formyl peptide	Zymosan	Formyl peptide	Zymosan
0 (control group)	135 ± 21	141 ± 18	118 ± 5	124 ± 7
1	112 ± 15*	115 ± 23*	98 ± 4*	99 ± 9*
2	119 ± 16*	94 ± 23**	79 ± 3**	89 ± 8**
4	105 ± 31**	113 ± 9**	55 ± 4**	77 ± 3**
10	68 ± 18**	78 ± 24**	54 ± 6**	67 ± 3**

Compared with the control group: * $P < 0.01$, ** $P < 0.001$

macrophages. It has been found that Danshen can also inhibit the overactivity of alveolar macrophages of rats with acute respiratory distress syndrome, and the secretion of Inflammatory factors such as IL-1 and TNF-α [20].

11.1.1.4 Effects on Neutrophil Function

Polymorphonuclear Neutrophils (PMNs) are important immunocytes in the inflammatory process. Acute inflammation response can lead neutrophils and endothelial cells (ECs) to release cytokines (such as IL-1, TNF-α), intercellular adhesion molecule-1 (ICAM-1), and its ligands in PMNs (CD11b/CD18, the molecules mediating the adhesion of PMNs to ECs). These factors can stimulate ECs and PMNs so as to increase the expression of ICAM-1. The upregulation of the expression of the adhesion molecules on the surfaces of neutrophils (CD11b/CD18, ICAM-1) can facilitate the adhesion and activation of PMNs and ECs so that PMNs extend pseudopods and migrate across the endothelia, reach the inflammatory areas, produce a lot of active oxygen and proteolytic enzymes, lead to angiemphraxis and disturbance of microcirculation, and cause tissue damage around EC.

Using an in vitro chemotactic migration experimental system and indirect immunofluorescence, found that Danshen extract IH764-3 could significantly inhibit the migration of neutrophils across the polycarbonate (PC) membrane and human blood vessel endothelial cells induced by IL-8. This might be one of the mechanisms of IH764-3's anti-inflammatory effects. It was also found that IH764-3 could inhibit the expression of CD11b, and this might be the mechanism of its inhibition of neutrophil migration [21]. Wang Chaolin et al. found that Danshen injection could inhibit the chemotaxis of neutrophils in human peripheral blood [16] (Table 11.9). Gao Yiyuan et al. found that when human white blood cells were incubated with 50 ng/ml of tanshinone for 1 h, the chemotaxis of white blood cells was significantly inhibited, while the random migration of white blood cells was not affected. If the incubation was prolonged to 19 h, then tanshinone at a concentration of 5 ng/ml was enough to inhibit the chemotaxis and random migration of white blood cells significantly [22]. Li Hui et al. found that after administration of tanshinone at 10, 40, and 80 mg/kg, lysosome release, phagocytosis, and adhesion of neutrophils in the peripheral blood of rabbits with acute myocardial infarction were significantly inhibited [23], and tanshinone II_A also had significant dose-dependent anti-inflammatory effects on the chemotaxis and migration of human white blood cells in vitro. The IC_{50} was 0.46 μg/ml, significantly stronger than that of hydrocortisone [24] (Table 11.10).

11.1.2 Effects on Cytokines

11.1.2.1 Effects on Inflammatory Cytokines

Inflammatory cytokines mainly refer to IL-1, IL-6, TNF-α, and IL-8. Among them, IL-1, IL-6, and TNF-α are called pro-inflammatory factors, which can lead to the high expression of adhesion molecules by neutrophils or endothelial cells. IL-8 represents the chemotactic factor family; its target cells are neutrophils, it is strongly chemotactic toward granulocytes, it can cause the degranulation of granulocytes, produce

Table 11.10 The distance of WBC migration and the inhibition rate of drugs [24]

	Dosage (μg/ml)	Migration distance (times)				Average	*P* value	Inhibition rate (%)
		1	2	3	4			
fMLP		3	3	3	3	3.0 ± 0.00		
Tanshinone II_A	0.1	2	2	2	2	2.0 ± 0.00	<0.05	33.33
	1	1	1	2	1	1.25 ± 0.22	<0.01	58.33
	10	1	1	0	0	0.5 ± 0.26	<0.01	83.33

Table 11.11 Number of ICE positive cells in the cortices of gerbil after Danshen treatment ($\bar{X} \pm s$, $n = 12$, number/Field) [26]

Group	Time (d)			
	1	2	3	4
Sham-operation group	8.8 ± 1.8	9.3 ± 2.2	8.1 ± 2.7	10.3 ± 2.3
Test group	205.9 ± 11.4	296.3 ± 12.8	503.6 ± 16.1	634.1 ± 15.1
Danshen group	150.1 ± 14.6*	166.0 ± 13.6*	203.9 ± 16.0*	246.3 ± 14.0*

Compared with the experimental group: * $P < 0.01$

oxygen free radicals and proteases, increase granulocyte penetration of the endothelial layer and the permeability of blood vessel endoderm, and directly or indirectly make granulocytes enter the tissue space or inflammatory areas through chemotaxis.

More and more studies in recent years have shown that Alzheimer's disease (AD), cerebral ischemia, hypertension, and atherosclerosis are all closely related to inflammatory responses, so there are many reports about the effects of Danshen on the cytokines in these blood vessel diseases. Li Longxuan et al. [25] studied the effects of tanshinone on the mRNA expression of IL-1β and IL-6 in the hippocampi of rats with AD-like symptoms, and investigated the molecular mechanism of tanshinone's toxic effects on the β-starch like peptide neurons. Amyloid β-peptide 1–40 ($A\beta_{1-40}$) was injected into the dorsal cells in the gyrus dentatus of bilateral hippocampi of rats in order to establish an AD animal model. Light–dark box passive avoidance method and shuttle method were used to test learning and memory functions. RT-PCR was used to measure the mRNA levels of IL-1β and IL-6 in the hippocampus. Tanshinone (50 mg/kg) was applied to rats treated with $A\beta_{1-40}$ via intragastric administration for 14 days in order to observe the effects of intervention. The results showed that after 14 days' injection of Aβ1–40 into hippocampi, the learning and memory functions of rats were significantly impaired according to ethological tests, and the expressions of IL-1β and IL-6 mRNA in the hippocampus were significantly higher than those of the control group ($P < 0.01$). Tanshinone (50 mg/kg) could significantly inhibit the deterioration of the learning and memory functions of rats.

Zhang Jintao et al. found that the expression of interleukin 1β-converting enzyme (ICE) increased in the gerbil global cerebral ischemia-reperfusion model; if Danshen injection (10 g/kg) was administered 30 min before ischemia, the number of ICE positive cells decreased, and the morphology of neurocytes was improved. Therefore, Danshen has inhibiting actions on the expression of ICE and can alleviate the brain injury caused by IL-1β [26] (Table 11.11). Among the risk factors causing hypertension, the inflammation, and infection hypothesis of atherosclerosis has received attention again in recent years. The activation of cytokines involved in inflammation may indirectly or directly participate in the impairment of endothelial functions and the pathological changes of other vessels. Wan Lei et al. found that Danshen could inhibit the expression of IL-6 in rabbits with atherosclerosis [27]. Nie Guili et al. cultured the rat aorta smooth muscle cells in vitro to establish the cell model of TNF-α injury, and they found that salvianolic acid B and tanshinone II_A had the strongest inhibitory effects on the expression of IL-β [28].

In many pulmonary diseases, the level of IL-8 increases. Chlorine was used to produce pulmonary edema in rats, and radioimmunoassays were used to measure IL-8 and endothelin (ET). The results showed that Danshen injection could decrease IL-8 and ET levels in rats with pulmonary edema [29]. Wu et al. found that the level of IL-8 was significantly increased in pulmonary edema in rats. However, after Danshen was used, the level of IL-8 in the plasma of rats with pulmonary edema was significantly lowered [30]. Other studies showed that intra-articular injection of Danshen injection could significantly decrease the levels of IL-1 and TNF-α in the synovial fluid of the rabbit knee osteoarthritis model [31]. Tanshinone II_A could significantly lower the levels of IL-1, IL-6, and TNF-α in the rat arthritis model [32]. Xu et al. [33] found that Danshen and Tanshinol had significant inhibitory effects on the cytokines, such as IL-1, IL-6, IL-8, and TNF-α, secreted by the Kuffer's cells (KC) of rats under the stimulation of endotoxin, and the mechanisms of inhibition included the stimulation of the secretion of immunosuppression substance PGE_2 by activated KC and inhibition of the secretion of granulocyte macrophage colony stimulatory factor (GM-CSF). Not only does Danshen alone have inhibitory actions on inflammatory cytokine production, the compound formula containing astragalus root has these effects also. For example, astragalus root and Danshen injection can lower the levels of NO and TNF-α in the plasma of rats with pregnancy-induced hypertension [34]. Liu et al. found that Danshen injection could significantly reduce the production of IL-6 in the early stage of burns, and facilitate the production of IL-10 so as to exert anti-inflammatory effects [35].

11.1.2.2 Effects on Adhesion Molecules

Adhesion molecules are a type of molecule mediating the contact and binding between cells and matrix substances, which are mainly distributed in the cell surface and extracellular matrix, and are involved in the signal transduction and activation, cell extension and mitigation, and the biochemical activities and differentiation of cells. The mechanisms for the involvement in the generation of inflammation are mainly through the binding between adhesion molecules expressed in white blood cells, such as CD11a/CD18, CD11b/CD18, CD11c/CD18, VLA-4, L-selectin, CD15, CD15s, and the intercellular adhesion molecules expressed in endothelial cells, such as ICAM-1, ICAM-2, vascular cell adhesion molecule-1 (VCAM-1), E-selectin, and P-selectin. These molecules adhere to each other, facilitate the migration of white blood cells to the inflammatory areas, and aggravate local inflammatory reactions.

Wu et al. found that E-selectin levels increased significantly during pulmonary edema in rats, and the maximal value might exceed six times the level in normal tissues. The level of E-selectin in the plasma of rats with pulmonary Edema treated with Danshen decreased significantly, to about onefold lower than that in the plasma of rats with pulmonary edema [36]. The effects of Danshen on the inflammation in lungs after smoking were observed, and the results showed that when compared with the smoking group, Danshen injection could significantly lower the number of inflammatory cells in the bronchial lavage liquid, and reduce the expression of CD11b and ICAM-1 in the epithelial cells of the lungs and airway [37]. Compound Danshen preparations can significantly inhibit the expression of CD44 and CD54 in the surface of lung cancer cells with high metastasis, and have significant inhibitory effects on the adhesion between PG cells and the activated and unactivated blood vessel endothelial cells [38]. Using recombinant selectin to select its antagonist, Liu et al. [39] found that 25 mg/ml Danshen injection could block more than 95 % of the adhesions between target cells and P-selectin. The antagonistic action of tanshinone II_A sulfonate (10 μg/ml) was higher than 85 %, and the antagonistic action of total salvianolic acid (16 μg/ml) was near 95 %, indicating that the action site of Danshen and its extracts was P-selectin.

As for the treatment of atherosclerosis (AS), the adhesion between monocytes and endothelial cells plays an important role in AS formation. It has been found that Danshen and hawthorn extracts can inhibit the expression of P-selectin in the platelets of the thrombosis animal model so as to suppress thrombosis [40] (Table 11.12). Jiang et al. [41, 42] found that sodium tanshinone II_A sulfonate can inhibit the expression of adhesion molecules in human umbilical vein endothelial cells and human blood platelets. Tanshinol can inhibit the expression of CD11 in activated white blood cells and has inhibitory effects on the adhesion of white blood cells, the expression of VCAM-1 and E-selectin in human umbilical vein endothelial cells induced by TNF-α, and can protect blood vessel endothelia, reduce the adhesion of white blood cells, and exert anti-AS effects (Tables 11.13, 11.14, 11.15, 11.16 and 11.17). Both salvianolic acid B (SalB) and tanshinone II_A can reduce the expression of VCAM-1 in the aorta of AS rabbits, and the effect of SalB was more significant. Tanshinone II_A can significantly lower the level of IL-8 in the serum of AS rabbits, but SalB has no influence on IL-8 [43]. Jiang et al. found that sodium tanshinone II_A sulfonate could significantly inhibit adhesion between cultured human umbilical vein endothelial cells and platelets, and the expression of adhesion molecule P-selectin [44]. Chen et al. [45] studied human thoracic aorta endothelial cells, and they found that SalB and the alcohol extracts of Danshen could significantly inhibit the expression of VCAM-1 and ICAM-1 induced by TNF-α. The alcohol extract of Danshen (50 μg/ml) and SalB (5 μg/ml) could significantly inhibit the binding rate between human monocyte U937 and TNF-α activated human thoracic aorta endothelial cells, and significantly lower the activity of NF-κB induced by TNF-α. The results showed that SalB and the alcohol extract of Danshen had anti-inflammatory effects, and the anti-inflammatory activities and inhibitory effects on low-density lipoprotein (LDL) could account for the antiatherosclerosis effects.

Adhesion molecules also play an important role in the inflammatory reactions of cerebral ischemia. After cerebral ischemia, the cytokines that can cause inflammation are produced in the ischemic area which induce the expression of various adhesion molecules, further facilitate the infiltration of white blood cells, produce inflammatory reactions, and aggravate brain injury. Liu et al.

Table 11.12 Effects of WH505 (extract of Danshen and Chinese hawthorn fruit) on platelet *P*-selectin ($\bar{X} \pm s$) [40]

Group	Mean fluorescent intensity (Ig)
Control	1.93 ± 0.25
Thrombin (1 Kμ/L)	6.51 ± 0.37
Thrombin + WH505 0.6 g/L	2.35 ± 0.13
Thrombin + WH505 1.2 g/L	3.03 ± 0.37
Thrombin + WH505 2.4 g/L	2.31 ± 0.31
Thrombin + WH505 4.8 g/L	2.10 ± 0.08

Table 11.13 Effects of Tanshinol on the expression of CD11b induced by Neutrophil fMLP ($n = 5$–6, $\bar{X} \pm s$) [42]

Group	Mean fluorescent intensity (Ig)
Control	$2.31 \pm 0.53^{**}$
fMLP 1×10^{-7} mol/L	4.71 ± 0.368
fMLP 1×10^{-7} mol/L + Tanshinol 100 mg/L	$3.93 \pm 0.33^{**}$
fMLP 1×10^{-7} mol/L + Tanshinol 200 mg/L	$3.08 \pm 0.4^{**}$
fMLP 1×10^{-7} mol/L + Tanshinol 300 mg/L	$2.65 \pm 0.15^{**}$

Compared with fMLP 1×10^{-7} mol/L group: $^{**}P < 0.01$

Table 11.14 Effects of Tanshinol on the thrombin-induced expression of *P*-selectin in platelets ($n = 4$–6, $\bar{X} \pm s$) [42]

Group	Mean fluorescent intensity (Ig)
Control group	$1.93 \pm 0.25^{**}$
Thrombin 1 kU/L	6.51 ± 0.37
Thrombin + Tanshinol 100 mg/L	$5.92 \pm 0.55^{*}$
Thrombin + Tanshinol 200 mg/L	$6.42 \pm 0.65^{*}$
Thrombin + Tanshinol 300 mg/L	$5.67 \pm 0.61^{*}$

Compared with thrombin (1 kU/L) group: $^{*}P < 0.05$, $^{**}P < 0.01$

Table 11.15 Effect of Tanshinol on ICAM-1 expression in cultured human umbilical vein endothelial cells ($n = 4$, $\bar{X} \pm s$) [42]

Group	Mean fluorescent intensity (Ig)
TNF-α 5 μg/L	23.87 ± 0.89
TNF-α + Tanshinol 100 mg/L	25.63 ± 1.39*
TNF-α + Tanshinol 200 mg/L	20.98 ± 1.51*
TNF-α + Tanshinol 300 mg/L	22.22 ± 1.35*

Compared with TNF-α 5 μg/L group: $^{*}P < 0.05$

Table 11.16 Effects of Tanshinol on the expression of VCAM-1 in cultured human umbilical vein endothelial cells ($n = 4$, $\bar{X} \pm s$) [42]

Group	Mean fluorescent intensity (Ig)
TNF-α 5 μg/L	2.4 ± 0.62
TNF-α + Tanshinol 100 mg/L	1.83 ± 0.05**
TNF-α + Tanshinol 200 mg/L	1.38 ± 0.13**
TNF-α + Tanshinol 300 mg/L	1.43 ± 0.31**

Compared with TNF-α 5 μg/L group: $^{**}P < 0.01$

Table 11.17 Effects of Tanshinol on the expression of E-selectin in cultured human umbilical vein Endothelial cells ($n = 4$, $\bar{X} \pm s$) [42]

Group	Mean fluorescent intensity (Ig)
TNF-α 5 μg/L	11.85 ± 0.49
TNF-α + Tanshinol 100 mg/L	9.63 ± 2.1**
TNF-α + Tanshinol 200 mg/L	9.39 ± 1.07**
TNF-α + Tanshinol 300 mg/L	8.28 ± 2.9**

Compared with TNF-α 5 μg/L group: $^{**}P < 0.01$

[46] found that Danshen could significantly increase the blood flow in the middle cerebral artery area in the ischemic side, lower the number of CD18 and CD11b positive cells in the periphery blood, and inhibit the leukocyte infiltration and necrosis of neurons in the ischemic middle cerebral artery area. So, the mechanism for the inhibition of adhesion between white blood cells and blood vessel endothelial cells was related to the blockage of CD11B/CD18 (a member of integrin family), the ligand of the white blood cell adhesion molecules in peripheral blood [47]. The expression of ICAM-1 is in close relation with neutrophil infiltration during cerebral ischemia. Liu et al. found that Danshen could inhibit the expression of ICAM-1, prevent adhesion between blood vessel Endothelial cells and white blood cells, and inhibit neutrophil infiltration [48]. Gao et al. [49] found that Danshen could inhibit the adhesion between polymorphonuclear neutrophils (PMN) and cerebral capillary endothelial cells (CCEC) so as to mitigate inflammatory reactions in cerebral ischemic tissues. The number of CD18/CD11b positive white blood cells in peripheral blood in the Danshen group was significantly lower than that in the control group, indicating that Danshen could block CD18/CD11b antigens on the surfaces of white blood cells [10, 49]. Therefore, Danshen can protect neurons after cerebral ischemia by blocking the adhesion of blood vessel endothelial cells and leukocytes, reducing neutrophil infiltration in ischemic brain tissues, and mitigating the injury to neurocytes.

Wang et al. [50] found that Danshen could simultaneously act on red blood cells and endothelial cells, and reduce the number and strength of adhesions between red blood cells and endothelial cells to improve blood circulation. Danshen can reduce the adhesion of neutrophils and endothelial cells of rabbits with burns, and decrease the expression of CD11a/CD18 and CD11b/CD18 on the surfaces of neutrophils [51] (Figs. 11.1, 11.2 and 11.3). Danshen injection

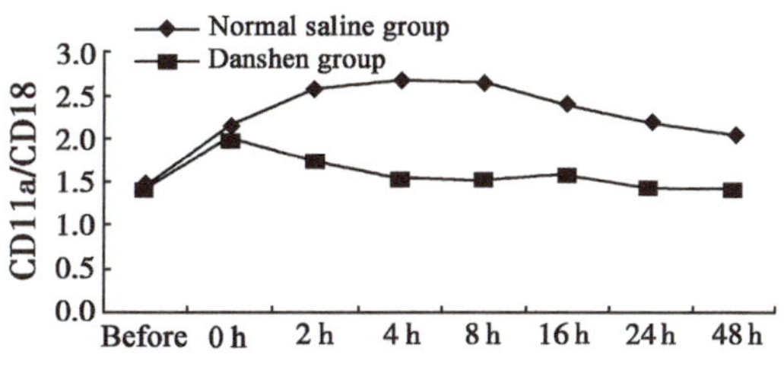

Fig. 11.1 Effects of Danshen on the CD11a/CD18 expression in rabbits after burning

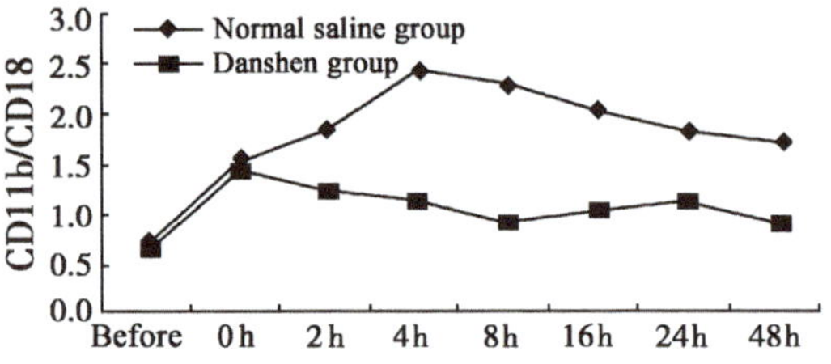

Fig. 11.2 Effects of Danshen on CD11b/CD18 expression in rabbits after burning

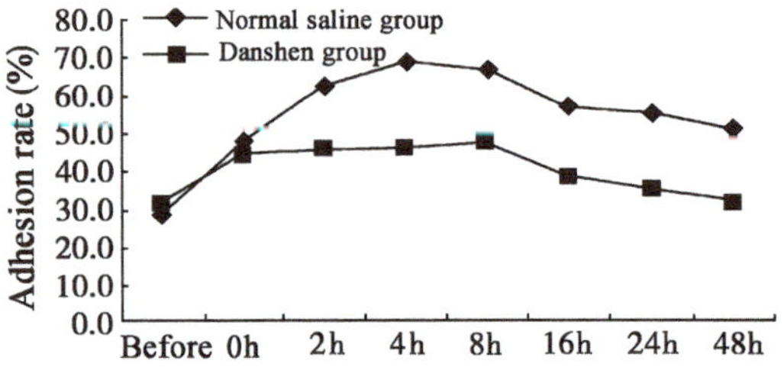

Fig. 11.3 Effects of Danshen on the adhesion between PMN and EC after burning

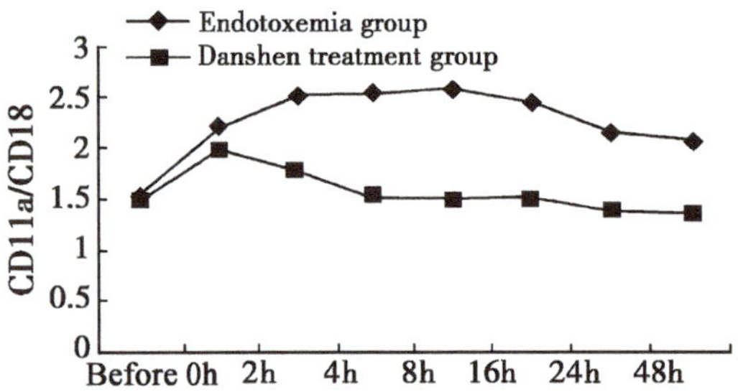

Fig. 11.4 Effects of Danshen on the CD11a/CD18 expression in rabbits with Endotoxemia

can decrease the expression of adhesion molecules CD11a/CD18 and CD11b/CD18 in rabbits with Endotoxemia, inhibit adhesion between neutrophils and endothelial cells [52] (Figs. 11.4, 11.5 and 11.6), and Danshen can inhibit adhesion between neutrophils and endothelial cells, and the expression of adhesion molecules in the early stage of acute pancreatitis in rabbits [53].

11.1.2.3 Others

Besides the studies on the effects of Danshen on immunocytes and cytokines, there are some studies on the other immunological pharmacological activities of Danshen, such as the actions

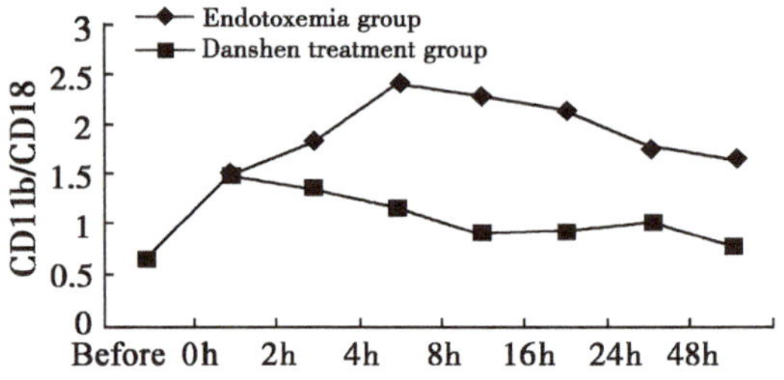

Fig. 11.5 Effects of Danshen on the CD11b/CD18 expression in rabbits with Endotoxemia

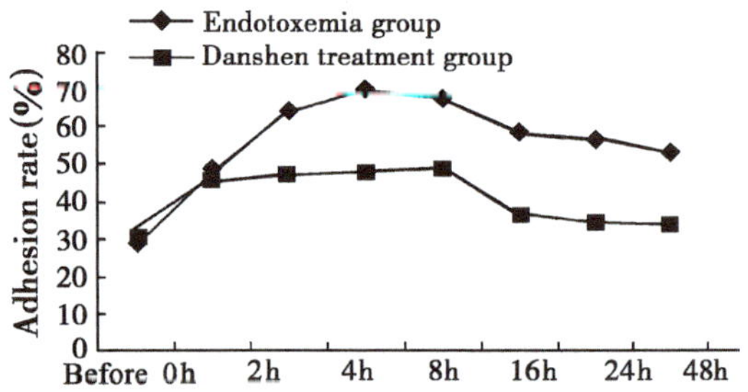

Fig. 11.6 Effects of Danshen on the adhesion between PMN and EC in rabbits with Endotoxemia

on factors related to the autoimmune disease antiphospholipid syndrome (APS). APS is a type of antibody against phospholipids. APS can induce thrombosis in arteries and veins for reasons currently unknown. It can also cause habitual abortion, neurological diseases, and symptoms such as thrombocytopenia. Its serological markers are antiphospholipid antibodies (aPL) such as lupus anticoagulant (LAC) and anticardiolipin antibodies (aCL). Chen Gang et al. used β2-glycoprotein to induce APS in mice, and found that 2.5 g/kg, 5 g/kg, and 10 g/kg of Danshen injection could significantly inhibit the production of aCL in mice, change the unbalanced T-Helper/T-Suppressor ratio back to normal, and lower the level of IL-2 [54, 55] (Tables 11.18 and 11.19).

It has been known that during cerebral ischemia, after the upregulated expression of basic fibroblast growth factor (bFGF), the survival of neurons and the growth of axons can be facilitated via high-affinity tyrosine kinase receptors. PDGF can stimulate the mitosis of fibroblast cells, osteoblasts, smooth muscle cells, and glial cells. It also can act on blood vessel

Table 11.18 Alteration of T cell subset of antiphospholipid mice ($n = 10$, $\bar{X} \pm s$)

Group	T-Helper cells	T-Suppressor cells	Th/Ts ratio
A	0.39 ± 0.03	0.26 ± 0.02	1.48 ± 0.03
B	0.36 ± 0.02*	0.27 ± 0.01*	1.33 ± 0.03**
C	0.33 ± 0.02**	0.29 ± 0.01**	1.16 ± 0.04**
D	0.33 ± 0.02	0.29 ± 0.02	1.15 ± 0.03
E	0.39 ± 0.03	0.26 ± 0.02	1.51 ± 0.05
F	0.33 ± 0.02	0.29 ± 0.02	1.15 ± 0.03

*,** Compared with group E: * $P < 0.05$, ** $P < 0.01$
The treatment of each group is as follows: A: Danshen injection 2.5 g/(kg d) + 150 μg of β_2-Glycoprotein; B: Danshen injection 15 g/(kg d) + 150 μg of β_2-Glycoprotein; C: Danshen injection 10 g/(kg d) + 150 μg of β_2-Glycoprotein; D: Danshen injection 15 g/(kg d); E: 150 μg of β_2-Glycoprotein; F: Blank control

Table 11.19 Comparison of IL-2 activity ($n = 10$, $\bar{X} \pm s$, A) [54]

Group	Dilution factor		
	1:2	1:4	1:8
A	0.85 ± 0.03	0.79 ± 0.03	0.72 ± 0.04
B	0.71 ± 0.05*	0.62 ± 0.04*	0.58 ± 0.02*
C	0.58 ± 0.03*	0.51 ± 0.04*	0.45 ± 0.02*
D	0.52 ± 0.04	0.45 ± 0.03	0.37 ± 0.04
E	0.87 ± 0.02	0.81 ± 0.05	0.74 ± 0.03
F	0.53 ± 0.03	0.46 ± 0.02	0.38 ± 0.02

Compared with group E: * $P < 0.01$; the treatment of each group was the same as shown in Table 11.18

endothelial cells and neurocytes, and lead to cellular responses such as proliferation, chemotaxis, recombination of actin, and flow of calcium. Kuang Genpei et al. found that after cerebral ischemia reperfusion in rats, Danshen could cause a bFGF-like immune response, and the degree of degeneration of neurocytes was milder, indicating Danshen could upregulate the expression of bFGF and intensify the reparative effects of bFGF during cerebral ischemia reperfusion [56]. Jiang et al. [57] found that Danshen could significantly reduce the expression of PDGF-A in rat brains with ischemic injury in the temporal lobe, improve the spatial cognitive impairment, and reduce infarct volume.

Miao et al. found that the water decoction of Danshen could inhibit the secretion level of transforming growth factor-β (TGF-β) in the lung tissues of mice with experimental pulmonary fibrosis caused by bleomycin. TGF-β is the most important cytokines in pulmonary fibrosis, and can cause the synthesis and deposit of extracellular matrix. Since the water decoction of Danshen could significantly inhibit the expression of TGF, it may be one of the important mechanisms of Danshen's mitigating effects on pulmonary fibrosis caused by bleomycin [58].

Interleukin-12 (IL-12) and interferon-γ (IFN-γ) are secreted by TH1 cells. Studies [59] showed that tanshinone could significantly inhibit the LPS-induced secretion of IL-12 and the keyhole limpet hemocyanin (KLM) stimulated secretion of interferon-γ in a dose-dependent manner. Further studies showed that tanshinone could inhibit the expression of IL-12 p40 mRNA and inhibit the binding of nuclear factor (NF)-κB to the κB site. Studies showed that Danshen could reduce the protein levels in urine by inhibiting the metabolism of arachidonic acid, decreasing the impairment of the glomerular filtrating membrane [60, 61], maintaining the dilatation of renal capillary vessels, lowering the viscosity of blood, and increasing renal blood flow, which could facilitate the absorption of immune complexes and the reparation of inflammatory damage [62, 63]. In addition, the anti-inflammatory effects of the magnesium salts of salvianolic acid B might be related to its inhibition of polymorphic nucleus leukotrienes and the activity of 5-lipooxidase [64]. Tang et al. proved that tanshinone II_A could significantly inhibit the activity of phospholipase A2 to achieve an anti-inflammatory effect [65].

To sum up, Danshen has strong anti-inflammatory activities, and the mechanisms of these actions include mainly the inhibition of inflammatory cytokine secretion by macrophages and the chemotaxis, migration and adhesion of neutrophils, and reducing the expression of adhesion molecules.

11.2 The Pharmacological Effects of Danshen in Antibacterial Activity

Li Zhang and Guanhua Du

Tanshinone has strong inhibitory effects on a large variety of Gram-positive bacteria and relatively weak inhibitory effects on Gram negative bacteria. It has been shown that tanshinone can produce free radicals during bacterial metabolic processes, which can damage bacterial components such as DNA and lead to the death of bacteria. One of the mechanisms of acne treatment with tanshinone preparations is based on the powerful antibacterial effects of tanshinone [66].

11.2.1 Effects on *Staphylococcus Aureus*

Gao et al. found that tanshinone had significant therapeutic effects on acute infections caused by *Staphylococcus Aureus*, including those caused by the drug-resistant strains, and it also had antibacterial effects on 2 *Trichophyton sp*. They found that tanshinone had strong antibacterial effects on *S. aureus* and inhibitory effects on *Hemolytic streptococcus*, and the minimum inhibitory concentration for *Bacillus tuberculosis* $H_{37}RV$ was <1.5 μg/ml. After oral administration or subcutaneous administration of total tanshinone, antibacterial substances can be detected in the tissues and excreta. In addition, it was found that total tanshinone had inhibitory actions on *Trichophyton ferrugineum*, and 100 μg/ml of tanshinone could inhibit the growth of *Trichophyton ferrugineum* and *Trichophyton rubrum* [3]. Wang Dingbang has found that total tanshinone contains 10 components, five of which have antibacterial effects. Among these, cryptotanshinone has the strongest antibacterial activity, and they mainly act on Gram-positive bacteria. The quinoid group is the essential group for bacteriostatic effects, and the bacteriostatic action of ortho-quinone is stronger than that of paraquinone. Both hydroxylation and dehydrogenation of ring A can lower bacteriostatic activity, which is also influenced by the different substitution of α-H in the furan ring [67]. Fang et al. performed in vitro bacteriostatic experiments on 10 components of Danshen, and found that cryptotanshinone, tanshinone II_B, methyltanshinonate, hydroxytanshinone II_A, and dihydrotanshinone had strong inhibitory effects on *S. aureus* and its drug-resistant strains, and tanshinone I, II_A, cryptotanshinone, and hydroxytanshinone II_A had strong inhibiting effects on *Bacillus tuberculosis* (H37RV) strains [68]. Sun et al. conducted a test using the Kirby-Bauer disk diffusion method, and found that 100 % Danshen extract had significant bacteriostatic effects on *Escherichia coli*, *S. aureus*, *Staphylococcus albus*, *Bacillus proteus,* and *Streptococci group* B [69] (Table 11.20).

Table 11.20 Bacteriostatic effects of Danshen

Drug concentration	Bacteria	Average diameter of bacteriostatic ring (d/cm)
Raw liquor	*Escherichia coli*	1.138
Raw liquor	*Staphylococcus aureus*	1.307
Raw liquor	*Staphylococcus albus*	0.909
Raw liquor	*Bacillus proteus*	1.123
Raw liquor	*Streptococci group* B	0.89

Table 11.21 In vitro bacteriostatic effects of Danshen injectable powder (double dilution method, sulphuric acid–ethanol medium, $n = 3$)

Tube number	Volume of bacterial liquid/ml	Concentration or dilution rate of drugs/mg ml^{-1}	Growth conditions of bacteria
1	0.1	100.00	No growth, completely clear
2	0.1	50.000	No growth, completely clear
3	0.1	25.000	No growth, completely clear
4	0.1	12.500	No growth, completely clear
5	0.1	6.250	No growth, completely clear
6	0.1	3.125	No growth, completely clear
7	0.1	1.563	No growth, completely clear
8	0.1	0.781	The growth was inhibited; a little turbid
9	0.1	0.391	The growth was inhibited; a little turbid
10	0.1	0.195	The growth was inhibited; a little turbid
11	0.1	0.098	The growth was inhibited; a little turbid
12	0.1	0.098	No growth, completely clear
Positive control	0.1		Lantern-like normal growth
Blank control			No growth, completely clear

Note With the increase of the dilution rate of the drugs, the inhibition of *Staphylococcus aureus* growth was decreased, and the turbidity increased. The concentration of the *Staphylococcus aureus* liquid added was 10^5 CFU/ml

Zhang et al. used in vitro antibacterial and bacteriostatic tests to determine the bactericidal effects of Danshen injection on *S. aureus*, and the results showed that in the test using the double dilution method, Danshen injection had strong killing effects on *S. aureus*; the minimum inhibitory concentration was 0.098 mg/ml, and the minimum bactericidal concentrations were 1.563 mg/ml (sulfuric acid–ethanol medium) and 0.195 mg/ml (nutritional broth medium). Using direct inoculation, the MIC and minimum bactericidal concentrations were 0.156 mg/ml and 0.311 mg/ml, respectively [70] (Tables 11.21, 11.22 and 11.23).

Zhu et al. [71] used a hydrophilic solvent instead of a hydrophobic solvent and agar diffusion to observe the antibacterial effects of tanshinone II_A. The results showed that the MIC of tanshinone II_A and tanshinone II_B for *E. coli* were 50 and 25 μg/ml, respectively; the MIC for *S. aureus* ATCC-25923 were 100 and 50 μg/ml, respectively; the MIC for *Pseudomonas aeruginosa* ATCC-27853 were 50 μg/ml and 25 μg/ml, respectively; the MIC to *Hemolytic streptococcus* were 12.5 and 25 μg/ml, respectively (Tables 11.24, 11.25 and 11.26).

In our opinion, the reason Gao Yugui et al. did not find the bacteriostatic effect of tanshinone II_A against *S. aureus* is because they used agar diffusion, which is not appropriate. The studies by Gao Yugui showed that among total Tanshinones, cryptotanshinone had the strongest bacteriostatic action. The analysis of the structures of cryptotanshinone and tanshinone II_A shows that tanshinone II_A has stronger lipophilicity than does cryptotanshinone, suggesting that tanshinone II_A in the CH_3Cl filter paper could not enter the agar to exert its effects. The pharmacophoric group of tanshinone II_A is identical to the active group responsible for oxidation–reduction in cryptotanshinone (ortho-quinone). The mechanisms of the bacteriostatic effects of cryptotanshinone are already known, i.e., by the formation of superoxide anions [72].

11.2.2 Effects on Helicobacter Pylori

Helicobacter pyloridis (HP) is in close relation with chronic gastritis and digestive tract ulcers. Xu et al. have conducted in vitro HP sensitivity

Table 11.22 In vitro bacteriostatic effects of Danshen injectable powder (double dilution method, nutritional broth culture medium, $n = 3$)

Tube number	Volume of bacterial liquid/ml	Concentration or dilution rate of drugs/mg ml^{-1}	Growth conditions of bacteria
1	0.1	100.00	No growth, completely clear
2	0.1	50.000	No growth, completely clear
3	0.1	25.000	No growth, completely clear
4	0.1	12.500	No growth, completely clear
5	0.1	6.250	No growth, completely clear
6	0.1	3.125	No growth, completely clear
7	0.1	1.563	No growth, completely clear
8	0.1	0.781	No growth, completely clear
9	0.1	0.391	No growth, completely clear
10	0.1	0.195	No growth, completely clear
11	0.1	0.098	The growth was inhibited; a little turbid
12	0.1	0.098	No growth, completely clear
Positive control	0.1		Lantern-like normal growth
Blank control			No growth, completely clear

Note With the increase of the dilution rate of the drugs, the inhibition of *Staphylococcus aureus* growth was decreased, and the turbidity increased. The concentration of the *Staphylococcus aureus* liquid added was 10^5 CFU/ml

Table 11.23 In vitro bacteriostatic effects of Danshen Injectable Powder (direct inoculation method, $n = 3$)

Tube number	Volume of bacterial liquid/ml	Concentration or dilution rate of drugs/mg ml^{-1}	Growth conditions of bacteria
1	0.1	15	The growth was inhibited; a little turbid
2	0.1	15	The growth was inhibited; a little turbid
3	0.1	15	No growth, completely clear
4	0.1	15	No growth, completely clear
5	0.1	15	No growth, completely clear
6	0.1	15	No growth, completely clear
7	0.1	15	No growth, completely clear
8	0.1	15	No growth, completely clear
9	0.1	15	No growth, completely clear
10	0.1	15	No growth, completely clear
Positive control	0.1	15	Lantern-like normal growth
Blank control		15	No growth, completely clear

Note Danshen injectable powder was dissolved with sterile normal saline to obtain 50 mg/ml drug solution

tests, determination of bacteriostatic and bactericidal concentrations of sensitive drugs, combined sensitivity tests, etc., and found that Danshen has a killing effect on HP, and the minimal bacteriostatic and bactericidal concentration of the water decoction of Danshen was 6.3 % [73].

Table 11.24 Bacteriostatic activities of tanshinone II_A and tanshinone II_B measured by agar diffusion

Compounds	*Escherichia coli* ATCC-25922	*Staphylococcus aureus* ATCC-25923	*Pseudomonas aeruginosa* ATCC-27853
Tanshinone II_A	11	16	17
	11	11	16
Tanshinone II_B	11	13	20
	11	15	20

Culture diameter (6 mm); control solvent; dimethyl formamide (DMF): bacteriostatic diameter 0

Table 11.25 Results of antibacterial sensitivity assay using agar diffusion and bacterial strains isolated from clinical patients

Compounds	Diameter of bacterial inhibition zone/mm															
	Escherichia coli						*Staphylococcus aureus*						*Streptococcus mutans*		*Hemolytic streptococcus*	
	Bacterial strain 1		Bacterial strain 2		Bacterial strain 3		Bacterial strain 1		Bacterial strain 2		Bacterial strain 3		Test 1	Test 2	Test 1	Test 2
Tanshinone II_A	20	17	13	4	5	13	30	0	20	12	15	20	10	0	10	0
Tanshinone II_B	19	19	15	14	7	14	30	0	10	17	22	20	11	0	16	0
Methenyl tanshinone	18	19	16	12	6	15	20	0	15	17	17	20				

Culture diameter (6 mm); control solvent; dimethyl formamide (DMF): bacteriostatic diameter 0

Table 11.26 Minimal inhibitory concentrations of tanshinone II_A and tanshinone II_B

Bacterial strain	*Escherichia coli* ATCC-25922					*Staphylococcus aureus* ATCC-25923				
	200	100	50	25	12.5	200	100	50	25	12.5
Tanshinone II_A	–	–	–	+		–	–	+	+	+
Tanshinone II_B	–	–	–	–	+	–	–	–	+	+
Bacterial strain	*Pseudomonas aeruginosa* ATCC-27853					*Hemolytic streptococcus*				
	200	100	50	25	12.5	200	100	50	25	12.5
Tanshinone II_A	–	–	–	|	+	–	–	–	–	–
Tanshinone II_B	–	–	–	–	+	–	–	–	–	++

–: clear; +: a little turbid; ++: turbid

11.2.3 Others

Luo Houwei et al. found that among the human *Mycobacterium*, *Mycobacterium tuberculosis* H37Rv was the most sensitive bacteria to the bacteriostatic action of tanshinone II_A, and the minimal inhibitory concentration (MIC) was 0.31 μg/ml [74]. It has been shown by other studies that compound formulas composed of heat-clearing drugs, such as Danshen and coptis, have significant inhibitory effects on anaerobic bacteria such as *Bacteroides fragilis*, Fusobacterium, *Fusobacterium sp.*, *Eubacterium sp.*, *Peptococcus sp.*, and *Peptostreptococcus sp.*). They also have bacteriostatic effects on aerobic bacteria such as *S. aureus*, *Pseudomonas aeruginosa*, *Klebsiella pneumoniae*, *Streptococcus viridans*, and *Enterococcus* group D [75]. The research by Sun Haitao et al. used microcalorimetry to prove that the water decoction of Danshen had antibacterial action on *Pseudomonas aeruginosa*, and the two kinds of bacteria could not grow when Danshen concentrations in the water decoction were 299.6 μg/10 ml and 41.5 μl/10 ml [76].

The research by Luo Houwei et al. also showed that Danshen and its components had various degrees of bacteriostatic effects on mycobacteria 607 and Mycobacterium ulcerans, Danshen had strong bacteriostatic action against *E. coli*, Bacillus proteus and dermatophyte, and some bacteriostatic action against Salmonella typhosa and Bacillus dysenteriae [1, 14, 15].

To sum up, Danshen has antibacterial effects. The antibacterial action against Gram-positive bacteria is strong, and the bacteriostatic action against mycobacteria is relatively strong. The mechanism of action is mainly by producing superoxide anions.

References

1. Zhang GC. Chin J Clin Pediatr. 2004;12(3):959.
2. Gao YG, Wong LZ, Tang JX. J Integr Tradit Chin W Med. 1983;3(5):300.
3. Gao YG, Song YM, Yang Y, et al. Acta Pharm Sinica. 1979;14(2):75.
4. Yan YZ, Chen XY, Zeng WD. J Chin Acad Med Sci. 1986;8(6):417.
5. Zhao ZB, Feng JY, Chen SQ, et al. J Sichuan Tradit Chin Med. 1993;11:13.
6. Lang XC, Li MX, Jia BY, et al. J Hebei Med Coll. 1991;12(3):140.
7. Ying ZZ, Han ZH, Wu YF. Shizhen J Tradit Chin Med Res. 1994;5(2):19.
8. Ye HJ, Wang XM, Zhang L, et al. Chin J Clin Hepatol. 1995;11(3):142.
9. Qu L, Wang D, Yang XL. J Microbiol. 2000;20(2): 26, 38.
10. Sun HM. Tradit Chin Med Res. 199;1(3):50.
11. Wang P, Wang W, Wang WL, et al. Chin Arch Tradit Chin Med. 2002;20(1):88.
12. Wang WJ, Wu XZ, Yao Z, et al. Chin J Immunol. 1995;11(6):370.
13. Zhang LX, Wang M, Qian Y, et al. Pharmacol Clin Chin Materia Medica. 1990;6(4):31.
14. Jia TY. Tradit Chin Med Res. 1995;8(4):17.
15. Miao J, Chen J, Gao L. J Tianjin Coll Tradit Chin Med. 1998;17(1):38.
16. Wang CL, Khara. A, Faber V. J Guangxi Med Univ. 1997;14(3):9.
17. Lv SJ, Huang HL. Chin J Exp Clin Immunol. 1992; 4(2):41.
18. Liu J, Wu WP, Kuang PG, et al. Chin J Neuroimmunol Neurol. 1998;3(4):225.
19. Liu J, Kuang PG, Wu WP, et al. Chin J Geriatr Cardiovasc s. 2002;4(2):127.
20. Lu Q, Sun ZJ. Chin J Crit Care Med. 1999;19(7):389.
21. Chen KQ, Zhou Y, Xing HY, et al. Chin J Pathophysiol. 1999;15(9):781.
22. Gao YY, Wang SF, Zhang ZJ, et al. Chin J Pathophysiol. 1986;2(2):93.
23. Li H, Tang RY. Acta Pharmacol Sin. 1991;12(3):269.
24. Zhou ZC, Zheng JR, Xu WY. J Chin Acad Med Sci. 1997;19(3):232.
25. Li LX, Xu ZE, Zhao B. Chin J Histochem Cytochem. 2004;13(2):168.
26. Zhang JT, Weiping Wu, Fengying Z, et al. Med J Chin People's Liberation Army. 2002;27(1):56.
27. Wan L, Xu T, Li SS. Chin J Gerontol. 2007;27(3): 249.
28. Nie GL, Guo MJ, Li YP, et al. Chin J Gerontol. 2007;27(9):823.
29. Wu S, Sheng YL, Jiang ZX, et al. Liaoning J Tradit Chin Med. 2003;30(10):796.
30. Wu S, Sheng YL, Jiang ZX, et al. China Clin Pract Med. 2007;1(1):4.
31. Wang XH, Su ZB LL, et al. J Tradit Chin Orthop Traumatol. 2003;15(10):1.
32. Yang DW, Meng B, Gao LZ, et al. Pract Clin J Integr Tradit Chin W Med. 2007;6:10.
33. Xu F, Lu BG, Yao Z, et al. Chin Crit Care Med. 1996;8(5):262.
34. He L, Wang RG, You ZL, et al. J Hunan Univ Tradit Chin Med. 2003;23(2):4.
35. Liu YL, Chen XY, Chen XL, et al. Tianjin J Tradit Chin Med. 2007;24(1):66.
36. Wu S, Sheng YL, Jiang ZX, et al. Liaoning J Tradit Chin Med. 2003;30(3):172.
37. Yuan WS, Xiong SD, Xu YJ, et al. Hubei J Tradit Chin Med. 2004;26(3):9.
38. Wang YY, Wu AQ, Xu N, et al. Chin J Mod Appl Pharm. 2003;20(5):343.
39. Liu JN, Jan H XB. Shenyang J Military Med. 2006;19 (1):10.
40. Jiang KY, Ruan CG, Gu ZL. Acta Acad Med Suzhou. 2000;20(3):203.
41. Jiang KY, Ruan CG, Gu ZL, et al. Acta Pharmacol Sin. 1998;19(1):47.
42. Jiang KY, Gu ZL, Ruan CG. Chin Pharmacol Bull. 2000;16(6):682.
43. Zhang FY, Hua SY, Fan YC. J Shandong Univ Tradit Chin Med. 2006;30(2):152.
44. Jiang KY, Ruan CG, Gu ZL, et al. Acta Pharmacol Sin. 1998;19(1):47.
45. Chen YH, Lin SJ, Ku HH, et al. J Cell Biochem. 2001;82(3):512.
46. Liu J, Kuang PG, Li B, et al. Chin J Neuroimmunol Neurol. 1998;5(2):77.
47. Liu J, Wu WP, Kuang PG, et al. Chin J Neuroimmunol Neurol. 1998;3(4):225.
48. Liu J, Kuang PG, Wu WP, et al. Chin J Nerv Mental Dis. 1999;25(4):198.
49. Gao CY, Zhou HD, Chen ME, et al. J Apoplexy Nerv Dis. 2001;18(2):113.
50. Wang L, Huang X, Ding ZH, et al. W China J Pharm Sci. 1995;10(4):193.

51. Chen X, Zhao Y, Sha JP, et al. J Chengdu Univ Tradit Chin Med. 2002;25(1):42.
52. Zhao Y, Yin Y, Tian J, et al. J Chengdu Univ Tradit Chin Med. 2003;26(4):33.
53. Sha JP, Zhu BD, Xu Y, et al. Chin J Integr Tradit W Med Intensive Crit Care. 2003;10(5):279.
54. Chen G, Liu RX, Chen XL, et al. J Xi'an Med Univ. 2001;22(5):429.
55. Chen G, Liu RX, Fu Jie, et al. J Fourth Mil Med Univ. 2003;24(11):997.
56. Kuang PG, Liu J, Wang FZ, et al. J Brain Nerv Dis. 1997;5(1):1.
57. Jiang SJ, Wu WP. Chin J Neurol. 1999;32(5):290.
58. Miao WN, Dong J, Liu ST, et al. Pharmacol Clin Chin Materia Med. 2003;19(5):24.
59. Kang BY, Chung SW, Kim SH, et al. Immunopharmacology. 2000;49(3):355.
60. Kim SY, Moon TC, Chang HW, et al. Phytother Res. 2002;16(7):616.
61. Ryu SY, Oak MH, Kim KM. Planta Med. 1999;65 (7):654.
62. Liu XR, Guan YJ, Chen KZ, et al. J Shandong Med Univ. 1992;30(1):85.
63. Sun W, Zhou Y. Chin J Integr Tradit W Nephrol. 2004;5(8):490.
64. Liu P, Goukou JH, Shenze CS. Pharmacol Clin Chin Materia Med. 1993;9(3):13.
65. Tang T, Guo WQ, Wang Y, et al. Lett Biotechnol. 2007;18(1):51.
66. Lee DS, Lee SH, Noh JG, et al. Biosci Biotechnol Biochem. 1999;63(12):2236.
67. Wang DB. J Tradit Chin Med. 1982;23(11):846.
68. Fang QN, Zhang PL, Xu ZP, et al. J Chem. 1976;34(3): 197.
69. Sun JL, Chang YT, Song HY, et al. Lishizhen Med Materia Med Res. 2003;14(12):725.
70. Zhang XM, Zhang KR, Hu CQ. Tradit Chin Drug Res Clin Pharmacol. 2003;14(5):321.
71. Zhu JR, Luo HW. J China Pharm Coll. 2004;35(4): 368.
72. Lee D, Lee S, Han S, et al. Bunge Biosci Biotechnol Biochem. 1999;63(12):2236.
73. Xu Z, Zhou DR, Wang Y, et al. China J Tradit Chin Med Pharm. 1993;8(5):25.
74. Luo HW, Gao JW, Zheng JR. J China Pharm Coll. 1988;19(4):258.
75. Cui NQ, Zhao EP, Qu C, et al. Chine J Integr Tradit W Med. 1995;1(1):1.
76. Sun HT, Zhang HG, Zhao XY, et al. Chin Pharm J. 1996;31(4):201.

Part II
Quality Control

The study of quality control in traditional Chinese medicine has always been a difficult and hot issue, and it is also the key to the modernization and internationalization of TCM. The quality control methods for herbal medicines in developed countries mainly adopt the simultaneous determination of multiple marker compounds based on the chemical constituents contained in the herb. Scientists in the United States and the European Union proposed the adoption of a fingerprint method for the quality control of herbal medicines to show that each constituent in the herb and the preparations and their ratios, would be consistent and stable to ensure the quality of the herbal medicine. In recent years, Chinese scientists have actively approached methodologies for the comprehensive and systemic quality control of TCM, including fingerprint techniques. Technical guidance for the fingerprint of TCM injections was established and computer software for evaluating fingerprint similarity was developed, which laid a solid foundation for the objective evaluation of TCM fingerprinting.

Currently, the generally accepted mode for the comprehensive quality control of TCM is to adopt conventional quality control methods including origin identification, morphological and microscopic identification, physio-chemical identification, pesticide residue and heavy metal detection, plus fingerprint techniques combined with multi-component quantitative analysis. In the last few years, Chinese scientists have conducted considerable research with this quality control mode to make the best use of the fingerprint technique to represent the characteristics of traditional Chinese medicine, to control the qualities of finished products, so as to control the preparation procedure and the qualities of the raw materials. By studying a large amount of TCM fingerprints, we have established a systemic determination approach, a set of quality control parameters, some mathematical models, and preliminarily realized the goal of operability, controllability, stability, and quantifiability of the fingerprint technique, and have thus raised the standard of TCM quality control.

Meanwhile, the application of both traditional chemical techniques of isolation and purification and modern techniques of liquid chromatography-mass spectrometry in the identification of the major peaks in the fingerprints has provided us with abundant constituent information, which makes the fingerprint technique more reliable and reasonable for TCM quality control. It has been proven in our practice that this mode of TCM quality control is feasible for control of TCM quality scientifically, reasonably, and comprehensively.

Danshen is one of the most commonly used and one of the most important TCMs. In recent years, scientists have conducted a large amount of research on Danshen and have made breakthroughs in the areas of active constituent identification and quality control, which are the solid bases for Danshen's scientific and rational application and global development. Because of the scattered distribution of these research data in publication, their reference value and utilization rate have been greatly limited. In view of this, this part is compiled to present the research results on quality control and other aspects achieved in recent years to provide convenient references for research, production, physicians, and patients. It also provides an exemplified study on the methodology of modern TCM quality control.

This part contains five chapters; Qualitative Research, General Quality Control Methods, Content Assaying, Fingerprinting, and Quality Control of Dantonic™, which introduces in detail, from the above-mentioned various aspects, the latest advancements in the quality research on Danshen and its preparations carried out by scientists in China and other countries.

The authors of this part are experts and scholars who have been engaged in the quality control studies or the related aspects of Danshen in China, including a number of young scientists. I am extremely grateful for their diligent and tireless work.

We are also thankful to Tianjin Tasly Co., Ltd. the organizer of this book, who has delivered her great support in the compilation process. Due to our limited knowledge and the scope of the book, mistakes and omissions are inevitable. We welcome any feedbacks and comments from the reader.

Shanghai, January 2008 Guo De-an

12 Qualitative Research

Guoqiang Fan, Guoqing Wu, Xiaoqian Zhang, Xian Zhang, Rixin Liang, Manling Li, Fenglan Cao, Ming Zhu, Zhangzhao Jin, Bilian Chen, Linke Ma, Shen Ji, Qing Gong, Zhengliang Ye and Jun Gao

There are five methods for the identification of medicinal Danshen: morphological identification, microscopic identification, physio-chemical identification, TLC (Thin-Layer Chromatography) identification, and spectral identification. These methods are simple and feasible, and capable of accurately identifying the authenticity and quality of Danshen. At present, there are a wide variety of Danshen preparations, the most representative preparations of which are Composite Danshen Dropping Pill (Dantonic™), Danshen Injection, Compound Danshen Tablet, Xiangdan Injection, and Guanxinning Injection. This chapter discusses the qualitative identification of these five preparations.

12.1 Identification of Medicinal Danshen

12.1.1 Morphological Identification

Guoqiang Fan

Danshen is the dried rhizome of plant *Salvia miltiorrhiza* Bge. (Family *Labiatae*). The plant is collected in spring and fall, cleaned, and dried.

The rhizome of *S. miltiorrhiza* is stubby; sometimes there are residual basal stems on the top. It contains several roots of long cylindrical shape and slightly bending; some roots have branches with fibrous roots, 10–20 cm long and 0.3–1 cm in diameter. The surface is red-brown or dark brown, rough, and with longitudinal wrinkles. The outer skin of the old root is loose, purple, often has scale-like flaking. The texture is hard and brittle, and the cross-section is loose with cracks, slightly flat, and compact. The bark part is red-brown; the xylem is yellow or brown, and the vascular bundle is yellow-white, showing a radial arrangement. It smells a little and tastes slightly bitter (Fig. 12.1).

12.1.2 Microscopic Identification

Guoqing Wu, Xiaoqian Zhang and Zhang Xian

12.1.2.1 Cross-Section

The cork layer of Danshen's roots includes 3–7 layers of cork cells which are square or rectangular, extending tangentially, containing a purple-brown substance, and sometimes with rhytidome tissue. The cortex is broad and composed of rectangular or elliptical parenchyma cells. The phloem is narrow, half-moon like, and composed of sieve tube groups and parenchyma cells. The cambium is obviously ring-like shaped. The xylem is broad, yellow, with 3–4 vascular bundles, and divided into 3–4 strands. The vessels are round, polygonal, or oval shaped, some

G. Fan · G. Wu · X. Zhang · X. Zhang · R. Liang · M. Li · F. Cao · M. Zhu · Z. Jin · B. Chen · L. Ma · S. Ji · Q. Gong · Z. Ye (✉) · J. Gao
Zhejiang Institute of Food and Drug Control, Hangzhou, China
e-mail: zhumingd@hotmail.com

X. Yan (ed.), *Dan Shen (Salvia miltiorrhiza) in Medicine*,
DOI 10.1007/978-94-017-9463-3_12,

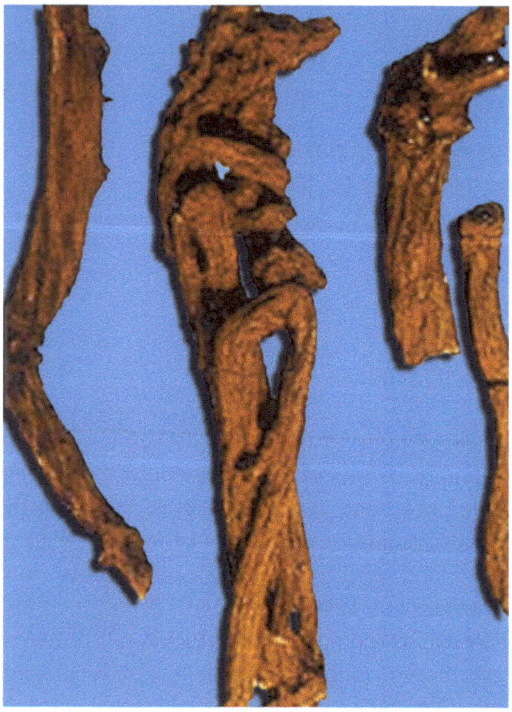

Fig. 12.1 The morphology of medicinal Danshen

extending radially, scattering individually or in a group of 2–12 vessels, and rich in wood fibers. The xylem ray is broad and composed of more than 20 lines of radially extending parenchyma cells. The ray cells are parallel to the tangential vascular bundle and thickened by lignification [1–3] (Fig. 12.2).

It was also reported that some radicular fibers are mainly distributed in the phloem; sometimes the stone cells can be seen. Cultivated Danshen (such as in Sichuan) usually has no fibers or stone cells. There are many vascular bundles, from nearly one dozen to more than 20 bundles. The vessels spread sparsely, arrayed along the radial direction with a single line; the cortex and phloem of a small number of roots have clerenchyma [2].

At present, Danshen in the market is primarily cultivated. Hu Cultivated Danshen under simulated wild conditions and compared them with the market varieties. The results showed that the sections of simulated wild cultivated species are relatively loose; the xylem presents as brown-red with fewer vascular bundles [4].

Yang and He conducted market surveys to compare the varieties commonly confused with Danshen, such as *S. bowleyana*, *S. przewalskii*, and *S. yunnanensis*. In the cross-section, the cork layer of Danshen has many breakages, and the cork cells are flat and scattered with stone cells.

Fig. 12.2 Cross section of Danshen root (× 6)

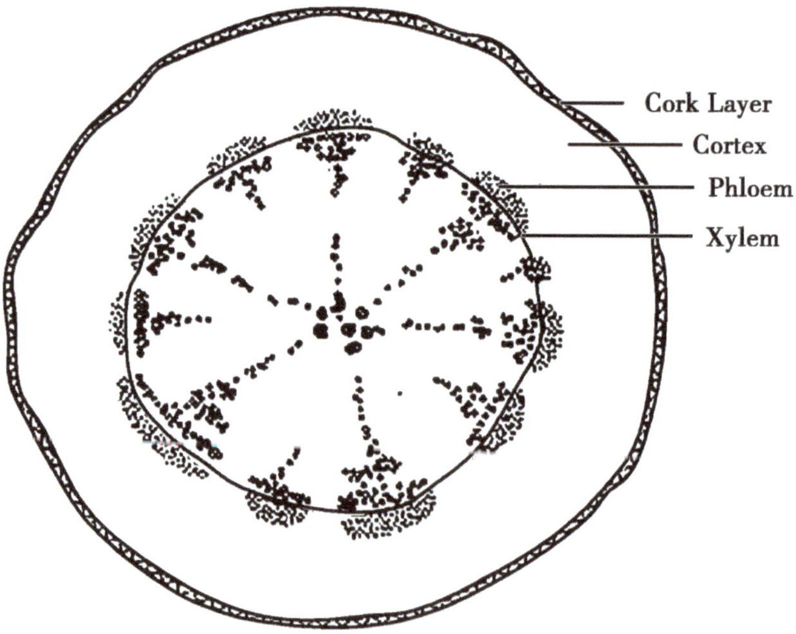

There is no obvious boundary between the cortex and the phloem. There are lots of stone cells and phloem fibers, which are scattered separately or in groups. The stone cells are primarily distributed in the external layer and the phloem fiber is primarily distributed in the internal side, and the xylem is distributed in the center of the old roots, forming a cell suberization ring with relatively more vascular bundles [5, 6].

The commonly counterfeit species of Danshen include *Przewalsk Sage* root, radix dipsaci, great burdock root, etc. The researchers compared the identified features of the sections with those of Danshen [7–10].

12.1.2.2 Features of Danshen Powder

The powder of Danshen is light pink in color. The stone cells are scattered separately or connected, nearly colorless or light yellow, and round-like, spindle-like, square, rectangle-like, or triangle-like or irregular in shape. The edge is not even, 14–70 μm in diameter, and about 190 μm in length. The thickness of the wall is 5–20 μm, with sparse pits, short, small or coarse and a long pit canal; some have branches, and some do not reach the edge. The lines are fewer and visible, some cells contain yellow-brown or brown substances.

The vascular bundles have lines and bordered pits, with diameters of 11–60 μm. The reticulate vessel molecule is long spindle-shaped. The ends are long and sharp or slightly tilted. The pits are narrow and some of the pit canals do not reach the edges; the holes are located at the side wall or end wall. A small number of vascular molecules contain brown substances; the vessels with bordered pits are relatively larger, with bordered pit-round, arranged compactly, some with tertiary reticulated thickness.

Wood fibers are primarily fiber tracheids, many of them in bundles or scattered separately, and are long spindle-shaped, blunt or sharp, or round at the end, 12–27 μm in diameter, and the wall thickness is 2–5 μm. With bordered pits, pits crack-shaped or cross-like, the pit canals are sparse, and relatively dense in local parts.

Phloem fibers are in bundles or scattered separately, long spindle-shaped, and round or blunt at the two ends. The edges are uneven, 60–170 μm long, 7–27 μm in diameter, and the wall thickness is 3–12 μm. The pit canals are relatively obvious; some of them have visible lines and pits.

The cork cells are connected, yellow-brown, rectangular, or polygonal forms. The wall is a little thick, with slight lignification [2] (Fig. 12.3).

12.1.3 Physical and Chemical Identification

Guoqing Wu, Xiaoqian Zhang, Rixin Liang and Li Manling

12.1.3.1 Color Reaction

Take 5 g of Danshen root powder, add 50 ml of water, boil for 15–20 min, cool down, filter, condense the filtrate to a viscous paste in a water bath, cool down and add 3–5 ml of ethanol to dissolve it, filter. Take 0.5 ml of the filtrate, add 1–2 drops of chloride ferric test solution, and it exhibits a dirty green color [1].

12.1.3.2 Fluorescence Reaction

Place Danshen powder on a white porcelain plate. Under an ultraviolet lamp (365 nm), the powder exhibits gray fluorescence [11].

Take several drops of the filtrate from 3.1, apply on a filter paper strip, dry, and observe under an ultraviolet lamp (365 nm); the spots exhibit brilliant slate fluorescence. Hang the filter paper strip in a concentrated ammonia solution flask (do not touch the liquid), remove after 20 min, and observe under the ultraviolet lamp. The spots exhibit light blue-greenish fluorescence [12].

12.1.4 Identification by Thin-Layer Chromatography (TLC)

Guoqing Wu, Xiaoqian Zhang, Rixin Liang and Manling Li

12.1.4.1 Identification of Liposoluble Constituents by TLC

Take 1 g of Danshen powder, add 5 ml of ethyl ether, place the mixture in a capped test tube, shake, incubate for 1 h, filtrate, volatilize, add 1 ml of acetidin to the residue to dissolve. This is

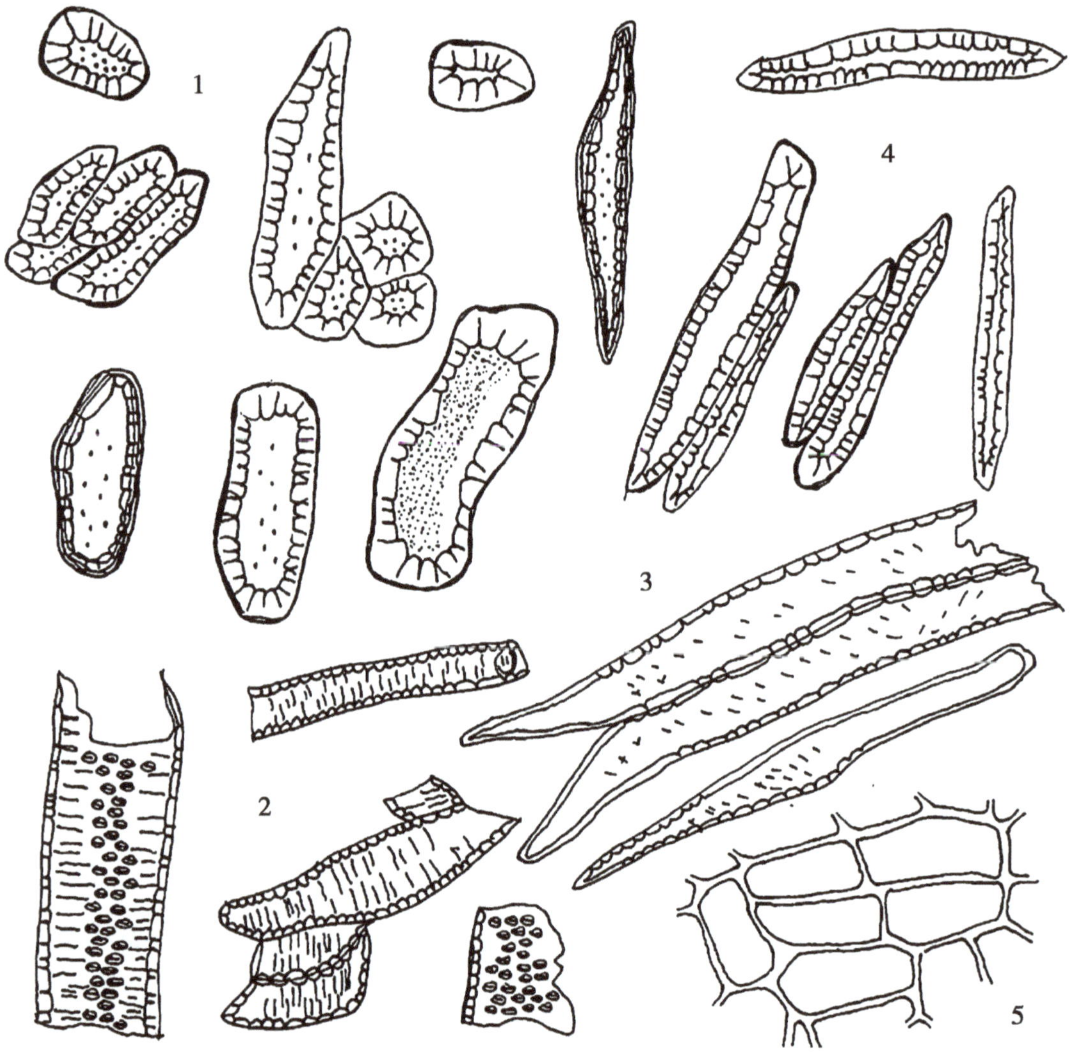

Fig. 12.3 Danshen Root Powder (× 220). *1* Stone cells; *2* vascular bundle; *3* wood fiber; *4* phloem fiber; *5* cork cells

the Simple Solution. Take 1 g of Danshen reference sample to make the Reference Herb Solution by the same method. Take tanshinone II_A, add acetidin to make a 2 mg/ml solution, and keep it as Reference Substance Solution. Run a test by TLC. Take 5 μl of each of the above solutions, apply on the same silica gel G thin-layer plate; develop in a solution of benzene–acetidine (19:1), take them out, and then dry in open air. At the corresponding location in the color spectrum of the Test Solution, the spot exhibits the same color as that of the reference herb solution; and at the corresponding location in the color spectrum of the reference substance solution, the spot exhibits the same dark red [3, 11] (Fig. 12.4).

Take 0.5 g of Danshen powder, add 10 ml of dichloromethane-methanol (5:1), and incubate for 2 h with constant shaking. Take the supernatant, concentrate to 1 ml, and apply 10 μl onto a TLC plate. Absorbent: high-performance silica gel GF_{254} thin-layer plate; chromogenic condition: ultra-violet lamp (254 nm). The developing solution of ligroin-cyclohexanone (8:1) can separate

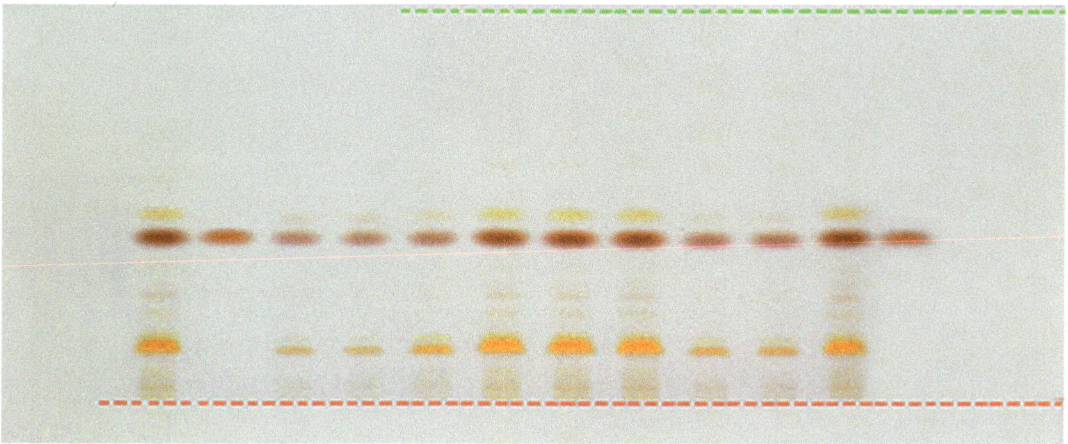

1 S 2 3 4 5 6 7 8 9 10 S

Fig. 12.4 Thin-layer Chromatogram of Tanshinone II_A. *2–9* Danshen herb samples; *1* and *10* Danshen herb reference; *S* tanshinone II_A reference

saprorthoquinone, tanshinone II_A, tanshinone I, cryptotanshinone I, and dihydrotanshinone; further developing with ligroin-absolute alcohol (7:1) can separate przewaqinone A and tanshinone II_B [1].

He studied the ethyl ether extract and ethanol extract of Danshen and *S. yunnanensis* roots: take 1 g of powder, add 5 ml of ethyl ether, mix well, incubate for 1 h, filtrate, volatilize the solution, add 1 ml of ethanol to the residue, forming the ethyl ether extract. Make the ethanol extract in the same way. Take tanshinone II_A reference substance, and add ethanol to make a 2 mg/ml solution. Apply 5 μl of each of the test solutions and reference substance solution on silica gel G thin-layer plates with benzene–acetidine (19:1) as the developing agent; inspect under sunlight [6].

Wang compared the chemical constituents of four different Danshen processed products: simple slices, simple stir-frying, soaking in alcohol and then drying, and soaking in wine and then stir-frying. Take 1 g of the powder, add 20 ml of ethyl ether, soak for 6 h, shake constantly, filtrate, and concentrate the filtrate to 1 ml. Take tanshinone II_A as the reference substance. Use silica gel G as the thin-layer plate, chloroform-acetidine (9:1) as the developing agent, and inspect under sunlight. Each sample showed 11 spots [20].

12.1.4.2 Identification of Water-Soluble Constituents by TLC

1. Take 0.2 g of Danshen powder, add 25 ml of 75 % methanol, heat under reflux for 1 h, filter, concentrate the filtrate to 1 ml, and use it as a test solution. Use 75 % methanol to make 1 mg/ml salvianolic acid B solution, and use it as a reference substance solution. Take 5 μl of each of the above solutions, apply to a silica gel GF_{254} thin-layer plate, develop in methyl benzene–chloroform–acetidin–methanol–formic acid solution (2:3:4:0.5:2), air dry the plate. Inspect under an ultraviolet lamp (254 nm or 365 nm). At the corresponding location in the spectrum of the test solution, the spot exhibits the same color as that of the reference substance solution [11] (Figs. 12.5 and 12.6).
2. Take 0.5 g of Danshen powder, add 30 ml of water, soak overnight, heat under reflux for 2 h, stand for a while, filter, adjust the filtrate to pH 2 with 10 % hydrochloric acid, extract with ethyl acetate (10 ml × 3 times), combine the acetidin extracts, concentrate, dissolve the residue with 2 ml of methanol, and use 5 μl for spotting. Absorbent: high-performance silica gel GF_{254} thin-layer plate; chromogenic conditions: under ultraviolet lamp (254 nm);

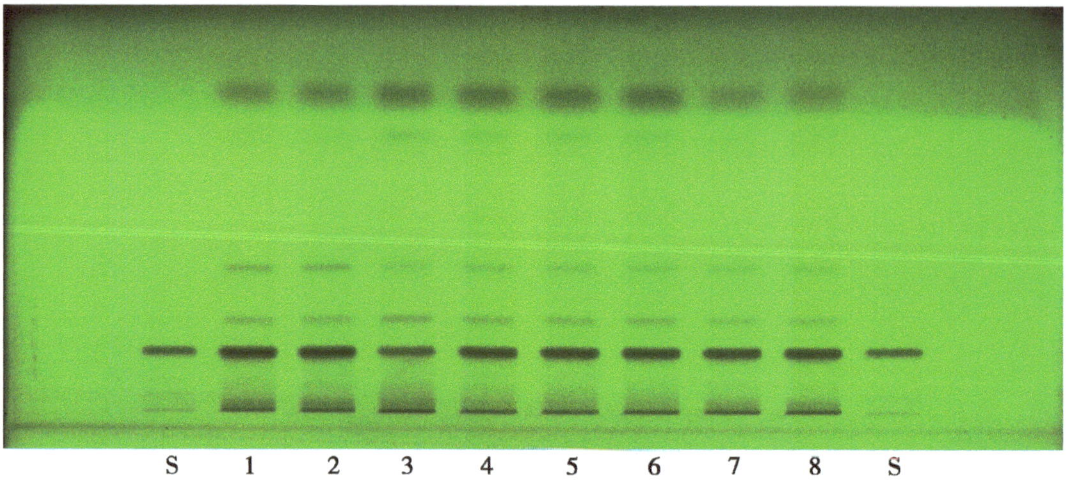

Fig. 12.5 Thin-layer chromatogram of salvianolic acid B in Danshen (254 nm). *S* Salvianolic acid B; *1–8* Danshen samples

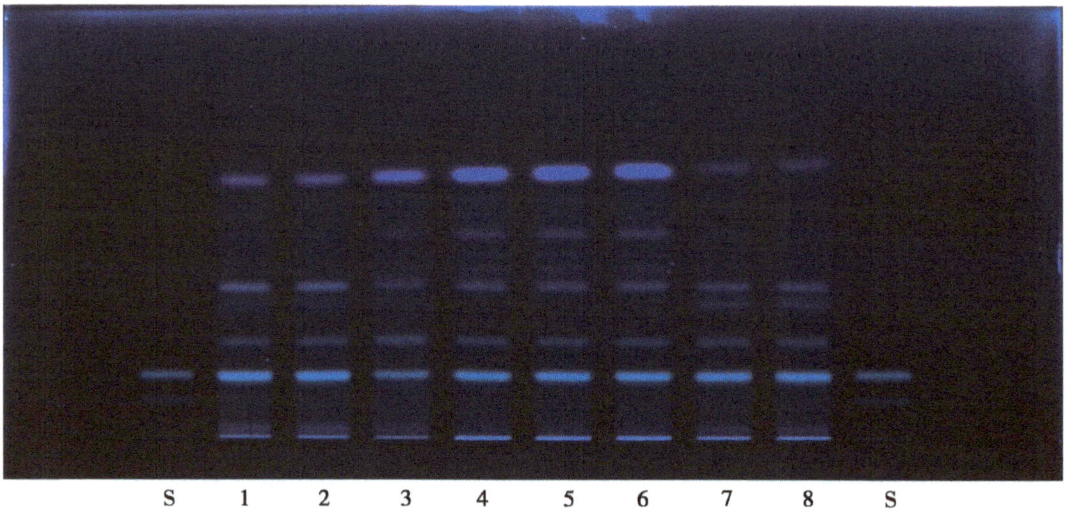

Fig. 12.6 Thin-layer chromatogram of salvianolic acid B in Danshen (365 nm). *S* Salvianolic acid B; *1–8* Danshen samples

developing solution: chloroform–acetidin–ben zoic–formic acid–methanol (1.5:2:1:1:3) can separate caffeic acid, methyl rosmarinate, salvianolic acid A, rosmarinic acid, salvianolic acid C, and salvianolic acid B; developing solution: chloroform–acetidin benzene–formic acid (2.4:2:1:0.6) can separate protocatechuic aldehyde [1, 3].

3. The water-soluble components of 17 different Danshen and Danshen-related medicinal herbs (including varieties) were compared with the following method. Using protocatechuic aldehyde, protocatechuic acid, and danshensu as reference substances [2]: take 1 g of Danshen powder (40 meshes), place in a 100 ml round bottom flask, add 50 ml of distilled water, extract on a boiling water bath for 1 h, decant the supernatant, add distilled water, extract twice more, half an hour per extraction. Combine the three water extracts,

concentrate to 5–10 ml, add ethanol to 80 %, stand overnight, filter, volatilize ethanol from the filtrate under decompression until no alcohol odor is left. Add 1 mol/L hydrochloric acid to adjust to a pH of 2, extract with ethyl ether five times, 15 ml per extraction, dehydrate the ethyl ether solution with anhydrous sodium sulfate, volatilize until dry, and dissolve the residue in 1 ml of ethanol. Develop with benzene–acetidin–formic acid (4:4:1). After the front moves 11 cm, take out, volatilize the solvent, and then develop with benzene–acetidin–formic acid (80:50:4), until the front moves 16.5 cm, take out and volatilize until dry. Observe under sunlight and under an ultraviolet lamp (254 nm), spray potassium ferricyanide-ferric chloride chromogenic agent, and then observe under sunlight.

4. Take 10 g of Danshen powder, add water twice, decoct for 30 min each time, filter, concentrate the filtrate to 20 ml, adjust to pH 2 with hydrochloric acid, extract with ethyl ether twice, combine the ethyl ether extracts, volatilize to 2 ml. With protocatechuic aldehyde and danshensu as the reference substances, silica gel G as the thin-layer plate, and benzene–acetidin–formic acid (16:10:1.6) as the developing agent, observe the fluorescence. Spread the plate with chromogenic agent ferric chloride-potassium ferricyanide and observe the color reaction. 10 fluorescence spots and eight chromogenic spots were obtained from Danshen slice and other processed products [20].
5. Liwei et al. used TLC to identify and differentiate medicinal Danshen and its related species, which were produced in different places, collected in different periods, and of different quality grades. By comparing the R_f values, colors, and sizes of TLC spots and fingerprints of the whole TLC, and quantifying the fingerprint characteristics, they could identify the authenticity and quality. Their method is as follows: take 1 g of the powder, add 50 ml of 2 % ammonia water, treat with ultrasound for 1 h, heat the solution on a boiling water bath for 45 min, adjust the extract to pH 3–4 with glacial acetic acid, filter, extract the filtrate with acetidin for three times, 20 ml each time, combine the acetidin solutions, volatilize, dissolve the residue with 1 ml of methanol, to create the water-soluble extract sample. Take 1 g of the powder, add 10 ml of methanol-dichloromethane (8:2), treat with ultrasound for 30 min, filter, and store the filtrate as the liposoluble extract sample. Take tanshinone II_A, tanshinone I, cryptotanshinone, protocatechuic aldehyde, protocatechuic acid, and danshensu as the reference substances; with high-performance silica gel GF_{254} (Merck) as the thin-layer plate, chloroform–acetidin–benzene–formic acid (24:20:10:6) as the water-soluble developing agent, and benzene–acetidin (19:1) as the liposoluble developing agent. Inspect under an ultraviolet lamp (254 nm) [21].
A new developing agent, ligroin (60–90 °C)-acetidin (8:2), could separate the spots of tanshinone II_A and tanshinone I, from those of other constituents [22].

12.1.5 Spectral Identification

Guoqing Wu, Xiaoqian Zhang, Rixin Liang and Manling Li

12.1.5.1 Ultraviolet Spectrum

Take 1 g of the sample powder, add 20 ml of water, shake, soak in cold water for 2 h, filter, dilute the filtrate to a suitable concentration, and determine the UV absorption spectrum. It is found that its maximum absorbance appears at 285 nm and the minimum absorbance appears at 266 nm [6].

Take 0.5 g of the sample powder, add 10 ml each of ligroin (60–90 °C), chloroform, and methanol, seal, shake, treat with ultrasound for 20 min, filter, dilute the filtrate to a suitable concentration, and determine the UV absorption zeroth-order and 2nd derivative spectra, with the corresponding reagent as a blank reference. The ligroin extract shows strong absorbance at 224 nm and a strong broad peak at 240–280 nm; the chloroform extract shows no obvious absorbance; and methanol extract exhibits an absorption peak at 281 nm [13].

Take 0.5 g of sample coarse powder, add 20 ml of ethanol, treat with ultrasound for 10 min, filter, dilute the filtrate, and determine; characteristic absorption peaks appear at 253 nm and 277 nm [7].

Take 1 g of the sample powder, place in a test tube with stopper, add 5 ml of ethyl ether, shake, stand for 1 h, filter, volatilize the filtrate, add acetidin to dissolve the residue, dilute, and determine the UV absorption spectrum; its maximum absorbance appears at 419 nm [14].

Liu [15] determined the ultraviolet absorption spectrum of Danshen extracted by ligroin, chloroform, dehydrated alcohol and water.

12.1.5.2 Near Infrared Spectrum

Take 2 g of sample coarse powder, extract with ligroin, ethyl ether, and water in order to obtain the corresponding extracts, concentrate and prepare the sample by potassium bromide window coating, and determine the infrared spectrum. The characteristic absorption peaks appear at 3,440.0, 2,925.5, 1,633.4, 1,411.7, and 1,062.6 cm^{-1} (width) [15].

Liu studied the Danshen products of nine habitats and nine species of the same genus by near infrared diffuse-reflectance spectrometry (NIRDRS) combined with cluster analysis and discriminant analysis. Dry the sample by baking at 60 °C for 12 h, pulverize, sift through a 80-mesh sieve, take about 2 g, place in a measuring glass, use gold foil as a reference in integrating sphere diffuse-reflectance, scan 64 times with a scanning scope of 3,700–12,000 cm^{-1} and resolution of 4 cm^{-1}, perform signature analysis on the NIRDRS of Danshen and another nine plants of the same genus by factor analysis, treat with rotation transformation etc., bring forward information with regularity, and perform identification by eigenvector projection. Danshen herbal materials from different habitats have very high similarity, while there are large differences between Danshen herbal materials and the materials of different species in the same genus [16].

Pulverize the herbal materials, sift through a 120-mesh sieve, dry at 60 °C to a constant weight, tablet with potassium bromide, determine the infrared absorption spectrum. The sample exhibits weak absorbance at 1,411.7 cm^{-1}, which is very different from that of fake Danshen [13].

Bake the herbal materials at 60 °C for 3 h, pulverize, sift through a 200-mesh sieve, tablet with potassium bromide, determine the infrared absorption spectrum. The spectrum of the discussed Danshen is very similar to that of cultivated Danshen [17].

Xiao [18] used convolution transform analysis of the near infrared spectrum of TCM drugs, and the obtained result was accordant with that of traditional plant taxonomy.

Li et al. [19] studied the phenolic acids of Danshen by ultraviolet spectroscopy, mass spectroscopy, nuclear magnetic resonance spectroscopy, and two dimensional NMR.

12.2 Identification of Danshen Preparations

12.2.1 Qualitative Identification of Dantonic™

Fengnan Cao

12.2.1.1 Identification of Danshen

Preparation of reference substance solution: Take an appropriate amount of salvianic acid A sodium reference substance, add acetidin to make a 1.0 mg/ml solution, and keep as the reference substance solution.

Preparation of test solution: Take 15 pills of the product, place in a centrifuge tube, add 1 ml of water and two drops of diluted hydrochloric acid, shake to dissolve, add 3 ml of acetidin, shake for 1 min and centrifugate for 2 min, and take the supernatant as the test solution.

Test procedure: The test is conducted according to the TLC method described in *Chinese Pharmacopoeia* 2005, volume 1, Appendix VI B. Take 10 μl of the test solution and 2 μl of the reference substance solution, apply on the same silica gel G thin-layer plate,

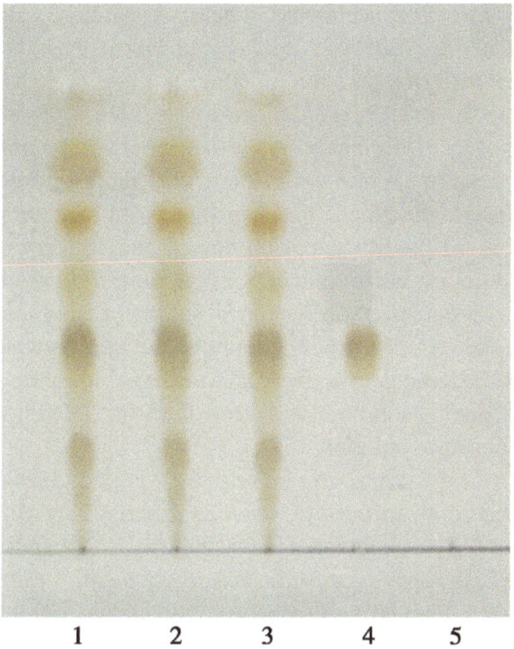

Fig. 12.7 Identification of danshensu in Dantonic™. *1–3* Sample solutions; *4* Salvianic acid A sodium reference substance solution; *5* Negative sample solution

develop with chloroform–acetone–formic acid (25:10:4) as the developing agent, take out and dry in open air; it exhibits light yellowish spots after being fumigated in ammonia vapor. Inspect the plate under an ultraviolet lamp (365 nm) after standing for 30 min. At the corresponding position in the spectrum of the reference substance solution, the spectra of the test solutions have fluorescent spots of the same color (Fig. 12.7).

12.2.1.2 Identification of Borneol

Chemical Identification

Take 15 pills of the product (if the pills have film coatings, crush to break the film), place in an evaporating dish, add 0.5 ml of water, cover the dish, and heat with fire for about 1 min. Drop 1 drop of freshly-prepared 1 % vanillin sulfuric acid onto the white sublimated substance adhered to the cover, and the drop will show a rose red color (Fig. 12.8).

TLC Identification

Preparation of test solution: Take an appropriate amount of the product and crush, add 10 ml of dehydrated alcohol to 1.0 g of the product, ultrasound for 10 min, filter, and keep the filtrate as the test solution.

Preparation of reference substance solution: Take an appropriate amount of borneolum syntheticum reference substance, add dehydrated alcohol to make a 1.0 mg/ml solution, and keep it as the reference substance solution.

Fig. 12.8 Color reaction of borneol

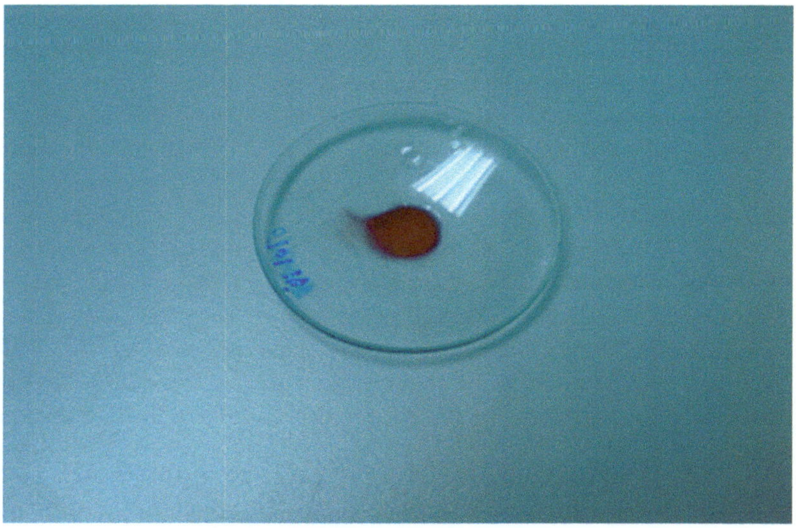

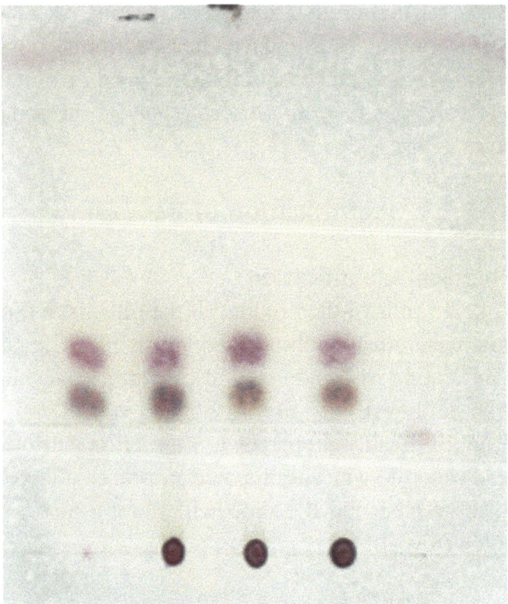

Fig. 12.9 TLC identification of Borneolum Syntheticum in Dantonic™. *1* Borneolum Syntheticum reference solution; *2–4* The test solutions of 3 different lots; *5* Negative reference solution

Test procedure: The test is conducted according to the TLC method described in *Chinese Pharmacopoeia* 2005, volume 1, Appendix VI B. Take 5 μl of each of the above solutions, apply on the same silica gel G thin-layer plate, develop with cyclohexane-acetidin (8.5:1.5), take out, dry in open air, spray with 1 % sulfuric acid-vanillin solution, and bake at 105 °C for 2–3 min. At the corresponding position in the spectrum of the reference substance solution, the spectra of the test solutions have spots of the same color (Fig. 12.9).

12.2.1.3 Identification of *Radix Notoginseng*

Preparation of test solution. Pretreatment method 1: Take 40 pills of Dantonic™, add 10 ml of methanol, shake for 10 min, place at 4 °C for 24 h to allow polyethylene glycol to be completely separated out, filter, and keep the filtrate as the test solution.

Pretreatment method 2: Take 20 pills of Dantonic™, place in a centrifuge tube, add 9 ml of diluted ammonia solution at about 40 °C (take 8 ml of concentrated ammonia solution, add water to 100 ml, mix well), shake to dissolve, centrifugate, take the supernatant, filter through a D101 macro-reticular resin column (60–80 meshes, with inner diameter of about 0.7 cm and length of about 5 cm), elute with 15 ml of water, discard the eluant, and then elute with methanol, discard the initial 0.4 ml eluant, and collect the following eluant (about 5 ml) as the test solution.

Preparation of reference substance solution. Take an appropriate amount of notoginsenoside R1 reference substance, add methanol to make a 2 mg/ml solution, and keep it as the reference substance solution.

Test procedure: The test is conducted according to the TLC method described in *Chinese Pharmacopoeia* 2005, volume 1, Appendix VI B. Take 15 μl of the test solution and 5 μl of the reference substance solution, apply on the same silica gel G thin player plate, develop with the upper phase of *n*-butanol–acetidin–water (4:1:5), take out, dry in open air, spray with 10 % sulfuric acid-ethanol solution, and heat at 105 °C for about 10 min. At the corresponding position in the spectrum of the reference substance solution, the spectra of the test solutions have spots of the same color (Figs. 12.10 and 12.11).

Notes on the Method

Pretreatment method 1 is convenient but time consuming. It can meet the basic requirements for the identification of *Radix Notoginseng* in Dantonic™. There are slight interferences in the identification of *Panax notoginseng* saponins from the other ingredients in Dantonic™.

Pretreatment method 2 uses a hot diluted ammonia solution and selectively absorbs saponin to resin, which is easy, time-saving, and has a perfect effect of eliminating the interferences by Danshen and the excipients in Dantonic™. Therefore, the spots of *P. notoginseng* saponins obtained by TLC are clearer and more specific.

The differences between the developing system of TLC of this method and that described in the *Chinese Pharmacopoeia*: the developing systems of *Chinese Pharmacopoeia* and ours were compared in the studies on quality standards, and both showed good separating results.

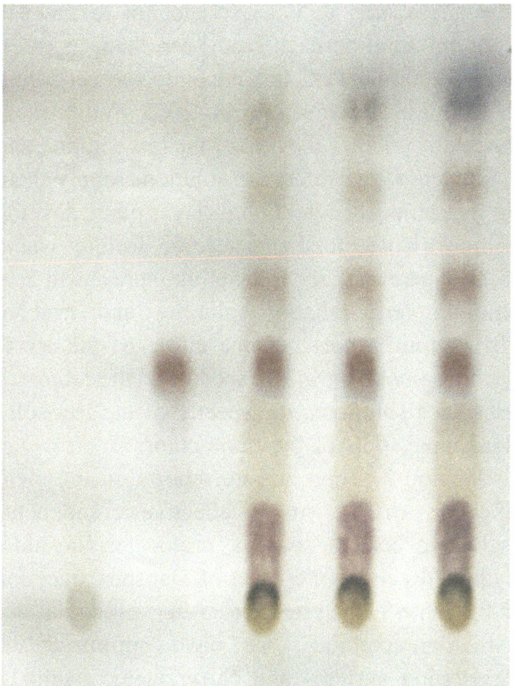

Fig. 12.10 Thin-layer chromatogram of notoginsenoside R_1 in Dantonic™ (Pretreatment method 1). *1* Negative sample solution; *2* Notoginsenoside R_1 reference substance solution; *3–5* The test solutions from 3 different lots of Dantonic™

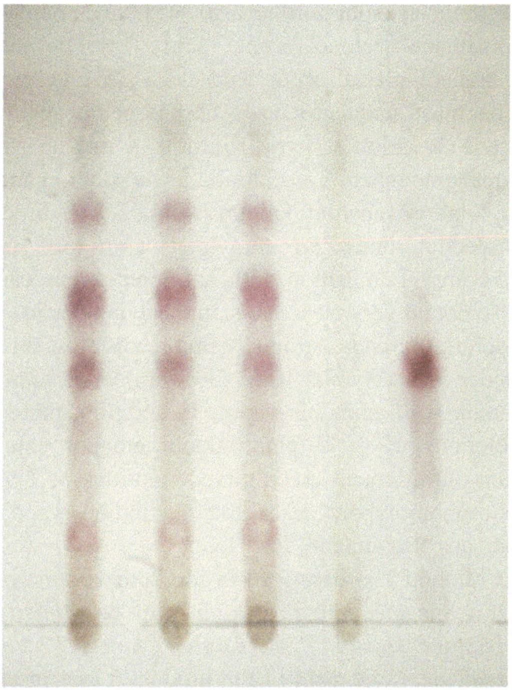

Fig. 12.11 Thin-Layer Chromatogram of *notoginsenoside* R_1 in Dantonic™ (Pretreatment method 2). *1–3* The test solutions from 3 different lots of Dantonic™; *4* negative sample solution; *5* Notoginsenoside R_1 reference substance solution

However, our system was easier to prepare and could be used without pre-refrigeration, and each constituent of Panax notoginseng saponins was well separated with round and intact spots, which is especially applicable to quality control during drug production.

12.2.2 Qualitative Identification of Danshen Injection

Ming Zhu, Zhangzhao Jin, Bilian Chen, Linke Ma and Qing Gong

12.2.2.1 Color Reaction

Take 2 drops of the product, add two drops of water, shake, add one drop of chloride ferric, and it shows a dirty green color.

Apply one drop of the product on a filter paper, dry in open air, inspect under an ultraviolet lamp (365 nm), and it exhibits blue fluorescence. It exhibits flavo-green fluorescence after fumigation with ammonia vapor for about 1 min.

12.2.2.2 Identification by Ultraviolet Absorption

Take 0.5 ml of the product, add water to 500 ml, mix well, and determine the UV absorbance according to the spectrophotometric method described in *Chinese Pharmacopoeia* 2005, Volume 1, Appendix A. The maximum absorbance appears at a wavelength of 281 ± 3 nm.

12.2.2.3 Identification by TLC

Overview. Danshen injection is an injection of Danshen root extracted with water and precipitated with alcohol; most of the compounds are water-soluble phenolic acids of Danshen roots, thus TLC is used primarily to identify these compounds. The most frequently used TLC developing agents for this purpose are benzene–acetidin–formic acid (8:5:0.8), trichloromethane–acetone–formic acid (8:1:1), and n-butanol–glacial acetic acid–water (5:1:4), etc. The most frequently used chromogenic agents are 2 % chloride ferric solution, a mixture of equal amounts of 2 % chloride ferric solution and 1 % ferricyanatum kalium solution, or direct inspection under an ultraviolet lamp (365 or 254 nm). Fumigation with ammonia vapor can also obtain very clear spots, and salvianic acid A sodium exhibits strong sapphire fluorescence under an ultraviolet lamp (365 nm) after fumigation with ammonia vapor. In addition, mixed solutions of 1 % phloroglucin ethanol solution–sulfuric acid (1:1), mixed solutions of 7 % phosphomolybdic acid–ethanol, iodine vapor, etc. are also suitable for TLC.

Method 1 is described in the 20th volume of *Drug Specifications Promulgated by the Ministry of Public Health, P. R. China, Chinese Patent Medicine*. Take 4 ml of the product, place in an evaporating dish, evaporate to dryness, add 1 ml of dehydrated alcohol to dissolve the product, and keep it as the test solution. Take the protocatechuic aldehyde reference substance, add dehydrated alcohol to make a 1 mg/ml solution, and keep it as the reference solution. TLC test is conducted according to *Chinese Pharmacopoeia* 2005, Volume 1, Appendix VI B. Take 2–5 μl of each of the above solutions, apply on the same silica gel G thin-layer plate, develop with benzene–acetidin–formic acid (8:5:0.8), take out, dry in open air, and spray with a mixture of 2 % chloride ferric and 1 % ferricyanatum kalium solution (1:1). At the corresponding position in the spectrum of the reference substance solution, the spectra of the test solutions have spots of the same color.

This method uses only protocatechuic aldehyde as the reference, and toxic benzene is used in the developing agent. So, it has much room for improvement.

Method 2 is based on method 1, using salvianic acid A sodium and protocatechuic aldehyde as references. Take 4 ml of the product, place in an evaporating dish, evaporate to dryness, add 1 ml of dehydrated alcohol to dissolve, and keep it as the test solution. Take salvianic acid A sodium and protocatechuic aldehyde reference substances, add dehydrated alcohol to make a 1 mg/ml solution each, and keep them as reference solutions. TLC test is conducted according to Chinese Pharmacopoeia 2005, Volume 1, Appendix VI B. Take 5 μl of the test solution and 2 μl each of the reference solutions, apply them on the same silica gel G thin-layer plate, develop with trichloromethane–acetone–formic acid (8:2:1), take out, dry in open air, spray with 2 % chloride ferric-ethanol solution, and heat at 105 °C until spots become clear. At the corresponding position in the spectra of the reference substance solutions, the spectra of the test solutions have spots of the same color.

Method 3. There are no reports on the identification of the major effective components (salvianic acid A sodium, protocatechuic aldehyde and salvianolic acid) of Danshen Injection in the same TLC system yet. After testing various extraction methods, we have optimized the developing systems and chromogenic methods, which are described below: Take 5 ml of the product, adjust the pH with diluted hydrochloric acid to 2.0, extract with acetidin twice, 10 ml each time, combine the acetidin extracts, evaporate to dryness, dissolve the residue with 2 ml of ethanol, and keep it as the test solution. Take 1 g of Danshen reference substance, add about 80 ml of water, decoct and keep boiling for 2 h, cool, centrifugate at 3,000 rpm for 10 min, take the supernatant, concentrate to about 25 ml, adjust the pH to 2.0 with diluted hydrochloric acid, extract with acetidin twice, 30 ml each time, combine the acetidin extracts, and evaporate to dryness. Dissolve the residue with 1 ml of ethanol, and keep it as the herb reference solution. Take an appropriate amount of salvianic acid A sodium, protocatechuic aldehyde and salvianolic acid B reference substances, add 20 % methanol to make a 1 mg/ml solution each, and keep them as the reference solutions. TLC tests are conducted according to Chinese Pharmacopoeia 2005, Volume 1, Appendix VI B. Take 2 μl each of the above solutions, apply on the same silica gel GF_{254} thin-layer plate, develop with trichloromethane–acetone–methanol–formic acid (8:2:1:1.5), take out, dry in open air, fumigate in ammonia vapor, and inspect under ultraviolet

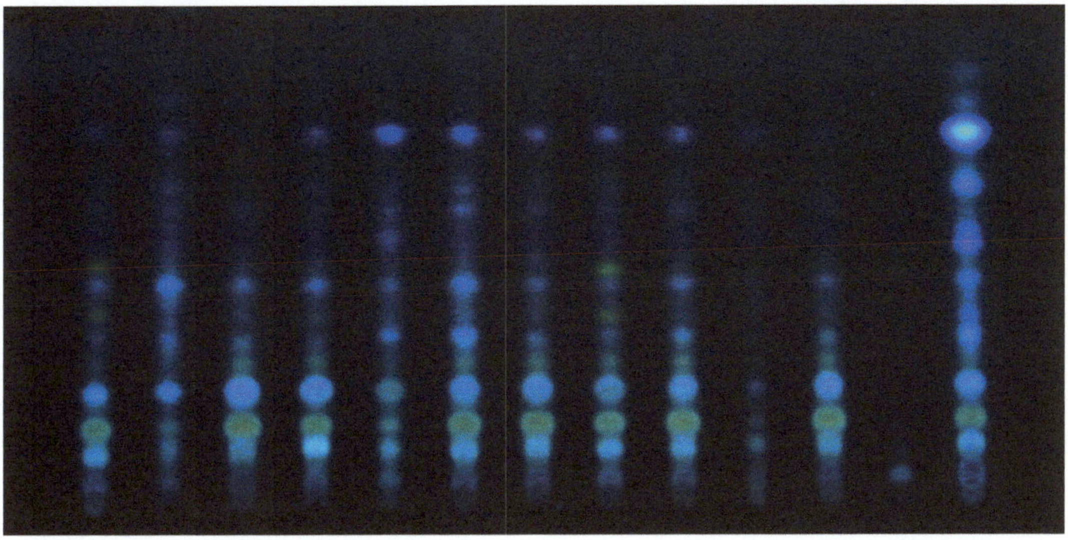

Fig. 12.12 Thin-layer chromatogram of Danshen Injection (un-fumigated, under UV 365 nm). *1–11* Danshen Injection; *12* Salvianolic acid B, salvianic acid A sodium, and protocatechuic aldehyde reference substances; *13* Danshen herb reference substance

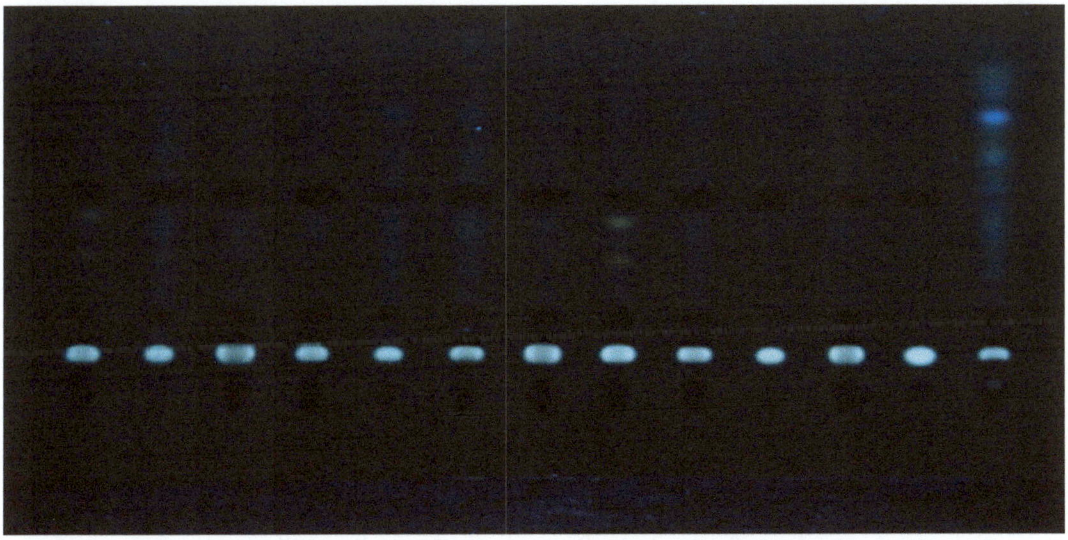

Fig. 12.13 Thin-layer chromatogram of Danshen Injection (fumigated with ammonia vapor, under UV 365 nm). *1–11* Danshen Injection; *12* Salvianolic acid B, salvianic acid A sodium, and protocatechuic aldehyde reference substances; *13* Danshen herb reference substance

lamps (365 and 254 nm). At the corresponding position in the spectra of the herb reference solutions, the spectra of the test solutions have spots of the same color. Spray with an equal proportion mixture of freshly prepared 1 % ferricyanatum kalium and 2 % chloride ferric, showing blue-black spots (Figs. 12.12, 12.13, 12.14 and 12.15).

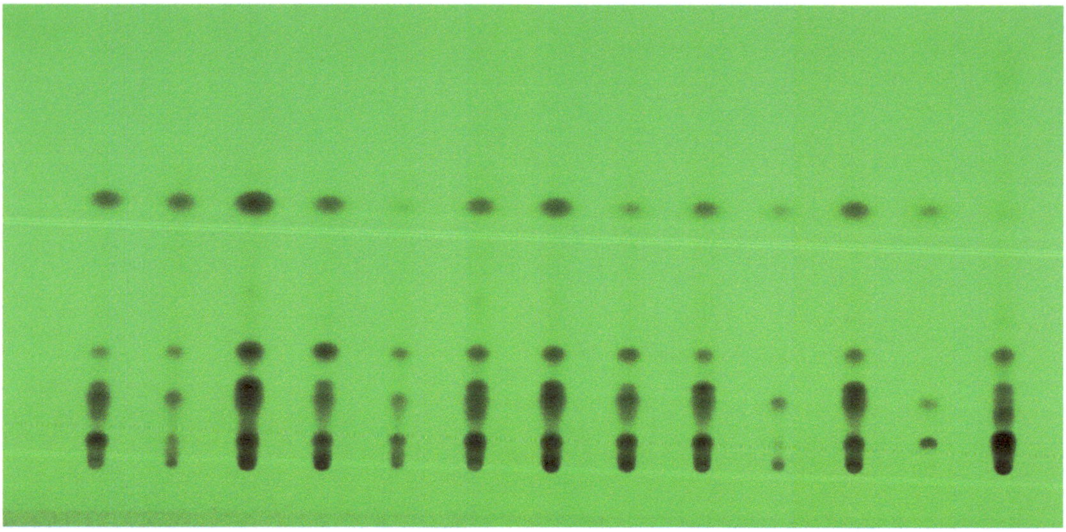

1 2 3 4 5 6 7 8 9 10 11 12 13

Fig. 12.14 Thin-layer chromatogram of Danshen Injection (fumigated with ammonia vapor, under UV 365 nm). *1–11* Danshen Injection; *12* Salvianolic acid B, salvianic acid A sodium, and protocatechuic aldehyde reference substances; *13* Danshen herb reference substance

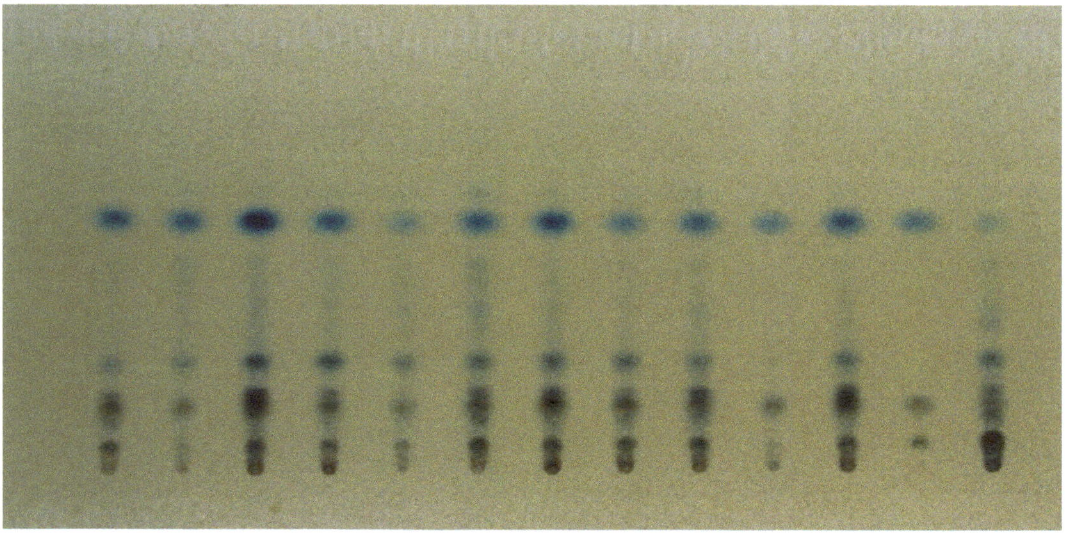

1 2 3 4 5 6 7 8 9 10 11 12 13

Fig. 12.15 Thin-layer chromatogram of Danshen Injection (Treated with potassium ferricyanide–ferric chloride mixed solution as chromogenic agent). *1–11* Danshen Injection; *12* Salvianolic acid B, salvianic acid A sodium, and protocatechuic aldehyde reference substances; *13* Danshen herb reference substance

Result analysis: judging directly from the above figures, the contents of the three major effective components in the Danshen Injection manufactured by different companies differ significantly. Digital photography can be used to extract and express TLC fingerprint characteristics. TLC has a particular advantage in the extraction and expression of the fingerprint features in Danshen Injection, but the technique is still in the development stage.

12.2.3 Qualitative Identification of Compound Danshen Tablet and Danshen Tablet by TLC

Shen Ji

12.2.3.1 Overview

Compound Danshen Tablet is made of Danshen roots, notoginseng roots, and broneolum syntheticum, and Danshen Tablet is made of Danshen roots. The effective constituents of Danshen can be grouped into the liposoluble part and the water-soluble part. The method is to identify liposoluble constituents, with ethyl ether as the extraction solvent, by ultrasonication and extraction after shaking, and with tanshinone II_A or Danshen herb as the reference.

12.2.3.2 Identification by TLC

Preparing Test Solution

Compound Danshen Tablet [12]. Take five tablets of the product, remove the coat, crush, add 10 ml of ethyl ether, treat with ultrasound for 5 min, filter, volatilize the filtrate to dryness, and dissolve the residue with 2 ml of acetidin.

Danshen Tablet [28]. Take 10 tablets of the product, remove the coat, crush, add 20 ml of ethyl ether, place in a test tube with stopper, shake, stand for 1 h, filter, volatilize the filtrate to dryness, and dissolve the residue with 1 ml of acetidin.

Preparing Reference Substance Solution

Compound Danshen Tablet [12]. Take tanshinone II_A reference substance, add acetidin to make a 0.5 mg/ml solution, and keep it as the reference substance solution;

Compound Danshen Tablet [28]. Take tanshinone II_A reference substance, add acetidin to make a 2 mg/ml solution, and keep it as reference substance solution. Take Danshen control herbal material, and make the reference herb solutions by the same method used to make the test solution.

Preparing Negative Blank Reference Substance Solution

Take the same amounts of ingredients other than Danshen to make a blank sample and to make a blank reference solution the same way the test solution is made.

Developing Condition of TLC

The plate: silica gel G plate with 0.5 % sodium carboxymethycellulose as the adhesive; developing agent: benzene–acetidin (19:1); inspection condition: under sunlight.

Method Validation

The blank reference solution shows no interference in the identification of Danshen in the preparations. So, the method is proven to be specific and feasible (Figs. 12.16 and 12.17).

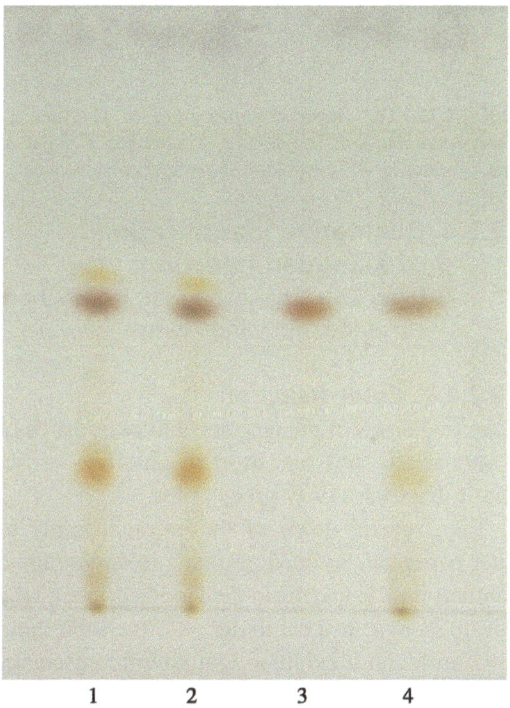

Fig. 12.16 Thin layer chromatogram of Danshen in Danshen Tablet. *1–2* Samples; *3* Tanshinone II_A reference substance; *4* Danshen herb reference substance

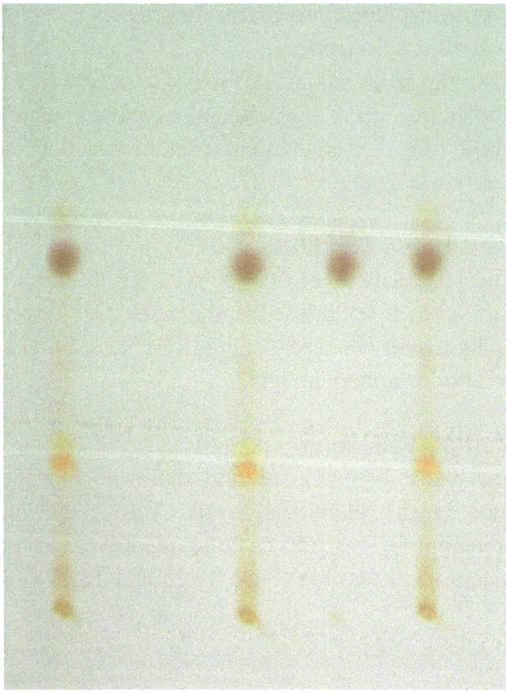

Fig. 12.17 Thin layer chromatogram of Danshen in Compound Danshen Tablet. *1, 3, 5* Samples; *2* Negative blank reference; *4* Tanshinone II_A reference substance

12.2.4 Qualitative Identification of Xiangdan Injection

Ming Zhu, Zhangzhao Jin, Bilian Chen, Linke Ma and Qing Gong

12.2.4.1 Color Reaction

Take two drops of the product, add twos drops of water, shake, add one drop of chloride ferrric, and it displays a dirty green color.

Take several drops of the product, apply to filter paper, dry in open air, hang in an ammonia solution flask (away from the liquid surface), take out in 20 min, inspect under an ultraviolet lamp (365 nm), and it exhibits light blue fluorescence.

12.2.4.2 Identification by UV Absorption

Take 0.5 ml of the product, add water to 500 ml, shake well, determine the UV absorbance according to the Spectrophotometric Method in *Chinese Pharmacopoeia* 2005, Volume 1, Appendix A, and its maximum absorbance appears at a wavelength of 281 ± 3 nm.

12.2.4.3 Identification by TLC

Overview. Xiangdan Injection is a compound injection made of Dalbergia aromatic water and the alcohol precipitate of the water extract of Danshen. The major effective components of Danshen are mainly the water-soluble phenolic acids, so the TLC test is primarily used to identify these phenolic acids. The most frequently used TLC developing agents for the identification of phenolic acids in Danshen preparations are benzene–acetidin–formic acid (8:5:0.8), trichloromethane–acetone–formic acid (8:1:1), and *n*-butanol–glacial acetic acid–water (5:1:4), etc. The most frequently used chromogenic agents are 2 % chloride ferric in ethanol, and an equiproportional mixture of 2 % chloride ferric and 1 % ferricyanatum kalium. Inspection can also be directly conducted under an ultraviolet lamp (365 nm or 254 nm). Clear spots can also be obtained by fumigation with ammonia vapor, and salvianic acid A sodium exhibits strong sapphire fluorescence under an ultraviolet lamp (365 nm) after fumigation. In addition, the following are also suitable for TLC: mixed solution of 1 % phloroglucin ethanol solution–sulfuric acid (1:1), mixed solution of 7 % phosphomolybdic acid–ethanol, iodine vapor, etc.

Method 1 is described in the 17th volume of *Drug Specifications Promulgated by the Ministry of Public Health, P. R. China, Chinese Patent Medicine*: Take 4 ml of the product, place it in an evaporating dish, evaporate on a water bath to dryness, add 1 ml of dehydrated alcohol to dissolve the residue, and keep the obtained solution as the test solution. Take the protocatechuic aldehyde reference substance, add dehydrated alcohol to make a 1 mg/ml solution, and keep it as the reference solution. TLC is performed according to *Chinese Pharmacopoeia* 2005, Volume 1, Appendix VI B. Take 2–5 μl of each of the above solutions, apply to the same silica gel G thin-layer plate, develop with benzene–acetidin–formic acid (8:5:0.8), take out, dry in open air, and spray 2 % chloride ferric–1 % ferricyanatum kalium solution (1:1). At the corresponding position in the spectrum of the test solution, the spot exhibits the same color as that of the reference solution.

This method uses only protocatechuic aldehyde as the reference, and toxic benzene is used in the developing agent. So, it has much room for improvement.

Method 2 is based on method 1, with salvianic acid A sodium and protocatechuic aldehyde used as reference substances. Take 4 ml of the product, place it in an evaporating dish, evaporate to dryness, add 1 ml of ethanol to dissolve, and keep it as the test solution. Take the salvianic acid A sodium and protocatechuic aldehyde reference substances, add dehydrated alcohol to make a 1 mg/ml solution for each, and keep them as the reference solutions. TLC is performed according to *Chinese Pharmacopoeia* 2005, Volume 1, Appendix VI B. Take 2–5 μl of each of the above solutions, apply to the same silica gel G thin-layer plate, develop with benzene–acetidin–formic acid (8:5:0.8) as the developing agent, take out, dry in open air, spray with 2 % chloride ferric-ethanol solution, and heat at 105 °C until spots become clear. At the corresponding location in the color spectrum of the test solution, the spot exhibits the same color as that of the reference solution.

Method 3. There are no reports on the identification of the major effective components (salvianic acid A sodium, protocatechuic aldehyde and salvianolic acid) in Xiangdan Injection in the same developing system. After optimizing the various extraction methods, the following developing systems and chromogenic methods have been developed. Take 5 ml of the product, adjust the pH with diluted hydrochloric acid to 2.0, extract with acetidin twice, 10 ml each time, combine the acetidin extracts, evaporate to dryness, dissolve the residue with 2 ml of ethanol, and keep it as the test solution. Take 1 g of Danshen herb material, add about 80 ml of water, decoct and keep boiling for 2 h, cool, centrifugate at 3,000 rpm for 10 min; take the supernatant, concentrate to about 25 ml, adjust the pH to 2.0 with diluted hydrochloric acid; extract with acetidin twice, 30 ml each time, combine the acetidin extracts, and evaporate to dryness. Take an appropriate amount of salvianic acid A sodium, protocatechuic aldehyde and salvianolic acid B reference substances, add 20 % methanol to make a 1 mg/ml solution of each, and keep them as the reference solutions. Conduct TLC. Take 2 μl of each of the above solutions, apply to the same silica gel GF_{254} thin-layer plate, develop with trichloromethane–acetone–methanol–formic acid (8:2:1:1.5), take out, dry in open air, fumigate in ammonia vapor, and inspect under ultraviolet lamps (254 nm). At the corresponding location in the color spectrum of the test solution, the spot exhibits the same color as that of the reference herb solution; spray with an equal proportion mixture of freshly prepared 1 % ferricyanatum kalium and 2 % chloride ferric, and the spots show a blue-black color (Figs. 12.18 and 12.19).

Result analysis: Analyzed directly from the figures, the content differences in the three compositions are significant. As TLC identification is undemanding in quantity, which method should be applied for quality control still remains a suspending problem for TLC identification.

12.2.5 Identification of Guanxinning Injection by TLC

Ming Zhu, Zhangzhao Jin, Chen Bilian, Linke Ma and Qing Gong

12.2.5.1 Color Reaction

Take two drops of the product, add two drops of water, shake, add one drop of chloride ferric, and it displays a dirty green color.

Apply one drop of the product on a filter paper, dry in open air, inspect under an ultraviolet lamp (365 nm), and it exhibits blue fluorescence. It exhibits flavo-green fluorescence after fumigation with ammonia vapor for about 1 min.

12.2.5.2 Identification by UV Absorption

Take 0.5 ml of the product, add water to 1000 ml, shake, determine the UV absorbance according to the Spectrophotometric Method in *Chinese Pharmacopoeia* 2005, Volume 1, Appendix A, and its maximum absorbance appears at a wavelength of 283 ± 2 nm.

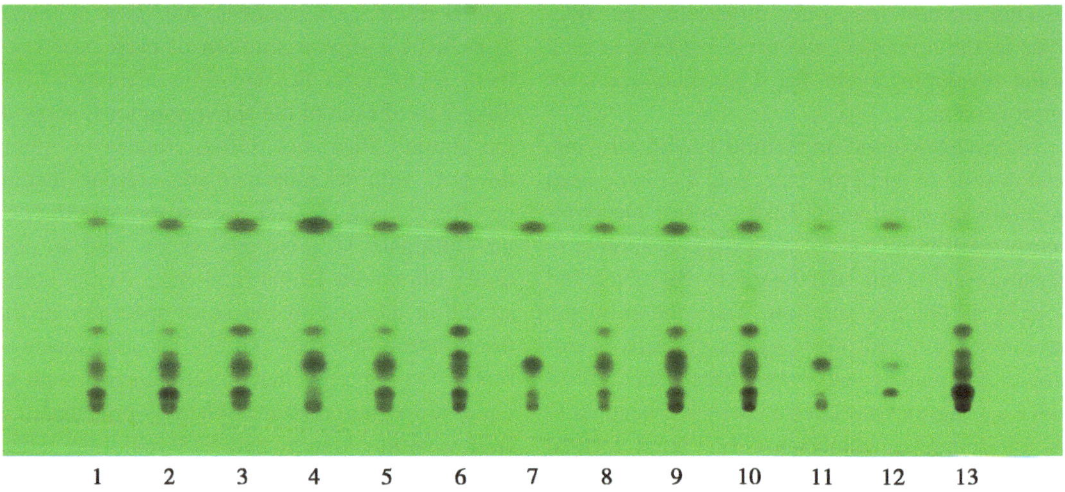

Fig. 12.18 Thin layer chromatogram of Xiangdan Injection (fumigated with ammonia vapor, under UV 254 nm). *1–11* Xiangdan Injection; *12* Salvianolic acid B, salvianic acid A sodium and protocatechuic aldehyde; *13* Danshen herb reference substance

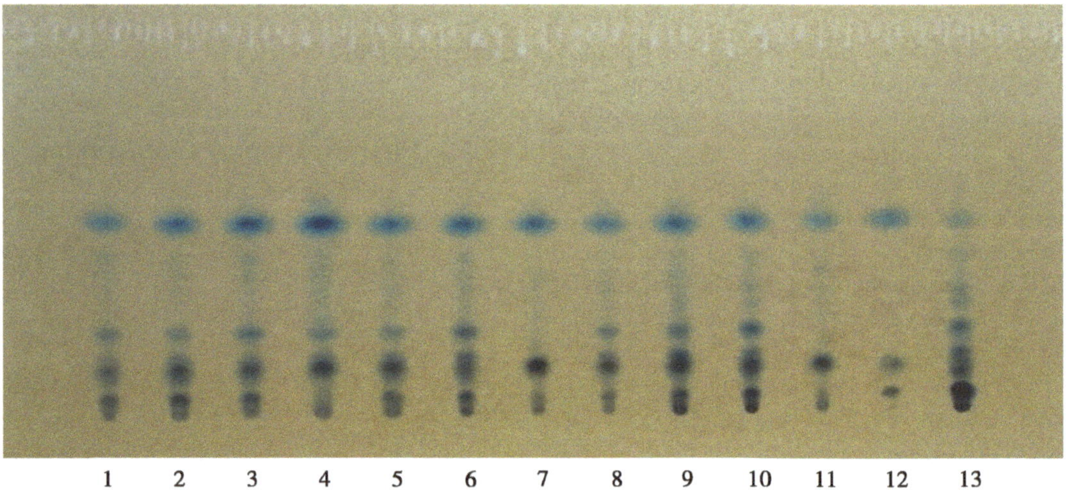

Fig. 12.19 Thin layer chromatogram of Xiangdan Injection (Treated with ferricyanatum kalium–chloride ferric solution as chromogenic agent). *1–11* Xiangdan Injection; *12* Salvianolic acid B, salvianic acid A sodium and protocatechuic aldehyde; *13* Danshen herb reference substance

12.2.5.3 Identification by TLC

Overview. Guanxinning Injection is an injection made of ethanol precipitates of the water extracts of Danshen and Sichuan lovage root. The effective components of Danshen in the preparation are mainly the water-soluble phenolic acids, so TLC is primarily used to identify these phenolic acids. The most frequently used TLC developing agents for the purpose are benzene–acetidin–formic acid (8:5:0.8), trichloromethane–acetone–formic acid (8:1:1), and *n*-butanol–glacial acetic acid–water (5:1:4) and so on. The most frequently used chromogenic agents are 2 % chloride ferric solution, and an equal proportion mixture of 2 % chloride ferric solution and 1 % ferricyanatum kalium solution. Inspection can also be directly conducted under an ultraviolet lamp (365 nm or 254 nm). Clear spots can also be obtained by fumigation with ammonia vapor, and salvianic acid A sodium exhibits strong sapphire fluorescence under an ultraviolet lamp (365 nm) after fumigation with ammonia vapor. In addition, mixed solution of 1 % phloroglucin ethanol solution–sulfuric acid (1:1), mixed solution of 7 % phosphomolybdic acid–ethanol, iodine vapor, etc., are also used.

Method 1. There are no reports on the identification of the major effective components (salvianic acid A sodium, protocatechuic aldehyde, and salvianolic acid) in Guanxinning Injection in the same developing system. After optimization of various extraction methods, developing systems and chromogenic methods, the following method is developed: Take 5 ml of the product, adjust the pH with diluted hydrochloric acid to 2.0, extract with acetidin twice, 10 ml each time, combine the acetidin extracts, evaporate to dryness, dissolve the residue with 2 ml of ethanol, and keep it as the test solution. Take 1 g of Danshen herb reference substance, add about 80 ml of water, decoct and keep boiling for 2 h, cool, centrifugate at 3,000 rpm for 10 min, take the supernatant, concentrate to about 25 ml, adjust the pH to 2.0 with diluted hydrochloric acid, extract with acetidin twice, 30 ml each time, combine the acetidin extracts, and evaporate to dryness. Take appropriate amounts of salvianic acid A sodium, protocatechuic aldehyde and salvianolic acid B reference substances, add 20 % methanol to make a 1 mg/ml solution of each, and keep them as the reference substance solutions. Take 2 μl of each of the above solutions, apply on the same silica gel GF_{254} thin-layer plate, develop with trichloromethane–acetone–methanol–formic acid (8:2:1:1.5), take out, dry in open air, fumigate in ammonia vapor until spots become clear, and inspect them under sunlight and under an ultraviolet lamp (254 nm), respectively. At the corresponding location in the spectrum of the test solution, the spot exhibits the same color as that of the reference herb solution (Figs. 12.20 and 12.21).

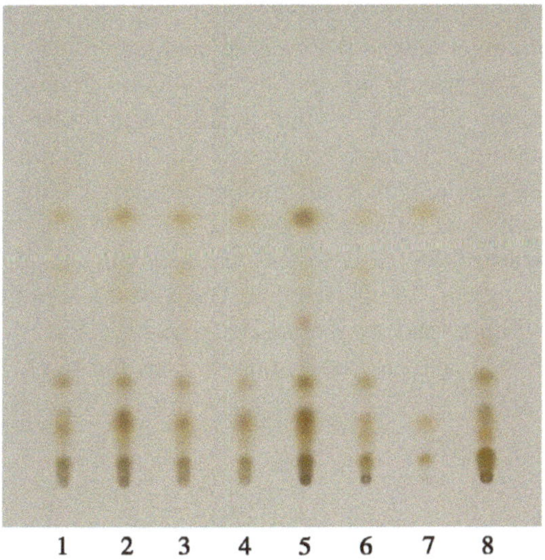

Fig. 12.20 Thin layer chromatogram of Guanxinning Injection (under sunlight). *1–6* Guanxinning Injection samples; *7* The reference substance solutions; *8* Danshen herb reference substance

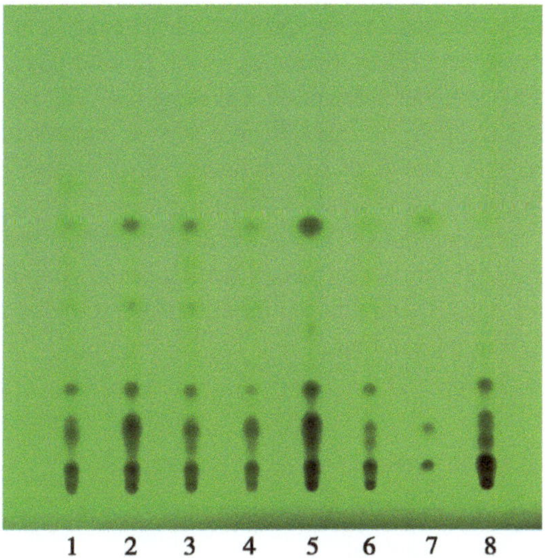

Fig. 12.21 Thin layer chromatogram of Guanxinning Injection (under an ultraviolet lamp, 254 nm). *1–6* Guanxinning Injection samples; *7* The reference substance solutions; *8* Danshen herb reference substance

12.2.6 Identification of Danshen Total Phenolic Acid Injection (Lypholized)

Zhengliang Ye and Jun Gao

12.2.6.1 Overview

Danshen Total Phenolic Acids Injection (Lypholized) is lypholized sterile powder for injection, manufactured from the active fraction of water-soluble components from the roots of Danshen. The main constituents are salvianolic acid B, E, rosmarinic acid, lithospermic acid, and other phenolic compounds. The identification tests are performed by means of phenolic hydroxyl group color reactions and the comparison of the HPLC retention time of salvianolic acid B and HPLC fingerprint.

12.2.6.2 Color Reaction

Take 10 mg of the product, dissolve with 1 ml of water, and add one drop of ferric chloride. The solution is dirty green in color.

12.2.6.3 HPLC

Chromatographic condition and system suitability test: Take octadecylsilane chemically bonded silica as the loading agent; acetonitrile–water–phos phoric acid (23.5:76.5:0.02) as the mobile phase; and test under a detection wavelength of 288 nm. The number of theoretical plates is calculated based on the peak of salvianolic acid B, and it should be no less than 5000.

Preparation of the reference solution: Precisely weigh the reference substance of salvianolic acid B, and add the mobile phase to make a 0.2 mg/ml solution.

Preparation of the test solution: Precisely weigh 20 mg of the product, transfer it into a 50 ml volumetric flask, add mobile phase, and dilute to scale. Shake well.

Assay: Precisely take 10 μl each of the reference solution and test solution, and inject into a liquid chromatograph. Run the machine. The test solution should present the chromatographic peaks which have the same retention time as those of the reference solution.

12.2.6.4 Fingerprint

Chromatographic condition and system suitability test: Take octadecylsilane chemically bonded silica as the loading agent (priority: phenomenex Luna150 mm × 4.6 mm, 5 μm column); acetonitrile–water–phosphoric acid (23.5:76.5:0.02) as the mobile phase; test under a detection wavelength of 288 nm, column temperature of 25 °C, and flow rate of 0.5 ml/min. The number of theoretical plates is calculated based on the peak of salvianolic acid B, and it should be no less than 5,000.

Preparation of the reference solution: Take an appropriate amount of salvianolic acid B reference substance, precisely weigh, and add mobile phase to make a 0.05 mg/ml solution.

Preparation of the test solution: Precisely weigh 250 mg of the product, add water to 100 ml, boil for 30 min, cool down, filter, wash the residue with small amount of water, combine with the filtrate, adjust the pH with 10 % hydrochloric acid to 2.0, extract with acetidin four times, 15 ml each time, combine the acetidin extracts, and evaporate to dryness. Add the mobile phase to dissolve the residue, transfer to a

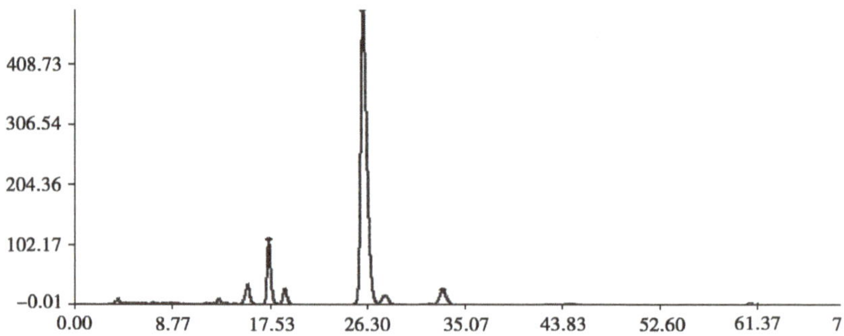

Fig. 12.22 Reference HPLC for fingerprint of Danshen total phenolic acid injection (Lypholized)

25 ml volumetric flask, dilute to scale, and shake well. Filter through a microporous membrane, and take the subsequent filtrate. This is the test solution.

Assay: Precisely take 10 μl each of the reference solution and test solution respectively, and inject into a liquid chromatograph. Run the machine.

Compare the fingerprints of the test solution and reference solution (Fig. 12.22), and calculate with simulation semblance computing software; the semblance should be 0.09–1.00.

References

1. Xiao PG. New book of traditional Chinese medicine. Beijing: Press of Chemical Industry; 2002.
2. Xu GJ. Collections and research on common Chinese crude drug. Fuzhou: Fujian Science and Technology Press; 2001.
3. Drug Administration of People's Republic of China. Modern Pragmatic Herbals. Beijing: People's Medical Publishing House; 1997.
4. Hu MH. Pharm J Chin People's Liberation Army. 2003;19(3):219.
5. Yang H. Shizhen Nat Med. 2004;15(7):417.
6. He GX. Chin Crude Drug. 1996;19(7):342.
7. Yan SQ. J Basic Trad Chin Med. 1999;13(1):27.
8. Yu LX. Brief Trad Chin Med Correspondence Course. 1994;3:46.
9. Zheng XK. J Experim Trad Chin Pharmacol. 2003;9(6):12.
10. Chen XF. Hunan J Trad Chin Med. 1996;2(3):40.
11. Wu ML. Identification of Chinese crude drug. Tianjin: Tianjin Science and Technology Press; 2001.
12. Committee of Chinese Pharmacopoeia. Chinese pharmacopoeia 2000, vol. 1. Beijing: Press of Chemical Industry; 2000.
13. Li HY. J Chin Med Mater. 1999;22(9):449.
14. Zhang H. J China Mod Pract Pharm. 2001;18(2):24.
15. Liu XH. Spectrum identification of Chinese crude drug. Shanghai: Press of Second Millitary Medical University; 2002.
16. Liu LL. Pharm Serv Res. 2002;2(1):23.
17. Yan JF. J Zaozhuang Normal Univ. 2000;17(5):53.
18. Xiao J. Analyt Chem. 2003;31(11):1295.
19. Institute of Materia Medica, Chinese Academy of Medical Sciences. Modern study on Chinese herbals. Vol 2. Beijing: Beijing Medical University/Peking Union Medical College United Press; 1996.
20. Wang WK. J Chin Med Mater. 1996;19(3):131.
21. Li W. World Sci Technol Modern Trad Chin Med. 2003;5(1):58.
22. Liu MY. Chin Patent Med. 2001;23(7):527.
23. Hong X. New Drug Clin Pharmacol Trad Chinese Med. 2000;11(5):286.
24. Huang TL. New Drug Clin Pharmacol Trad Chin Med. 2001;12(5):354.
25. Wang NS. New Drug Clin Pharmacol Trad Chinese Med. 2000;11(3):141.
26. Hong X. New Drug Clinical Pharmacol Trad Chin Med. 2002;13(4):24.
27. Huang YC. New Drug Clinical Pharmacol Trad Chin Med. 2002;13(3):174.
28. Committee of Chinese Pharmacopeia. Standards for Chinese patent medicines specified by Ministry of Health. Vol 10.

13 General Quality Control Methods

Xiaojian Zhang, Guoqing Wu, Rixin Liang, Manling Li, Xiaoqian Zhang and Shen Ji

13.1 Determination of Water Content

Zhang Xiaojian, Wu Guoqing, Liang Rixin, Li Manling and Zhang Xiaoqian

Take about 2 g of Danshen granules or chips (with diameter not more than 3 mm), precisely weighed, and place in a weighing bottle that has been dried to a constant weight; then dry it at 105 °C for 5 h before transferring to a desiccator to cool for 30 min. Precisely weigh and then dry at the above temperature for 1 h; cool, and weigh until the difference between two continuous weights is not more than 5 mg. Calculate the moisture content in the test product based on the lost weight. The moisture contents of the Danshen samples from some production regions are shown in Table 13.1, and the moisture contents of the Danshen samples from the Tasly Shaanxi Shangluo Danshen Base are shown in Fig. 13.1.

X. Zhang · G. Wu · R. Liang · M. Li · X. Zhang ·
S. Ji (✉)
Shanghai Municipal Food and Drug Administration,
Shanghai, China
e-mail: ji_shen@hotmail.com

13.2 Determination of Ash Content

Zhang Xiaojian, Wu Guoqing, Liang Rixin, Li Manling and Zhang Xiaoqian

13.2.1 Total Ash

Pulverize Danshen roots, sift through a 2# sieve, then place 4 g of Danshen powder in a crucible that has been burned to constant weight; weigh (accurate to 0.01 g), burn slowly till it is completely carbonized, place in a muffle furnace, gradually increase the temperature to 500 °C, and burn for 2 h until it is completely ashen to a constant weight. Calculate the percentage of the total ash in the test product based on the residue weight.

13.2.2 Acid-Insoluble Ash

Take the total ash obtained as above and place it in a crucible, add about 10 ml of diluted hydrochloric acid; cover the crucible with a watch glass, heat on a water bath for 10 min, rinse the watch glass with 5 ml of hot water which is added in the crucible, and then filter it through an ashless filter paper. Then transfer the residue to the filter paper from the crucible with water, and rinse until the obtained solution shows no chlorination reaction. Transfer the residue combined with the filter paper into a crucible, dry, burn slowly till it is completely carbonized; place in a muffle furnace,

X. Yan (ed.), *Dan Shen (Salvia miltiorrhiza) in Medicine*,
DOI 10.1007/978-94-017-9463-3_13,

Table 13.1 The water content in Danshen produced in different regions

No.	Place of origin	Moisture content (%)	No.	Place of origin	Moisture content (%)
1	Hebei	10.2	6	Drug market of Guangzhou	13.2
2	Zhengzhou, Henan	10.1	7	Drug market of Xi'an	10.4
3	Drug market of Chongqing	9.67	8	Linyi, Shandong	10.7
4	Anhui	10.0	9	Yangzhou, Jiangsu	10.4
5	Linxi, Neimenggu	10.9	10	Wujiang, Shanghai	10.9

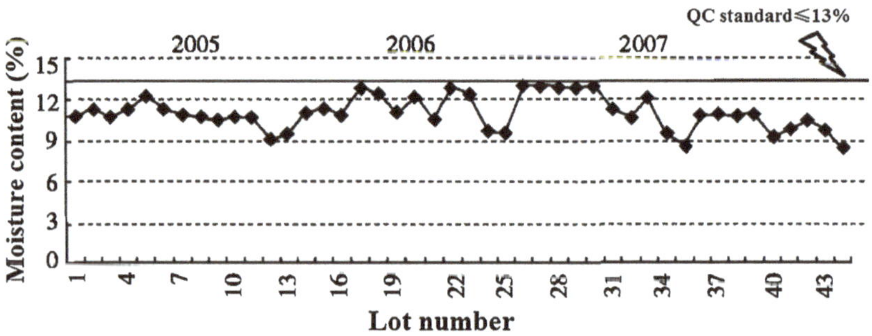

Fig. 13.1 The water content in Danshen produced in the Tasly Shaanxi Shangluo Danshen Base

Table 13.2 The content of total ash and acid-insoluble ash in Danshen produced in different regions

No.	Place of origin	Total ash (%)	Acid-insoluble ash (%)
1	Anhui	6.2	0.8
2	Chongqing	5.9	0.5
3	Linyi, Shandong	6.4	3.0
4	Shanghai	8.7	2.0
5	Zhengzhou, Henan	7.0	1.7
6	Yangzhou, Jiangsu	8.2	1.2
7	Hebei	8.5	2.4
8	Guangzhou	5.2	0.5

gradually increase the temperature to 500 °C, and burn for 1 h to a constant weight. Calculate the percentage of acid-insoluble ash in the test product based on the residue weight.

The test results show that the total ash content in Danshen is 5.2–8.7 %, with a mean of 7.0 %; acid-insoluble ash content is 0.5–3.0 %, with a mean of 1.5 % (Table 13.2).

The total ash content of Danshen produced in the Tasly Shaanxi Shangluo Danshen Base is 4.7–9.4 %, and the acid-insoluble ash content is 0.4–3.0 % (Figs. 13.2 and 13.3).

13.3 Determination of Extractives

Xiaojian Zhang, Guoqing Wu, Rixin Liang, Manling Li and Xiaoqian Zhang

13.3.1 Water-Soluble Extractives

Pulverize Danshen, sift through a $2^{\#}$ sieve, weigh 4 g and place in a 250–300 ml conical flask; precisely add 100 ml of water, seal, soak in cold water, shake constantly in the initial 6 h, and then let it stand for 18 h. Filter through a dry filter quickly, precisely measure 20 ml of filtrate, and place in an evaporating dish that has been dried to ca constant weight, evaporate on a water bath to dryness, then dry at 105 °C for 3 h, place in a desiccator, cool for 30 min, quickly and precisely

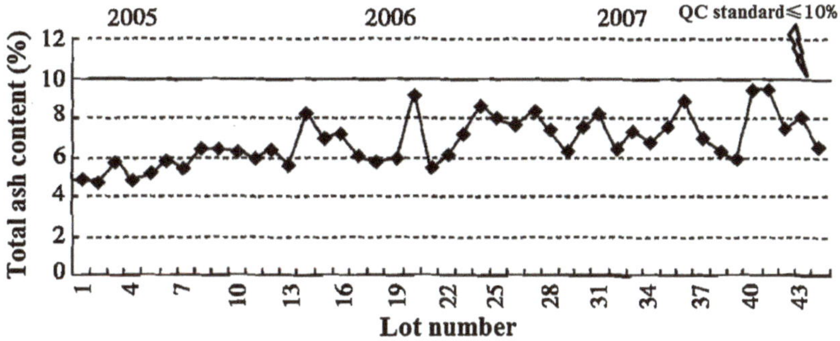

Fig. 13.2 The total ash content in Danshen produced in the Tasly Shaanxi Shangluo Danshen Base

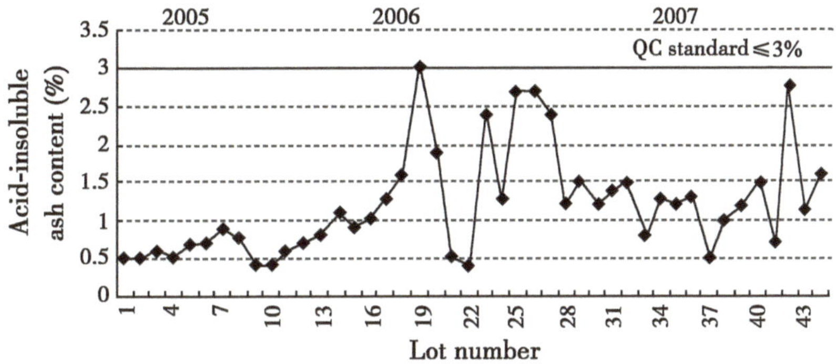

Fig. 13.3 Acid-insoluble ash content in Danshen produced in the Tasly Shaanxi Shangluo Danshen Base

Table 13.3 The content of water-soluble extractives in Danshen produced in different regions

No.	Place of origin	Water-soluble extractives content (%)	No.	Place of origin	Water-soluble extractives content (%)
1	Zhengzhou, Henan	36.6	6	Drug market of Xi'an	55.4
2	Drug market of Chongqing	41.5	7	Hebei	50.4
3	Anhui	43.9	8	Linyi, Shandong	55.3
4	Linxi, Neimenggu	52.8	9	Yangzhou, Jiangsu	59.5
5	Drug market of Guangzhou	40.4	10	Drug market of Hefei	56.7

weigh and calculate the water-soluble extractives content (%) in the test product based on the dried product (Table 13.3 and Fig. 13.4).

13.3.2 Alcohol-Soluble Extractives

Pulverize Danshen, sift through a 2# sieve, take about 2–4 g, weigh and place in a 100–250 ml conical flask; precisely add 50–100 ml of 95 % ethanol, seal, weigh, stand for 1 h, connect to reflux condensing tube, heat until boiling, and keep boiling for 1 h. Cool, take the conical flask off, seal, and weigh again; add 95 % ethanol to make up for the lost weight, shake, filter through a dry filter; then precisely measure 25 ml of filtrate, place in an evaporating dish that has been dried to constant weight, evaporate on a water bath to dryness, dry at 105 °C for 3 h, place in a desiccator, cool for 30 min, quickly and precisely weigh, and calculate the ethanol-soluble extractives content (%) in the test product based on the dried product (Table 13.4 and Fig. 13.5).

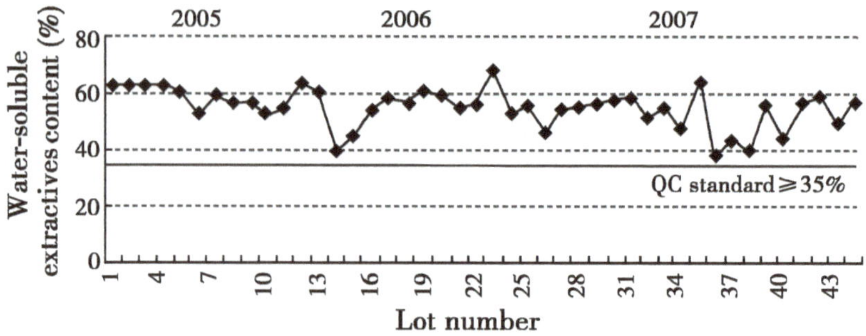

Fig. 13.4 The water-soluble extractives content in Danshen produced in the Tasly Shaanxi Shangluo Danshen Base

Table 13.4 The content of alcohol-soluble extractives in Danshen produced in different regions

No.	Place of origin	Alcohol-soluble extractives content (%)	No.	Place of origin	Alcohol-soluble extractives content (%)
1	Hebei	29.8	6	Drug market of Guangzhou	36.2
2	Zhengzhou, Henan	25.7	7	Drug market of Xi'an	30.8
3	Drug market of Chongqing	36.1	8	Linyi, Shandong	37.7
4	Anhui	36.0	9	Yangzhou, Jiangsu	32.7
5	Linxi, Neimenggu	23.3	10	Wujiang, Shanghai	33.9

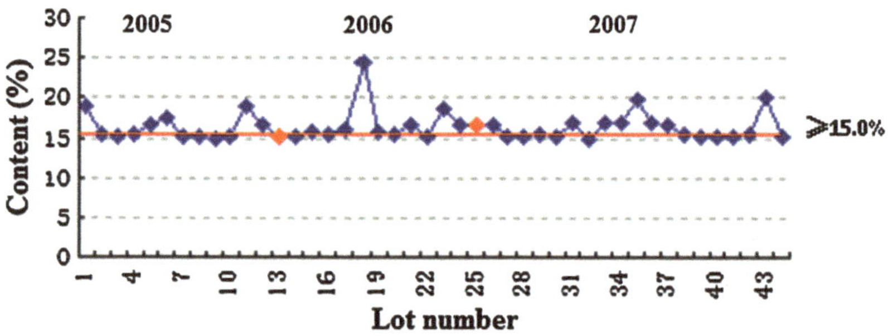

Fig. 13.5 The content of alcohol-soluble extractives in Danshen produced in the Tasly Shaanxi Shangluo Danshen Base

13.4 Heavy Metals

Shen Ji

13.4.1 Overview

The harmful elements possibly contained in Danshen also need to be strictly controlled. Here, the harmful elements primarily are copper, arsenic, chromium, mercury, lead, etc. At present, the methods for detecting these elements are primarily spectrophotometry, flame and graphite furnace atomic absorption spectroscopy, atomic fluorescence spectrometry, and inductively coupled plasma mass spectrometry (ICP-MS), etc. These methods have their own limitations: some need complicated sample pre-treatment, such as extraction, concentration and enrichment, or suppressing interference; some are time consuming and unable to implement multi-component or multi-element analysis; some fail to meet the requirements for detectability or sensitivity, etc. While compared with the traditional analytical techniques, the latest ICP-MS technique has many advantages, including high sensitivity, wide linear dynamic range, little interference, high accuracy, and rapid analysis speed.

The technique has become a universal method for element detection [1].

This section mainly introduces the methods for detecting harmful elements in Danshen by ICP-MS [2], which has been included in the appendix of *Chinese Pharmacopoeia* 2005, Volume 1.

13.4.2 Detection and Determination of Heavy Metals

13.4.2.1 Instruments and Reagents

ICP-MS, microwave digestion system, nitric acid (G.R.), water (deionized water), single element standard solutions (Cu, As, Cd, Pb, and Hg, 100 μg/ml each), and internal standard solutions (Li, Sc, Ge, In, and Bi, 10 μg/ml each).

13.4.2.2 Preparation of the Test Solution

Take about 0.5 g of the test product and add 5–10 ml nitric acid; seal and digest the sample according to the program; cool, transfer the obtained solution into a 50 ml volumetric flask, rinse the digestion tank with a little water for 3 times, and combine the rinse solutions into the volumetric flask; add 200 μl of 1 μg/ml standard gold solution and dilute with water to the scale; mix well. This is the test solution.

13.4.2.3 Preparation of the Reference Solution

Precisely measure appropriate amounts of standard arsenic, chromium, lead and copper single element standard solutions respectively, and dilute with 10 % nitric acid to make standard solutions containing 0, 1, 5, 10 and 20 ng/ml of arsenic and lead, 0, 0.5, 2.5, 5 and 10 ng/ml of chromium, and 0, 50, 100, 200 and 500 ng/ml of copper, respectively. Precisely take an appropriate amount of mercury single element standard solution and dilute with 10 % nitric acid to make solutions containing 0, 0.2, 0.5, 1, 2 and 5 ng/ml of mercury. These are the reference solutions (they should be prepared just before use).

13.4.2.4 Assaying Method

The isotopes selected for the detection are ^{63}Cu, ^{75}As, ^{114}Cd, ^{202}Hg and ^{208}Pb. Among these isotopes, ^{72}Ge is taken as the internal standard of ^{63}Cu and ^{75}As; ^{115}In as the internal standard of ^{114}Cd, and ^{209}Bi as the internal standard of ^{202}Hg and ^{208}Pb. A suitable correction equation is selected to correct the results according to the requirements of different instruments.

Determination by standard curve method. With the ratio of the readings of each concentration of the reference solution to those of the internal standard as the *Y*-axis and the corresponding concentration as *X*-axis, plot the standard curve of each element respectively. Measure the ratio of the readings of the test product to those of the internal standard, calculate based on the standard curve, and obtain the corresponding concentration. Run a blank test simultaneously, deduct the corresponding concentration of the blank solution from that of the sample, and obtain the content of each element respectively.

13.4.3 Detection of Heavy Metals in Danshen

13.4.3.1 Danshen Herb

Collect 10 batches of Danshen roots from different sources, and assay for heavy metal contents according to the methods described above. The results are as are shown in Table 13.5, Figs. 13.6 and 13.7.

As revealed by the results, the contents of copper, arsenic, chromium, mercury and lead in Danshen from different sources differ significantly, but all of them are under the standard limits.

13.4.3.2 Danshen Preparations

The test results of the Compound Danshen Tablet manufactured by Nantong Jinghua Pharmaceutics Co., Ltd, (Lot No. 0309022): ^{63}Cu: 2.2 μg/tablet, ^{75}As: 0.16 μg/tablet, ^{114}Cd: 0.0043 μg/tablet, ^{202}Hg: 0.043 μg/tablet, ^{208}Pb: 0.068 μg/tablet.

Table 13.5 Heavy Metal Content in Danshen (μg/kg)

Place of origin	^{63}Cu	^{75}As	^{114}Cd	^{202}Hg	^{208}Pb
Sanshangou Village, Tongjing Town, Yinan, Shandong	10776.0	650.8	105.3	30.0	1649.6
Bought from Guo'an Market	7623.1	302.7	25.4	4.9	320.9
Mengyang Country, Linyi, Shangdong	5478.5	167.5	13.2	0.4	286.5
Henan	19038.3	1322.4	116.7	79.3	3084.4
Shiquan 2nd Village, Zhongjiang, Siquan	6986.3	263.4	70.7	3.6	326.4
Dazhangzhuang Village, Tongjing Town, Yinan, Shandong	9574.7	254.7	32.1	2.6	689.4
Pingyi Country, Linyi, Shangdong	10943.7	1847.5	43.7	81.2	802.8
Dongyuezhuang Village, zhuanbu Town, Yinan Country, Linyi, Shandong	11481.5	427.5	33.0	10.2	726.3
Lunan, Henan	12364.1	409.2	143.6	8.2	1201.6
Shiquan 4th Village, Zhongjiang, Siquan	9783.0	290.4	102.5	1.3	387.2

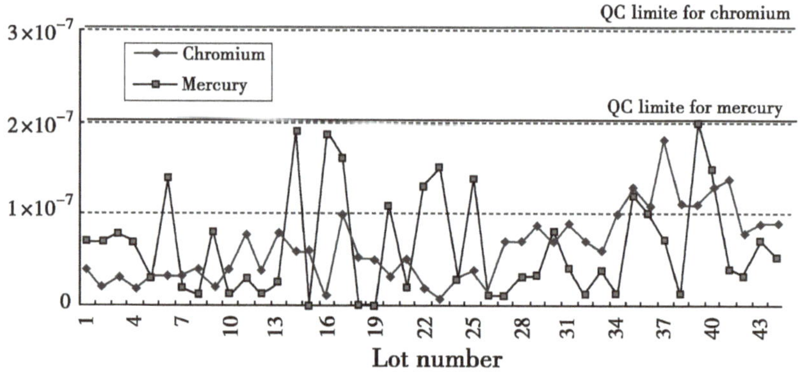

Fig. 13.6 The contents of mercury and cadmium in Danshen produced in the Tasly Shaanxi Shangluo Danshen Base

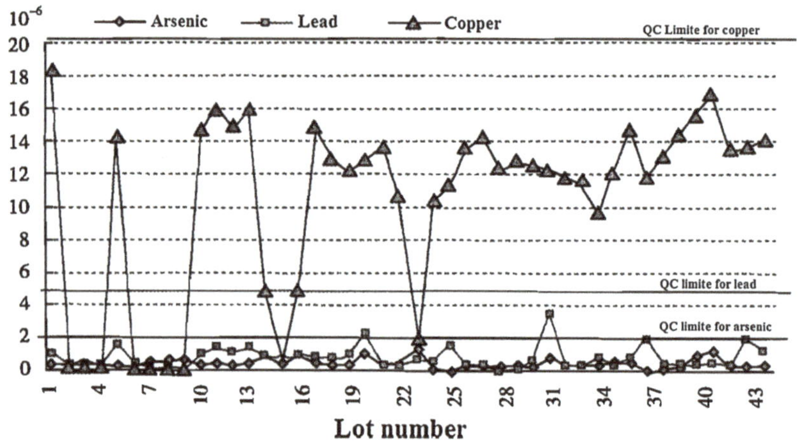

Fig. 13.7 The contents of arsenic, lead, and copper in Danshen produced in the Tasly Shaanxi Shangluo Danshen Base

13.4.4 Method Validation

1. Pre-treatment: The sample digestion programs, sampling quantities, nitric acid dosages, etc., were investigated, and the results showed that the above method could digest the samples completely, and the sensitivity met the requirement. The method is easy and safe.

2. Linearity: The standard curves of the five elements drawn after determination had good linear relationships, and the correlation coefficients were in the range of 0.9990 and 0.9999.
3. Repeatability: Under the requirement of trace analysis, the repeatability result of the method was very good and could meet the test requirements, and the relative standard deviations were all less than 10 %.
4. Recovery: The required trace analysis recovery rates were usually between 70 and 120 %. Recovery tests were run on each element at 3 concentration levels, and the recovery rates of copper, arsenic, chromium, mercury and lead were 118, 104, 107, 120, and 104 % ($n = 6$), respectively.
5. Stability: The test results showed that the stability of all the elements except for mercury was good in the solutions for 3 days. The mercury content showed a large deviation in the first 4 days, which was related to its unstable characteristic. Therefore, it is suggested that when running routine detections, especially the detection of mercury, prepare the test solution just before use.

13.5 Pesticide Residues

Shen Ji

13.5.1 Overview

Pesticide residue is the general term for the trace amounts of pesticides, toxic metabolites and foreign matter that remain in organisms, agricultural byproducts and the environment after pesticide usage. It is expressed in mg (or μg, ng, etc.) per 1 kg sample.

TCM drug manufacturing is a fast-growing branch of the pharmaceutical industry. TCM drugs come from the natural world, and pesticides have been widely used in Chinese crude drug cultivation for a long time. Its pollution of the natural environment is prone to cause excessive pesticide residues in these drugs. The constituents of Chinese herbs are complex, and many of them have similar physico-chemical properties to those of pesticides. On the other hand, the pesticide residuals in the herbs are relatively low, in the ultratrace range. Furthermore, TCM drugs have a great variety of preparations, which make the extraction, separation, purification and enrichment of the pesticides difficult. All of these factors have negative effects on the detection and determination of pesticide residuals, making the work more complicated than in foods and farm products. So far, no universalized detection methods and standards for pesticide residues in TCM herbs have been established at home or abroad. However, western developed countries such as the European Union, the United States, Japan, and Canada have been proposing more and more strict requirements for many indicators, including pesticide residues, in imported goods to protect the economic interests and the health of their people. Therefore, to protect people's health and meet the needs of foreign trade, it is extremely important to make an utmost effort to study the techniques for detecting pesticide residues.

In recent years, China has made great progress in the research and development of the technique to detect pesticide residues. In *Chinese Pharmacopoeia* 2000, Volume 1 appendix, "Assay Method for Organochlorine Pesticide Residues" was included for the first time. Even though only 2 herbs, namely astragalus root and licorice root, were stipulated for the assay, it nonetheless filled the gap in the area in *Chinese Pharmacopoeia*, and that was of great significance. In the appendix of Volume 1, *Chinese Pharmacopoeia* 2005, "Assay Method for Orgnaophosphorus Pesticide Residues" and "Assay Method for Pyrethroid Pesticide Residues" were included. Although significant progress has been made in pesticide residue assaying techniques, when compared with the international standards, China is still falling behind. For example, in *US Pharmacopoeia*, assays of 24 organochlorine pesticides, 17 orgnaophosphorus pesticides and 5 pyrethroid pesticides are specified, while in *Chinese Pharmacopoeia*, only 9 organochlorine pesticides, 12 organophosphorus pesticides and 3 pyrethroid pesticides are specified. Also, in the aspect of the validation of positive results, the *US Pharmacopoeia* specifies that 2 chromatographic columns with different polarities or gas chromatography mass spectrometry (GC-MS) be used, but there is still no concrete regulation in *Chinese Pharmacopoeia* for result validation.

This section is to emphasize the introduction of the methods developed in recent years for the detection of organochlorine, organophosphorus, and pyrethroid pesticides in Danshen herb and Danshen preparations. These methods have all been included in the appendix of Volume 1, *Chinese Pharmacopoeia* 2005.

13.5.2 Organochlorine Pesticide Residues

Organochlorine pesticide residues in Danshen roots are mainly detected according to the "Assay Method for Organochlorine Pesticide Residue" in Appendix IX Q of Volume 1, *Chinese Pharmacopoeia* 2005 [3].

13.5.2.1 Instruments and Reagents

Gas chromatograph (with ^{63}Ni electron capture detector), rotary evaporator, ultrasonic cleaner, centrifuge, nitrogen blowing meter, sodium chloride (A.R.), anhydrous sodium sulfate (A.R.), dichloromethane (P.R.), acetone (P.R.), ligroin (60–90 °C) (P.R.), and sulphuric acid (A.R.).

13.5.2.2 Preparation of the Test Solution

Dry the test sample at 60 °C for 4 h, and pulverize it into a fine powder. Precisely weigh about 2 g, place in a 100 ml conical flask with stopper, and add 20 ml of water, soaking overnight. Add 40 ml of acetone, weigh and treat with ultrasound for 30 min; cool, make up for the lost weight with acetone, add about 6 g of sodium chloride, then precisely add 30 ml of dichloromethane. Weigh, treat with ultrasound for 15 min, and make up for the lost weight with dichloromethane. Let stand to let the phases separate. Rapidly transfer the organic phase into a 100 ml conical flask with a stopper that is filled with an appropriate amount of anhydrous sodium sulfate, and stand for 4 h. Precisely measure 35 ml of this, concentrate on a 40 °C water bath under reduced pressure to approximate dryness, and add a little ligroin. Repeat the above procedures till dichloromethane and acetone are completely eliminated, and then dissolve with ligroin. Then transfer to a 10 ml graduated centrifuge tube with stopper, and add ligroin to 5 ml. Carefully add 1 ml of sulphuric acid, shake for 1 min, and centrifugate at 3,000 rpm for 10 min. Precisely measure 2 ml of the supernatant, place in a concentration flask with scales, connect a rotatory evaporator, concentrate the solution at 40 °C (or with nitrogen gas) to an appropriate amount, and precisely dilute to 1 ml. This is the test solution.

13.5.2.3 Preparation of the Reference Solutions

Precisely weigh appropriate amounts of benzene hexachloride (BHC) [α-BHC, β-BHC, γ-BHC, δ-BHC), dichloro-diphenyl-trichloroethane (DDT) [p,p′-DDE, p,p′-DDD, o,p′-DDT, p,p′-DDT), and quintozene (PCNB) pesticide reference substances, make with ligroin (with boiling point ranging from 60 to 90 °C) into solutions with concentrations of 0, 1, 5, 10, 50, 100 and 500 μg/L, respectively. These are the reference solutions.

13.5.2.4 Assay Method

Instruments: Fused silica capillary column SE-54 (30 m × 0.32 mm × 0.25 μm) (or DB-1701), and ^{63}Ni electron capture detector. Temperature of sample injection port: 230 °C; temperature of the detector: 300 °C. Splitless injection. Temperature programming: initial temperature 100 °C, increase the temperature 10 °C/min–220 °C, and then 8 °C/min–250 °C, and maintain for 10 min. As per α-BHC peak, the amount of theoretical plate should not be less than 10^6, and the separating degree of 2 adjacent peaks should be more than 1.5.

Precisely take 1 μl of each the test solution and the mixed reference solution with the corresponding concentrations, respectively, continuously inject 3 times, take the mean value of the 3, and calculate the 9 pesticide residues in the test product by external standard method.

13.5.2.5 The Results

Cong Xiaodong et al. used p-nitrotoluene as the internal standard, 3 m × 3 mm ID OV-17 packed column, sample injection port and detector temperature at 320 °C, ECD as the detector, and highly pure nitrogen as the carrier gas, and assayed the residuals of organochlorine pesticides (benhexachlor and dichloro-diphenyl-trichloroethane) in 10 Danshen herb samples. The samples were extracted with ligroin under ultrasound, and sulfonated and purified by sulfuric acid. The recovery rate of each isomer was 85.21–103.30 %, linear range 5–900 ng, and the detected isomers were in the range of 5.879×10^{-3}–0.4644 μg/g [4].

Wang Minjuan et al. determined the residues of benzene hexachloride and DDT in compound Danshen Tablet by high pressure gas chromatography. The results showed that when measuring the isomers and metabolites of benzene hexachloride and DDT, the detection limit was 1.8×10^{-4}–8.0×10^{-3} μg/g. The average recoveries were 91.43 % ± 1.7 and 90.88 ± 2.84 %, respectively, and relative standard deviations were 0.29–9.93 %. The residual amount of organochlorine pesticides in compound Danshen tablet was lower than the limit standards of all countries in the world [5].

13.5.3 Organophosphorus Pesticide Residues

13.5.3.1 Instruments and Reagents

Gas chromatograph (with nitrogen phosphorous detector), rotary evaporator, multifunction vacuum sample processor, activated carbon cartridge (120/400-mesh graphite carbon filler 0.25 g, inner diameter 0.9 cm cartridge, 3 ml), nitrogen evaporator, anhydrous sodium sulfate (A.P.), acetidin (P.R.), and N-hexane (P.R.).

13.5.3.2 Preparation of the Test Solution

Take 5.0 g of the test product powder, sift through a 20-mesh sieve, precisely weigh, add 5 g of anhydrous sodium sulfate and 50–100 ml of acetidin, place on an ice bath and treat with ultrasound for 3 min, let it stand for a while, take the upper layer liquid, filter, add 30–50 ml of acetidin into the gruff and place on ice bath and treat with ultrasound for 2 min. Place for a while, filter, and collect the 2nd filtrate; then rinse the filter paper and residue with a little acetidin, and combine the solution into the above filtrate. Recover the solvent in the entire filtrate below 40 °C and under reduced pressure near dryness, transfer with acetidin into a 5 ml volumetric flask. Dilute to the scale, precisely place 1 ml on an activated carbon cartridge (120/400-mesh, 0.25 g, inner diameter 0.9 cm, if it is Supelclean ENVI-Carb SPE Tubes, 3 ml activated carbon cartridge, pre-rinse the cartridge with 5 ml of acetidin); elute with 5 ml of multifunctional N-hexane–acetidin (1:1) mixture, collect all the eluent, place it on a nitrogen evaporator; concentrate to approximate dryness, and precisely add 1 ml of acetidin to dissolve. This is the test solution.

13.5.3.3 Preparation of the Reference Solution

Precisely weigh appropriate amounts of parathion, methyl parathion, dantox, folimat, methamidophos, monocrotophos, dimpylate, dicthion, carbofos, methidathion, dichlorovos, and acephate pesticide reference substances; use acetidin to make reference stock solutions. Then use acetidin to make the above 12 reference stock solutions into a mixed reference solution, and keep as the mixed reference stock solution. Precisely draw the above mixed reference stock solution, and dilute with acetidin to make a serial of solutions with concentrations of 0.1, 0.5, 1, 2 and 5 μg/ml. These are the reference solutions.

13.5.3.4 Assay Method

Using fused silica capillary column DB-17MS (30 m × 0.32 mm × 25 μm) (or HP-5), and nitrogen phosphorous detector. Temperature of sample injection port: 220 °C; temperature of

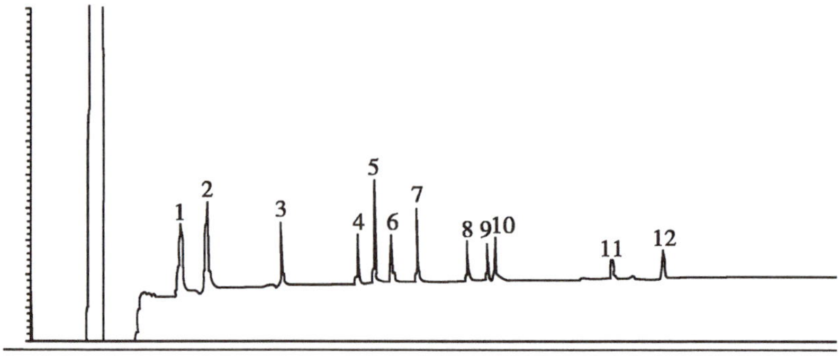

Fig. 13.8 Gas chromatogram of 12 organophosphorus pesticide reference substances *1* Dichlorvos; *2* Methamidophos; *3* Acephate; *4* Omethoate; *5* Diazinon; *6* Monoc rotophos; *7* Dimethoate; *8* Parathion—methyl; *9* Malathion; *10* Parathion; *11* Methidathion; *12* Ethion

detector: 300 °C. Splitless injection. Temperature programming: initial temperature 120 °C, increase the temperature 10 °C/min–200 °C; then 5 °C/min–240 °C, maintain for 2 min; then 20 °C/min–270 °C, and maintain for 0.5 min. As per dichlorovos peak, the amount of theoretical plate should not be less than 6,000, and the separating degree of 2 adjacent peaks should be more than 1.5.

Precisely draw 1 μl of each of the test solution and mixed reference solution with the corresponding concentrations, respectively, continuously inject 3 times, and calculate the 12 pesticide residues in the test samples by external standard method (Fig. 13.8).

13.5.3.5 Assay Results

The "Assaying Method for Organophosphorus pesticide Residues" in the appendix of Volume 1, *Chinese Pharmacopoeia* 2005, was written by Shanghai Institute for Food and Drug Control. They tested 10 Danshen samples from different sources while writing the protocol, and found none of the 12 organophosphorus pesticide residues in these samples.

13.5.3.6 Method Validation

1. Pretreatment of the samples: Various extraction methods were tested and the ultrasonic extraction method was finally chosen. The extraction times, solvent amount, and ultrasonication time, as well as the SPE cartridge used for purification and the eluent were tested. The results showed that under the above preparative method and conditions, organophosphorus pesticide residues in the samples could be extracted completely.
2. Linearity: The mixed standard solution with different concentrations of the 12 organophosphorus pesticides were tested, and a standard curve with the sample size on the *X*-axis and peak area on the *Y*-axis was plotted, and the regression equation was calculated. The results showed that the 12 pesticides exhibited good linearity, and the correlation coefficients were between 0.9959 and 0.9999.
3. Reproducibility: The reproducibility of the results using this method met the requirements for residue analysis, and were completely in line with the requirements for the test. The RSDs of all the test results were less than 15 %.
4. Recovery: The recovery tests were conducted on 10 medicinal herbs at 3 concentration levels. The recovery rates of the 12 organophosphorus pesticides were all between 86.37 and 98.25 %, which meets the residue analysis requirement (70–120 %).
5. Stability: The test solutions were tested in 13 h, and the results showed that the 12 pesticides were stable in 13 h.

13.5.4 Pyrethroid Pesticide Residues

13.5.4.1 Instruments and Reagents

Gas chromatograph (with ^{63}Ni electron capture detector), rotary evaporator, ultrasonic cleaner, anhydrous sodium sulfate (A.R.), alumina (A.R.), microcrystalline cellulose (A.R.), Florisil silica (A.R.), acetone (P.R.), ligroin (with boiling point ranging from 60–90 °C, P.R.), and ethylether (P.R.).

13.5.4.2 Preparation of the Test Solution

Dry the sample at 60 °C for 4 h, pulverize into fine powder, precisely weigh 1–2 g, and place in a 100 ml conical flask with stopper; then add 30 ml of ligroin–acetone (4:1) mixed solution, treat with ultrasound for 15 min, filter, and place the filtrate in another 100 ml conical flask. Repeat the above procedures another 2 times, and combine the filtrates. Add an appropriate amount of anhydrous sodium sulfate into the filtrate to dehydrate, place in a 100 ml round bottom flask, and concentrate at 40–45 °C and under reduced pressure to approximate dryness. Repeat the procedures with a little ligroin until acetone is completely eliminated, dissolve the residue with an appropriate amount of ligroin, and transfer into a treated glass chromatographic column [mixed columella, filled with, from top to bottom, 2 g of anhydrous sodium sulfate, 1 g of alumina, 1 g of microcrystalline cellulose, 4 g of florisil silica and 2 g of anhydrous sodium sulfate, and pre-washed with 20 ml of ligroin–ethylether (4:1) mixed solution], elute with 90 ml of ligroin–ethylether (4:1) mixed solution, collect the eluent, concentrate at 40–45 °C and under reduced pressure to approximate dryness. Repeat the above procedure with 3–4 ml of ligroin until the ethyl ether in the sample is completely eliminated, dissolve with ligroin, transfer into a 5 ml volumetric flask, dilute to the scale, and shake well. This is the test solution.

13.5.4.3 Preparation of the Reference Solution

Precisely weigh appropriate amounts of alphamethrin, fenvalerate and deltamethrin pesticide reference substances. Then prepare the respective reference stock solutions by using ligroin. Precisely measure the above reference stock solutions, dilute with ligroin to the scale, shake, and make a mixed reference stock solution. Precisely weigh the above mixed reference stock solution, and make serial solutions with concentrations of 0, 2, 10, 40, and 200 μg/L with ligroin. These are the reference solutions.

13.5.4.4 Assay Method

Instruments: Fused silica capillary column (30 m × 0.32 mm × 0.25 μm) SE-54 (or DB-5), and ^{63}Ni electron capture detector. Temperature of sample injection port: 270 °C; temperature of detector: 330 °C. Split ratio: 20:1 or 5:1 (or select the optimal split ratio based on the instrument setting). Temperature programming: initial temperature 160 °C, maintain for 1 min, increase the temperature 10 °C/min–278 °C, and maintain for 0.5 min; then 1 °C/min–290 °C, and maintain for 5 min. As per deltamethrin peak, the amount of theoretical plate should not be less than 10^5, and the separating degree of 2 adjacent peaks should be more than 1.5 (Fig. 13.9).

Precisely draw 1 μl of each of the test solution and the mixed reference solution with the corresponding concentration, respectively, continuously inject 3 times, take the mean value of the 3, and calculate the 3 pyrethroid pesticide residues in the test sample by external standard method.

13.5.4.5 Assay Results

Three batches of Danshen samples were tested with the method described above, and the results are shown in Table 13.6.

13.5.4.6 Method Validation

1. Pretreatment of samples: The extraction solvents, the packing materials for the purifying chromatographic column, and the eluting solvents were tested. The results showed that under the above preparative method and conditions, pyrethroid pesticides could be completely extracted from the samples.
2. Linearity: We assayed the mixed standard solutions with different concentrations of the

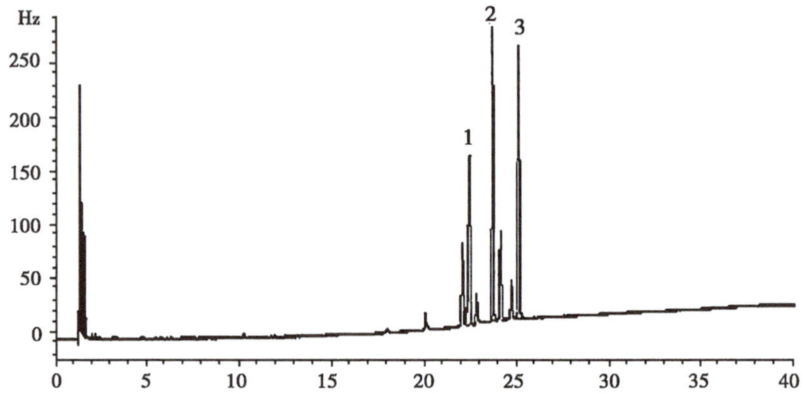

Fig. 13.9 Gas chromatogram of 3 pyrethroid pesticide reference substances *1* Cypermethrin; *2* Fenvalerate; *3* Deltamethrin

Table 13.6 Detection of pyrethroid pesticides in Danshen samples

Danshen samples	Alphamethrin (mg/kg)	Fenvalerate	Deltamethrin
Danshen 1	1.63×10^{-2}	Undetected	Undetected
Danshen 2	Undetected	Undetected	Undetected
Danshen 3	Undetected	Undetected	Undetected

3 pyrethroid pesticides, plotted a standard curve with sample size on the *X*-axis and peak area on the *Y*-axis, and calculated the regression equation. The results showed that the 3 pesticides exhibited good linearity and the correlation coefficients were all between 0.9950 and 0.9991.

3. Reproducibility: When tests ($n = 5$) were conducted with the method, the reproducible RSD values of alphamethrin, fenvalerate and deltamethrin were 3.57, 3.91 and 5.30 %, respectively, which means the reproducibility of the method is good.
4. Recovery: Using the following medicinal herbs and preparations, codonopsis root, notoginseng root, astragalus root, white peony root, licorice root, American ginseng, Chinese angelica, Danshen, honeysuckle flower, Xinhuang Tablet (新癀片), and Babaodan (八宝丹), we investigated the recovery rate of the method, and the results showed that the recovery rates were all between 80 and 110 %, which indicates that the method is accurate and reliable.

13.6 Aflatoxin

Shen Ji

13.6.1 Overview

Aflatoxin is a carcinogenic substance produced by the toxigenic strains of *Aspergillus flavus* and *Aspergillus parasiticus*, and is the most toxic mycotoxin discovered so far. Its toxicity is equivalent to 10 times of that of potassium cyanide, and 68 times of that of white arsenic, and its ability to induce liver cancer is 75 times greater than that of dimethyl nitrosamine. There are 4 types of aflatoxin, namely G2, G1, B2 and B1. Among these, B1 is generally acknowledged as the primary toxic substance, and mainly exists in agricultural products, animal feeds, traditional Chinese herbs, etc. Currently, the detection of aflatoxin at home and abroad is mainly focused on various kinds of foods, peanuts, and their processed products. The standards for the detection of aflatoxin in TCM herbs is still absent. However, TCM herbs are prone to mildew because of long storage times and poor storage conditions, so aflatoxin detection in TCM herbs is urgently needed.

Currently, the major detection methods for aflatoxin are thin-layer chromatographic scanning (TLCS), enzyme linked immunosorbent assay (ELISA), high performance liquid chromatography (HPLC), etc. TLCS detects and measures aflatoxin in samples by thin layer development after suitable pretreatments. The method is insensitive, and it is difficult to separate different samples well under unitary chromatographic conditions. In terms of thin layer scanning, the practice of locating the spot only by the R_f value cannot be further validated, meaning that this scanning is of poor specificity. When using ELISA, the assaying result is greatly affected by many conditions such as differences in kits, experimental temperature, and instrument sensitivity. The method has poor repeatability, a high false positive rate, and it is difficult to meet the related technical requirements. Considering the above reasons, we do not recommend TLCS or ELISA methods to be used in detecting content in TCM herbs. Although conventional HPLC has high sensitivity, it requires complicated sample pretreatment and skillful operation. We introduce here a HPLC—immunoaffinity column—post-column derivatization, which is specific, sensitive, easy to operate, and complies with the current state food inspection standards and the food and drug inspection standards of Hong Kong Government Laboratory, and it is an advanced method around the world. Its specificity is better than that of the thin layer chromatography currently adopted by United States Pharmacopeia [6].

13.6.2 Determination of Aflatoxin

13.6.2.1 Instruments and Reagents

High performance liquid chromatograph (with fluorescence detector), post-column derivatization system, high speed homogenizer, aflatoxin immuno-affinity column, methanol (A.R.), acetonitrile (A.R.), deionized water, iodine (A.R.), and sodium chloride (A.R.).

13.6.2.2 Preparation of the Test Solution

Extract with methanol–water system, filter and dilute the extract, purify the filtrate by an immunoaffinity chromatographic column containing aflatoxin specific antibody that has specificity to aflatoxin B1, B2, G1 and G2, to let aflatoxin cross-link to the antibody conjugated to the chromatographic media, remove the impurities from the immunoaffinity chromatographic column with water, elute the aflatoxin with methanol, adjust the volume of the eluent, and derivatize and assay the solution with the post-column iodine solution on a HPLC with a fluorescence detector.

13.6.2.3 Preparation of the Reference Solution

Take aflatoxin G2, G1, B2 and B1 mixed standard solutions, and add 70 % methanol to make 0.45–20 ng/ml solutions.

13.6.2.4 Assay Method

Use octadecylsilanized silica gel as a filler; methanol–acetonitrile–water (40:18:42) as the mobile phase, with flow rate of 0.8 ml/min; use post-column derivatization method with derivatization solution of 0.05 % iodine solution, flow rate of 0.3 ml/min, and derivatization reaction temperature of 70 °C; and detect by a fluorescence detector, with excitation wavelength of 360 nm and emission wavelength of 450 nm, and the separating degree between each chromatographic peak should be in accordance with the requirements (Fig. 13.10).

Precisely take the above standard solutions and inject into a liquid chromatograph respectively, measure and draw a standard curve. Determine the contents of aflatoxin G2, G1, B2 and B1 from the standard curve.

13.6.3 Assay Results

13.6.3.1 Danshen Herb

3 batches of Danshen samples collected from different production regions were assayed according to the method described above, and the results are shown in Table 13.7.

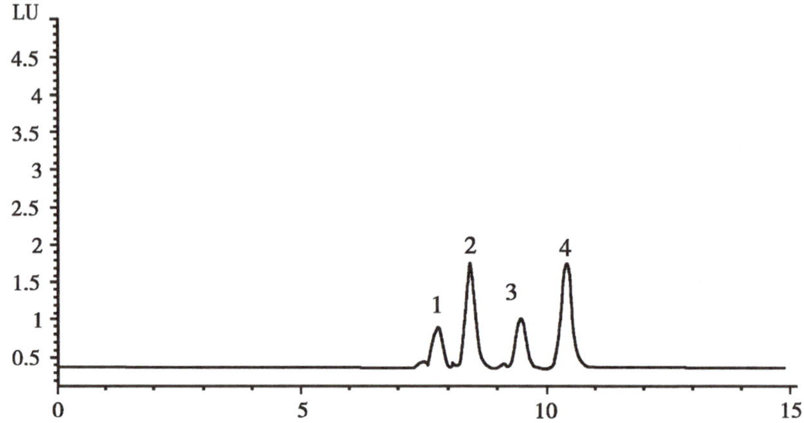

Fig. 13.10 HPLC profile of aflatoxin reference solutions *1* Aflatoxin G2; *2* Aflatoxin G1; *3* Aflatoxin B2; *4* Aflatoxin B1

Table 13.7 Detection of aflatoxin in Danshen samples (*Unit* ppb)

The origin of Danshen samples	G2	G1	B2	B1	Total amount
Dazhangzhuang Village, Tongjing Town, Yinan, Shandong	Undetected	Undetected	Undetected	Undetected	Undetected
Dongyuezhuang Village, Zhuanbu Town, Yinan Country, Linyi, Shandong	Undetected	Undetected	Undetected	Undetected	Undetected
Lunan, Henan	Undetected	Undetected	Undetected	1.54	1.54

13.6.3.2 Assaying of Danshen Preparations

Danshen Tablet and Compound Danshen Tablet manufactured by different companies were assayed for aflatoxin according to the method described above, and the results are shown in Table 13.8.

13.6.3.3 Method Validation

1. Linearity: Aflatoxin G2 and B2 had good linearities in the range of 2.25–150 pg; aflatoxin G1 and B1 had good linearities in the range of 7.5–500 pg, and all the correlation coefficients were >0.9999.
2. Limit of detection: The qualitative detection limits of aflatoxin G2, G1, B2 and B1 were 0.9, 1.5, 0.9 and 1.5 pg, respectively, and the quantitative detection limits were 2.25, 7.5, 2.25 and 7.5 pg, respectively, which are the lowest detection levels currently reported at home and abroad.
3. Precision: The RSD of the precision tests on aflatoxin G2, G1, B2 and B1 were 6.5, 3.1, 4.5 and 2.5 % ($n = 5$), respectively, indicating that the assay had good precision.
4. Repeatability: Add a certain amount of reference substances into the samples, then perform the assay ($n = 5$). The results showed that the RSD of aflatoxin G2, G1, B2 and B1 were 3.1, 1.6, 2.7 and 2.6 %, respectively, indicating that the assay had good repeatability.
5. Stability: Add a certain amount of reference substances to the samples, then perform the assay at different times. The results showed that the RSD of aflatoxin G2, G1, B2 and B1 in 0–24 h were 2.8, 2.1, 1.9 and 1.3 %, respectively, indicating that all subjects were basically stable during that time frame.
6. Recovery: The recovery tests were conducted on 3 concentration levels respectively; the results reveal the recovery rates of aflatoxin G2, G1, B2, B1, and the total recovery rates

Table 13.8 Detection of aflatoxin in different Danshen preparations (*Unit* ppb)

Samples	Manufacturer	Batch No.	G2	G1	B2	B1	Total amount
Danshen Tablet 1	Shanghai Ley's Pharmaceutical Co., Ltd	030116	Undetected	Undetected	0.49	Undetected	0.49
Danshen Tablet 2	Shanghai Huanghai Pharmaceutical Co., Ltd	030514	Undetected	Undetected	Undetected	Undetected	Undetected
Danshen Tablet 3	Shanghai Xinyijiahua Pharmaceutical Co., Ltd	20021206	Undetected	Undetected	2.43	Undetected	2.43
Compound Danshen Tablet 1	Shanghai Ley's Pharmaceutical Co., Ltd	0309033	Undetected	Undetected	Undetected	Undetected	Undetected
Compound Danshen Tablet 2	Shanghai Huanghai Pharmaceutical Co., Ltd	030920	Undetected	Undetected	Undetected	Undetected	Undetected
Compound Danshen Tablet 3	Guangzhou Baiyunshan Pharmaceutical Co., Ltd	04021016	Undetected	Undetected	Undetected	1.21	1.21

of aflatoxin were 66.4–94.2, 65.6–108.1, 77.5–109.5, 73.5–98.0 and 70.7–97.7 %, respectively.

References

1. Li JY, Guo DF, Yao JJ, et al. J Chin Mass Spectrom Soc. 2002;23(3):164.
2. Xia BF, Lu ZQ, Wang XM, et al. Mod Instrum. 2004;10(1):17.
3. Committee of Chinese Pharmacopoeia, Chinese Pharmacopoeia 2005, Volume I. Beijing: Press of Chemical Industry; 2005. Appendix 52–54.
4. Cong XD, Fang W. J Chin Pharm Univ. 1999;30(2):103.
5. Wang MJ, Cheng HX. Chinese Patent Medicine. 1994; vol 16(12):32.
6. Huang QC, Li J, Zhang XM, et al. Insp Quar Sci. 2003;13(3):53.

Content Determination

14

Wanying Wu, Rongxia Liu, Dean Guo, Xiaoqian Zhang, Guoqing Wu, Rixin Liang, Manling Li, Fengnan Cao, Ming Zu, Zhangzhao Jin, Bilian Cheng, Linke Ma, Qing Gong, Shen Ji, Ming Zhu, Zhengliang Ye, Jun Gao and Aihua Liu

14.1 Determination of Salvianolic Acid Content

14.1.1 Determination of Total Phenolic Acid Content in Compound Danshen Tablet

Wanying Wu, Rongxia Liu and Dean Guo

14.1.1.1 Selection of Methods and Reference Substances

Compound Danshen Tablet contains mainly phenolic acids, compounds which have strong absorptions in the ultraviolet region. Thus, ultraviolet spectrophotometry was selected to assay total phenolic acid content.

Salvianolic acid B is one of the major effective constituents of Compound Danshen Tablet, and is also the constituent with the highest content; thus salvianolic acid B was selected as the reference substance for content determination.

14.1.1.2 Methodological Studies

1. Standard Curve: See the results in Table 14.1. The regression equation was $Y = 0.0516X + 0.000171$, and the correlation coefficient was $r = 0.9993$. The linear range of salvianolic acid B was 13–65 μg.
2. Precision Study: See the results in Table 14.2, which shows that the precision of the instruments was relatively good.
3. Stability Study: See the results in Table 14.3, which shows that the sample's water solution was stable for at least 4 h.
4. Reproducibility Study: See the results in Table 14.4, which reveals that the method had good reproducibility.
5. Recovery Study: Because the content of salvianolic acid B in the sample is relatively high, even if taking 1/2 amount of the sample as described above to run the test would waste a large amount of reference substance. To save reference substance, the sample amount and dilution multiple were reduced. The results showed that this method was feasible (Table 14.5).

14.1.1.3 Total Phenolic Acid Content Calculation

Using UV spectrophotometry to determine the total phenolic acid content in Danshen, and using salvianolic acid B as the reference substance, the calculated results are usually higher than the actual content because the molecular weights of the other phenolic acids in Danshen are lower than that of salvianolic acid B. The HPLC results of 6 batches of samples show that besides the main constituent salvianolic acid B, Danshen also contains other phenolic acids such as

W. Wu · R. Liu · D. Guo (✉) · X. Zhang · G. Wu · R. Liang · M. Li · F. Cao · M. Zu · Z. Jin · B. Cheng · L. Ma · Q. Gong · S. Ji · M. Zhu · Z. Ye · J. Gao · A. Liu
Shanghai Institute of Materia Medica, Shanghai, China
e-mail: daguo@mail.shcnc.ac.cn

X. Yan (ed.), *Dan Shen (Salvia miltiorrhiza) in Medicine*,
DOI 10.1007/978-94-017-9463-3_14,

Table 14.1 Experimental data on the standard curve of total phenolic acid in Danshen roots

Sample (μg)	13	26	39	52	65
Absorbance (A)	0.257	0.498	0.743	0.992	1.268

Table 14.2 Precision test results

Serial no.	1	2	3	4	5
Concentration (μg/ml)	26.0	26.0	26.0	26.0	26.0
Absorbance (A)	0.499	0.500	0.500	0.503	0.504
Mean value	0.50				
RSD (%)	0.43				

Table 14.3 Stability test results

Time (h)	Absorbance (A)	Total phenolic acid content (mg/g)	Mean value (mg/g)	RSD (%)
0	0.742	874.6	875.9	0.3
1	0.744	872.5		
2	0.744	876.8		
3	0.745	877.7		
4	0.742	877.7		

Table 14.4 Reproducibility test results

Serial no.	Sample weight (mg)	Concentration (mg/ml)	Absorbance (A)	Total phenolic acid content (mg/g)
1	5.34	0.0559	1.078	850.7
2	4.43	0.0475	0.916	891.4
3	3.58	0.0381	0.735	876.8
4	3.85	0.0418	0.806	881.0
5	3.62	0.0381	0.735	853.0
Mean content	870.6			
RSD (%)	2.06			

Table 14.5 Recovery test results

Test no.	1	2	3	4	5
Sample weight (mg)	0.1243	0.1228	0.1403	0.1557	0.1354
Sample content (mg)	0.108	0.107	0.125	0.138	0.120
Reference substance added (mg)	0.138	0.138	0.138	0.138	1.138
Total weight (mg)	0.252	0.251	0.270	0.272	0.259
Detected weight (mg)	0.143	0.144	0.145	0.135	0.139
Recovery rate (%)	103.47	104.19	104.48	97.33	100.51
Average recovery (%)	101.99				
RSD (%)	2.99				

protocatechuic aldehyde, salvianolic acid E, rosmarinic acid, lithospermic acid and salvianolic acid A. Their molecular weights are very different from each other, thus the ratios of the average molecular weight of protocatechuic aldehyde, salvianolic acid E, rosmaric acid, lithospermic acid and salvianolic acid A to the molecular weight of salvianolic acid B is taken as a correction factor to calculate the contents of total phenolic acids. See the average molecular weight of the above five phenolic acids in Table 14.6.

Correction factor f = average molecular weight of the 5 phenolic acids/molecular weight of salvianolic acid B = 449.6/740 = 0.608.

Therefore, the calculation formula for total phenolic acid content (%) determined by UV spectrophotometry is:

$$\text{Total phenolic acid content in Danshen}\,(\%) = f(A - B) + B$$

where f is the correction factor 0.608; A is the total phenolic acid content determined by UV spectrophotometry with salvianolic acid B as the reference; B is the salvianolic acid B content in the sample determined by HPLC.

14.1.1.4 Determination of the Total Phenolic Acids Content in Compound Danshen Tablets from 10 Different Manufacturers

The results of total phenolic acids in Compound Danshen Tablet manufactured by 10 different companies are shown in Tables 14.7, 14.8 and 14.9.

14.1.2 Determination of Salvianolic Acid Content in Danshen Herb

Xiaoqian Zhang, Guoqing Wu, Rixin Liang and Manling Li

14.1.2.1 Determination of Salvianolic Acid B Content

1. Instruments and Materials

   Instruments: HP1100 HPLC, G1315A Diode Array Detector, G1311A Quat Pump,

Table 14.6 The molecular weights of phenolic acids

Phenolic acid	Protocatechuic aldehyde	Rosmarinci acid	Lithospermic acid	Salvianolic acid A	Salvianolic acid E	Mean value	Salvianolic acid B
Molecular formula	$C_3H_6O_3$	$C_{18}H_{15}O_8$	$C_{27}H_{22}O_{12}$	$C_{26}H_{22}O_{10}$	$C_{36}H_{30}O_{16}$	449.6	$C_{36}H_{28}MgO_{16}$
Molecular weight	138	360	538	494	718		740

Table 14.7 Total phenolic acid content determined by UV spectrophotometry

Manufacturer	Sample weight (g)	Dilution multiple	Absorbance (A)	Content (μg/ml)	Content A (%)
Shanghai	0.2002	100	0.611	46.957	11.733
Guangzhou	0.2001	100	0.456	35.152	8.779
Tianjin	0.2002	100	0.426	32.87	8.213
Hangzhou	0.2003	100	0.454	35	8.737
Beijing	0.2003	100	0.273	21.228	5.299
Henan	0.2003	100	0.506	38.597	9.635
Shanxi	0.2002	100	0.682	52.62	26.284
Xinjiang	0.2003	100	0.416	32.109	8.015
Xinjiang	0.2003	100	0.300	23.283	5.812
Guangdong	0.2002	100	0.506	38.597	9.635

Table 14.8 Salvianolic acid B content determined by HPLC

Manufacturer	Sample weight (g)	Dilution multiple	Peak area	Concentration (μg/ml)	Content B (%)
Shanghai	0.2002	10	1108.9	1.184	0.592
Guangzhou	0.2001	10	918.1	0.98	0.490
Tianjin	0.2002	10	74.6	0.08	0.040
Hangzhou	0.2003	10	686.6	0.732	0.366
Beijing	0.2003	10	240.5	0.256	0.128
Henan	0.2003	10	616.6	0.657	0.329
Shanxi	0.2002	10	80.8	0.086	0.043
Xinjiang	0.2003	10	537.9	0.573	0.286
Xinjiang	0.2003	10	286.3	0.284	0.142
Guangdong	0.2002	10	379.6	0.404	0.202

Table 14.9 Corrected total phenolic acid content

Manufacturer	Sample weight (g)	Content A (%)	Content B (%)	Total phenolic acid content (%) [0.608 (A − B) + B]
Shanghai	0.2002	11.733	0.592	7.37
Guangzhou	0.2001	8.779	0.49	5.53
Tianjin	0.2002	8.213	0.04	5.01
Hangzhou	0.2003	8.737	0.366	5.14
Beijing	0.2003	5.299	0.128	3.27
Henan	0.2003	9.635	0.329	5.99
Shanxi	0.2002	13.142	0.043	8.01
Xinjiang	0.2003	8.015	0.287	4.99
Xinjiang	0.2003	5.812	0.142	3.59
Guangdong	0.2002	9.635	0.202	5.94

G1313A HPLC Autosampler, G1316A Thermostatted Column Compartment, G1322A Degasser, Agilent ChemStation.

Reagents: Acetonitrile (L.C.), methanol (G. R.), water (self-made, high purity), and other chemicals (A.R.).

Reference substances: Salvianolic acid B (batch No. 200201) for content assay, purchased from the National Institutes for Food and Drug Control.

2. Chromatographic Condition

Chromatographic column. Inertsil ODS 3 column (4.6 mm × 150 mm, 4 μm). Column temperature: 35 °C. Mobile phase: Methanol–acetonitrile–formic acid–water (30:10:1:59). Flow rate: 1 ml/min.

Detection wavelength: 286 nm. See the UV absorption spectrum of salvianolic acid B reference substance in Fig. 14.1.

3. Preparation of the Reference Solution

Precisely weigh an appropriate amount of salvianolic acid B reference substance, dissolve in 75 % methanol, and make a solution of 0.14 mg/ml.

4. Preparation of the Test Solution

Sift Danshen powder through a #3 sieve, precisely weigh about 0.2 g, place it in a conical flask with stopper, precisely add 50 ml of 75 % methanol, seal, weigh, heat under reflux for 1 h, take out, cool, weigh again, add 75 % methanol to make up for the lost weight, shake well, filter through a micro-porous membrane (0.45 μm), and take the subsequent filtrate. This is the test solution.

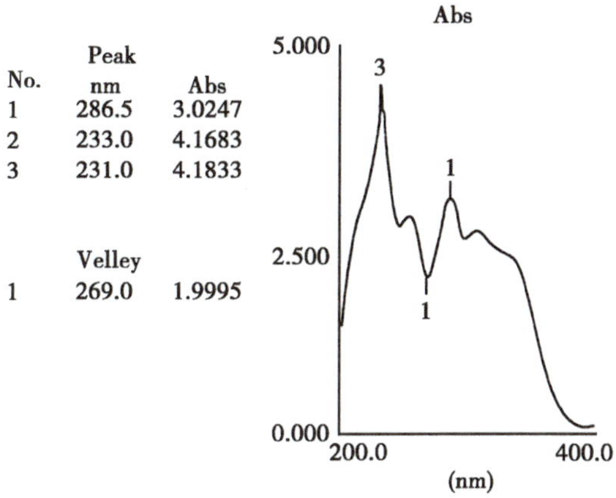

Fig. 14.1 UV absorption spectrum of salvianolic acid B reference substance

5. Sample Assay
   Assay the salvianolic acid contents in Danshen samples from 10 different production regions according to the above method. The results are shown in Table 14.10, Figs. 14.2 and 14.3.
6. Methodology Validation
   When the injection amount of salvianolic acid B was in the range of 0.56–2.8 μg, the curve showed good a linear relationship. The precision test showed that the RSD was 0.3 %. The stability test showed that the test solutions were stable for 24 h, with a RSD of 1.7 %. The reproducibility test showed that the RSD was less than 2.0 %. The standard recovery test showed that the average recovery of salvianolic acid B was 100.5 %, and the RSD was 2.0 %.

Table 14.10 The content of salvianolic acid B in Danshen samples from different regions

No.	Production region	Salvianolic acid B content (%)
1	Zhengzhou, Henan	1.19
2	Drug market of Chongqing	5.33
3	Anhui	3.01
4	Linxi, Neimenggu	4.35
5	Drug market of Guangzhou	0.09
6	Drug market of Xi'an	3.79
7	Hebei	4.95
8	Linyi, Shandong	3.59
9	Yangzhou, Jiangsu	6.81
10	Drug market of Hefei	5.71

Fig. 14.2 HPLC profile of salvianolic acid B reference substance

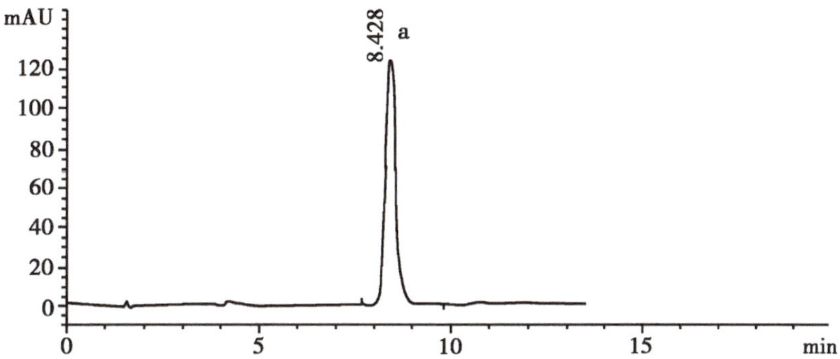

Fig. 14.3 HPLC profile of Danshen extract (the chromatographic peak at Rf 8.424 min is salvianolic acid B)

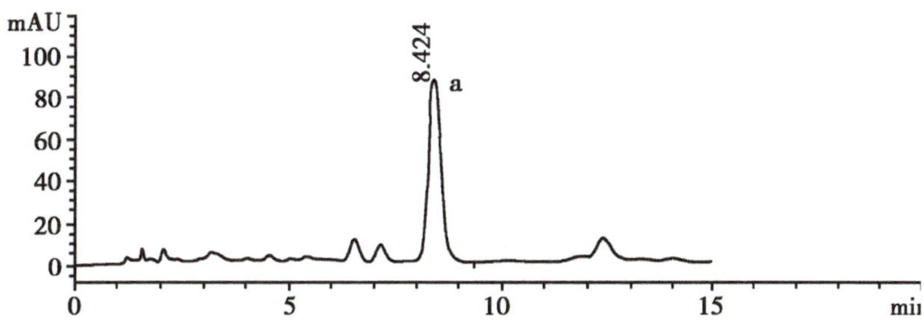

14.1.2.2 Determination of Danshensu and Protocatechuic Aldehyde

1. Instruments and Materials
   Instruments: HP1100 HPLC, G1315A Diode Array Detector, G1311A Quat Pump, G1313A HPLC Autosampler, G1316A Thermostatted Column Compartment, G1322A Degasser, Agilent ChemStation.
   Reagents: Acetonitrile (L.C.), methanol (G. R.), water (self-made, high purity), and other chemicals (A.R.).
   Reference substances: Sodium danshensu (batch no. 855-200102) for content assay, and protocatechuic aldehyde (batch no. 0801-9402) for identification, were purchased from the National Institutes for Food and Drug Control.
2. Chromatographic Condition
   Chromatographic column: Inertsil ODS-3 column (4.6 mm × 150 mm, 4 μm). Column temperature: 35 °C. Mobile phase: Methanol–glacial acetic acid–water (20:1:80). Flow rate: 1 ml/min. Detection wavelength: 280 nm.
3. Preparation of Reference Solution
   Precisely weigh appropriate amounts of sodium danshensu and protocatechuic aldehyde reference substances, dissolve in methanol, and make a sodium danshensu solution of 0.08 mg/ml (equivalent to 0.07 mg of danshensu) and a protocatechuic aldehyde solution of 0.012 mg/ml, respectively.
4. Preparation of Test Solution
   Sift Danshen powder through a #3 sieve, precisely weigh about 0.4 g, place it in a conical flask with stopper, precisely add 25 ml of water, seal, weigh, treat with ultrasound (power 250 W, and frequency 33 kHz) for 30 min, take out, cool, heat under reflux for 4 h, take out, cool, weigh again, add water to make up for the lost weight, shake well, filter through a micro-porous membrane (0.45 μm), and save the filtrate. This is the test solution.
5. Sample Assay
   Assay the contents of danshensu and protocatechuic aldehyde in Danshen samples from 10 different regions according to the above method. The results are shown in Table 14.11, Figs. 14.4 and 14.5.
6. Methodology Validation
   When the injected sodium danshensu and protocatechuic aldehyde were in the range from 0.216 to 1.081 μg and 0.036 to 0.180 μg, respectively, the curve showed good linearity. In the precision test, RSDs for both were less than 2.0 %. The stability test showed that the test solution was stable for 24 h, with RSD less than 2.0 %. In the reproducibility test, RSDs for both were less than 2.0 %; and in the standard recovery test, the average recovery of danshensu was

Table 14.11 The contents of danshensu and protocatechuic aldehyde in Danshen samples from different regions

No.	Production region	Danshensu (%)	Protocatechuic aldehyde (%)
1	Hebei	0.33	0.01
2	Zhengzhou, Henan	0.22	0.03
3	Drug market of Chongqing	0.16	0.05
4	Anhui	0.42	0.06
5	Linxi, Neimenggu	0.26	0.02
6	Drug market of Guangzhou	0.04	0.01
7	Drug market of Xi'an	0.36	0.02
8	Linyi, Shandong	0.39	0.02
9	Yangzhou, Jiangsu	0.68	0.09
10	Drug market of Hefei	0.50	0.03

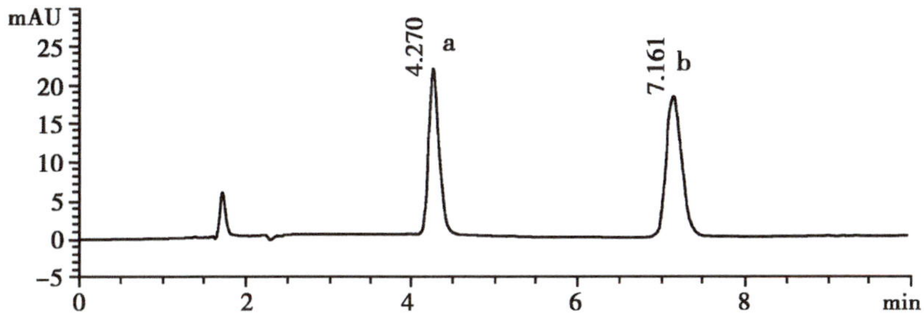

Fig. 14.4 HPLC profile of sodium danshensu and protocatechuic aldehyde reference substances. **a** Sodium danshensu; **b** Protocatechuic aldehyde

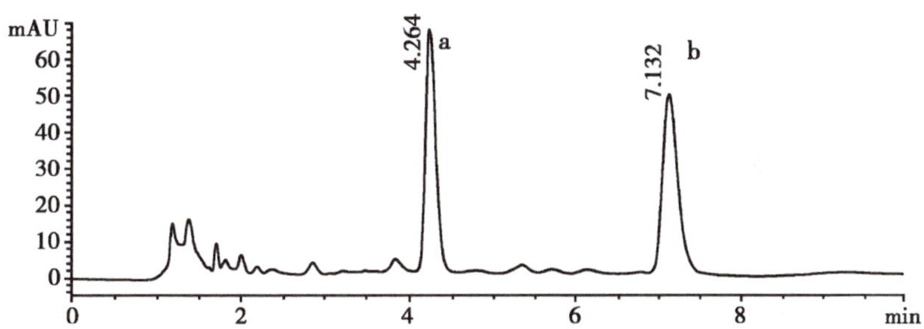

Fig. 14.5 HPLC profile of Danshen herb. **a** Sodium danshensu; **b** Protocatechuic aldehyde

98.94 % with RSD of 1.5 %; and the average recovery of protocatechuic aldehyde was 98.94 % with RSD of 1.7 %.

14.1.2.3 Determination of Other Constituents

Li et al. used TLC-scanning to detect 7 Danshen constituents (including protocatechuic aldehyde, caffeic acid, methhyl rosmarinate, salvianolic acid A, rosmarinic acid, salvianolic acid C, and salvianolic acid B), and compared the phenolic acid contents of Danshen samples collected from different regions and different times, as well as the contents of these constituents in plant species closely related to Danshen [1].

Their sample preparation method was: soak 0.5 g of Danshen powder in water overnight, extract in hot water for 0.5 h (extract for 2 h when detecting protocatechuic aldehyde), filter, adjust the filtrate with 10 % hydrochloric acid to pH 2, extract with acetidin for 3 times, 10 ml each time, combine the extracts, concentrate, and transfer into a 2 ml volumetric tube with methanol. Protocatechuic aldehyde could be separated by applying to a high performance silica gel GF_{254} thin layer plate and using chloroform–acetidine–benzene–formic acid (2.4:2:1:0.6) as the developing agent; the other 6 constituents could be separated by chloroform–acetidine–benzene–formic acid–methanol (1.5:2:1:1:0.1). The detection wavelength was 300 nm and reference wavelength was 240 nm [1].

Sun and He used HPLC to detect protocatechuic aldehyde, salvianolic acid A, salvianolic acid B, salvianolic acid E, lithospermic acid, and rosmarinic acid in Danshen and in other plants of the same genus. Protocatechuic aldehyde could be separated by using YWG-C18 as the stationary phase, methanol–water–formic acid (18:81.5:0.5) as the mobile phase, and PHBA as the internal standard; in the same chromatographic column, the other 5 constituents could be separated by using methanol–acetonitrile–water (41.5:1:57.5) (adjust to pH 2.2 with formic acid) as the mobile phase and cinnamic acid as the internal standard. The flow rate was 1.0 ml/min and the detection wavelength was 281 nm.

Jian et al. determined the contents of protocatechuic aldehyde and dihydroxy benzoic acid by thin-layer chromatography. They used silica gel H-CMC-Na thin layer plate and benzene–acetidin–formic acid (80:50:4) as the developing agent. The detection wavelength was 270 nm and the reference wavelength was 350 nm.

Cao et al. established a method for detecting salvianolic acid B in the water decoction of Danshen and Panax notoginseng. Chromatographic condition: Dalian Elite Spherisorb C18 chromatographic column (4.6 mm × 200 mm), with column temperature of 25 °C; mobile phase of methanol–formic acid solution (41.5:58.5) (pH 2.5), with flow rate of 0.8 ml/min; and detection wavelength of 280 nm [2].

Pan et al. assayed 4 water-soluble constituents in Danshen Injection (salvianolic acid B, rosmarinic acid, danshensu, and protocatechuic aldehyde) with HPLC. Chromatographic condition: Dalian Elite Hypersil C18 chromatographic column (4.6 mm × 250 mm), with column temperature of 35 °C; mobile phase: A and B gradient elution systems. A: water–dimethylformamide–glacial acetic acid (94:4:2), and B: methanol–dimethyl formamide–glacial acetic acid (94:4:2). Gradient elution was performed according to Table 14.12. Detection wavelength: 280 nm [3].

Wang et al. used HPLC to detect protocatechuic aldehyde in the cortex and xylem of Danshen. Chromatographic condition: C18 chromatographic column; mobile phase: methanol–1 % glacial acetic acid (8:92); and detection wavelength: 280 nm. The results showed that the protocatechuic aldehyde content in the cortex was lower than that in the xylem [4].

Table 14.12 HPLC mobile phase elution gradient

t/min	A (%)	B (%)
0	95	5
8	95	5
13	75	25
35	74	26
45	68	32
52	68	32

Yuan et al. used HPLC to determine the contents of danshensu and protocatechuic aldehyde in Compound Danshen Tablet. Chromatographic condition: ODS chromatographic column (4.6 mm × 250 mm), with column temperature of 35 °C; mobile phase: methanol–0.5 % glacial acetic acid (12:88); detection wavelength: 281 nm [5].

14.1.3 Determination of the Water-Soluble Constituents in Dantonic™

Fengnan Cao

14.1.3.1 HPLC-External Standard Method for the Determination of Danshensu and Protocatechuic Aldehyde Contents

1. Method 1
   (a) Chromatographic condition: Chromatographic column: YWG-C18, 10 μm (4.6 mm × 250 mm); mobile phase: methanol-water (3:97) with IPR-B7 sodium heptanesulfonate (2.5/100 ml); flow rate: 1 ml/min; column temperature: 30 °C; detection wavelength: 280 nm.
   (b) Preparation of the test solution: Precisely weigh 5 pills of Dantonic™, place them in a 10 ml volumetric flask, add an appropriate amount of methanol, treat with ultrasound for 2 h, take out, cool, dilute with methanol to the scale, shake well, filter through a 0.45 μm micro-porous membrane, and the filtrate is the test solution.
   (c) Preparation of the reference solution: Precisely weigh an appropriate amount of sodium danshensu and add methanol to make a 0.16 mg/ml solution (1 mg sodium danshensu is equivalent to 0.900 mg danshensu).
   (d) Assay: Precisely draw 5 μl of the reference solution and 5–10 μl of the test solution, respectively, inject into a liquid chromatograph, assay, calculate, and obtain the content (Fig. 14.6).

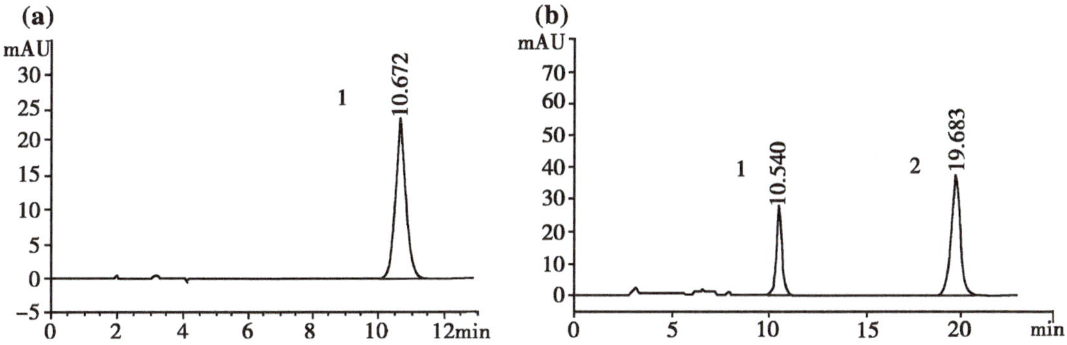

Fig. 14.6 HPLC profiles of Dantonic™ (**a** reference substance; **b** drug) *1* Danshensu; *2* Protocatechuic aldehyde

(e) Explanations: ① HPLC mobile phase separating system: it can separate danshensu and protocatechuic aldehyde in Dantonic™, and generated good peak shapes; however, the system uses a large amount of ion-pairing reagents which are apt to damage the chromatograph and chromatographic column, therefore, after the assay is completed, the chromatographic system should be washed with plenty of water to decrease the damage to the system. ② When the chromatographic mobile phase separating system was used, and when the sample size of danshensu was 0.14–0.7 μg, the linearity was good, and the regression equation was $Y = 0.000128C–0.0412$ (correlation coefficient r = 9989). ③ With decreasing chromatographic column efficiency, the use of the mobile phase separating system may result in bad peak shapes, or even failure to achieve separation. Thus, the chromatographic column efficiency needs to be relatively high.

2. Method 2

(a) Chromatographic condition: Filler: octadecylsilane chemically bonded silica; column temperature: 30 °C; mobile phase: methanol–water–glacial acetic acid (8:91:1); flow rate: 1 ml/min; detection wavelength: 281 nm. As per danshensu peak, the amount of theoretical plates should not be less than 2,000.

(b) Preparation of the test solution: Precisely weigh 30 pills of the product, place 12 of them in a 25 ml volumetric flask, (if it is dropping pill with film coat, break the coat, rinse the milky body with an appropriate amount of methanol, and combine the rinse solution into the volumetric flask), add about 15 ml of methanol, treat with ultrasound (50 W, 50 Hz, and initial bath temperature of 25 °C) for 10 min to dissolve, cool, add methanol to the scale, shake well, centrifugate for 5 min at 2,000 rpm, and the supernatant is the test solution.

(c) Preparation of the reference solution: Precisely weigh an appropriate amount of sodium danshensu and use methanol to make a 0.16 mg/ml solution (1 ml of the solution contains 0.144 mg of danshensu).

(d) Assay: Precisely draw 5 μl of each of the reference solution and the test solution, inject into a liquid chromatograph in order, assay, calculate, and obtain the content (Fig. 14.7).

(e) Explanations: ① The chromatographic mobile phase separating system did not use ion-pairing reagents and used glacial acetic acid to adjust the pH. It could separate danshensu and protocatechuic aldehyde in Dantonic™ well, be used for a long time, and generate a chromatogram with good peak shape and stable baseline. Its requirements for the chromatographic

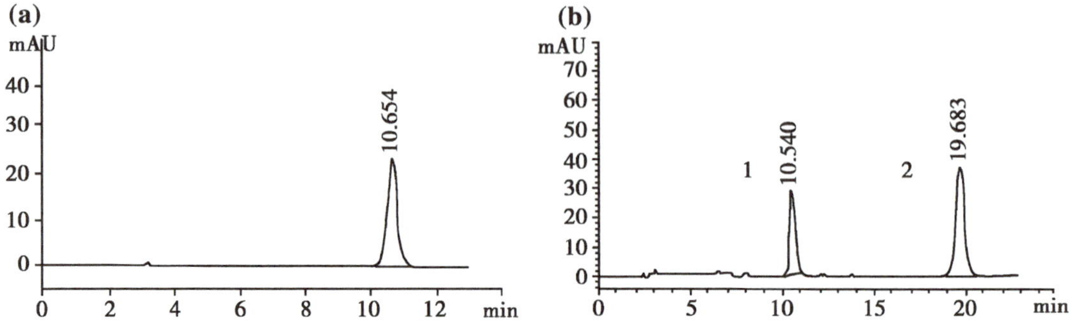

Fig. 14.7 HPLC profiles of Dantonic™ (**a** reference substance; **b** drug) *1* Danshensu; *2* Protocatechuic aldehyde

column efficiency and the maintenance of the chromatographic column and chromatography were less demanding. ② When assaying danshensu by this method, the linear range was 0.42–1.89 μg, and the regression equation was $Y = -272745.40X + 10071.68$ (correlation coefficient $r = 0.9997$), indicating that the method had good linearity. ③ The method had good selectivity, the other ingredients in the preparation had no interference in the determination of danshensu, and the other phenolic acids were well separated from danshensu. ④ The method had good precision, stability, repetitiveness, and reproducibility, and the RSDs were all less than 2.0 %. ⑤ In the recovery tests on danshensu of Dantonic™, the average recovery was 100.8 %, and RSD was 1.3 %, indicating that the method has good reliability.

14.1.3.2 HPLC: Internal Standard Method for the Determination of Danshensu and Protocatechuic Aldehyde Contents

1. Method 1
   (a) Chromatographic condition: using stainless steel column with octadecylsilane chemically bonded silica as filler (4.6 mm × 250 mm), and column temperature of 30 °C, applying methanol–1 % glacial acetic acid (9:91, containing appropriate amount of ion pairs) as the mobile phase, with flow rate of 1 ml/min, and detection wavelength of 281 nm.
   (b) Preparation of the test solution: Precisely weigh 5 Dantonic™ pills, place them in a 10 ml volumetric flask, precisely add 0.3 ml of PABA methanol solution (1 mg/ml), add an appropriate amount of methanol, shake well, ultrasonic treatment for 10 min until the sample is dissolved, add methanol to the scale, shake well, take the solution, centrifuge for 5 min, take the supernatant, and keep for HPLC assays.
   (c) Preparation of the reference solution: Precisely draw 2.0 ml of sodium danshensu reference solution (1 mg/ml), 0.4 ml of protocatechuic aldehyde reference solution (1 mg/ml), and 0.3 ml of PABA internal standard solution (1 mg/ml), place them in a 10 ml volumetric flask, add methanol to the scale, shake well, and keep for HPLC assays.
   (d) Assay: Precisely draw 5 μl of the test solution and reference solution, inject into the HPLC in order, assay, and obtain the sample content by internal standard method. See the chromatogram in Fig. 14.8.
   (e) Explanations: ① The selection of mobile phase and internal standard: Considering the constituents to be tested were phenolic acids, methanol–1.0 % acetic acid (9:91) was selected as the basic composition of mobile phase, and after studying various phenols and aromatic acids, para-amino benzoic acid (PABA) was selected as the

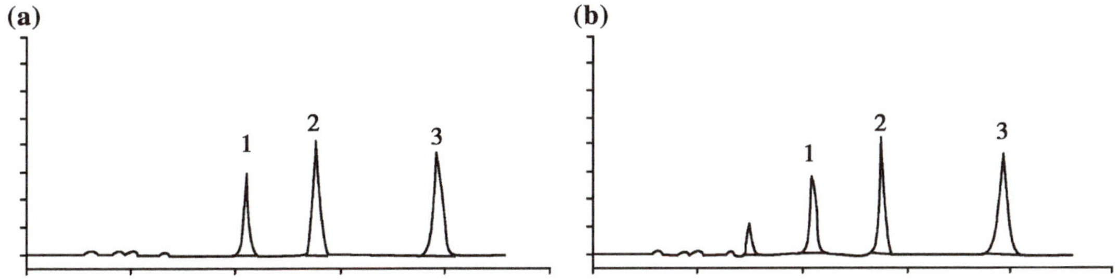

Fig. 14.8 HPLC profiles of Dantonic™ (**a** reference substance; **b** sample) *1* Danshensu; *2* PABA; *3* Protocatechuic aldehyde

best internal standard, because its retention time was between those of danshensu and protocatechuic aldehyde, and the chromatographic response met the requirements. ② The supplement of 0.38 ml of IPR-B7 ion-pairing reagent into 500 ml of mobile phase not only separated the constituents of PABA and the samples well and generated good chromatographic peak shapes, but also strengthened the acidity of the mobile phase (approximate pH 3.0) which was propitious to the useful life of the chromatographic column. System suitability tests and sample analysis also showed that the system is good. ③ The method was used to assay Dantonic™; the linearity ranges of danshensu and protocatechuic aldehyde were 0.3–1.5 μg and 0.05–0.6 μg, respectively, and the regression equations of danshensu and protocatechuic aldehyde were $A_s/A_{is} = 0.493\ W - 0.043$ ($r = 0.9994$) and $A_s/A_{is} = 5.45\ W - 0.045$ ($r = 0.9998$), respectively, indicating the linearity was good. ④ The sample treatment was simple. As shown by methodological studies, the method was sensitive, accurate, and simple, and had good precision, stability, and reproducibility, with all RSDs less than 1.5 %. The average recoveries were 99.2 and 99.7 %, respectively, and RSDs were 0.8 and 0.7 %, respectively. ⑤ The method can be used not only for Dantonic™, but also for Danshen and its other preparations.

2. Method 2
   (a) Chromatographic condition: Chromatographic column: Nucleosil C18 (4.6 mm × 250 mm, 10 μm); mobile phase: methanol–water–glacial acetic acid (19:80:1). Flow rate: 1 ml/min; and detection wavelength: 279 nm.
   (b) Preparation of the reference solution: Precisely weigh 20 mg of sodium danshensu, and 4 mg of protocatechuic aldehyde reference substances, place in a 200 ml volumetric flask, add 1 % acetic acid solution to dilute to the scale, shake well, and keep as the reference substance stock solution. Precisely draw 1.0–10.0 ml of the reference substance stock solutions (7 samples in all), place in a 10 ml volumetric flasks, respectively, precisely add 1.0 ml of internal standard solution into each, add 1 % acetic acid solution to the scale, shake well, and keep them as the reference solutions.
   (c) Preparation of internal standard solution: Precisely weigh 10 mg of PHBA, place in a 100 ml volumetric flask, add 1 % acetic acid solution to the scale, shake well, and keep it as the internal standard solution.
   (d) Preparation of the test solution: Precisely weigh about 125 mg of Dantonic™, place it in a 25 ml volumetric flask, precisely add 2.5 ml of the internal standard solution, and then add 1 % acetic acid solution, treat with ultrasound to dissolve, cool to room temperature, dilute with 1 % acetic acid

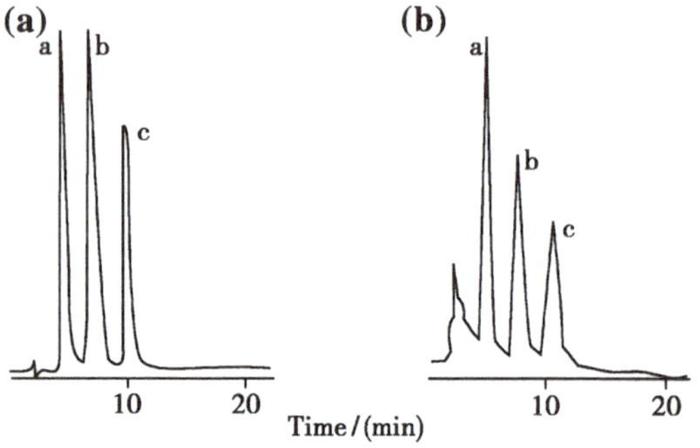

Fig. 14.9 HPLC profiles of Dantonic™ (**a** reference substances; **b** sample). *a* Sodium danshensu; *b* Protocatechuic aldehyde; *c* PHBA

solution to the scale, mix, filter through a 0.45 μm filter membrane, abandon the initial filtrate, and keep the subsequent filtrate as the test solution.

(e) Assay: Precisely take 10 μl of each of the reference solution and the test solution, inject into a liquid chromatograph in order, assay, calculate, and obtain the content by internal standard method (Fig. 14.9).

(f) Explanations: ① The method is convenient and fast, but the peak shapes and the extent of danshensu and protocatechuic aldehyde separation are not very good. Nonetheless, it can generally meet the QC requirements. ② Under the given chromatographic conditions, danshensu and protocatechuic aldehyde have good linearity in the ranges of 10–80 mg/L and 2–16 mg/L, respectively.

14.1.4 Determination of Salvianolic Acids in Danshen Injection

Ming Zu, Zhangzhao Jin, Bilian Cheng, Linke Ma and Qing Gong

14.1.4.1 Ultraviolet and Visible Light Spectrophotometry

1. Visible Light Spectrophotometry

Danshen Injection contains mainly water-soluble phenolic acids, so the total phenolic acid content is an important QC indicator. Phenolic acids can react with potassium ferricyanide-chloride ferric, and the reaction is chromogenic and sensitive, therefore the total water-soluble phenolic acids can be determined based on the reaction followed by visible light spectrophotometry, using protocatechuic aldehyde as the reference substance [6].

(a) Preparation of the reference solution: Precisely an weigh appropriate amount of protocatechuic aldehyde reference substance and add ethanol to make a 5 μg/ml solution.

(b) Preparation of the standard curve: Precisely draw 0, 0.5, 1, 2, 3, 4 and 5 ml of the reference solution, place in a 25 ml volumetric flask, respectively, add an appropriate amount of ethanol to make up the total volume to 5.0 ml, and add 2 ml of 0.3 % sodium dodecylsulfate and 1 ml of potassium ferricyanide-chloride ferric chromogenic agent. Place in a dark place for 5 min, add 0.1 mol hydrochloric acid solution to the scale, shake well, stand for 20 min. Use the first solution as blank, determine the absorbance at 720 nm according to ultraviolet-visible light spectrophotometry described in *Chinese Pharmacopoeia* 2005, Volume 1, Appendix VA. Plot a standard curve with absorbance on the Y-axis and concentration on the X-axis.

(c) Assay: Precisely measure an appropriate amount of Danshen Injection (containing about 1 mg of total phenolic acids), place in a separatory funnel, add 10 ml of acidic water and 1.5 g of sodium chloride, shake well and extract with ethyl ether twice, 20 ml each time, take the ethyl ether solutions, volatilize to dryness, dissolve the residue with ethanol, quantitate, transfer into a 50 ml volumetric flask, dilute with ethanol to the scale, shake well, filter, take 1 ml of the subsequent filtrate, place in a 25 ml volumetric flask, add 4.0 ml of ethanol, assay the absorbance according to the method described above by adding 2 ml of 0.3 % sodium dodecylsulfate. Read the total phenolic acid content (μg) in the test solution from the standard curve.

2. Methodological Study and Content Determination
   (a) Sensitivity test: The sensitivities of the sodium nitrite-sodium molybdate colorimetric method, phloroglucinol colorimetric method and potassium ferricyanide-ferric chloride colorimetric method were compared. The results showed that the potassium ferricyanide-ferric chloride colorimetric method was the most sensitive.
   (b) Standard curve: Using protocatechuic aldehyde as the reference substance, the colorimetric determination of the solutions at different concentrations was conducted according to the method described above, and a standard curve was plotted based on the results obtained. The curve exhibited good linearity when the amount of protocatechuic aldehyde was less than 35 μg; and when the amount of protocatechuic aldehyde was up to 45 μg, the curve bent downwards slightly. So, when conducting the actual assay, the amount of protocatechuic aldehyde should not exceed the limit.
   (c) Recovery test: The average recovery was 97.9 %, $n = 8$, and RSD = 1.0 %.
   (d) Results: See the results in Table 14.13.
3. UV Spectrophotometry (Standard Curve Method)

Table 14.13 The content of total phenolic acids in Danshen Injection manufactured by different companies

Batch no.	Total phenolic acid content % (counted based on protocatechuic aldehyde)
1 (780404)	0.0826
2 (800719)	0.0188
3 (801014)	0.209
4 (780113)	0.0481
5 (78653)	0.0548
6 (801110)	0.0057
7 (800107)	0.0171

The water-soluble constituents in Danshen Injection, such as danshensu, protocatechuic acid, protocatechuic aldehyde, salvianolic acid B, and salvianolic acid A, all have absorptions in the ultraviolet region, and both sodium danshensu and test solution of Danshen Injection have maximum absorptions at 280 ± 2 nm (Figs. 14.10 and 14.11). Therefore, the total phenolic acid content of Danshen Injection could be determined at the wavelength of 280 nm by standard curve method, with sodium danshensu as the reference substance. Methodological validation studies have shown that the method is stable, reliable, and simple.

(a) Method
   (i) Water-soluble phenolic acids: Precisely weigh 15 mg of sodium danshensu reference substance, place

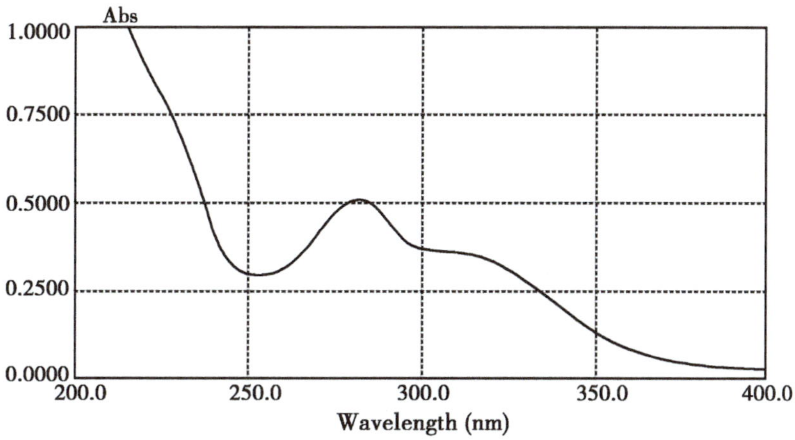

Fig. 14.10 Ultraviolet absorption spectrum of the test solution

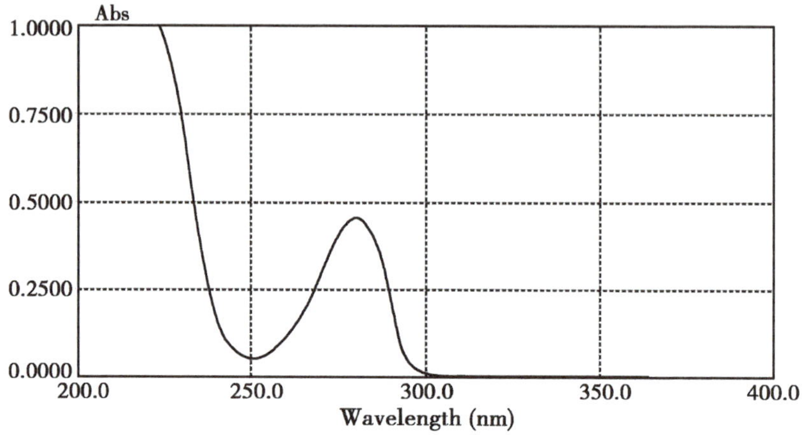

Fig. 14.11 Ultraviolet absorption spectrum of the reference solution

in a 100 ml volumetric flask, add water to the scale, shake well, and keep it as the reference solution.

(ii) Preparation of standard curve: Precisely measure 1, 2, 3, 4 and 5 ml of the reference solution, place in a 10 ml volumetric flask, respectively, add water to the scale, shake well, test according to Ultraviolet-visible Light Spectrophotometry in *Chinese Pharmacopeia* 2005, Volume I, Appendix VA. Use water as the blank, assay the absorbance at 280 nm, and plot a standard curve with absorbance on the Y-axis and concentration on the X-axis.

(iii) Assay: Precisely measure 1 ml of Danshen Injection, place in a 100 ml volumetric flask, add water to dilute to the scale, shake well, precisely measure 1 ml and place in a 5 ml volumetric flask, add water to dilute to the scale, shake well, assay the absorbance according to the method, read the weight of sodium danshensu in Danshen Injection from the standard curve.

(b) Methodological Study and Content Determination

(i) Chromatographic condition and wavelength selection: The ultraviolet-visible light spectrophotometry of sodium danshensu reference substance (supplied by the National Institutes for Food and Drug Control) and the test samples showed that they have maximum absorptions at 280 + 2 nm (Figs. 14.10 and 14.11); therefore, 280 nm was selected as the wavelength for content determination.

(ii) Related studies: When the concentrations of sodium danshensu reference solutions are in the range of 0.01301 mg/ml to 0.1301 mg/ml, the concentration and absorbance exhibit good linear relationship, with $Y = 12.013X - 0.009$ and $r = 0.9999$. The interference test showed that the ingredient-deficient sample (without Danshen) had no interference in the determination of water-soluble phenolic acid content in Danshen Injection. In reproducibility tests, RSD was 0.5 %. In recovery tests, the average recovery of nine samples was 98.5 %, and RSD was 1.1 %. The stability test showed that the test solution is stable in 6 h, with RSD of 0.4 % ($n = 7$).

(iii) Assay: The total phenolic acid contents in Danshen injection samples from different manufacturers are shown in Table 14.14.

14.1.4.2 Thin-Layer Chromatography

TLC-scanning has been used to determine the water-soluble effective constituents of Danshen Injection, i.e., danshensu, protocatechuic aldehyde and protocatechuic acid [7].

Table 14.14 The total phenolic acid contents in Danshen Injection samples from different manufacturers

Manufacturer	Batch no.	Total phenolic acid content (%)
Manufacturer of Zhejiang	040403	44.61
Manufacturer of Zhejiang	0401191	9.87
Manufacturer of Shanghai	030805	15.39
Manufacturer of Shanghai	040301	12.92
Manufacturer of Sichuan	010319	45.16
Manufacturer of Jiangxi	040905	17.99
Manufacturer of Jiangxi	040901	66.16
Manufacturer of Shanxi	040312	72.32
Manufacturer of Shanxi	040329	46.35
Manufacturer of Guangxi	040202	64.15
Manufacturer of Jiangsu	040401	69.78

1. Method
   TLC-scanning was conducted according to the method described in *Chinese Pharmacopoeia* 2005, Volume 1, Appendix VI B. Danshen Injection was used as the test solution. Danshensu, protocatechuic aldehyde and protocatechuic acid were used as the reference substances, and they were dissolved in methanol to make 4 mg/ml, 1 mg/ml and 1 mg/ml solutions, respectively. Take 4 μl of the test solution, and 2 and 4 μl of the reference solutions, apply on the same silica gel GF_{254} thin layer plate, respectively, develop with chloroform acctidin–benzene–formic acid (5:6:3:1), take out, dry in open air, scan at wavelengths: $\lambda_s = 280$ nm and $\lambda = 400$ nm, assay the test substance absorbance score and the reference substance absorbance score, and calculate the content.
2. Methodological Study and the Test Results
   (a) After developing the product, 7–8 spots were shown under ultraviolet light (254 nm). The spots in the locations with R_f 0.11, 0.47 and 0.55 were danshensu, protocatechuic acid and protocatechuic aldehyde, respectively. Under these conditions, danshensu, protocatechuic acid and protocatechuic aldehyde were completely separated from the neighboring spots, and the separation degrees were all greater than 1.5.
   (b) When danshensu content was 4–20 μg, the linearity was good; when protocatechuic aldehyde and protocatechuic acid contents were in the ranges of 1–5 μg, the linearity was good. The results of recovery tests were 99.9 % (n = 8), 99.9 % (n = 8) and 99.6 % (n = 8), respectively. The results of stability tests showed that danshensu was stable for 100 min in the solution, while the contents of protocatechuic acid and protocatechuic aldehyde gradually decreased 30 min after the preparation. The co-eluting reference standards had little influence on the assay results.
   (c) The temperature of plate activation has some influence on the separation, because the polarities of danshensu, protocatechuic acid and protocatechuic aldehyde are all relatively large, and if the plate activating temperature is too high, the adsorption force will be too strong and consequently cause tailing. Thus, activation of the plate under 80 °C for 1 h is appropriate.
   (d) See the results in Table 14.15.

14.1.4.3 HPLC Method

The primary water-soluble phenolic acids of Danshen Injection, such as danshensu, protocatechuic aldehyde, protocatechuic acid, salvianolic acid B, and salvianolic acid A, all have absorptions in the ultraviolet region, so their contents can be determined by HPLC specifically.

1. Determination of Danshensu, Protocatechuic Aldehyde by HPLC
   (a) Method
      (i) Sodium danshensu and protocatechuic aldehyde: Assay was conducted according to the HPLC protocol described in *Chinese Pharmacopoeia* 2005, Volume 1, Appendix VI D.
      (ii) Chromatographic condition and system suitability test: Use octadecylsilane chemically bonded silica as the

Table 14.15 The content of phenolic acids in Danshen Injection samples from different manufacturers

Manufacturer	Batch no.	Danshensu (mg/ml)	Protocatechuic acid (mg/ml)	Protocatechuic aldehyde (mg/ml)
Manufacturer of Shanghai	831004	10.2	0.146	0.706
	840206	9.62	0.261	0.707
	840905	9.62	0.144	0.908
	850303	6.48	0.258	1.32
	840302	4.78	0.177	0.769
Manufacturer of Jiangsu	840307	6.32	0.212	0.638
	830806	5.41	0.174	0.747
	8309219	6.71	0.166	1.06
Manufacturer of Jiangsu	8404041	5.90	0.158	1.12
	831004	10.2	0.146	0.706

filler; methanol–0.5 % glacial acetic acid solution (17:83) as the mobile phase; detection wavelength of 280 nm; and as per protocatechuic aldehyde peak, the amount of theoretical plates should not be less than 3,000.

(iii) Preparation of the reference solutions: Precisely weigh appropriate amounts of sodium danshensu and protocatechuic aldehyde reference substances that have been dried under reduced pressure by drying agent phosphori pentoxidum for 24 h, and add 5 % methanol to make a mixed solution containing 0.12 mg/ml and 14 μg/ml of the compounds, respectively.

(iv) Preparation of the test solution: Precisely measure 2 ml of the product, transfer into a 50 ml volumetric flask, dilute with water to the scale, and shake well.

(v) Assay: Draw 10 μl of the reference solutions and the test solution, respectively, inject into a liquid chromatograph, and run the machine.

(b) Methodological Study and the Test Results

(i) This method is a modification of the original ministerial standards, supplementing with the determination of sodium danshensu and changing the mobile phase to methanol–0.5 % glacial acetic acid (17:83), which decreases the acidity. The detection wavelength is 280 nm; column temperature: 25 °C; flow rate: 1 ml/min. Under these conditions, sodium danshensu and protocatechuic aldehyde peaks can be separated very well from the impurity peaks (Figs. 14.12, 14.13 and 14.14).

The results of methodological studies showed that when the sample size of sodium danshensu was 0.0629–5.2100 μg, the curve had good linearity with $Y = 520.17X + 11.83$ and $r = 0.9999$, and when the sample size of protocatechuic aldehyde reference solution was 0.01574–0.47205 μg, the curve had good linearity with $Y = 3921.5X-5.1805$ and $r = 0.9999$. In precision testing, the RSD of sodium danshensu was 1.0 % ($n = 5$), and the RSD of protocatechuic aldehyde was 1.3 % ($n = 5$). In reproducibility tests of sodium danshensu, the RSD was 0.8 % ($n = 6$), and the recovery rate was 101.8 % ($n = 9$); and in reproducibility tests of protocatechuic aldehyde, the RSD was 0.4 % ($n = 6$), and the recovery rate was 100.1 %

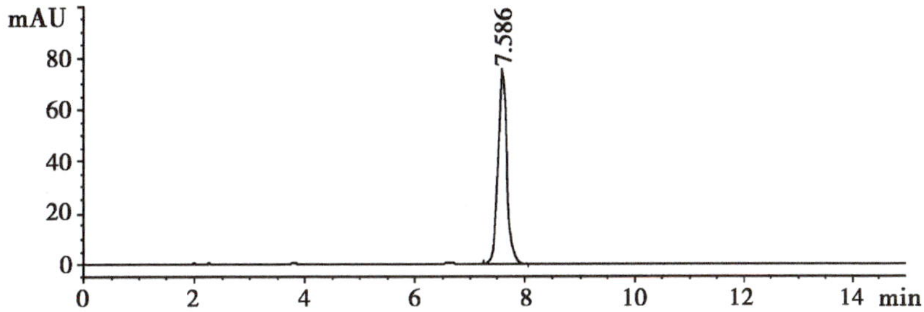

Fig. 14.12 HPLC profile of sodium danshensu reference substance

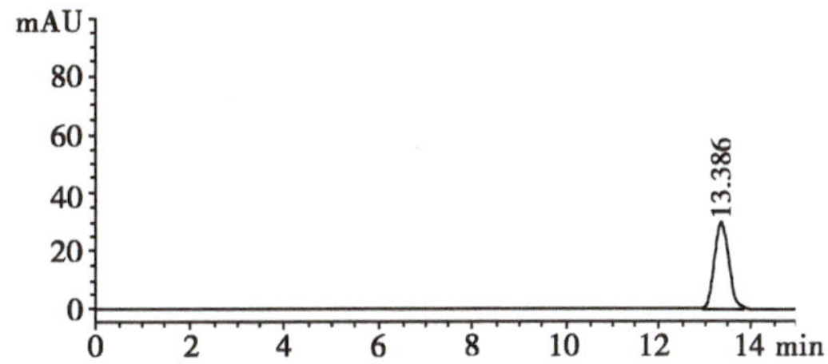

Fig. 14.13 HPLC profile of protocatechuic aldehyde reference substance

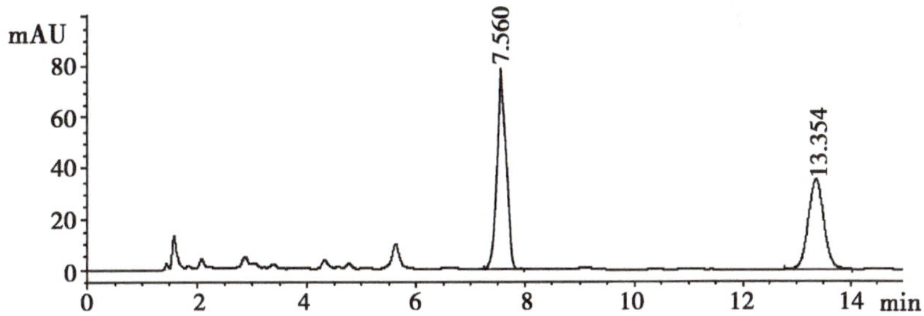

Fig. 14.14 HPLC profile of Danshen injection

($n = 9$). The stability tests showed that the test solution was stable for 29 h; sodium danshensu RSD was 1.0 % ($n = 5$), and protocatechuic aldehyde RSD was 1.3 % ($n = 5$). These results showed that the method has good reproducibility, and can effectively control the quality of the preparation.

(ii) Results: See the danshensu and protocatechuic aldehyde contents in Danshen Injection samples from different manufacturers in Table 14.16.

(c) Related Reports

(i) Protocatechuic aldehyde: Precisely measure 2 ml of the product, place in a 25 ml volumetric flask, and dilute with water to the scale. Chromatographic column: C_{18} column; mobile phase: methanol–0.5 % ammonium acetate (12:88) [adjust to pH 2.2 with sulphuric acid]; detection wavelength: 280 nm.

Table 14.16 The contents of danshensu and protocatechuic aldehyde in Danshen Injection samples from different manufacturers

Manufacturer	Batch no.	Danshensu (mg/ml)	Protocatechuic aldehyde (mg/ml)
Zhejiang	030116	3.82	0.41
Zhejiang	0104241	0.30	0.22
Sichuan	020601	3.21	0.41
Jiangsu	0109271	2.65	0.52
Chengdu	020801	2.23	0.31
Huaxi	021002	2.93	0.22
Sichuan	010402	4.32	1.11
Guangdong	0208009	1.48	0.16

(ii) Sodium danshensu and protocatechuic aldehyde [8]: Precisely measure 2 ml of the product, place in a 5 ml volumetric flask, add 0.3 ml of internal standard solution (PHBA 10 mg/ml), and dilute with methanol to the scale. Chromatographic column: RP-18 column (4 mm × 200 mm, 5 μm); mobile phase: methanol–0.5 % acetic acid (10:70); detection wavelength: 281 nm; column temperature: 35 °C; flow rate: 1.2 ml/min.

(iii) Sodium danshensu and protocatechuic aldehyde [9]: Precisely measure 1 ml of the product, place in a 10 ml volumetric flask, and dilute with water to the scale. Chromatographic column: RP-18 column (4 mm × 20 cm, 5 μm); mobile phase: methanol–dimethyl formamide–glacial acetic acid–water (4:4:2:90); detection wavelength: 281 nm; flow rate: 1.2 ml/min.

(iv) Sodium danshensu, protocatechuic aldehyde and protocatechuic acid [10]: Precisely measure 0.5 ml of the product, place in a 5 ml volumetric flask, and dilute with methanol to the scale. Chromatographic column: Bondapak C_{18} column; mobile phase: methanol–water–glacial acetic acid (19:80:1); flow rate: 1.5 ml/min; sodium danshensu and protocatechuic aldehyde were detected at wavelength of 280 nm, and protocatechuic acid was detected at wavelength of 262 nm.

2. Determination of Salvianolic Acid B by HPLC

The effective constituents of Danshen Injection are phenolic acids. Of them, danshensu and protocatechuic aldehyde generally exhibit peaks in the same mobile phase at approximately the same time during HPLC, so they can be determined simultaneously. However, there is a significant difference in polarity between the two compounds mentioned above and the other phenolic acids such as salvianolic acid B and salvianolic acid A. Thus the latter can only be determined by gradient elution.

(a) Method

(i) Sodium danshensu, protocatechuic aldehyde and salvianolic acid B: determined according to the HPLC protocol described in *Chinese Pharmacopoeia* 2005, Volume 1, Appendix VI D.

(ii) Chromatographic condition and system suitability test: Filler: octadecylsilane chemically bonded silica; mobile phase: methyl cyanide–0.0265 % phosphoric acid solution; gradient elution: in the 0–65th minute, methyl cyanide linearly increases from 2 to 67 %, and in the 65–66th minute, methyl cyanide linearly increases from 67 % to 98 %; column temperature: 30 °C; flow rate: 0.8 ml/min; detection wavelength: 288 nm. As per protocatechuic aldehyde peak, the amount of theoretical plates should not be less than 3000.

(iii) Preparation of the reference solution: Precisely weigh appropriate amounts of sodium danshensu, protocatechuic aldehyde and salvianolic acid B reference substances, add 5 % methanol to make a mixed solution containing 0.32 mg/ml of sodium danshensu, 0.10 mg/ml of protocatechuic aldehyde, and 0.12 mg/ml of salvianolic acid B.

(iv) Preparation of the test solution: Precisely measure 2 ml of the sample, transfer into a 10 ml volumetric flask, dilute with water to the scale, and shake well.

(v) Assay: Take 10 μl of the reference solution and the test solution, respectively, inject into a liquid chromatograph, start, and collect data.

(b) Methodological Study and Content Determination

(i) There is a big difference between the polarities of salvianolic acids A & B and those of danshensu and protocatechuic aldehyde; thus, the former can only be determined simultaneously after gradient elution. After different mobile phases and different chromatographic columns had been tested, methyl cyanide–phosphoric acid water solution was selected as the mobile phase for gradient elution; Discovery C_{18} (4.6 mm × 250 mm, 5 μm) was selected as the chromatographic column; and based on the Diode-Array Detection 3D graphics of Danshen Injection, the detection wavelength was set at 280 nm. Under these conditions, the peaks of sodium danshensu, protocatechuic aldehyde and salvianolic acid B could be well separated from the impurity peaks (Fig. 14.15).

(ii) The test results showed that when the sample sizes of sodium danshensu, protocatechuic aldehyde, and salvianolic acid B were 0.16075–9.645 μg, 0.0584–3.504 μg, and 0.06245–3.747 μg, respectively, the curves had good linearity. In precision tests, all RSDs were less than 2.0 %. The stability test showed that the test solution was stable for 29 h, with RSD less than 2.0 %. In reproducibility tests, all RSDs were less than 1.0 %. In recovery tests, the average recovery of sodium danshensu of 9 samples was 100.9 %, and RSD was 1.7 %; the average recovery of protocatechuic aldehyde was 98.9 %, and RSD was 1.5 %; and the recovery rate of salvianolic acid B was 100.1 %, and RSD was 2.5 %.

(iii) See the content assay results of Danshen Injection samples from different manufacturers in Table 14.17.

(c) Related reports: Sodium danshensu, protocatechuic acid, protocatechuic aldehyde and salvianolic acid [11]: Precisely measure 1 ml of the product, place in a 10 ml volumetric flask, and dilute with water to the scale. Chromatographic column: Hypersil C_{18} (4.6 mm × 250 mm, 5 μm); mobile phase: methanol–glacial acetic acid; gradient elution: in the 0–5th minute, methanol content was 10 %, in the 5–10th minute, methanol content linearly increased from 10 to 35 %, and in the 10–20th minute, methanol content kept stable at 35 %; detection wavelength: 281 nm; column temperature: 30 °C; flow rate: 1 ml/min.

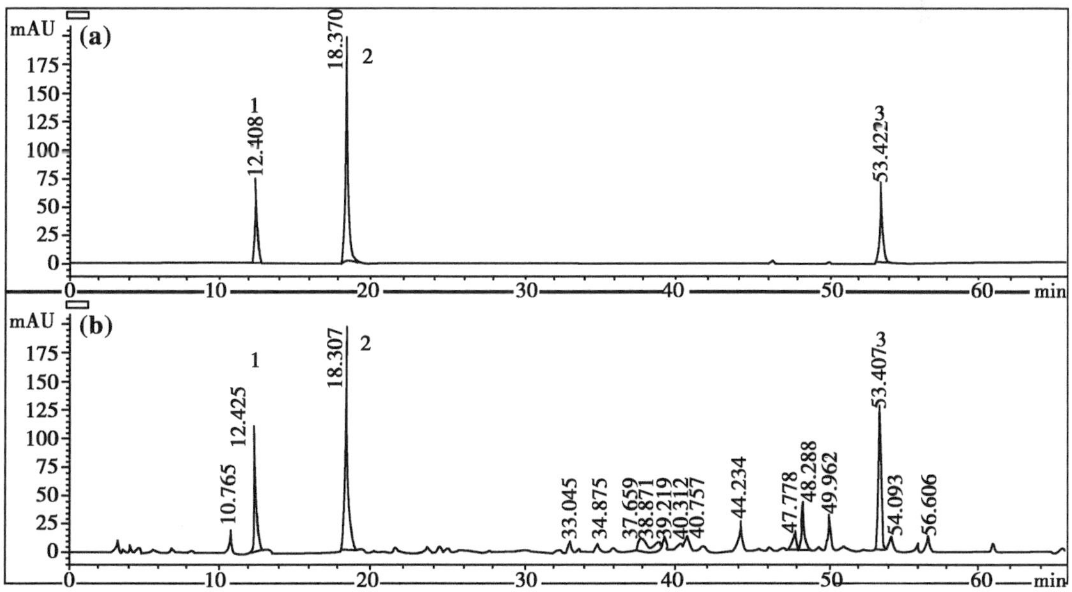

Fig. 14.15 **a** HPLC profile of the reference substances; **b** HPLC profile of Danshen injection. *1* Sodium danshensu; *2* Protocatechuic aldehyde; *3* Salvianolic acid B

Table 14.17 The contents of danshensu, protocatechuic aldehyde and salvianolic acid B in Danshen Injection samples from different manufacturers

Manufacturer	Batch no.	Danshensu (mg/ml)	Protocatechuic aldehyde (mg/ml)	Salvianolic acid B (mg/ml)
Manufacturer of Siquan	040210	2.62	1.13	1.03
Manufacturer of Shanxi	040329	2.752	0.31	1.62
Manufacturer of Yunnan	040205	2.89	0.54	1.10
Manufacturer of Shanghai	030203	5.31	0.77	0.52
Manufacturer of Guangdong	0404010	4.58	0.37	2.18
Manufacturer of Siquan	040601	1.17	0.23	0.80
Manufacturer of Jiangsu	040404	4.73	0.83	0.47
Manufacturer of Shanghai	040301	1.04	0.26	0.00
Manufacturer of Shanghai	040305	1.98	0.57	0.10
Manufacturer of Jiangxi	040905	0.61	0.05	0.34
Manufacturer of Zhejiang	0401183	0.55	0.24	0.05
Manufacturer of Zhejiang				

14.1.5 Determination of the Phenolic Components in Compound Danshen Tablet and Danshen Tablet

Shen Ji

14.1.5.1 Overview

The active constituents of Danshen can be roughly classified into 2 categories: one of liposoluble constituents, represented by tanshinones, and the other of water-soluble constituents, represented by salvianolic acids. The manufacture of Compound Danshen Tablet uses the techniques of ethanol reflux and water decocting, so the drug contains both the liposoluble and water-soluble constituents of Danshen. The water-soluble constituents of Danshen are primarily danshensu, salvianolic acid A, salvianolic acid B, salvianolic acid C, protocatechuic aldehyde, protocatechuic acid, etc. The commonly used methods for determination of these constituents are colorimetric reactions, TLC-scanning, and HPLC [12]. Among these constituents, salvianolic acid B has a relatively high content and relatively stable chemical properties; therefore, by determining salvianolic acid B content, the quality of the preparation can be controlled. Because of the complexity of constituents of Chinese herbs, HPLC is usually used to determine salvianolic acid B content at present. The method described below has been adopted by *Chinese Pharmacopoeia* 2005.

The manufacture of Danshen tablet also uses the techniques of ethanol refluxing and water decocting, so the drug contains both the liposoluble and water-soluble constituents of Danshen. Similar to Compound Danshen Tablet, Danshen Tablet's salvianolic acid B content is also determined by HPLC.

14.1.5.2 Instruments and Reagents

Salvianolic acid B reference substance (supplied by the National Institutes for Food and Drug Control), HPLC, methanol (A.R.), and methyl cyanide (A.R.).

14.1.5.3 Determination of Phenolic Components in Compound Danshen Tablet

1. Preparation of the Test Solution: Take 10 tablets of the product, remove the coat, precisely weigh, pulverize, take about 0.15 g of the powder, precisely weigh, place in a 50 ml volumetric flask, add an appropriate amount of water, treat with ultrasound for 30 min, cool, add water to the scale, shake well, centrifuge, take the supernatant.

2. Preparation of the Reference Solution: Take an appropriate amount of salvianolic acid B reference substance, precisely weigh, and add water to make a 60 μg/ml solution.
3. Assay: Filler: octadecylsilane chemically bonded silica; mobile phase: methanol–acetonitrile–formic acid–water (30:10:1:59); detection wavelength: 286 nm. As per salvianolic acid B peak, the amount of theoretical plates should not be less than 4000. Use 10 μl of the reference solution and the test solution, respectively, for HPLC.
4. Results: 12 batches of Compound Danshen Tablet samples from 5 manufacturers were tested with the method described above, and the results are shown in Table 14.18.
5. Method Validation
   (a) The detection wavelength: By diode array detector, the detection wavelength was set at 286 nm.
   (b) The linearity: When the sample size of salvianolic acid B was 0.114–5.70 μg, the peak area and the sample size exhibited good linearity, and the correlation coefficient was $r = 0.9999$.
   (c) The precision test: The RSD of the method was 0.1 % ($n = 5$), indicating that the precision was good.
   (d) The extraction solvent, and extraction method: Salvianolic acid B is a water-soluble constituent, so, based on its properties, water was used for extraction, followed by ultrasonic treatment. The length of ultrasonic treatment was also tested.
   (e) The blank test: Blank samples free of Danshen were prepared according to the prescription, and they were assayed with the same method. The results showed that the blank samples did not interfere with salvianolic acid B determination.
   (f) The reproducibility test: The RSD was 0.4 % ($n = 6$), indicating the method had good reproducibility.
   (g) The stability test: The test solutions were tested at different times, and the results showed that the solutions were substantially stable for 12 h.
   (h) The recovery test: The recovery test was run at 3 concentration levels, and the results showed that the average recovery was 98.4 % and RSD was 1.2 %, indicating that the method had a good recovery rate and was feasible.

14.1.5.4 Determination of Salvianolic AcidS in Danshen Tablet

1. Preparation of the test solution: Take 10 tablets of the product, remove the coat, precisely weigh, pulverize, take about 0.2 g of the powder, precisely weigh, place in a 50 ml

Table 14.18 Salvianolic acid B content in compound Danshen tablet samples from different manufacturers

No.	Manufacturer	Batch no.	Salvianolic acid B content (mg/tablet)
1	Shanghai Ley's Pharmaceutical Co., Ltd	040233	6.9
2	Shanghai Ley's Pharmaceutical Co., Ltd	040234	6.4
3	Shanghai Ley's Pharmaceutical Co., Ltd	040239	5.8
4	Shanghai Zhongxi Pharmaceutical (Group) Co., Ltd	030101	8.0
5	Shanghai Zhongxi Pharmaceutical (Group) Co., Ltd	030201	8.1
6	Guangzhou Baiyunshan Chinese Medicine Factory	04021012	4.3
7	Guangzhou Baiyunshan Chinese Medicine Factory	04021004	3.3
8	Guangzhou Baiyunshan Chinese Medicine Factory	04021016	4.4
9	Sanjin Group Guilin Chinese Medicine Factory	030502	9.3
10	Shanghai Huanghai Pharmaceutical Co., Ltd	031036	2.7
11	Shanghai Huanghai Pharmaceutical Co., Ltd	031054	2.7
12	Shanghai Huanghai Pharmaceutical Co., Ltd	031169	2.2

volumetric flask, add appropriate amount of water, treat with ultrasound for 20 min, cool, add water to the scale, shake well, filter, precisely measure 1 ml of the filtrate, place in a 25 ml volumetric flask, add water to the scale, and shake well. This is the test solution.

2. Preparation of the reference solution: Take appropriate amount of salvianolic acid B reference substance, precisely weigh, add water to make a 10 μg/ml solution, and this is the reference solution.
3. Assay: Filler: octadecylsilane chemically bonded silica; mobile phase: methanol–acetonitrile–formic acid–water (30:10:1:59); detection wavelength: 286 nm. As per salvianolic acid B peak, the amount of theoretical plates should not be less than 4000. Use 20 μl of the reference and the test solutions, respectively, for HPLC.
4. Results: 10 batches of samples from 3 manufacturers were tested using the method described above, and the results are shown in Table 14.19.
5. Method Validation
   (a) Detection wavelength: By diode array detector, the detection wavelength was set at 286 nm.
   (b) The linearity: When the sample size of salvianolic acid B was between 3.944–19.720 μg, the peak area and the sample size exhibited good linearity, and the correlation coefficient was $r = 0.9999$.
   (c) The precision test: The RSD of the method was 8 % ($n = 5$), indicating the precision was good.
   (d) The extraction solvent and extraction method: Salvianolic acid B is a water-soluble constituent, so it was extracted with water, followed by ultrasonication. The length of ultrasonication was tested.
   (e) The blank test: Blank samples free of Danshen were prepared according to the prescription, and they were assayed with the same method. The results showed that the blank samples did not interfere with the salvianolic acid B determination.
   (f) The reproducibility test: The RSD of the method was 0.4 % ($n = 6$), indicating that the method had good reproducibility.
   (g) The stability test: The test solutions were tested at different times, and the results showed that they were substantially stable for 15 h.
   (h) The recovery test: The recovery test was run at 3 concentration levels, and the results showed that the average recovery was 99.9 %, and RSD was 0.2 %, indicating that the method had a good recovery rate and was feasible.

Table 14.19 Salvianolic acid B content in Danshen tablet

Manufacturer	Batch no.	Salvianolic acid B content (mg/tablet)
Shanghai Ley's Pharmaceutical Co., Ltd	030201	10.9
Shanghai Ley's Pharmaceutical Co., Ltd	030114	11.5
Shanghai Ley's Pharmaceutical Co., Ltd	030125	11.3
Shanghai Ley's Pharmaceutical Co., Ltd	030924	10.6
Shanghai Ley's Pharmaceutical Co., Ltd	030118	9.6
Shanghai CTDT Pharmaceutical Co., Ltd	20030404	15.5
Shanghai CTDT Pharmaceutical Co., Ltd	20030505	14.1
Shanghai Huanghai Pharmaceutical Co., Ltd	030412	3.0
Shanghai Huanghai Pharmaceutical Co., Ltd	030513	3.7
Shanghai Huanghai Pharmaceutical Co., Ltd	030514	3.0

14.1.6 Determination of Salvianolic Acids in Xiangdan Injection

Ming Zhu

14.1.6.1 UV-Visible Light Spectrophotometry

1. Visible Light Spectrophotometry: All of the water-soluble constituents of Danshen are ortho-dihydroxybenzene compounds, so visible light spectrophotometry can be used after the chromogenic reactions [13].
   (a) Method
      (i) Preparation of the reference solution: precisely weigh an appropriate amount of 3,4-dihydroxybenzaldehyde reference substance, and add water to make a 12.85 μg/ml solution.
      (ii) Preparation of the standard curve: Precisely measure 0, 1, 2, 3 and 4 ml of the reference solution, place in test tubes with stoppers, add appropriate amounts of water to make up the total volume to 5.0 ml, add 0.3 ml of $NaNO_2$ solution (1:20), stand for 5 min, add 0.3 ml of $Al(NO_3)_3$ solution (1:10), 5 min later, add 5.0 ml of 1 mol/L NaOH solution, and stand for 15 min. Use the first solution as a blank, determine the solutions' absorbances at 525 nm, and plot a standard curve with absorbance on the Y-axis and concentration on the X-axis.
      (iii) Assay: Precisely measure 2 ml of Xiangdan Injection, place in a 250 ml volumetric flask, add water to the scale, take 1 ml of the subsequent filtrate, place in a test tube with stopper, add water to make the total volume to 5.0 ml, add 0.3 ml of $NaNO_2$ solution (1:20), stand for 5 min, add 0.3 ml of $Al(NO_3)_3$ solution (1:10), 5 min later, add 5.0 ml of 1 mol/L NaOH solution, stand for 15 min. Determine the absorbance, read the ortho-dihydroxybenzene compound content in the test solution out from the standard curve, and calculate the content.
   (b) Methodological Study and Content Determination
   When 3,4-dihydroxybenzaldehyde concentration was between 0.01301–0.1301 mg/ml, the concentration and the absorbance had good linearity, $C = 157.14A - 0.916$, and $r = 0.9999$. The assay result of ortho-dihydroxybenzene compound contents in Xiangdan Injection were: the samples from a Shanghai manufacturer (920117): 6.71 mg/ml, the samples from a Jiangsu manufacturer (891218-2): 1.56 mg/ml, the samples from a Jiangsu manufacturer (900215-2): 2.03 mg/ml; and the samples from a Jiangsu manufacturer (841217): 3.08 mg/ml.
2. UV Spectrophotometry (Standard Curve Method)
   (a) Xiangdan Injection is made of Danshen and rosewood, but rosewood volatile oil has no absorption in the UV region, while the water-soluble phenolic acids of Danshen such as danshensu, protocatechuic aldehyde, protocatechuic acid, salvianolic acid B and salvianolic acid A all absorb in the UV region, with sodium danshensu and Xiangdan Injection having maximum absorption at 280 ± 2 nm. Therefore, UV spectrophotometry can be used to determine total phenolic acids in Xiangdan Injection and assay the total phenolic acid content by standard curve method at 280 nm, with sodium danshensu as the reference substance.
   (b) The test results of phenolic acid contents in Xiangdan Injection samples from different manufacturers are shown in Table 14.20.
   (c) Related reports: Determining the content of protocatechuic aldehyde in Xiangdan Injection by first-derivative spectroscopy [14]: Precisely an weigh appropriate

Table 14.20 The total phenolic acid contents in Xiangdan Injection samples from different manufacturers

Manufacturer	Batch no.	Total phenolic acid content (%)
Manufacturer of Zhejiang	040403	33.98
Manufacturer of Zhejiang	0404133	32.71
Manufacturer of Jiangsu	040110	39.98
Manufacturer of Jiangsu	040301	29.16
Manufacturer of Guangdong	0403007	33.53
Manufacturer of Guangxi	040416	94.60
Manufacturer of Jilin	040209	29.89
Manufacturer of Siquan	040412	59.15
Manufacturer of Shanxi	040329	22.26
Manufacturer of Guangxi	040201	42.80
Manufacturer of Jiangsu	040401	69.78

amount of protocatechuic aldehyde reference substance, add ethanol to make a 54 μg/ml solution, and keep it as the reference solution. Precisely measure an appropriate amount of Xiangdan Injection, add an appropriate amount of saturated sodium chloride solution, extract with ethyl ether 5 times, 30 ml each time, combine the ethyl ether layers, volatilize to dryness, add an appropriate amount of ethanol to dissolve the residue, quantitatively transfer into a 25 ml volumetric flask, add ethanol to the scale, shake well, precisely transfer 5 ml to a 10 ml volumetric flask, add ethanol to the scale, shake well, and keep it as the test solution. Determine the semi-amplitudes (D value, distance of peak from 0) of the first derivative spectrums of the above 2 solutions at 168 nm, with ethanol as the blank, and calculate the contents. The results showed that when the protocatechuic aldehyde concentration was between 2.16 and 19.44 μg/ml, the concentration and D value exhibited very good linearity. The average recovery was 98.1 %, and RSD was 1.3 %. The test results of the samples from a manufacturer were: 0.3245 mg/ml (990901), 0.4577 mg/ml (990902), and 0.2846 mg/ml (990903), respectively.

14.1.6.2 HPLC

Xiangdan Injection is made of Danshen and rosewood, but rosewood volatile oil has no absorption in the UV region, while the water-soluble phenolic acids of Danshen such as danshensu, protocatechuic aldehyde, protocatechuic acid, salvianolic acid B and salvianolic acid A all absorb in the UV region, so HPLC can be used to determine single or multiple constituents.

1. Determination of Danshensu and Protocatechuic Aldehyde by HPLC [15]
   (a) Method
      (i) HPLC: According to *Chinese Pharmacopoeia* 2005, Volume 1, Appendix VI D.
      (ii) Chromatographic condition and system suitability test: Filler: octadecylsilane chemically bonded silica; mobile phase: methanol—methyl cyanide—1 % glacial acetic acid (containing 0.04 % sodium heptanesulfonate reagent) (10:2:88); column temperature: 40 °C; detection wavelength: 280 nm. As per protocatechuic aldehyde peak, the amount of theoretical plates should not be less than 3000.
      (iii) Preparation of the reference solutions: Precisely weigh appropriate amounts of sodium danshensu and protocatechuic aldehyde reference substances that have been dried under reduced pressure with desiccant phosphori pentoxidum for 24 h, and add methanol to make 56 μg/ml and 10 μg/ml mixed solutions, respectively. These are the reference solutions.
      (iv) Preparation of the test solution: Precisely measure 2 ml of the product, transfer into a 50 ml volumetric flask, dilute with water to the scale, and shake well. This is the test solution.

(v) Assay: Inject 10 μl of the reference solution and the test solution, respectively, into a HPLC and collect data.

(b) Methodological Study and Content Determination

(i) When the concentration of sodium danshensu reference solution was between 0.07 and 1.40 μg/ml, the curve showed good linearity with $Y = 0.711C + 1.861$ and $r = 0.9997$; when the concentration of protocatechuic aldehyde reference solution was between 024 and 0.48 μg/ml, the curve showed good linearity with $Y = 5.011C–3.944$ and $r = 0.9995$. In the precision test, the RSD for sodium danshensu was 1.6 % ($n = 6$), and the RSD for protocatechuic aldehyde was 2.4 % ($n = 6$). The recovery rate of sodium danshensu was 99.69 %, with RSD of 0.9 %; and the recovery rate of protocatechuic aldehyde was 98.54 %, with RSD of 2.1 %.

(ii) The contents of the two compounds in Xiangdan Injection samples from different manufacturers are shown in Table 14.21.

(c) Related Reports

(i) Protocatechuic aldehyde: Precisely measure 5 ml of the product, place in a 100 ml volumetric flask, and dilute with water to the scale. Chromatographic column: C_{18} column; mobile phase: methanol—1 % glacial acetic acid solution; and detection wavelength: 280 nm.

(ii) Danshensu [16]: Precisely measure 0.5 ml of the product, place in a 25 ml volumetric flask, and dilute with water to the scale. Chromatographic column: C_{18} column (HEWLETT 4.6 mm × 200 mm, 5 μm); mobile phase: methanol—0.4 % acetic acid solution (5:95); flow rate: 1.2 ml/min; detection wavelength: 281 nm. The results: the average recovery of the samples was 99.71 %, and RSD was 1.0 %. The results showed significant differences (Table 14.22).

(iii) Sodium danshensu and protocatechuic aldehyde [17]: Precisely measure 5 ml of the product, place in a 100 ml volumetric flask, add 5 ml of internal standard solution (PHBA 1.2 mg/ml), and dilute with 50 % methanol to the scale. Chromatographic column: Shim-Pack VP-ODS (4.6 mm × 15 cm); mobile phase: methanol—1 % acetic acid solution (12.5:87.5); detection wavelength: 280 nm; column temperature: 25 °C; and flow rate: 1.0 ml/min.

(iv) Sodium danshensu and protocatechuic aldehyde [18]: Precisely measure 2 ml of the product, place in a 50 ml volumetric flask, and dilute with methanol to the scale. Chromatographic column:

Table 14.21 The contents of protocatechuic aldehyde and danshensu in Xiangdan Injection samples from different manufacturers

Manufacturer	Batch no.	Danshensu (mg/ml)	Protocatechuic aldehyde (mg/ml)
A	981214	2.35	0.482
B	980438	3.07	0.462
B	970632	2.14	0.567
B	970633	0.37	0.091
C	990404	3.63	0.384
C	990615	3.11	0.156
C	990605	2.69	0.298
C	990506	4.44	0.122

Table 14.22 The danshensu content in Compound Danshen Injection samples from different manufacturers

Manufacturer	Batch no.	Danshensu (mg/ml)
A	981204	4.225
B	981240	2.947
C	980340	4.043
D	990128	4.194
E	981101	2.340
F	980829	2.595

Table 14.23 The contents of protocatechuic aldehyde and sodium danshensu in Xiangdan Injection samples from different manufactruers

Manufacturer	Batch no.	Sodium danshensu (mg/ml)	Protocatechuic aldehyde (mg/ml)
Drug manufacturer A of Shanghai	010913	2.65	0.39
Drug manufacturer A of Shanghai	02013904	2.16	0.44
Drug manufacturer B of Shanghai	0201071	2.45	0.33
Drug manufacturer C of Shanghai	020423	3.67	0.45
Drug manufacturer of Kunming	020302	1.88	0.24
Drug Manufacturer of Guangdong	020203	2.57	0.18

YWG C_{18} (10 mm × 300 mm, 10 μm); mobile phase: methanol—water—glacial acetic acid (1:4:0.25); flow rate: 1.1 ml/min; and detection wavelength: 280 nm. See the results in Table 14.23.

2. Determination of Salvianolic Acid a by HPLC [19]
   (a) Method
   The HPLC protocol was described in *Chinese Pharmacopoeia* 2005, Volume 1, Appendix VI D.
   (i) Chromatographic condition and system suitability test: Filler: octadecylsilane chemically bonded silica; mobile phase: methanol–water–formic acid (41:59:0.8); column temperature: 25 °C; detection wavelength: 285 nm. As per salvianolic acid A peak, the amount of theoretical plates should not be less than 3000.
   (ii) Preparation of the reference solution: Precisely weigh an appropriate amount of salvianolic acid A reference substance, and add mobile phase to make a 79.64 μg/ml solution; this is the reference solution.
   (iii) Preparation of the test solution: Precisely measure 2 ml of the product, transfer into a 25 ml volumetric flask, dilute with mobile phase to the scale, and shake well; this is the test solution.
   (iv) Assay: Inject 10 μl of the reference and the test solutions, respectively, into HPLC, and collect data.
   (b) Methodological Study and Content Determination
   (i) The methodological test showed that salvianolic acid A reference solution had a good linearity when the sample size was between 0.24 and 5.6 μg: $Y = 1.00 \times 10^6X - 5.656 \times 10^4$, $r = 0.9997$; RSD of the intraday precision test was 1.8 % ($n = 5$), and RSD of reproducibility test was 2.4 % ($n = 6$). The average recovery was 101.3 %, and RSD was 0.9 %. The stability test showed that the test solution was stable for 4 days.
   (ii) The contents of salvianolic acid A in the samples from different manufacturers: Salvianolic acid A was undetectable in the samples from manufacturers T and J; contents in the samples from manufacturers H, S, and Y were 1.148 mg/ml, 0.457 mg/ml, and 0.479 mg/ml, respectively; the RSDs were 1.9, 1.4, and 0.9 %, respectively ($n = 3$).
   (c) Determination of Sodium Danshensu, Protocatechuic Aldehyde, and Salvianolic Acid B by HPLC
   (i) Method
   Xiangdan Injection is made of Danshen and rosewood, but rosewood volatile oil has no absorption in the UV region, therefore, the contents of sodium danshensu, protocatechuic aldehyde, and salvianolic acid B can be determined by HPLC with a detection

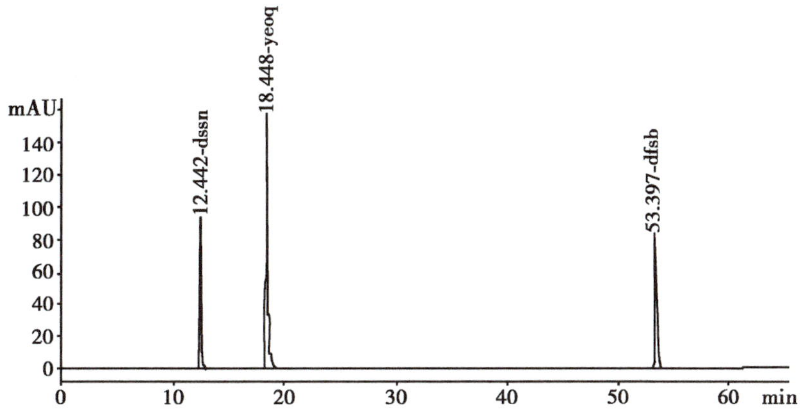

Fig. 14.16 HPLC profile of sodium danshensu, protocatechuic aldehyde and salvianolic acid B reference substances

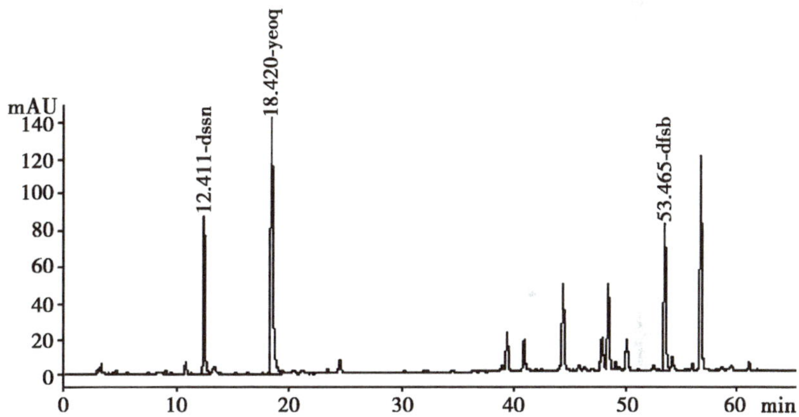

Fig. 14.17 HPLC profile of Xiangdan injection

wavelength of 280 nm and with methyl cyanide–0.03 % phosphoric acid solution as the mobile phase, by gradient elution of sodium danshensu, protocatechuic aldehyde and salvianolic acid B (Figs. 14.16 and 14.17).

(ii) The results of the tests are shown in Table 14.24.

14.1.6.3 High Performance Capillary Electrophoresis (HPCE) [20]

1. Method

   Determining the contents of protocatechuic aldehyde and danshensu according to the High Performance Capillary Electrophoresis described in the *Chinese Pharmacopoeia* 2005, volume 2, Appendix VI D.

   (a) Electrophoretic separation condition: Uncoated capillary 27 cm × 75 μm; running buffer 20 mmol/L borax—3 mmol/L β-CD (pH 9.0); sample introduction under high pressure for 5 s, operating voltage of 6 kV(+)→(−); detection wavelength: 209 nm; column temperature: 25 °C.

   (b) Preparation of the reference solutions: Precisely weigh appropriate amounts of sodium danshensu and protocatechuic aldehyde reference substances which have been dried under reduced pressure with

Table 14.24 The contents of danshensu, protocatechuic aldehyd and salvianolic acid B in Xiangdan Injection samples from different manufacturers

Manufacturer	Batch no.	Danshensu (mg/ml)	Protocatechuic aldehyde (mg/ml)	Salvianolic acid B (mg/ml)
Manufacturer of Siquan	040210	0.58	0.174	0.46
Manufacturer of Shanxi	040314	3.61	0.792	2.99
Manufacturer of Yunnan	040201	1.55	0.274	1.33
Manufacturer of Jiangsu	040301	1.07	0.124	0.62
Manufacturer of Guangdong	0403007	2.03	0.161	0
Manufacturer of Hebei	0404001	1.34	0.283	0.57
Manufacturer of Jiangsu	040110	2.54	0.45	0.40
Manufacturer of Guangdong	0403007	2.03	0.161	0.0
Manufacturer of Kunming	20040304	2.09	0.392	0.0
Manufacturer of Zhejiang	040101	2.17	0.263	1.17
Manufacturer of Zhejiang	040106	1.71	0.291	0.96
Manufacturer of Zhejiang	0404133	1.52	0.338	0.71

desiccant phosphori pentoxidum for 24 h, and add water to make a mixed solution containing 0.4 mg/ml sodium danshensu and 0.1 mg/ml protocatechuic aldehyde.

(c) Preparation of the internal standard solution: Precisely weigh an appropriate amount of benzoic acid, dissolve with water and dilute to the volume to make a 0.1 mg/ml solution.

(d) Correction factor determination: Precisely weigh 0.6 ml of the reference solution, place in a 10 ml volumetric flask, precisely add 0.6 ml of the internal standard solution, dilute with water to the scale, shake well, inject the sample, and calculate the correction factor.

(e) Assay: Precisely measure 2 ml of Xiangdan Injection and 0.6 ml of internal standard solution, place in a 10 ml volumetric flask, dilute with water to the scale, shake well, and inject the samples.

2. Methodological Study

The linear range of danshensu was 4.06–48.72 μg/ml, and the correlation coefficient was 0.9994; the linear range of protocatechuic aldehyde was 3.59–43.08 μg/ml, and the correction coefficient was 0.9978. The recovery rate of danshensu was 98.91 %, and the recovery rate of protocatechuic aldehyde was 98.96 %. In the reproducibility tests, the RSD for danshensu was 1.2 %, and for protocatechuic aldehyde was 1.5 %.

3. Result

See the test results of danshensu and protocatechuic aldehyde contents in Xiangdan Injection in Table 14.25.

Table 14.25 The contents of danshensu and protocatechuic aldehyde in Xiangdan injection

Manufacturer	Batch no.	Danshensu (mg/ml)	Protocatechuic aldehyde (mg/ml)
A	990716	0.5623	0.1437
A	990718	0.5765	0.1475
A	990720	0.5879	0.1428

14.1.7 Determination of Salvianolic Acids in Lyophilized Danshen Total Phenolic Acid Injection

Zhengliang Ye and Jun Gao

14.1.7.1 Salvianolic Acid B

1. Method

(a) Chromatographic condition and system suitability test: Filler: octadecylsilane chemically bonded silica; mobile phase: acetonitrile–water–phosphoric acid (23.5:76.5:0.02); detection wavelength: 288 nm. As per salvianolic acid B peak, the amount of theoretical plates should not be less than 5000.

(b) Preparation of the reference solution: Precisely weigh salvianolic acid B reference substance and add mobile phase to make a 0.2 mg/ml solution.

(c) Preparation of the test solution: Precisely weigh approximately 20 mg of the product, transfer into a 50 ml volumetric flask, add mobile phase to the scale, and shake well.

(d) Assay: Inject 10 μl of the reference and the test solutions, respectively, into HPLC and collect data.

2. Methodological Study and Content Determination

(a) Standard curve: Precisely weigh an appropriate amount of salvianolic acid B reference substance, add mobile phase to make a 0.88 mg/ml solution, and keep it as the stock solution. Take 0.25, 0.5, 1.0, 2.0, 3.0 and 4.0 ml the stock solution, place in a 10 ml volumetric flask, respectively, add mobile phase to the scale, shake well, precisely take 10 μl of each to inject into HPLC, and calculate, to get the regression equation $Y = 4821.6X + 9764.6$ and correlation coefficient $r = 0.9993$. The linear range of salvianolic acid B was 220–3520 ng.

(b) The precision test: Take the same reference solutions and test by HPLC 5 times. The RSD of peak area was 1.2 %.

(c) The blank test: Prepare excipients blank accessories according to the prescription, precisely weigh about 9 mg of it, dissolve, and assay using the same method as for the samples. The result showed no interference peaks appearing at the retention time of salvianolic acid B.

(d) The stability test: Precisely weigh about 35 mg of Lyophilized Danshen Total phenolic acid Injection, place in a 25 ml volumetric flask, add mobile phase to the scale, shake well, precisely transfer 5 ml into a 25 ml volumetric flask, add mobile phase to the scale, shake well, and precisely inject 10 μl into HPLC. The result showed that the solution was stable for 4 h.

(e) The reproducibility test: Take 5 samples, prepare the test solutions, respectively, measure the peak areas, and calculate contents. The RSD for content was 1.3 %.

(f) The recovery test: Precisely weigh 10 mg from 5 samples, place in a 50 ml volumetric flask, respectively, add about 3 mg of salvianolic acid B reference substance, add mobile phase to the scale, shake well, inject 10 μl of the sample into HPLC, determine the peak area, and calculate the recovery rate. The results are shown in Table 14.26.

Table 14.26 The recovery rates of salvianolic acid B in Lyophilized Danshen Total phenolic acid Injection determined by HPLC

Sample weight (mg)	Sample content (mg)	Added standard substance weight (mg)	Total weight (mg)	Detected weight (mg)	Recovery rate (%)	Average recovery (%)	RSD (%)
9.36	3.60	3.52	7.13	3.53	100.3	97.4	2.2
9.36	3.60	3.52	6.96	3.36	95.5		
9.41	3.62	2.78	6.32	2.70	95.2		
9.41	3.62	2.78	6.38	2.76	99.0		
9.41	3.62	2.78	6.38	2.76	96.9		

Table 14.27 The content of salvianolic acid B in Lyophilized Danshen Total phenolic acid Injection

Batch no.	20030901	20030902	20030903	20030830	20040405	20040412	20040501	20050101	20050102	20050103
Salvianolic acid B (mg/ampule)	56.2	55.4	55.7	51.5	49.9	51.3	52.5	50.6	50.3	52.0

(g) Sample Determination: Precisely weigh approximately 20 mg of the product, transfer into a 50 ml volumetric flask, add mobile phase to the scale, and shake well. Precisely inject 10 µl into HPLC. The results are shown in Table 14.27.

14.1.7.2 Total Salvianolic Acids

1. Method
   (a) Preparation of the reference solution: Precisely weigh salvianolic acid B reference substance and add methyl cyanide–water–phosphoric acid (23.5:76.5: 0.02) mixture solution to make a 20 µg/ml solution; this is the reference solution.
   (b) Preparation of the test solution: Precisely weigh 35 mg of the product, place in a 50 ml volumetric flask, add methyl cyanide–water–phosphoric acid (23.5:76.5: 0.02) mixture solution to dissolve and dilute to the scale, shake well, precisely transfer 2 ml to another 50 ml volumetric flask, add the above mixture solution to the scale, and shake well; this is the test solution.
   (c) Assay: Take the reference solution and test solution, respectively, using the mixture solution as a blank, measure the absorbance at 288 nm, and calculate according to the formula below:

$$\text{The content of total salvianolic acids (mg/ampule)} = f(A - B) + B$$

where f is the correction factor 0.626; A is total salvianolic acid content assayed by spectrophotometry and calculated with salvianolic acid B as reference; and B is salvianolic acid B content assayed by HPLC.

2. Methodological Study and Content Determination
   (a) Determination of the Maximum Absorption Wavelength
      (i) UV absorption spectrum of salvianolic acid B: Take the reference substance salvianolic acid B, add mobile phase to make a 32 µg/ml solution, and determine the absorption curve at 200–400 nm on a UV spectrophotometer. The results showed that the maximum absorption wavelength was 288 nm; therefore, 288 nm was selected as the detection wavelength.
      (ii) UV absorption spectrum of Danshen Total Phenolic Acid Injection: Use the mobile phase to make a solution containing approximately 20 µg/ml of the product, weigh excipient based on the formula proportion, and prepare another solution with mobile phase complying with the same method; determine the absorption curve at 200–400 nm on a UV spectrophotometer. The result showed that the maximum absorption wavelength was the same as that of salvianolic acid B, 288 nm, and the excipient had no interference.
   (b) Standard curve: Precisely weigh an appropriate amount of salvianolic acid B reference substance, add mobile phase to make a 0.084 mg/ml solution, and keep it as the stock solution. Precisely transfer 0.5, 1.0, 2.0, 4.0, 8.0 and 4.0 ml to a 10 ml volumetric flask, respectively, add mobile phase to the scale, and shake well. Using the mobile phase as the blank, determine the UV absorbance at 288 nm. The calculating regression equation was $Y = 0.0198X + 0.0098$, the correlation

coefficient was $r = 0.9999$, and the linear range was 4.2–67.2 μg/ml.

(c) The precision test: Precisely take 3.0 ml of the reference stock solution, place in a 10 ml volumetric flask, add mobile phase to the scale, shake well, determine the absorbance ($n = 5$), and calculate the precision. The RSD was 0.1 %.

(d) Calculation explanation and calculation formula: The calculated total phenolic acid contents in Danshen Total Phenolic Acid Injection, based on UV spectrophotometry and with salvianolic acid B as the reference, were a little higher than expected. The reason for is that the molecular weights of the other phenolic acids are lower than that of salvianolic acid B. Besides the main constituent salvianolic acid B, the total phenolic acids also include protocatechuic aldehyde, salvianolic acid E, rosmarinic acid, lithospermic acid, and salvianolic acid A. Their molecular weights are very different from each other, thus we take the ratio of the average molecular weight of protocatechuic aldehyde, rosmarinic acid, lithospermic acid, salvianolic acid A, and salvianolic acid E to the molecular weight of salvianolic acid B as the correction factor to calculate the contents of the other phenolic acids. The molecular weights of these five phenolic acids are shown in Table 14.28.

Correction factor f = average molecular weight of 5 phenolic acids/molecular weight of salvianolic acid B = 449.6/718 = 0.626

Therefore, the calculation formula for total phenolic acid content (%) assayed by ultraviolet spectrophotometry should be: $f(A - B) + B$

Where f = correction factor; A = total salvianolic acid content assayed by ultraviolet spectrophotometry and calculated with salvianolic acid B as reference; B = salvianolic acid B content determined by HPLC.

(e) The stability test: Precisely weigh about 35 mg of Danshen Total Phenolic Acid Injection, place in a 50 ml volumetric flask, add mobile phase to dissolve and dilute to the scale, shake well, precisely transfer 2 ml into another 50 ml volumetric flask, add mobile phase to the scale, shake well, determine UV absorbance every 1 h, and calculate the content. The results showed that the test solution was stable for 4 h.

(f) The reproducibility test: Take 5 samples, prepare the test solutions using the above method, determine the absorbance, and calculate the contents. The RSD was 0.1 %.

(g) The recovery test: Precisely weigh 17.5 mg of the product from 5 samples, place in a 50 ml volumetric flask, respectively, add mobile phase to dissolve and dilute to the scale, and shake well. Precisely transfer 2 ml to another 50 ml volumetric flask, precisely add 1 ml of salvianolic acid B reference solution (0.498 mg/ml) to each flask, dilute with mobile phase to the scale, shake well, assay, and calculate the recovery rate. The results are shown in Table 14.29.

(h) Sample determination: Precisely weigh approximately 35 mg of the product, transfer into a 50 ml volumetric flask, add mobile phase to the scale, and shake well. Precisely transfer 2 ml to another 50 ml volumetric flask, dilute with mobile phase to the scale, shake well, with the excipient

Table 14.28 The molecular weights of 5 phenolic acids

Phenolic acid	Protocatechuic aldehyde	Rosmarinic acid	Lithospermic acid	Salvianolic acid A	Salvianolic acid E	Salvianolic acid B	Mean value
Molecular formula	$C_7H_6O_3$	$C_{18}H_{16}O_8$	$C_{27}H_{22}O_{12}$	$C_{26}H_{22}O_{10}$	$C_{36}H_{30}O_{16}$	$C_{36}H_{30}O_{16}$	
Molecular weight	138	360	538	494	718	718	449.6

Table 14.29 The recovery Rates of Salvianolic Acid B in Danshen Total Phenolic Acid Injection

Sample weight (mg)	Sample content (mg)	Added standard substance weight (mg)	Total weight (mg)	Detected weight (mg)	Recovery rate (%)	Average recovery (%)	RSD (%)
20.24	12.54	9.96	22.40	9.86	99.00		
20.17	12.53	9.96	22.34	9.81	98.49		
20.09	12.48	9.96	22.29	9.81	98.49		
20.14	12.51	9.96	22.34	9.83	98.69		
20.11	12.49	9.96	22.29	9.80	98.39		

Table 14.30 The contents of total phenolic acids in Danshen Total Phenolic Acid Injection determined by UV Spectrophotometry

Batch no.	20030901	20030902	20030903	20030830	20040405	20040412	20040501	20050101	20050102	20050103
Salvianolic acid B (mg/ampule)	78.4	76.4	78.4	88.6	72.8	74.0	81.2	70.6	70.5	72.7

prepared by the same method as blank. Determine the absorbance at 288 nm and calculate the contents. See the results in Table 14.30.

14.1.8 Determination of the 6 Major Phenolic Acids in Danshen and Its Preparations

Aihua Liu and Dean Guo

14.1.8.1 Materials and Method

1. Instruments, Reagents and Drugs
   High performance liquid chromatograph: Agilent 1100 HPLC, equipped with Quat Pump, Autosampler, Thermostatted Column Compartment, and Diode Array Detector. Sartorius electronic balance; methanol and methyl cyanide (L.C.), double distilled water, the other reagents were all A.R. grade.
   The 6 phenolic acid reference substances, danshensu (1), protocatechuic aldehyde (2), rosmarinic acid (3), lithospermic acid (4), salvianolic acid B (5) and salvianolic acid A (6), were isolated from Danshen. The purities of the 6 compounds were more than 98 %, as determined by HPLC. Their structures are shown in Fig. 14.18. All medicinal materials and preparations were bought from local drug markets or drugstores in China. The properties of different preparations are shown in Table 14.31.
2. HPLC condition: Stationary phase: Zorbax Extend-C_{18} chromatographic column (4.6 mm × 250 mm, 5 μm); mobile phase: methyl cyanide (A)—0.026 % phosphoric acid water solution (B); run gradient elution as follows:

Time (min)	A (%)	B (%)
0	2	98
20	23	77
35	28.5	71.5

   Flow rate of mobile phase: 1.0 ml/min, detection wavelength: 288 nm; column temperature: 20 °C; and sample size: 10 μl.
3. Preparation of the reference solutions: Precisely weigh appropriate amounts of the 6 phenolic acid, place in 5 ml volumetric flasks, and dissolve with 50 % methanol to make reference stock solutions with known concentrations. Take appropriate amounts of the reference stock solutions, respectively, place in another 5 ml volumetric flask, make serial dilutions with 50 % methanol, and determine the UV absorbance of each concentration to generate a standard curve. Keep the reference solutions in a 4 °C refrigerator for further use.

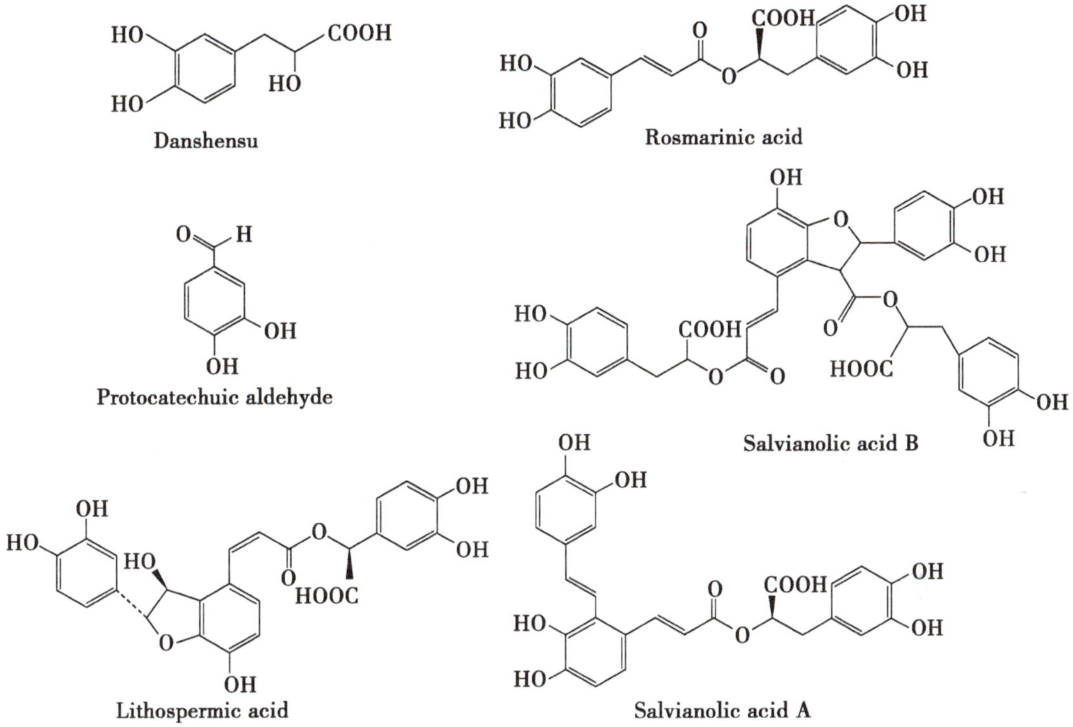

Fig. 14.18 The chemical structure of the 6 phenolic acids in Danshen

Table 14.31 Traditional Chinese medicine preparations

Drug name	Medicament form	Drug formulation	Indication
Compound Danshen Tablet	Tablet	Danshen (450 g), notoginseng (141 g), broneolum syntheticum (8 g)	For chest distress, angina pectoris, etc.
Dantonic™	Pill	Danshen, notoginseng, and broneolum syntheticum	For chest distress, angina pectoris, etc.
Danshen Injection	Injection	Danshen	For coronary artery disease, chest distress, etc.
Xiangdan Injection	Injection	Danshen (1,000 g) Rosewood (1,000 g)	For heart infarction, angina pectoris, etc.

4. Preparation of the test solutions: Pretreatment: Dry Danshen to constant weight, pulverize, and sift through a 100-mesh sieve; remove the coating or film of Compound Danshen Tablet, pulverize, and sift through a 100-mesh sieve; and grind Dantonic™ to a powder. Treatment of sample: Precisely weigh 0.300 g of the treated solid sample, add 10 ml of 70 % methanol solvent, extract under reflux for 1 h, cool, weigh, make up for the lost weight, filter through a 0.45 μm filter membrane, and repeat to extract and assay each sample 3 times. Or, take 1 ml of liquid sample (Danshen Injection and Xiangdan Injection), dilute to 5 ml, mix, and filter through a 0.45 μm filter membrane. Keep all samples in a 4 °C refrigerator.
5. System suitability test: Continuously and repeatedly test the mixed reference solutions 5 times to evaluate the system suitability; and as for all the test samples, the tailing factor

should be less than 1.2, separating degree greater than 1.5, and number of theoretical plates more than 10,000.

14.1.8.2 Results and Discussion

1. Optimization of chromatographic condition: A perfect chromatographic condition should be able to reach baseline separation of the adjacent peaks in the shortest possible time. To optimize the chromatographic condition to achieve as good as possible of a separating effect, we investigated the primary influencing factors such as stationary phase, mobile phase, column temperature, detection wavelength, and flow rate.
   First, we investigated the separating effects of different stationary phases on phenolic acids and compared the target peaks' separating degree, shape, stability, scope of use. The results showed that Zorbax Extend C_{18} was more applicable to phenolic acids than were BDS–Hypersil C_{18}, YMC–Pack ODS–A C_{18} and Luna C_{18} columns. Then, we investigated different mobile phases, and the results showed different proportions of water-methyl cyanide were more applicable to phenolic acids separation than was water-methanol. Although the peak forms and degree of separation obtained from water–methyl cyanide were not very good, based on the references [21–23], it was clear that the mobile phase supplemented with an appropriate amount of acid could depress the ionization of compounds containing phenolic hydroxyl groups or carboxyl groups, and could consequently improve the degree of separation and peak form. Therefore, we investigated different proportions of phosphoric acid, formic acid and acetic acid in the test, and found that 0.026 % phosphoric acid could separate phenolic acids well. In addition, we also investigated the separating effects at different column temperatures, including 15, 20, 30 and 40 °C. The results showed that the separating effect was the best at a column temperature of 20 °C. Flow at 1.0 ml/min could achieve a good degree of separation and shorten the assaying time. We also detected detection wavelengths ranging from 200 to 400 nm, and found that 288 nm was the most suitable.
2. Optimization of extraction conditions: To get the best extraction result, we investigated the factors influencing extraction. The first factor investigated was solvent. With 6 chromatographic peaks as the object of analysis, we compared the extraction efficiencies of methanol, 70 % methanol, 50 % methanol, 30 % methanol, 10 % methanol, and water. Results showed that water could maximize danshensu and protocatechuic aldehyde extraction, but could not completely extract salvianolic acids B and *A*. On the other hand, pure methanol could extract salvianolic acids B and A completely, but its extraction of danshensu and protocatechuic aldehyde was incomplete. 70 % methanol could extract all 6 constituents to thc maximum extent (Fig. 14.19). We then investigated extraction modes such as ultrasound, soaking, and reflux, and the results showed that reflux extraction was the best extraction mode. Finally, we investigated the extraction time, and results showed that extraction for 30 min could completely extract phenolic acids (Fig. 14.20). In summary, the best way of extracting phenolic acids from Danshen was reflux extraction with 10 ml of 70 % methanol for 30 min.
3. Method Validation: Based on the guiding principles of method validation in Method Validation ICH (the International Conference on Harmonization) and pertinent literature and reports [24, 25], the linearity, precision, accuracy, stability, etc. were investigated.
   (a) Standard curve establishment and limit of detection (LOD): By external standard method, 7 different concentrations of each analyte were tested, and each reference solution concentration was tested 3 times. The standard curve was established by plotting the ratio of peak area to sample concentration, and the final standard curve was obtained after averaging the 3 results (Table 14.32). The results showed that all 6 compounds displayed excellent linearity in the range of the standard curve

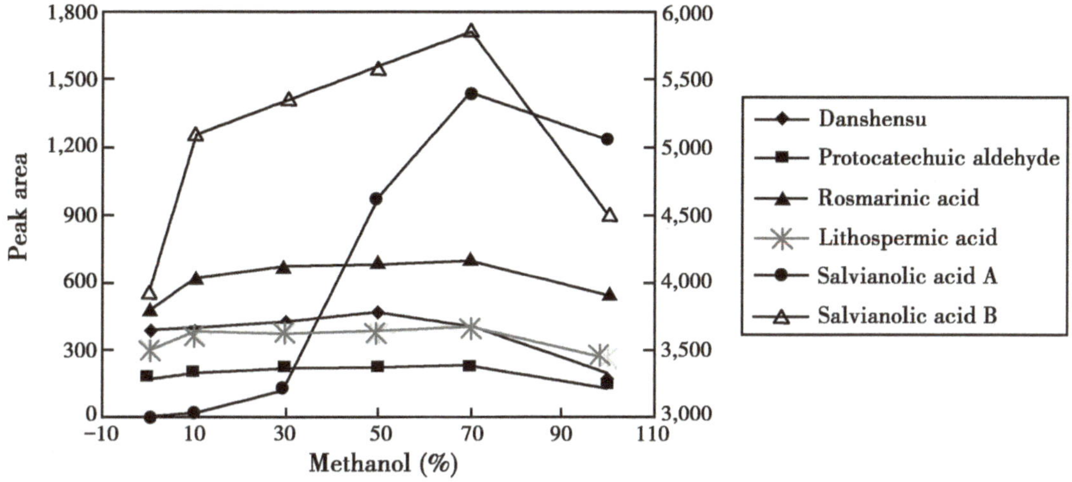

Fig. 14.19 The extraction efficiencies of different solvents

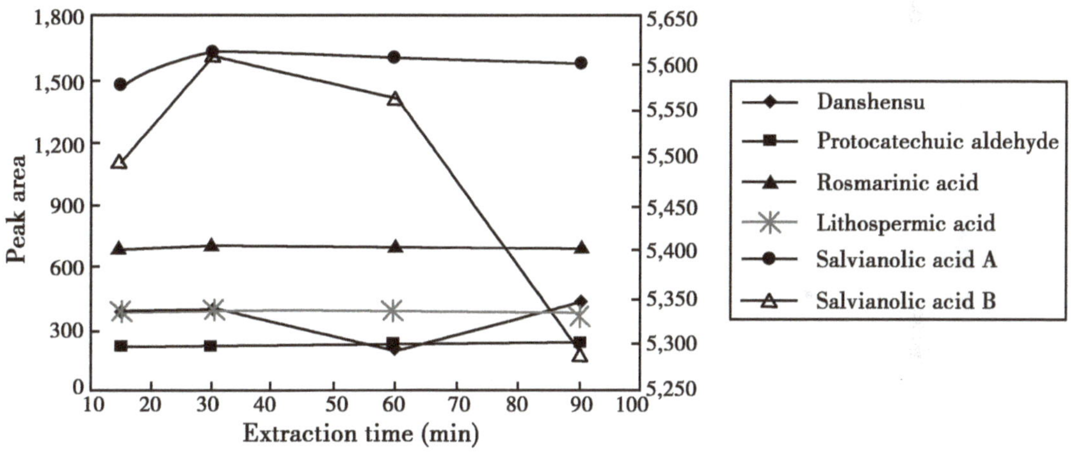

Fig. 14.20 The efficiencies of different extraction time

Table 14.32 The calibration curves of 6 phenolic acids (n = 4)

Compound	Retention time (min)	Standard curve	r^2	Test range (μg/ml)	LOD (μg/ml)
Danshensu	10.30	Y = 5.2530X-1.1146	0.9998	10.84–346.9	0.15
Protocatechuic aldehyde	14.62	Y = 39.061X-4.3277	0.9998	0.690–24.15	0.04
Rosmarinci acid	27.50	Y = 19.156X-8.4844	0.9998	1.115–78.05	0.15
Lithospermic acid	28.38	Y = 11.272X-7.9138	0.9997	2.163–108.1	0.32
Salvianolic acid B	30.97	Y = 9.7908X-30.393	0.9999	13.50–810.0	0.43
Salvianolic acid A	33.73	Y = 33.616X + 15.493	0.9997	2.530–88.55	0.18

Y Peak area; *X* Analyte concentration (μg/ml); *LOD S/n* = 3

($r^2 > 0.999$).

According to ICH, LOD is defined as the lowest concentration of the assayed substance that can be distinguished from background interference, namely the sample concentration when signal-noise ratio is 3. The LOD of each assayed substance is shown in Table 14.32.

(b) The precision test: The evaluation of precision is achieved by determining the intra-day and inter-day values of a reference substance, during which, 3 concentration levels, namely, low, middle and high concentrations, of each reference substance are selected. The intra-day precision is estimated by determining 6 samples that have been treated by the same method in a day; and the inter-day precision is completed in three consecutive days. The reproducibility of the method is estimated by calculating the relative standard deviation (RSD). As revealed by the results, both intra- and inter-day RSDs of the 6 compounds were less than 5 %, indicating that the method had good repeatability (Table 14.33).

Table 14.33 The intra-day and inter-day precision of the method

Concentration (μg ml^{-1})	Intra-day (n = 5)			Inter-day (n = 9)		
	Found	RSD[a] (%)	Accuracy (%)	Found	RSD (%)	Accuracy (%)
Danshensu						
21.68	21.51 ± 0.02	0.10	99.22	21.42 ± 0.10	0.46	98.80
216.80	214.81 ± 0.17	0.08	99.08	215.44 ± 1.53	0.71	99.37
325.20	320.48 ± 4.60	1.44	98.55	324.29 ± 3.06	0.94	99.72
Protocatechuic aldehyde						
0.79	0.77 ± 0.00	0.23	97.47	0.77 ± 0.01	0.65	97.09
10.35	10.18 ± 0.01	0.07	98.36	10.29 ± 0.13	1.27	99.40
10.35	19.02 ± 0.03	0.17	98.45	18.92 ± 0.15	0.79	97.93
Rosmarinic acid						
2.23	2.35 ± 0.00	0.00	105.38	2.34 ± 0.04	1.75	105.11
33.45	32.04 ± 0.04	0.12	95.8	33.62 ± 1.28	3.81	100.50
62.44	60.30 ± 0.06	0.10	96.57	63.08 ± 1.34	2.12	101.03
Lithospermic acid						
4.33	4.51 ± 0.01	0.22	104.16	4.49 ± 0.17	3.83	103.63
4.33	42.91 ± 0.08	0.19	99.21	42.61 ± 2.01	4.73	98.51
90.83	93.90 ± 0.05	0.06	103.38	94.38 ± 2.87	3.04	103.91
Salvianolic acid B						
27.00	26.95 ± 0.06	0.24	99.81	27.49 ± 0.60	2.17	101.80
405.00	391.17 ± 0.39	0.10	96.59	395.30 ± 2.88	0.73	97.60
756.00	722.97 ± 0.78	0.11	95.63	757.04 ± 23.36	3.04	100.14
Salvianolic acid A						
2.53	2.42 ± 0.01	0.58	95.66	2.55 ± 0.11	4.31	100.79
2.53	37.53 ± 0.08	0.22	98.89	36.79 ± 0.86	2.34	96.94
70.84	69.15 ± 0.23	0.34	97.61	67.73 ± 1.46	2.16	95.66

[a] Relative standard deviation (%) = (standard deviation/mean) × 100 %

[b] Recovery rate (%) = (mean of the measured amount/added amount) × 100 %

(c) The stability test: The test solutions were assayed in the interval of 2, 4, 6, 9 and 12 h, respectively, and the RSD of the contents of the 6 test substances were less than 1.9, 1.3, 2.8, 3.5, 3.2 and 4.3 %, respectively, indicating that the samples were stable at room temperature for at least 12 h.

(d) The reproducibility test: Precisely weigh Danshen sample, prepare 3 test solutions according to the above method, and assay the samples 5 times under the same assaying condition. The results showed that the RSD of the contents of the 6 substances were 1.4, 1.2, 2.4, 3.3, 3.4 and 4.2 %, respectively, indicating that the method had good reproducibility and was feasible.

(e) The recovery test: To a Danshen sample with known content, add a known amount of mixed reference stock solution, perform extraction and HPLC determination, and calculate the recovery rate based on the ratio of the measured theoretical value to actual addition amount. In the recovery test, we investigated 3 different concentration levels, namely high, middle, and low concentrations, in the standard curve. Repeat each test for 3 times and calculate average recovery; the results can be seen in Table 14.34. The results indicated that the test had good accuracy.

4. Sample determination: Test the reference solutions and test solutions with the above

Table 14.34 The recovery test results (n = 4)

Phenolic acid spiked (μg/ml)	Found (μg/ml)	RSD[a] (%)	Recovery[b] (%)
Danshensu			
21.68	20.61 ± 0.53	2.57	95.05
130.08	129.84 ± 1.53	1.18	99.81
220.37	217.90 ± 1.77	0.81	98.88
Protocatechuic aldehyde			
0.69	0.68 ± 0.01	0.84	98.99
6.90	6.75 ± 0.06	1.48	97.83
13.80	13.12 ± 0.33	2.52	95.07
Rosmarinic acid			
2.23	2.31 ± 0.07	2.88	103.74
22.30	23.30 ± 0.23	0.98	104.48
55.75	54.42 ± 0.69	1.26	97.61
Lithospermic acid			
4.33	4.39 ± 0.09	2.09	101.44
43.25	42.85 ± 1.30	3.04	99.08
69.20	71.9 ± 0.84	1.17	103.90
Salvianolic acid B			
27.00	28.30 ± 0.24	0.84	104.78
216	219.99 ± 1.77	0.80	101.85
351	364.80 ± 4.34	1.19	103.93
Salvianolic acid A			
5.06	5.07 ± 0.05	0.93	100.26
25.30	24.74 ± 0.78	3.13	97.80
50.60	51.79 ± 0.58	1.11	102.36

[a] Relative standard deviation (%) = (standard deviation/mean) × 100 %

[b] Recovery rate (%) = (mean of the measured amount/added amount) × 100 %

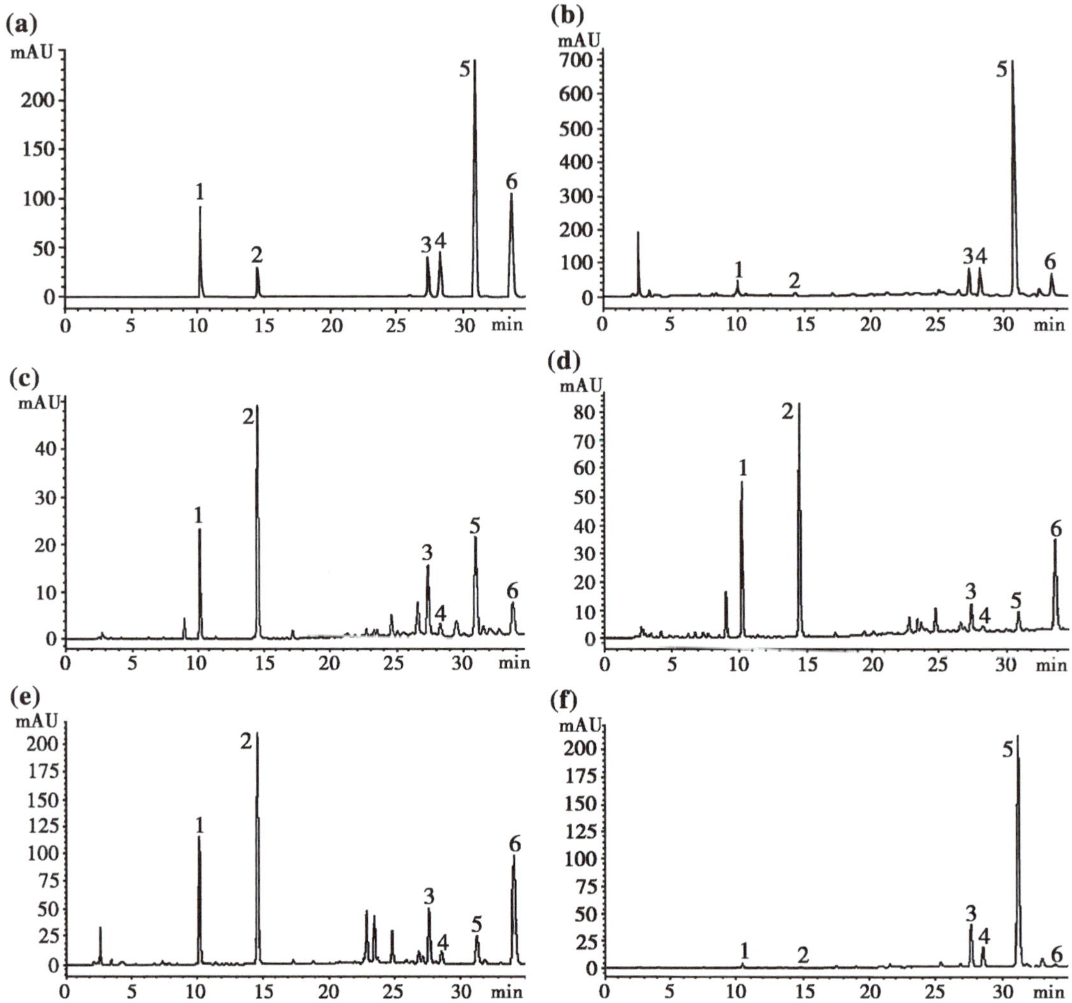

Fig. 14.21 Representative HPLC profiles Panels: **a** High concentration of standard solution; **b** Compound Danshen Tablet (Hebei, China, 9919); **c** Danshen Injection (Guangzhou, China, 040811); **d** Xiangdan Injection (Shanghai, China, 200408214); **e** Dantonic™ (Tianjing, China, 2004218; **f** Danshen (Henan, China, 2003). Peaks: *1* Danshensu; *2* Protocatechuic aldehyde; *3* Rosmarinic acid; *4* Lithospermic acid; *5* Salvianolic acid B; *6* Salvianolic acid A

established method; the typical chromatograms are shown in Fig. 14.21. From the chromatograms, the retention times of danshensu, protocatechuic aldehyde, rosmarinic acid, lithospermic acid, salvianolic acid B and salvianolic acid A were 10.30, 14.62, 27.50, 28.38, 30.97 and 33.73 min, respectively. The indicator peaks and other peaks had baseline separation, and no interference peaks appeared. We tested a total of 28 compound Danshen samples; among them, 20 were from different manufacturers, and 8 were products from the same manufacturer but with different batch numbers. The contents of the 6 phenolic acids in the Danshen sample were calculated based on the standard curve equation, which was for the first time the 6 primary water-soluble constituents in Danshen and its preparations were determined, and the results are shown in Table 14.35. It can be seen from Table 14.35 that the phenolic acid contents in the samples

Table 14.35 The contents of 6 phenolic acids in Danshen and its preparations ($n = 3$)

No.	Batch no.	Place of collection	Content (%) [f]($n = 3$)					
			Danshensu	Protocatechuic aldehyde	Rosmarinic acid	Lithospermic acid	Salvianolic acid B	Salvianolic acid A
1[a]	30902	Guangdong	1.03 ± 0.00	0.08 ± 0.00	1.56 ± 0.00	1.05 ± 0.02	17.70 ± 0.16	0.22 ± 0.00
2[a]	31207	Beijing	10.00 ± 0.08	5.24 ± 0.01	1.18 ± 0.02	0.95 ± 0.02	19.66 ± 0.06	1.16 ± 0.01
3[a]	40210	Henna	5.21 ± 0.08	0.60 ± 0.00	1.64 ± 0.01	0.87 ± 0.01	19.68 ± 0.09	1.99 ± 0.06
4[a]	30207	Zhejiang	2.01 ± 0.04	0.12 ± 0.00	0.83 ± 0.00	1.81 ± 0.03	21.71 ± 0.30	1.03 ± 0.03
5[a]	20040561	Yunnan	23.67 ± 0.03	0.36 ± 0.00	1.17 ± 0.01	1.29 ± 0.03	24.16 ± 0.15	1.10 ± 0.02
6[a]	2003110	Heilongjiang	4.24 ± 0.04	0.36 ± 0.00	1.11 ± 0.01	1.69 ± 0.03	23.22 ± 0.23	1.67 ± 0.04
7[a]	40401	Jiangxi	1.93 ± 0.00	0.15 ± 0.00	0.82 ± 0.02	0.81 ± 0.03	12.49 ± 0.18	1.01 ± 0.03
8[a]	20301	Jiangxi	0.73 ± 0.03	0.09 ± 0.00	0.75 ± 0.01	1.27 ± 0.02	21.14 ± 0.43	0.11 ± 0.00
9[a]	40107	Sichuan	1.00 ± 0.02	0.02 ± 0.00	1.07 ± 0.01	1.46 ± 0.02	23.82 ± 0.48	0.17 ± 0.00
10[a]	10427	Beijing	8.30 ± 0.01	0.44 ± 0.00	1.30 ± 0.02	0.90 ± 0.02	15.36 ± 0.29	2.19 ± 0.03
11[a]	40301	Guangdong	2.96 ± 0.03	0.10 ± 0.00	1.50 ± 0.02	0.89 ± 0.02	11.51 ± 0.17	0.84 ± 0.02
12[a]	30901	Guangdong	0.77 ± 0.01	0.11 ± 0.00	1.97 ± 0.01	0.84 ± 0.02	13.62 ± 0.20	0.13 ± 0.00
13[a]	31211	Guangdong	1.30 ± 0.03	0.06 ± 0.00	0.71 ± 0.02	1.44 ± 0.02	18.67 ± 0.30	0.53 ± 0.00
14[a]	40309	Guangxi	0.45 ± 0.01	nd[g]	0.71 ± 0.02	1.17 ± 0.00	19.33 ± 0.18	nd[g]
15[a]	40403	Jiangxi	1.84 ± 0.01	0.19 ± 0.00	0.95 ± 0.02	0.86 ± 0.00	14.84 ± 0.18	0.77 ± 0.01
16[a]	40501	Guangdong	0.48 ± 0.01	0.05 ± 0.00	0.46 ± 0.00	0.52 ± 0.02	11.03 ± 0.15	0.03 ± 0.00
17 [a*]	4051005	Guangdong	2.57 ± 0.02	0.19 ± 0.00	1.13 ± 0.18	1.02 ± 0.03	16.24 ± 0.23	1.26 ± 0.03
18 [a*]	4011008	Guangdong	2.70 ± 0.02	0.18 ± 0.00	0.92 ± 0.02	1.07 ± 0.01	15.89 ± 0.26	1.57 ± 0.03
19 [a*]	4041012	Guangdong	2.70 ± 0.02	0.17 ± 0.00	0.97 ± 0.00	0.93 ± 0.00	14.99 ± 0.28	1.16 ± 0.01
20 [a*]	4021008	Guangdong	2.76 ± 0.02	0.17 ± 0.00	0.92 ± 0.02	0.88 ± 0.00	13.80 ± 0.20	1.23 ± 0.03
21 [a*]	4031018	Guangdong	2.74 ± 0.02	0.19 ± 0.00	0.96 ± 0.01	1.02 ± 0.02	16.06 ± 0.16	1.27 ± 0.03
22 [a*]	3111005	Guangdong	2.73 ± 0.02	0.17 ± 0.00	1.04 ± 0.01	1.11 ± 0.01	16.00 ± 0.12	1.48 ± 0.02

(continued)

Table 14.35 (continued)

No.	Batch no.	Place of collection	Content (%) [f]($n = 3$)					
			Danshensu	Protocatechuic aldehyde	Rosmarinic acid	Lithospermic acid	Salvianolic acid B	Salvianolic acid A
23 [a*]	4081001	Guangdong	2.97 ± 0.02	0.17 ± 0.00	1.01 ± 0.01	1.07 ± 0.01	15.82 ± 0.24	1.26 ± 0.02
24 [a*]	3121020	Guangdong	2.87 ± 0.01	0.18 ± 0.00	1.00 ± 0.01	1.04 ± 0.03	15.23 ± 0.26	1.31 ± 0.01
25[a]	20040407	Liaoning	4.38 ± 0.06	0.54 ± 0.00	1.26 ± 0.03	1.70 ± 0.03	26.71 ± 0.36	2.05 ± 0.04
26[a]	403006	Jiangsu	6.81 ± 0.03	0.33 ± 0.00	1.58 ± 0.02	1.40 ± 0.00	22.81 ± 0.31	3.34 ± 0.05
27[a]	403006	Xinjiang	0.97 ± 0.03	0.06 ± 0.00	0.61 ± 0.01	1.24 ± 0.01	16.35 ± 0.13	0.31 ± 0.00
28[a]	20090903	Shanci	3.38 ± 0.02	0.24 ± 0.00	1.38 ± 0.01	1.04 ± 0.03	16.87 ± 0.22	2.42 ± 0.06
29[b]	40915	Guangdong	18.91 ± 0.10	6.41 ± 0.07	5.13 ± 0.02	1.94 ± 0.04	19.76 ± 0.37	1.55 ± 0.01
30[c]	40710	Shanghai	22.61 ± 0.13	5.47 ± 0.07	1.70 ± 0.07	0.85 ± 0.01	3.41 ± 0.05	3.65 ± 0.03
31[d]	20030618	Tianjin	6.89 ± 0.04	2.02 ± 0.03	1.16 ± 0.01	0.69 ± 0.00	1.67 ± 0.02	1.79 ± 0.05
32[e]	040913	Henna	0.59 ± 0.01	0.03 ± 0.00	2.37 ± 0.01	2.00 ± 0.03	30.30 ± 0.13	0.25 ± 0.00

[*] Samples from the same manufacturer
[a] Compound Danshen Tablet
[b] Danshen Injection
[c] Xiangdan Injection
[d] Dantonic™
[e] Danshen herb
[f] Content = mean ± SD ($n = 3$). Unit for solid samples: mg/g drug, and unit for liquid samples: mg/10 ml drug
[g] *nd* not detected

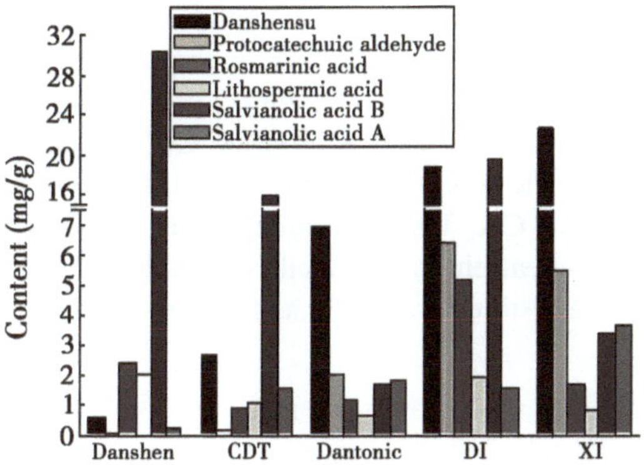

Fig. 14.22 Contents of the 6 phenolic acids in Danshen and Danshen preparations *CDT* compound Danshen tablet; *DI* Danshen injection; *XI* Xiangdan injection

of different manufacturers differed greatly, which could have been caused by the differences in the original medicinal herb, processes of production, storage, and transportations. It can also be seen from Table 14.35 that the qualities of the 8 products from the same manufacturer with different batch numbers were relatively stable.

Among Danshen and 4 Danshen preparations, the determined proportions of the 6 phenolic acids are very different (Table 14.35 and Fig. 14.22). Salvianolic acid B is richest in Danshen, and among the 4 preparations, only Compound Danshen Tablet contains the approximate proportion of salvianolic acid B, while in Dantonic™, Danshen Injection and Xiangdan Injection, the proportions of the 6 phenolic acids have changed a lot, which is caused mainly by the different preparative technologies during the process of production.

14.1.8.3 Conclusion

We have successfully determined the contents of 6 phenolic acids in Danshen and 4 Danshen preparations by HPLC, and the results indicate that the method is simple, rapid, accurate, and applicable to the quality evaluation of Danshen and Danshen preparations.

14.2 Determination of Tanshinone Contents

Aihua Liu and Dean Guo

14.2.1 Determination of Tanshinones in Danshen

14.2.1.1 Tanshinone II_A

According to the High Performance Liquid Chromatography method described in *Chinese Pharmacopoeia* 2005, volume 1, Appendix VI D.

1. Chromatographic condition and system suitability test: Filler: octadecylsilane chemically bonded silica; mobile phase: methanol–water (75:25); detection wavelength: 270 nm. As per tanshinone II_A peak, the amount of theoretical plates should not be less than 2,000.
2. Preparation of the reference sPrecisely weigh 10 mg of tanshinone II_A reference substance, place in a 50 ml brown volumetric flask, add methanol to the scale, shake well, precisely transfer 2 ml to a 25 ml brown volumetric flask, add methanol to the scale, and shake well; this is the reference solution (16 μg/ml).
3. Preparation of the test solution: Take 0.3 g of Danshen powder (sift through a #3 sieve), precisely weigh, place in a conical flask with stopper, precisely add 50 ml of methanol, weigh, heat under reflux for 1 h, cool, precisely weigh again, add methanol to make up for the lost weight, shake well, filter, and take the subsequent filtrate; this is the test solution.
4. Assay: Precisely inject 5 μl of the reference solution and the test solution, respectively, into the liquid chromatograph.

14.2.1.2 Total Tanshinones

According to the Ultraviolet Spectrophotometry method described in *Chinese Pharmacopoeia* 2005, Volume 1, Appendix V A.

1. Chromatographic condition and instrument: Shimadzu UV-2450; detection wavelength: 270 nm.

2. Preparation of the reference solution: Take an appropriate amount of tanshinone II_A reference substance, and add methanol to make a 5 μg/ml solution; this is the reference solution.
3. Preparation of the test solution: Precisely weigh about 10 mg of total tanshinone extract powder, place in a 25 ml volumetric flask, add methanol to the scale, shake well, precisely transfer 0.5 ml to another 25 ml volumetric flask, add methanol to the scale, shake well, filter, and take the subsequent filtrate; this is the test solution.

14.2.2 Determination of 4 Tanshinones in Danshen and Danshen Preparations

14.2.2.1 Materials and Method

1. Instruments, reagents and drugs: High performance liquid chromatograph, equipped with Quat Pump, Autosampler, Thermostatted Column Compartment, and Diode Array Detector. Satorius electronic balance; SB3500 ultrasonic instrument (50 kHz, 350 W); FW100 pulverizer (24,000 rpm, 460 W). Methanol and methyl cyanide (L.C.); double distilled water; and other A.R. grade reagents. The 4 tanshinone compound reference substances, namely dihydrotanshinone I, cryptotanshinone, tanshinone I, and tanshinone II_A, were all purchased from National Institutes for Food and Drug Control. Their purities were all greater than 98 % as assayed by HPLC, and the constitutional formulas can be seen in Fig. 14.23. All medical materials and preparations were purchased from local drug markets or drugstores in China.
2. Liquid-chromatographic condition: Stationary phase: Zorbax Extend C_{18} chromatographic column (250 mm × 4.6 mm, 5 μm) (Palo Alto, CA, USA); Zorbax Extend C_{18} chromatographic column (20 mm × 4 mm, 5 μm) (Palo Alto, CA, USA); mobile phase: implement gradient elution by methyl cyanide (A)–double distilled water (B) as follows:

Time (min)	A (%)	B (%)
0	45	55
3	60	40
14	60	40
15	80	20
20	82	18

The flow rate of mobile phase: 1.2 ml/min; detection wavelength: 270 nm; column temperature: 20 °C; and sample size: 10 μl.
3. Preparation of the reference solution: Precisely weigh appropriate amounts of the 4 tanshinones, respectively, place in 5 ml volumetric flasks, and add methyl cyanide to the scale to make reference stock solutions with known concentrations. Take appropriate amounts of the reference stock solutions, place in another 5 ml volumetric flask, respectively, and adjust the volume with methyl cyanide to make solutions with the concentrations for the development of the standard curve. Prepare and test the solutions 4 times. Keep the reference solutions in a 4 °C refrigerator for further use.
4. Preparation of the test solution: Pretreatment: Dry Danshen to constant weight, pulverize, and sift through a 100-mesh sieve; remove the coating or film of Compound Danshen Tablet, pulverize, and sift through a 100-mesh sieve;

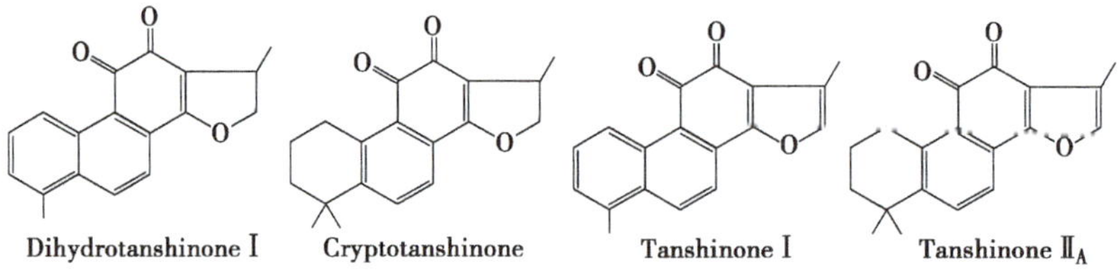

Fig. 14.23 The chemical structures of 4 tanshinones

grind Dantonic™ to powder. Treatment of sample: Precisely weigh 0.300 g of treated solid sample, add 10 ml of methanol:chloroform (7:3, v/v) solvent, extract under ultrasonic for 0.5 h, cool, filter, transfer the filtrate into a 25 ml flask with stopper, recover under reduced pressure at 28 °C, and adjust the volume to 10 ml. Repeat extraction of each sample 3 times. Or, take 1 ml of liquid sample (Danshen Injection and Xiangdan Injection), dilute to 5 ml, filter through a 0.45 μm filter membrane, and store in a 4 °C refrigerator for further use.

5. System suitability test: Continuously and repeatedly test the mixed reference solution for 5 times to evaluate the system suitability, and as for all the analytes, the tailing factor should be less than 1.2, separating degree more than 1.5, and number of theoretical plates not less than 10,000.

14.2.2.2 Results and Discussion

1. Optimization of extraction conditions: To get the best extraction efficiencies, we investigated the factors influencing extraction. Based on references [26–28], we tested the extraction efficiencies of a series of solvents, including methanol, methanol–chloroform (7:3, 5:5, 3:7, v/v) and chloroform, and the results showed that the solvent methanol–chloroform (7:3, v/v) could effectively extract the 4 target compounds (Table 14.36).

Table 14.36 The extraction efficiencies of different solvents

Extraction solvent	Tanshinone peak area			
	1	2	3	4
Methanol	194.8	392.6	614.2	2016.1
Methanol–chloroform (7:3)	287.7	539.0	863.1	2747.3
Methanol–chloroform (5:5)	165.9	324.3	504.7	1655.2
Methanol–chloroform (3:7)	140.7	276.8	433.6	1426.8
Chloroform	177.3	127.0	418.2	826.7

Table 14.37 The extraction efficiencies of the extraction duration

Extraction time (min)	Tanshinone peak area			
	1	2	3	4
10	347.5	1100.1	498.5	1130.5
20	464.4	1752.0	1012.9	2459.7
30	452.9	1811.2	1071.3	2522.4
45	464.8	1797.8	1063.0	2527.8
60	436.9	1778.3	1052.5	2457.9

We compared the extraction efficiencies of soaking, reflux, and ultrasound, and the results showed that ultrasound was the most effective extraction method. Finally, we analyzed and compared the extraction efficiencies of time on 4 target tanshinones, and the results showed that 30 min was the best extraction time within which tanshinones could be completely extracted (Table 14.37).

2. Optimization of chromatographic condition: A perfect chromatographic condition should be able to reach baseline separation of the adjacent peaks in the shortest possible time. To optimize the chromatographic condition to achieve as good a separation effect as possible, we investigated the primary influencing factors such as stationary phase, mobile phase, column temperature, detection wavelength, and flow rate.

First, we investigated the separating factors of different stationary phases on tanshinones. The effects of different stationary phases on the target peaks' separating degree, shape, stability, scope of use, etc., were compared, and the results showed that Zorbax Extend C_{18} was more applicable to tanshinones than were BDS-Hypersil C_{18}, YMC-Pack ODS-A C_{18} and Luna C_{18} columns. Then, we investigated different mobile phases, and the results showed that different proportions of water–methyl cyanide were more suitable for tanshinone separation than were water–methanol solutions, and that adding acid could not improve the separating effect, so no acid was added in the aqueous phase. Because the retention behaviors of cryptotanshinone and tanshinone I were very similar, 60 % methyl cyanide was maintained from the 3rd minute to the 14th minute to separate them. In addition, we also investigated the effects of column temperature on tanshinone separation, and the results showed that 20 °C was the best. Flow rate at 1.2 ml/min could achieve both a good degree of separation and shortened assay time. We also tested the detection wavelengths ranging from 200 to 400 nm, and found 270 nm was the most suitable.

3. Method validation: Based on the guiding principles of method validation in Method Validation the international conference on harmonization (ICH) and the pertinent literature and reports, we examined the linearity, precision, accuracy, stability, etc.
   (a) The development of standard curve: By external standard method, 7 different concentrations of each analyte were determined, and the reference solution at each concentration level was tested 3 times. The standard curve was developed by plotting the peak areas against the sample concentrations, and the final standard curves were obtained after averaging 3 test results (Table 14.38). The results showed that the 4 tanshinones displayed excellent linearity in the range of the standard curve ($r^2 > 0.999$).
   (b) LOD and limit of quantitation (LOQ): According to ICH, LOD means the lowest concentration of the substance that can be distinguished from the background interference, namely the sample concentration when signal-noise ratio (S/N) is 3. LOQ is the assayed substance concentration when S/N is 10, which means the lowest detectable concentration of the assayed substance in a sample based on the premise that certain reliability is ensured. The samples were tested 5 times, and the mean values (RSD < 5 %) were calculated, and the results showed that the LOD of the 4 compounds were 0.05, 0.03, 0.03 and 0.02 μg/ml, respectively, and the LOQ of the 4 compounds were 0.13, 0.08, 0.06 and 0.05 μg/ml, respectively (Table 14.38).
   (c) The precision test: The evaluation of precision is achieved by determining the intra-day and inter-day values of a reference substance, during which 3 concentration levels, namely, low, middle and high concentrations, of each reference substance are selected. The intra-day precision is estimated by determining 6 samples that have been treated by the same method in a day; and the inter-day precision is completed over three consecutive days. The reproducibility of the method is estimated by calculating the RSD. Both the intra-day and inter-day RSDs of the 4 compounds were less than 5 %, indicating that the method had good repeatability (Table 14.39).
   (d) The stability test: The Danshen test solutions were assayed in the interval of 2, 4, 6, 9, 12 and 24 h. The results showed that the RSDs of the 4 assayed substances were 2.7, 3.4, 3.2 and 4.2 %, respectively, indicating that the samples were stable at room temperature for at least 24 h.

Table 14.38 The standard curves of the 4 tanshinones

Analyte	Retention time	Standard curve	r^2	Test range (μg/ml)	LOD (μg/ml)	LOQ (μg/ml)
Dihydrotanshinone I	8.02	$Y = 24.32XX + 6.525$	0.9998	1.4–64.1	0.05	0.13
Cryptotanshinone	11.95	$Y = 26.835X + 11.065$	0.9998	6.9–171.9	0.03	0.08
Tanshinone I	13.23	$Y = 30.049X + 0.850$	0.9998	4.1–82.70	0.03	0.06
Tanshinone II_A	18.13	$Y = 41.220X + 23.234$	0.9999	30.86–185.14	0.02	0.05

Y Peak area; *X* Analyte concentration (μg/ml)

Table 14.39 The intra-day and inter-day precision of the method for Tanshinone determination

Concentration (μg ml^{-1})	Inter-day (n = 9)			Intra-day (n = 5)		
	Found	RSD[a] (%)	Accuracy[b] (%)	Found	RSD (%)	Accuracy (%)
Dihydrotanshinone I						
5.83	5.69 ± 0.19	3.3	97.6	5.53 ± 0.05	0.85	94.9
34.95	13.00 ± 0.17	1.3	104.0	35.10 ± 0.06	0.18	100.4
58.25	60.49 ± 0.73	1.2	103.8	60.20 ± 0.63	1.04	102.9
Cryptotanshinone						
13.75	14.04 ± 0.53	3.8	102.1	13.24 ± 0.02	0.12	96.3
82.50	85.3 ± 1.71	2.0	96.7	83.22 ± 0.39	0.47	100.9
158.13	165.75 ± 1.85	1.1	104.8	164.81 ± 0.35	0.21	104.2
Tanshinone I						
8.27	8.15 ± 0.16	2.0	98.5	8.03 ± 0.02	0.26	97.1
49.60	50.05 ± 1.4	2.7	100.9	49.53 ± 0.31	0.63	99.9
74.40	75.49 ± 1.5	2.0	101.5	75.94 ± 0.23	0.30	102.1
Tanshinone IIA						
38.6	37.04 ± 1.04	1.0	95.9	37.23 ± 0.07	0.18	96.4
92.6	89.44 ± 2.01	2.3	96.6	90.71 ± 0.22	0.24	98.0
177.4	172.31 ± 4.70	2.7	97.1	178.90 ± 0.45	0.25	100.8

[a] RSD (%) = (SD/mean) × 100 %

[b] Recovery (%) = (mean of measured concentration/spiked concentration) × 100 %

(e) The reproducibility test: Precisely weigh Danshen sample, prepare 3 sample solutions according to the above method, and test the solutions 5 times. The results showed that the RSDs of the contents of the four assayed substances were 2.7, 3.3, 3.0, and 4.0 %, respectively, indicating that the reproducibility of the method was good.

(f) The recovery test: Add a known amount of mixed reference stock solution into a Danshen sample with known content, perform extraction and HPLC assay, and calculate the recovery rate based on the ratio of the measured theoretical value to actual added amount. In the recovery test, we investigated three different concentration levels, namely high, medium and low concentrations, in the standard curve range, repeated each test three times, and calculated average recovery. The results showed that the method had good accuracy (Table 14.40).

4. Sample determination: Take the reference and sample test solutions, test by the above method, and the typical chromatogram can be seen in Fig. 14.23. From the chromatogram, the retention times of the 4 compounds were 8.02, 11.95, 13.23 and 18.13 min, respectively, the chromatographic peaks of the indicator constituents had baseline separation from other chromatographic peaks, and no interference peaks appeared. After obtaining the chromatograms of all samples, the peaks were identified by comparing their retention times and UV absorptions with the peaks of the reference substances. The contents of the four compounds were calculated with the curve equations. The results are shown in Table 14.41 (Fig. 14.24).

   We reported for the first time the determination of the 4 primary liposoluble constituents in Danshen and 4 Danshen containing preparations. We tested a total of 40 samples; among them were 20 Compound Danshen Tablet samples, 14 samples of Danshen from

Table 14.40 The results of the recovery tests on the 4 tanshinones ($n = 4$)

Tanshinone spiked (μg/ml)	Found (μg/ml)	RSD[a] (%)	Recovery[b] (%)
Dihydrotanshinone I			
3.0	2.94 ± 0.10	3.3	98.1
36.2	35.41 ± 0.57	1.6	97.8
48.6	48.37 ± 1.35	2.8	98.8
Cryptotanshinone			
8.8	9.07 ± 0.13	1.5	103.1
96.8	95.07 ± 1.38	1.5	98.2
142.1	138.28 ± 2.35	1.7	97.3
Tanshinone I			
8.9	9.24 ± 0.31	3.4	103.9
48.9	47.53 ± 1.5	3.2	97.2
63.4	60.45 ± 3.5	5.8	95.4
Tanshinone IIA			
38.5	40.20 ± 2.39	5.9	104.4
84.7	81.13 ± 2.81	3.5	95.8
126.6	117.26 ± 4.0	3.4	95.6

[a] RSD (%) = (SD/mean) × 100 %

[b] Recovery (%) = (mean of measured concentration/spiked concentration) × 100 %

14 different regions, 1 Danshen Injection, 1 Xiangdan Injection, and 4 Dantonic™ samples. As revealed by the results in Table 14.41, there were large differences in the contents of the 4 tanshinones in Danshen samples from different regions, which may be caused by different climates, soils, environments, and harvesting seasons. The quality differences among crude drugs have been the primary reason for unstable traditional Chinese drug quality and unstable clinical effects; therefore, the quality control of tanshinone constituents in Danshen is of significant importance. Among the 4 Danshen preparations, only Compound Danshen Tablet contained tanshinones, while in Danshen Injection, Xiangdan Injection, and Dantonic™, tanshinones were almost undetectable. The difference among the preparations is mainly caused by different production processes.

As required by *Chinese Pharmacopoeia* 2005, the contents of tanshinone II_A in Danshen and Compound Danshen Tablet should be not less than 0.2 % and 200 μg/tablet, respectively. Among the 14 tested Danshen samples, 11 samples (No. 21–23 and No. 27–34) met the requirement. While all Compound Danshen Tablet samples met the requirement, the contents of tanshinone I, cryptotanshinone, and dihydrotanshinone I were very different. Considering that the 3 constituents have strong pharmacologic actions, it is suggested that they should be included in the quality control standards of Danshen and Compound Danshen Tablet.

Table 14.41 The contents of 4 tanshinones in Danshen and its preparations

No.	Batch no. or harvest time	Place of collection	Content[f] ($n = 3$)			
			Tanshinone II_A	Tanshinone I	Cryptotanshinone	Dihydrotanshinone I
2[a]	31207	Beijing	531.3 ± 9.0	201.5 ± 9.3	528.5 ± 21.4	123.0 ± 6.1
3[a]	40703	Shenzhen	534.6 ± 6.3	144.3 ± 6.7	291.6 ± 5.2	93.9 ± 3.2
4[a]	30113	Guangxi	534.5 ± 21.2	87.2 ± 3.3	192.6 ± 3.6	42.5 ± 3.4
5[a]	40210	Henna	714.4 ± 18.6	477.4 ± 3.4	768.7 ± 6.2	270.3 ± 21.6
6[a]	9919	Hebei	411.6 ± 6.7	210.5 ± 12.7	273.5 ± 3.1	84.2 ± 3.2
7[a]	20040342	Anhui	696.6 ± 9.5	297.2 ± 3.5	834.6 ± 2.5	159.5 ± 3.5
8[a]	40237	Shanghai	621.5 ± 39.2	249.6 ± 6.6	777.4 ± 24.6	183.6 ± 9.7
9[a]	4120369	Beijing	483.3 ± 9.4	285.3 ± 9.5	222.2 ± 3.3	141.3 ± 6.4
10[a]	20040561	Yunnan	666.1 ± 9.3	309.7 ± 6.2	744.4 ± 3.5	207.5 ± 6.7
11[a]	40614	Sichuan	609.4 ± 3.5	354.3 ± 9.3	507.3 ± 9.7	135.7 ± 3.2
12[a]	20040307	Hebei	447.2 ± 12.5	240.7 ± 3.4	453.5 ± 3.2	111.2 ± 3.7
13[a]	40110	Fujian	429.3 ± 33.5	141.8 ± 6.5	279.3 ± 6.3	75.4 ± 3.3
14[a]	40505	Guangxi	642.6 ± 15.3	129.3 ± 6.3	354.4 ± 6.1	78.6 ± 2.8
15[a]	20301	Jiangxi	507.2 ± 6.4	237.5 ± 3.2	369.6 ± 3.5	81.8 ± 3.1
16[a]	31101	Guangdong	675.5 ± 2.6	144.3 ± 6.3	300.3 ± 6.3	87.3 ± 6.2
17[a]	40343	Guangdong	576.2 ± 15.3	141.6 ± 3.2	72.1 ± 3.0	ND[g]
18[a]	40208	Guangdong	777.4 ± 24.5	264.8 ± 6.3	426.3 ± 6.3	135.5 ± 3.6
19[a]	301900	Guangdong	615.2 ± 3.3	408.3 ± 6.6	1863.5 ± 27.6	399.7 ± 6.4
20[a]	40427	Beijing	528.4 ± 27.5	252.2 ± 6.2	510.3 ± 9.7	114.5 ± 6.2
21[b]	200406	Shandong	2.01 ± 0.09	1.82 ± 9.7	2.63 ± 0.03	1.37 ± 0.01
22[b]	200407	Sichuan	5.03 ± 0.27	1.22 ± 9.4	4.08 ± 0.05	3.72 ± 0.04
23[b]	200308	Shanxi	7.45 ± 0.32	2.01 ± 0.03	5.49 ± 0.05	3.51 ± 0.03
24[b]	200309	Liaoning	0.44 ± 0.02	0.17 ± 0.02	0.23 ± 0.00	0.07 ± 0.00
25[b]	200409	Henna	1.79 ± 0.08	1.16 ± 0.03	0.98 ± 0.01	0.48 ± 0.01
26[b]	200408	Zhejiang	0.40 ± 0.02	0.36 ± 0.01	0.17 ± 0.00	0.10 ± 0.00
27[b]	200107	Shanxi	2.23 ± 0.11	2.03 ± 0.02	1.81 ± 0.02	3.95 ± 0.04
28[b]	200309	Liaoning	2.37 ± 0.10	1.27 ± 0.01	2.59 ± 0.03	2.13 ± 0.02
29[b]	200402	Henna	2.45 ± 0.12	1.09 ± 0.04	1.53 ± 0.01	1.34 ± 0.01
30[b]	200408	Shanxi	2.03 ± 0.09	1.41 ± 0.03	0.75 ± 0.01	0.98 ± 0.01
31[b]	200407	Hebei	2.48 ± 0.09	0.98 ± 0.01	1.29 ± 0.01	0.82 ± 0.01
32[b]	200403	Shanxi	2.27 ± 0.08	1.20 ± 0.01	1.39 ± 0.01	0.94 ± 0.01
33[b]	200301	Shanxi	2.60 ± 0.09	1.95 ± 0.03	1.81 ± 0.02	2.05 ± 0.02
34[b]	200303	Shandong	5.11 ± 0.14	3.29 ± 0.05	3.09 ± 0.04	2.41 ± 0.02
35[c]	040710	Shanghai	ND[g]	ND[g]	ND[g]	ND[g]
36[d]	040828	Guangdong	ND[g]	ND[g]	ND[g]	ND[g]
37[e]	20031001	Tianjin	ND[g]	2.8 ± 0.3	5.4 ± 0.4	ND[g]

(continued)

Table 14.41 (continued)

No.	Batch no. or harvest time	Place of collection	Content[f] ($n = 3$)			
			Tanshinone II_A	Tanshinone I	Cryptotanshinone	Dihydrotanshinone I
38[e]	20030618	Tianjin	ND[g]	2.9 ± 0.2	5.5 ± 0.5	ND[g]
39[e]	20040218	Tianjin	ND[g]	2.7 ± 0.2	5.4 ± 0.4	5.3 ± 0.4
40[e]	20040508	Tianjin	2.7 ± 0.2	2.7 ± 0.3	5.3 ± 0.3	5.2 ± 0.3

[a] Compound Danshen Tablet
[b] Danshen
[c] Danshen Injection
[d] Xiangdan Injection
[e] Dantonic™
[f] Content = mean ± SD ($n = 3$). Unit for Danshen: mg/g; unit for Compound Danshen Tablet: μg (tanshinone)/300 mg tablet; unit for Dantonic™: μg (tanshinone)/27 mg pill; and unit for Danshen Injection and Xiangdan Injection: mg/10 ml ampoule
[g] Not detected
[h] Batch number used by the manufacturer, the harvesting time of Danshen

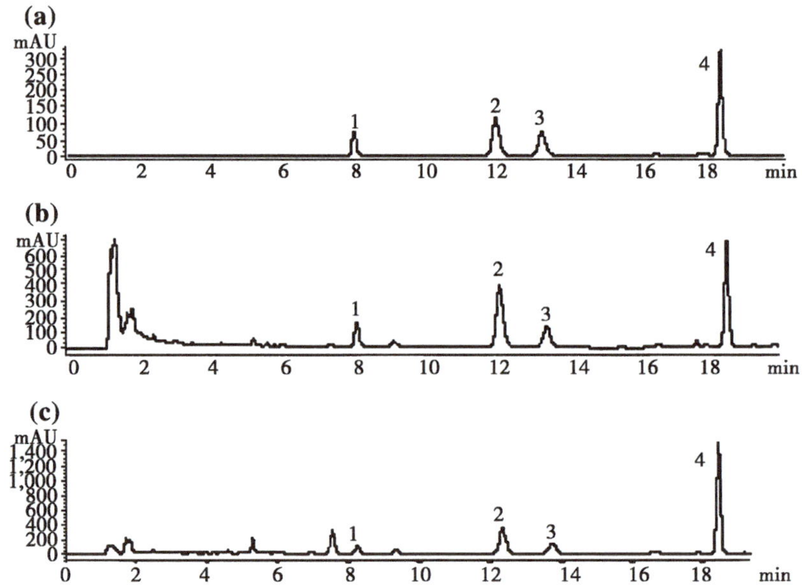

Fig. 14.24 Representative HPLC profiles panels: **a** Standard solution with middle concentration; **b** Compound Danshen tablet (Liaoning, China); **c** Danshen (Shanxi, China). Peaks: *1* Dihydrotanshinone I; *2* Cryptotanshinone; *3* Tanshinone I; *4* Tanshinone II_A

14.2.2.3 Conclusion

The HPLC method is simple, rapid, accurate, and applicable to the quality evaluation of tanshinones in Danshen and Danshen preparations. Using this method, we have successfully determined 4 primary tanshinones in Danshen and 4 Danshen preparations (Compound Danshen Tablet, Dantonic™, Danshen Injection and Xiangdan Injection) for the first time, and the results revealed that there was a large difference in the tanshinone contents of Danshen from different production regions and Compound Danshen Tablets from different manufacturers. The tanshinones were almost undetectable in Dantonic™, Danshen Injection and Xiangdan Injection.

References

1. Li J, He LY, Song WZ. Acta Pharm Sin. 1993;28 (7):543.
2. Cao F, Ma CH, Qiao YJ. J Beijing Univ Tradit Chin Med. 2003;26(6):45.
3. Pan YN, Yuan D, Fu WW, et al. J Shenyang Pharm Univ. 2004;21(3):196.
4. Wang YZ, Lie WS. J Pediatr Pharm. 2004;10(4):18.
5. Yuan JX, Zhu LZ. Chin Patent Med. 1996;18(1):15.
6. Liu RL. Chin J Pharm Anal. 1983;3(3):163.
7. Ni KY. Pharm J Nanjing. 1986;17(4):265.
8. Sheng JF. Study Chin Patent Med. 1987;7:14.
9. Sun W, Liu ZT. Heilongjiang Med Sci. 1987;12(1):17.
10. Ni KY, Zhang GQ. Acta Pharm Sin. 1987;23(4):293.
11. You YJ, Cheng GQ, Chen MH. Chin J Pharm Anal. 2004;24(1):46.
12. Feng GQ. J Pharm Pract. 1998;16(4):225.
13. Yuan L, Su GL, Hu GS. Chin Herbal Med. 1994;25 (6):299.
14. He JG, Yang LQ, Zhu JH, et al. Lishizhen Med Mater Med Res. 2002;13(10):590.
15. Shi YL, Li MY, Zhu Q, et al. Chin Pharm Sci. 2002;37(4):300.
16. Dai CQ, Yuan J. Pharm J Chin People's Liberation Army. 2000;16(5):277.
17. Ye SM, Xiong XC. Primary J Chin Mater Med. 2001;15(4):7.
18. Huang YL, Sun GH, Sun WG, et al. Chin Pharm. 2002;11(12):36.
19. Huang X, Zhu YN, Huang RB. Chin J Pharm Anal. 2002;22(4):308.
20. Cai M, Zhao LH, Hu CQ, et al. J Chin Pharm Univ. 2001;32(2):130.
21. Mulinacci N, Bardazzi C, Romani A, et al. Chromatographia. 1999;49:197.
22. Escarpa A, Oonzalez MC. J Chromatogr A. 1999;830:301.
23. Cui H, He CX, Zhao GW. J Chromatogr A. 1999;855:171.
24. Sandor G. J Pharm Biomed Anal. 2005;36:931.
25. de Oliveira AM, Lowen TC, Cabral LM, et al. J Pharm Biomed Anal. 2005;4:751.
26. Li HB, Chen F. J Chromatogr A. 2001;(1–2):109.
27. Pan X, Niu G, Liu H. J Chromatogr A. 2001; (1–2):371.
28. Qiu ZY. Chin Trad Patent Med. 2001;1:18.

Fingerprint Study

15

Dean Guo, Jinlan Zhang, Ming Zhu, Xiaohui Fan, Yongjiang Wu, Aihua Liu, Min Yang and Haibing Qu

Danshen is a complicated system containing multiple components, which are mainly liposoluble diterpenoid quinones and water-soluble phenolic acids. These compounds are the material foundation of Danshen's pharmaceutical functions. At present, the methods for Danshen's quality control are based upon the use of salvianolic acid B and tanshinone II_A as the quantitative and qualitative indicators, which of course could not indicate the overall quality of the product. Also, these approaches neglect the synergism among the various components, which is not in line with the holistic ideal of TCM in clinical application. Fingerprinting technology can solve the above problems and can objectively and comprehensively reflect the quality of Danshen. It is of great significance to apply fingerprinting technology to illuminate the mechanisms of TCM functions, to improve the quality of TCM, and to promote the modernization and internationalization of TCM.

D. Guo (✉) · J. Zhang · M. Zhu · X. Fan · Y. Wu · A. Liu · M. Yang · H. Qu
Shanghai Institute of Materia Medica, Shanghai, China
e-mail: daguo@mail.shcnc.ac.cn

15.1 Study of Chromatographic Fingerprinting

15.1.1 Study of the Chromatographic Fingerprinting of Salvianolic Acids

15.1.1.1 The Chromatographic Fingerprinting of Salvianolic Acids in Danshen

Jinlan Zhang and Ming Zhu

1. Instruments and Reagent

   Agilent 1100 series HPLC (equipped with diode array detector), PE SCIEX QSTAR MASS analyzer; chromatographic column: Zorbax Extend C_{18} (4.6 × 250 mm, 5 μm). Methanol, hydrochloric acid and acetidin were the products of Beijing Chemical Factory; acetonitrile was the product of Caledon, a Canadian company. Danshensu, protocatechuic aldehyde, tanshinone, cryptotanshinone and tanshinone II_A were purchased from National Institutes for Food and Drug Control; salvianolic acid B and C were self-made, each with a purity of over 97 % (tested by HPLC). Wild and cultivated Danshen herbs were collected and purchased nation-wide in China, and they were identified by the Modern Research Centre for TCM at Beijing University (see Table 15.1).

X. Yan (ed.), *Dan Shen (Salvia miltiorrhiza) in Medicine*,
DOI 10.1007/978-94-017-9463-3_15,

Table 15.1 Information about the Danshen samples

No.	Places of collection or purchase	Growth pattern	Place of origin
1	Luxi, Yunnan	Wild	Luxi, Yunnan
2	Qujing, Yunnan	Wild	Qujing, Yunnan
3	Lingbao, Henan	Wild	Lingbao, Henan
4	Yishui, Shandong	Wild	Yishui, Shandong
5	Lushi, Henan	Wild	Lushi, Henan
6	Hengqu, Shanxi	Wild	Hengqu, Shanxi
7	Liaoning Zhaoyang City Herb Company	Wild	
8	Shanxi Chinese Crude Drug Market	Wild	
9	Shanxi Chinese Crude Drug Supplying Station	Wild	
10	Shangluo, Shaanxi	Wild	Shangluo, Shaanxi
11	Funiushan Moutain, Lushi County, Henan	Wild	Funiushan Moutain, Lushi County, Henan
12	Qiao Village, Gujiang Town, Jiang County, Shanxi	Wild	Qiao Village, Gujiang Town, Jiang County, Shanxi
13	Shanyang County, Shaanxi	Wild	Shanyang County, Shaanxi
14	Yishui, Shandong	Cultivation	Shandong
15	Hengqu, Shanxi	Cultivation	Hengqu, Shanxi
16	Yuzhou, Henan	Cultivation	Yuzhou, Henan
17	Guangxi Pingnan Herb Company	Cultivation	Guangxi Pingnan Herb Company
18	Guiyang Herb Company	Cultivation	Hebei
19	Sichuan Zhongjiang Herb Company (first level)	Cultivation	Sichuan Zhongjiang
20	Sichuan Zhongjiang Herb Company (second level)	Cultivation	Sichuan Zhongjiang
21	Lingbao,Henan	Cultivation	Lingbao, Henan
22	Pinglu, Shanxi	Cultivation	Pinglu, Shanxi
23	Zhejiang Qin'an Herb Company	Cultivation	Qin'an, Zhejiang
24	Drug market of Chongqing	Cultivation	Chongqing
25	Shandong Pingyi Herb Company	Cultivation	Pingyi, Shandong
26	Anguo, Hebei	Cultivation	Anguo, Hebei
27	Zhejiang Yongkang Herb Company A	Cultivation	Zhejiang
28	Hai city, Anguo, Hebei	Cultivation	Anguo, Hebei
29	Gansu Gangu Herb Company	Cultivation	Gansu Gangu
30	Qiugou Village, Yongluo Town, Bing County, Shaanxi	Cultivation	Qiugou Village, Yongluo Town, Bing County, Shaanxi
31	Gaobadian Town, Shanyang County, Shaanxi	Cultivation	Gaobadian Town, Shanyang County, Shaanxi
32	East Gaobadian Town, Shanyang County, Shaanxi	Cultivation	East Gaobadian Town, Shanyang County, Shaanxi
33	Tongshun Herb Wholesale Station of Shenyang, Liaoning	Cultivation	Tongshun, Shenyang, Liaoning

(continued)

Table 15.1 (continued)

No.	Places of collection or purchase	Growth pattern	Place of origin
34	Liaoning Yingkou Herb Company	Cultivation	Hebei
35	Market of Fushun, Liaoning	Cultivation	Fushun, Liaoning
36	Harbin, Heilongjiang	Cultivation	Hebei
37	Shanyang County, Shaanxi	Cultivation	Shanyang County, Shaanxi
38	Majie Town, Shangluo, Shaanxi	Cultivation	Majie Town, Shangluo, Shaanxi
39	Minxing Country, Gaoluo Country, Hengqu Country, Shanxi	Cultivation	Goushu Village, Minxing Coungtry, Gaoluo Country, Hengqu County
40	Zhejiang Yongkang Herb Company B	Cultivation	Zhejiang

2. The Optimization and Determination of the Extraction Method for Water-Soluble Phenolic Acids
   The effects of extraction method, extraction time, and extraction solvent on the results of extraction were studied. Accurately weigh about 250 mg of samples of Danshen powder which have been sifted through a 60-mesh sieve; a total of 6 samples were taken. Add 30 ml of water, weigh, soak in cold water overnight; extract three samples by heating them under reflux for 1.5, 2.0 and 2.5 h, respectively; extract one sample with ultrasound for half an hour. All the above samples, after extraction, should be made up for the lost weight and then be filtered. Adjust the filtrate to pH 2. Precisely take 20 ml of the filtrate and extract with 60 ml of acetidin; concentrate the acetidin layer and evaporate to dryness. Add 5 ml of methanol to dissolve the residue, and filter through a 0.45 μm microporous membrane before testing by HPLC. Take another two samples for extraction by heating under reflux for 2 h, and then extract them with 60 ml of acetidin and 20 ml of n-butanol, respectively. Compare the chromatographic peak areas of the major phenolic acids. The results show that the best extraction method was the following: Accurately weigh 250 mg of Danshen powder that has been filtered through a 60-mesh screening, add 30 ml of water, weigh, soak in cold water overnight, extract by heating under reflux for 2.0 h, place at room temperature, make up for the lost weight, filter, adjust the filtrate to pH 2; accurately take 20 ml of filtrate, extract with 60 ml of acetidin, concentrate the acetidin layer and evaporate to dryness. Add 5 ml of methanol to dissolve the residue, and filter through a 0.45 μm microporous membrane before testing by HPLC (Table 15.2).
3. The Establishment of Analytical Methods for Water-Soluble Constituents
   Using Agilent 1100 HPLC-DAD and Zorbax Extend C_{18} chromatographic columns (4.6 × 250 mm, 5 μm), we compared the effects of mobile phases, such as methanol–water, acetonitrile–water, acetonitrile-acetic acid water, and acetonitrile-phosphoric acid water, on the separation. The results showed that acetonitrile-phosphoric acid water was better suited for the analysis of the water-soluble constituents of Danshen. Using the acetonitrile (A)–0.026 % phosphoric acid in water (B) system, we compared constant composition and gradient elution, and finally determined the following gradient elution process: A/B(v/v) = 12:88 (0–5 min) → A/B = 15:85 (20 min) → A/B = 20:80 (35 min) → A/B = 23:77 (45 min) → A/B = 30:70 (55 min) → A/B = 12:88 (65 min). (See chromatogram in Fig. 15.1).
4. HPLC-MS Analysis and Identification of Water-Soluble Constituents
   In order to gather more structural information on the various constituents, HPLC-MS analysis was performed.
   (a) Analytical Condition: Dissolve an appropriate amount of Danshen total phenolic acids in methanol, filter through a 0.45 μm

Table 15.2 Comparison of extraction methods for the water-soluble constituents of Danshen

Retention time	T_R (min)										
	4.8	9.4	11.9	35.6	36.5	41.8	44.8	49.8	50.9	53.4	53.9
Hot water reflux	*Area*										
1.5 h	160.9	466.2	639.9	773.8	314.7	1,227	875.3	9,500	2,088	1,029	1,120
2.0 h	316	907.9	723.7	1,504	516.1	1,937	1,583	14,395	2,232	1,260	1,565
2.5 h	667.2	614.4	614.4	1,367	829.9	1,686	1,399	12,473	2,040	9,278	1,865
Ultrasound extraction	*Area*										
0.5 h	174.5	83.5	540.6	810.9	259.6	335	1,263	17,266	2,441	732.9	624.3
Different solvents	*Area*										
EtOAc (1:1)	316	907.9	723.7	1,504	516.1	1,938	1,583	14,395	2,232	1,260	1,565
EtOAc (3:1)	1,056	883.6	792.5	1,612	764.4	2,318	1,916	17,325	2,363	1,305	1,584
n-BuOH (1:1)	1,023	728.8	935	1,597	859.7	2,240	1,685	17,246	2,123	1,175	2,094

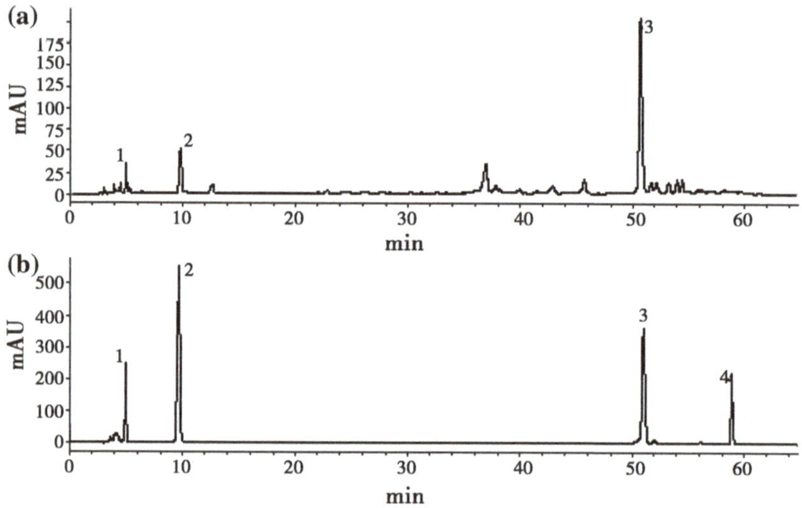

Fig. 15.1 The chromatogram of water-soluble constituents. Panels: **a** Danshen; **b** Reference substances. Peaks: *1* Danshensu; *2* Protocatechuic aldehyde; *3* Salvianolic acid B; *4* Salvianolic acid C

microporous membrane, and take the filtrate for HPLC-MS analysis. With the same liquid phase condition as above, the condition for mass spectrometric detection was as follows: Ion Source: Turbo Ionspray; ionization mode: ESI negative ionization; ionization voltage: 3,300 eV; MS range: 100–1,500 *m/z* amu; GSl: 60 psi; GS2: 40 psi; CUR: 35 psi; ion source temperature: 400 °C; DPl: −40 eV; FP: −215 eV; DP2: −15 eV.

(b) HPLC-MS Results: By selective ion monitoring (SIM) and extraction of the mass spectrogram (MS) of the major chromatographic peaks, the molecular weights of six constituents were collected. By referring to the published literature and reports [1] and comparing with the reference substances, a total of 10 constituents in the chromatogram were identified (Figs. 15.2 and 15.3, and Table 15.3).

5. Methodological Study

The precision, repeatability, and stability of the methods described above were investigated.

(a) Precision

Follow the methods described above to extract a Danshen sample and analyze the

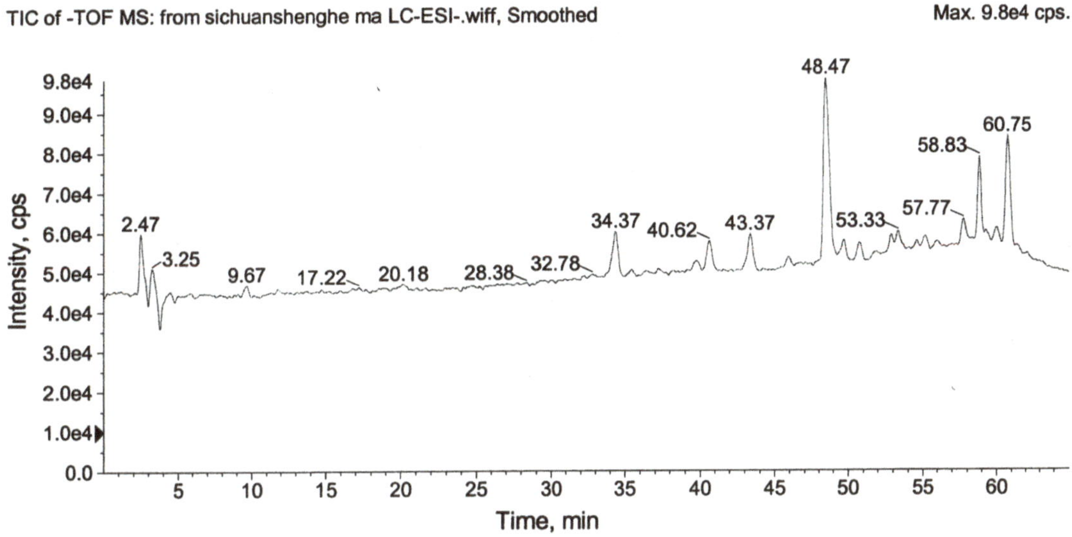

Fig. 15.2 The total ion chromatogram (TIC) of total phenolic acids in Danshen

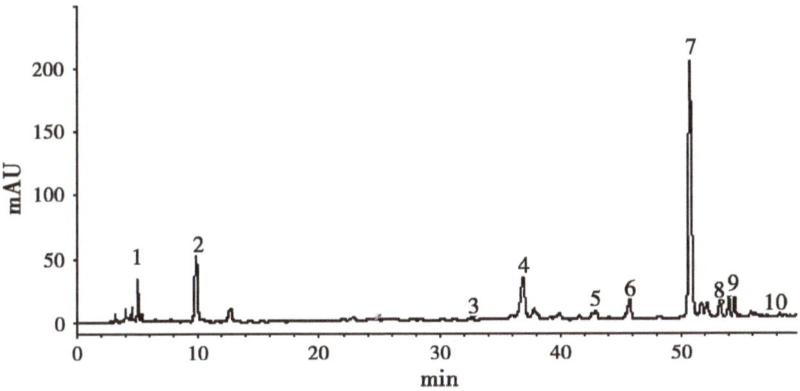

Fig. 15.3 HPLC-DAD analysis and peak identification of total phenolic acids in Danshen. *1* Danshensu, *2* Protocatechuic aldehyde; *3* Salvianolic acid G; *4* Salvianolic acid D; *5* Rosmarinic acid; *6* New compound (294); *8* Salvianolic acid E; *9* Salvianolic acid A; *10* Salvianolic acid C

Table 15.3 Identification of phenolic acids in the HPLC fingerprint of Danshen

Peak no.	Retention time (min)	Mass number	Possible constituents
1	4.98	198	Danshensu
2	9.83	138	Protocatechuic aldehyde
3	32.64	340	Salvianolic acid G
4	36.97	418	Salvianolic acid D
5	42.99	360	Rosmarinci acid
6	45.79	494	New chemical compound
7	5,075	718	Salvianolic acid B
8	53.32	718	Salvianolic acid E
9	54.52	494	Salvianolic acid A
10	58.32	492	Salvianolic acid C

extract with HPLC continuously for 6 times. The collected data was processed by "*The Similarity Evaluation System for Chromatographic Fingerprint of Chinese Crude Drug*" (2004A). The results showed that the precision similarity was over 0.95 (Table 15.4).

(b) Repeatability: Follow the established extraction methods to extract 6 samples, and analyze the extracts with HPLC. The data collected was processed by "*The Similarity Evaluation System for Chromatographic Fingerprint of Chinese Crude Drug*" (2004A). The results showed that the reproducibility similarity was over 0.95 (Table 15.5).

(c) Stability
Follow the above method to extract a Danshen sample, and dissolve the extract in methanol. Place it at room temperature, and conduct the HPLC analysis at different intervals. The data collected was then processed by "*The Similarity Evaluation System for Chromatographic Fingerprint of Chinese Crude Drug*" (2004A). The results showed that the sample was stable for 12 h (Table 15.6).

6. HPLC Fingerprint Analysis of the Water-Soluble Constituents in Wild and Cultivated Danshen
Danshen samples collected from different habitats, either wild or cultivated, were extracted and analyzed using the established methods. The results and similarity calculations are shown in Tables 15.7 and 15.8, Figs. 15.4 and 15.5.
The major water-soluble constituent in Danshen is salvianolic acid B, and its content varies with the wild or cultivated types of Danshen and with the habitats. For example, the similarities of HPLC fingerprints of wild Danshen from Yishui of Shandong, Lingbao and Lushi of Henan, Shanyang of Shaanxi, Yi County and Hunyuan County of Shanxi were over 0.9, while those of wild Danshen from other habitats were lower. The similarities of HPLC fingerprints of cultivated Danshen from Yuzhou of Henan, Yuanqu of Shanxi, Yishui of Shandong, Anguo of

Table 15.4 The precision of the HPLC method

	S1	S2	S3	S4	S5	S6	Fingerprint comparison
S1	1.000	1.000	1.000	1.000	1.000	1.000	1.000
S2	1.000	1.000	1.000	1.000	1.000	1.000	1.000
S3	1.000	1.000	1.000	1.000	1.000	1.000	1.000
S4	1.000	1.000	1.000	1.000	1.000	1.000	1.000
S5	1.000	1.000	1.000	1.000	1.000	1.000	1.000
S6	1.000	1.000	1.000	1.000	1.000	1.000	1.000
Fingerprint comparison	1.000	1.000	1.000	1.000	1.000	1.000	1.000

Table 15.5 The repeatability of the extraction method

	S1	S2	S3	S4	S5	S6	Fingerprint comparison
S1	1.000	1.000	1.000	0.999	0.999	0.999	1.000
S2	1.000	1.000	0.999	0.999	0.999	1.000	1.000
S3	1.000	0.999	1.000	0.999	0.999	0.999	0.999
S4	0.999	0.999	0.999	1.000	1.000	1.000	1.000
S5	0.999	0.999	0.999	1.000	1.000	1.000	0.999
S6	0.999	1.000	0.999	1.000	1.000	1.000	1.000
Fingerprint comparison	1.000	1.000	1.000	1.000	0.999	1.000	1.000

Table 15.6 The stability of the test samples

Time (h)	S1	S2	S3	S4	S5	S6	Fingerprint comparison
0	1.000	1.000	1.000	1.000	1.000	1.000	1.000
1.5	1.000	1.000	1.000	1.000	1.000	1.000	1.000
3	1.000	1.000	1.000	1.000	1.000	1.000	1.000
6	1.000	1.000	1.000	1.000	1.000	1.000	1.000
9	1.000	1.000	1.000	1.000	1.000	1.000	1.000
12	1.000	1.000	1.000	1.000	1.000	1.000	1.000
Fingerprint comparison	1.000	1.000	1.000	1.000	1.000	1.000	1.000

Table 15.7 The similarity evaluation of the water-soluble constituents in wild Danshen

No.	1	2	3	4	5	6	7	8	9	10	11	12	13
Similarity	0.711	0.712	0.941	0.928	0.917	0.278	0.892	0.949	0.931	0.893	0.771	0.587	0.934

Table 15.8 The similarity evaluation of the water-soluble constituents in cultivated Danshen

No.	14	15	16	17	18	19	20	21	22	23	24	25	26
Similarity	0.979	0.874	0.84	0.96	0.965	0.887	0.891	0.768	0.624	0.22	0.358	0.405	0.209
No.	27	28	29	30	31	32	33	34	35	36	37	38	39
Similarity	0.096	0.864	0.117	0.463	0.075	0.939	0.917	0.936	0.956	0.876	0.15	0.874	0.761

Hebei, and Zhongjiang of Sichuan were over 0.9, while the similarities of cultivated Danshen from other habitats were lower.

15.1.1.2 The Chromatogram and Fingerprinting of Salvianolic Acids in Dantonic™

Xiaohui Fan and Yongjiang Wu

Dantonic™ is a compound preparation composed of Danshen, notoginseng, broneolum syntheticum. It has the function of activating blood circulation and dissipating blood stasis, regulating qi, and alleviating pain. Ever since its entry into the market, it has become an important medicine in treating cerebrovascular diseases, and has been one of the emergency prerequisite Chinese patent medicines in all TCM hospitals nationwide. However, the current quality standards for Dantonic™ specify only the identification of saponin R_1 from notoginseng by TLC, and the determination of danshensu, which cannot objectively reflect the overall quality of Dantonic™. Therefore, it is necessary to study and establish the chemical fingerprint of Dantonic™ to enhance its quality control.

While working on the project "Demonstration Research on the Fingerprint Application", one of the Tenth Five-year State Plan of National Science and Technology Research Programs, we found that the major effective constituents of Dantonic™ [2], the Danshen phenolic acids (or called water-soluble constituents) and notoginseng saponins, cannot be accurately reflected in the same chromatogram simultaneously. Taking the UV detector of HPLC for example, the water-soluble constituents of Danshen have a maximum UV absorption at a wavelength around 280 nm, with end absorption at around 200 nm. However, notoginseng saponins only have end absorption at around 200 nm (in publications, 203 nm is usually taken as the detection wavelength). Due to the fact that Dantonic™ contains a relatively low level of notoginseng saponins, it will be very difficult to completely reflect the chemical composition characteristics of the pill, even if 203 nm is selected as the detection wavelength. As shown in Fig. 15.6, notoginseng saponin R_1 and other

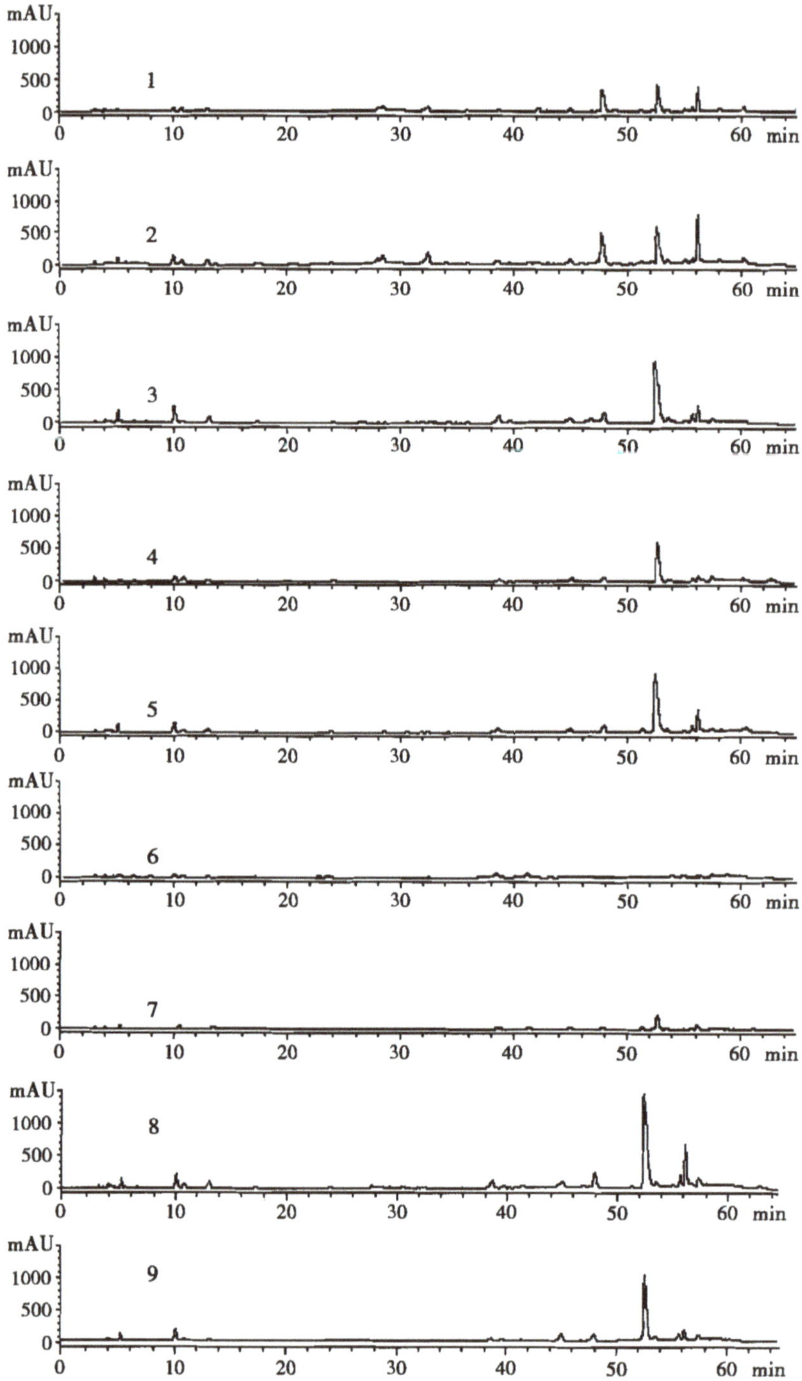

Fig. 15.4 HPLC fingerprint of water-soluble constituents in wild Danshen from different habitats

indicator components are not reflected in the 203 nm wavelength chromatogram.

Therefore, we created the technique of multiple chemical fingerprinting, based on which we successfully established the multiple chemical fingerprint of Dantonic™: we used two fingerprints (one was the chromatographic fingerprint of Danshen phenolic acids, and the other was the chromatographic fingerprint of notoginseng saponins) to characterize the two major constituents

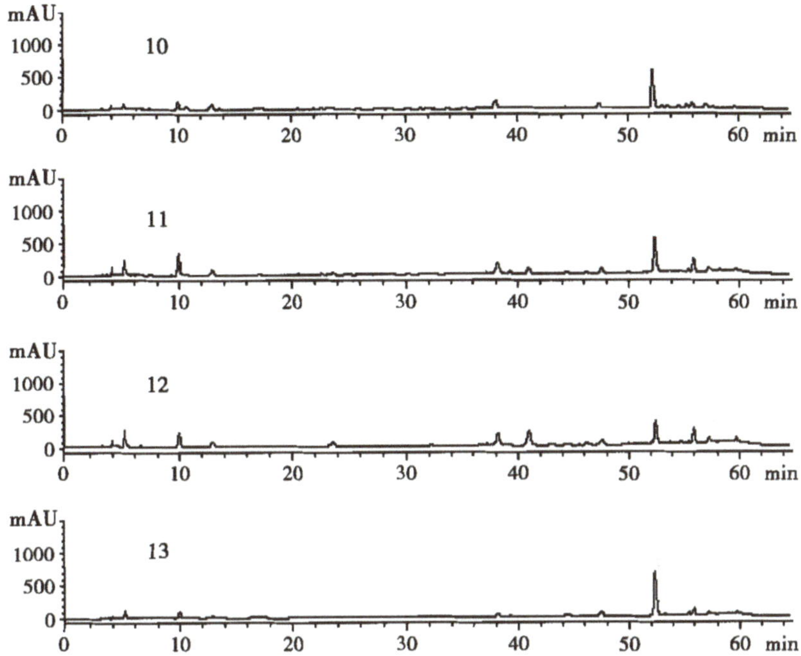

Fig. 15.4 (continued)

of Dantonic™. In this way, the overall chemical composition of Dantonic™ would be well characterized. Due to space limitations, this section will only discuss the chromatographic fingerprinting of Danshen phenolic acids in Dantonic™. For the chromatographic fingerprinting of notoginseng saponins in Dantonic™, please see Chap. 16 and the related literature [3–5].

1. Study of the Constituents of Danshen Phenolic Acids in Dantonic™

   Sample preparation: Weigh 148.4 mg of Dantonic™; add 6 ml of water, and dissolve by ultrasound for 15 min. Filter through a 0.45 μm nylon membrane to get a yellow solution sample.

   Instrument and reagents: Agilent 1100 HPLC; Diode Array Detector; Autosampler; Thermostatted Column Compartment; Degasser; HP Instrument chromatographic work station; 1100 Series LC-MSD/Trap Mass Spectrometer (Bruker); ESI ion source; Ion trap detector; Extract—Clean C_{18} column; chromatographic pure acetonitrile; redistilled water; analytical pure acetic acid.

   Testing conditions: Use Agilent Zorbax SB-C_{18} chromatographic column (5 μm, 4.6 mm × 25 cm, Agilent, SN: USCL009296) for HPLC analysis; the methods of gradient elution and the conditions for mass spectrometric detection for all samples are shown in Tables 15.9 and 15.10. The HPLC and MS-TIC chromatograms of water-soluble constituents of Danshen in Dantonic™ are shown in Fig. 15.7. The characterization result and explanation is shown in Tables 15.11 and 15.12. The molecular structures for all chemical components are shown in Fig. 15.8. By comparing with the reference, Peaks 1, 2, 7 and 8 were identified as danshensu, protocatechuic aldehyde, rosmarinic acid and salvianolic acid B, respectively. HPLC/MS^n analysis was conducted to collect structural information. HPLC/MS^n data is shown in Table 15.12.

   Judging by the MS^n cleavage of the reference substances, it could be concluded that the collision induced dissociation (CID) fragment ions of salvianolic acid were mainly produced by a-cleavage and neutral losses of danshensu. Therefore, by referring to the MS^n data of

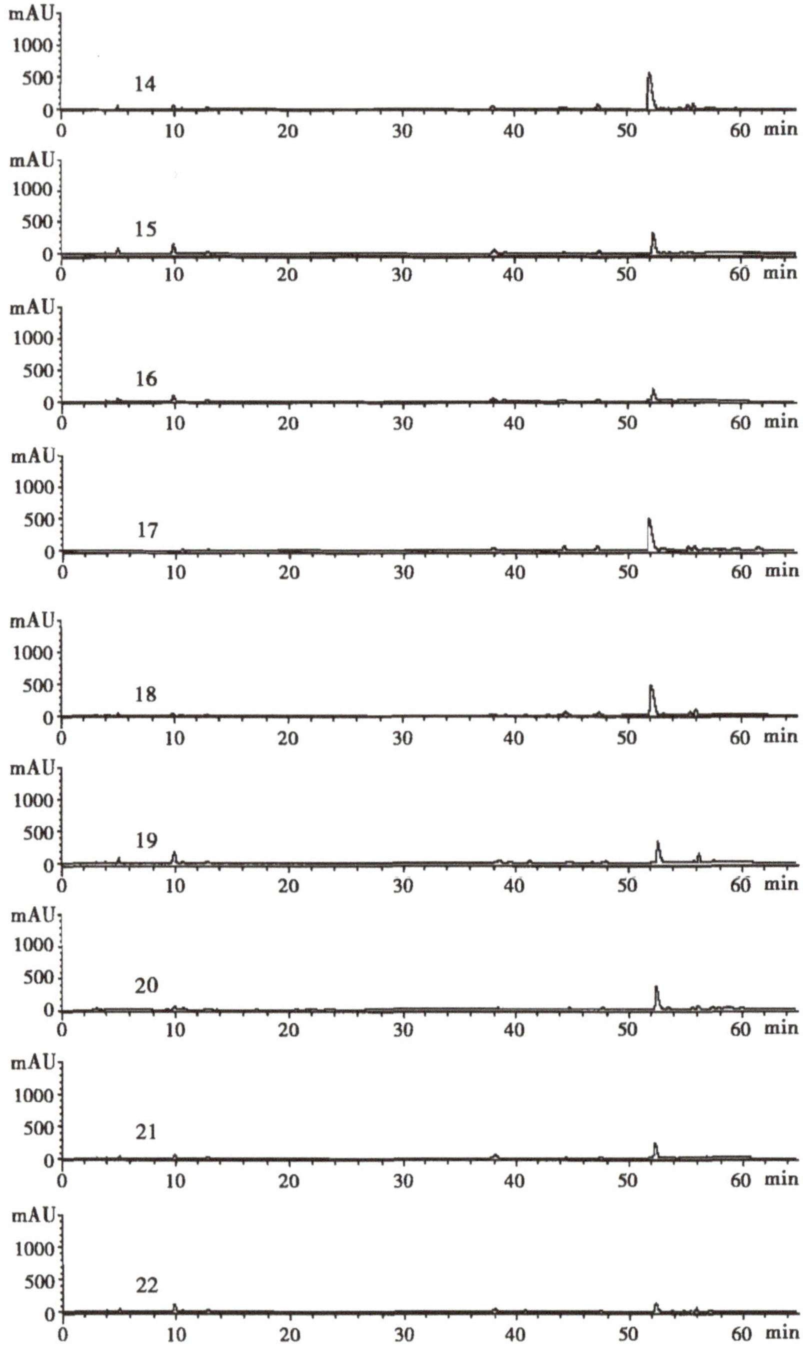

Fig. 15.5 HPLC fingerprint of water-soluble constituents in cultivated Danshen from different habitats

various peaks elaborated in literature, we were able to have a relatively accurate characterization of all chemical compounds. After neutral loss of 198, Peak 10 had a neutral loss of 136, which was in line with the structure of salvianolic acid A. Also, the ultraviolet absorption of Peak 10 was almost identical to that of salvianolic acid A, so it was identified as salvianolic acid A. Similarly, we could identify Peak 5 and Peak 9 as salvianolic acid

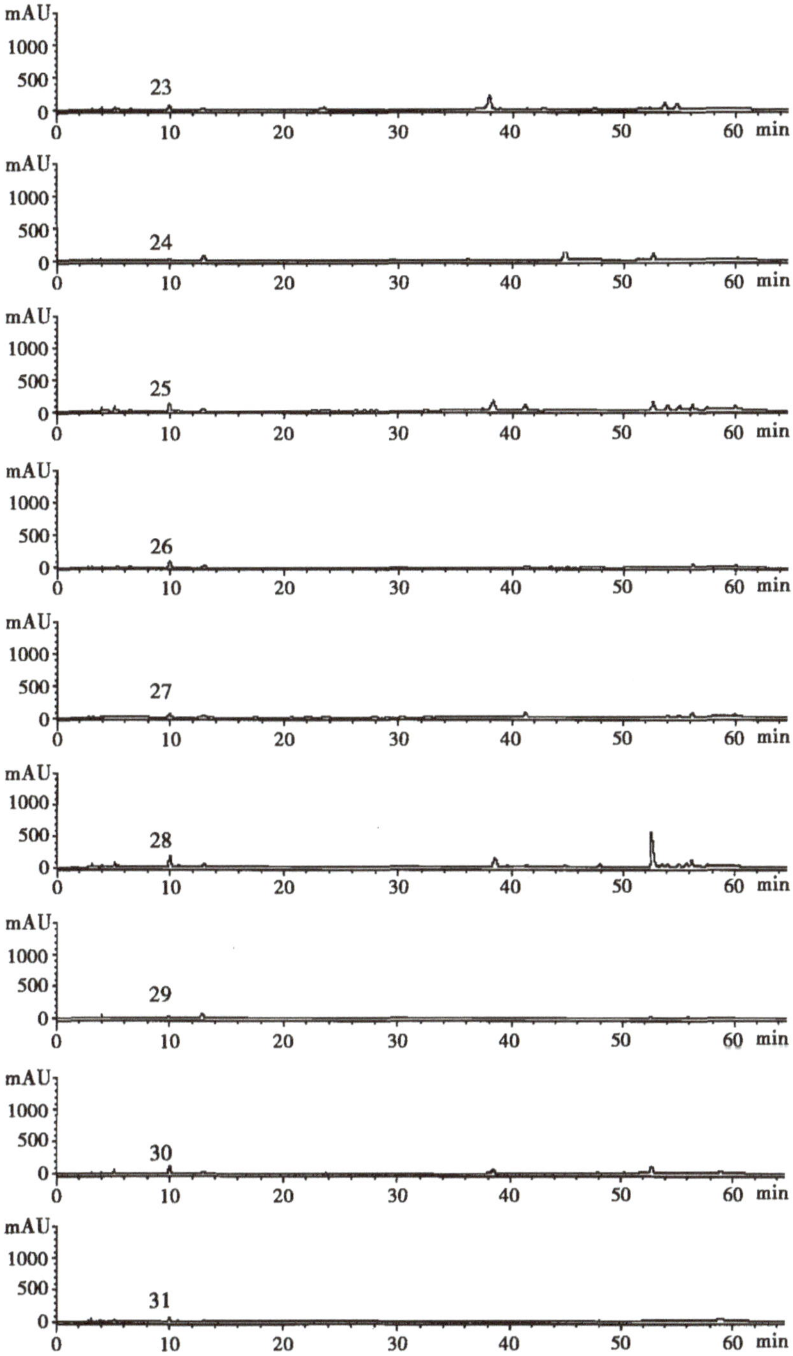

Fig. 15.5 (continued)

D and salvianolic acid G, respectively. The MS^n of Peak 6 was basically the same as that of salvianolic acid B, but their ultraviolet absorptions were different. Due to its phenyl benzene-dihydrofuran structure, salvianolic acid B had a relatively strong absorption around 253 nm, while Peak 6 did not. By comparing with the literature, we found that it was quite similar to salvianolic acid E, and therefore we identified Peak 6 as salvianolic

Fig. 15.5 (continued)

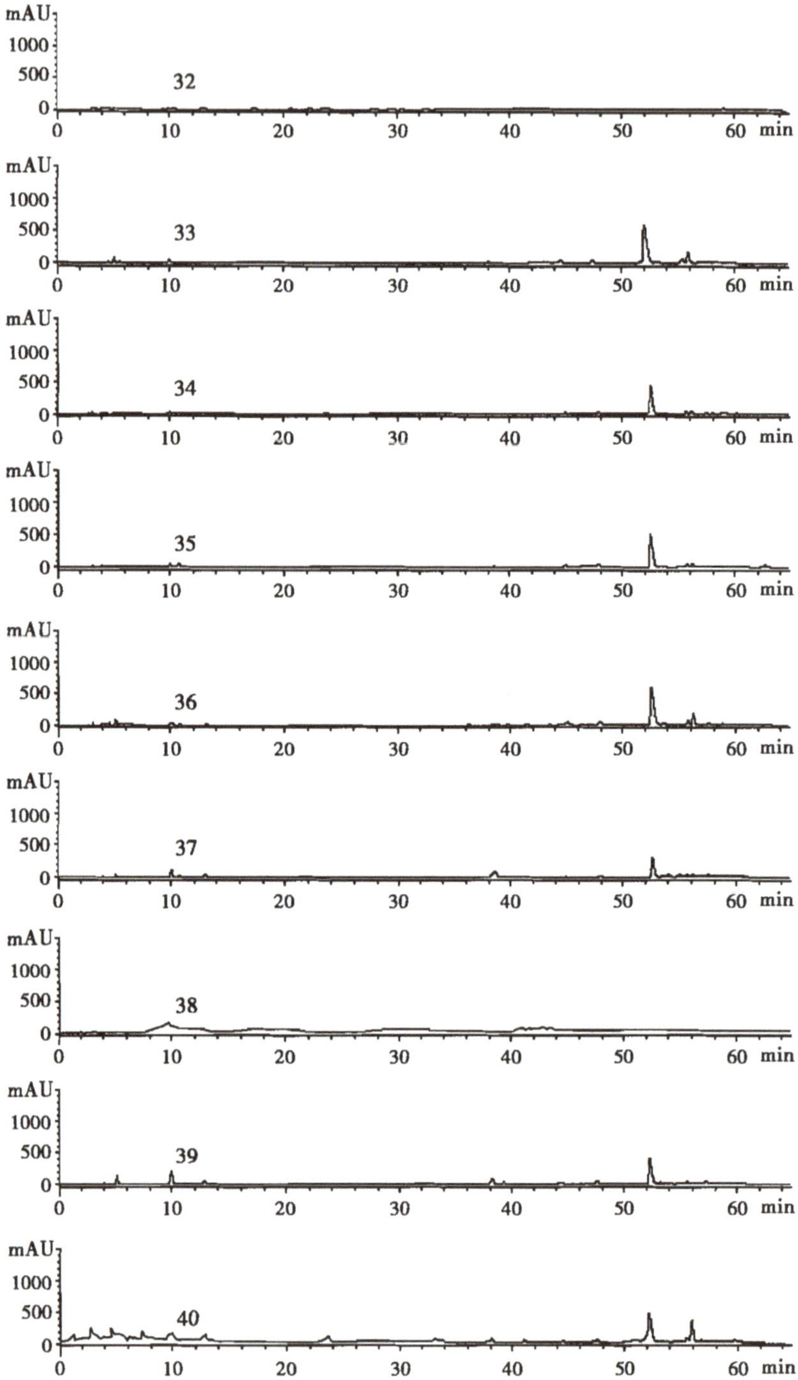

acid E. Peak 3 and Peak 4 had the same molecular weight and MS^n profiles, and their structures showed features similar to those of lithospermic acid, as described in the literature. However, their ultraviolet absorption around 253 nm was quite different from that of lithospermic acid, which was the same case as Peak 6 and salvianolic acid B. We

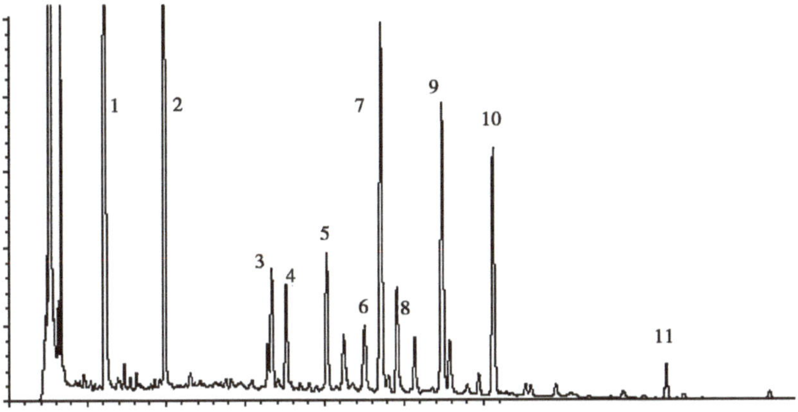

Fig. 15.6 HPLC fingerprint of Dantonic™ at 203 nm. Peaks: *1* Danshensu; *2* Protocatechuic aldehyde; *3, 4* Chemical compounds I/II; *5* Salvianolic acid D; *6* Salvianolic acid E; *7* Rosmarinci acid; *8* Ginsenoside Rg1; *9* Salvianolic acid B; *10* Salvianolic acid A; *11* Ginsenoside Rb_1

Table 15.9 The gradient elution conditions for LC/MSn analysis of water-soluble constituents of Danshen in Dantonic™

Time (min)	Mobile phase *A* (%)	Mobile phase *B* (%)	
0	95	5	Mobile phase *A*: glacial acetic acid:water = 0.01:100; mobile phase *B*: glacial acetic acid: acetonitrile = 0.01:100; flow rate: 0.5 ml/min; temperature: 30 °C; detection wavelength: multi-wavelength detection (showing wavelength of 281 mm)
15	78.3	21.7	
33	78.3	21.7	
38	65	35	

Table 15.10 The mass spectrum conditions for LC/MSn analysis of water-soluble constituents of Danshen in Dantonic™

Ion detecting method	Negative ion detection	
Dry gas flow rate (L/min)	10	10
Nebulizer pressure (psi)	60	60
Dry temprature (°C)	350	350
Capillary voltage (v)	3,500	3,500
Mass scan range (*m/z*)	100–1,200	100–800
Fragment ampl.	/	1.5–3.0

presumed that these two compounds had the same backbone as salvianolic acid E, carboxylic diphenyl ethylene. As there were no available reports about the two compounds, we tentatively named them Compound I and II. Their identities require further separation and NMR characterization.

2. Collection of the Chromatographic Fingerprint of Salvianolic Acids in Dantonic™

(a) Instrument, Reagents and Materials

Agilent 1100 series HPLC (equipped with Diode Array Detector), acetonitrile (chromatographic pure, Fisher Company, USA), phosphoric acid (analytical pure), Millipore ultrapure water. Dantonic™ (final product, intermediate and crude drugs) was provided by Tianjin Tasly Group Co. Ltd.

(b) Preparation of the test solution

(i) Dantonic™: Weigh 10 pills and put them in a 10 ml volumetric flask; add an appropriate amount of distilled water, and treat with ultrasound for 15 minutes before adding water to the scale.

(ii) Dantonic™ Intermediate: Weigh 0.1 g of the sample and put it in a 10 ml volumetric flask. Then follow the same preparation process as above.

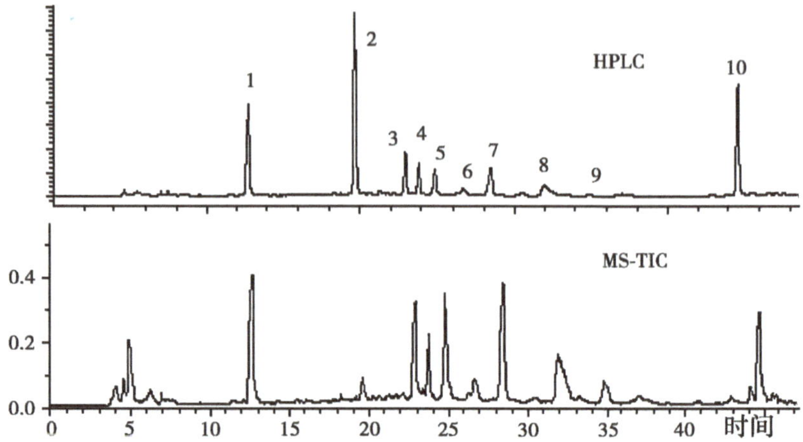

Fig. 15.7 The HPLC and MS-TIC chromatograms of Danshen water-soluble constituents in Dantonic™

Table 15.11 The HPLC/MS data of water-soluble constituents of Danshen in Dantonic™, and the identification results

Peak no.	Retention time	Identity	$[M-H]^-$	$[2M-H]^-$	$[M-2H]^{2-}$	λ_{max}
1	12.73	Danshensu	197	395		280
2	19.69	Protocatechuic aldehyde	137	275		231, 280, 310
3	22.99	Chemical compound I or II	537			327
4	23.83	Chemical compound I or II	537			327
5	24.89	Salvianolic acid D	417			247, 321
6	26.7	Salvianolic acid E	717		358	330
7	28.51	Rosmarinic acid	359	719		329
8	31.93	Salvianolic acid B	717		358	254, 286, 309
9	34.86	Salvianolic acid G	339			395
10	44.64	Salvianolic acid A	493			288

(iii) Danshen: Weigh 5 g of Danshen and put it in a round bottom flask; follow the production process to extract the product and put it in a 100 ml volumetric flask, add water to the scale.

All test solutions should be filtered through a 0.45 μm microporous membrane before analysis.

(c) Optimization of Chromatographic Conditions

(i) Selection of Mobile Phase: After comparing different systems, such as methanol–water, methanol-acetic acid–water, acetonitrile-acetic acid–water, and acetonitrile-phosphoric acid–water, we finally chose acetonitrile–0.02 % phosphoric acid–water as the mobile phase, because it had the best separation results with little baseline drift.

(ii) Selection of Gradient Elution: The above mobile phase at the same gradient (30, 50 %B) and 30 °C was used for elution, and the results are shown in Fig. 15.9a, b, respectively. The elution results using 0–100 % gradient within 0–30 min are shown in Fig. 15.9c. It could be concluded that gradient elution could effectively enhance the chromatographic separation. After trying many gradients, the following gradients was selected

Table 15.12 The HPLC-MSn data of water-soluble constituents of Danshen in Dantonic™ and explanation

Peak no.	Identity	Fragment ion *m/z*
3	Chemical compound I or II	MS^2 (537):493[M–H–CO_2]$^-$,295[M–H–CO_2–ROH]$^-$
		MS^3: (295):159, 109
4	Chemical compound I or II	MS^2 (537):493[M–H–CO_2]$^-$,295[M–H–CO_2–ROH]$^-$
		MS^3 (295):159, 109
5	Salvianolic acid D	MS^2 (417):373[M–H–CO_2]$^-$,175[M–H–CO_2–ROH]$^-$
		MS^3 (175):147, 157, 133
6	Salvianolic acid E	MS^2 (717):519[M–H–ROH]$^-$,321[M–H–2ROH]$^-$
		MS^3 (519):321[M–H–2ROH]$^-$,339[M–R–ROH]$^-$
		Grade3 (321):279, 293, 249, 223, 185
7	Rosmarinci acid	MS^2 (359):179[M–R]$^-$,161[M–H–ROH]$^-$
8	Salvianolic acid B	MS^2 (717):519[M–H–ROH]$^-$,321[M–H–2ROH]$^-$
		MS^3 (519):321[M–H–2ROH]$^-$,339[M–R–ROH]$^-$
		MS^3 (321):279, 293, 249, 223, 185
9	Salvianolic acid G	MS^2 (339):321[M–H–H_2O]$^-$,295[M–H– CO_2]$^-$
		MS^3 (295):279, 267
		MS^4 (279):251
10	Salvianolic acid A	MS^2 (493):295[M–H–ROH]$^-$
		MS^3 (295):159, 109

(Table 15.13) and the results derived from the system are shown in Fig. 15.9d, which indicates that the major constituents of Danshen in Dantonic™ were well separated.

(iii) Selection of Column Temperature: Under the selected mobile phase gradient, set the column temperature at 20, 25, 30, 35, and 40 °C, and inspect the influence of column temperature on the fingerprints of Dantonic™. The results showed that the fingerprints displayed certain regularity with the change in column temperature. Figure 15.10 shows the fingerprints obtained at column temperatures of 20, 30 and 40 °C, from which it can be seen that with the increase of column temperature, the retention time of each chromatographic peak becomes shorter. Through a comprehensive consideration of the chromatographic peak's separation degree, peak shape, analytical time and other factors, a column temperature of 30 °C was finally selected.

(iv) Selection of Detection Wavelength and Band Width: Due to the features of the production process and of Danshen's composition, the major components extracted are phenolic acids which contain a conjugated system structure and thus a relatively strong ultraviolet absorption. In the literature, 280 nm is usually taken as the detection wavelength to quantify Danshen phenolic acids. The components of the extracted notoginseng crude drug are mainly saponins, and except for the end absorption at around 200 nm, they are difficult to detect at other wavelengths. In the literature, 203 nm is generally taken as the detection wavelength. Figure 15.11 shows the spectrograms of the 14 major peaks

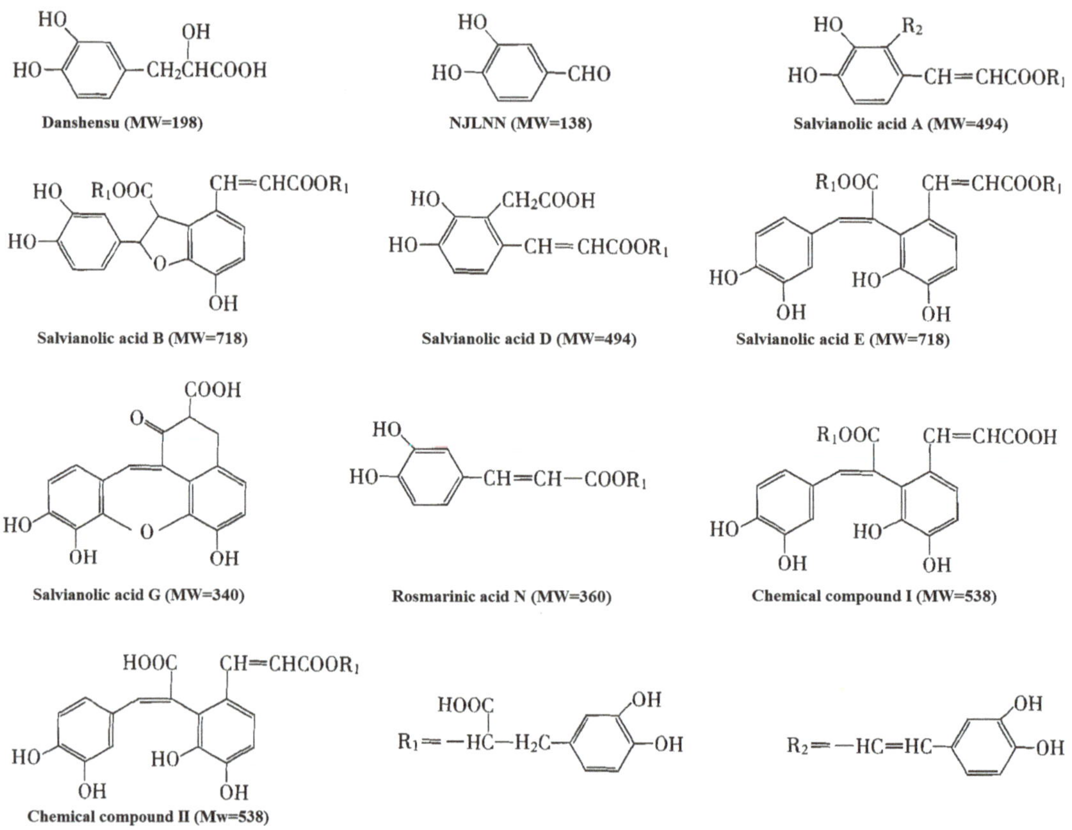

Fig. 15.8 The molecular structures of Danshen phenolic acids in Dantonic™

from the chromatogram of Dantonic™. For a better comparison, we standardized each spectrum, setting the maximum absorption of each spectrum as 1, and the other response values were then adjusted accordingly. From Fig. 15.11, we can see that Danshen phenolic acids have a relatively large absorption at 280 nm, which can better reflect the overall situation of phenolic acids and eliminate the influence on salvianolic acid fingerprints by notoginseng saponins. Besides selecting the detection wavelength, the diode array detector of Agilent 1100 HPLC needs bandwidth selection. With the other chromatographic conditions unchanged, we inspected the chromatograms under bandwidths of 4, 8, 16 and 24 nm. The results showed that the chromatograms obtained from different bandwidths overlapped together, which indicates that bandwidth has little influence on the chromatogram, nor on retention time or peak width. Through an overall consideration, the parameters of DAD used for the analysis of Danshen were set as: assaying wavelength of 280 nm and band width of 16 nm (Fig. 15.12).

(v) Selection of Sample Size: Take the test solution of Dantonic™ and inject 5, 10 and 20 μl of samples, respectively; after comparing their chromatograms directly, it is found that different sample amounts had little influence on the overall separation effect. Thus, 10 μl was finally selected as the sample size for HPLC analysis.

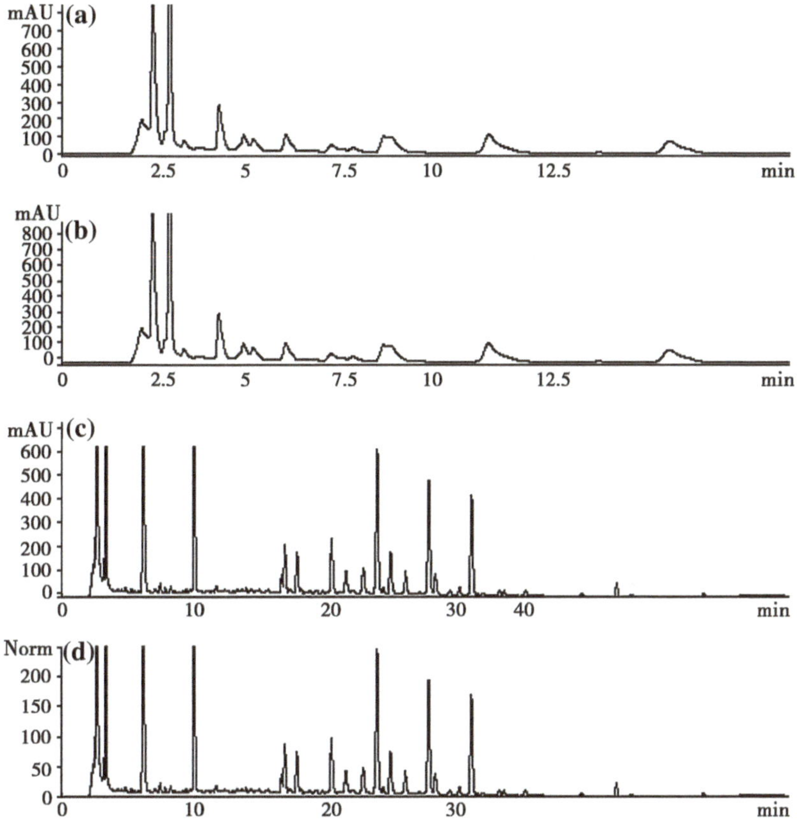

Fig. 15.9 Influence of mobile phase gradient on the chromatographic fingerprint

Table 15.13 The chromatographic mobile phase gradient program

Time (min)	Mobile phase A (water + 0.02 % phosphoric acid) (%)	Mobile phase B (80 % acetonitrile + 0.02 % phosphoric acid) (%)
0	90	10
8	78	22
15	74	26
55	48	52

(vi) Selection of Preparation Methods for the Test Solution of Dantonic™: Dantonic™ is a pill preparation, which differs from injection preparations, and its test solution requires a specified preparation process. We tested the different solvents and the length of ultrasound treatment.

Water, 50 % methanol, methanol, 50 % acetonitrile, acetonitrile, 50 % ethanol, and ethanol were used to prepare samples by following the method described above. Results showed that methanol, acetonitrile, and ethanol could not completely dissolve the sample, while other solvents could. Judging by the chromatograms, samples prepared with water, 50 % methanol and 50 % acetonitrile produced almost the same results. For the sake of convenience, water was selected as the solvent for preparing the test solution.

Using water as the solvent to prepare samples according to the above-mentioned method, we treated the samples with ultrasound for 5, 10, 15, 20 and 30 min. The result showed that all samples were dissolved and the chromatograms were almost the same, which indicates that to completely dissolve Dantonic™ samples, the length of ultrasonic treatment has no

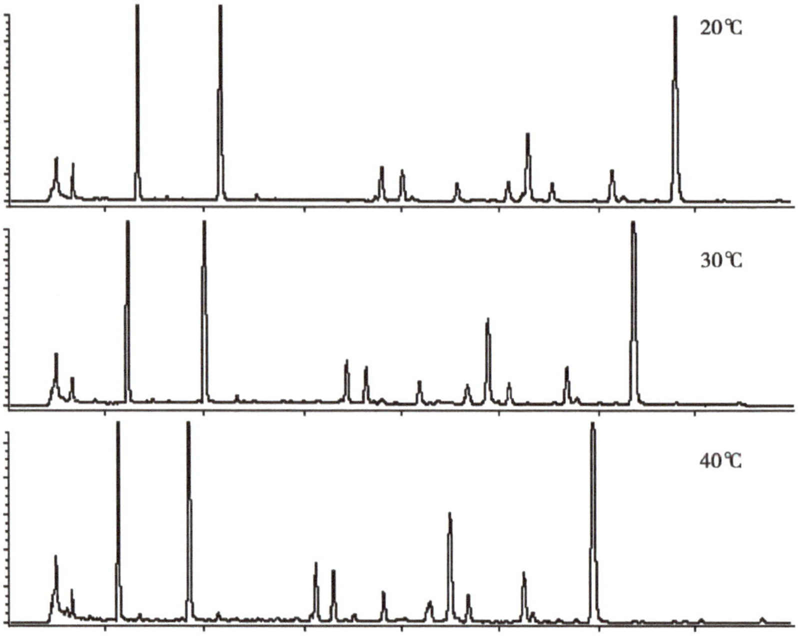

Fig. 15.10 Influence of column temperature on the fingerprints of salvianolic acids in Dantonic™ (280 nm)

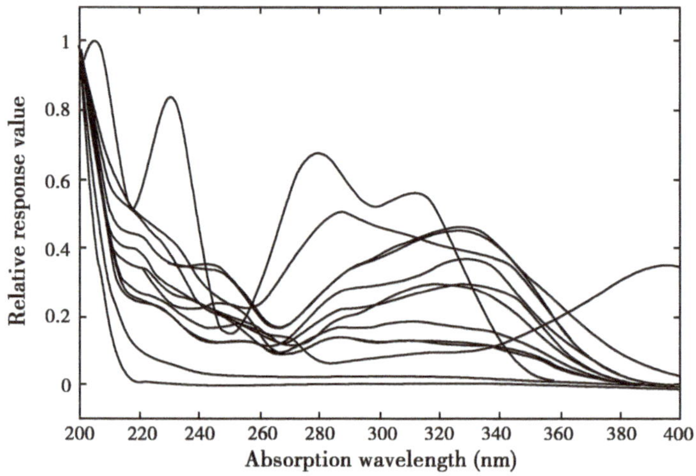

Fig. 15.11 Ultraviolet spectrograms of the major constituents in Dantonic™

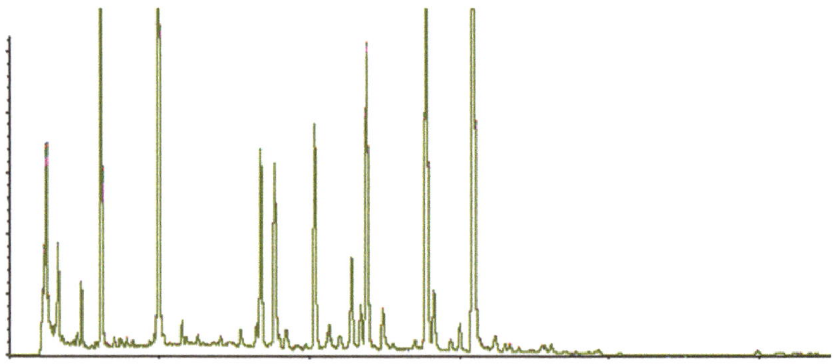

Fig. 15.12 The chromatograms at various bandwidths

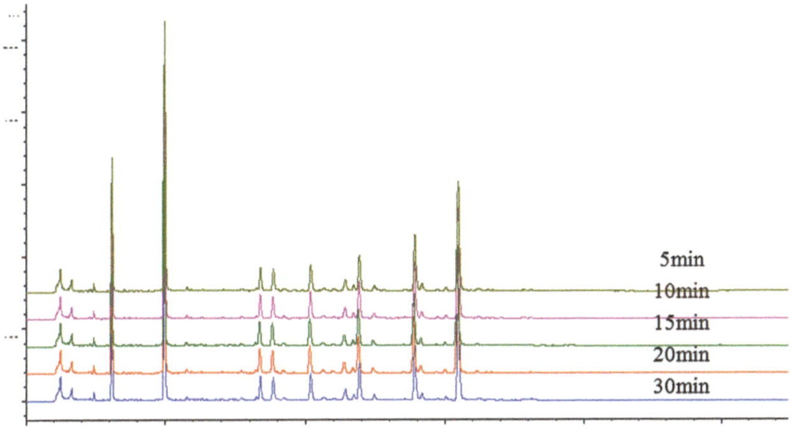

Fig. 15.13 HPLC chromatograms of samples treated with ultrasonic for different lengths of time

negative effect on the samples. We chose 15 min arbitrarily as the length of ultrasound treatment (Fig. 15.13).

After the above optimization of chromatographic conditions, the final testing conditions for the chromatographic fingerprint of Dantonic™ were set as: Agilent Zorbax SB-C_{18} analytical column (4.6 × 250 mm); mobile phase: phase A was 0.02 % (V/V) phosphoric acid water solution; phase B was 80 % acetonitrile water solution containing 0.02 % phosphoric acid; linear gradient elution process: 0 min (90:10), 8 min (78:22), 15 min (74:26), 40 min (57.75:42.25); flow rate: 1 ml/min; detection wavelength: 280 nm; column temperature: 30 °C; and sample size: 10 µl.

3. The Identification of Danshen Phenolic Acids in the Chromatographic Fingerprint and Methodological Study

   Not every analysis spectrum (image) of a TCM drug can be regarded as its chemical fingerprint. It is generally required to investigate whether the obtained spectrogram reflects the chemical composition of the sample, so as to answer the question of "why it is the fingerprint" and to prove that the obtained spectrum (image) can be used as the chemical fingerprint of this sample for its quality control.

   (a) The correlation of chromatographic fingerprints of salvianolic acid content in Dantonic™

   Inspecting the corresponding HPLC chromatographic fingerprints of Danshen, the intermediate of Dantonic™, and the final product, under the above-mentioned chromatographic conditions, we found that the fingerprints of the intermediate and final product had an excellent correlation, while Danshen differs from the final product and the intermediate in certain peaks. For example, Peaks A and B in the water extract of Danshen were relatively smaller than those in the intermediate and the final product, and Peak C (salvianolic acid B) was relatively bigger in the water extract of Danshen than in the intermediate and final product (Fig. 15.14).

   To find the cause of these differences, it is necessary to further inspect the impact of different extraction processes on the water-soluble constituents of Danshen. It was found that the pH of the extraction solvent would significantly affect the sizes of Peaks A, B, and C. With the increase in pH value, the sizes of Peaks A and B would also increase, but Peak C would decrease. When extracted by water (with a pH of around 7) or acidic solution (pH = 4), the sizes of Peaks A and B were very small, but they would increase notably if extracted by alkaline buffered salt solution (e.g., pH = 8.2 or 9.2). An increase in pH would reduce the content of salvianolic acid B (Fig. 15.15).

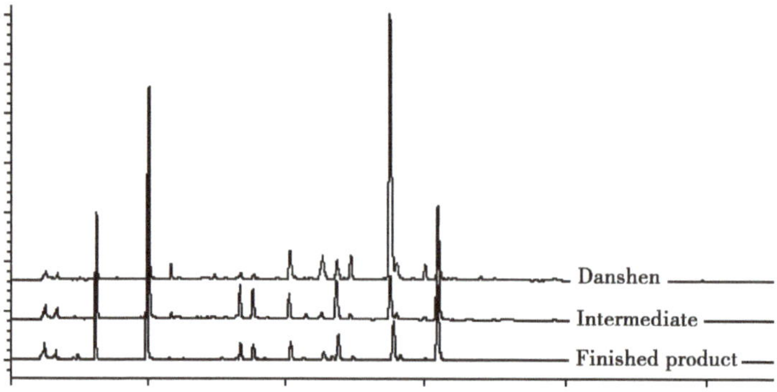

Fig. 15.14 Correlation study of HPLC chromatographic fingerprints of salvianolic acids

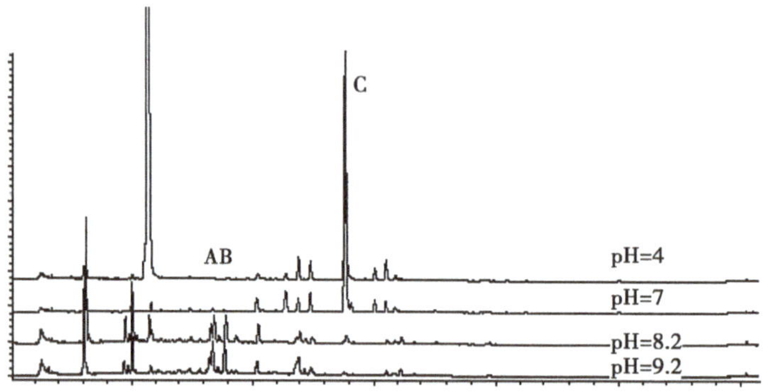

Fig. 15.15 The effect of pH values of extraction solvents on the HPLC fingerprints of Danshen

(b) Identification of the Chromatographic Peaks in the Fingerprints of Salvianolic Acids in Dantonic™

In order to verify that the obtained fingerprint was a real reflection of the salvianolic acids in Dantonic™, an HPLC-DAD-MS combination method was used to identify the peaks' chemical natures. The results showed that major chromatographic peaks corresponded to the following constituents: danshensu, protocatechuic aldehyde, salvianolic acid D, salvianolic acid E, rosmarinic acid, salvianolic acid G, salvianolic acid B and salvianolic acid A. Therefore, it is concluded that the fingerprint is a real reflection of the major water-soluble constituents of Dantonic™ (Fig. 15.16).

(c) Method Validation

(i) Precision: The same test solution was tested continuously for 6 times to examine the precision of the fingerprint. The results are shown in Fig. 15.17. Take the retention time and peak area of Peak 1 as a reference (set as 1) and calculate the relative peak area as well as the RSD of each peak. The results showed that the relative peak areas for each peak were relatively stable (with RSD within 1 %) (Table 15.14).

(ii) Repeatability: Take the same batch of Dantonic™ (Batch No.:20011213) and prepare six samples of test solution. Examine the variation of their HPLC profiles. The results showed

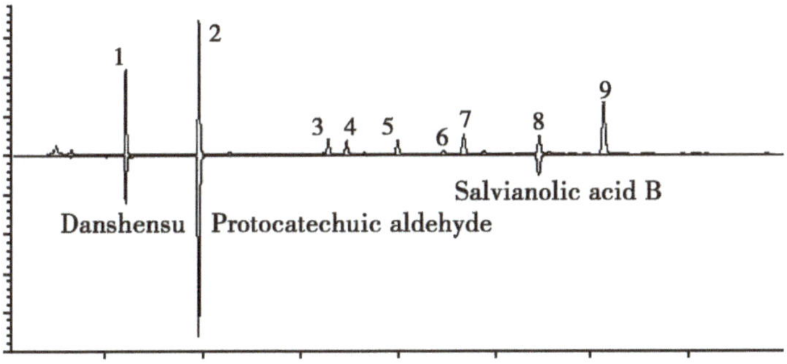

Fig. 15.16 The fingerprint of salvianolic acids in Dantonic™. Peaks: *1* Danshensu; *2* Protocatechuic aldehyde; *3, 4* Compound I/II; *5* Salvianolic acid D; *6* Salvianolic acid E; *7* Rosmarinic acid; *8* Salvianolic acid B; *9* Salvianolic acid A

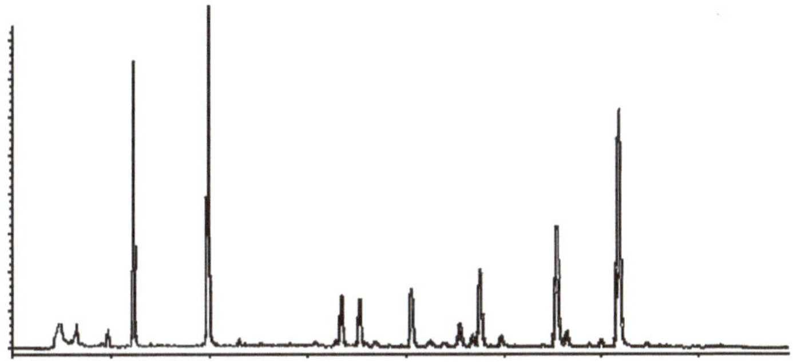

Fig. 15.17 The precision test results of Dantonic™ HPLC fingerprint

Table 15.14 The precision test results

Peak no.	Mean relative retention time	RSD (%)	Mean relative peak area	RSD
1	1	0.00	1	0.00
2	1.61	0.05	2.43	0.21
3	2.73	0.16	0.30	0.19
4	2.87	0.23	0.31	0.32
5	3.31	0.38	0.36	0.06
6	3.71	0.41	0.17	0.07
7	3.871	0.46	0.53	0.06
8	4.05	0.43	0.08	0.08
9	4.51	0.50	0.85	0.07
10	5.03	0.69	1.71	0.38

that the relative peak area of each chromatographic peak was relatively stable, with RSD within 2 % (Table 15.15, Fig. 15.18).

(iii) Stability: Store one Dantonic™ test solution at room temperature, and over 2 days, conduct multiple tests to examine the stability of the fingerprint. The corresponding HPLC fingerprints and the specific similarities (cosine method) are shown in Table 15.16 and Fig. 15.19. The results showed that the test solution of Dantonic™ was stable during the 2 days.

(iv) Tolerance

- Chromatographic Column's Influence on the Fingerprints: The chromatographic column is one of the major factors affecting the chromatographic fingerprints. We examined 6 chromatographic columns from different manufacturers or from different batches of the same manufacturer. By applying the 6 chromatographic columns on

Table 15.15 The repeatability test results

Peak no.	Relative peak area					Mean value	RSD (%)
	1	2	3	4	5		
1	1	1	1	1	1	1	0.00
2	1.75	1.75	1.75	1.77	1.75	1.75	1.02
3	0.27	0.28	0.28	0.28	0.28	0.28	0.38
4	0.25	0.26	0.26	0.24	0.26	0.25	0.60
5	0.26	0.26	0.26	0.26	0.26	0.26	0.16
6	0.19	0.19	0.18	0.18	0.18	0.19	0.57
7	0.69	0.69	0.67	0.68	0.69	0.68	0.80
8	0.17	0.17	0.17	0.15	0.17	0.17	0.63
9	0.56	0.56	0.55	0.55	0.55	0.56	0.56
10	1.48	1.51	1.48	1.52	1.5	1.50	1.51

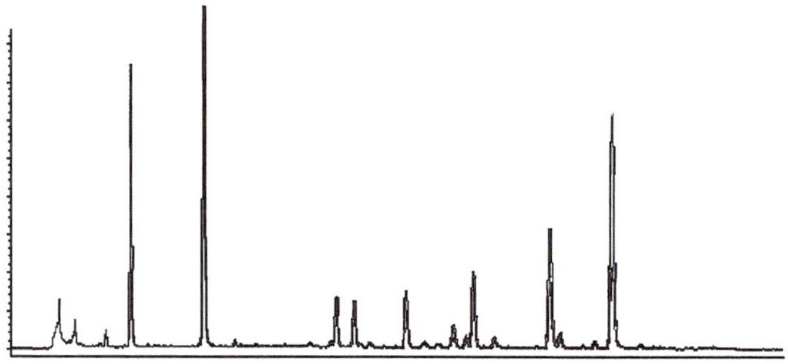

Fig. 15.18 The reproducibility test results of Dantonic™ HPLC fingerprint

Table 15.16 The stability test results of Dantonic™ HPLC fingerprint

Detection time (h)	0	1.5	3.0	6.0	9.0	12.0	15.0
Similarity	0.9996	0.9996	0.9999	0.9999	0.9999	0.9999	0.9999
Detection time (h)	18.0	21.0	24.0	30.0	36.0	42.0	48.0
Similarity	0.9999	1.0000	1.0000	1.0000	1.0000	1.0000	0.9999

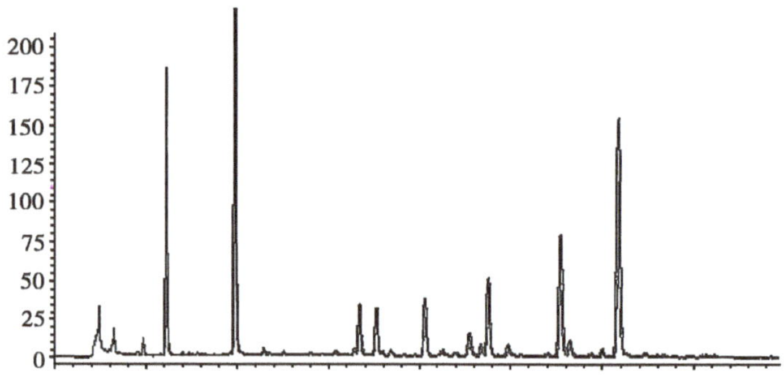

Fig. 15.19 The stability test results of Dantonic™ HPLC fingerprint

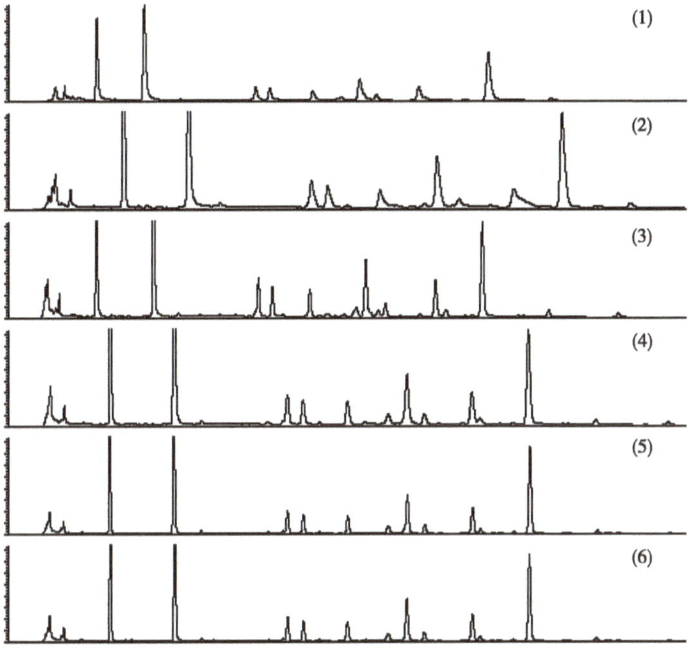

Fig. 15.20 The effect of chromatographic columns on the HPLC fingerprint of Dantonic™

Table 15.17 The chromatographic columns and models in the study

No.	Specifications	Filler	Manufacturer	Batch no.
1	250 × 4.6 mm, 5 μm	ODS C_{18}	Dalian Elite Analytical Instruments Co, Ltd.	
2	250 × 4.6 mm, 5 μm	Lichrospher C_{18}	Jiangsu Hanbang	01111604
3	250 × 4.6 mm, 5 μm	Extend C_{18}	Agilent Zorbax	USHR002173
4	250 × 4.6 mm, 5 μm	SB C_{18}	Agilent Zorbax	USCL01309
5	250 × 4.6 mm, 5 μm	SB C_{18}	Agilent Zorbax	USCL010304
6	250 × 4.6 mm, 5 μm	SB C_{18}	Agilent Zorbax	USCL013107

the same HPLC device, we tested 5 Dantonic™ samples. Figure 15.20 lists the 6 corresponding chromatograms of a sample. It is obvious that due to the different fillers, the fingerprints obtained from the chromatographic columns of different manufacturers showed great disparity in peak shape, retention value of the peak, etc., while those columns from the same manufacturer's different batches showed little discrepancy (Table 15.17 and Fig. 15.20).

The results of calculated similarities among the HPLC fingerprints obtained from different chromatographic columns are shown in Table 15.18. The fingerprint from the fifth chromatographic column was taken as the reference for calculation.

- Instrument Influence on the Fingerprints: Under the same conditions, different chromatographs may pose some influence on the analytical results of the samples. In this study, the test solutions from 5 batches of Dantonic™ were used to examine the influence of different instruments on the fingerprints, with the other experiment conditions identical. For a specified batch, the HPLC fingerprints obtained from different

Table 15.18 The similarities of HPLC fingerprints obtained from different chromatographic columns

Batch no. of the sample	20011202	20011208	20011116	20011120	20020124
Column 1	0.9817	0.9911	0.9947	0.9891	0.9974
Column 2	0.9871	0.9960	0.9970	0.9947	0.9978
Column 3	0.9866	0.9980	0.9978	0.9935	0.9985
Column 4	0.9859	0.9969	0.9969	0.9934	0.9982
Column 5	1	1	1	1	1
Column 6	0.9860	0.9940	0.9979	0.9934	0.9980

Table 15.19 The similarities of fingerprints obtained from different chromatographs

Instrument	20011116	20011120	20011202	20011208	20020124
Agilent 1100 (1)	0.9969	0.9939	0.9854	0.9954	0.9982
Agilent 1100 (2)	0.9982	0.9948	0.9850	0.9964	0.9982
Waters 2695	0.9973	0.9947	0.9870	0.9961	0.9984

chromatographs demonstrated that the two systems, Agilent and Waters, havc little difference on the separation of peaks and the overall fingerprint quality. The results are presented in Table 15.19 and Fig. 15.21.

- Influence of the Chromatographic Column Efficiency on Dantonic™ Fingerprints: In the process of use, changes in column efficiency will also affect the fingerprints. In order to check the effect of column efficiency on the fingerprints, we took samples from the same batch of product and obtained their HPLC fingerprints at different times. During the process, the column

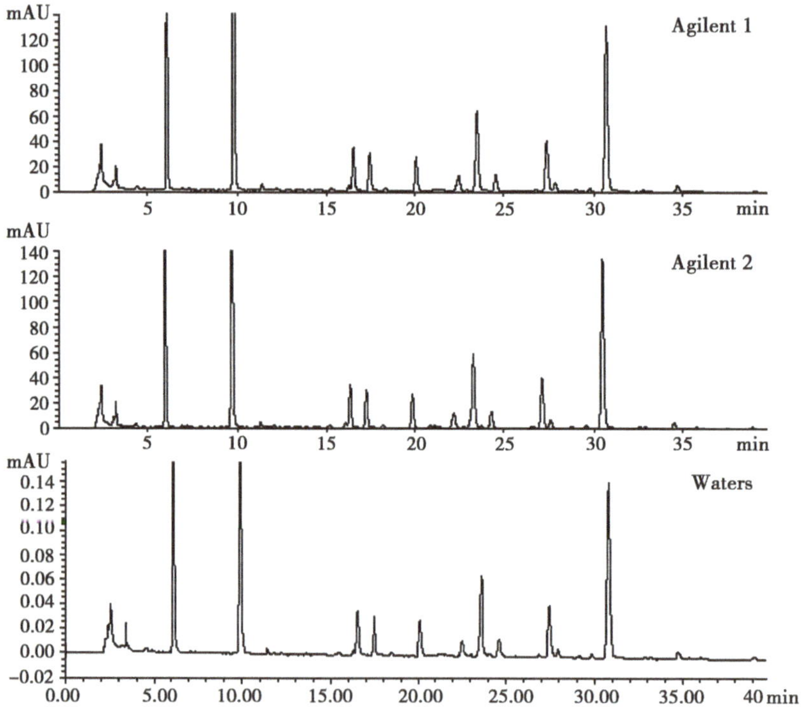

Fig. 15.21 The effect of chromatographs on HPLC fingerprints of Dantonic™

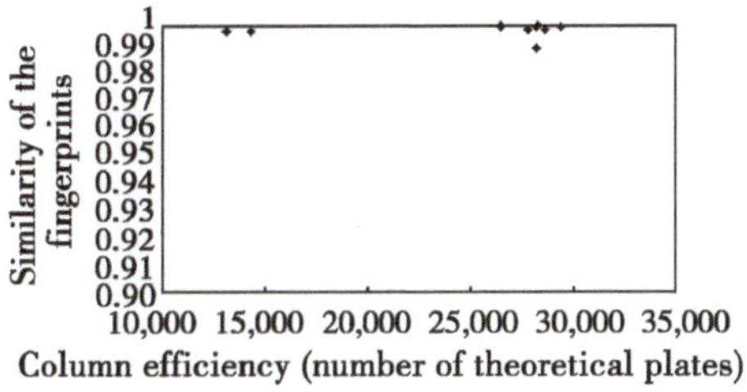

Fig. 15.22 The effect of column efficiency on chromatographic fingerprints

efficiency was recorded and the similarities of the fingerprints were calculated to check the effect of column efficiency on the fingerprints. As the results show, under different column efficiency, the fingerprints were very similar, suggesting that even though the column efficiency fluctuated within a certain range, it posed no obvious impact on the fingerprints (Fig. 15.22).

The above results show that the HPLC profiles obtained by the above methods can well reflect the composition of the Danshen water-soluble constituents in Dantonic™. The methods are stable and reliable, and the HPLC profiles can be taken as the chromatographic fingerprints for the above water-soluble constituents. However, the profiles do not show the notoginseng saponins in Dantonic™. That is why we need to establish another fingerprinting technique for notoginseng saponins in order to completely reflect the chemical composition of Dantonic™ (see references for the relevant methods).

15.1.1.3 The Chromatographic Fingerprinting of Salvianolic Acids in Danshen Injection

Jinlan Zhang and Dean Guo

1. Instruments and Reagents
   Agilent 1100 series HPLC (equipped with Diode Array Detector); PE SCIEX QSTAR MASS analyzer; Chromatographic column: Zorbax Extend C_{18} (4.6 × 250 mm, 5 μm).
   Methanol (GR), hydrochloric acid (AR), and acetidin (AR) were produced by Beijing Chemical Factory; acetonitrile was produced by Caledon, a Canadian company. Danshen Injection was produced by Ya'An Sanjiu (Batch No.:020603), specification: 10 ml, 1 g Danshen/ml. Danshensu and protocatechuic aldehyde were purchased from National Institutes for Food and Drug Control. Salvianolic acid B and C were self-made, with each at a purity of over 97 % (determined by HPLC).
2. Sample Preparation and Analysis
   The Danshen Injection was diluted with double distilled water at a ratio of 1:3, and then underwent HPLC analysis directly. The Danshen roots used for producing Danshen Injection were treated in the same way as the samples for HPLC fingerprinting of water-soluble constituents. The conditions for HPLC were also the same. See Fig. 15.23 for the HPLC chromatogram. From the chromatogram, it can be found that after Danshen was processed into Danshen Injection, the proportion of the constituents changes significantly. Danshensu and protocatechuic aldehyde were the major constituents of Danshen Injection, with a greater variety of poly-phenolic acids than in Danshen.
3. HPLC-MS Analysis on the Water-Soluble Constituents in Danshen Injection
   The analytical conditions for HPLC-MS were the same as those for Danshen analysis. The total ion current (TIC) of Danshen Injection is shown in Fig. 15.24. Through selected ion recording (SIR) and extracting the MS of the major chromatographic peaks, the molecular weights of 11 components were obtained (Table 15.20). By referencing the published literature [1] and comparing with the reference substances, a total of 12 constituents were identified, and their positions in the chromatogram were confirmed (Fig. 15.25).
4. Methodological Study
   The precision, repeatability, and stability of the HPLC method for the analysis of Danshen Injection was studied.

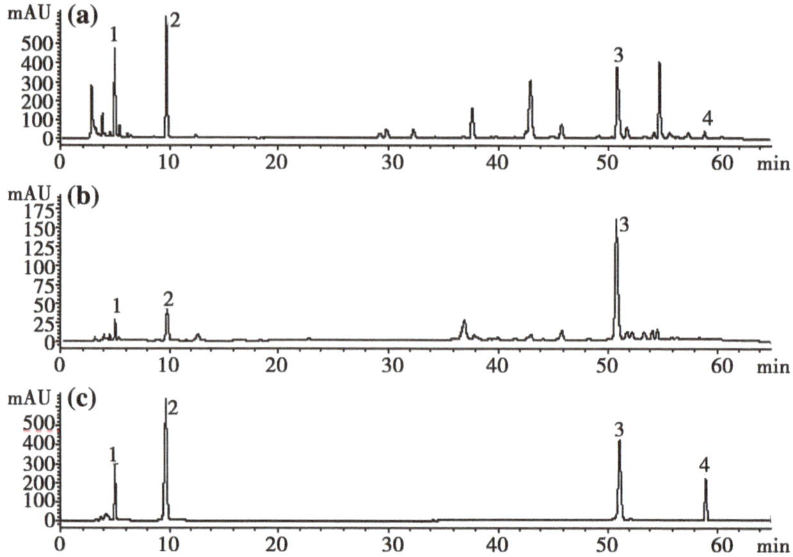

Fig. 15.23 The HPLC fingerprints of water-soluble constituents in Danshen. Panels: **a** Danshen Injection; **b** Danshen; **c** References. Peaks: *1* Danshensu; *2* Protocatechuic aldehyde; *3* Salvianolic acid B; *4* Salvianolic acid C

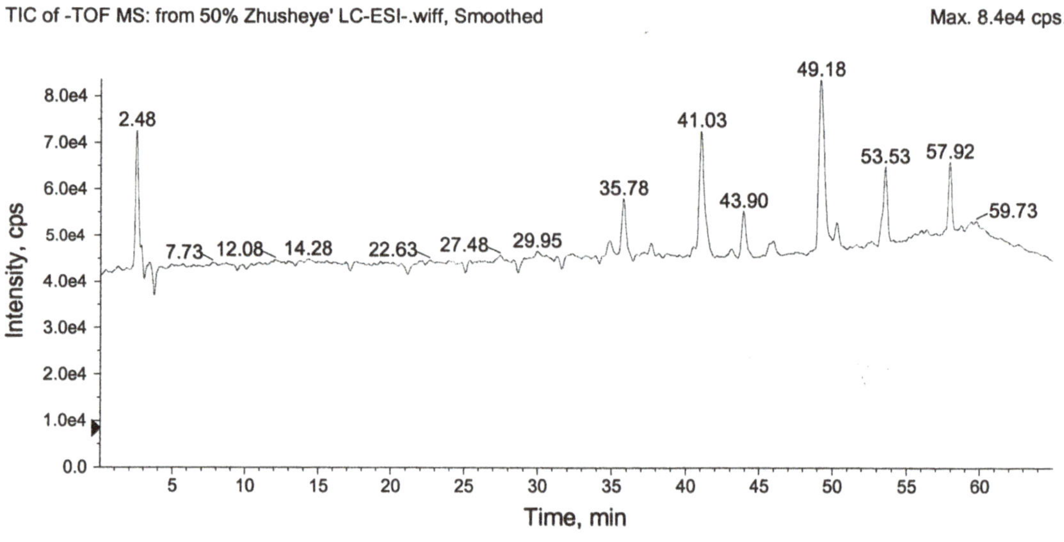

Fig. 15.24 The total ion current of Danshen injection

Table 15.20 Identification of the phenolic acids in the HPLC fingerprint of Danshen Injection

Peak no.	Retention time (min)	Mass number	Possible identity	Peak no.	Retention time (min)	Mass number	Possible identity
1	4.88	198	Danshensu	7	43.90	494	New compound
2	9.48	138	Protocatechuic aldehyde	8	45.97	538	lithospermic acid
3	27.40	538	Salvianolic acid H or I	9	49.18	718	Salvianolic acid B
4	34.88	340	Salvianolic acid G	10	53.28	718	Salvianolic acid E
5	35.78	418	Salvianolic acid D	11	53.55	494	Salvianolic acid A
6	41.03	360	Rosmarinci acid	12	57.92	492	Salvianolic acid C

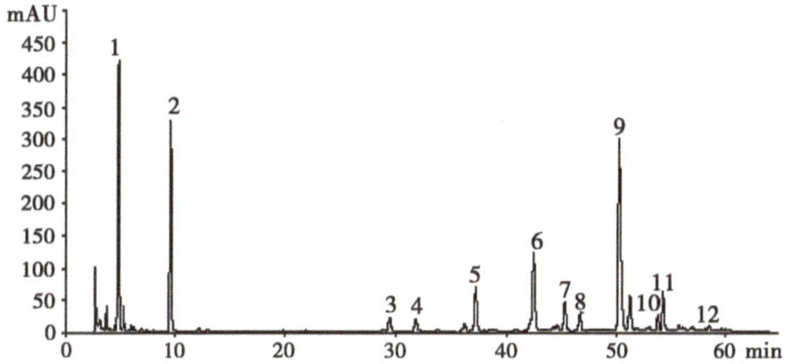

Fig. 15.25 Identification of the chemical components in the HPLC chromatogram of Danshen Injection. *1* Danshensu; *2* Protocatechuic aldehyde; *3* Salvianolic acid H or I; *4* Salvianolic acid G; *5* Salvianolic acid D; *6* Rosmarinic acid; *7* New Compound; *8* lithospermic acid; *9* Salvianolic acid B; *10* Salvianolic acid E; *11* Salvianolic acid A; *12* Salvianolic acid C

(a) Precision: After dilution of Danshen Injection, the solution was tested 6 times repeatedly by the HPLC method. The obtained data was then processed by "*Similarity Evaluation System for Chromatographic Fingerprint of Chinese Crude Drug*" (2004A). The results showed that the similarities of the results were over 0.95 (Table 15.21).

(b) Repeatability: Following the sample preparation method for Danshen Injection, 6 samples were prepared and tested by HPLC to check the repeatability of the method. The results showed that the method had good repeatability (Table 15.22).

(c) Stability: After dilution of Danshen Injection, the samples were stored at room temperature, and tested by HPLC at

Table 15.21 The precision test results

	S1	S2	S3	S4	S5	S6	Fingerprint comparison
S1	1.000	0.999	0.999	0.998	0.997	0.995	0.999
S2	0.999	1.000	0.999	0.999	0.999	0.998	1.000
S3	0.999	0.999	1.000	0.999	0.998	0.998	1.000
S4	0.998	0.999	0.999	1.000	1.000	0.999	1.000
S5	0.997	0.999	0.998	1.000	1.000	1.000	1.000
S6	0.995	0.998	0.998	0.999	1.000	1.000	0.999
Fingerprint comparison	0.999	1.000	1.000	1.000	1.000	0.999	1.000

Table 15.22 The repeatability test results

	S1	S2	S3	S4	S5	S6	Fingerprint comparison
S1	1.000	1.000	0.999	1.000	0.999	0.996	0.999
S2	1.000	1.000	0.998	0.999	0.999	0.996	0.999
S3	0.999	0.998	1.000	0.999	1.000	0.999	1.000
S4	1.000	0.999	0.999	1.000	1.000	0.998	1.000
S5	0.999	0.999	1.000	1.000	1.000	0.999	1.000
S6	0.996	0.996	0.999	0.998	0.999	1.000	0.999
Fingerprint comparison	0.999	0.999	1.000	1.000	1.000	0.999	1.000

Table 15.23 The stability test results

Time (h)	S1	S2	S3	S4	S5	S6	S7	S8	Fingerprint comparison
0	1.000	1.000	1.000	1.000	0.999	0.998	0.946	0.987	0.998
1.5	1.000	1.000	1.000	1.000	0.999	0.998	0.948	0.988	0.998
3.0	1.000	1.000	1.000	1.000	0.999	0.999	0.948	0.989	0.998
6.0	1.000	1.000	1.000	1.000	1.000	0.999	0.95	0.991	0.999
9.0	0.999	0.999	0.999	1.000	1.000	1.000	0.952	0.993	0.999
12.0	0.998	0.998	0.999	0.999	1.000	1.000	0.953	0.995	0.999
24.0	0.946	0.948	0.948	0.95	0.952	0.953	1.000	0.956	0.963
48.0	0.987	0.988	0.989	0.991	0.993	0.995	0.956	1.000	0.994

different times. The results showed that the samples were stable within 48 h (Table 15.23).

5. The HPLC Fingerprints of Water-Soluble Constituents in Danshen Injection Samples from Different Manufacturers

   The previously established HPLC analytical method was followed to analyze the samples provided by different manufacturers. The results are shown in Fig. 15.26.

6. Summary

   (a) The HPLC-MS analysis results showed that the major constituents of Danshen Injection were water-soluble phenolic acids from Danshen. Therefore, Danshen that contains less tanshinone II_A than the pharmacopoeia's standard can still be used to produce qualified Danshen Injection, as long as its content of salvianolic acid B is at a high level. Therefore, based on our study results, it is suggested that new quality control standards for Danshen should be established.

   (b) There exist discrepancies in the fingerprints of Danshen Injections from different companies. All samples contained danshensu and protocatechuic aldehyde, and the major discrepancy was represented in the variety and content of salvianolic acids. This may be caused by the different sources of Danshen and production processes by different manufacturers.

   (c) It is imperative to unify the production process of Danshen Injection and to establish specified quality standards for Danshen Injection.

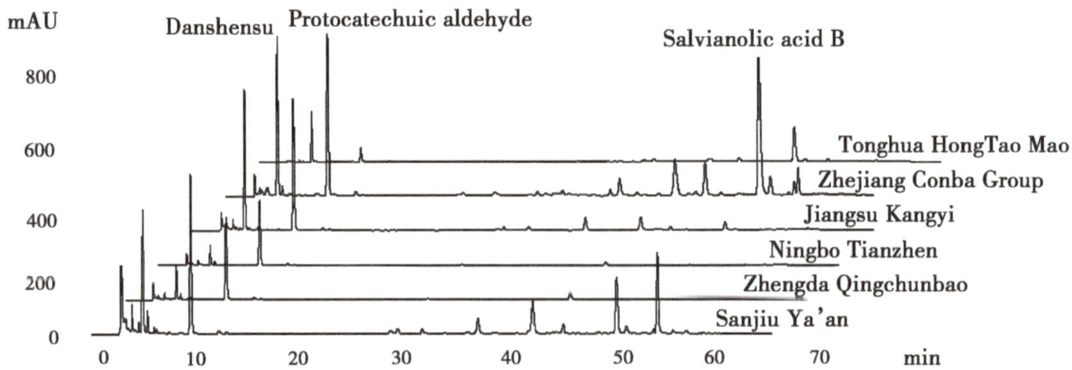

Fig. 15.26 The HPLC fingerprints of the water-soluble constituents in Danshen Injection samples from different companies

15.1.1.4 The Chromatographic Fingerprinting of Salvianolic Acids in Guanxinning Injection

Ming Zhu

1. The Establishment of Chromatographic Fingerprint of Salvianolic Acids in Guanxinning Injection

Guanxinning Injection is made of Danshen and Sichuan lovage root. Its indications are for "angina and coronary heart disease by activating and promoting blood circulation and dissipating blood stasis." According to the literature, the active components of Danshen are mainly the liposoluble diterpenoid quinones and water-soluble phenolic acids. Because the production of Guanxinning Injection uses water extraction and alcohol precipitation techniques, the emphasis of QC should be on water-soluble constituents. The water-soluble phenolic acids from Danshen mainly include: rosmarinic acid, methylrosmarinata, dimethyl lithospermate, ethyllithospermate, lithospermic acid B, salvianolic acid A, B, C, D, E, F, G, H, I, isosalvianolic acid C, protocatechuic aldehyde, protocatechuic acid, caffeic acid, isoferulicacid, etc. [1, 6, 7]. Among them, salvianolic acids have the functions of anti-hepatic fibrosis, antiplatelet agglutination, anti-oxidation, and have protective effects on the heart and cell injury of brain tissue [8, 9].

As reported in literature [10], Sichuan lovage root contains volatile oil, phthalide derivatives, alkaloid, and salvianolic acids. Because Guanxinning Injection is produced by water extraction and alcohol precipitation techniques, assays should be conducted on the water-soluble constituents of Sichuan lovage root. Phthalide derivatives include ligustilide, 3-butylidenephthalide, senkyunolide, etc. Alkaloids include chuanxiongzine, etc. Phenolic acids include ferulaic acid, vanillic acid, protocatechuic acid, sedanic acid, 4-(Hydroxymethyl) benzoic acid, etc. Other compounds include vanillin, etc. Among these compounds, ferulaic acid, senkyunolide H, and senkyunolide I could reduce the deformability and aggregation of erythrocytes [11, 12], while chuanxiongzine, vanillin, and chrysophanic acid could affect the membrane receptors of myocardial cell membranes [13].

Guanxinning Injection is an injection of Danshen and Sichuan lovage root decocted with water and precipitated with alcohol. The major water-soluble constituents of Danshen are salvianolic acids, and the water-soluble constituents of Sichuan lovage root include phenolic acids, phthalide derivatives, etc. As most of these constituents have ultraviolet absorption, HPLC/UV method is thus selected for the fingerprint of this drug.

(a) Selection of Extraction Method: To select a method for the preparation of test solutions, the following two were tried: direct dilution and extraction with acetidine. The results showed that the major water-soluble constituents from Danshen and Sichuan lovage root could be detected in the test solution prepared by the direct dilution method. Therefore, direct dilution was selected for preparing the test solution.

(i) The Effect of Solvents and pH on the Stability of the Major Components: The influence of different diluent solvents on the stability of test solutions was studied. The tested solvents were 10 % methanol solution, 10 % methanol (containing 0.2 % glacial acetic acid) solution, 10 % methanol (pH = 9), 10 % methanol (pH = 8). The results showed that salvianolic acid A was unstable in neutral or alkaline conditions, while the major chromatographic peaks were relatively stable in acidic conditions. Therefore, 10 % methanol containing 0.2 % glacial acetic acid was chosen as the solvent.

(ii) Extraction with Acetidine: The major constituents detected by gas chromatography in the acetidin extract were protocatechuic aldehyde, caffeic acid, ferulaic acid, and phthalide derivatives, which were mainly from Sichuan lovage root. All of them could be easily detected in the HPLC chromatograms of samples prepared by direct dilution, and the number and quantity of detected peaks from directly diluted solutions were more than that of acetidine extraction. So, the direct dilution method was chosen for the preparation of the test solution: precisely measure 1 ml sample, place in a 5 ml volumetric flask, add 10 % methanol (containing 0.2 % glacial acetic acid) to the scale, and shake well.

(b) Selection of the Mobile Phase

(i) Elution mode: The constant gradient and program gradient of acetonitrile–0.03 % phosphoric acid solution was used as the mobile phase for elution. The results showed that in a constant gradient system, if the proportion of acetonitrile was too low, compounds like salvianolic acid B could not be eluted, even though danshensu, protocatechuic aldehyde and other constituents could be separated very well. If the concentration of acetonitrile was too high, compounds like salvianolic acid B could be eluted and separated, but other components like danshensu and protocatechuic aldehyde were eluted too quickly to get a satisfactory separation. Yet, program gradient elution was able to quickly and satisfactorily separate most components. Therefore, program gradient elution was selected for fingerprinting (Fig. 15.27).

(ii) Mobile Phase: The following mobile phases were tested: methanol–0.1 % TFA solution, methanol–0.5 % phosphoric acid, acetonitrile–0.5 % phosphoric acid, and acetonitrile–0.03 % phosphoric acid. Agilent Zorbax Extend C_{18} (4.6 × 250 mm, 5 μm) was used as the chromatographic column. The elution mode was program gradient as shown in Table 15.24.

In order to unify the fingerprint methods for Danshen Injection and simplify the fingerprinting conditions, several research institutions which were engaging in research in this area agreed on using acetonitrile–0.03 % phosphoric acid solution as the mobile phase for the fingerprinting of Danshen Injections (Fig. 15.28).

(iii) Acidity's Influence on Chromatogram: Using the above-mentioned test solution and elution method, and the same chromatographic column on the same chromatograph, the following mobile phases were tested to study their influence on chromatogram: acctonitrile–0.02 % phosphoric acid, acetonitrile–0.0265 % phosphoric acid, acetonitrile–0.03 % phosphoric acid, acetonitrile–0.04 % phosphoric acid. The results showed that when phosphoric acid was 0.02 %, caffeic acid could not be separated from other chromatographic peaks. When phosphoric acid concentration was over 0.0265 %, the major chromatographic peaks could be well separated, and the variation in the phosphoric acid concentration had little influence on the retention times of the analytes. Therefore, acetonitrile–0.03 % phosphoric acid solution was selected as the mobile phase (Fig. 15.28).

(c) Determination of Wavelength: Under the above-mentioned chromatographic conditions, use HPLC-DAD and a combination of HPLC-PDA-MS/MS techniques to determine the chromatogram of Guanxinning Injection (wavelength scanning: 200–400 nm) and the total ion chromatogram (mass scanning: 100–800). Based on the three dimensional chromatogram (Fig. 15.29) detected by HPLC-DAD and the equal-absorption figure (Fig. 15.30), it can be concluded that at the wavelength of 288 nm and under the same chromatographic conditions, Danshen's major salvianolic acids, Sichuan lovage root's phenolic acids and phthalide derivatives can all be detected simultaneously, and the results could reflect the water-soluble

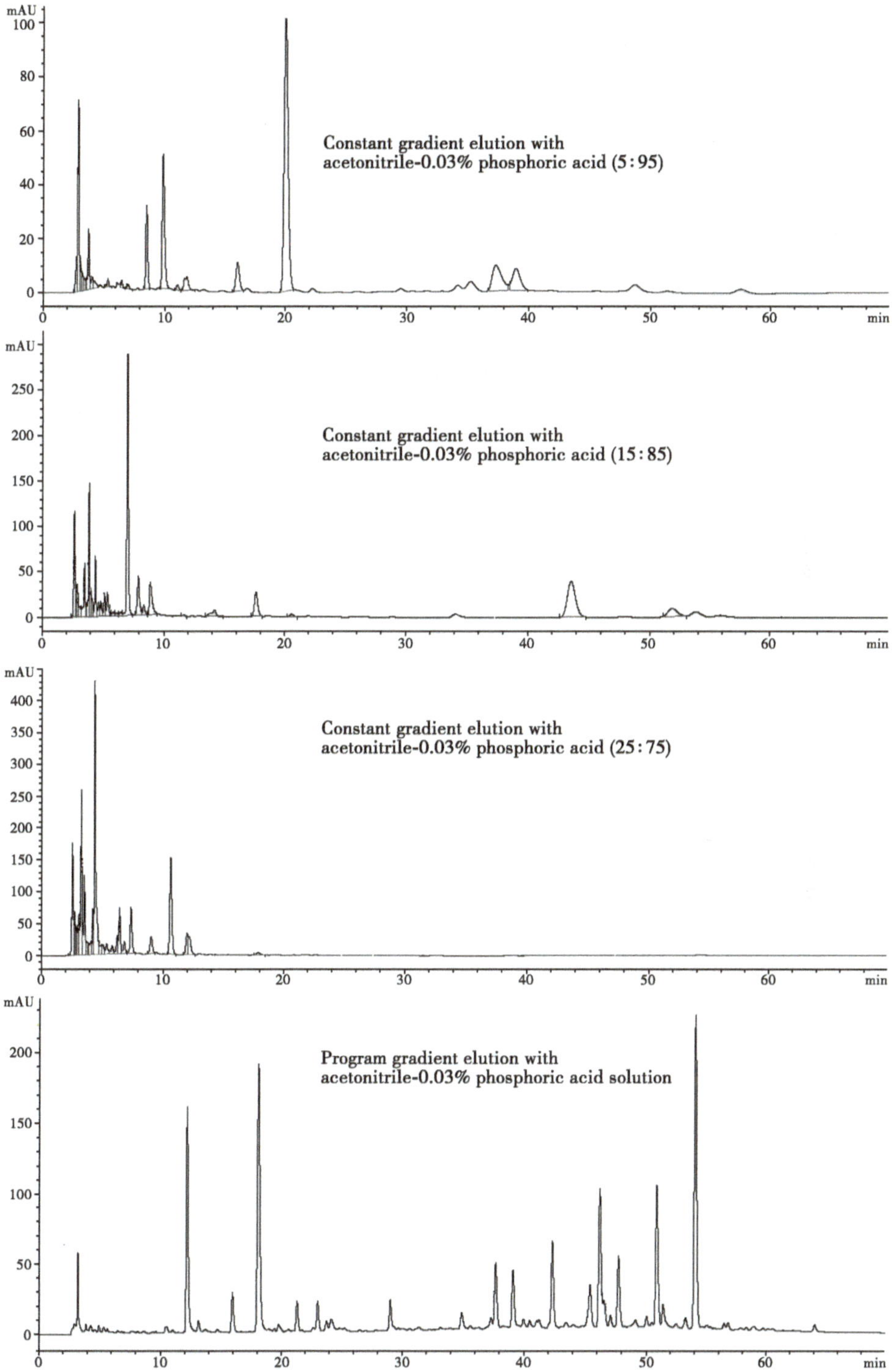

Fig. 15.27 The HPLC chromatograms generated by different elution modes

Table 15.24 Program gradient elution table

Time (min)	A (%)	B (%)
0 → 65	2 → 33	98 → 67
65 → 70	33 → 2	67 → 98

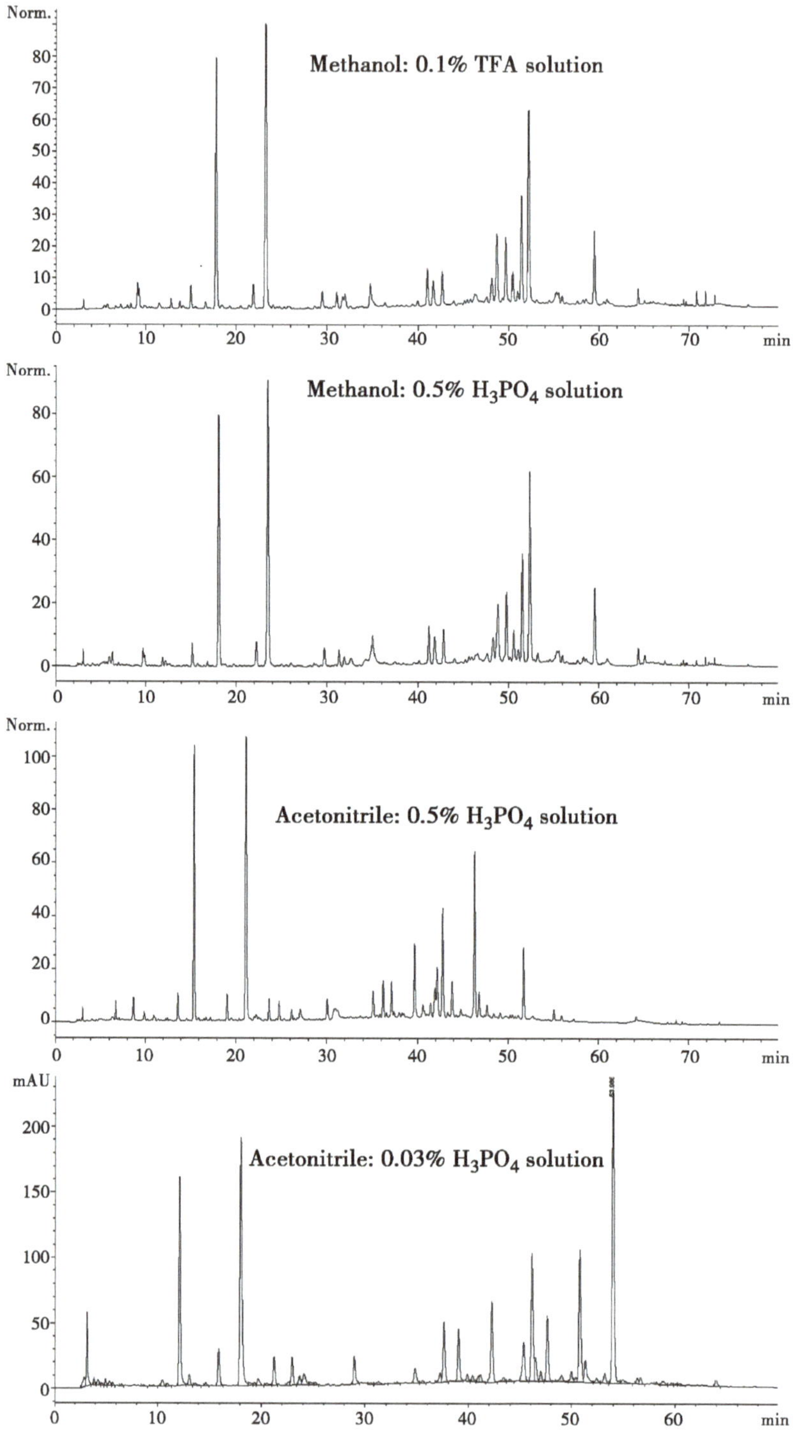

Fig. 15.28 The effect of mobile phase on fingerprinting

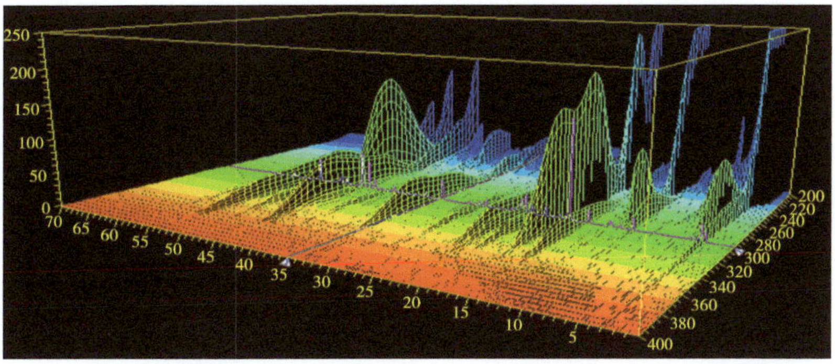

Fig. 15.29 3D fingerprint of Guanxinning Injection generated by HPLC-DAD

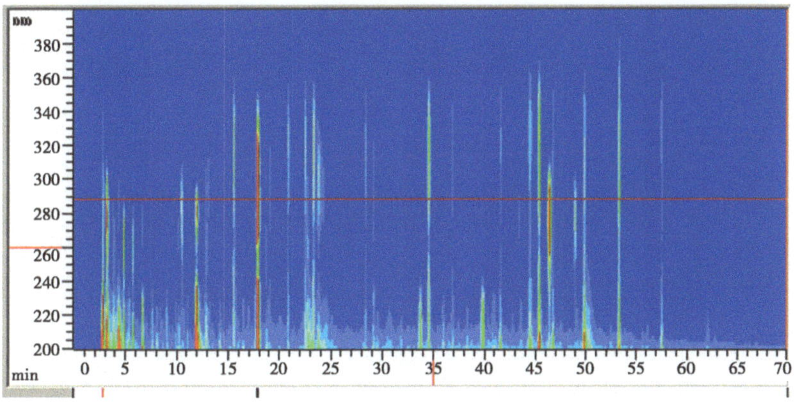

Fig. 15.30 The equal-absorption spectrum of Guanxinning Injection generated by HPLC-DAD

constituents of Danshen and Sichuan lovage root in Guanxinning Injection.

(d) Selection of Chromatographic Column: The above-mentioned test solutions and elution mode were used to select chromatographic columns made by different manufacturers: Agilent Zorbax extend C_{18} (4.6 × 250 mm, 5 μm), Discovery C_{18} (4.6 × 250 mm, 5 μm), Alhech C_{18} (4.6 × 250 mm, 5 μm), Shimadzu shim-pack (4.6 × 150 mm, 5 μm) and Dikma-C_{18} (4.6 ram × 250 mm, 5 μm). The results showed that all chromatographic columns from the above five manufacturers could separate well the major chromatographic peaks of Guanxinning Injection, but the fingerprints from columns of different companies display quite significant discrepancies. Therefore, it is necessary to choose a chromatographic column. In order to unify the fingerprint methods for Danshen Injections and simplify the fingerprint conditions, several research institutions which were engaging in research in this area agreed on the Agilent Zorbax extend C_{18} (4.6 × 250 mm, 5 μm) column for the fingerprinting of Danshen Injections (Fig. 15.31).

(e) Determination of Elution Time: Under the conditions stated above, a 140 min-chromatogram of Guanxinning Injection was generated. The gradient elution process is described in Table 15.25.
The result showed that the constituents detected at the wavelength of 288 nm could be eluted completely within 70 min. So, the elution time was set to be around 70 min (Fig. 15.32).

(f) Identification of the Chromatographic Peaks: The fingerprints of Guanxinning Injection were primarily used to determine the salvianolic acids of Danshen and the phenolic acids and water-soluble phthalide derivatives of Sichuan lovage root. By using the reference substances,

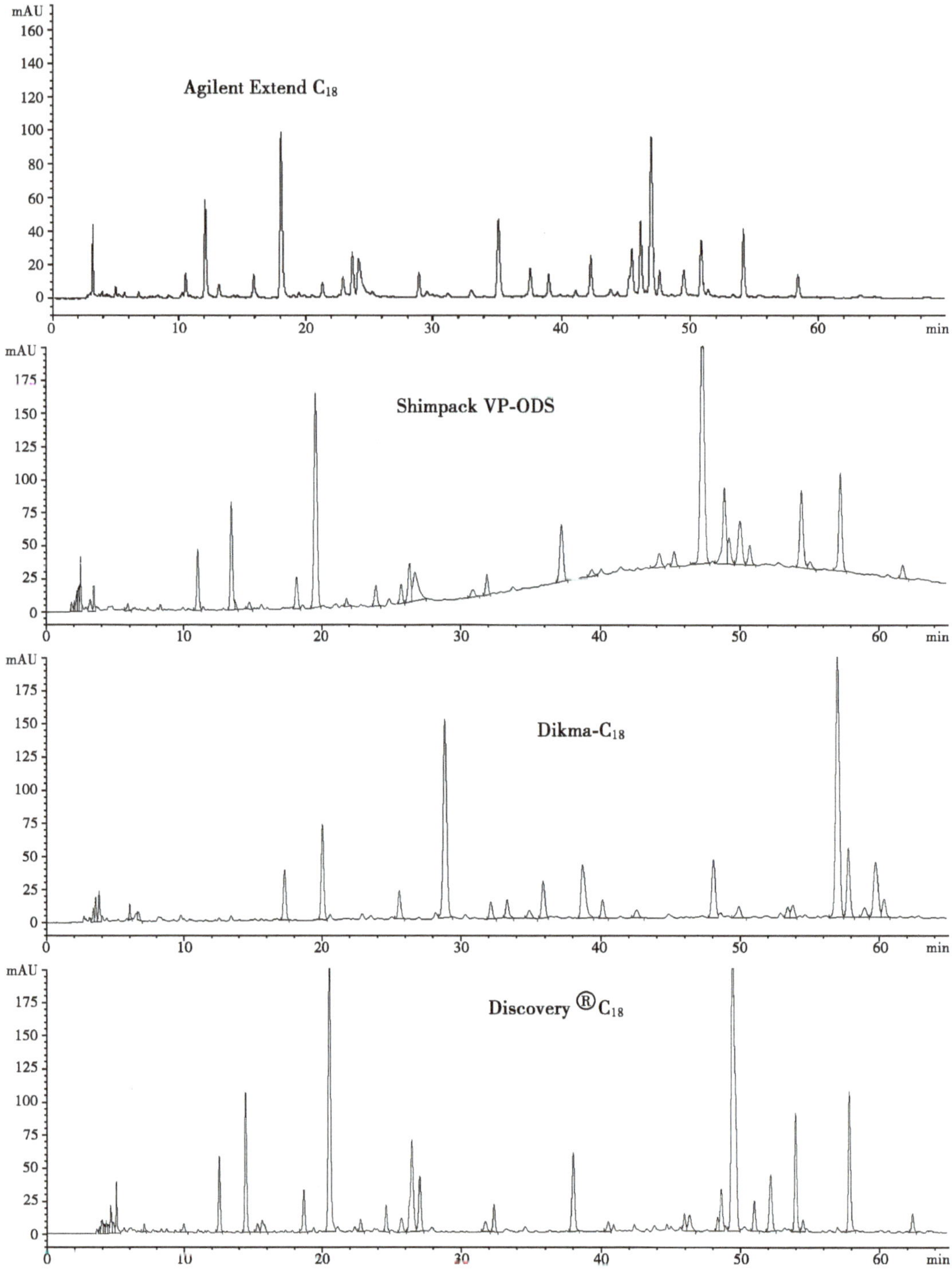

Fig. 15.31 The fingerprints of Guanxinning Injection generated by different chromatographic columns

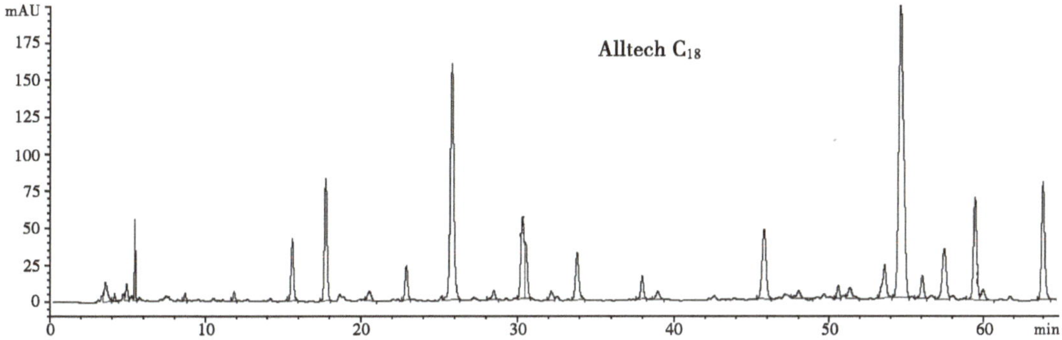

Fig. 15.31 (continued)

Table 15.25 Program gradient elution table

Time (min)	Acetonitrile (%)	0.03 Phosphoric acid solution (%)
0 → 65	2 → 33	98 → 67
65 → 140	33	67

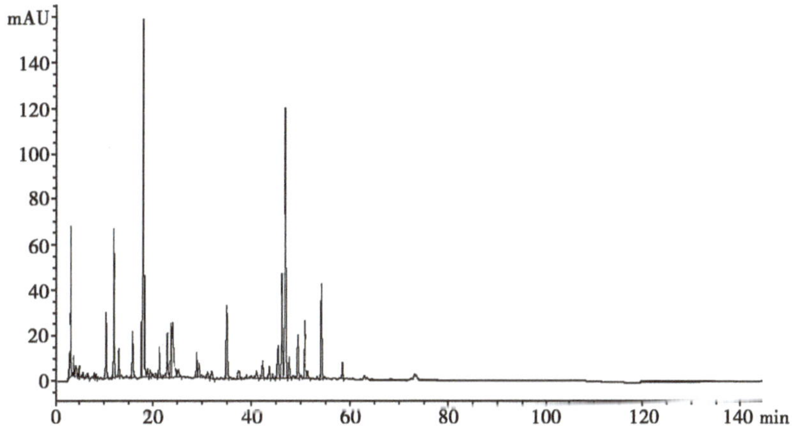

Fig. 15.32 The chromatogram of Guanxinning Injection eluted for 140 min

comparing the retention times and spectra, the structures of six major chromatographic peaks were identified: danshensu, protocatechuic aldehyde, caffeic acid, vanillin, ferulaic acid and salvianolic acid B, all from Danshen. As for the water-soluble constituents of Sichuan lovage root, the techniques of liquid phase isolation and preparation, GC/MS, and HPLC-PDA-MS/MS were used to identify the components, and the results showed it contained senkyunolide I and its two isomers. As for Danshen's other salvianolic acids, HPLC-PDA-MS was used to determine their identities, and the results showed they were salvianolic acid A, salvianolic acid C, lithospermic acid, and rosmarinic acid.

Based on the result, 13 chromatographic peaks in the fingerprint of Guanxinning Injection were identified (Fig. 15.33).

(g) Methodological Study
   (i) The Precision Test: Precisely follow the above-mentioned chromatographic conditions, and continuously determine the test solution 6 times. The similarities in the resulting chromatograms were evaluated by adopting "*Similarity*

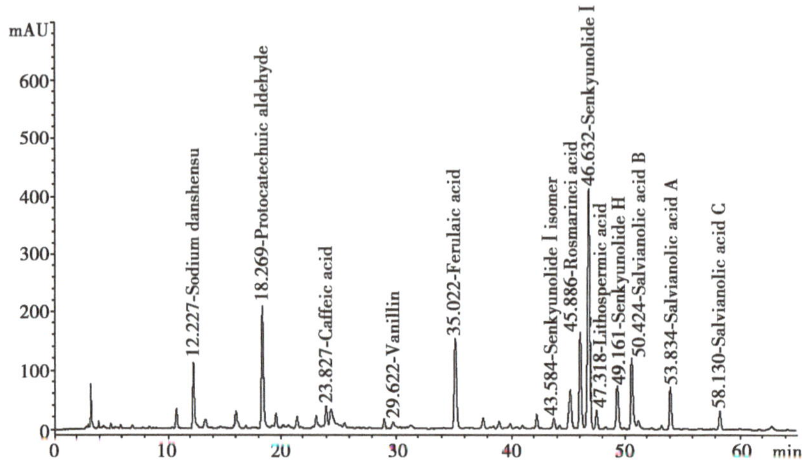

Fig. 15.33 Identification of the chromatographic peaks

Evaluation System for Chromatographic Fingerprint of Chinese Crude Drug" (Edition A) provided by the Committee of *Chinese Pharmacopoeia*, with calculations based on the first chromatogram as the reference. The results showed that the method had good precision (Fig. 15.34 and Table 15.26).

(ii) The Repeatability Test: Precisely weigh 6 samples, and perform extractions and assays by following the above-mentioned methods. The similarities in the results were evaluated by "*Similarity Evaluation System for Chromatographic Fingerprint of Chinese Crude Drug*" provided by the Committee of *Chinese Pharmacopoeia*, with calculations based on the first chromatogram as the reference. The results showed that this method had relatively good repeatability (Fig. 15.35 and Table 15.27).

(iii) The Stability Test: Take the same test solution used in the precision test, and test it at a certain interval. Then evaluate the similarities in the results by adopting "*Similarity Evaluation System for Chromatographic Fingerprint of Chinese Crude Drug*", with

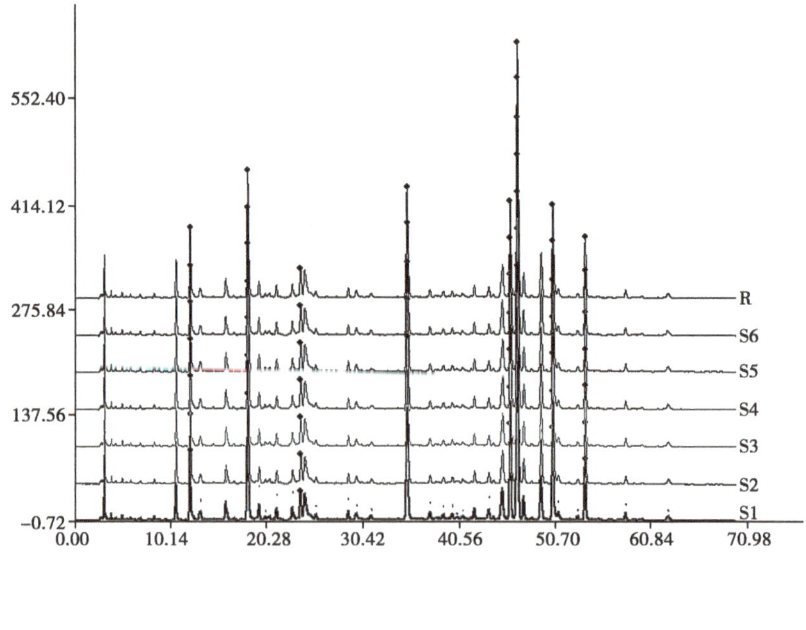

Fig. 15.34 The precision test chromatograms

Table 15.26 The calculated similarities among the results from the precision tests

Serial no.	1	2	3	4	5	6
Similarity	1.000	1.000	1.000	1.000	1.000	1.000

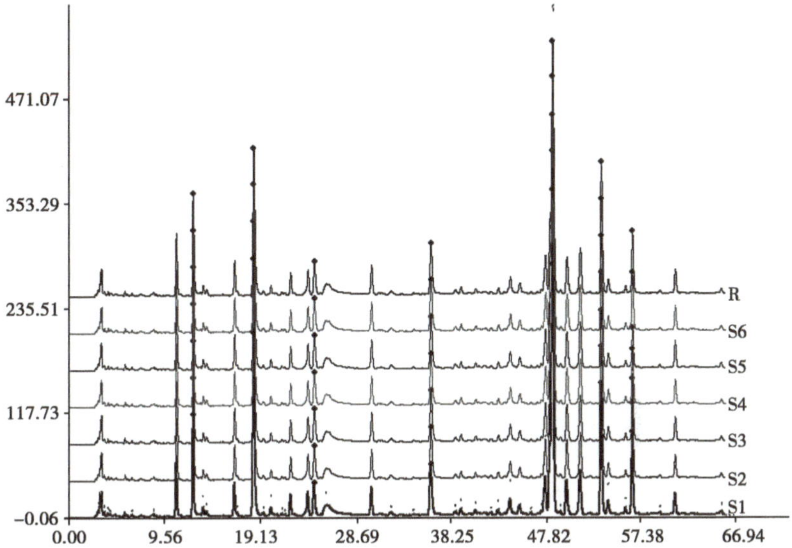

Fig. 15.35 The repeatability test chromatograms

Table 15.27 The calculated similarities among the results of the repeatability tests

Serial no.	1	2	3	4	5	6
Similarity	1.000	1.000	0.986	0.985	0.995	0.986

calculations based on the first chromatogram as the reference. The results showed that this method remained stable for 49 h (Fig. 15.36 and Table 15.28).

(h) The method for the establishment of the reference fingerprint template for Guanxinning Injection
   (i) Take the samples prepared by the adjusted production process, compare them with other samples made by the same manufacturer, select the samples with a similarity value above a certain threshold, and use them as the reference samples to generate fingerprint templates.
   (ii) Take the reference templates from different manufacturers and synthesize them into a unified reference fingerprint template.
   (iii) Use "*Similarity Evaluation System for Chromatographic Fingerprint of Chinese Crude Drug*" (2004A) provided by the Committee of *Chinese Pharmacopoeia* to establish the reference fingerprint template.

2. Correlation Study
   (a) The correlations among the raw materials, the intermediate, and the final product of Guanxinning Injection are shown in Fig. 15.37.
   (b) The Chromatographic Fingerprints from Different Manufacturers' Products: We found that the major cause for the variation of fingerprints of this product was the production process. After adjustment of the production process, the differences among the fingerprints were indeed reduced to achieve basic similar levels (Figs. 15.38 and 15.39).

Fig. 15.36 The stability test chromatograms

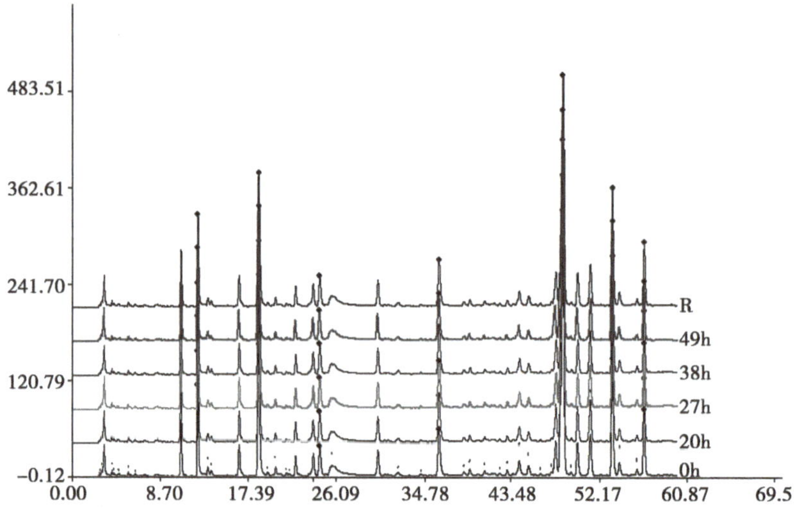

Table 15.28 The calculated similarities among the results of the stability tests

Time (h)	0	20.0	27.0	38.0	49.0
Similarity	1.000	1.000	1.000	1.000	1.000

15.1.1.5 The Chromatographic Fingerprinting of Compound Danshen Tablet

Aihua Liu and Dean Guo

1. Collection of Samples
   In this study, we collected 58 samples of Compound Danshen Tablet (CDT), among which 49 samples came from 49 different manufacturers and the other 9 came from different batches of the same company. The details of these samples are shown in Table 15.29.
2. Methodological Study
   (a) Extraction Method: Different extraction methods, different extraction times, and different extraction solvents were tried to study their effects on the final results. The results showed that the best extraction method was reflux extraction for 1 h with 70 % methanol as the extraction solvent. The experiments are described below.
   (b) Determination of the Extraction Solvent: Carefully remove the coat of the CDT, and grind the tablets with a mortar before sifting through a 100-mesh sieve to obtain the powder. Weigh 8 samples of this powder, 0.300 g each, and add 5 ml of different solvents to each sample: water, methanol, chloroform, or methanol–water and methanol–chloroform of different proportions. The extraction results showed that 70 % methanol gave the best extraction, which was able to completely extract all of the water-soluble constituents and liposoluble constituents of Danshen in CDT (Table 15.30).
      (i) Extraction Method: Carefully remove the coat of CDT and grind the tablets with a mortar before sifting through a 100-mesh sieve to obtain the powder. Weigh 3 samples of this powder, 0.300 g each, add 10 ml of 70 % methanol, extract by ultrasound, reflux, and soak overnight. The results showed that the extract obtained by reflux displayed the best extraction efficiency (Table 15.31).
      (ii) Extraction Time: Carefully remove the coat of CDT and grind the tablets with a mortar before sifting through a 100-mesh sieve to obtain the powder. Weigh 4 samples of this powder, 0.300 g each, and add 10 ml of 70 % methanol. Compare the extraction effects of extracting for 30, 60, 120,

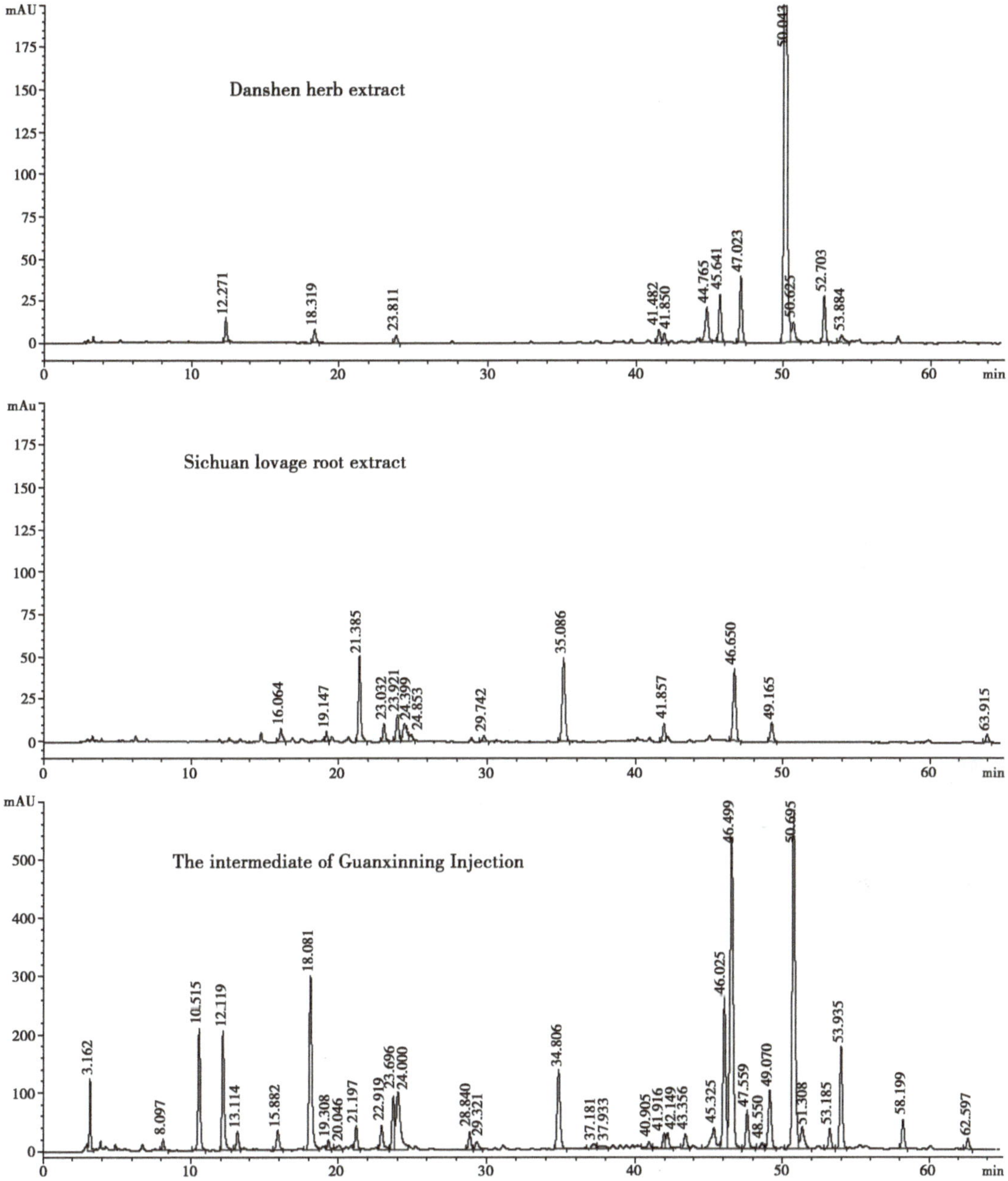

Fig. 15.37 The chromatograms showing the correlations among the raw materials, intermediate, and the final product of Guanxinning Injection

and 180 min, respectively. The results showed that extracting for 120 min displayed the best extraction efficiency (Table 15.32).

(c) Selection of Mobile Phase: Apply the Zorbax Extend C_{18} chromatographic column (4.6 × 250 mm, 5 μm) in the Agilent 1100 series HPLC and compare the following mobile phase systems: methanol–water, methanol-acid water, acetonitrile–water, acetonitrile-acetic acid water, and acetonitrile-phosphoric acid water. The results showed that the system of acetonitrile-phosphoric acid water was better suited for

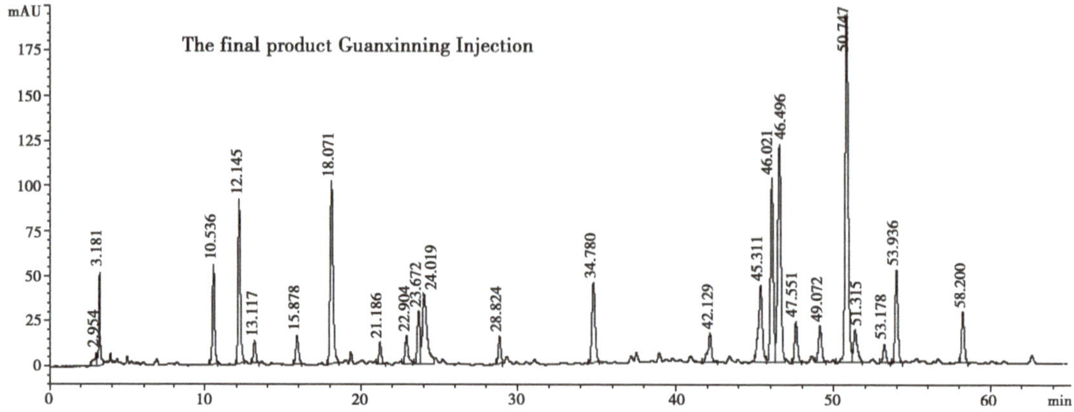

Fig. 15.37 (continued)

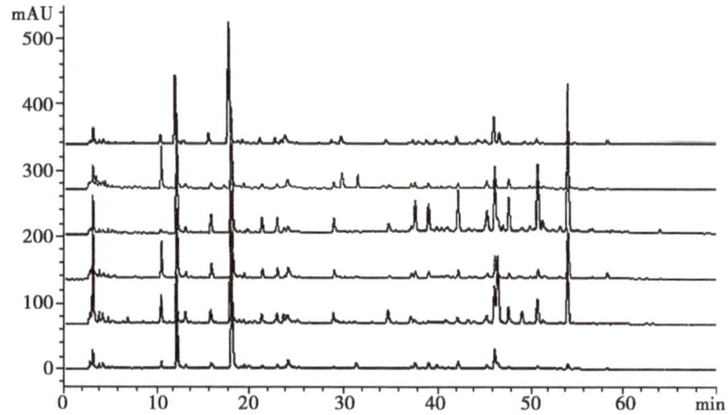

Fig. 15.38 The fingerprints of Guanxinning Injection samples from 6 manufacturers before the production process adjustment

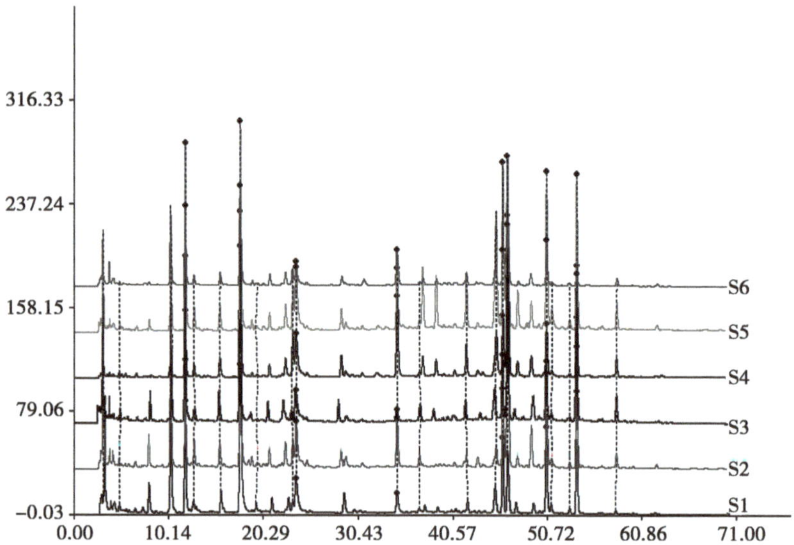

Fig. 15.39 The fingerprints of Guanxinning Injection samples from 6 manufacturers after the production process adjustment

Table 15.29 Information about the samples of compound Danshen tablet (CDT)

Product no.	Region of production	Batch no.	Product no.	Region of production	Batch no.
CDT-1	Guangdong	030902	CDT-30	Guangdong	031001
CDT-2	Guangdong	040404	CDT-31	Beijing	040427
CDT-3	Beijing	031207	CDT-32	Guangxi	0405231
CDT-4	Shenzhen	040703	CDT-33	Guangdong	040301
CDT-5	Guangxi	030113	CDT-34	Guangdong	030902
CDT-6	Henan	040210	CDT-35	Guangdong	030901
CDT-7	Liaoning	03040111	CDT-36	Shanghai	031211
CDT-8	Hebei	9919	CDT-37	Guangxi	030311
CDT-9	Sichuan	20040315	CDT-38	Guangxi	040309
CDT-10	Anhui	20040342	CDT-39	Jiangxi	040403
CDT-11	Sichuan	030203	CDT-40	Guangxi	040426
CDT-12	Zhejiang	030207	CDT-41	Guangdong	040501
CDT-13	Shanghai	040237	CDT-42	Guangdong	030206
CDT-14	Beijing	4120369	CDT-43[a]	Guangdong	04051005
CDT-15	Yunnan	20040561	CDT-44[a]	Guangdong	04011008
CDT-16	Heilongjiang	20030802054	CDT-45[a]	Guangdong	04041012
CDT-17	Sichuan	040614	CDT-46[a]	Guangdong	04021008
CDT-18	Jiangxi	040401	CDT-47[a]	Guangdong	04031018
CDT-19	Hebei	20040307	CDT-48	Liaoning	20040407
CDT-20	Fujian	040110	CDT-49[a]	Guangdong	03111005
CDT-21	Guangxi	040505	CDT-50[a]	Guangdong	04081001
CDT-22	Jiangxi	020301	CDT-51[a]	Guangdong	03121020
CDT-23	Guangdong	040602	CDT-52	Jiangsu	0403006
CDT-24	Guangdong	031101	CDT-53	Xinjiang	20040604
CDT-25	Guangdong	030703	CDT-54	Xinjiang	0403006
CDT-26	Jiangxi	040343	CDT-55	Shaanxi	20030903
CDT-27	Guangdong	040208	CDT-56	Shanxi	20040409
CDT-28	Sichuan	040107	CDT-57	Shanghai	040201
CDT-29	Guangdong	0301900	CDT-58[a]	Guangdong	04051006

Note [a] Represents the different batches from the same company

the analysis of Danshen constituents in CDT. The suitable concentration of phosphoric acid was 0.026 % (V/V). The study also revealed that gradient elution could more effectively separate Danshen constituents than constant gradient elution. Therefore, gradient elution with acetonitrile and 0.026 % phosphoric acid water was selected.

(d) Determination of Wavelength: Studies have shown that the two major fractions of Danshen are the water-soluble phenolic acids and the liposoluble diterpenoid quinones. As phenolic acids have the largest absorption at 288 nm and diterpenoid quinones have the largest absorption at 270 nm, we selected 280 nm as the detection wavelength (Fig. 15.40).

(e) Selection of Chromatographic Column: The following chromatographic columns were tested: Zorbax Extend C_{18} (4.6 × 250 mm, 5 μm), BDS-Hypersil C_{18}

Table 15.30 The effects of solvents on the extract of compound Danshen tablet

Peak (retention time)	11.51	16.12	30.03	31.37	37.84	57.95	63.96	65.11
30 % methanol	768.5	259.0	653.3	5,701.8	442.6	55.5	177.1	0.0
50 % methanol	728.7	261.2	671	5,933.6	1,461.9	448.5	57.7	332.7
70 % methanol	758.4	267.5	696.6	6,027.4	1,612.1	613.3	795.6	1,229.6
Methanol	477.1	170	530.1	5,834.2	1,316.9	571.8	775.8	1,056.6
Chloroform	0	4.1	0	0	0	295.9	368.6	573.8
Chloroform-methanol (1:1)	132.1	75.7	218.5	772.4	492.3	420.2	567.5	900.5
Chloroform-methanol (1:2)	217.6	98.3	303.8	1,822.9	746.9	426	575.2	910
Chloroform-methanol (2:1)	53.2	46.7	105.9	258.4	223.9	318.1	420.8	668.2

Table 15.31 The effects of different extraction methods on the extract of compound Danshen tablet

Peak (retention time)	11.52	16.13	30.03	31.38	37.84	57.95	63.96	65.11	74.52
Reflux	634.3	238.6	587.9	4,614.9	1,411	556.0	730.4	1,206.6	1,543.7
Ultrasound	477.7	170.0	530.1	4,434.2	1,316.9	571.8	775.8	1,229.6	1,631.9
Soaking	441.8	185.2	369.5	3,625.1	1,219.6	506.9	926.8	846.9	1,863.9

Table 15.32 The effects of extraction time on the extract of compound Danshen tablet

Peak (retention time)	11.52	16.13	30.03	31.38	37.84	57.95	63.96	65.11	74.52
0.5 (h)	558.2	213.3	439.9	3,811.8	1,253.2	515.2	938.7	857.8	1,748.3
1.0 (h)	531.6	198.8	408.2	3,326.1	1,180.3	443.4	862.69	785.3	1,648.1
2.0 (h)	606.9	212.3	438.6	3,664.63	1,304.5	488.9	891.5	862.9	1,745.2
3.0 (h)	576.8	201.3	408.4	3,261.2	1,205.3	457.7	846.1	817.5	1,617.4

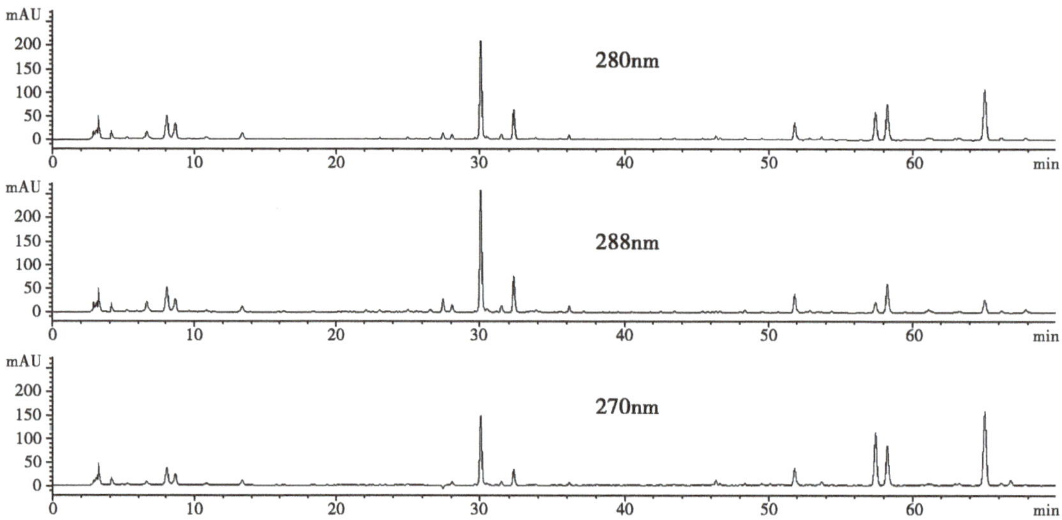

Fig. 15.40 Chromatograms of CDT samples at different wavelengths

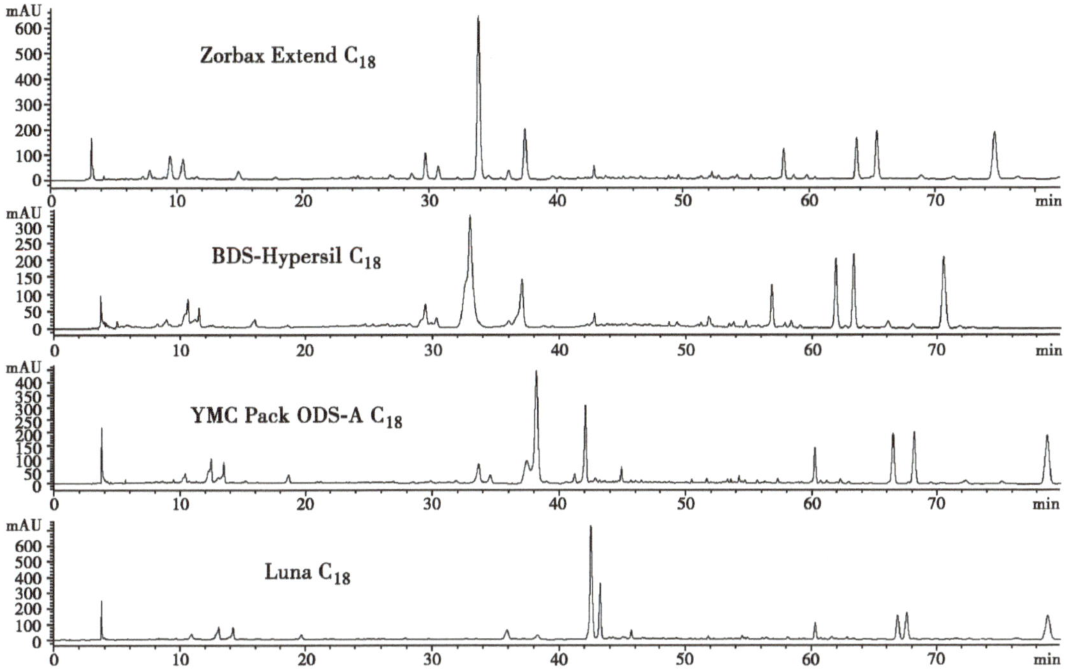

Fig. 15.41 The chromatograms of CDT sample with different chromatographic columns

(4.6 × 250 mm, 5 μm), YMC Pack ODS-A C_{18} (4.6 × 250 mm, 5 μm) and Luna C_{18} (4.6 × 250 mm, 5 μm). The results showed that only Zorbax Extend C_{18} column could separate the constituents of Danshen well and generate peaks with good shape. Therefore, Zorbax Extend C_{18} column was selected (Fig. 15.41).

(f) Selection of Column Temperature: The influence of column temperatures, namely, 15, 20, 25, 30 and 40 °C, on the separation of chemical constituents in Danshen was studied. The results revealed that 20 °C was the best temperature (Fig. 15.42).

3. Assay Method
   (a) Pretreatment: Carefully remove the coat of CDT, grind the tablets with a mortar; sift through a 100–200 mesh sieve to get the powder; dry the powder in a vacuum dryer for 12 h; store the powder in the vacuum dryer for further use.
   (b) Sample Treatment: Precisely weigh 0.300 g of the pretreated sample powder; transfer into a 50 ml volumetric flask; add 10 ml of methanol; extract by boiling water-bath reflux for 1 h; cool the solution to room temperature; add methanol to make up the lost volume; shake well, and filter through a 0.45 μm microporous membrane. The filtrate was used directly for HPLC analysis: Chromatographic column: Zobax Extend C_{18} (4.6 × 250 mm, 5 μm); Pre-column: Zobax Extend C_{18} (4.6 × 250 mm, 5 μm); Column incubator temperature: 20 °C; Flow rate: 0.8 ml/min; Detection wavelength: 280 nm; Sample volume: 10 μl; Mobile phase: acetonitrile-phosphoric acid water, gradient elution.
4. Methodological Study
   On the basis of the established extraction method and HPLC conditions, the precision, repeatability and stability of the method were further studied.
   (a) Precision: A sample was extracted and then assayed continuously 6 times. With the obtained chromatograms, the RSDs of the area of the common peaks were calculated to evaluate the precision of the method. The results showed that the

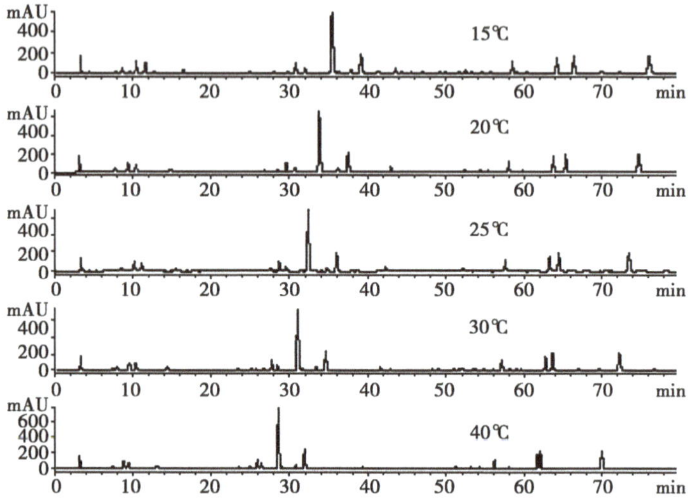

Fig. 15.42 The chromatograms of CDT samples at different column temperatures

RSD were all less than 5 %, indicating that this method had an excellent precision. Besides, the analysis of the 6 obtained chromatograms with "*Similarity Evaluation System for Chromatographic Fingerprint of Chinese Crude Drug*" (2004A) showed the similarities were over 0.990, which further proved that this method had good precision (Tables 15.33 and 15.34).

(b) Repeatability: The same batch of CDT was divided into 6 parts and extracted and tested individually. The data was processed in the same way as mentioned above. The results revealed that the method had a good repeatability (Tables 15.35 and 15.36).

(c) Stability: Extraction was conducted on a sample of CDT, which was then filtered through a 0.45 μm microporous membrane and stored at room temperature. Analysis by HPLC was carried out at different intervals. The data was processed in the same way as mentioned above. The results revealed that the sample was stable within 24 h (Tables 15.37 and 15.38).

(d) Chromatographic Fingerprinting of CDT: With optimization of chromatographic conditions and the validation of the methodology, the chromatographic

Table 15.33 The precision test results

No.	Peak (retention time)									
	11.420	16.010	29.780	30.815	34.072	37.699	57.952	63.760	65.104	74.509
1	947.6	332.2	825.6	404.2	6,550.9	2,278.1	878.4	1,305.8	1,901.2	2,370.5
2	914.4	331.3	805.7	403.5	6,533.6	2,275.5	884.1	1,302.8	1,905	2,321.2
3	920.4	337.1	812.7	402.2	6,533.3	2,277.8	886.0	1,301.5	1,895.9	2,316.4
4	944.5	354.5	806.4	404.4	6,540.6	2,275.0	885.0	1,302.4	1,872.1	2,321.3
5	920	338.6	809.1	405.3	6,544.3	2,277.5	886.7	1,304.8	1,889.0	2,331.7
6	921.8	341.7	816.8	405.2	6,562.9	2,278.0	881.4	1,307.3	1,902.8	2,377.5
SD	14.15	8.44	7.55	1.16	11.30	1.37	3.15	2.23	12.31	27.07
Average	928.1	339.23	812.72	404.13	6,544.27	2,277.0	883.6	1,304.1	1,894.33	2,339.77
RSD (%)	1.63	2.49	0.93	0.29	0.17	0.20	0.36	0.17	0.65	1.16

Table 15.34 The similarities of the chromatograms from the precision tests

	S1	S2	S3	S4	S5	S6	Reference chromatogram
S1	1.000	1.000	0.999	1.000	0.999	1.000	1.000
S2	1.000	1.000	0.999	1.000	0.999	1.000	1.000
S3	0.999	0.999	1.000	0.999	1.000	0.999	0.999
S4	1.000	1.000	0.999	1.000	0.999	1.000	1.000
S5	0.999	0.999	1.000	0.999	1.000	0.999	0.999
S6	1.000	1.000	0.999	1.000	0.999	1.000	1.000
Reference chromatogram	1.000	1.000	0.999	1.000	0.999	1.000	1.000

Table 15.35 The repeatability test results

No.	Peak (retention time)									
	11.532	16.183	30.069	31.160	34.467	38.045	58.020	63.827	65.256	74.689
1	889.1	334.1	782	397.3	6,351.8	2,165.9	851.7	1,268.0	1,872.8	2,318
2	911.0	328.5	778	411.4	6,522.5	2,133.5	840.9	1,202.8	1,895	2,325.7
3	886.8	328.5	774.3	396.1	6,347.2	2,127.2	836.4	1,210.5	1,889.6	2,348.3
4	925.0	345.0	820.3	403	6,584.1	2,285.3	893.3	1,222.6	1,941.5	2,404
5	903.6	336.0	804.7	394.8	6,436.3	2,233.2	861.1	1,283	1,902.8	2,345.6
6	893.6	320.0	800.7	400.1	6,437.9	2,224.2	889.8	1,232.5	1,903.2	2,411.5
SD	14.7	8.5	18.1	6.1	94.1	62.8	24.3	37.6	34.4	40.4
Average	901.5	332.0	793.3	400.5	6,446.9	2,194.9	862.2	1,244.1	1,905.7	2,365.5
RSD (%)	1.63	2.55	2.28	1.53	1.46	2.86	2.82	2.60	1.81	1.71

Table 15.36 The similarities of the chromatograms from the repeatability tests

	S1	S2	S3	S4	S5	S6	Reference chromatogram
S1	1.000	1.000	0.999	1.000	1.000	1.000	1.000
S2	1.000	1.000	1.000	1.000	1.000	1.000	1.000
S3	0.999	1.000	1.000	1.000	1.000	0.999	1.000
S4	1.000	1.000	1.000	1.000	1.000	1.000	1.000
S5	1.000	1.000	1.000	1.000	1.000	1.000	1.000
S6	1.000	1.000	0.999	1.000	1.000	1.000	1.000
Reference chromatogram	1.000	1.000	1.000	1.000	1.000	1.000	1.000

fingerprinting of CDT was successfully established.

5. Data Analysis

   In this study, a combination of data comparison and software analysis was adopted to analyze the data obtained.

   (a) Selection of the Reference Sample: Due to the fact that products from different companies vary significantly in quality, only by collecting a substantial number of samples for comprehensive analysis can the fingerprint features of CDT be fully revealed. We select the needed reference samples based on these features to ensure that they are representative.

   After analysis and content determination of the collected CDTs, samples from 9 manufacturers were selected, and they were no. 3, 6, 7, 10, 12, 31, 33, 34, and 57 (Table 15.29). These samples contained

Table 15.37 The stability test results

Time	Peak (retention time)									
	11.532	16.183	30.069	31.160	34.467	38.045	58.020	63.827	65.256	74.689
0 (h)	947.6	332.2	825.6	404.2	6,550.9	2,278.1	878.4	1,368.2	1,901.2	2,370.5
1.5 (h)	914.4	331.3	805.7	403.5	6,533.6	2,275.5	884.1	1,342.5	1,905.0	2,321.2
3 (h)	920.4	337.1	812.7	402.2	6,533.3	2,277.8	886.0	1,310.6	1,895.9	2,316.4
6 (h)	902.7	323.3	806.4	395.6	6,143.6	2,232.9	869.5	1,322.2	1,872.1	2,281.9
9 (h)	915.3	365.9	801.1	395.9	6,423.3	2,277.5	865.4	1,383.0	1,889.2	2,299.8
12 (h)	903.6	336.0	804.7	394.8	6,436.3	2,232.9	861.1	1,332.9	1,902.8	2,345.6
18 (h)	925.0	345.0	820.3	403.0	6,586.1	2,233.2	893.3	1,314.1	1,941.5	2,404.0
24 (h)	939.8	362.4	804.1	407.1	6,550.2	2,285.3	911.7	1,312.7	1,897.0	2,343.8
SD	15.99	15.19	8.70	4.67	143.61	23.89	16.50	16.50	19.55	39.29
Average	921.1	341.65	810.08	400.79	6,469.66	2,261.65	881.19	1,335.8	1,900.59	2,335.4
RSD (%)	1.74	4.45	1.07	1.16	2.22	1.00	1.87	0.98	1.03	1.68

Table 15.38 The similarities of the chromatograms from the stability tests

	0	1.5	3	6	9	12	18	24	0	Reference chromatogram
0 (h)	0.999	1.000	0.999	1.000	1.000	1.000	1.000	1.000	0.999	1.000
1.5 (h)	1.000	0.999	1.000	0.999	0.999	0.999	0.999	0.999	1.000	0.999
3 (h)	0.999	1.000	0.999	1.000	1.000	1.000	1.000	1.000	0.999	1.000
6 (h)	0.999	1.000	0.999	1.000	1.000	1.000	1.000	1.000	0.999	1.000
9 (h)	0.999	1.000	0.999	1.000	1.000	1.000	1.000	1.000	0.999	1.000
12 (h)	0.999	1.000	0.999	1.000	1.000	1.000	1.000	1.000	0.999	1.000
18 (h)	0.999	1.000	0.999	1.000	1.000	1.000	1.000	1.000	0.999	1.000
Reference chromatogram	1.000	0.999	1.000	0.999	0.999	0.999	0.999	0.999	1.000	0.999
	0.999	1.000	0.999	1.000	1.000	1.000	1.000	1.000	0.999	

the comprehensive constituents of Danshen with appropriate contents. In the meantime, 9 samples from different batches of the products by a manufacturer, which has the largest sales volume nationwide and well-regulated production management, were selected. They were samples no. 43, 44, 45, 46, 47, 49, 50, 51, and 58, because our test results showed that these samples contained the comprehensive constituents of Danshen with appropriate contents, as well as stable quality. Therefore, these 18 samples were taken as the reference samples for establishing the standard fingerprints.

(b) The Establishment of Standard Chromatographic Fingerprints: By following the established extraction and HPLC methods, the 18 reference samples of CDT were determined by HPLC. The results were processed with the "*Similarity Evaluation System for Chromatographic Fingerprint of Chinese Crude Drug*" (2004A) to get the standard fingerprints.

(c) Establishment of Chromatographic Fingerprint for Test Samples: The other samples were tested in the same way as were the reference samples (see Fig. 15.43).

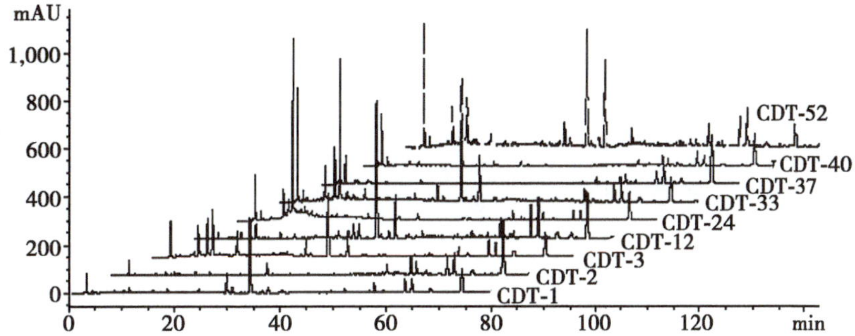

Fig. 15.43 The fingerprints of the test samples

(d) Determination of the Common Peak: The so-called common peak refers to the chromatographic peaks that share the same relative retention time within a certain comparison window. CDT had 12 common peaks (Table 15.39).

(e) Selection of Reference Peak: Currently, the extrapolated peak is used in most analytical work for reference. However, the chromatographic fingerprinting of TCM samples requires higher standards for separation, because there are many compounds with similar properties. Besides, the separation of extrapolation references should also be considered. These high requirements left rather limited choices for the reference peaks. Therefore, the internal reference peak method was used in our study. The selected internal reference peak must comply with the following conditions: first, it must separate well from the nearby peaks, with its position in the middle; second, it should be present in both the standard and the test samples. For the CDT, even though our analysis revealed a great discrepancy in the constituents contained in the samples from different manufacturers, tanshinone II_A was present in every sample and it was the only constituent that met the requirements as an internal reference peak. Since Chinese Pharmacopoeia specified the content of tanshinone II_A in CDT, we chose this compound (Peak 12) as the internal reference peak.

(f) Data Processing: A combination of data comparison and software analysis was adopted to analyze the data obtained from Danshen and CDT samples.

(i) Data Comparison: In the chromatographic fingerprints, the peak area is positively proportional to the content of the corresponding constituent, therefore, comparison of the peak areas of the same peaks from different samples could reveal the difference in their content. There are two ways to express the normalized value of a peak area, and the way we chose was to set the internal reference peak area as the benchmark, and compare all the other peak areas against the benchmark, thus obtaining a series of normalized area values, which fully reveal the quantitative relationships among the constituents in the samples. For example, the closeness degree of the normalized values directly shows the closeness degree of the ratio of contained constituents in the sample. Therefore, the normalized value is a quantized display of the similarity between samples. In our study, the area and normalized value of every common peak in the samples were calculated (Table 15.39).

(ii) Software Analysis: After collection of all of the chromatograms and data, "*Similarity Evaluation System for Chromatographic Fingerprint of Chinese Crude Drug*" (2004A) was

Table 15.39 The peak areas of common peaks and their normalized values

No.	A/NA	1	2	3	4	5	6	7	8	9	10	11	12
Standard chromatogram	A	1,189.790	1,139.118	547.750	999.706	426.458	6,966.499	344.369	2,956.021	893.517	1,473.375	1,836.503	2,767.858
	NA	0.430	0.412	0.198	0.365	0.154	2.517	0.124	1.068	0.323	0.532	0.664	1.000
CDT-1	A	77.921	200.948	111.365	1,008.947	288.716	4,944.634	131.302	282.279	412.437	745.751	802.180	2,095.018
	NA	0.037	0.096	0.053	0.482	0.138	2.360	0.063	0.135	0.197	0.356	0.383	1.000
CDT-2	A	0.000	148.072	0.000	646.376	64.932	0.000	0.000	0.000	550.944	996.668	1,217.432	4,806.360
	NA	0.000	0.031	0.000	0.134	0.014	0.000	0.000	0.000	0.115	0.207	0.253	1.000
CDT-3	A	1,788.258	2,210.674	0	659.019	196.263	4,381.875	96.567	1,317.480	431.816	932.346	784.363	3,376.997
	NA	0.530	0.655	0	0.206	0.058	1.298	0.029	0.390	0.128	0.276	0.232	1.000
CDT-4	A	421.410	198.292	49.748	706.473	92.842	1,555.897	0.000	169.808	367.140	546.834	639.257	2,106.969
	NA	0.200	0.094	0.024	0.335	0.044	0.738	0.000	0.081	0.174	0.260	0.303	1.000
CDT-5	A	0.000	0.000	0.000	200.049	0.000	0.000	0.000	0.000	250.508	444.962	434.984	2,257.090
	NA	0.000	0.000	0.000	0.089	0.000	0.000	0.000	0.000	0.111	0.197	0.193	1.000
CDT-6	A	244.239	1,388.238	1,206.031	1,132.231	391.693	5,200.334	260.627	3,072.587	920.668	1,429.2550	1,981.145	3,074.742
	NA	0.079	0.451	0.392	0.368	0.127	1.691	0.085	0.999	0.299	0.465	0.644	1.000
CDT-7	A	78.879	1,397.975	879.935	1,645.762	655.704	9,481.560	550.606	5,591.303	1,962.580	3,774.621	2,790.320	5,611.179
	NA	0.014	0.249	0.157	0.293	0.117	1.690	0.098	0.996	0.350	0.673	0.480	1.000
CDT-8	A	0.000	673.039	155.745	1,258.872	971.750	11,062.200	315.319	1,360.128	366.780	656.235	1,019.558	2,132.141
	NA	0.000	0.316	0.073	0.590	0.456	5.188	0.148	0.638	0.172	0.308	0.478	1.000
CDT-9	A	352.727	0.000	0.000	57.631	0.000	183.319	0.000	0.000	582.504	680.716	1,656.285	1,731.839
	NA	0.204	0.000	0.000	0.033	0.000	0.106	0.000	0.000	0.336	0.393	0.956	1.000
CDT-10	A	45.321	195.909	1,887.912	1,092.691	202.123	3,848.633	127.104	3,035.368	623.176	1,782.792	1,531.435	3,294.998
	NA	0.014	0.593	0.573	0.332	0.061	1.168	0.039	0.921	0.189	0.541	0.465	1.000
CDT-11	A	0.000	445.558	242.823	721.955	308.222	4,530.683	113.703	858.435	119.095	401.109	261.887	1,993.986
	NA	0.000	0.022	0.122	0.362	0.154	2.272	0.057	0.431	0.060	0.201	0.131	1.000
CDT-12	A	179.665	740.457	4,245.673	712.175	790.442	9,467.446	127.104	1,887.068	905.990	1,889.956	2,415.896	4,125.414
	NA	0.181	0.361	0.068	0.525	0.140	1.926	0.058	0.600	0.184	0.275	0.368	1.000
CDT-13	A	48.108	811.755	509.423	508.016	71.975	2,051.771	83.561	1,752.771	83.561	1,505.548	999.190	2,480.373
	NA	0.019	0.327	0.205	0.205	0.029	0.827	0.003	0.707	0.275	0.607	0.403	1.000

(continued)

Table 15.39 (continued)

No.	A/NA	1	2	3	4	5	6	7	8	9	10	11	12
CDT-14	A	98.697	355.868	66.685	1,315.143	173.068	2,725.112	76.478	707.888	561.702	476.837	1,178.701	2,044.728
	NA	0.048	0.174	0.033	0.643	0.085	1.333	0.037	0.346	0.275	0.233	0.576	1.000
CDT-15	A	0.000	735.012	631.444	816.508	342.631	6,996.887	187.147	1,766.306	766.300	1,493.946	1,387.056	2,927.415
	NA	0.000	0.251	0.216	0.279	0.117	2.390	0.064	0.603	0.262	0.510	0.474	1.000
CDT-16	A	0.000	1,353.662	0.000	944.429	645.194	9,118.844	0.000	3,077.164	381.493	774.544	888.853	1,874.071
	NA	0.000	0.722	0.000	0.504	0.071	4.866	0.000	1.666	0.204	0.413	0.474	1.000
CDT-17	A	1,935.232	248.280	0.000	1,180.246	749.252	14,858.120	360.592	163.793	694.356	930.807	1,676.287	2,409.466
	NA	0.803	0.103	0.000	0.490	0.311	6.167	0.150	0.068	0.288	0.386	0.696	1.000
CDT-18	A	905.111	691.408	261.023	685.742	465.693	5,669.813	312.430	1,874.680	884.680	2,416.463	2,501.194	5,521.771
	NA	0.164	0.125	0.047	0.124	0.084	1.027	0.057	0.339	0.160	0.438	0.463	1.000
CDT-19	A	2,008.162	361.092	270.126	278.702	197.311	2,851.519	149.064	1,030.550	412.208	590.430	1,120.903	1,928.988
	NA	1.041	0.187	0.140	0.144	0.102	1.478	0.077	0.534	0.214	0.462	0.581	1.000
CDT-20	A	16.824	36.381	0.000	769.001	157.010	1,898.221	0.000	0.000	333.111	575.129	647.196	1,929.597
	NA	0.009	0.019	0.000	0.398	0.081	0.984	0.000	0.000	0.173	0.298	0.335	1.000
CDT-21	A	0.000	0.000	0.000	645.521	0.000	176.483	0.000	0.000	326.803	735.590	436.670	2,882.210
	NA	0.000	0.000	0.000	0.224	0.000	0.061	0.000	0.000	0.113	0.261	0.186	1.000
CDT-22	A	144.868	137.428	0.000	650.252	491.111	8,624.178	193.759	175.784	313.572	758.468	1,071.760	2,152.361
	NA	0.068	0.064	0.000	0.302	0.228	4.007	0.090	0.082	0.146	0.352	0.498	1.000
CDT-23	A	111.197	86.932	47.890	1,411.965	182.877	1,604.072	0.000	0.000	153.130	267.449	316.364	1,322.916
	NA	0.084	0.657	0.036	1.067	0.138	1.212	0.000	0.000	0.116	0.202	0.239	1.000
CDT-24	A	7,163.34	17.988	0.000	1,100.964	0.000	354.110	0.000	0.000	323.201	546.868	484.253	2,526.983
	NA	2.835	0.008	0.000	0.436	0.000	0.140	0.000	0.000	0.128	0.216	0.192	1.000
CDT-25	A	303.153	552.578	298.526	518.433	235.463	2,834.708	0.000	866.076	266.686	422.440	504.402	696.353
	NA	0.313	0.570	0.308	0.535	0.243	2.924	0.000	0.893	0.275	0.436	0.520	1.000
CDT-26	A	435.81	933.296	246.820	82.558	102.551	1,207.398	0.000	1,327.068	135.779	160.720	653.062	2,738.162
	NA	0.159	0.341	0.090	0.030	0.037	0.441	0.000	0.594	0.050	0.059	0.239	1.000
CDT-27	A	47.888	0.000	0.000	52.560	142.772	0.000	0.000	0.000	769.555	1,278.453	1,557.838	4,107.393
	NA	0.012	0.000	0.000	0.013	0.035	0.000	0.000	0.000	0.187	0.311	0.379	1.000

(continued)

Table 15.39 (continued)

No.	A/NA	1	2	3	4	5	6	7	8	9	10	11	12
CDT-28	A	95.781	933.296	246.820	82.558	102.551	1,207.398	0.000	1,627.068	135.779	160.720	653.062	2,738.162
	NA	0.036	0.058	0.000	0.249	0.152	2.571	0.038	0.058	0.616	1.319	0.660	1.000
CDT-29	A	0.000	608.913	82.538	1,103.903	1,247.772	14,362.980	629.484	599.170	1,616.155	3,458.588	1,731.193	2,621.787
	NA	0.000	0.232	0.031	0.421	0.476	5.478	0.240	0.228	0.616	1.319	0.660	1.000
CDT-30	A	1,640.157	0.000	0.000	0.000	0.000	0.000	0.000	0.000	41.017	60.523	220.970	1,483.011
	NA	1.106	0.000	0.000	0.000	0.000	0.000	0.000	0.000	0.028	0.041	0.149	1.000
CDT-31	A	469.580	935.900	176.362	1,362.656	363.512	4,998.805	150.121	1,557.156	478.262	714.966	955.366	2,595.350
	NA	0.181	0.361	0.068	0.525	0.140	1.926	0.058	0.600	0.184	0.275	0.368	1.000
CDT-32	A	487.916	227.932	173.577	1,804.230	294.883	3,903.087	135.900	536.054	407.742	883.663	889.731	2,856.822
	NA	0.171	0.080	0.061	0.632	0.103	0.1366	0.046	0.188	0.143	0.309	0.311	1.000
CDT-33	A	2,512.231	2,009.532	556.213	880.224	248.196	4,576.802	264.992	3,027.679	544.731	1,030.175	1,048.194	2,199.726
	NA	1.142	0.914	0.253	0.400	0.113	2.081	0.120	1.376	0.248	0.468	0.477	1.000
CDT-34	A	314.430	668.462	307.872	478.578	369.951	4,370.296	219.363	1,891.391	349.391	776.291	775.477	1,645.810
	NA	0.191	0.406	0.187	0.291	0.225	2.655	0.133	1.149	0.212	0.472	0.471	1.000
CDT-35	A	492.790	164.786	140.196	1,634.737	274.209	4,665.848	125.058	208.063	512.346	1,021.560	1,004.583	2,805.809
	NA	0.176	0.059	0.050	0.583	0.100	1.663	0.045	0.074	0.183	0.364	0.358	1.000
CDT-36	A	423.524	399.007	84.136	632.937	655.888	8,640.085	315.890	998.139	1,338.013	2,524.375	2,802.052	5,039.172
	NA	0.084	0.079	0.017	0.126	0.130	1.715	0.063	0.198	0.266	0.501	0.556	1.000
CDT-37	A	0.000	0.000	0.000	0.000	0.000	0.000	0.000	0.000	401.442	705.060	1,343.396	4,374.442
	NA	0.000	0.000	0.000	0.000	0.000	0.000	0.000	0.000	0.092	0.161	0.307	1.000
CDT-38	A	201.435	64.035	0.000	436.791	303.967	303.937	5,274.148	70.627	492.919	1,205.349	881.845	2,394.989
	NA	0.084	0.027	0.000	0.182	0.127	2.202	0.000	0.029	0.206	0.503	0.368	1.000
CDT-39	A	740.520	131.638	42.587	392.821	221.453	6,644.261	299.235	1,827.585	398.814	1,023.721	796.027	2,408.641
	NA	0.307	0.277	0.171	0.325	0.148	2.759	0.124	0.759	0.166	0.425	0.330	1.000
CDT-40	A	152.981	33.071	0.000	197.861	0.000	0.000	0.000	0.000	258.966	817.506	469.569	2,809.636
	NA	0.054	0.001	0.000	0.070	0.000	0.000	0.000	0.000	0.092	0.291	0.167	1.000
CDT-41	A	740.520	131.638	42.578	392.821	221.453	4,973.733	76.408	0.000	511.784	1,203.954	942.513	3,424.367
	NA	0.216	0.038	0.012	0.115	0.065	1.452	0.022	0.000	0.149	0.035	0.275	1.000

(continued)

Table 15.39 (continued)

No.	A/NA	1	2	3	4	5	6	7	8	9	10	11	12
CDT-42	A	63.828	0.000	0.000	476.753	0.000	136.915	0.000	0.000	425.786	937.916	755.073	3,193.342
	NA	0.016	0.000	0.000	0.122	0.000	0.035	0.000	0.000	0.109	0.240	0.193	1.000
CDT-43	A	657.580	760.665	349.672	816.926	379.000	6,583.882	289.518	2,062.641	779.132	1,207.310	1,959.846	2,626.437
	NA	0.250	0.290	0.133	0.311	0.144	2.507	0.110	0.785	0.297	0.460	0.746	1.000
CDT-44	A	452.859	920.307	392.670	816.237	379.000	6,583.882	289.518	2,062.640	779.132	1,207.310	1,959.846	2,626.437
	NA	0.250	0.290	0.133	0.311	0.144	2.507	0.110	0.785	0.297	0.460	0.746	1.000
CDT-45	A	998.729	777.237	314.320	827.450	383.023	6,724.818	300.259	2,085.455	737.878	1,127.799	1,842.198	2,492.901
	NA	0.401	0.312	0.126	0.332	0.154	2.670	0.120	0.837	0.276	0.452	0.739	1.000
CDT-46	A	1,380.849	805.703	326.377	792.037	364.528	6,205.287	267.828	2,159.616	789.145	1,234.198	1,876.014	2,434.991
	NA	0.567	0.372	0.134	0.325	0.150	2.548	0.110	0.887	0.324	0.507	0.770	1.000
CDT-47	A	834.127	857.204	339.524	825.849	423.549	7,166.891	308.618	2,201.432	845.921	1,265.101	1,893.613	2,419.821
	NA	0.345	0.354	0.140	0.341	0.175	2.962	0.128	0.910	0.350	0,523	0.783	1.000
CDT-48	A	577.707	900.051	827.833	685.315	402.771	6,633.624	253.612	2,455.785	869.016	1,200.698	1,935.524	2,375.678
	NA	0.243	0.379	0.348	0.288	0.170	2.792	0.107	1.034	0.366	0.505	0.815	1.000
CDT-49	A	466.213	869.771	315.560	894.034	461.900	7,076.680	335.099	2,583.890	870.454	1,409.781	1,809.531	2,234.424
	NA	0.209	0.390	0.141	0.400	0.207	3.167	0.150	1.156	0.390	0.631	0.846	1.000
CDT-50	A	1,019.169	1,032.057	356.472	866.619	502.716	7,528.644	324.3740	2,216.227	727.169	1,096.074	1,447.850	2,044.086
	NA	0.499	0.505	0.174	0.424	0.246	3.683	0.159	1.804	0.356	0.536	0.708	1.000
CDT-51	A	1,458.042	981.046	341.544	886.275	443.666	7,067.704	303.724	2,342.499	843.735	1,355.302	1,626.561	2,122.128
	NA	0.687	0.462	0.161	0.418	0.209	3.330	0.143	1.104	0.398	0.639	0.766	1.000
CDT-52	A	3,011.272	2,160.789	556.885	1,381.401	460.010	8,157.718	504.325	5,827.531	1,236.385	1,689.127	2,409.354	2,491.120
	NA	1.209	0.867	0.224	0.555	0.185	3.275	0.242	2.339	0.496	0.678	0.967	1.000
CDT-53	A	37.266	337.547	72.253	721.607	349.558	6,599.350	221.380	749.704	515.091	1,042.278	1,202.023	2,193.772
	NA	0.017	0.154	0.033	0.329	0.159	3.008	0.101	0.342	0.235	0.475	0.548	1.000
CDT-54	A	32.042	304.000	100.309	545.784	527.1541	70,259.146	0.000	665.179	304.404	745.709	465.846	1,070.145
	NA	0.030	0.284	0.093	0.510	0.493	3.568	0.000	0.622	0.284	0.697	0.435	1.000
CDT-55	A	141.619	929.762	375.935	1,032.851	317.894	5,445.319	313.566	3,974.669	893.066	1,224.914	1,961.021	2,489.969
	NA	0.057	0.373	0.151	0.415	0.128	2.187	0.126	1.596	0.359	0.492	0.788	1.000

(continued)

Table 15.39 (continued)

No.	A/NA	1	2	3	4	5	6	7	8	9	10	11	12
CDT-56	A	0.000	963.204	405.694	1,672.775	553.760	7,634.091	402.493	2,756.210	994.259	837.792	639.257	1,758.626
	NA	0.000	0.548	0.231	0.951	0.315	4.341	0.229	1.567	0.565	0.476	0.363	1.000
CDT-57	A	4,627.619	3,268.125	1,277.113	1,548.079	551.571	13,085.480	1,045.279	7,067.701	1,714.660	1,916.991	2,916.232	3,376.997
	NA	1.370	0.968	0.378	0.458	0.163	3.875	0.310	2.093	0.508	0.568	0.864	1.000
CDT-58	A	1,188.790	934.533	328.235	819.177	399.539	6,525.101	293.686	2,261.908	842.017	1,301.737	1,887.203	2,355.601
	NA	0.505	0.397	0.139	0.348	0.170	2.770	0.125	0.960	0.357	0.553	0.801	1.000

used to process the data and obtain the similarities between the chromatogram of each sample and its standard chromatogram (Table 15.40).

(g) The Correlation of Chromatographic Fingerprints of CDT From Different Manufacturers: The Danshen extraction process specified by the *Pharmacopeia* for the manufacture of CDT (extracting first by reflux with ethanol; then by reflux with 50 % ethanol, and finally by extracting with water), can guarantee that the two effective fractions of Danshen are completely extracted. However, the quality of Danshen herb, the actual extraction process, and the input of the raw materials by different companies may vary, which, accordingly, leads to the variation of the extracted constituents and their contents. As a result, the quality of CDT cannot be guaranteed. That is why the quality of CDTs on the market varies significantly, and why the clinical dosages of CDTs from different brands are different.

Among the 58 tested samples in this study, the following samples had a similarity value over 0.900 (compared with the standard chromatogram): No. 6, 7, 11, 12, 15, 16, 31, 33, 34, 36, 38, 39, 43, 44, 45, 46, 47, 48, 49, 50, 51, 52, 53, 55, 56, 57, and 58; the following samples had a similarity value between 0.8 and 0.9: No. 1, 3, 8, 10, 14, 17, 18, 19, 22, 25, 28, 29, 32, 35, 41, and 54; the following samples had a similarity value between 0.5 and 0.8: No. 4, 13, 20, and 23; the following samples had a similarity value less than 0.5: No. 2, 5, 9, 21, 24, 26, 27, 30, 37, 40, and 42. A series of similarities were generated after processing the data with "*Similarity Evaluation System for Chromatographic Fingerprint of Chinese Crude Drug*" (2004A). However, how to use these numbers to evaluate the quality of the samples needs careful analysis before a relatively rational conclusion can be reached. In the following, we attempt

Table 15.40 The similarity between the sample chromatogram and standard chromatogram

Sample no.	1	2	3	4	5	6	7	8	9
Similarity	0.871	0.377	0.874	0.727	0.358	0.954	0.938	0.888	0.498
Sample no.	10	11	12	13	14	15	16	17	18
Similarity	0.864	0.929	0.973	0.711	0.821	0.931	0.908	0.849	0.868
Sample no.	19	20	21	22	23	24	25	26	27
Similarity	0.871	0.767	0.390	0.884	0.703	0.311	0.871	0.499	0.357
Sample no.	28	29	30	31	32	33	34	35	36
Similarity	0.861	0.850	0.57	0.967	0.839	0.901	0.936	0.854	0.911
Sample no.	37	38	39	40	41	42	43	44	45
Similarity	0.397	0.903	0.937	0.412	0.854	0.443	0.996	0.992	0.997
Sample no.	46	47	48	49	50	51	52	53	54
Similarity	0.998	0.997	0.962	0.991	0.990	0.997	0.943	0.938	0.836
Sample no.	55	56	57	58					
Similarity	0.970	0.954	0.967	0.964					

to evaluate a sample's quality by combining the similarity value with the specific constituents.

(i) The Samples with Similarity Value Over 0.9: From the data displayed in Table 15.35, it can be seen that samples with similarity over 0.900 are similar to the standard chromatogram, indicating that the sample contains a complete set of constituents. The water-soluble constituents are mainly represented by salvianolic acid B (Peak 6) with its integral areas in the range of 4,370–13,085. The other constituents are listed as follows by ranking their integral area from largest to smallest: salvianolic acid A (Peak 8), danshensu (Peak 2), rosmarinic acid (Peak 4), protocatechuic aldehyde (Peak 3), lithospermic acid (Peak 5). The liposoluble constituents include mainly the following four compounds: dihydrotanshinone (Peak 9), tanshinone I (Peak 10), cryptotanshinone (Peak 11) and tanshinone II_A (Peak 12). Tanshinone II_A has a relatively higher UV absorption value, with integral areas of 1645–5,611. Water-soluble salvianolic acid B and liposoluble tanshinone II_A have their relative integral areas within 1.690–4.866. Judging from these chromatograms and data, it can be inferred that the manufacturers of these samples have followed the *Chinese Pharmacopoeia* to extract Danshen, so the effective dissolution of the water-soluble and liposoluble constituents was guaranteed.

(ii) The Samples with Similarity Value Over 0.8 but less than 0.9: They also contain a relatively complete set of constituents. Water-soluble constituents mainly include salvianolic acid B, but the content of salvianolic acid B varies significantly among different samples, with the smallest being 2,851 and the largest reaching 14,858. The proportion of other constituents is not completely in line with the standard chromatogram. As for liposoluble constituents, they display little deviation from the standard chromatogram, mainly including the following four: dihydrotanshinone, tanshinone I, cryptotanshinone and tanshinone II_A. Tanshinone II_A has a relatively higher UV absorption, with its integral areas within 969–5,521. Water-soluble salvianolic acid B and liposoluble tanshinone II_A have their relative integral areas within 1.027–6.568. All of these results show that CDTs from these manufacturers contain relatively complete constituents, but that the constituents differ significantly in contents among the samples from different manufacturers. The proportion of some constituents is not totally in line with the standard chromatogram. Therefore, it can be inferred that the manufacturers of these samples have basically followed the *Chinese Pharmacopoeia* to extract Danshen, and the variation in the contents of the constituents may be caused by the difference in the quality of the crude material.

(iii) The Samples with Similarity Value Over 0.5 but less than 0.8: The water-soluble constituents in these samples mainly include salvianolic acid B and salvianolic acid A, but their contents are relatively low, the integral areas of salvianolic acid B are only around 1,555–2,051. The proportion of the four major liposoluble constituents displays little deviation from the standard chromatogram, but their contents are low, with tanshinone II_A having integral areas within 1,322–2,480. The relative integral areas of salvianolic acid B are within 0.738–1.212, which, evidently, are less than that of the standard chromatogram, 2.517. These samples, though containing water-soluble and liposoluble constituents, have a relatively low content of water-soluble constituents. The reason may be that the Danshen they purchased was of poor quality, or insufficient raw material was used for the production, or the extraction time or number of extraction processes was not sufficient.

(iv) The Samples with Similarity Value of less than 0.5: Judging by the chromatogram, these samples contain only a few or even no water-soluble constituents. There is almost no salvianolic acid B, in particular, and only a little rosmarinic acid exists. The proportion of liposoluble constituents shows little deviation, but has low quantity. Based on these results, it can be inferred that these companies did not follow the *Chinese Pharmacopoeia* to extract Danshen, but added the intermediates of liposoluble constituents during the production process in order to pass the quality check required by *Chinese Pharmacopoeia* (check on the content of tanshinone II_A) before finally putting them into the market.

(v) Discussion: The test samples were processed by following the optimized extraction method and HPLC conditions, and the data obtained were compared with standard fingerprints by using "*Similarity Evaluation System for Chromatographic Fingerprint of Chinese Crude Drug*" (2004A). Among the 50 samples from different manufacturers, 20 samples had a similarity value over 0.900, suggesting that these products contained a complete set of the constituents from Danshen, with appropriate contents. Their extraction process was in line with that from *Pharmacopoeia*. There were 16 samples with a similarity value between 0.8 and 0.9. These samples contained a relatively complete set of the constituents, with the extraction process basically in line with *Pharmacopoeia*. However, the proportion of some constituents was not in line with the standard chromatogram. Four samples had a similarity value between 0.5 and 0.8; they contained a low quantity of water-soluble constituents. The reason may be that the Danshen herb was of poor quality, or the raw material used for production was insufficient, or the extraction time or the number of extraction processes was not enough. There were 11 samples with a similarity value less than 0.5. These samples contained almost no water-soluble constituent, which revealed that the manufacturers did not follow the extraction process of the *Pharmacopoeia*. In order to make the content of tanshinone II_A meet the requirement of *Pharmacopoeia*, adulteration or cheating may be involved. Of all the products from 50 different manufacturers, 40 % of them produce products with a similarity over 0.9; 72 % with a similarity of over 0.8. From this we can see that CDT from different manufacturers varies greatly in quality. Therefore, it is recommended that a comprehensive quality standard for CDT should be established so as to ensure the effectiveness and safety of the product.

15.1.1.6 Identification of Phenolic Acids in Danshen

HPLC-DAD-ESI-MS^n analysis of phenolic acids of Danshen.

1. Reagents and the Drug

   Acetonitrile and methanol were chromatographic pure; double distilled water was purified by Milli-Q system; formic acid and trifluoroacetic acid (analytical pure) were purchased from Beijing Chemical Reagents Company; solid phase extraction column was purchased from Waters Company (OASIS MAX, 6 cc, 0.2 g).

   Danshen was purchased from Liaoning Chaoyang City Herb Company (collected from around Chaoyang City), and then verified by Professor Dean Guo from Beijing University. The following 11 reference products were isolated by the authors of this article: danshensu, caffeic acid, protocatechuic acid, protocatechuic aldehyde, rosmarinic acid, salvianolic acid D, salvianolic acid C, lithospermic acid, salvianolic acid A, salvianolic acid B, and salvianolic acid E.

2. Chromatographic Condition
   (a) HPLC Condition: Agilent 1100 series HPLC (equipped with Diode Array Detector); Zorbax SB-C_{18} chromatographic column (250 × 4.6 mm, 5 μm); Zorbax SB-C_{18} Pre-column (20 × 4 mm, 5 μm). Mobile phase was acetonitrile acid water system. Detection wavelength: 288 nm. Full wavelength scanner range: 190–400 nm. Sample size: 15 μl. Column temperature: 20 °C.

   Mass Spectrum Conditions
   (b) Mass Spectrum Conditions: LCQ liquid chromatography-mass spectrometry analyzer (Finnigan MAT, San Jose, CA), equipped with ESI source and LCQ data processing system. ESI ion source: negative ion scanning mode.

3. Preparation of the Sample Solutions
   (a) Preparation of the Reference Solution: Weight 1 mg each of the 11 reference substances, respectively, dissolve them in 1 ml of 50 % acetonitrile, and dilute them to 0.1 mg/ml. Inject 250 μl for continuous mass spectrometry analysis.
   (b) Preparation of the Test Solution: Take 0.2 g of Danshen (100-mesh) and place it in a 25 ml conical flask, add 10 ml of 70 % methanol solution, extract by water-bath reflux for 1 h, filter, and put the filtrate in a rotary evaporator at 28 °C for recovery. The residue is then dissolved in 1 ml of methanol solution.

   Activate the solid phase extraction column by 5 ml of water and methanol; load 1 ml of extract into the column; wash with 5 ml of 2 % NH_4OH and methanol solution to remove alkaline and neutral impurities; elute the target phenolic acids with 5 ml of 2 % TFA methanol solution. The eluent was then put in a rotary evaporator at 28 °C to recover; the residue was dissolved in 1 ml of 50 % acetonitrile and then filtered through a 0.20 μm microporous membrane before directly undergoing HPLC analysis.

4. Result and Discussion
   (a) Multiple-stage mass spectrometry of reference substances Inject the 11 reference substances (0.1 mg/ml) into the MS for analysis. Based on the results, the patterns of mass spectral fragmentation of these references are summarized. During the experiment, we tried both positive and negative ion scanning modes, and the results showed that the negative ion mode was better suited for phenolic acids. Then, the molecular ion $[M\text{–}H]^-$ was selected as the target of CID (the impact energy of CID was 35–50 %) to produce secondary daughter ions, and the ion with the largest abundance was selected for 3–5 stage fragmentation.

   The phenolic acids in Danshen can be divided into monomers and multimers. So far, the multimers isolated from Danshen are dimers, trimers and tetramers, which are polymerized by two, three, and four monomers (i.e., danshensu, and caffeic acid) through ester bonds or ether bonds, respectively. The 11 reference substances in this study include 4 monomers, 2 dimers, 3 trimers and 2 tetramers. The 4 monomers contain carboxyl, methyl, hydroxyl or carbonyl group, respectively, and their $[M\text{–}H]^-$ ions can easily lose CO_2, CO, or H_2O. For the 7 multimers, their $[M\text{–}H]^-$ ions can easily break at the *a* or *b* position in the ester bond, thus losing danshensu $[M\text{–}H\text{-}198]^-$ or caffeic acid $[M\text{–}H\text{-}180]^-$ (see Fig. 15.44).

   (i) The Fragmentation Behavior of danshensu, Caffeic Acid, Protocatechuic Acid, and Protocatechuic Aldehyde (The Monomers): With the impact energy fixed at 30 %, danshensu's molecular ion peak was *m/z* 179. When affected by CID, this molecular ion would first lose one water molecule and produce a daughter ion *m/z* 179, which would soon lose a CO_2 to produce a third-stage daughter ion *m/z* 135. Caffeic acid's molecular ion peak was *m/z* 179, which had fragmentation behavior similar to that of danshensu, i.e., it could easily lose a CO_2 to produce a daughter ion *m/z* 135. Protocatechuic acid and protocatechuic aldehyde shared a similar structure, with the former connecting to a carboxyl in its Position 1 and the latter to a carbonyl. Their molecular ion peaks were *m/z* 153 and *m/z* 137, respectively, and when affected by CID, the former would easily lose a CO_2 while the latter easily loses a CO.

   In summary, the monomers in Danshen are mostly the derivatives of

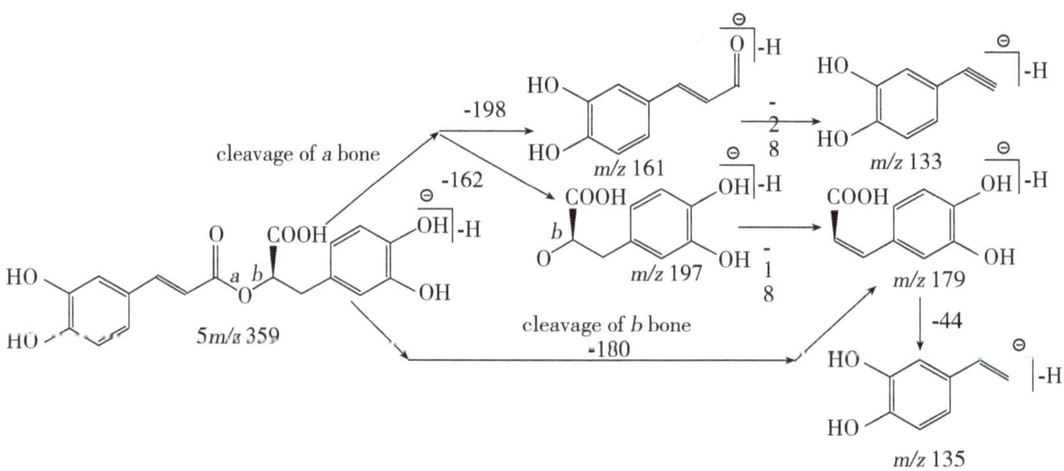

Fig. 15.44 The fragmentation pathway of rosmarinic acid

3,4-dihydroxybenzene. These types of compounds can easily have neutral loss in Position 1 and lose the substituent group of this position.

(ii) The Fragmentation Behavior of Rosmarinic Acid and Salvianolic Acid D (The Dimers): With the impact energy set at 30 %, the molecular ion of rosmarinic acid was *m/z* 359. When affected by CID, this compound could easily break its *a* bond and lose danshensu, producing the daughter ions $[M–H-198]^-$ (*m/z* 161, 100 %) and $[M–H-162]^-$ (*m/z* 197, 25 %). On the other hand, the molecular ion peak of rosmarinic acid could break its *b* bond and lose caffeic acid, producing a daughter ion $[M–H-180]^-$ (*m/z* 179, 28 %). By comparing the ion abundance of the second-stage daughter ions, it can be seen that it is easier to break the *a* bond and lose danshensu than to break the *b* bond and lose caffeic acid. Second-stage daughter ions *m/z* 161, 179 and 197, when affected by CID, would lose CO, CO_2 and H_2O, respectively, to produce the third-stage daughter ions $[M–H-198-28]^-$ (*m/z* 133), $[M–H-180-44]^-$ (*m/z* 135) and $[M–H-162-18]^-$ (*m/z* 179), accordingly. The fragmentation behavior of second-stage daughter ion *m/z* 161 was in accordance with that of protocatechuic aldehyde, the monomer, *m/z* 179 with caffeic acid, and *m/z* 197 with danshensu (Fig. 15.44).

With the impact energy set at 30 %, the molecular ion peak of salvianolic acid D was *m/z* 417. This molecular ion under the effect of CID had many fragmentation paths: First, it could easily lose a CO_2 and produce a second-stage daughter ion $[M–H–CO_2]^-$ (*m/z* 373, 100 %), which is caused by the loss of carboxyl from the carbon at Position 2. The second-stage daughter ion (*m/z* 373) could easily break the *a* bond and produce *m/z* 175 (100 %) and *m/z* 197 (23 %), with no daughter ion observed from the cleavage of the *b* bond. Second, the molecular ion could also cleave at *a* bond and *b* bond to produce daughter ions $[M–H-220]^-$ (*m/z* 197, 19 %) and $[M–H-180]^-$ (*m/z* 237, 24 %). It can be seen from the relative abundance of daughter ions that the cleavage of *b* bond is easier than that of *a* bond. This is different from the regularity of previous compounds, which may be caused by the substitution of Position 2 by a carboxyl group. Lastly, the molecular ion of salvianolic acid D could also directly lose a water molecule and produce a second-stage daughter ion *m/z* 399 (78 %) which could undergo *a* and *b* bond cleavage to lose caffeic acid and danshensu, respectively, thus producing third-stage daughter ions *m/z* 219 (100 %) and *m/z* 201 (42 %), accordingly. By a comprehensive comparison of the three fragmentation paths of salvianolic acid D, the following three patterns can be concluded: first, when a molecular ion loses the substituent at Position 2, it is easier to have *a* bond cleavage, which is in line with the previous fragmentation regularity. Second, when a substituent is present at Position 2, *b* bond fragments are more likely than *a* bond, which reveals that the substituent at Position 2 can inhibit *a* bond cleavage and the loss of danshensu. Third, based on the daughter ion *m/z* 399, it is clear that the two o-hydroxyls in 3,4-dihydroxybenzene can easily dehydrate, which is different from the previous monomers. Therefore, Position 2, when substituted with carboxyl, will facilitate the dehydration of the two o-hydroxys in 3,4-dihydroxy benzene.

(iii) The Fragmentation Behavior of Salvianolic Acid C, Lithospermic Acid, and Salvianolic Acid A (The Trimers):

With the impact energy fixed at 30 %, salvianolic acid C produced a molecular ion *m/z* 491, which could easily undergo *a* and *b* bond cleavage to lose danshensu and caffeic acid, respectively, producing second-stage daughter ions $[M–H-198]^-$ (*m/z* 293, 100 %) and $[M–H-180]^-$ (*m/z* 311, 35 %), respectively. These second-stage daughter ions, affected by CID, would display the same fragmentation behavior as those of monomers and dimers.

The molecular ion of lithospermic acid was *m/z* 537, whose molecular ion peak at the impact energy of 30 % was not the highest abundant. The highest abundant peak was $[M–H–CO_2]^-$ (*m/z* 493, 100 %). This meant that the molecular ion was not stable and could easily lose a carboxyl to produce second-stage daughter ions of greater stability. A comparison of its structure and that of salvianolic acid C revealed that its benzo-furan ring contained one more carboxyl than salvianolic acid C. Therefore, it could be inferred that this carboxyl could easily break away from the benzo-furan ring, which also meant that lithospermic acid could easily lose a carboxyl to form a more stable second-stage daughter ion: $[M–H–CO_2]^-$ (*m/z* 493, 100 %). This daughter ion had fragmentation behavior similar to that of salvianolic acid C; it could easily break *a* and *b* bond and lose danshensu and caffeic acid, producing third-stage daughter ions: $[M–H-198]^-$ (*m/z* 295, 100 %) and $[M–H-180]^-$ (*m/z* 313, 15 %), respectively.

The molecular ion of salvianolic acid A was *m/z* 493 which could easily lose danshensu and caffeic acid to produce second-stage daughter ions *m/z* 295 (100 %) and 311 (15 %). In addition, salvianolic acid A could also easily lose 3,4-dihydroxybenzene and produce a second-stage daughter ion $[M–H-110]^-$ (*m/z* 383, 10 %). A comparison of the structures of salvianolic acid A and salvianolic acid C revealed that the benzo-furan ring of salvianolic acid C is replaced by ethenyl in salvianolic acid A, from which we could infer that the production of the daughter ion $[M–H-110]^-$ was related to 3,4-dihydroxystyryl.

In summary, mass spectral fragmentation of trimers in salvianolic acids has the following patterns. First, from the abundance of the second-stage daughter ions of all trimers, it can be seen that danshensu is more easily lost than caffeic acid. Second, from lithospermic acid, we can see that in phenolic acids, the carboxyl connected to benzo-furan can be easily lost. Third, from salvianolic acid A, we can see that when 3,4-dihydroxy-styryl is present in phenolic acids, the fragmentation of phenolic acids will produce the daughter ion $[M–H-110]^-$.

(iv) The Fragmentation Behavior of Salvianolic Acid B and Salvianolic Acid E (The Tetramers): A comparison of the structures of salvianolic acid B and E revealed that salvianolic acid B contained a benzo-tetrahydrofuran ring, while in salvianolic acid E, this benzo-tetrahydrofuran ring was replaced by an open loop straight chain.

With the impact energy set at 30 %, the molecular ion of salvianolic acid B *m/z* 717 would first lose a danshensu molecule and produce a second-stage daughter ion $[M–H-198]^-$ (*m/z* 519). Therefore, in the molecular ion of salvianolic acid B, the loss of danshensu was easier than that of caffeic acid. There are two ways to produce a second-stage daughter ion *m/z* 519. One is cleavage at *a* bond; the other is cleavage at *a'* position. This molecular ion's third-stage ions mainly included *m/z* 321 and *m/z* 339, with the former caused by the continual loss of two danshensus $[M–H-198-198]^-$ (100 %), and the latter caused by losing another caffeic acid [M–H-198-18]$^-$ (22 %). The abundance of these two third-stage ions shows that danshensu was more easily fragmented than was caffeic acid. In addition, its fourth-stage daughter ion shared fragmentation behavior similar to that of the monomers (see Fig. 15.45).

Similar to salvianolic acid B, the molecular ion of salvianolic acid E also lost a danshensu at first and produced a second-stage daughter ion [M–H-198]$^-$ (*m/z* 519). However, its abundance was only 27 %, which suggests that the loss of the first danshensu in salvianolic acid E was harder than in salvianolic acid B. There are also two paths of losing the first danshensu for salvianolic acid E: cleavage at *a* bond or *a′* bond, with the latter in the same environment as that of salvianolic acid B while the former is not. Thus, it can be inferred that the loss of the first danshensu is due to the cleavage of *a* bond. This explains that the substituent on benzo-tetrahydrofuran ring can be easily dissociated. This pattern is in line with that of lithospermate.

In summary, the loss of the first danshensu in salvianolic acid B and E is caused by the cleavage of *a* bond, while the loss of the first caffeic acid is by the cleavage of *b* bond. The loss of the second danshensu and caffeic acid is caused by *a* and *b* bonds, respectively. In addition, the fragmentation of salvianolic acid B shows that the substituent on benzo-tetrahydrofuran ring can be easily dissociated.

(b) HPLC-DAD-ESI-MS^n Analyses of the Salvianolic Acids in Danshen Extract: The HPLC-UV and TIC results of Danshen extract are shown in Fig. 15.46. A comparison of the effect of different solvents on the extraction revealed that 70 % methanol could extract the phenolic acids almost completely [14], while simultaneously extracting other constituents. In order to purify phenolic acids, we tested solid phase extraction columns, and by comparing the effects of different types of SPE columns (Waters OIASIS HLB, MAX and MCX) in purifying phenolic acids, we selected MAX column which could effectively enrich and purify phenolic acids.

Adding acid to the mobile phase in the HPLC-ESI-MS^n analysis of Danshen extract could improve the deionization of phenolic acids, and thus improve the peak shape. By comparing different concentrations of formic acid and acetic acid, we found that 0.4 % formic acid was more appropriate.

By applying HPLC-ESI-MS^n analysis, 55 constituents of phenolic acids from Danshen extract were detected, and 42 were identified. Among them, 16 were new compounds (see Table 15.41).

(i) Identification of Monomers: Apart from the reference substances 1–4, three more monomers in Danshen extract were identified. After CID, the molecular ion (*m/z* 117) of Peak 1 (t_R 4.91 min) lost a CO_2 and produced a second-stage daughter ion *m/z* 73. So, Peak 1 was identified as succinic acid [15]. The molecular ion (*m/z* 197) of Peak 2 (t_R 8.97 min) had the same fragmentation behavior as that of compound 1, so Peak 2 was identified as the stereoisomer of compound 1, namely, (S)-3-(3,4-Dihydroxyphenyl) lactic acid [16]. The molecular ion (*m/z* 193) of Peak 7 (t_R 20.11 min) had the following second-stage daughter ions: *m/z* 150 [M–HC0$_2$]$^-$ and *m/z* 135 [M–HC0$_2$–CH$_3$]$^-$. Peak 7 had the same fragmentation behavior as that of ferulic acid. Therefore, it was identified as ferulic acid [17].

(ii) Identification of Dimers: Based on the fragmentation behavior of the compounds, besides compound 5 and 6 in Danshen, 9 dimers were identified,

Fig. 15.45 The fragmentation behavior of the tetrameric salvianolic acids B and E. **a** MS^2 of salvianolic acid B; **b** MS^3 of salvianolic acid G; **c**, **d** MS^4 of salvianolic acid G; **e** MS^2 of salvianolic acid E; **f** MS^3 of salvianolic acid E; **g**, **h** MS^4 of salvianolic acid E

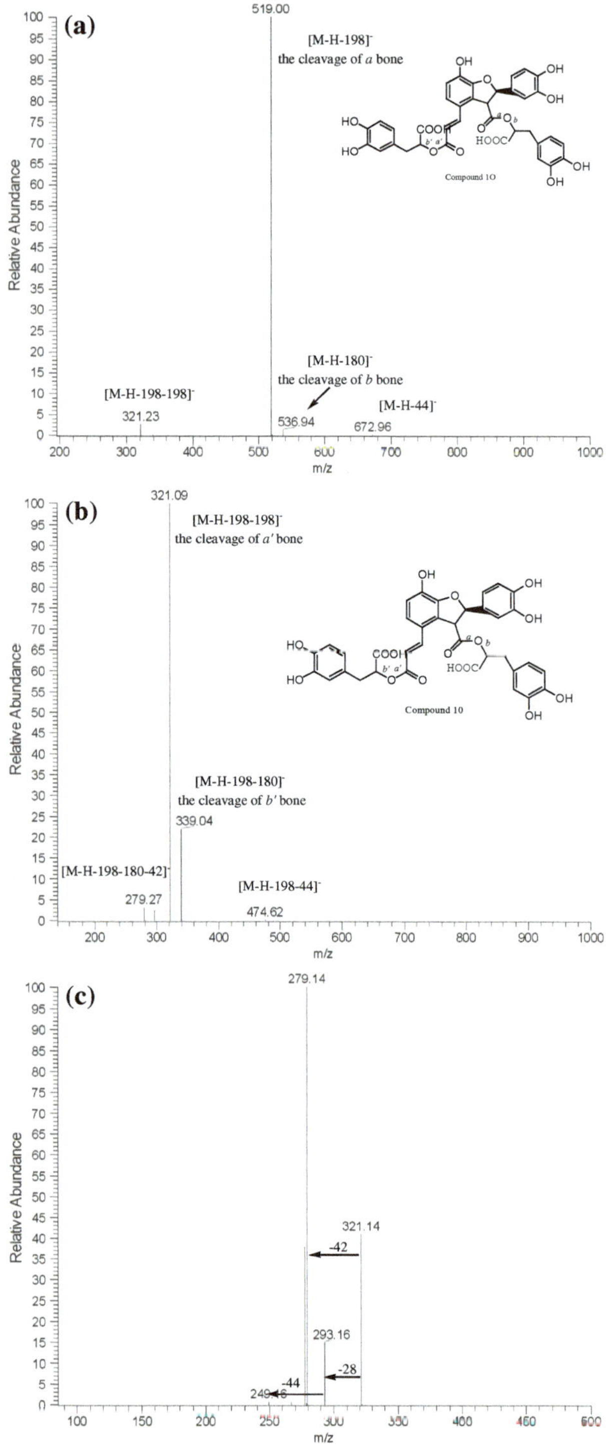

Fig. 15.45 (continued)

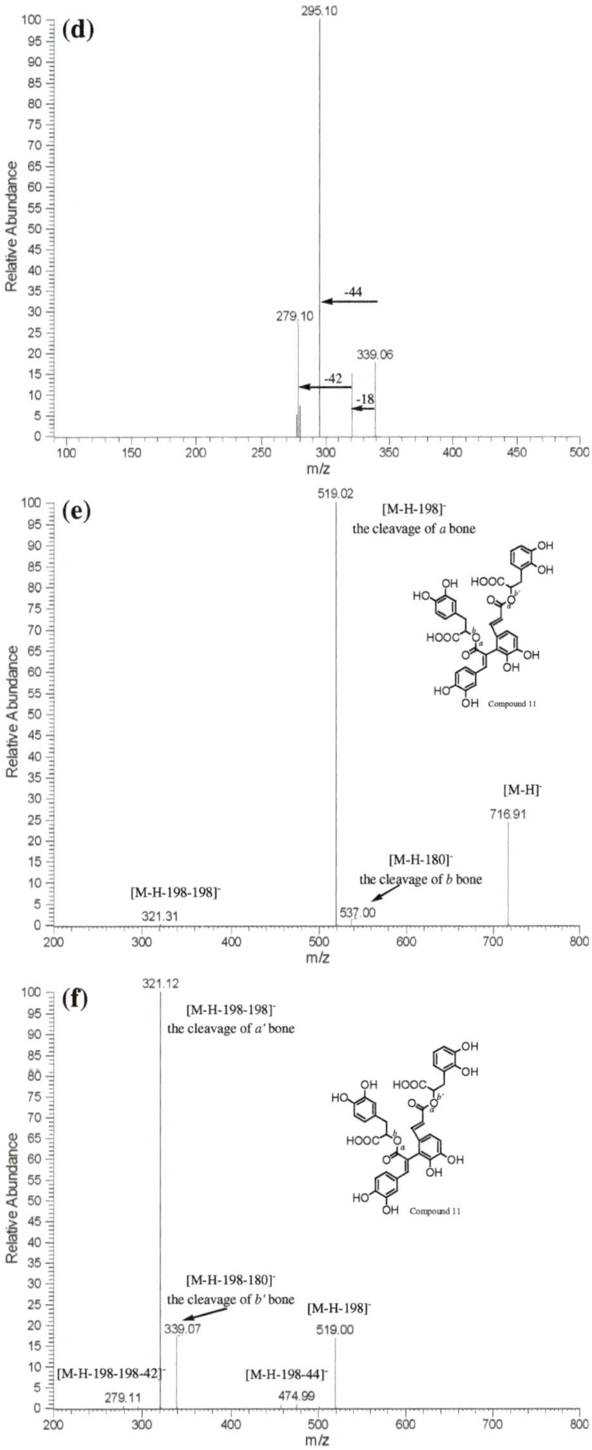

Fig. 15.45 (continued)

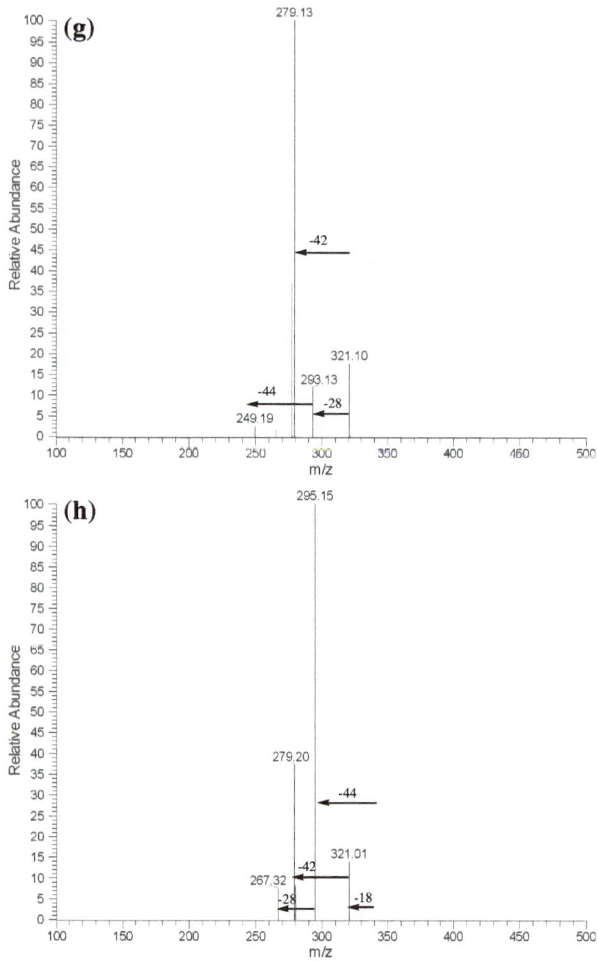

among which 4 had been reported before from *Salvia* plants and the rest were new compounds.

The molecular ion (*m/z* 333) of Peak 8 (t_R 15.83 min) produced the following second-stage daughter ions after CID: *m/z* 197, 179, 315 and 153. These ions were produced during the loss of protocatechuic acid and danshensu by the molecular ion. The molecular ion *m/z* 197 could produce the third-stage ion *m/z* 135, which further proved that this compound contained the structural unit of danshensu. The above information proved that the structure of Peak 8 contained danshensu and protocatechuic acid. Therefore, Peak 8 was identified as 9-oxo-[(3,4-dihydroxyphenyl) methanoyl]-3,4; di-hydroxybenzenepropanoic acid.

The molecular ion (*m/z* 325) of Peak 9 (t_R 20.14 min), affected by CID, could easily produce the second-stage daughter ions *m/z* 193 and *m/z* 131. These two ions showed the same molecular ion peaks as those of ferulaic acid and benzo-furan. Therefore, Peak 9 was polymerized by ferulaic acid and benzo-furan, and thus identified as 2-(7-dihydroxyl)-benzofuranyl-, and ferulic acid.

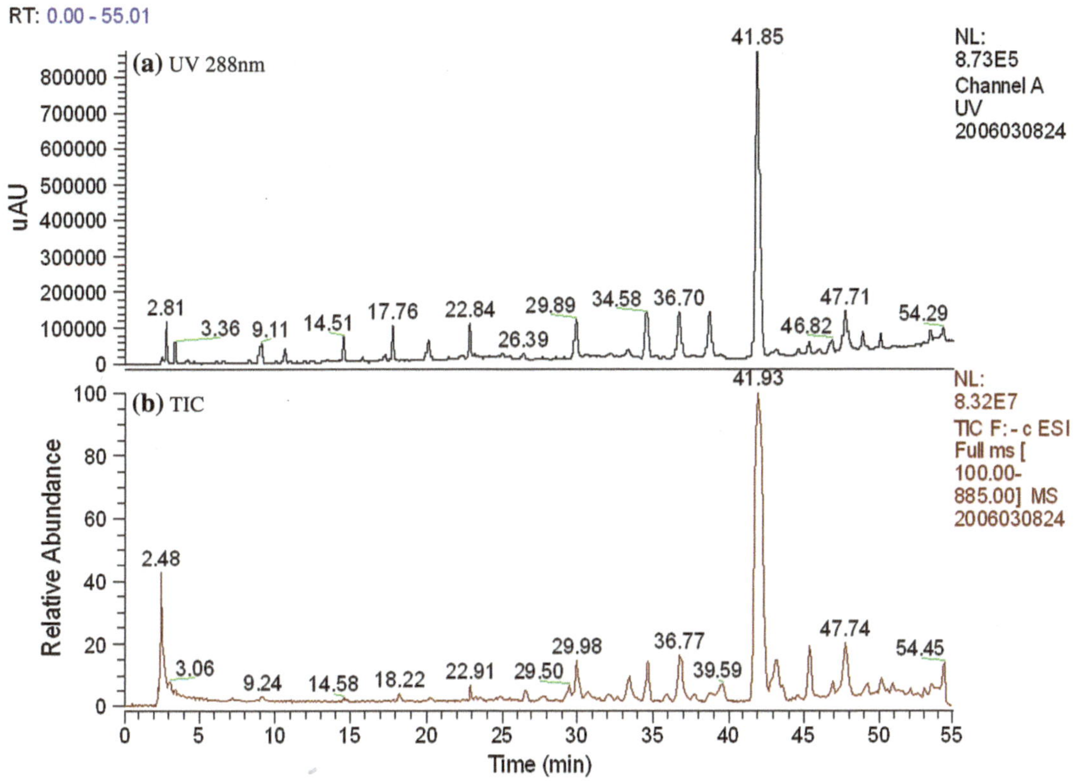

Fig. 15.46 HPLC-DAD-ESI-MS^n analysis of Danshen extract. **a** HPLC-UV spectrum under 288 nm; **b** Total ion current spectrum of LC-negtive ion ESI MS

The molecular ion (*m/z* 313) of Peak 10 (t_R 22.96 min) could easily lose CO_2 first to produce a second-stage daughter ion *m/z* 269, which then lost 3,4-dihydroxybenzene to produce a third-stage ion *m/z* 159. These fragmentation behaviors were similar to those of salvianolic acid F. Therefore, Peak 10 was identified as salvianolic acid F [18]. The molecular ion of Peak 18 (t_R 38.33 min) was *m/z* 327 whose molecular weight was 328. It was 14 Da more than salvianolic acid F. That meant it might contain one more methyl than Peak 10. The molecular ion of Peak 18 produced two second-stage daughter ions: *m/z* 295 and *m/z* 217. The former was produced by losing methanol (the presence of methanol proves the existence of methoxy), while the latter was produced by losing 3,4-dihydroxybenzene (similar to salvianolic acid F). The third-stage daughter ion of Peak 18 was produced also by the loss of 3,4-dihydroxybenzene. Therefore, comparing Peak 18 with Peak 10, we could preliminarily identify Peak 18 as 2-(3,4-dihydroxyphenyl)-7-hydroxy-5 benzenepropanoic acid.

Both Peak 11 (t_R 23.52 min) and Peak 15 (t_R 32.29 min) had the molecular ion *m/z* 371, which were isomers. The molecular ion of Peak 11, after loss of methanol, produced a second-stage daughter ion *m/z* 339, which was similar to Peak 18. Therefore, Peak 11 contained a methoxy group. After losing H_2O and CO_2, *m/z* 339 produced *m/z* 295 and 321. All of the above fragmentation patterns were in line with a compound in literature; therefore,

Table 15.41 The phenolic acids detected in Danshen extracts by HPLC-ESI-MS^n

Peak no.	Retention time (t_R, min)	$[M-H]^-$ m/z	HPLC-ESI-MS^n m/z (% of base peak)	Identification
1▼	4.91	117	MS^2 [117]:73(100)	Succinic acid
2▼	8.97	197	MS^2 [197]:179(100);MS^3 [179]:135 (100)	3,4-trihydroxy-,(S)-benzenepropanoic acid
3★,▼	9.25	197	MS^2 [197]:179(100);MS^3 [179]:135 (100)	Danshensu(1)
4★,▼	10.86	153	MS^2 [153]:109(100)	Protocatechuic acid(3)
5★,▼	14.60	137	MS^2 [137]:109(100)	Protocatechuic aldehyde(4)
6★,▼	17.86	179	MS^2 [179]:135(100)	Caffeic acid(2)
7▼	20.11	193	MS^2 [193]:150(100), 179(46);MS^3 [150]:135(100);	Ferulic acid
			MS^3 [179]:135(100)	
Dimmer				
8	15.83	333	MS^2 [333]:197(100), 179(15), 315 (10), 223(10), 153(10);MS2 E1971:179(100);MS^3 [179]:135(100)	9-oxo[(3,4 dihydroxyphenyl) methanoyl]-3,4;dihydroxy benzenepropanoic acid
9	20.14	325	MS^2 [325]:193(100), 131(10), 113(8), 178(1);MS^3 [193]:178(100), 134(18), 149(10)	2-(7-ihydroxyl)-benzofuranyl, ferulic acid
10▼	22.96	313	MS^2 [313]:269(100), 313(10);MS^3 [269]:159(100), 109(100)	Salvianolic acid F
11▼	23.52	371	MS^2 [371]:339(100), 321(2), 295(2); MS^3 [339]:321(85), 295(100), 280(10)	4-(2-carboxyethenyl)-2-(3,4-dihydroxyphenyl)-2,3-dihydro-7-hydroxy-,3-methyl ester,[2,3,4(E)]-3_benzofurancarboxylic acid,
12▼	23.89	539	MS^2 [539]:297(100), 359(40), 341 (10), 315(10), 401(8), 279(8)	6-[(1E)-3-[1 carboxy 2(3,4 dihydroxyphenyl)ethoxyl-3-OXO-1-propenyl]-3-(3,4 dihydroxyphenyl)8 hydroxy-7-OXO-,bicyclo[2.2.2]oct-5-ene-2 carboxylic acid
			MS^3 [297]:279(100), 161(40)	
13★,▼	29.72	417	MS^2 [417]:175(100), 373(60), 197(20), 237(18), 339 (10), 179(10);MS^3 [175]:175(100), 157(20), 147(18), 131 (5)	Salvianolic acid D(6)
14▼	30.73	339	MS^3 [321]:293(100), 265(10);MS^4 [280]:252(100)	Salvanolic arid G
15	32.29	371	MS^2 [371]:327(100), 295(60), 217(8); MS^3 [327]:295(100)	3-carboxyl-2-(3,4-dihydroxyphenyl)-2,3-dihydro-7-hydroxy-,4-methyl ester,-4-benzofurancarboxyethenylic acid
16★,▼	34.42	359	MS^2 [359]:161(100), 179(25), 197 (25);MS^3 [161]:133(100)	Rosmarinic acid(5)
17	35.16	311	MS^2 [311]:293(100), 283(20), 267 (65);MS^2 [293]:249(100)	2(3,4-dihydroxyphenyl)-7-hydroxy-5-benzenepropanoic acid
18	38.33	327	MS^2 [327]:217(100), 295(20), 185(20)	Methyl salvianolate F

(continued)

Table 15.41 (continued)

Peak no.	Retention time (t_R, min)	[M–H]$^-$ m/z	HPLC-ESI-MS^n m/z (% of base peak)	Identification
Trimer				
19	21.62	571	MS^2 [571]:483(100), 527,439,329; MS^3 [527]:483(100), 329 MS^4 [483]:439(100)	Yunnaneic acid E
20	22.51	683	MS^2 [683]:665(100), 621(20), 485(1), 441(1);MS^3 [665]:621(100).423(80)	3,4 dihydroxy(1,3,4,5)-l-carboxy 4 hydroxy l, 3,5-cyclohex anetriyl ester-*benzcnepropanoic* acid
21	28.64	585	MS^2 [585]:497(100), 387(90), 343 (30), 299(20);MS^3 [497]:	Methyl yunnaneate E
22	28.54	535	MS^2 [535]:373(100);MS^3 [373]:I35 (100), 31 l(20), 1 79(10)	5-(2-car})oxyethcnyl)-2-(3,4-dihydroxyphenyI)-7-hydroxy-,3-[1-carboxy-2(3,4-dihydroxyphenyl)ethyl] ester,(E)-3-benzo- furancarboxylic acid
23▼	29.86	537	MS^2 [537]:339(100), 295(30), 493(5); MS^3 [339]:295(100), 321(28);MS^4 [295]:280(100), 295(80), 277(45), 267 (42),l85(10), 253(5)	Salvanolic acid H/I
24▼	31.73	537	MS^2 [537]:493(100);MS^3 [493]:295 (100)	Salvanolic acid J
25▼	32.90	537	MS^2 [537]:339(100), 295(42), 493(1), 321(1); MS^3 [339]:295(100), 32 l(1); MS^4 [295]:280(50).295(100), 277(25), 267(20), 185(10), 253(5)	Salvanolic acid H/I
26	31.13	495	MS^2 [495]:297(100);MS^3 [2,973:253 (100), 269(25)	7,7′,8,8′-tetrahydrogen-salvanolic acid C
27★,▼	36.93	5.7	MS^2 [537]:493(100), 295(20);MS^3 [493]:295(100), 159(18), 109(8), 277 (5)	Lithospermic acid(8)
28	37.21	551	MS^2 [551]:321(100), 519(80), 353 (85), 309(40), 507(10);MS^3 [321]:277 (100), 249(45)	Methyl salviabolate H/I
29	44.80	551	MS^2 [551]:507(100), 309(50), 327(5), 197(5);MS^3 [507]:327(100), 197(32)	9″-methyl lithospermate
30	47.04	551	MS^2 [551]:353(100), 321(42), 519 (40), 231(1), 339(2), 371(1);MS^2 [353]:335(100), 294(90), 321(60), 353 (20), 279(20), 243(20), 325(18), 217 (5);MS^3 [335]:335(100), 276(50), 307 (40), 253(5)	Methyl salvianolate H/I
31★,▼	46.14	493	MS^2 [493]:295(100), 313(18), 203(5), 383(1), 159(1), 185(1) MS^3 [295]:159 (38), 109(25), 251(2), 267(2), 277(18), 175(10), 185(5), 202(1)	Salvanolic acid A(9)

(continued)

Table 15.41 (continued)

Peak no.	Retention time (t_R, min)	[M–H]⁻ m/z	HPLC-ESI-MSⁿ m/z (% of base peak)	Identification
32	49.10	493	MS² [493]:295(100);MS³ [295]:295 (100), 267(85), 277(60), 239(18), 280 (10)	EE(2E)-3-[(2R)2(3,4-dihydroxyphenyl)2,3-dihydr0-7 hy-droxy-4-benzofuranyl]-1-oxo-2-propenyl]oxy]-3,4-dihydroxy-ben-zenepropanoic acid
33▼	50.10	491	MS² [491]:293(100), 311(65), 249 (10), 265(8);MS³ [293]:249(100), 265 (70), 293(20), 276(15);MS⁴ [249]:249 (100), 221(18)	Isosalvianolic acid C
34★,▼	53.38	491	MS² [491]:293(100), 311(20), 265(1); MS³ [293]:29(100), 247(5), 265(42), 221(5), 276(5), 237(2)	Salvanolic acid C(7)
Tetramers				
35★,▼	33.30	717	MS² [717]:519(100), 321(18), 537(8), 339(1), 295(1); MS³ [519]:321(100), 339(30), 279(5), 277(1), 295(1); MS⁴ [321]:321(100), 279(95), 277(50), 293 (10);MS⁵ [279]:251(15)	Salvanolic acid E(11)
36	37.11	749	MS² [749]:551(100), 519(40), 321 (20), 572(5):MS³ [551]:321(100), 519 (50), 339(20), 295(1), 371(1); MS⁴ [321]:279(100)	8-hydroxy-9″-methyl-,salvianolate B
37	39.16	749	MS² [749]:519(100), 551(95), 321 (25), 717(18), 493(1), 569(91), 249(1); MS³ [519]:321(100), 339(25), 279(1), 371(1) MS⁴ [321]:279(95), 293(20), 303(5)	8-hydroxy-9″-methyl-,salvianolate B
38	41.70	717	MS² [717]:519(100), 339(18), 537(2), 321(1), 279(1), 673(1), 475(1);MS³ [519]:321(100), 339(25), 295(5), 279 (5);MS⁴ [321]:279(100), 293(18), 249 (2), 277(50)	Salvanolic acid B(10)
39	43.09	717	MS² [717]:519(100), 321(25), 537(1), 279(1).339(2) MS³ [519]:321(100), 339(28), 295(2), 279(5), 457(1) MS⁴ [321]:279(100), 321(85), 293(22), 249 (5).303(5)	Salvanolic acid L
40	45.66	715	MS² [715]:321(100), 519(80), 339 (85), 671(5), 339(28)	dedihydr0,salvanolic acid B/isomer
41	47.54	731	MS² [731]:533(100), 55 l(5), 335(5), MS³ [333]:335(100), 353(32), 309(5), 320(4) MS³ [335]:320(100), 307(6), 291(4), 276(5), 335(10)	4-methoxyl-salvianolic acid B
42	52.00	731	MS² [731]:533(100), 551(2), 335(10), 353(2) MS³ [533]:321(100), 339(30), 280(3), MS³ [335]:321(100), 279(58), 277(28)	9″ methyl salvianolate B

★ Compared with references; ▼ Reported in sages

Peak 11 was tentatively identified as 4-(2-carboxyethenyl)-2-(3,4-dihydroxyphenyl)-2,3-di-hydro-7-hydroxy-,3-methyl ester, [2,3,4(E)]-3-benzofurancarboxylic acid [19]. Compared with Peak 11, the second-stage daughter ion of Peak 15 (*m/z* 327) indicated that Peak 15 contained a carboxyl group which could easily produce CO_2, similar to Compound 8. Therefore, Peak 15 contained the carboxyl that was linked to the furan ring. In conclusion, Peak 15 was tentatively identified as 3-carboxyl-2-(3,4-di-hydroxyphenyl)-2,3-dihydr0-7-hydroxy-,4-methylester,-4-benz of urancarboxyethenylic acid.

The molecular ion of Peak 14 (t_R 30.73 min) was *m/z* 339. Its second-stage daughter ions were *m/z* 295 and 321. According to the literature [20], the compound was identified as salvianolic acid G.

(iii) Identification of Trimers: Based on the molecular ions and multi-stage daughter ions, besides compound 7 and 8 in Danshen, 14 trimers were identified, among which 7 were known compounds and the rest were new compounds.

The molecular ion of Peak 19 (t_R 21.62 min) was *m/z* 571, and its second-stage daughter ions included *m/z* 527, 483, 439, and 329, which were produced by losing one CO_2, two CO_2, three CO_2, and danshensu, respectively. Therefore, this compound contained at least three carboxyl groups and one danshensu unit. The above information was in line with yunnaneic acid E, and therefore Peak 19 was identified as yunnaneic acid E [21]. The molecular ion of Peak 21 (t_R 28.64 min) was *m/z* 585, and we could infer that its molecular weight was 586, which was 14 Da more than that of Peak 19. Considering its second-stage daughter ions (*m/z* 497, 387, 343, and 299), we could preliminarily identify this compound as the methyl ester product of Peak 19. Therefore, Peak 21 was tentatively identified as methyl yunnaneate E.

The molecular ions of four peaks, including Compound 7, were *m/z* 537. The second-stage and third-stage daughter ions of Peak 24 (t_R 31.73 min) were *m/z* 493 and 295, respectively, which was similar to Compound 7. Therefore, Peak 24 must be the isomer of Compound 7, and thus was identified as salvianolic acid J. As for Peak 23 (t_R 29.86 min) and Peak 25 (t_R 32.90 min), their second-stage daughter ions included *m/z* 339 (100 %), 295, and 493, indicating that danshensu in Peak 23 and Peak 25 could be more easily lost than carboxyl. Therefore, it could be inferred that Peak 23 and Peak 25 did not contain a benzo-furan structure. The third-stage daughter ions of Peak 23 and Peak 25 were almost the same. So, they are stereoisomers, and thus tentatively identified as salvianolic acid H/I.

The molecular ion of Peak 28 (t_R 37.21 min), Peak 29 (t_R 44.80 min) and Peak 30 (t_R 47.04 min) were *m/z* 551. Therefore, they contained one more methyl than *m/z* 537. The second-stage daughter ions of Peak 28, Peak 29, and Peak 30 were *m/z* 321, 507, and 353, respectively. *m/z* 507 of Peak 29 was produced when the molecular ion lost one carboxyl; from this fact it could be inferred that this carboxyl was most likely to be linked to the benzo-furan ring, and the methyl was linked to the other carboxyl. In addition, Peak 29 could also lose danshensu and produce daughter ion *m/z* 309. Therefore, Peak 29 was tentatively identified as 9″-methyl lithospermate. *m/z* 321 of Peak 28 was produced when the molecular ion lost methanol and danshensu, while the daughter ion (*m/z* 353) of Peak 30 was produced when the molecular ion lost danshensu. The rest of the second-stage and third-stage daughter ions of

Peak 28 and Peak 30 were similar to those of salvianolic acid H/I. Therefore, Peak 28 and Peak 30 were identified as methyl salvianolate H/I [22].

The molecular ion of Peak 33 (t_R 50.10 min) was *m/z* 491, indicating that Peak 33 was the isomer of Compound 9. Its second-stage daughter ions (*m/z* 293,311 and 265) were similar to those of Compound 9, so Peak 33 was identified as isosalvianolic acid C [23]. The molecular ion of Peak n26 (t_R 31.13 min) was *m/z* 495, which was similar to Compound 9. It produced daughter ions *m/z* 297 and 253. Therefore, Peak 33 should be the tetrahydrogenated product of Compound 9, thus identified as 7, 7′, 8, 8′ Ltetrahydrogen-salvanolic acid C.

The base peak of the second-stage daughter ion produced from molecular ion (*m/z* 493) of Peak 32 (t_R 49.10 min) was *m/z* 295, which was similar to Compound 8. But unlike Compound 8, its third-stage daughter ion was *m/z* 267, indicating that Peak 32 could lose H_2O easily. By comparing with Compound 6 and 8, Peak 32 was thus identified as [(2E)-3-[(2R)-2-(3,4-dihydroxy-phenyl)-2,3-dihydro-7-hydr-oxy-4-benzofuranyl]-1-OXO-2-propenyl] oxy]-3,4-dihydroxy-, -benzeneprop anoic acid [24].

The molecular ion of Peak 20 (t_R 22.51 min) was *m/z* 683, whose second-stage daughter ions included *m/z* 665, 621, 485, and 441—they were produced by losing H_2O, CO_2, danshensu and two CO_2, respectively. The above ion fragments were consistent with 3,4-dihydroxy (1,3,4,5) -1-carboxy-4-hydroxyl-l,3,5-cyclohexanetriyl ester-benzenepropanoic acid. Therefore, Peak 20 was identified as that compound [25].

(iv) Identification of Tetramers: Apart from Compounds 10 and 11, 6 tetramers from the Danshen extract were identified; among them 2 compounds had been reported in the species plant, and the rest were new compounds.

On the HPLC chromatogram, besides Compounds 10 and 11, the molecular ion of Peak 39 (t_R 43.09 min) was also *m/z* 717. The second-stage ions of Peak 39 included *m/z* 519, 321, and 537, which was similar to Compound 11. This showed that the compound did not contain a benzo-furan ring. By referring to the third-stage ion of Peak 39, it was identified as salvianolic acid L [26].

The molecular ion of Peak 40 (t_R 45.66 min) was *m/z* 715, its second-stage daughter ions included *m/z* 321, 519, 339, and 671, which were produced by losing danshensu, dedihydrodanshensu, caffeic acid, and CO_2, respectively. By comparing with Peak 39 and Compound 10, Peak 40 could be identified as dedihydro-, salvanolic acid B/isomer.

The molecular ions of both Peak 41 (t_R 47.54 min) and Peak 42 (t_R 52.00 min) were *m/z* 731. The second-stage, third-stage and fourth-stage daughter ions of Peak 41 were *m/z* 533, 335, and 320, respectively, which were produced by losing one danshensu, two danshensus, and a methyl. On the other hand, the second-stage and third-stage daughter ions of Peak 42 were *m/z* 533 and 321, respectively, which were produced by losing a danshensu and a methyl danshensu. Based on the information, Peak 41 and Peak 42 were identified as 4 methoxyl, salvianolic acid B and 9″-methyl salvianolate B.

The molecular ions of both Peak 36 (t_R 37.11 min) and Peak 37 (t_R 39.16 min) were *m/z* 749. The second-stage daughter ions of Peak 36 included *m/z* 551(100 %), 519 (40 %), and 321 (20 %), which were produced by losing a danshensu, a danshensu methyl ester, and a second danshensu, respectively. The second-stage daughter ions of Peak 37 included *m/z* 519 (100 %), 551 (95 %), and 321 (25 %), which were produced by losing a danshensu methyl ester and a danshensu, respectively. The above information indicates that Peak 36 could lose danshensu more easily, while Peak 37 could lose danshensu methyl ester more easily. By comparing with Compound 10 which breaks its *a* bond first, Peak 36 and Peak 37 were identified as 8 hydroxyl, 9″-methyl-salvinalate B, and 8-hydroxyl, 9″-methyl-salvian0-late B, respectively.

15.1.2 Chromatographic Fingerprints of Diterpene Quinone Constituents

15.1.2.1 Chromatographic Fingerprints of Diterpene Quinone Constituents in Danshen

Jinlan Zhang and Dean Guo

1. Equipment and Reagents (Omitted)
2. Optimization of the Method for Extraction of Liposoluble Diterpene Quinone in Danshen

The diterpene quinone compounds are thermally unstable, so the ultrasonic method was used to extract the compounds from Danshen. The effects of treatment length and solvents on the extraction were investigated. Method: Precisely weigh 50 mg of Danshen powder and sift through a 60-mesh sieve. Add 10 ml of solvent and weigh. Soak overnight at a cold temperature. Perform ultrasonic extraction for 0.5 h. Make up for the lost weight after extraction, and then filter. Concentrate and evaporate the filtration to dryness. Precisely add 5 ml of methanol to dissolve the residue. Pass through a 0.45 μm microporous membrane and perform HPLC. Take another three Danshen powder samples, add methylene chloride: methanol (8:2). Perform ultrasonic extraction for 10, 20, 30, and 40 min, respectively. Follow the previous steps. Compare the peak areas of the main constituents: cryptotanshinone, tanshinone I, tanshinone II_A, and choose the best extraction method (Table 15.42).

Based on the above comparison, the best extraction method was confirmed: Precisely weigh 50 mg of Danshen powder and pass through a 60-mesh sieve. Add 10 ml of solvent and weigh. Soak overnight at a cold temperature. Perform ultrasonic extraction for 0.5 h. Make up the weight loss after extraction, and filter. Concentrate and evaporate the filtration to dryness. Precisely add 5 ml of

Table 15.42 Comparison of the methods for extracting liposoluble constituents from Danshen

	Extraction method	Peak area		
		Cryptotanshinone	Tanshinone I	Tanshinone II_A
Extraction solvent	Methanol	3,621.9	2,189.4	4,040.2
	Methylene chloride	4,592.9	2,346.2	1,057.6
	Methylene chloride: methanol (8:2)	4,177.1	2,591.3	1,221.9
	Methylene chloride: methanol (3:2)	3,469.9	2,284.4	4,352.2
	Methylene chloride: methanol (2:8)	2,748.7	1,757.2	832.40
Extraction time	10 min	3,590.9	2,119.6	970.10
	20 min	4,200.9	2,652.0	1,244.0
	30 min	4,177.1	2,591.3	1,221.9
	40 min	3,948.9	2,533.6	1,086.5

methanol to dissolve the residue. Pass through a 0.45 μm microporous membrane and perform HPLC.

3. The Establishment of the Analytic Method of Liposoluble Constituents and Their Identification
   Aglient 1100 HPLC-DAD and ZORBAX-EXTEND C_{18} chromatographic column were used for the analysis, and the different mobile phases methanol–water, acetonitrile–water, and acetonitrile–acetic acid were tested. The results showed that acetonitrile–water was more suitable for analyzing the liposoluble constituents in Danshen than other systems. The effects of constant composition and gradient composition of the system on elution were also compared, and the following gradient elution program was selected: acetonitrile: water (v/v) = 45:55 (0 min) → 50:50 (10 min) → 60:40 (40 min) → 70:30 (40–60 min) → 45:55 (70 min); collect HPLC chromatogram; compare with chromatograms of the reference products under the same chromatographic conditions. Three main constituents in the fingerprint of liposoluble constituents were identified: cryptotanshinone, tanshinone I, and tanshinone II_A (Fig. 15.47).
4. Methodological Validation
   The precision, repeatability and stability of the method for the extraction and analysis of the liposoluble constituents in Danshen were tested.
   (a) Precision: Following the above method, extract one part of Danshen and, test the samples 5 consecutive times. The result similarities were greater than 0.95, meeting the requirements (Table 15.43).
   (b) Repeatability: Using the established extraction method, extract 5 Danshen samples, then perform HPLC analysis. The result similarities were greater than 0.95, meeting the requirements (Table 15.44).
   (c) Stability: Extract one Danshen sample, dissolve the extract into methanol, and place at room temperature. Perform

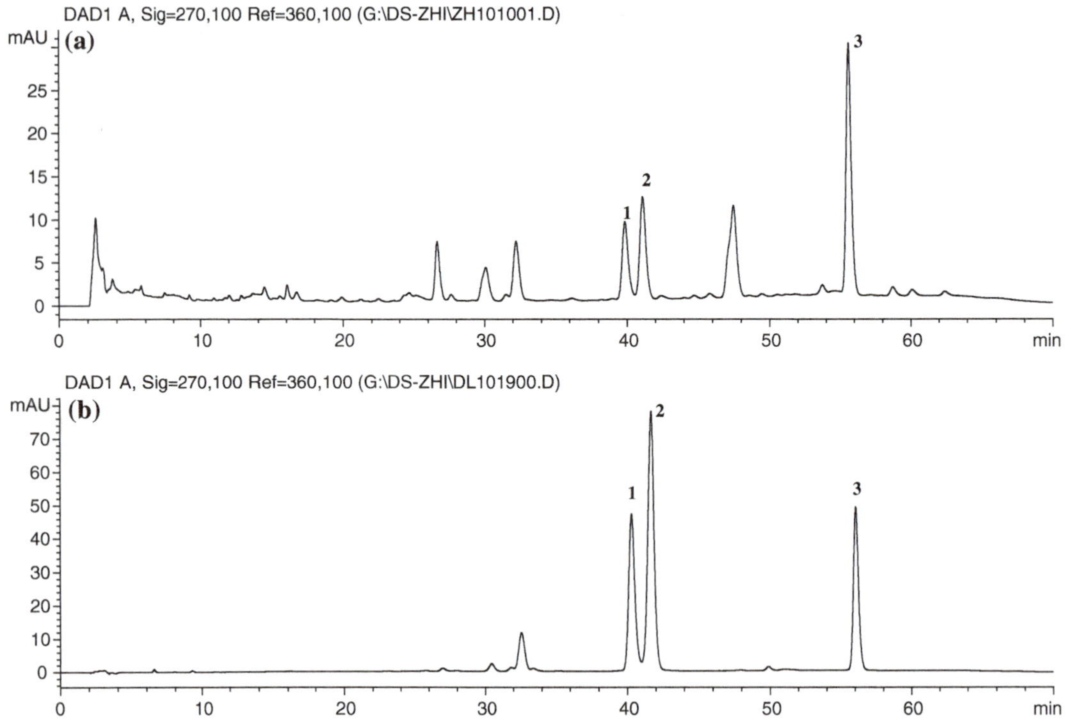

Fig. 15.47 The HPLC chromatograms of Danshen's liposoluble constituents (**a**) and the reference substances (**b**). Peaks *1*, *2*, and *3* are cryptotanshinone, tanshinone I, and tanshinone II_A, respectively

Table 15.43 The precision test results

	S1	S2	S3	S4	S5	S6	Reference fingerprint
S1	1.000	1.000	1.000	1.000	1.000	1.000	1.000
S2	1.000	1.000	1.000	1.000	1.000	1.000	1.000
S3	1.000	1.000	1.000	1.000	1.000	1.000	1.000
S4	1.000	1.000	1.000	1.000	1.000	1.000	1.000
S5	1.000	1.000	1.000	1.000	1.000	1.000	1.000
S6	1.000	1.000	1.000	1.000	1.000	1.000	1.000
Reference fingerprint	1.000	1.000	1.000	1.000	1.000	1.000	1.000

Table 15.44 The repeatability test results

	S1	S2	S3	S4	S5	S6	Reference fingerprint
S1	1.000	1.000	0.997	1.000	0.992	0.983	0.999
S2	1.000	1.000	0.998	0.999	0.991	0.983	0.998
S3	0.997	0.998	1.000	0.996	0.99	0.984	0.997
S4	1.000	0.999	0.996	1.000	0.993	0.984	0.999
S5	0.992	0.991	0.990	0.993	1.000	0.987	0.995
S6	0.983	0.983	0.984	0.984	0.987	1.000	0.990
Reference fingerprint	0.999	0.998	0.997	0.999	0.995	0.990	1.000

Table 15.45 The stability test results

Time (h)	S1	S2	S3	S4	S5	S6	S7	Reference fingerprint
0	1.000	1.000	1.000	1.000	1.000	1.000	1.000	1.000
1	1.000	1.000	1.000	1.000	1.000	1.000	1.000	1.000
2	1.000	1.000	1.000	1.000	1.000	1.000	1.000	1.000
3	1.000	1.000	1.000	1.000	1.000	1.000	1.000	1.000
4	1.000	1.000	1.000	1.000	1.000	1.000	1.000	1.000
6	1.000	1.000	1.000	1.000	1.000	1.000	1.000	1.000
8	1.000	1.000	1.000	1.000	1.000	1.000	1.000	1.000
Reference fingerprint	1.000	1.000	1.000	1.000	1.000	1.000	1.000	1.000

HPLC at different time points. The results showed that the sample was stable for at least 8 h (Table 15.45).

5. HPLC Fingerprinting of the Liposoluble Constituents in Wild and Cultivated Danshen from Different Regions
   Collect Danshen samples from different regions, and analyze them by the methods described above. The results are shown in Tables 15.46 and 15.47, and Figs. 15.48 and 15.49.

Except for the wild Danshen from Yunnan, the similarities of the HPLC fingerprints of liposoluble constituents in Danshen from different regions all exceeded 0.9. Except for the Danshen cultivated in Shanxi's Pinglu, Guizhou's Guiyang and the one purchased from Hebei, the similarities of the HPLC fingerprints of liposoluble constituents in Danshen from other regions all exceeded 0.9.

Table 15.46 The similarities of the test results of the liposoluble constituent in wild Danshen

No.	1	2	3	4	5	6	7	8	10
Similarity	0.579	0.756	0.972	0.961	0.97	0.976	0.977	0.964	0.968

Table 15.47 The similarities of the test results of the liposoluble constituent in cultivated Danshen

No.	14	15	16	17	18	19	20	21	23	24
Similarity	0.958	0.983	0.936	0.989	0.747	0.954	0.958	0.985	0.847	0.923
No.	25	26	29	30	31	32	37	38	39	40
Similarity	0.969	0.586	0.959	0.958	0.981	0.93	0.935	0.93	0.94	

15.1.2.2 Identification of the Diterpenoid Quinone Components in Danshen

Dean Guo and Min Yang

The object of the study was to establish the HPLC-ESI-MS^n qualitative analysis method so that the comprehensive chemical constituents in Danshen could be evaluated.

1. Materials and Methods
   (a) Drugs and Reagents: Tanshinone II_A, cryptotanshinone, tanshinone I,15,16-cryptotanshinone I, tanshinone II_B, purple danshensu, methyl tanshinonate, tanshinonediol B, and tanshinonediol C were isolated from Danshen extracts by the authors; 1,2-didehydro-miltionone and miltionone were isolated from Danshen extract. The purities of these compounds all exceeded 95 %. The Danshen sample was collected from the Danshen Cultivation Base located in Sichuan's Zhongjiang. HPLC grade acetonitrile was purchased from Fisher Company. Water was ultrapure. Chloroform and methanol used for extraction and methanoic acid used for HPLC mobile phase were analytical reagents and purchased from Beijing Chemical Factory.
   (b) HPLC Analysis: Agilent 1100 series HPLC (equipped with Diode Array Detector), chromatographic column Zorbax Extend C_{18} column (4.6 × 250 mm, 5 μm). Mobile phase: acetonitrile (A) and 0.03 % (v/v) methanoic acid (B); gradient elution: 0–5 min, acetonitrile linearly increased from 27 % (V/V) to 45 %, and to 70 % at 20 min, and to 100 % at the last 10 min. Flow rate was 1.0 ml/min and column temperature was room temperature. The DAD ultraviolet spectrum recording range was 190–400 nm, and the detection wavelength was 270 nm.
   (c) MS Analysis: LCQ Advantage Ion-Trap Mass Spectrometer (ThermoFinnigan, San Jose, CA) connected with Agilent 1100 HPLC through electrospray ionization (ESI). The LC effluent entered ionization source at the split ratio of 2:1. Collision gas was super high pure helium (He); atomized gas was high pure nitrogen gas (N_2). MS conditions: Ion spray voltage 4.5 kV; sheath gas (N_2) 50 u; auxiliary gas (N_2) 10 u; capillary tube temperature 340 °C; capillary tube voltage 30 V. Full scanning mass spectrographic range *m/z* 100–500. When applying data-dependent software during HPLC/MS procedure, the ion of the highest abundance at every scanning would be picked as the parent ion for mass chromatographic analysis of the next class. Impact energy was 45 %.
   (d) HPLC/MS Analysis: Weigh 1 g of Danshen powder. Reflux with 10 ml of chloroform–methanol (3:7) mixture solution on a water bath for 60 min. Filter, evaporate the filtrate to dryness, and then dissolve in 5 ml of methanol. Filter through a 0.22 μm microporous

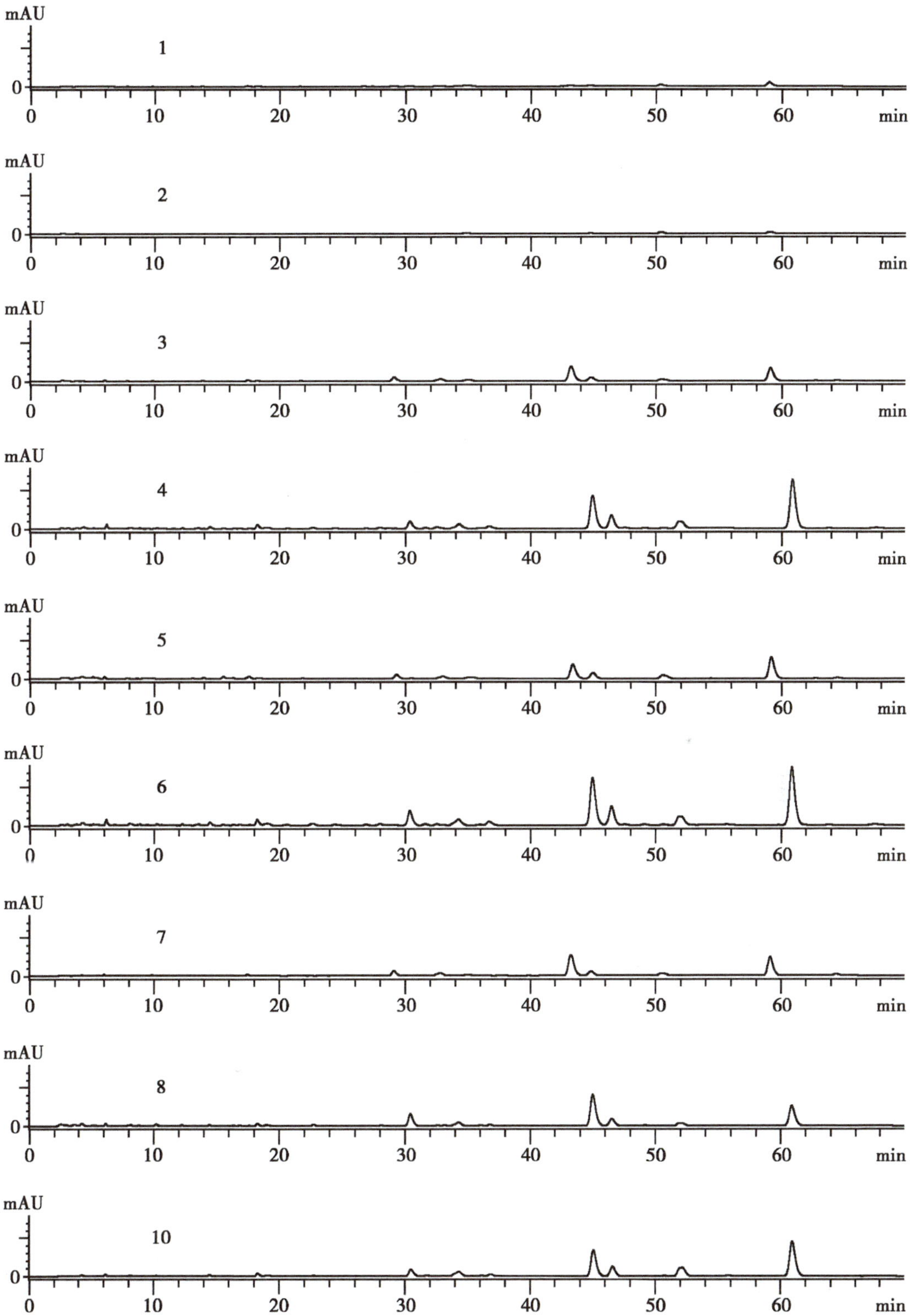

Fig. 15.48 The HPLC fingerprints of the liposoluble constituents in wild Danshen from different regions

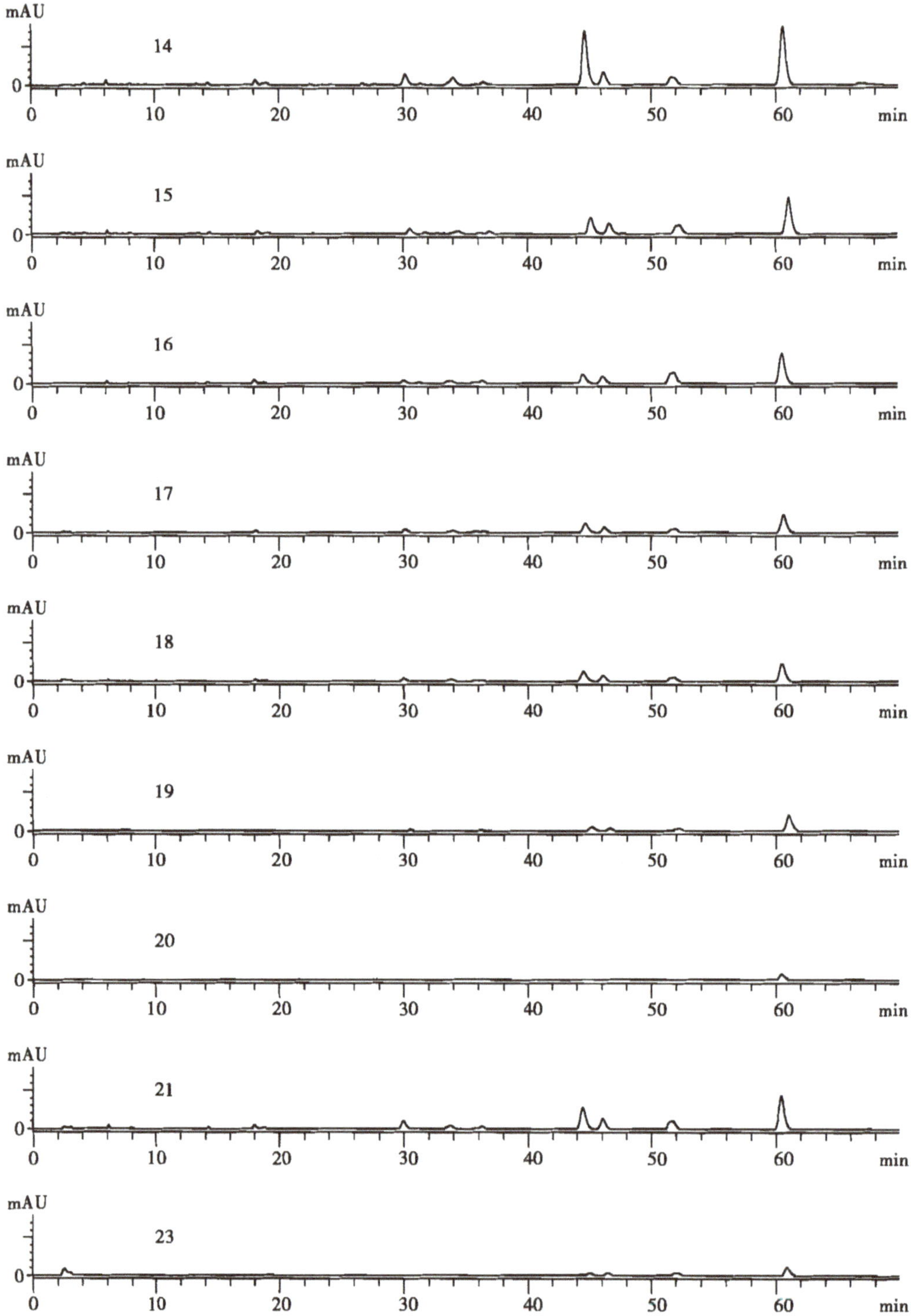

Fig. 15.49 The HPLC fingerprints of the liposoluble constituents in cultivated Danshen from different regions

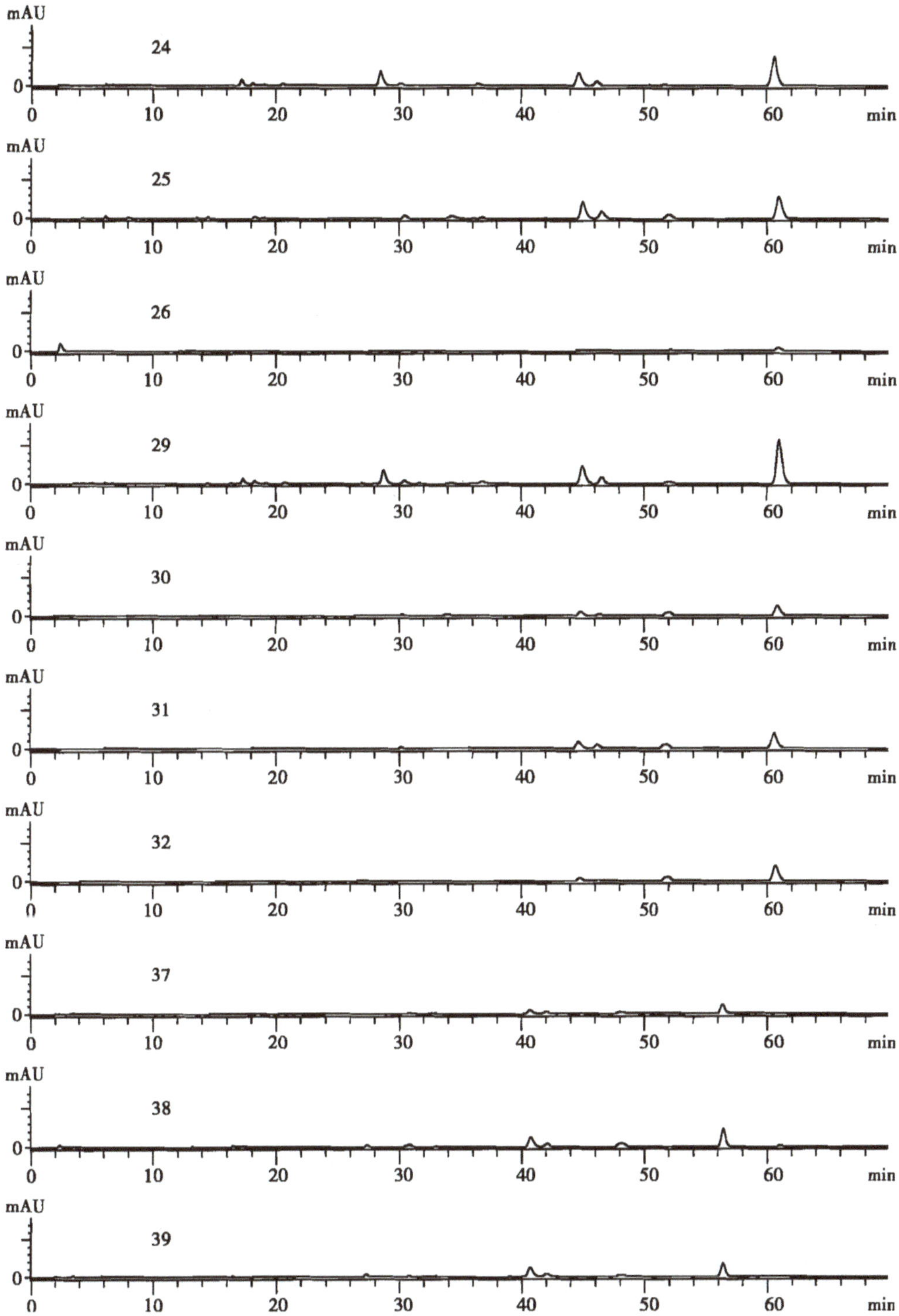

Fig. 15.49 (continued)

membrane before use. The sample size was 5 μl.

2. Results and Discussion
   (a) Optimization of HPLC Conditions: In order to achieve a better separating effect, different solid phases, mobile phases, column temperatures, wavelengths and flow rates were tested [15]. When analyzing tanshinone compounds, we found that Zorbax Extend C_{18} column was better than BDS hypersil C_{18}, YMC Pack ODS-A C_{18}, Luna C_{18}, and other columns. An acetonitrile/water system was significantly better than a methanol/water system. Addition of a little acid in the mobile phase could help ionization under positive ion mode. Flow rate of mobile phase was 1.0 ml/min, column temperature was room temperature, and detection wavelength is 270 nm.
   (b) The Analysis of Tanshinone Compounds' ESI-MS Fragmentation Pattern: Four modes were compared: $APCI^+$, $APCI^-$, ESI^+, and ESI^-, and the results showed that ESI positive ion mode gave the best response in analyzing tanshinone compounds.

   We analyzed 11 tanshinone compounds individually by multistage mass spectrometry. $[M + H]^+$ was analyzed by CID MS/MS, and the predominant ions were picked as the parent ion of next stage. Perform MS^n (n = 3–6) spallation stage by stage. The range of impact energy was 25–50 %. Based on the chemical structures and spallation routes (see Figs. 15.50 and 15.51), these compounds could be divided into 6 groups for discussion (Table 15.48).
      (i) Group I (Tanshinone I): The ESI-MS of tanshinone I showed an obvious $[M + H]^+$ peak *m/z* 277. After CID, the MS/MS spectrum showed a predominant ion peak *m/z* 249, which was generated by $[M + H]^+$ ion losing one CO (28 Da) (Fig. 15.50a). The other weaker ions, *m/z* 259 and 231, were generated by dehydration (18 Da) and dehydration plus the loss of CO. In the MS^3 spectrum, *m/z* 249, 221, 193 were abundant, with *m/z* 221 as the base peak (Fig. 15.50b). The existence of these ions indicated that there was a continuous CO loss process under the action of CID. Meanwhile, some peaks related to CH_3 loss (15 Da) could also be observed: *m/z* 234 (249-15) and *m/z* 178 (193-15). The continuous loss of CO from $[M + H]^+$ ions and the similar abundance of *m/z* 249 and 193 to that of the base peak in MS^3 spectrum are the features of the compounds with a structure similar to tanshinone I.
      (ii) Group II (15,16-dihydrotanshione I): The 15th and 16th positions of 15,16-dihydrotanshinone I are saturated, which results in a significant difference in fragmentation mode from that of tanshinone I. In MS/MS, the $[M + H]^+$ ion of *m/z* 279 lost one H_2O instead of CO, apparently caused by the saturation at the 15th and 16th positions. The obtained ion *m/z* 261 lost one CO in MS^3 and another CO in MS^4, generating ions *m/z* 233 and 205.
      (iii) Group III (Tanshinone II_A, Tanshinone II_B, Przewaquinone A, Methyl tanshinonate): The compounds in this group have a backbone similar to that of tanshinone II_A. With an impact energy of 41 %, the $[M + H]^+$ ion of tanshinone II_A, *m/z* 295, lost one H_2O (18 Da), generating ion *m/z* 277. In addition, except for $[M + H–H_2O–CO]^+$ ion *m/z* 249, there was a very weak ion *m/z* 253 possibly generated by losing one propene (CH_2=$CHCH_3$, 42 Da). The dehydrated ion *m/z* 277 lost one CO in MS^3 and another in MS^4, generating ions *m/z* 249 and *m/z* 221, respectively. In MS^4, there were demethylation signals with medium intensity, *m/z* 234 (249-15) and 206(221-15).

      Both tanshinone II_B and przewaquinone A are tanshinone II_A with a single hydroxyl group substitution, whose

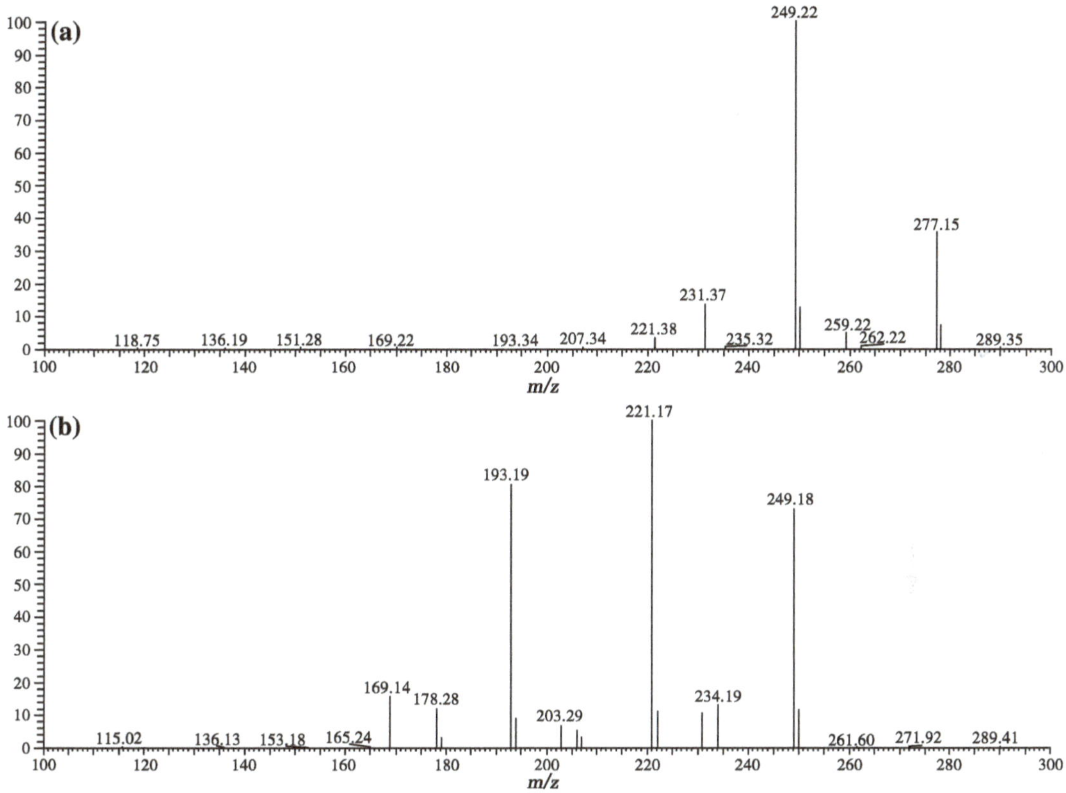

Fig. 15.50 The ESI-MS spectrum of tanshinone I. **a** The 2nd stage mass spectrum $[M + H]^+$ *m/z* 277; **b** The 3rd mass spectrum *m/z* 249(277-249)

$[M + H]^+$ ion is *m/z* 311. Both base peaks in MS/MS and MS^3 were dehydrated peaks *m/z* 293 ($[M + H\text{-}H_2O]^+$) and 275 ($[M + H\text{-}H_2O]^+$). There is a big difference in their third-stage mass spectra. The signal of propene loss (42 Da) by tanshinone II_B, *m/z* 251, was much stronger than that of przewaquinone A, while its abundance was almost the same as that of the base peak. When compared with tanshinone II_A, the methyl group at the fourth position of methyl tanshinonate is transformed into a carboxyl group, resulting in a different fragmentation mode. Its $[M + H]^+$ ion *m/z* 339 first lost one methyl formate ($HCOOCH_3$, 60 Da). The resulting ion *m/z* 279 lost one H_2O and then two CO, generating *m/z* 261, 233, and 205, respectively, just like tanshinone II_A.

(iv) Group IV (Tanshindiol B and Tanshindiol C): The difference of these two compounds is the configuration of their hydroxyl groups at the third position. The $[M + H]^+$ ion was *m/z* 313. Their MS^n spectra were almost the same. The $[M + H]^+$ ion *m/z* 313 lost one water (313 → 295) first, followed immediately by losing one CO (295 → 267), instead of losing two water molecules as predicted. The process might be like this: the dehydration occurred at the fourth position hydroxyl group, resulting in the formation of an enolate with the hydroxyl group at the third position. The third position carbonyl group was then rearranged, and the net result was the loss of one CO instead of H_2O. The fragmentation of the obtained ion *m/z* 267 was very similar

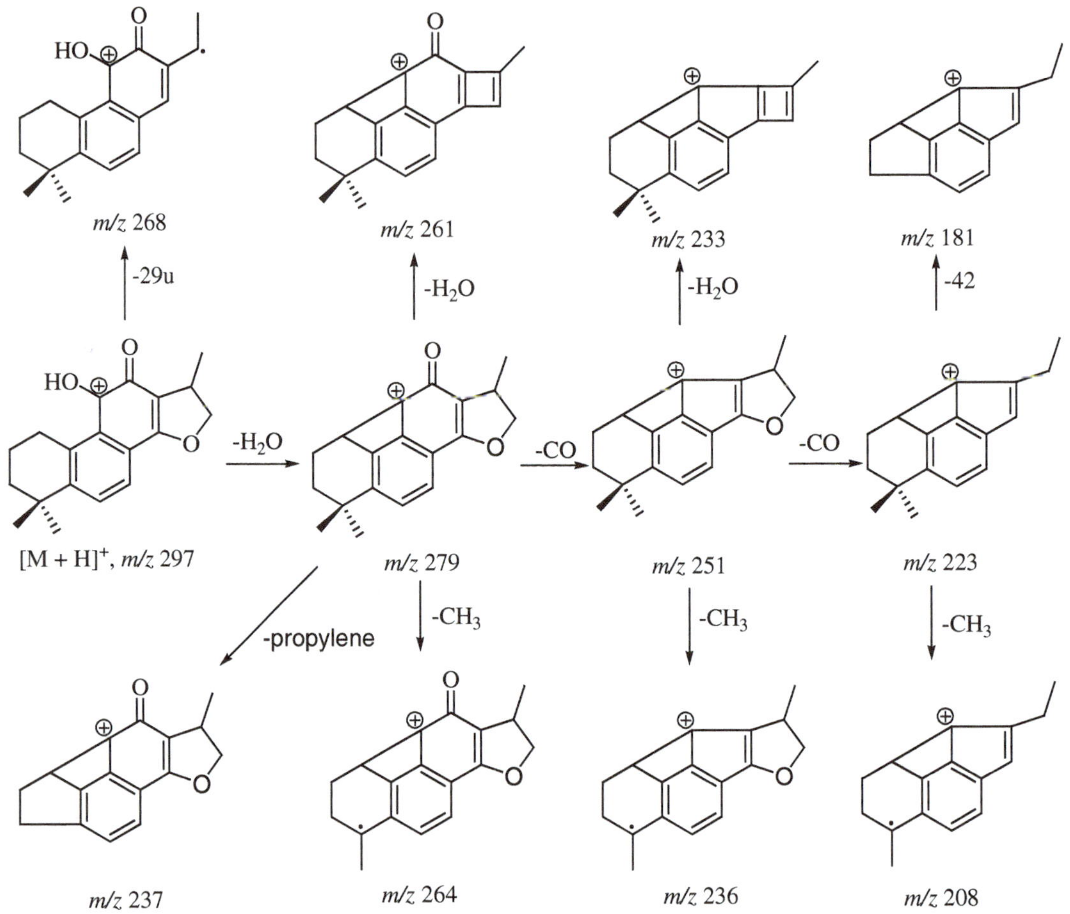

Fig. 15.51 The possible fragmentation pathways of cryptotanshinone

to that of tanshinone II_A, generating *m/z* 249, 221, 193, respectively. ESI-MS could not distinguish between these two compounds; however, their retention times were different, so they could be distinguished in a chromatogram.

(v) Group V (Cryptotanshinone): Compared with tanshinone II_A, the 15th and 16th positions of cryptotanshinone are saturated bonds; therefore its $[M + H]^+$ ion, *m/z* 297, was 2 Da heavier than that of tanshinone II_A. In MS^n ($n = 2$–5), the main ions were *m/z* 279, 251, 223, and 208. They had the same fragmentation pathways as tanshinone II_A (Fig. 15.51). However, in MS^n spectra, every signal of losing 42 Da ($CH_2{=}CHCH_3$), *m/z* 251, 237, 209, was relatively strong. These must be caused by its different conjugated system from that of tanshinone II_A.

(vi) Group VI (Miltirone and 1,2-didehyd romihirone): The 14th position of the compounds in this group does not have a hydroxyl group, and the isopropyl group at the 13th position does not form a furan ring with its neighbors. The difference between them is that the first and second positions of 1,2-didehydromiltirone are double bonds, which also results in the difference in the fragmentation pattern. The MS/MS spectrum of miltirone showed the main fragment ion was

Table 15.48 The tanshinone compounds identified in Danshen extracts by HPLC-ESI-MS^n

No.	Name	R1	R2	R3	R4	Double bond	Mass number
1	15,16-Dihydrotanshindiol B/C	β-OH	α-CH_3	OH	CH_3	–	314
2	Tanshindiol C	β-OH	α-CH_3	β-OH	CH_3	Δ15,16	312
3	Danshenxinkun A	–	–	–	–	–	296
4	Tanshinone VI	–	–	–	–	–	296
5	17-Hydroxycryptotanshinone	CH_3	CH_3	H	β-CH_2OH	–	312
6	3α-Hydroxymethylenetanshinquinone	=CH_2	–	α-OH	CH_3	Δ15,16	294
7	Tanshinol B	OH	CH_3	H	CH_3	Δ15,16	296
8	Methylenetanshinquinone	=CH_2	–	H	CH_3	Δ15,16	278
9	Trijuganone A	–	–	–	–	–	294
10	1,2-Didehydrotanshinone V	CH_2OH	OH	–	–	Δ1,2	312
11	Tanshinone II_B	β-CH_2OH	α-CH_3	H	CH_3	Δ15,16	310
12	1-Ketoisocryptotanshinone	–	–	–	–	–	310
13	Trijuganone C	β-$COOCH_3$	α-CH_3	H	CH_3	–	340
15	1,2-Didehydrotanshinone II_A	CH_3	CH_3	H	CH_3	Δ1,2,Δ15,16	292
16	15,16-Dihydrotanshinone I	–	–	–	–	Δ1,2	278
17	Sibiriquinone A	–	–	–	–	–	280
18	Trijuganone B	–	–	–	–	–	280
19	Methyl tanshinonate	β-$COOCH_3$	α-CH_3	H	CH_3	Δ15,16	338
20	1,2-Didehydrocryptotanshi none	CH_3	CH_3	H	CH_3	Δ1,2	294
21	Cryptotanshinone	CH_3	CH_3	H	β-CH_3	–	296
22	Tanshinone I	–	–	–	–	Δ1,2,Δ15,16	276
23	1,2-Dihydrotanshinone I	–	–	–	–	Δ15,16	278
24	1,2-Didehvdromiltirone	CH_3	H	–	–	Δ1,2	280
25	Tanshinone II_A	CH_3	CH_3	H	CH_3	Δ15,16	294
26	2,3-Didehydrotanshinone II_A	CH_3	CH_3	H	CH_3	Δ2,3,Δ15,16	292
27	Miltirone	CH_3	H	–	–	–	282

dehydrated peak *m/z* 265 (Fig. 15.52a). The $[M + H–H_2O]^+$ ion in MS^3 spectrum mainly lost one propene (42 Da), generating *m/z* 223. Compared to these, the predicted peak from losing CO (*m/z* 237) was much less abundant. In other words, in the fragmentation process of these compounds, the free isopropyl group at the 13th position was much easier to lose than was the carbonyl group at the 12th position. In the MS/MS spectrum, *m/z* 241 which lost one propene also had a peak with medium intensity, and *m/z* 223 lost one CO to generate a peak of *m/z* 195 (MS^4). Meanwhile, peaks related to the loss of a methyl group were also observed (*m/z* 283 → 268, 265 → 250, 223 → 208, 195 → 180).

In the MS/MS spectrum of 1,2-didehydromiltirone, the base peak was *m/z* 221, which was generated by dehydration and loss of a propene. The CO-losing ion *m/z* 253 was also very strong, and its abundance was almost the same as that of the base peak, while the dehydrated peak *m/z* 263 had a weaker intensity since it could readily lose propene (Fig. 15.52b). Therefore, the transition to a double bond at the first and the second positions, which

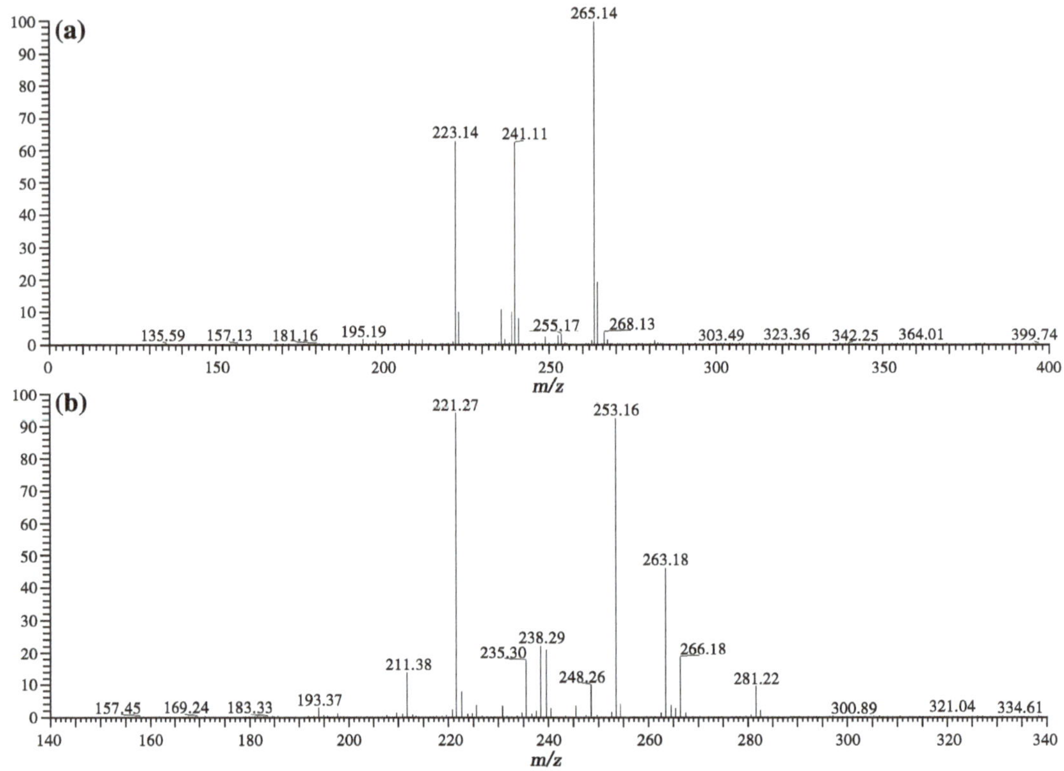

Fig. 15.52 The ESI-MS spectra of the compounds in group VI. **a** The second stage mass spectrum of miltirone, $[M + H]^+$ at *m/z* 283; **b** The second stage mass spectrum of 1,2-didehydro-miltirone, $[M + H]^+$ at *m/z* 281

extends the conjugated system, has a great influence on fragmentation and enhances the trend of losing propene and CO. Since the fragmentation of tanshinone I $[M + H]^+$ ion starts from the direct loss of one CO while the fragmentation of tanshinone II_A $[M + H]^+$ ion starts from the loss of one H_2O, it is obvious that the extension of the whole conjugated system in ring A has a significant effect on the fragmentation pathway.

(c) HPLC-ESI-MS^n Analysis of Danshen Extracts: Fig. 15.53 shows the HPLC/UV and total ion-current (TIC) of the chloroform-methanol (3:7) extract of Danshen. A total of 27 compounds were identified in Danshen extracts; among them, 9 compounds were identified by comparing with the reference substances and 5 were new compounds. The structures of unknown compounds were mainly based on the MS^n spectra and deduced from the fragmentation patterns summarized from the above studies in Table 15.49.

Tanshinone I (t_R = 24.27 min) was identified by comparing with the reference substances. In its TIC spectrum, three peaks generated $[M + H]^+$ ion of *m/z* 279. Among them, the 18.95 min peak was identified as 15,16-dihydrotanshinone I by comparing with the reference substances. The MS^n (n = 2–5) spectra of the 25.92 min peak were almost identical to those of methyl tanshinonate after the loss of methyl formate (60 Da), so the compound was identified primarily as 1,2-dihydro-tanshinone I. Similarly, the 12.08 min peak was primarily identified as methylenetanshinquinone.

Three $[M + H]^+$ ions with *m/z* 281 peaks were identified. The 27.85 min peak was

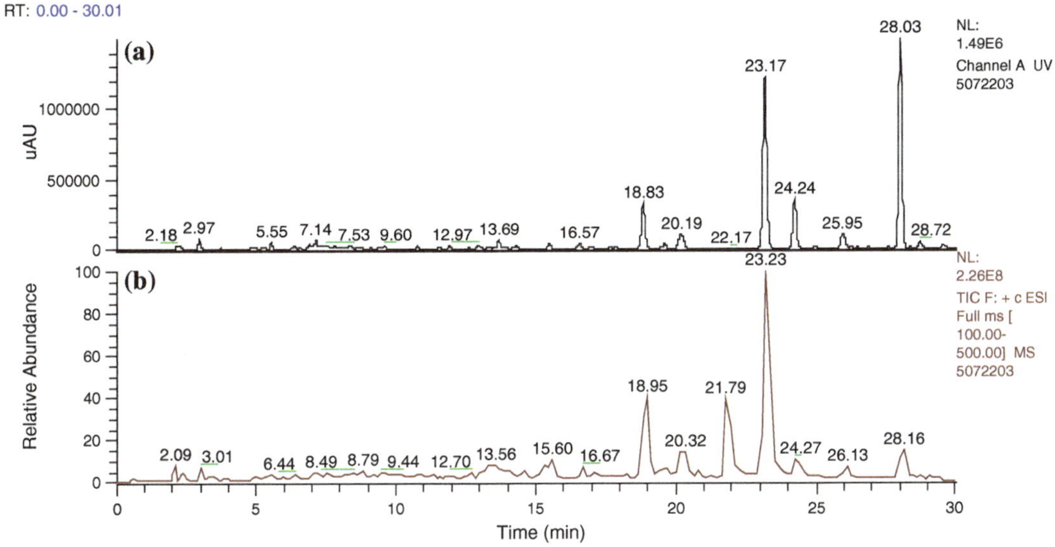

Fig. 15.53 The HPLC/DAD/ESI-MS^n spectra of the chloroform-methanol extract of Danshen. **a** HPLC/UV spectrum, detection wavelength was 270 nm; **b** ESI-MStotal ion current

identified as 1,2-didehydromiltirone by comparing to the reference substances. The mass spectrum of the 19.61 min peak was almost identical to that of 1,2-didehydromiltirone, therefore it was identified as its isomer, para-quinone compound sibiriquinone A, which was isolated from *Veronicastrum sibiricum* [27]. Another $[M + H]^+$ ion with the *m/z* 281 had a retention time of 20.19 min, and its MS^n spectra were similar to those of 15,16-dihydrotanshinone I, generating major ion peaks *m/z* 263, 235, 207, and 192. The corresponding peaks were all 2 Da heavier, indicating that there was hydrogenation at some positions. Therefore, the compound was tentatively identified as 1,2,15,16-tetrahydrotanshinone I, a compound also known as trijuganone B.

Miltirone (t_R 28.73 min) was identified by comparing to reference substances. In its TIC spectrum, there were three peaks whose $[M + H]^+$ ions were *m/z* 293. Its MS/MS first lost one CO, generating *m/z* 265, which was somewhat similar to tanshinone I, only 16 Da heavier. The MS^3 spectrum of the ion showed a trend of continuously losing CO; besides the $[M + H\text{-}2CO]^+$ *m/z* 237, ions *m/z* 265 ($[M + H–CO]^+$) and *m/z* 209 ($[M + H–3CO]^+$) were relatively strong as well, indicating that its structure was similar to that of tanshinone I, probably a monohydroxy-substituted tanshinone I. However, its substitution location could not be determined, and there are no compounds isolated from Danshen with such a structure.

The MS^n spectra of the other two peaks were very similar to those of tanshinone II$_A$, with major ion fragments of *m/z* 275 ($[M + H\text{-}18]^+$), 247 ($[M + H\text{-}18\text{-}28]^+$), 219 ($[M + H\text{-}18\text{-}56]^+$), and 204 ($[M + H\text{-}18\text{-}56\text{-}15]^+$), All were 2 Da lighter than tanshinone II$_A$, indicating they were probably didehydrotanshinone compounds. The difference between miltirone and 1,2-didehydromiltirone was that the retention times of the compounds with double dehydrogenations at 1,2 positions were a little shorter, therefore the 17.69 min peak was tentatively identified as 1,2-didehydrotanshinone II$_A$, and the 28.60 min peak was named as 2,3-didehydrotanshinone II$_A$, speculating the dehydrogenation positions

Table 15.49 The HPLC/ESI-MS^n characteristics of tanshinone compound in Danshen (*m/z*, relative abundance)

Peak no.	Retention time (min)	Identification of constituents	UV λ_{max} (nm)	$[M + H]^+$ *m/z*	HPLC/ESI-MS^n *m/z* (% base peak)
1†	5.8	15,16-Dihydrotanshindiol B/15,16-Dihydrotanshindiol C	264	315	MS^2 [315]:297(100), 269(10)
					MS^3 [315 → 297]:297(4), 269(100)251(1), 223(8)
					MS^4 [315 → 297 → 269]:251(100), 2417(4), 223(62), 209(6)
					MS^5 [315 → 297 → 269 → 251]:251(12), 236(14), 233(48), 223(10), 209 (50), 205(20), 195(8), 181(6)
					MS^6 [315 → 297 → 269 → 251 → 223]:205(10), 195(100), 169(12)
2*	7.1	Tanshindiol C	–	313	MS^2 [313]:295(100), 285(4), 267(16)
					MS^3 [313 → 295]:277(2), 267(100)
					MS^4 [313 → 295 → 267]:252(1), 249(100), 221(18)
					MS^5 [313 → 295 → 267 → 249]:249(8), 231(12), 221(100), 193(6)
					MS^6 [313 → 295 → 267 → 249 → 221]:203(12), 193(100)
3	8.3	Danshenxinkun A	264	297	MS^2 [297]:279(100), 269(20), 261(10)255(10), 251(50), 233(4), 223(20), 209(4)
					MS^3 [297 → 279]:261(100), 252(12), 251(30), 237(10), 233(32), 209(12), 182(4)
					MS^4 [297 → 279 → 261]:261(22), 233(100), 219(3), 205(4), 190(6)
					MS^5 [297 → 279 → 261 → 233]:218(6), 205(100), 191(10)
4	9.5	Tanshinoe VI	258	297	MS^2 [297]:279(100), 269(6), 251(42), 223(2)
					MS^3 [297 → 279]:278(3), 261(100), 251(11), 237(4), 233(6), 209(4)
					MS^4 [297 → 279 → 261]:233(100), 215(9), 205(4)
					MS^5 [297 → 279 → 261 → 233]:233(36), 232(22), 205(100), 190(30)
5	9.7	17-Hydroxycryptotanshinone	264	313	MS^2 [313]:295(100), 277(24), 269(44), 267(32), 255(18), 253(10), 249(4)
					MS^3 [313 → 295]:295(4), 280(30), 277(100), 267(45), 253(22), 249(20)
					MS^4 [313 → 295 → 277]:277(100), 262(2), 249945), 235(8), 221(4), 195 (2)

(continued)

Table 15.49 (continued)

Peak no.	Retention time (min)	Identification of constituents	UV λ_{max} (nm)	$[M + H]^+$ m/z	HPLC/ESI-MS^n m/z (% base peak)
6	11.3	3α-Hydroxymethylenetans hinquinone	280	295	MS^2 [295]:277(100), 267(50), 264(12), 251(20), 249(60), 221(3)
					MS^3 [295 → 277]:277(22), 259(62), 249(100), 231(10), 221(4)
					MS^4 [295 → 277 → 249]:249(96), 221(100), 193(60), 178(45)
					MS^2 [297]:279(100), 269(20), 261(55)
					MS^3 [297 → 279]:261(100)
7	11.88	Tanshin01 B	270	297	MS^4 [297 → 279 → 261]:233(100), 205(3)
					MS^5 [297 → 279 → 261 → 233]:218(4), 205(100), 190(20), 179(2)
					MS^6 [297 → 279 → 261 → 233 → 205]:205(97), 190(100)
8	12.08	Methylenetanshinquinone	–	279	MS^2 [279]:261(100).251(3), 233(12), 205(10)
					MS^3 [279 → 261]:233(100), 205(6)
					MS^4 [279 → 261 → 233]:233(5), 218(32), 205(100), 190(28)
					MS^2 [295]:277(100), 267(4), 249(10), 225(4)
					MS^3 [295 → 277]:249(100), 22 l(14)
9	12.70	Trijuganone A	–	295	MS^4 [295 → 277 → 249]:249(13), 234(3), 231(2), 221(100), 206(6), 193 (10), 178(2)
					MS^5 [295 → 277 → 249 → 221]:221(100), 206(70), 203(12), 193(40), 179(10)
					MS^2 [313]:295(100), 267(65)
					MS^3 [313 → 295]:280(15), 277(100), 267(40), 262(5), 253(24), 249(50), 235(10), 225(30)
10†	12.96	1,2 Didchydrotanshinone V	242	313	MS^4 [313 → 295 → 277]:277(16), 262(50), 259(4), 249(100), 235(30), 231(10), 221(4)
					MS^5 [313 → 295 → 277 → 249]:249(12), 234(75), 221(100), 207(30), 206(15), 193(20)
					MS^6 [313 → 295 → 277 → 249 → 221]:206

(continued)

Table 15.49 (continued)

Peak no.	Retention time (min)	Identification of constituents	UV λ_{max} (nm)	$[M + H]^+$ *m/z*	HPLC/ESI-MS^n *m/z* (% base peak)
11*	13.68	Tanshinone IIB	270	311	MS^2 [311]:293(100), 283(4), 275(6), 267(10), 251(3)
					MS^3 [311 → 293]:278(15), 275(100), 265(18), 251(84), 247(5), 205(1)
					MS^4 [311 → 293 → 275]:275(100), 247(30), 229(10), 219(4), 204(2)
					MS^2 [311]:293(16), 283(100), 275(14), 267(3)
					MS^3 [311 → 283]:268(4), 265(100), 254(7), 237(70), 223(4)
12	14.07	l-Ketoisocryptotanshinone	248	311	MS^4 [311 → 283 → 265]:265(10), 250(10), 247(30), 237(100), 223(50), 219(12), 195(5)
					MS^5 [311 → 283 → 265 → 237]:223(12), 219(5), 209(100),195(10), 167 (9)
					MS^2 [341]:313(4), 281(100), 263(46), 235(12), 220(1)
					MS^3 [341 → 281]:263(100), 235(32), 211(1)
13	15.60	Trijuganone C	264	341	MS^4 [341 → 281 → 263]:248(3), 245(2), 235(100), 221 8), 207(3)
					MS^5 [341 → 281 → 263 → 235]:235(15), 220(73), 217(24), 207(100), 193(10)
					MS^6 [341 → 281 → 263 → 235 → 207]:207(34), 192(100), 179(10), 164 (6)
					MS^2 [293]:293(10), 275(8), 266(14), 265(100), 247(6)
14†	16.58	Monohydroxytanshinone I	–	293	MS^3 [293-265]:265(55), 250(20), 247(12), 237(100), 219(8), 209(30), 194 (4), 181(3)
					MS^4 [293 → 265 → 237]:237(8), 222(25), 219(12), 209(100), 191(10), 181(30)
					MS^5 [293 → 265 → 237 → 209]:209(60), 191(90), 181(100)
15	17.69	1,2-Didehydrotanshinone IIA	–	293	MS^2 [293]:293(2), 278(8), 276(18), 275(100), 265(10), 251(8), 247(40), 229(2).205(1)
					MS^3 [293 → 275]:275(10), 260(4), 247(100), 229(4), 205(1)
					MS^4 [293 → 275 → 247]:247(5), 232(14), 219(100), 204(44), 193(3)
					MS^5 [293 → 275 → 247 → 219]:219(100), 204(92)

(continued)

Table 15.49 (continued)

Peak no.	Retention time (min)	Identification of constituents	UV λ_{max} (nm)	$[M + H]^+$ m/z	HPLC/ESI-MS^n m/z (% base peak)
16*	18.95	15,16-Djhydrotanshinone I	244	279	MS^2 [279]:26 l(100), 251(2), 233(8), 209(1)
					MS^3 [279 → 261]:233(100), 215(3), 205(8), 169(1)
					MS^4 [279 → 261 → 233]:233(6), 218(8), 205(100),l90(4), 179(3)
					MS^5 [279 → 261 → 233-205]:205(80), 190(100)
					MS^2 [281]:281(1), 266(22), 263(58), 253(98), 248(12), 239(20), 238(20), 221(100), 211(22)
17	19.61	Sibiriquinone A	–	281	MS^3 [281 → 221]:221(2), 206(14), 193(100)
					MS^3 [281 → 253]:238(48), 235(36), 225(45), 224(28), 21 l(100), 197(8), 183(4)
					MS^4 [281 → 221 → 193]:193(42), 178(100), 165(4)
					MS^4 [281 → 253 → 2 ll]:196(100), 195(20), 183(30), 169(70)
					MS^2 [281]:263(100), 235(30), 22l(2), 193(1)
					MS^3 [281 → 263]:235(100).221(9), 207(4)
18	20.19	Trijuganone B	276	281	MS^4 [281 → 263 → 235]:235(12).220(80), 217(26).207(100), 202(8), 193 (8), 192(15)
					MS^5 [281 → 263 → 235 → 207]:207(28), 192(100), 179(25), 165(4)
19	20.47	Methyl tanshinonate	–	339	MS^2 [391:311(4), 279(100), 26 l(1 5)
					MS^3 [339 → 279]:261(100), 233(2)
					MS^4 [339 → 279 → 261]:233(100), 205(2)
					MS^5 [339 → 279 → 261 → 233]:233(3), 218(8), 205(100), 190(20),l79(2)
					MS^6 [339 → 279 → 261—233 → 205]:205(70),l90(100)
					MS^2 [295]:280(12), 278(13), 277(100), 267(13), 249(50), 235(8), 225(30)

(continued)

Table 15.49 (continued)

Peak no.	Retention time (min)	Identification of constituents	UV λ_{max} (nm)	$[M + H]^+$ m/z	HPLC/ESI-MS^n m/z (% base peak)
20†	20.80	1,2-Didehydrocryptotanshinone	–	295	MS^3 [295 → 277]:277(18), 262(57), 259(5), 249(100), 235(40), 231(5), 207(2)
					MS^4 [295 → 277 → 249]:249(13), 234(80), 23 l(10), 221(100), 220(13), 217(8), 207(21), 206(22), 193(3)
					MS^5 [295 → 277 → 249 → 221]:221(16), 206(100)
					MS^2 [297]:282(11), 280(12), 279(100), 269(4), 268(20), 255(5), 254 (28).251(75).237(10)
21*	23.23	Cryptotanshinone	264	297	MS^3 [297 → 279]:279(40), 264(38), 261(30), 251(100), 237(72), 233(10), 223(15), 209(10), 185(3)
					MS^4 [297 → 279 → 251]:251(20), 250(6), 236(46), 235(12), 233(20), 223 (100), 209(46),197(34), 181(1 5)
					MS^5 [297 → 279 → 251—223:222(10), 208(100), 105(88), 194(22), 1 93 (30),181(74), 167(32)
					MS^6 [297 → 279 → 251 → 223 → 208]:193(100)
					MS^2 [277]:277(4), 259(3), 249(100), 23 l(12), 22 l(4), 207(1)
22	24.27	Tanshinone I	246	277	MS^3 [277 → 249]:249(82), 234(10), 231(8), 221(100),193(80).178(10), 169(14)
					MS^4 [277 → ,249 → 221]:221(10), 206(20), 203(4), 193(100), 178(8)
					MS^5 [277 → 249 → 221—193]:193(5), 178(100)
					MS^2 [279]:261(100), 233(9)
23	25.92	1,2-Dihydrotanshinone I	290	279	MS^3 [279–261]:261(4), 233(100), 205(2), 169(1)
					MS^4 [279 → 7261—233]:233(14), 218(18), 205(100), 190(24), 179(1)
					MS^5 [279 → 261 → 233 → 205]:205(100), 190(60)
					MS^2 [281]:281(8), 266(35), 263(54), 248(10), 239(22), 238(18), 235(17), 221(100), 211(30),193(4)

(continued)

Table 15.49 (continued)

Peak no.	Retention time (min)	Identification of constituents	UV λ_{max} (nm)	$[M + H]^+$ m/z	HPLC/ESI-MS^n m/z (% base peak)
24	27.85	1,2 Didehydromihirone	276	281	MS^3 [281 → 221]:221(2), 206(7), 193(100)
					MS^3 [281 → 253]:238(38), 235(24), 225(30), 211(100), 183(5)
					MS^4 [281 → 221 → 193]:193(54), 178(100)
25	28.16	Tanshinone IIA	268	295	MS^2 [295]:277(100), 253(4), 249(14), 235(4), 207(1)
					MS^3 [295 → 277]:277(12).262(28), 259(3), 249(100), 235(3), 231(8), 221 (6), 206(2),179(1)
					MS^4 [295 → 277 → 249]:249(10), 234(66), 231(4), 221(100), 206(50), 193(7), 179(2) MS^5
					[295 → 277 → 249 → 221]:221(22), 206(100), 193(20)
					MS^6 [295 → 277 → 249 → 221 → 206]:206(68), 191(100)
26	28.60	2,3-DidehydrotanshinoneIIA	258	293	MS^2 [293]:293(3), 278(10), 276(15), 275(100), 265(8), 251(3), 247(36), 229(2)
					MS^3 [293 → 275]:275(8), 247(100), 229(2), 204(2)
					MS^4 [293 → 275 → 247]:232(12), 219(100), 204(20), 193(22)
					MS^5 [293 → 275 → 247 → 219]:219(26), 204(100)
					MS^2 [283]:265(100), 241(50), 237(10), 223(60)
					MS^3 [283 → 265]:265(20), 250(26), 247(14), 237(75), 236(20), 223(100), 209(9), 195(12)
27	28.73	Mihirone	258	283	MS^4 [283 → 265 → 223]:223(2), 208(63), 207(12), 205(32), 195(100), 181(20), 167(6)
					MS^5 [283 → 265 → 223 → 195]:180(100)

⁻ The signal was too weak, therefore the maximum UV absorption wavelength was not detected

* The structure was confirmed by comparing with the reference substances

† First reported

were at 2,3. This structure has never been reported.

Tanshinone II_A (t_R 28.16 min) was identified by comparing with the reference substances. The 20.80 min peak also generated $[M + H]^+$ ion *m/z* 295. Its MS^n (n = 2–5) spectra showed main ion peaks *m/z* 277, 249, 221, and 206, losing H_2O, CO, CO, and CH_3 one after another, which was similar to tanshinone II_A. However, the abundances of ions *m/z* 235 (277-42) and 207 (249-42) were much larger than those of tanshinone II_A, indicating that it probably has a structure similar to crypto-tanshinone. Also, in its MS/MS spectrum, ions *m/z* 267 ($[M + H–CO]^+$) and 225 ($[M + H–CO-42]^+$) were relatively strong, suggesting that a double bond might be formed at its 1, 2 positions. Therefore, this compound was tentatively identified as 1,2-didehydro-cryptotanshinone; the structure has never been reported.

The $[M + H]^+$ ion of the peak at 11.53 min lost one H_2O in the MS/MS spectrum, producing *m/z* 277 ion, which lost one CO, generating *m/z* 249 (MS^3). Ion *m/z* 249 MS^4 showed a fragmentation pattern similar to tanshinone I, consecutively losing CO to generate *m/z* 221 and 193. Therefore, the conjugated system of $[M + H]^+$ ion after losing one water was similar to that of tanshinone I, but not exactly the same; hence this compound was tentatively identified as 3α-hydroxymethylenetanshinquinone.

The peak at 12.70 min also showed a $[M + H]^+$ ion of *m/z* 295. Its MS^n (n = 2–4) were similar to those of 15,16-dihydrotanshinone I, with the major ions being *m/z* 277 (295-18), 249 (295-18-28), and 221 (295-18-56), which were 16 Da heavier than those of the latter. In MS^5, except for the very strong de-methyl ion peak *m/z* 206, there was a relatively strong dehydrated peak (*m/z* 203) and de-CO peak (*m/z* 193), indicating the possible existence of a phenolic hydroxyl group. The known compound trijuganone, reported as a component of *Salvia trijuga* [28], meets the above conditions.

In the TIC spectrum, there were four peaks with a $[M + H]^+$ ion of *m/z* 297, including cryptotanshinone (t_R 23.23 min). Both peaks of 8.33 and 9.05 min in MS/MS first lost one water molecule, generating ions *m/z* 279, which showed an almost identical fragmentation pathway to that of 15,16-dihydrotanshinone I, indicating that the two compounds, after losing a water molecule, generated the same fragment as did 15,16-dihydrotanshinone I or 15,16-dihydroisotanhinone I. It was found that tanshinone VI and neotanshinone A were consistent with this conditions. In addition, a strong *m/z* 251 ion peak appeared in the MS/MS spectrum, which was very likely generated from the t_R 23.23 min peak by losing one ethanol (46 Da) after the rearrangement of the position 13 1-hydroxy isopropyl group. This phenomenon provided further support to our conjecture. According to the retention time relationship, isotanshinone is generally eluted first on a reversed-phase column; therefore the 8.33 min peak was tentatively identified as danshenxinkun A, while the 9.05 min peak was identified as tanshinone VI.

The peak at 11.88 min also generated a $[M + H]^+$ ion of *m/z* 297, which, in MS/MS, lost one water first, generating *m/z* 279. Ion *m/z* 279 had MS^n (n = 3–6) spectra almost the same as those of methyltanshinonate, demonstrating that they had the same ion fragments after MS/MS spallation. Therefore, it was conjectured that the difference between the 11.88 min peak and methyltanshinonate was in the 4th position substituent, i.e., a hydroxyl group replaced a methoxy formyl group (-$COOCH_3$). It was initially identified as danshensu B.

Tanshinone II_B (t_R 13.68 min) was identified by comparing with the reference substances. A peak at 14.07 min had the same $[M + H]^+$ ion of *m/z* 311, which, in MS/MS, lost one CO (*m/z* 311 → 283) first, and the resulting ion's MS^n ($n = 3$–5) spectra were similar to those of cryptotanshinone, producing the base peaks of *m/z* 265 (283-18), 237 (283-18-28), and 209 (283-18-56), respectively. Also, the peaks involved in propene loss (*m/z* 265 → 223, 237 → 195, 209 → 167) were very strong. Therefore, this compound must be a derivative of cryptotanshinone or isocryptotanshinone, with certain carbon in its non-aromatic portion oxidized to a carbonyl group. The known compound 1-ketoisocryptotanshinone is consistent with the above conditions.

There were three peaks which generated $[M + H]^+$ ions of *m/z* 313. The 7.21 min peak was identified as tanshindiol C by comparing with the reference substances. The MS^n ($n = 2$–4) spectra of the 9.57 min peak were very similar to those of przewaquinone A, only 2 Da heavier than the corresponding ions. It first lost one water molecule, generating *m/z* 295 and 277. However, the peak of *m/z* 253 (295-42) was not very high, which differed from tanshinone II_B. We thought that this compound was similar to przewaquinone A, and tentatively identified it as 17-hydroxycryptotanshinone.

The peak at 12.96 min also generated a $[M + H]^+$ ion of *m/z* 313. Its MS/MS spectrum showed a base peak ion of *m/z* 295 (313-18) and a strong ion of *m/z* 267 (313-46), indicating that this compound had structural fragments similar to those of danshenxinkun A or tanshinone VI (i.e., the open ring structure formed from tanshinone benzofuran after hydration). Also, *m/z* 295 showed a very similar fragmentation pattern to 1,2-didehydrocryptotanshinone (t_R 20.80 min), suggesting that the fragment after dehydration was 1,2-didehydrocryptotanshinone; therefore this compound was tentatively identified as 1,2-didehydrotanshinone V.

The peak at 5.18 min had a $[M + H]^+$ ion of *m/z* 315. Its MS/MS and MS^3 were similar to tanshindiol B and tanshindiol C, first losing one water molecule (*m/z* 315 → 297), followed by losing one CO (*m/z* 297 → 269), and the resulting ion *m/z* 269 showed a similar fragmentation pattern to cryptotanshinone, indicating this compound was probably tanshindiol B or a structure hydrogenated at C15 and C16. It was tentatively identified as 15,16-dihydrotanshindiol B/15,16-dihydrotanshindiol C. Its configuration could be determined based on ESI-MS alone. The structure of this compound has not been reported.

The peak at 20.47 min was identified as methyltanshinonate by comparison with reference substances.

The peak at 15.60 min showed a $[M + H]^+$ ion of *m/z* 341, which was 2 Da heavier than methyltanshinonate. Its fragmentation pathway was similar to that of methyl tanshinonate as well, suggesting that it was probably methyl tanshinonate hydrogenated at the 15th and the 16th positions. Therefore, it was identified as trijuganone C.

We have performed ESI-MS^n multistage mass fragmentation on tanshinone compounds, and we found that the major fragments lost by these compounds were water and CO, accompanied by losses of methyl groups and propene. Different conjugated systems have great influence on the fragmentation pathways. For example, tanshinone I loses one CO first, while tanshinone II_A loses one H_2O first. For the compounds with a saturated bond at 15th and 16th positions, their fragmentation begins with the loss of one water molecule. Applying this rule, we have performed LC/MS analysis on Danshen extracts and identified 27 compounds, 5 of which were discovered for the first time. The method can be used rapidly and conveniently for Danshen's qualitative analysis, and for the studies of in vivo metabolism of Danshen and its compound formulas.

15.2 Study on Near Infrared Spectral Fingerprint

Haibing Qu

15.2.1 Introduction to Near Infrared Spectroscopy

Near infrared (NIR) spectroscopy is a new and efficient analysis technique. The method has a number of advantages over HPLC. It is fast and nondestructive, with easy sample preparation, and is environmentally friendly. The technique offers stable spectral characteristics, rich information, and good reproducibility, and it has been widely used in the fields of agricultural, food, petrochemical, and pharmaceutical industries [29].

NIR spectroscopy uses the electromagnetic spectrum in the range from 780 to 2,526 nm (12,820–3,959 cm^{-1}), which mainly reflects the frequency doubling and co-frequency absorption of the hydrogen-containing groups of organic molecules, such as C–H, N–H, O–H, etc. The spectral characteristics are: The signal frequency is higher than mid-infrared region, between mid-infrared region and visible light region, and similar to visible light, so it is easy to obtain and process the signals. The spectral information is similar to the mid-infrared region of vibrational spectrum, contains abundant structural information of hydrogen-containing groups, and the same group usually has absorptions at multiple locations in the NIR region. The intensity of spectral information is lower than mid-infrared region, with wider spectral peaks, weaker absorption, complicated background, and overlapping absorption peaks.

The characteristics of NIR spectra determine that the NIR spectroscopy analysis technique has significant advantages over routine analysis methods. First, a wide range of objects can be analyzed with the method. The substances and occasions which are suitable to be studied by NIR spectroscopy are extensive, and the technique is applicable to almost all physico-chemical property analyses of hydrogen-containing samples. It not only can reveal the compositional and structural information of most organic compounds, but also can indirectly reveal information on some substances (such as some inorganic compounds) which have no absorptions in the NIR region via the spectral changes caused by coexisting substances. The absorption coefficient of substances in the NIR region is small, and the analytical procedure is simple. As absorption spectra produced by molecular vibration energy level transitions, the frequency doubling or co-frequency absorption coefficients in the NIR region are very small, generally about 1–3 orders of magnitude lower than those of mid-infrared absorption, so samples can be directly measured without dilution. Although the small absorption coefficients hinder the detection of impurities in trace amounts, it nonetheless ensures that the impurities or constituents with weak absorptions in the NIR region will not interfere with the analysis. Glass or quartz media do not absorb NIR light, so they can be used as sample cell materials, and are cheap in cost and convenient to use. Sometimes samples can be assayed in glass containers directly, without need to open sealed containers, thus avoiding sample transfer and contamination. It is applicable to diffuse-reflectance technique. The light scattering in NIR region has strong effects and large penetration depth, and the penetration depth of the shortwave NIR region (700–1,100 nm) on solid samples can reach some centimeters while the absorbed light intensity is small, making NIR spectroscopy capable of determining samples directly by diffuse-reflectance techniques. Assaying is rapid and convenient; in most cases, it takes less than 1 min to assay a sample and obtain the results, which is very efficient. Meanwhile, the highly automated instruments decrease the requirements of operators' skills. It does not damage samples, does not consume solvents, and does not pollute the environment.

In NIR spectroscopy analysis, the only thing that needs to be obtained is the sample's spectral signals, and sometimes, even without any other solvent, samples can be directly assayed in the primary containers and then be sent back to the places where they are manufactured or the containers after the assay. No contamination is caused during the process. The analytical results are reliable. Since there is no need for sample pretreatment, the experimental errors are reduced to a minimum, so NIR spectroscopy analytical results are more reliable than those obtained by other analysis techniques. The technique can be used to assay the sample's non-chemical properties. NIR spectra can not only reflect the chemical composition and structural information of a sample, but also reflect the physical properties of the sample. Based on the influence of various physical parameters on NIR spectra, it can also be used to measure the compactness and particle size of a solid sample, the density of a liquid sample, the viscosity of a sticky sample, etc.; thus, such spectra contain a large amount of information.

NIR spectroscopy mainly reflects the frequency doubling and co-frequency absorption of hydric groups of the assayed sample, with severely overlapping absorption peaks and multifarious information, but as long as the chemical composition and proportion of the components in the test sample are stable and operated under uniform conditions, the sample's NIR spectrum will be constant. In other words, NIR spectra can represent the entirety of chemical features, meeting the two basic properties of chemical fingerprint, "entirety" and "fuzziness", and can be used to identify the quality of TCM drugs.

15.2.2 The Near Infrared Spectral Fingerprints of Danshen

Danshen is a frequently used TCM drug and is produced in many regions of China, but only one species, *Salviae miltiorrhiza* Bge in *Labiatae* family, is adopted in *Chinese Pharmacopoeia*. In practice, however, another 9 plants in the same genus, such as *Salvia przewalskii* Maxim., *S. digitaloides* Diels and *S. yunnanensis* C. H. Wright, are usually used as Danshen's substitutes. Influenced by many external factors including geographical environment and growing conditions, Danshen from different habitats have somewhat different chemical constituents and the differences in chemical constituents between Danshen and its substitutes are even larger, and most substitutes are not suitable as Danshen crude drug [30].

Danshen and the plants belonging to the same genus are all rhizome herbs. Besides abundant cellulose and starch, the other chemical constituents of their roots are primarily lipid-soluble conjugated quinones (tanshinone I, tanshinone II_A, tanshinone II_B, cryptotanshinone, isotanshinone, etc.) and ketones (diterpene perylene ketone, danshenspiroketallactone, etc.) as well as water-soluble compounds (danshensu, protocatechuic aldehyde, salvianolic acid B, etc.). These compounds are rich in hydric groups such as C–H, N–H and O–H; therefore, the quality of Danshen can be determined by NIR spectroscopy [31].

15.2.2.1 Instruments and Samples

Antaris Fourier Transform Near-Infrared Spectrometer (Thermo Nicolet Company, USA), equipped with Integrating Sphere for reflectance analyses, RESULT™ software and TQ Analyst™ software.

Sixty-two Danshen samples were collected from 14 different regions. Among these, 2 samples were from Anhui (#1–2), 1 from Gansu (#3), 2 from Anguo, Hebei (#4–5), 6 from Henan (#6–11), 6 from Lotus Pond Drug Market (#12–17), 5 from Shandong (#18–22), 4 from Shanxi (#23–26), 5 from Sichuan (#27–31), 3 from Chongqing (#32–34), 3 from Hubei (#35–37), 1 from Jiangxi (#38), 3 from Guangdong (#39–41), 2 from Yunnan (#42–43), and 19 from Danshen GAP Base, Shangluo, Shanxi Province (China) of Tasly Company (#44–62).

15.2.2.2 Analysis of NIR Spectrum Fingerprints and HPLC Fingerprints

Pulverize Danshen, sift through a 100-mesh sieve, place in a sample cup, and collect the NIR spectrum of the sample by the Integrating Sphere sampling system. Spectrum collection conditions were as follows: the internally installed background of the spectroscopy as the reference, wavelength ranging from 10,000 to 4,000 cm^{-1}, scanning times of 32, and resolution of 4 cm^{-1}. Run duplicate tests on 3 samples of each batch, and calculate the average spectrum as the NIR fingerprint.

Take 5 g of the herb, place in a reflux flask, extract the herb by a certain procedure, make the extract to volume in a 100 ml volumetric flask, centrifuge, filter, take the supernatant, run duplicate tests on 2 samples of each batch, use 10 μl for HPLC analysis. The chromatographic analytical conditions: Agilent Zorbax SB-C_{18} analytical column (4.6 × 250 mm, 5 μm); mobile phase A: 0.02 % (v/v) phosphoric acid water solution, phase B: 80 % methyl cyanide water solution containing 0.02 % phosphoric acid, and linear gradient elution (phase A concentration variation: At 0 min, 90 %; at 8 min, 78 %; at 15 min, 74 %; and at 55 min, 48 %); flow rate: 1 ml/min, detection wavelength: 280 nm, column temperature: 30 °C, and record time: 55 min.

Figure 15.54a is the raw NIR spectra, and it is difficult to identify the differences between herbs from different habitats or different batches from the same habitat. This results from the severe self-overlap of NIR spectral bands on one hand, and from the complex chemical constituents of the TCM drug on the other hand.

To reduce the influence of the instrument, sample particle size and homogeneity, etc., on the NIR spectra, multiplicative scatter correction (MSC) or standard normalization variate (SNV) was usually used to pretreat the raw NIR spectra. The spectral baseline drift can be corrected by differential treatment. First-order differential treatment can eliminate the constant background, and second order differential treatment can eliminate linear background. Because the noise signals are enlarged during the differential treatments, spectral data smoothing is necessary. Figure 15.54b is the spectra obtained after MSC, second-order differential treatment, and smoothing filter by Norris derivative. It can be seen from the figure that the after the treatments, not only the baseline drift in the NIR spectra was eliminated, but also the samples' spectral characteristics were more exquisitely reflected.

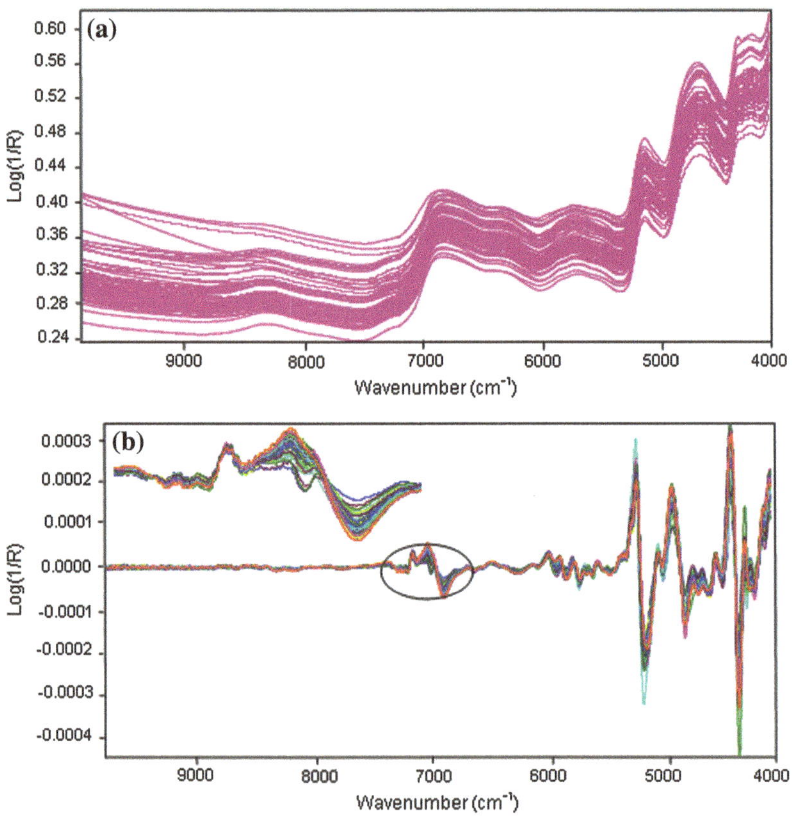

Fig. 15.54 The original NIR spectrogram **a** and the second-order differential spectrogram **b** of 62 Danshen samples

15.2.2.3 Investigation of Sample Distribution

After the pretreatment, the principal components of the raw spectral data were analyzed, and the first two major components represented 99.94 % of the information. Figure 15.55 shows the plot of the scores of the second major component against the scores of the first major component. There were significant differences among Danshen samples from different production regions and the GAP Base, and the samples from GAP Base clustered in a small range. Meanwhile, from the distribution of the NIR spectral information of each sample's major component, the difference among different batches of herbs could be distinguished.

15.2.2.4 The Method for Calculating Spectral Similarity

In recent years, chemical fingerprint analysis techniques have been gradually applied to the quality control of TCM drugs. Chemical fingerprinting can reflect the overall chemical composition characteristics of a TCM drug, while fingerprint similarity can quantitatively assess the differences in chemical composition.

To evaluate drug quality by NIR spectroscopy, a reference NIR spectral library should be established first; that is, to collect the NIR spectra from qualified samples (samples which have been tested by other classical methods) as reference NIR spectra, compare the NIR spectrum of a test sample with the reference NIR spectra, and judge whether the sample is qualified or to which quality grade the sample belongs. This method has been used in western drug quality control, however, the chemical composition of TCM crude drugs is much more complicated, the raw NIR spectra look very similar, and the spectra in the NIR spectral library are a group of reference spectra rather than a single spectrum. Therefore, the routine similarity measurement is not applicable to NIR spectral similarity calculation, and a new computing method is needed to measure the similarities between an individual NIR spectrum and a group of NIR spectra.

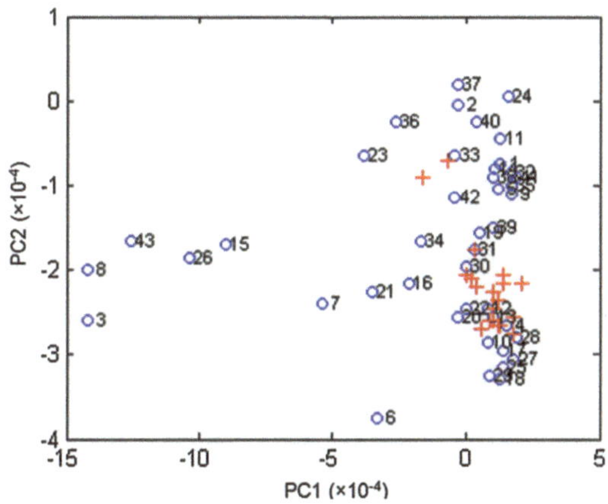

Fig. 15.55 The phase map of the second major component vs. the first major component. +: Samples from GAP Base; O: Samples from the other habitats

Suppose a reference NIR spectral library is composed of NIR spectra of m samples, and the number of scanning points in the spectrum of each sample is n, then the standard spectral library can be represented in matrix X:

$$X = \begin{bmatrix} x_{11} & x_{12} & \cdots & x_{1n} \\ x_{21} & x_{22} & \cdots & x_{2n} \\ \vdots & \vdots & \ddots & \vdots \\ x_{m1} & x_{m2} & \cdots & x_{mn} \end{bmatrix}$$

If the rank of the matrix is r, then it can be orthogonally decomposed to scoring matrix T and loading matrix P:

$$X = TP = \begin{bmatrix} t_{11} & t_{12} & \cdots & t_{1r} \\ t_{21} & t_{22} & \cdots & t_{2r} \\ \vdots & \vdots & \ddots & \vdots \\ t_{m1} & t_{m2} & \cdots & t_{mr} \end{bmatrix} \begin{bmatrix} p_{11} & p_{12} & \cdots & p_{1n} \\ p_{21} & p_{22} & \cdots & p_{2n} \\ \vdots & \vdots & \ddots & \vdots \\ p_{r1} & p_{r2} & \cdots & p_{rn} \end{bmatrix}$$

where $P_j = [P_{j1}, P_{j2}, \ldots, P_{jn}]$ ($j = 1, 2, \ldots, r$) is a group of mutually orthogonal vectors. Because P is composed of orthogonal vectors, therefore, the following equation holds:

$$T = XP^T$$

Suppose the NIR spectral data of a test sample to be evaluated is x_{new}; based on the above equation, the scoring vector t_{new} can be calculated:

$$t_{\text{new}} = x_{\text{new}} P^T = [s_1, s_2, \ldots, s_r]$$

When $j = 1$, check $e_{i1} = |s_1 - t_{i1}|^2$ ($i = 1, 2, \ldots, m$), find the minimum of $e_{11}, e_{21} \ldots, e_{m1}$, record it as e_{k1}, and record the corresponding t_{k1} as t'_1.

By the same method, when $j = 2, 3, \ldots, r$, get $t'_2, t'_3, \ldots, t'_r$, and calculate the residual spectrum e according to the following formula:

$$e = [s_1 - t'_1, s_2 - t'_2, \ldots, s_r - t'_r] \begin{bmatrix} p_{11} & p_{12} & \cdots & p_{1n} \\ p_{21} & p_{22} & \cdots & p_{2n} \\ \vdots & \vdots & \ddots & \vdots \\ p_{r1} & p_{r2} & \cdots & p_{rn} \end{bmatrix}$$

The residual spectrum e reflects the difference in the spectra between the sample to be evaluated and the reference sample. Based on the residual spectrum e, calculate the similarity match value similarity for short below (SMV) of the spectrums between the to-be-evaluated sample and reference sample:

$$\mathrm{SMV} = \left(1 - \frac{||e||}{||x_{\mathrm{new}}||}\right) \times 100$$

The similarity ranges from 0 to 100, and when the value is higher, the spectrum of the sample to be evaluated is more similar to that of the reference sample.

15.2.2.5 The Construction of a Reference NIR Spectral Library and the Results of Similarity Calculations

Randomly select 10 Danshen samples (#45–48, 50, 53, 55, 57, 59, 62) produced in the GAP Base to form a set of reference samples, and take the corresponding NIR spectra as the reference sample spectral data. To sufficiently use the NIR spectral information, use the whole spectral range (9,879.54 – 4,119.21 cm^{-1}) to build the model.

Input the NIR spectrums of the remaining samples into the model built above, calculate the similarities between them and the reference NIR spectra, and compare the results with the similarities of HPLC fingerprints. Figure 15.56 shows the similarities of fingerprints of the other 9 Danshen root samples from the GAP Base obtained by NIR spectra and HPLC, respectively. It can be seen from the figure that the similarities of NIR spectral fingerprints were basically the same as those of HPLC fingerprints, with the similarities of sample #58 being the lowest; 98.59 and 98.48, respectively. The high similarities among the samples from the GAP Base indicate that the quality of Danshen from the Base is comparatively stable and constant.

Figure 15.57 shows the NIR spectral and HPLC fingerprint similarities among Danshen samples from other habitats. Influenced by growing conditions, harvesting seasons, etc., the stabilities and homogeneities of the herb from the other habitats were poorer than those from the GAP Base. By comparing the similarities calculated from the two kinds of fingerprints, it was found that HPLC fingerprint similarities, except for those of sample #3, #17, #26, #39 and #43, all were greater than 97. Therefore, it is difficult to tell the difference in quality between samples from other regions and

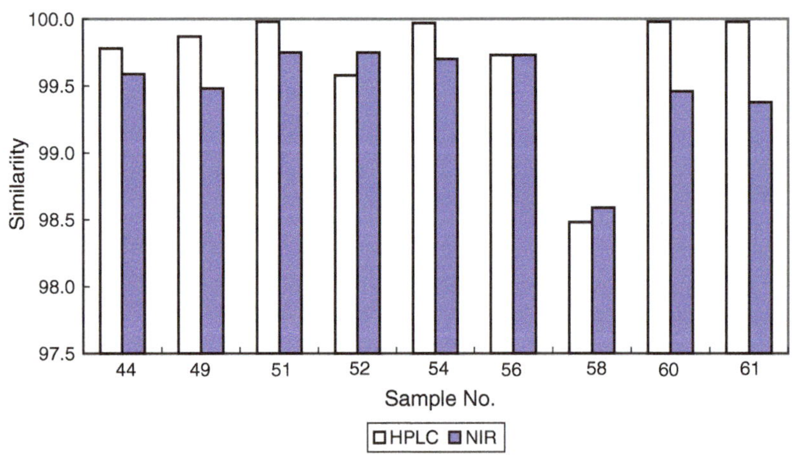

Fig. 15.56 The comparison of NIR spectral and HPLC fingerprint similarities among Danshen samples from the GAP Base

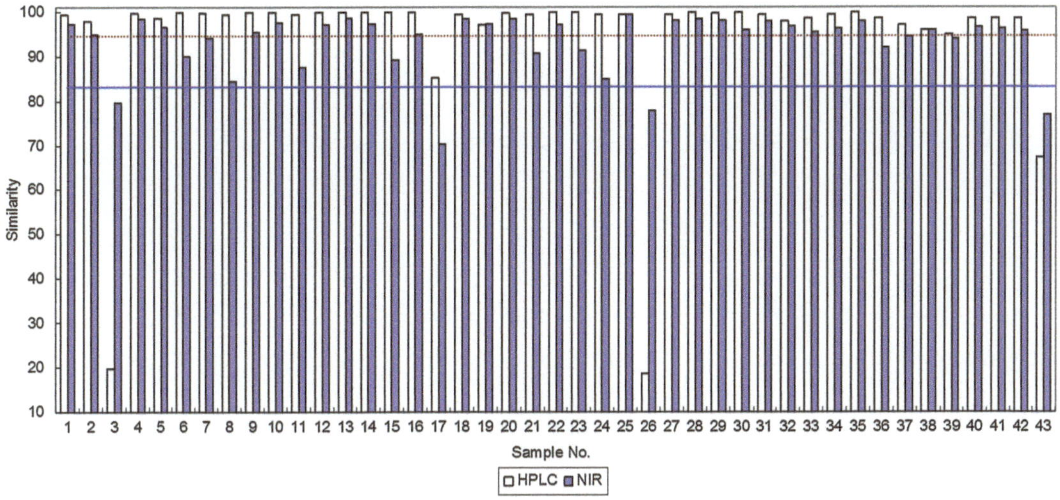

Fig. 15.57 The comparison of NIR spectral and HPLC fingerprint similarities among Danshen samples. *Dashed line* HPLC similarity threshold line; solid line: NIR similarity threshold line

samples from the GAP Base. Also, the similarity values of #3 and #26 were only 19.74 and 18.75, respectively; they belonged to abnormal results and may be caused by experimental errors or incorrect operation. While NIR spectral similarity values distributed from 70.16 to 99.36, the range was wide, and there were no abnormal values. Among the similarity values, those of sample #3, #17, #26, and #43 were below 80.

15.2.2.6 Distinguish the Quality of Medical Materials

1. Distinguish Danshen Samples from the GAP Base and from Other Habitats
   Using the above similarity results and selecting an appropriate threshold, the quality of herbal drugs can be distinguished. Based on the similarities of Danshen samples from the GAP Base (Fig. 15.57), the threshold could be set at 98.5 to distinguish samples from the GAP Base and those from the other habitats, and only samples #13 and #25 from the other habitats were judged as from the GAP Base.
2. Distinguish the Quality of Danshen
   Using the above similarity results and selecting an appropriate threshold, the qualities of herbal drugs can be evaluated. When setting the HPLC fingerprint similarity threshold at 97 and NIR spectral similarity threshold at 84, the herbs with similarities lower than the thresholds were evaluated as unqualified. The evaluation results of the tested samples can be seen in Table 15.50. Except for sample #39 which was judged as unqualified by HPLC but qualified by NIR spectroscopy, the results of the other samples by HPLC and NIR spectral fingerprints were consistent.

15.2.2.7 The Reproducibility of NIR Spectral and HPLC Fingerprints

Test a Danshen sample 5 times by HPLC and NIR spectroscopy and calculate the similarities, respectively; the results can be seen in Table 15.51.

It can be seen from Table 15.51 that NIR spectral analysis had better reproducibility than HPLC analysis, which was mainly because some pretreatments such as extraction were not essential in NIR spectral analysis, resulting in reduced experimental errors.

The above study showed that NIR spectral analysis technique can be used for the rapid determination of Danshen quality. It is simple to operate and convenient to use, requires no

Table 15.50 HPLC fingerprints (Threshold 97) and NIR spectral fingerprints (Threshold 84) of Danshen samples

Serial no.	HPLC		NIR		Serial no.	HPLC		NIR	
	Similarity value	Judgment result	Similarity value	Judgment result		Similarity value	Judgment result	Similarity value	Judgment result
1	99.43	√	97.23	√	23	99.85	√	91.25	√
2	97.94	√	94.92	√	24	99.34	√	84.95	√
3	19.74	×	79.59	×	25	99.5	√	99.36	√
4	99.55	√	98.48	√	26	18.75	×	77.79	×
5	98.61	√	96.43	√	27	99.44	√	98.23	√
6	99.92	√	89.92	√	28	99.9	√	98.49	√
7	99.55	√	94.08	√	29	99.57	√	97.98	√
8	99.32	√	84.44	√	30	99.92	√	95.94	√
9	99.84	√	95.34	√	31	99.36	√	97.75	√
10	99.94	√	97.69	√	32	97.88	√	96.86	√
11	99.3	√	87.43	√	33	98.66	√	95.57	√
12	99.99	√	97.05	√	34	99.32	√	96.29	√
13	99.94	√	98.7	√	35	99.93	√	97.88	√
14	99.88	√	97.2	√	36	98.75	√	92.03	√
15	99.85	√	89	√	37	97.02	√	94.35	√
16	99.82	√	94.83	√	38	97.01	√	96.11	√
17	85.18	×	70.16	×	39	94.95	×	93.91	√
18	99.51	√	98.46	√	40	98.6	√	96.53	√
19	97.11	√	97.31	√	41	98.51	√	96.36	√
20	99.58	√	98.3	√	42	98.74	√	95.79	√
21	99.39	√	90.74	√	43	67.27	×	76.66	×
22	99.92	√	96.98	√					

Note √ Qualified; × Unqualified

Table 15.51 The reproducibility comparison between NIR spectral fingerprints and HPLC fingerprints

Method	Similarity					Mean value	Standard deviation	RSD (%)
	1	2	3	4	5			
HPLC	99.32	96.01	98.34	96.01	98.12	97.56	1.33	1.36
NIR	95.32	95.41	94.75	95.64	96.06	95.44	0.43	0.45

sample preparation procedures such as extraction and separation, or organic solvents, and does not damage samples. By the construction of a Danshen reference NIR spectral fingerprint library and using a specified similarity computing method, Danshen from GAP Base and from other habitats can be distinguished rapidly. By NIR spectral fingerprinting, the quality of Danshen can also be evaluated. With HPLC fingerprint analysis as a reference standard, and an appropriate threshold, the results of NIR spectral fingerprints and those of HPLC fingerprints are basically the same.

15.2.3 NIR Spectral Fingerprint Based on Wavelet Transform

Wavelet transform (WT) is a time-frequency analytical method for signal processing. It is characterized by multi resolution analysis and is capable of showing the local features of signals in both time and frequency fields; thus it is called "a mathematical microscope" and has become a powerful signal processing tool that has been successfully used in chemical signal analysis including noise smoothing, data compression, signal analysis and feature extraction. In this section, we use Danshen as a study object, take advantage of the characteristics of WT multi-resolution analysis and combine WT with digital information visualization methods, analyze the NIR spectra of Danshen by wavelet decomposition, extract fingerprint information from them, and visually represent the fingerprint information so as to intuitively reflect the overall quality mode differences.

15.2.3.1 Fundamental Principles

1. Wavelet Decomposition of NIR Spectrum
The NIR spectrum can be regarded as the time sequence signal. When wavelet is used for multi-resolution analysis, we only need to use the wavelength axis as the time axis. The multi-resolution signal decomposition algorithm proposed by Stephane G. Mallat is the most widely-applied discrete WT algorithm, with its basic principles already published. In this section, this algorithm is used to conduct wavelet decomposition of the NIR spectral data of the tested drugs, so as to inspect the change in wavelet coefficients under different decomposition scales. By setting a certain threshold to select the wavelet coefficients, information on the characteristics of chemical models corresponding to the NIR spectra is obtained.
Figure 15.58 displays the obtained high-frequency wavelet coefficients under 9 different wavelet decomposition scales for NIR spectra of Danshen, where the wavelet basis function is the commonly-used *Daubechies* wavelet ($N = 2$). In the figure, a is the high-frequency wavelet coefficient (d_1) at scale 2^{-1} in the original spectrum, which mainly reflects the noise part of the achieved data; e is the high-frequency wavelet coefficient (d_5) at scale 2^{-5} in the original spectrum, which mainly reflects evident local characteristics; the wavelet coefficients under the decomposition scale 5 and 6 have shown features similar to the HPLC "fingerprint region". Therefore, this section selects decomposition scale 5 for wavelet decomposition. By carrying out the

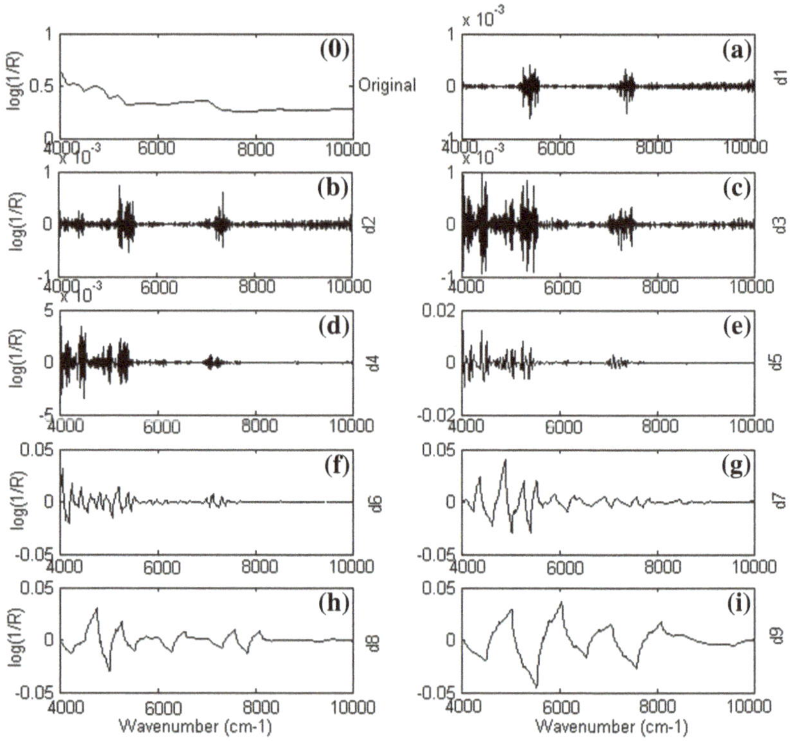

Fig. 15.58 Wavelet decomposition under 9 different wavelet decomposition scales for NIR spectra. **a**–**i**: Decomposition scale *1*–*1*; *0* stands for original NIR signal

above wavelet decomposition on the NIR spectrum, characteristic information of the chemical models, namely the fingerprint information, of Danshen is obtained.

2. Visual Expression of the NIR Digitalized Characteristic Information

The fingerprint information obtained from the wavelet decomposition of NIR spectra is usually abstract data, which should be visually expressed as some sort of image before it can be identified and understood by readers. There are many ways to visually display the information of chemical characteristics in NIR spectra in the wavelet-basis model. By considering that chromatographic fingerprints have become familiar mainstream spectra, the following method for visualized data expression in this section will transform the above digital information into a series of chromatographic peaks similar to those of chromatographic fingerprints.

Supposing the virtual spectral peaks are normal peaks, with the half band width and interval between the peak positions set as a constant, the mathematical model is:

$$h(t) = \sum_{i=1}^{N} \frac{A_i}{\sqrt{2\pi}\sigma} \exp\left(-\frac{(t - t_i)^2}{2\sigma^2}\right)$$

where A_i is the peak height of the virtual spectral peak; t_i is the retention time of the peak; σ is half band width; N is the number of peaks.

With the original NIR spectra being WTed as above, the characteristic information obtained is a wavelet basis eigenvector. In the visualized expression, each element of this vector corresponds to a virtual peak, with the element value representing the peak height and element sequence representing the sequence of the virtual peak. Then, the value of each element is put into the above formula to obtain the virtual peaks, the result of which is the visual expression of the wavelet basis eigenvector as a series of peaks. Thus, the NIR spectrum is turned into the NIR spectral fingerprint based on wavelet transformation.

15.2.3.2 Experiment

1. HPLC Analysis
   Instrument: Agilent 1100 HPLC equipped with quaternary gradient pump, online degasser, automatic sampler, column incubator, and ultraviolet and diode detectors.
   Reagents: Methyl cyanide (chromatographic pure, Fisher company of USA), phosphoric acid (A.R.), and Wahaha purified water.
   Samples: 40 batches of Danshen samples in all; among these, 17 were from GAP Base, Tasly, Shangluo, and Shanxi, and 23 from other habitats.
   Preparation of test product: Weigh 2.5 g of the herb powder, add 50 ml of distilled water, reflux for 1.5 h, filter, add 50 ml of distilled water into the residue, reflux for 1 h, combine the filtrates, bring volume to 100 ml, centrifuge, filter, take the supernatant, inject, and run duplicate tests on 2 samples.
   The chromatographic analytical condition: Agilent SB-C_{18} analytical column (4.6 × 250 mm, Zorbax SB USCL010304); Mobile phase: phase A: 0.02 % (v/v) phosphoric acid water solution; phase B: 0.02 % (v/v) phosphoric acid-methyl cyanide solution. Flow rate: 1 ml/min; DAD detection wavelength: 280 nm; column temperature: 30 °C; and sample size: 10 μl. The mobile phase gradient conditions are shown in Table 15.52.
2. NIR Spectral Analysis
   Instrument: Antaris Fourier Transform Near-Infrared Spectrometer (Thermo Nicolet Company, USA), equipped with Integrating Sphere for reflectance analyses, RESULT™ software and TQ Analyst™ software.
   Samples: the same as above.
   NIR spectrum collection: Take dry Danshen, pulverize, sift through a 100-mesh sieve, place a little in a quartz specimens cup (with sample height about 2.5 cm), and collect the NIR spectrum of the sample by the integrating sphere diffuse-reflectance spectrum sampling system of the spectroscopy. Spectrum collection condition: With the internally installed background of the spectroscope as the reference, wavelength: 10,000–4,000 cm^{-1}; scan times: 32; resolution: 4 cm^{-1}. Run duplicate tests on 3 samples, and take the average values. The NIR spectra of Danshen are shown in Fig. 15.59. Please note that the NIR spectra of Danshen samples from 2 different regions were very similar, and it is difficult to distinguish them directly.
3. NIR Spectral Fingerprint Based on Wavelet Transformation
   The HPLC fingerprints of Danshen are shown in Fig. 15.60.

Table 15.52 The gradient program of the mobile phase for HPLC analysis

Time (min)	A (%, v/v)	B (%, v/v)
0	92	8
8	82	18
15	79	21
40	66	34

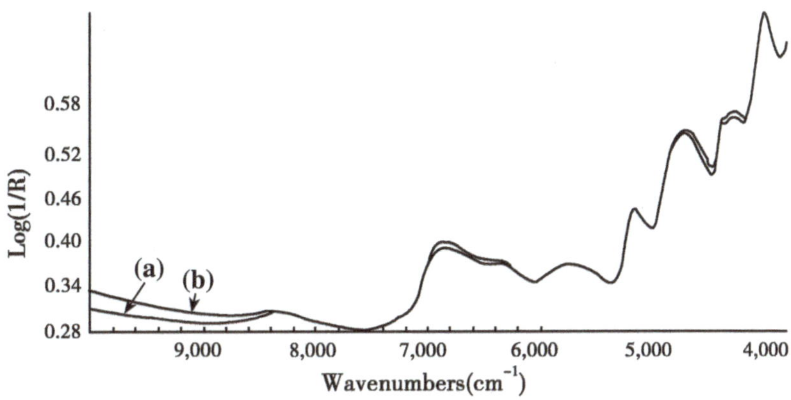

Fig. 15.59 The NIR diffuse-reflectance spectra of Danshen powder. **a** Danshen sample from Yunnan; **b** Danshen sample from the GAP base

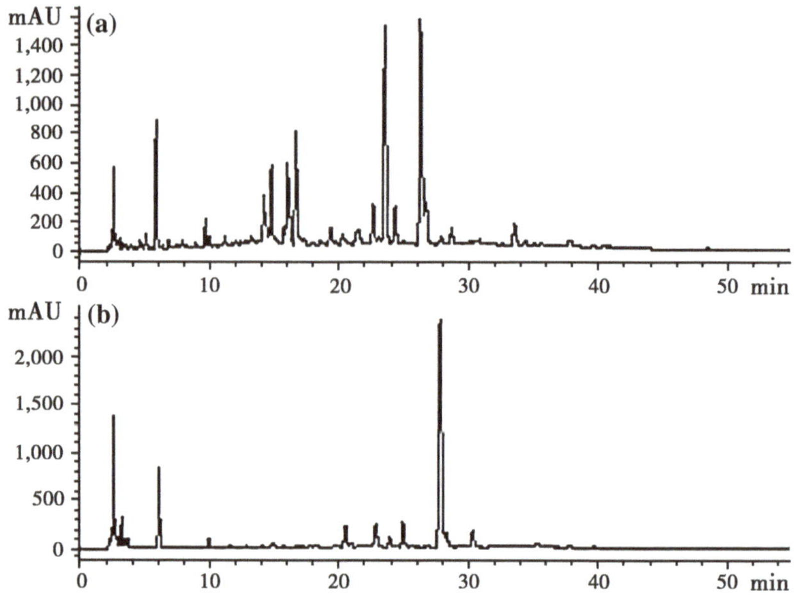

Fig. 15.60 The HPLC fingerprints of Danshen. **a** Danshen samples from Yunnan; **b** Danshen samples from the GAP Base

The HPLC fingerprints of Danshen samples from the GAP Base were similar to each other, and the primary peak shapes were basically the same; only the retention times were drifted a little. These fingerprints had a bigger difference from those of Danshen samples from Chrysanthemum Garden, Yunnan.

With D*b*2 as the wavelet basis function, decompose the two NIR spectra, obtain the high frequency wavelet coefficient at scale 5, and set the wavelet threshold value at 0.005. Implement visual representation of the obtained digitalized characteristic vectors by the method described above, and express them as NIR spectral fingerprints in the wavelet basis characteristic space, as shown in Fig. 15.61, in which the x-axis is the NIR spectral fingerprint virtual retention time and the y-axis is the corresponding response value. Figure 15.61a is the NIR fingerprint of Danshen from Chrysanthemum Garden, Yunnan; and Fig. 15.61b is the NIR fingerprint of Danshen from the GAP Base, Shangluo, Shanxi. The NIR fingerprints of Danshen roots from the GAP Base were very similar, and the peak shapes were basically the same. There were some differences in the NIR fingerprints between GAP Base Danshen and Danshen from Chrysanthemum Garden, Yunnan.

To compare the actual performance of NIR fingerprints to that of HPLC fingerprints, the similarities of these data from 40 Danshen samples were calculated, and the results are shown in Fig. 15.62.

Figure 15.62a shows the similarities of HPLC and NIR fingerprints of Danshen samples of GAP Base. It can be seen from the figure that the similarities of the HPLC and NIR fingerprints of each sample were high, which shows that the quality of Danshen from the GAP Base is stable. Figure 15.62b shows the similarities of HPLC and NIR fingerprints of Danshen samples from other habitats, and it can be seen from the figure that the change trends of the similarities in the two chemical fingerprints were nearly the same. Expressing the above two groups of similarity values into two groups of multidimensional vectors and calculating the correlation coefficient R of the two groups of vectors yields 0.98, which further confirms the concordance of the analytical results of the NIR spectral fingerprints based on wavelet transformation and HPLC fingerprints.

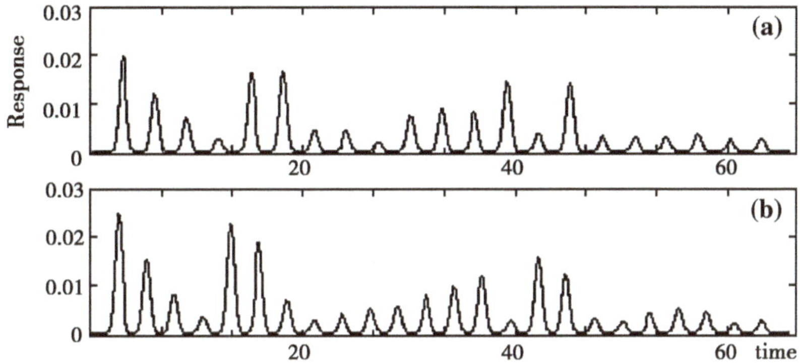

Fig. 15.61 NIR fingerprints of Danshen. **a** Danshen samples from Yunnan; **b** Danshen samples from the GAP Base

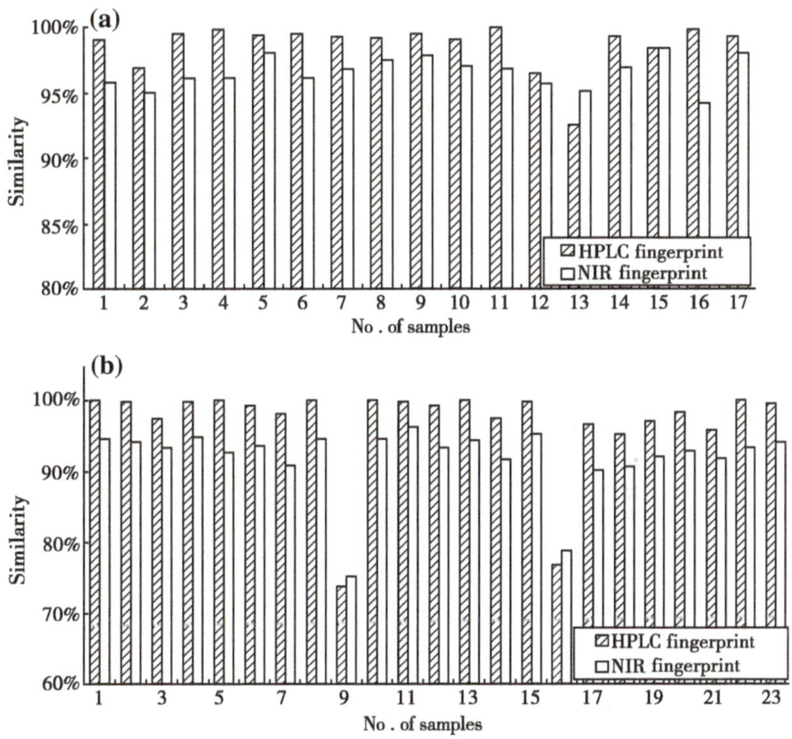

Fig. 15.62 Similarity values obtained from HPLC fingerprints and NIR fingerprints. **a** Samples from the GAP base; **b** Samples from other habitats

References

1. Xiao GP, editor. Modern Chinese materia medica, vol. 1. Beijing: Chem. Industrial Publishing House; 2002.
2. Ma JW, Ye ZL. World Sci Technol Modern Trad Chin Med Mater. 2000;2(2):44–7.
3. Fan XH, Ye ZL, Cheng YY. Chem J Chin Univ. 2006;27(1):26–9.
4. Fan XH, Ye ZL, Cheng YY. Chin J Pharm Anal. 2005;26(9):7–10.
5. Fan XH, Cheng YY, Ye ZL, et al. Anal Chim Acta. 2006;555(2):217–24.
6. Du GH et al. Basic Med Sci Clin. 2000;6:10.
7. Zhou CX, et al. J China Pharm Univ, 1999;30(6):411.
8. Ge SR, et al. J Chin Med Mater. 2002;25(9):683.
9. Gao XM, et al. Gianjin J. Trad Chin Med. 2002;19 (2):74.
10. Xiao YZ, et al. China J Chin Mater Med. 2002;27 (9):579.
11. Min Hong, et al. Lishizhen Med Med Res. 2003;14 (12):739.
12. Ding MY, et al. World Sci Technol/Modern Trad Chin Med. 2002;4(1):36.

13. Zhang YN, et al. China J Chin Mater Med. 2004;29 (7):660.
14. Liu AH, et al. J Pharm Biomed Anal. 2006;11:48.
15. Verweel HJ, MacGillavry CH, et al. Zeitschrift fuer Kristallographie. 1939;102:60.
16. Rao NN et al. Archiv der Pharmazie (Weinheim, Germany), 1988;321(3):179.
17. Packert M, Steinhart H. J Chromatogra Sci. 1995;11:631.
18. Dalla V, Cotelle P. Tetrahedron. 1999;22:6923.
19. Wada H, Kido T, Tanaka N, et al. Chem Pharm Bull. 1992;8:2099.
20. Ai CB. LI LN. Chin Chem Lett. 1991;1:17.
21. Tanaka T, Nishimura A, Kouno LO et al. Chem Pharm Bull. 1997;10:1596.
22. Zhang HJ, Li LN. Planta Med. 1994;1:70.
23. Qian TX, Li LN. Phytochemistry. 1992;3:1068.
24. De B, Hendrieus BA, Raggers RJ. Eur Pat Appl. 2003;32:16.
25. Zidorn C, Petersen Bo, Udovicic V, et al. Tetrahedron Lett. 2005;8:1291.
26. Ai CB, Li LN, et al. Planta Med., 1992, 2:197.
27. Gao W, Zhang R, Jia W. Chem Pharm Bull. 2004;52:136.
28. Lu X, Luo H, Niwa M. Planta Med. 1990;56:87.
29. Lu WZ, Yuan HF, Xu GT, et al. Modern near infrared spectrum analytical technique. Beijing: China Petrochemical Press; 2000.
30. Qu HB, Ye YY, Cheng YY. Chin J Pharm Anal. 2006;26(10):1413.
31. Ye Zhengliang Yu, Ke Cheng Yiyu. Chem J Chin Univ. 2007;28(3):441.

Quality Control of Dantonic™ 16

Shunhang Liu, Jun Gao, Yan Liu, Shunnan Zhang, Haiou Dong, Xueming Zhang, Jianping Lin, Junquan Wang, Xuesong Liu, Haibin Qu and Xiaohui Fan

16.1 Quality Control of Raw Material Medicinals

Shunhang Liu and Jun Gao

With the rapid modernization of TCM and the expansion of TCM corporations, unprecedented attention has been paid to the quality of raw material medicinals, which has been the bottleneck to the growth of Chinese pharmaceutical enterprises. To solve the problem, Tasly Group has begun construction of a medicinal herbs resource base since the end of last century. After exploration for many years, Tasly has obtained some outcomes in the quality control of TCM raw materials, and accumulated abundant experience.

16.1.1 Overview

16.1.1.1 A Medicinal Base is Fundamental to a TCM Corporation

It has been proven that to control the quality of raw material at the source, it is necessary to build ones own medicinal base; otherwise, the modernization of TCM is out of the question. Tasly Group realized the importance of such a base, and they began their construction of a Danshen base in 1998. Based on extensive research and discussion, the Shangluo region of Shaanxi Province was chosen as Tasly's Danshen production base, and Shaanxi Tasly Plant Pharmaceutical Co., Ltd. was established. There are mainly four reasons that Shangluo region was chosen, and they are elaborated upon below.

1. The Historical Factors
   In TCM, the origin of an herb, or its production place, has always been stressed. The reason for this idea is inherited from our ancestors. For example, the so-called Si Da Huai Yao, or the Big Four Herbs of Huai River (四大淮药) (rehmannia root, common yam rhizome, two-toothed achyranthes root, safflower) are produced in Henan, and the quality of the so-called Zhe Ba Wei, or Zhejiang's Eight Drugs (浙八味) (fritillary bulb, dwarf lilyturf tuber, turmeric root tuber, chrysanthemum flower, white atractylodes rhizome, corydalis tuber, corydalis rhizome) is highest if they are produced in Zhejiang. Similarly, the coptis rhizome produced in Chongqing's Shizhu and Hubei's Lichuan; the monkshood produced in Sichuan's Jiangyou; the Chinese angelica produced in Gansu's Langxian; the rhubarb root and rhizome produced in Gansu's Liquan; the astragalus root produced in Shanxi's Datong; the codonopsis root produced in Shanxi's Wutai; the poria produced in Hubei's Yingshan and Luotian and Anhui's Jinzhai; the notoginseng root produced in Guangxi's Bose; the

S. Liu · J. Gao · Y. Liu · S. Zhang (✉) · H. Dong · X. Zhang · J. Lin · J. Wang · X. Liu · H. Qu · X. Fan
Tasly Holding Group, Tianjin, China
e-mail: zsn@tasly.com

X. Yan (ed.), *Dan Shen (Salvia miltiorrhiza) in Medicine*,
DOI 10.1007/978-94-017-9463-3_16,

momordica fruit produced in Guangxi's Lingui and Yongfu; the eupatorium produced in Guangxi's Liuzhou, etc.

The location of Tasly's Danshen base, Shaanxi's Shangluo, has historically been the major Danshen production site. According to *Shen Nong's Classic of the Materia Medica*, the first monograph of materia medica (about the first century B.C.), Danshen was growing in Tongbai Mountains and rivers as well as in Tai Mountain. These areas are today's Shaanxi's Shangluo region.

Since 1949, farmers in the Shangluo region have grown more and more Danshen. Between 1974 and 1980, Professor Qin Guanshu of Xi'an Botanical Garden and his associates conducted the study of the "Purification of Danshen's varieties and propagation of Danshen's seeds" in Shangluo. They trained many farmer technicians and accumulated a large quantity of research data, and promoted the rapid growth of Danshen cultivation in the area. At that time, Danshen's cultivation area was over 10,000 μ (1 μ = 1/6 acre).

2. Quality Factors

   Danshen has a wide distribution in Northern China, such as in Shandong's Linyi, Sichuan's Zhongjiang, and Anhui's Bozhou, which are all the main production areas of Danshen. Tasly invited several experts from Shenyang Pharmaceutical University and China Pharmaceutical University to inspect Zhongjiang (Sichuan Province), Suizhou and Hengshui (Hubei Province), Bozhou (Anhui Province), Lushi (Henan Province), Taiyuan (Shanxi Province), Shangluo, Hanzhong and Chang'an (Shaanxi Province), as well as some places in Shandong and Gansu provinces. The chemical composition of the Danshen samples from these areas were also analyzed. The results showed that Danshen in Shangluo had the most superior quality, which was in agreement with the experts' conclusions based on field visits. Therefore, Shangluo is believed one of the most suitable places for Danshen plantations.

3. Environmental Factors

   Danshen has wide distribution and strong adaptability. Danshen generally grows in the edges of forests, hillside fields, beside brooks, in underbrush, and on roadsides that are characterized by sufficient sunshine, humid air and moist soil. It likes a mild climate, is relatively resistant to cold, and can resist low temperatures above −15 °C. The optimal temperature for Danshen's growth is 20–26 °C, with an optimal relative humidity of 80 %, annual average temperature of 11–17 °C, altitude greater than 500 m, and annual precipitation greater than 500 mm. The Shangluo region is located in the south of Qinling, a special climate zone between the warm temperature and subtropical zones, where the Yangtze River and the Yellow River meet. In this region, the air is clean, the soil and water are free from pollution, and the air quality reaches the first class in national standards. The annual rainfall is 800–900 mm, the soil is loose, and the heavy metal and "agrochemical residue" are both below the international standard.

4. Human Factors

   In early 1998, Shangluo government made Chinese medicinal cultivation their most important strategy to revitalize the economy, and preferential policies were offered to make TCM medicinal cultivation the backbone of their economic development. Moreover, the farmers in Shangluo mainly engage in plantation, and they are industrious and kind-hearted. They have abundant experience and traditions in planting Danshen.

16.1.1.2 Software, Hardware, and Related Supporting Facilities are the Solid Foundation of Quality Control

Shaanxi Tasly Plant Pharmaceutical Co., Ltd. was founded in 1998. Since then, it has developed from a small company with assets of 9 million RMB into a comprehensive corporation which engages in plantation, storage, and processing of medicinals, with total assets exceeding 50 million RMB. To realize the goal of providing stable, homogenous, and controllable Danshen herb, the company has invested a great deal of capital in software and hardware construction.

1. Hardware Construction
   Since 1999, the company has invested a total of 30 million RMB to establish 50 μ of scientific research and experimental gardens, 200 μ of quality seed parks, a 2,000 m^2 herbal pieces processing workshop, 500 m^2 packaging production line, and 4,000 m^2 elevated warehouse. They have also purchased forklift trucks; medicinal washing, drying, and packaging machines, farmland ploughing, sowing, and fertilizing machines; and established standardized laboratories, such as tissue culture and drug quality laboratories, equipped with instruments such as HPLC and ultraviolet spectrometers.
2. Management System
   To effectively control the entire process of Danshen production within the base, standard operating procedures (SOP) for planting Danshen were developed. They include: (1) SOP for selecting Danshen's resource base; (2) SOP for producing Danshen's stock seeds; (3) SOP for determining the quality of Danshen seeds; (4) SOP for Danshen cultivation techniques and management; (5) SOP for Danshen pest control; (6) SOP for Danshen harvest and primary processing techniques; (7) SOP for package, storage, and transport; (8) SOP for quality assurance; (9) SOP for personnel and equipment allocation; (10) SOP for document management.
3. Institutional Construction
   The company was one of the first corporations to pass GAP authentication by the Chinese government in Nov. 2003, and successively passed ISO9001 in December 2005, ISO14001 in December 2006, and national GMP certification for TCM herbal pieces in January 2007. We are also the first TCM plantation company to pass through a four-system authentication. The company became the first TCM plantation enterprise in the nation to pass all four-system certifications. In 2007, the company integrated the four systems, which improved the management to a new level.
4. Human Resources Team Construction
   At present, the company has 112 staff; 15 of them have a bachelor's degree, 25 of them have college education, 5 people have senior professional titles, and 13 people have intermediate professional titles.

16.1.1.3 Scientific Research Provides Powerful Support to Quality Control

From the very beginning, the company made scientific research their most important priority. In cooperation with the Institute of TCM Crude Drugs at Nanjing Agriculture University, the College of Life Science at Northwest Agriculture and Forestry University of Science and Technology, and China Pharmaceutical University, they conducted a series of research studies on Danshen, such as on germplasm, pest control, effective component formation, etc., and achieved significant results.

1. Main Research Projects
   The project titles are: The effects of breeding materials on the yield and quality of Danshen; The optimal plantation timing for Danshen; The effects of planting density on the yield and quality of Danshen; Study on the demand of irrigation and fertilization by Danshen; The effects of flower bud removing on the yield and quality of Danshen; Study on the optimal timing for harvesting Danshen; Study on the breeding conditions for 6-2-22 tetraploid Danshen; The effects of ridging on the yield and quality of Danshen; The fertilizer efficiency trial on Fuwanjia, a foliar potassium fertilizer; Danshen-specific fertilizer demonstration trial; Study on the regulation of secondary metabolites and its mechanism in Danshen; Study on Danshen's germplasm resources and DNA fingerprints; The effects of various ecological factors in Shangluo on the yield and quality of Danshen; The effects

of microelements on the growth, development, and accumulation of the effective components in Danshen.

2. Results Achieved
The company formulated and compiled a standardized Danshen-planting SOP, developed a specific fertilizer for Danshen growth, established a systemic, precise, and scientific Danshen identification system, set up Danshen standardized production theory and the key technologies quantitative system, determined the effects of various ecological factors on the yield and quality of Shangluo Danshen, and successfully bred a fine strain of Shangluo Danshen.

16.1.1.4 Strict Monitoring of Production is an Important Safeguard for Quality Control

In addition to the strong software and hardware foundation, and the powerful scientific research support, strict monitoring of each production step is also an important assurance of Danshen quality. Tasly strictly carries out the SOP of Danshen plantation and production, and maintains complete records for each step, including base selection, seed breeding, farmland management, and production process.

The quality of Danshen crude drug material is largely dependent on the compliance of Danshen farmland to the demand of the GAP base. If there is excessive heavy metal and pesticide residue in the field, the quality of the crude drug material must inevitably be disqualified. Thus, the selection of a breeding base is of ultimate importance. Tasly strictly complies to the demands of Chinese GAP by monitoring the soil, atmosphere, and water quality of each base. The COAs are achieved, and those bases which do not meet the criteria would be resolutely closed.

Every year during harvesting season, researchers would screen good seeds according to the company's seed breeding regime at each site, and cultivate good seedings in seedbeds to ensure that these seedings be used in Danshen bases to maintain the purity and characteristics of Shangluo Danshen. During the period of Danshen growth and planting, the technicians would usually visit Danshen-planting farmers, be informed of Danshen growth status, and organize the farmers to fertilize and control weeds and pests according to the SOP, so as to ensure that sufficient nutrients are acquired by the plants and to minimize damage from pests. In the process, information on fertilizer and agricultural chemical usage is recorded.

In addition to monitoring the planting process of Danshen crude material, the steps of production and processing are equally important. There are several working procedures, including washing, cutting, drying, and selecting, and in each step the quality assurance personnel would monitor and record according to the in-house SOP, by which the entire production process could be effectively controlled and the stabilization and consistency of Danshen crude drug quality could be achieved.

16.1.2 Summary

Through the above-mentioned measures, the yield of Danshen in Shangluo base increased from an initial 250–300 kg/μ, and the content of tashinone II increased from 0.18–1.2 to 1.7 ± 0.1 %. The similarities of eight major effective components monitored by multivariant fingerprinting were up to 0.99. While helping more than 500,000 farmers in the Shangluo region to prosper by planting Danshen, Tasly's Shangluo Danshen Base has become a well-known Danshen crude drug production base in China.

16.2 Quality Control in the Extraction Process

Shunhang Liu
and Yan Liu

Today, economic globalization is promoting the idea that standardization is the core of business management, and more and more Chinese enterprises are entering into the global market. Tasly has made standardized management its important entry point, and in accordance with the characteristics of pharmaceutical production and extraction processes, Tasly integrated industry standards with international standards to realize a diversified management, leading to the complete merging of enterprise quality management with international standards. In 2001, Tasly lead the nation in proposing, establishing, and implementing the Current Good Extraction Practice (CGEP), which has become the key link between TCM medicinal production (GAP) and drug manufacture (GMP). In 2003, Tasly became the first pharmaceutical corporation in China to pass the authentications of ISO9001/ISO14001/OH-SAS18001/Integration Management System, ISO10012 Calculation-Detection System, GMP of TCM extraction, as well as Australia TGA authentication. In Tasly, every step and practice has been standardized: extraction, documentation management, technological procedures, production, inspection, work field clearance, etc.

16.2.1 Strict Quality Standard for Crude Drug Materials

Danshen crude drug is produced in the GAP Base in Shangluo. The quality of the crude drug material not only needs to meet the standards specified in the *Chinese Phamacopeia*, but its contents of heavy metals and pesticidal residues, as well as its physical size and granularity and mid-infrared standard pattern, must also meet certain specifications. Some crude drug materials are required to meet a multivariant fingerprinting index; for instance, the similarities of eight major effective components controlled by multivariant fingerprinting must be greater than 0.99.

16.2.2 Advanced Techniques and Equipment

A dripping pill refers to a kind of preparation made by appropriately mixing crude drug extract with excipients and dropping into an unmixable condensation liquid, and finalized by concentration and condensation. The whole process is in fact a dispersal of the solid materials followed by formation of the dripping pills. Dripping pills are accepted in the global market; the preparation process is comparatively more advanced, however, the preparation requires a large unit drug loading capacity, higher extractum yield, well-known chemical composition, and easily conducted quality control and stabilization studies. All of these requirements necessitate that the preparation depend on excellent processes and techniques in extraction, refinement, and separation. On the other hand, the contents of moisture and macromolecular impurities in the extractum directly influence the quality of the dripping pill, which makes it necessary to control the entire process with advanced processes, equipment, analytical instruments, and control techniques to ensure the quality of the product. The extraction process should integrate traditional extraction techniques with modern ones, and be capable of realizing "three lows" (low temperature extraction, low temperature condensation, and low temperature drying), "three highs" (high purity, high yield, and high efficiency), and meet the demands of environmental protection, meanwhile ensuring the homogeneity and stability of dripping pill quality. Tasly has invested more than 10 million RMB in constructing a TCM extraction and separation pilot test center which consists of five modules: extraction, purification, condensation, refinement, and drying. The center can realize the automatization of process operations, real-time quality control, and provide a bridge for the transition of TCM extraction techniques from the laboratory to industrialization, and ensure the quality stabilization of TCM extraction products.

16.2.3 Quality Control of Extraction Process: Implementation of CGEP Management

The quality control of the extraction process includes raw material source control, transportation control, process control, and quality control. It requires that the extraction process meet four standards: management standardization, process specification, streamlined process control, and quality standardization, so as to ensure that the procedures of extraction and separation are normalized, standardized, controllable, and retraceable. CGEP require the establishment of SOPs for purchase, transportation, reception, storage, maintenance, warehouse departure, and for each step of extraction, including mixing, feeding, extracting, condensating, solid–liquid separating, alcohol precipitating, drying, packaging, and inspecting.

Tasly has constructed elevated warehouses with automated and informationalized management, realizing digital management for storage and maintenance conditions, such as temperature, humidity, and ventilation, so as to meet the demand for the storage of various crude drug materials.

Both the external quality of the crude drugs, such as particle size and fineness, and the internal quality of the crude drugs, such as the contents of effective components and the level of extractives, are strictly specified.

The process of material feeding is tightly closed to ensure the reliability of material entry.

The transportation of medicinal liquids in each step is also conducted in tightly sealed containers to ensure their cleanliness.

The production of extractum is carried out in a workshop with purity up to 100,000 grade to ensure the cleanliness of the products.

The precision requirement for the parameters in the extraction process is very high, and there are internal control parameters in each step, such as for solvent volume, pH, temperature, pressure, time, specific density of concentrated solution, solid material content, etc. The control of the parameters is quantified, with the fluctuation range of key parameters smaller than 0.002.

Through the application of statistical process control (SPC) methods, each step in the process is tested during production and thousands of data points are accumulated, which can be used in determining the optimal production parameters, tightening the in-process quality control, and meeting the high demand of dripping pill for extraction stability.

The solid content method is used to replace the previous specific gravity method for determining the endpoints of concentrations, which can strictly control the water content in the extractum within specified limits.

The fingerprinting technology is applied in each step of the process to ensure that the process is stable and controllable.

The extraction process has a strict requirement for the filtration and impurity removal system; thus, advanced filtration equipment plus self-made product-specific filters are used. Each step requires stepwise filtration, which ultimately produces a smooth and transparent paste without impurities, and the quality of the paste meets the international standards for plant soft gel exportation.

The extraction process has strict requirements for field cleaning, and the cleaning methods, conditions, frequency, and effectiveness are inspected.

Visualization training of the entire staff includes establishing a color herbs library, making digital videos of the entire extraction process, periodically training employees with the visual materials, and qualifying each worker by examination.

16.2.4 High Quality Standards for Extractum

To ensure the quality of the preparation, strict specifications of the extractum are established, which include appearance, moisture content, relative density, color, effective component contents, TLC identification, mid-infrared standard atlas, and limits on microbes, heavy metals, pesticide residues, and aflatoxins. Furthermore, the criteria of multivariant fingerprints are established, and the similarities of eight major effective components of Danshen controlled by multivariant fingerprinting are up to 0.99.

16.3 Quality Control in the Preparation Process

Shunnan Zhang, Haiou Dong, Junquan Wang, Xuesong Liu and Haibin Qu

In recent years, with the rapid growth of the TCM industry in China, the research, development, and production of dripping pill preparations have achieved significant success. The dripping pills of proprietary Chinese medicines are the products of Chinese patent medicine manufacturing and a modern pharmaceutical industry, which enrich the types and markets of Chinese patent medicines. The application of modern technology in TCM makes the production of various kinds of dripping pills possible. Annual output of dripping pills from each machine could reach more than hundred million pills. Thus, further development of dripping pill preparations can play to our strengths, and is an important component of TCM modernization and internationalization.

16.3.1 Overview

Shunnan Zhang and Haiou Dong

Dripping pill is a kind of preparation which is formed by appropriately mixing crude drug extractives with excipients and dropping into unmixable condensation liquid.

Dripping pill has the common advantages of other solid oral preparations, such as convenience of use, but it can also overcome the shortcomings of the latter, such as slow onset. It can increase bioavailability, enhance the therapeutic effects, and speed up onset, which makes it an ideal preparation type with a fast effect, other than injections. Therefore, dripping pill is a very promising preparation. The theoretical study and practical technology of dripping pill are still in the developing stage, and the studies are mainly concentrated on the following aspects: (1) the study of dripping pill formation theory; (2) the study of the solid dispersion state of the effective components; (3) the study of drop process stability.

16.3.1.1 The Theory of Dripping Pill Formation

After medicinal ingredients are melted with the matrix to form a solid solution (i.e., stock in dripping pill production), the mixture is dripped into unmixable condensation liquid. When the cohesive force of the stock is greater than the adhesive force between the stock and condensing agent, the formation force of dripping pills appears. The cohesive force of the stock depends on the difference between the surface tensions of the stock and condensing agent, i.e.:

$$\begin{aligned}\text{Forming force} &= \mathrm{W}c - \mathrm{W}a \\ &= 2_{\gamma \mathrm{A}} - (\gamma_{\mathrm{A}} + \gamma_{\mathrm{B}} - \gamma_{\mathrm{AB}}) \\ &= \gamma_{\mathrm{A}} + \gamma_{\mathrm{AB}} - \gamma_{\mathrm{B}}\end{aligned}$$

where Wc is the cohesive force of the stock, Wa is the adhesive force between the stock and condensing agent; γ_{A} is the surface tension of the stock; γ_{B} is the surface tension between the stock and condensing agent; γ_{AB} is the surface tension of the condensing agent.

The forming force is the so-called "spreading coefficient" in interface chemistry [1], which is generally accepted as a comparatively mature theory for dripping pill formation. However, the theory has its limitations, which makes it difficult to be applied in large-scale production. First, the surface tensions of stock and condensating agents are hard to find, and even if such data are located, it is difficult to use them directly as parameters for dripping pill production control because the data are influenced by all kinds of factors, such as temperature or interference from impurities in condensing agents. Second, the theory does not consider other factors which might influence the formation of dripping pills. For instance, when the temperature of the condensating agent is too low, even though the forming force meets the conditions, the roundness of the pills has difficulty meeting the requirements because the condensating agent cools down too quickly to give sufficient time for shrinking. On the other hand, if the temperature of the condensing agent is too high and the condensation process is insufficient, or the heat dissipation of the pills is slow, the pills are still not solidified at the end of the dripping process, which will cause the pills to stick together. The speed of heat dissipation in the pills depends on the specific heat values of the matrix and medicinal ingredients.

Therefore, it is required to control indirect parameters during the production of dripping pills to stabilize the forming force and solidification rate of the pills. Those parameters that can influence the forming force and solidification rate include: the properties of the raw materials, excipients and matrix, the proportions of the ingredients, the viscosity and relative density of the stock, the dripping temperature, the height and dripping rate of the stock, the relative density, viscosity, and temperature of the condensing agent, the height of the condensation column, as well as the inner diameter of the dripper and wall thickness. All of these parameters should be strictly controlled during production.

1. Formulation
   The proportion of extractum is the main influencing factor: if too much extractum is added, the surface tension of the stock would be affected, which could make contraction difficult and impair the roundness of the pills. Moreover, the pills would be soft and adhesions would occur during the dripping process.
2. The Stock Conditions
   The viscosity, relative density, temperature, and the height of the stock have large effects on the quality of dripping pills. Stock viscosity is dependent on the properties of medicinal ingredients and the matrix, as well as on the temperature. If the stock temperature is high and the viscosity is low, the dripping pills are apt to deform when they enter the condensating liquid, and if the temperature of the condensating liquid is low, the product would form "flat pills." On the other hand, if the stock temperature is low and viscosity is high, the dripper would draw and produce "little pills," which are hard to contract completely, and their roundness is compromised. The height of the stock can affect the dripping rate and pill weight. The change in height is the major cause of pill weight variation. When the dripping rate is fast, the pills may adhere together; if the dripping rate is slow, the production efficiency may decrease.
3. The Condensing Agent Condition
   The viscosity, relative density, temperature, and repeat use cycles, as well as the height of the condensating column, would influence the quality of the pills. When the pills enter the condensing agent, they generate a concentrating force because of the difference in surface tension, and they quickly lose heat and solidify. The viscosity, relative density, and temperature of the condensing agent, as well as the relative density of the stock, determine the movement rate of the pills in the condensing agent; so the faster the pills sink, the more force they meet, and as a result, the appearance of the pills become flatter. When the temperature of the condensing agent is too low, the pills solidify too quickly to form spherical pills; if the temperature of the condensing agent is too high, the solidification of the pills would be slow. If the height of the condensating column is not high enough, it is easy to cause a conglutination of the pills.
   If used repeatedly, the condensing agents may contain trace quantities of water and impurities, which can influence the surface tension between the stock and the condensing agent, which would decrease the contractibility of the pills. Thus, the condensing agent should be monitored and replaced in a timely manner. In short, we can improve the product quality and production process stability by analysis, quantized control, and real-time monitoring of the parameters related to the production process.

16.3.1.2 The Solid Dispersion State of the Drug Components in the Dripping PillS

The reason that dripping pill is a fast and potent preparation is that the drug components are in a state of solid dispersion in the pills. For instance, griseofulvin dripping pills manufactured in China contain PEG 6000 as the matrix. Under the microscope, only 5.39 % of the crystal grains are the size of 2–3 μm, and all the rest are smaller than 2 μm. When compared with the tablet in a dog model, the blood drug level of dripping pill was at least 1-fold higher than that of the tablet. In a clinical trial with 237 patients, the dosage of dripping pill was only a half that of the tablet, and it caused milder side effects than did the tablet. The required dosage of biphenyldicarboxylate pill was only 1/3 that of the corresponding tablet [2].

When the melting matrix is added to the soluble drug ingredients, the latter will distribute in the matrix in its molecular state to form the stock. When the stock is dripped into the condensing agent, because of quenching, the viscosity of the stock solution increases quickly and the stock solidifies. The drug molecules cannot aggregate into large crystals; rather, they form microcrystal grains. When the pills disintegrate in the body, the microcrystal grains can dissolve into the molecular state quickly in the body fluid and be absorbed; thus the bioavailability of the effective ingredients is improved.

For those insoluble ingredients, the particle size of the medicine can be minimized by micronization technology. When these materials are added to the melted matrix, the drug particles would be encapsulated by a high viscosity matrix which avoids the aggregation of drug ingredients. When the pill enters and disintegrates in the body, drug ingredients would be wetted with the dissolution of surrounding soluble matrix, which also can promote the bioavailability of the active ingredients.

It has been reported that drug ingredients and matrix could form a eutectic system and produce solid dispersion in the pills, i.e., the drug ingredients and matrix, in a certain proportion, could uniformly precipitate in a microcrystal state. For example, 40 % ibuprofen and 60 % PEG 2000 could form a eutectic system [5]. Also, when drug ingredients and matrix melt and solidify, they could form complexes, which have similar effects to a solid dispersion. Examples are narceine and amber acid, berberine hydrochloride and sulfanilamide, etc. [5]. Although these studies have reached a similar effect as solid dispersion, because of the strict demands for preparation formulation and matrix properties, it is difficult to be applied to the production of dripping pills.

At present, the study and application of solid dispersion in the production of dripping pill preparation are insufficient, and there are two main reasons:

First, the means of research on solid dispersion are limited. Currently, the only method that can be used for direct analysis of solid dispersion are X-ray diffraction and differential thermal analysis, and they are mainly used in R&D. Tasly Group has successfully used x-ray diffraction to analyze the solid dispersion state of borneol in Dantonic™. The indirect method for the study of solid dispersion is a dissolution rate assay. The solid dispersion state of drug ingredients can be reflected in the dissolution rate of the pills. However, this method is not suitable for all drugs.

Second, there are no specific equipments used for solid dispersion production. Because the history of dripping pill development is short, there are almost no specific equipments designed for its solid dispersion production, so the only substitutes are micronized equipment and colloid mill equipment, and their feasibility must be studied and confirmed.

Therefore, the keys to controlling solid dispersion pill production are the micronized treatment of drug ingredients and the monitoring of granularity during the dripping process.

16.3.1.3 The Stability of the Dripping Process

The stability of process dripping refers to the uniform dispersion of drug ingredients in the matrix, and the maintenance of stabilization. Documents related to this field are rare, which reflects the fact that the problem has not been paid enough attention.

When drug ingredients are added as a powder, they are hard to disperse in the matrix uniformly because of the high viscosity of melted matrix, which may decrease the homogeneity of the pills and affect the product quality.

When the drug ingredients are in the form of extractum, it mixes with the matrix as a suspension with a tendency to separate gradually into layers, which not only can affect the homogeneity of the pills and cause chromatic aberration, but also might affect the progress of the dripping process. Therefore, the stability of the dripping process is one of the major problems in the production of TCM pill preparations.

The extractum contains mainly three types of contents: (1) The effective constituents, which are the components that have remarkable biological activities or pharmacologic actions, such as alkaloids, glucosides, and volatile oils. (2) The adjuvant constituents, which are the components that have secondary biological activities or pharmacologic actions, and these components do occasionally have some application value in clinical practice. Some adjuvant constituents can promote the absorption of the effective constituents and enhance the curative effects; for example, digitonin can increase the absorption of digitalis and enhance the cardiotonic action of digitalis. (3) The ineffective constituents, which are the components with no biological activities, which have no medical effects in clinical practice, but exist widely in TCM herbs. For example, carbohydrates, proteins, pigments, resins, and inorganic salts.

The properties of extractum, such as viscosity and related density, are determined by the inactive macromolecules, and these properties determine the sedimentation velocity of the extractum in the melted matrix. Also, the fluidity of the stock is also the main factor which affects pill formation. The contributing factors to stock fluidity include matrix viscosity, extract viscosity, water content, stock temperature, etc. If the fluidity of the stock fluctuates over a wide range during the dripping process, it may have great negative effects on the dripping process.

Therefore, it is critical to maintain the stability of the dripping process by controlling the process. Control involves the following aspects: the particle size of the drug ingredient powder, and the viscosity and relative density of the extractum. Before dripping begins, the homogeneity and fluidity of the stock should be monitored, and the stability of flow ability can be realized by adjusting the dripping temperature and water content of the stock.

16.3.2 Quality Control of the Production Process of Dantonic™

Shunnan Zhang, Xueming Zhang and Jianping Lin

As the representative product of modern TCM drugs, Tianjin Tasly Group's Dantonic™ has the characteristics of good roundness, uniform pill weight, homogeneous content, and good solid dispersion, which provide a guarantee of clinical efficacy. During the period between research and development and entering the markets, they obtained 19 patents for Dantonic™, which include process, preparation, detection methods, etc., demonstrating their progress in the control of the production process.

The production of Dantonic™ uses the fourth generation automatic production line, developed independently by Tasly Group, which is the biggest dripping pill production line in the world with the highest degree of automation. It is equipped with digital electronic platforms and has the capacity of controlling and monitoring more than 80 data collection points through online gathering of parameters for rotation speed, temperature, pressure, flow rate, liquid level and viscosity, application of near infrared and particle size distribution instruments, real-time determination and closed-loop control, realizing the entire digital control of the production process, including material melting, transportation, dripping, pill formation, oil and pill separation.

16.3.2.1 Raw and Excipient Material Production Process Control

Polyethylene glycol (PEG): use near infrared methods for rapid assays of quality, so that stability of PEG quality is ensured.

Extractum: use near infrared methods for rapid determination of danshensu content and several other salvianolic acid components, and ensure the quality of the extractum by controlling its viscosity and relative density.

Borneol: Treat with micronizing equipment, and use real-time particle size measurement instruments to monitor the size distribution and ensure the quality of the powder.

16.3.2.2 Material Melting Process Control

Material liquid temperature: the material temperature is monitored by an independent closed-loop control system, which is a part of the material melting instrument, and it can make feedback adjustments to the temperature of the stock with a precision of ±1 °C, a precision three times higher than that of the traditional temperature control system.

Stock viscosity: multiple real-time viscosity meters are installed in the material melting device to monitor the viscosity of the stock. By adjusting the temperature and water content of the stock, the viscosity can be controlled to meet the requirements for the dripping process.

Homogeneity: a real-time near infrared detector is used for rapid determination of the distribution and the contents of danshensu and other salvianolic acid components in the stock to ensure the homogeneity of these effective constituents. The stability of the stock and the solid dispersion status of borneol are analyzed through a real-time particle size measurement instrument to ensure the stability and homogeneity of the pills during the production process.

16.3.2.3 Dripping Process Control

Dripping height: the height of the stock column is monitored and controlled by an alarm system, so that the dripping height will be maintained within a normal range.

Temperature of condensing liquid: an independent closed-loop control system temperature monitoring is used for feedback control of the condensation temperature.

Condensation height: by determining the height of condensation for feedback adjustment of the flow rate and pressure of condensation.

Life cycle of condensation: use a real-time near infrared detector to determine the purity of liquid paraffin to ensure its quality is up to the demand for the dripping process.

Pill quality: use a real-time near infrared detector to monitor the contents of the effective constituents and water, and the roundness and intactness of the pills.

16.3.2.4 Production Process Control Model

1. Temperature Control Model for the Melting and Mixing Process

   The stock is heated while being stirred in the melting and mixing tank (Fig. 16.1); the temperature of the stock is the key factor influencing the property of the stock and meeting the requirement for the dripping process. Suppose the density of the stock is constant during the course of heating, i.e., $\rho \equiv \rho_0$,

   Definition:

ρ	stock (extractum) density
h	stock height
r	radius of the bottom of melting and mixing tank
S	area of the bottom of melting and mixing tank
t	heating time
T_0	initial temperature of the stock
T	stock temperature at time t
T_{st}	tank wall heating temperature
u	total heat transfer coefficient
C_p	heat capacity of stock

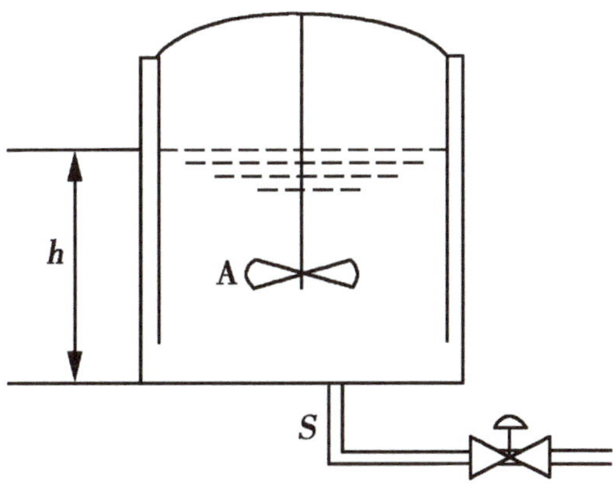

Fig. 16.1 Brief schematic diagram of a melting and mixing tank

Because the tank is kept still, the potential energy of stock in the tank is unchanged, in terms of liquid, and the internal energy is approximately equal to the enthalpy, i.e.:

$$U \approx H$$

where U is internal energy of the stock and H is the enthalpy of the stock. The following conclusion can be drawn:

If the heat is only provided through the tank wall and there is no water bath at the bottom of the tank, assuming the temperature of the tank wall is constant, then the stock temperature changing rate is related to the radius of the bottom of the melting and mixing tank, heat transfer coefficient, heating temperature of the tank wall, stock density, heat capacity of the stock, and stock temperature at a certain time, but not related to the quantity of the extractum:

$$\frac{\mathrm{d}T}{\mathrm{d}t} = \frac{u2\pi rh(T_{st} - T)}{\rho\pi r^2 hC_p} = \frac{2u(T_{st} - T)}{\rho rC_p}$$

The characterization of the temperature changing rate can be derived:

$$T = T_{st} - (T_{st} - T_0)\mathrm{e}^{\frac{-2ut}{\rho rC_p}}$$

Figure 16.2 is the heating–warming curve diagram of the melting and mixing tank, which complies to the exponential function.

During the melting and mixing course, the temperature rising time from time T_1 to T_2:

$$t_{12} = \frac{\rho rC_p}{2u}(\ln^{T_{st}-T_1} - \ln^{T_{st}-T_2})$$

2. Dripping Rate Control Model

   Characterization of dripping course: natural dripping process includes free dripping and charging dripping. It is derived that the dripping rate change during the free dripping course is:

$$v = \frac{\rho gH_h}{nV_dR}\mathrm{e}^{-\frac{\rho gt}{SR}}$$

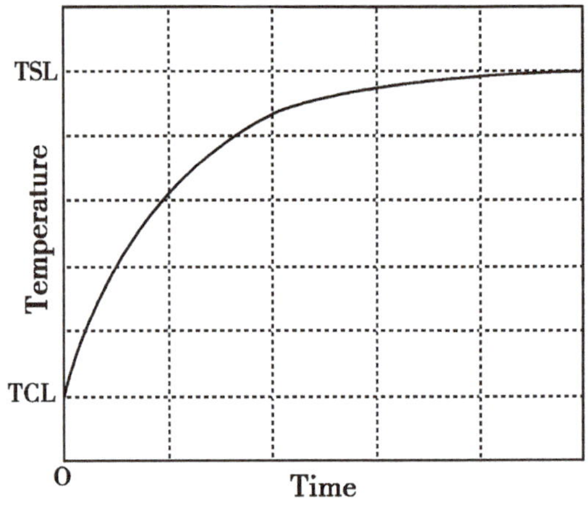

Fig. 16.2 The temperature rising curve of a melting and mixing tank

H_h is the highest initial level of the stock.

The dripping rate change of the charging dripping course is:

$$\upsilon = \frac{1}{nV_dR}\left[q_1R + (\rho gH_1 - q_1R)\mathrm{e}^{\frac{-\rho gt}{ST}}\right]$$

H_l is the lowest initial level of the stock. Where:

h	the height of stock in dripping tank at time t
T	temperature of the dripping tank at time t (even distribution)
S	bottom area of dripping tank
ρ	density of the stock
g	gravity acceleration
q_l	flow rate of charging valve
R	coefficient of flow resistance
n	number of dripping holes
υ	dripping speed at time t
V_d	dimension of the pills

(a) The nonlinear dripping rate control model based on fuzzy reasoning: using Danton-ic™ production data samples and fuzzy reasoning to build a systemic dripping rate model, the structure of the model is shown in Fig. 16.3, and the fitting curve in Fig. 16.4. The membership function of antecedent input variance is a standard triangular function, whereas the function of consequent output is a first-order linear equation. The sample training data is utilized to optimize the model which

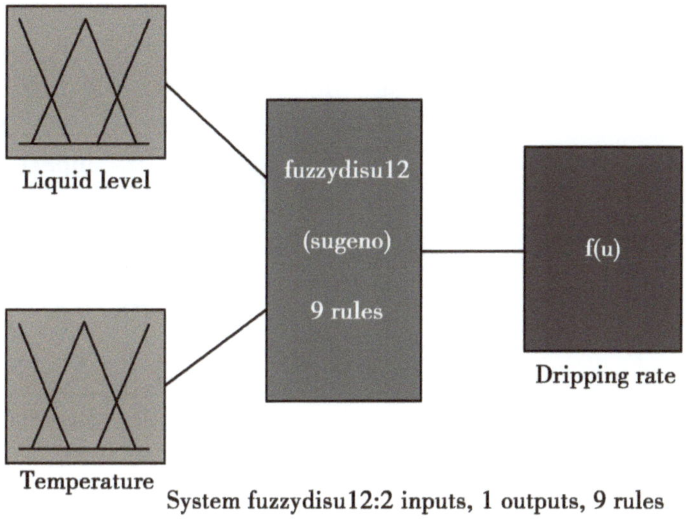

Fig. 16.3 The structural diagram of fuzzy dripping rate model

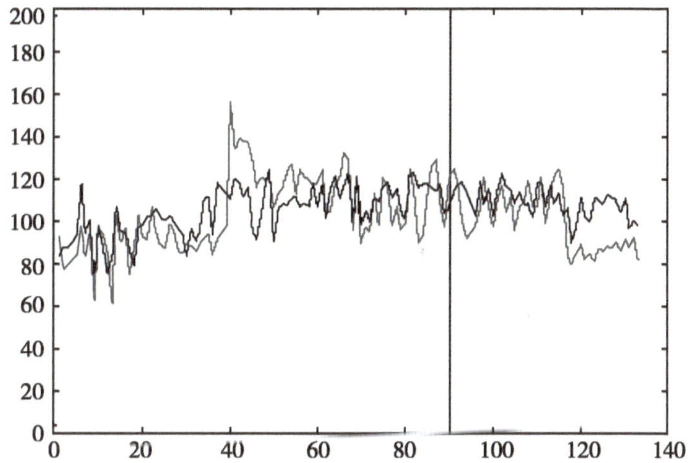

Fig. 16.4 The fitting curve of the fuzzy dripping rate model

employs a genetic algorithm, and the optimal variant is used as the constant of the consequent linear Equation.

(b) Principal component analysis (PCA)-FUZZY modeling: Fig. 16.5 shows the dripping rate model based on PCA and fuzzy reasoning, and its fitting curve is shown in Fig. 16.6.

3. The Yield-Targeting Partial Process Parameter Optimization Model

First, build a neural network model with partial process parameters as the input and yield as the output to describe the relationship between the process parameters and the yield. Based on the analysis of the process parameters, four critical parameters are selected as the input of the neural network model, and the output of the model is the pill yield. The Back-Propagation network has one input layer, one hidden layer, and one output layer. The input layer is composed of four nodes, the output layer is a single node, and the excitation function of the hidden layer (f_1) is a hyperbolic function. The excitation function of the output layer (f_2) is an s-type function. The structure of the model is shown in Fig. 16.7.

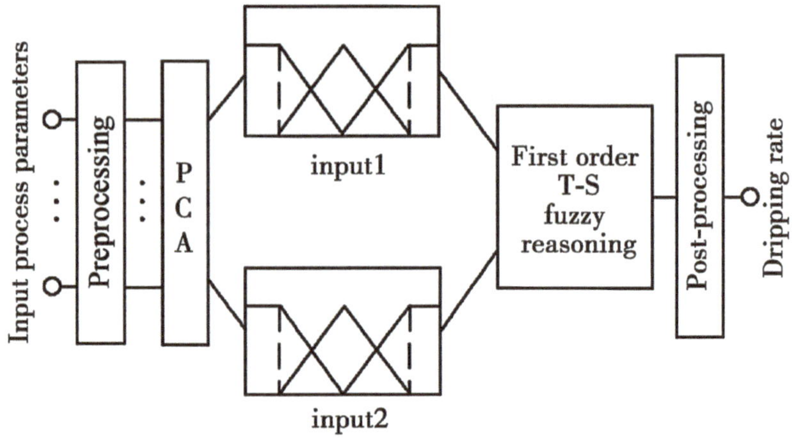

Fig. 16.5 The PCA-FUZZY model for the dripping rate assessment

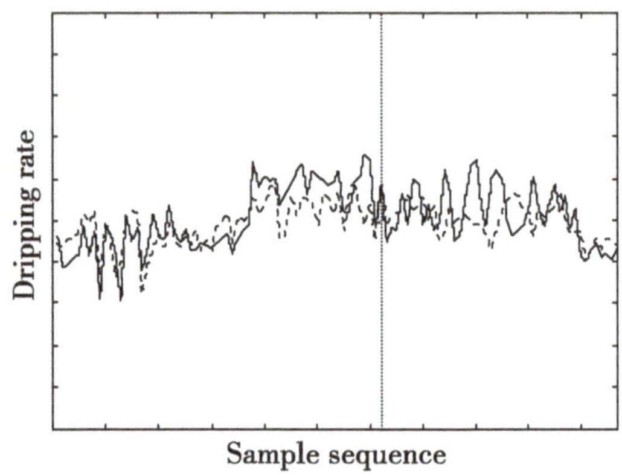

Fig. 16.6 The fitting curve of the PCA-FUZZY dripping rate model

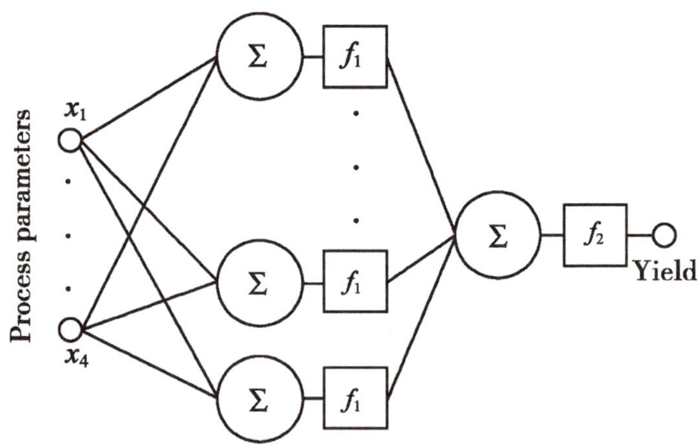

Fig. 16.7 The structural diagram of preparation process neural network model

In practice, by artificial immunity algorithm to optimize the aforementioned neural network model, five sets of parameter optimization control values can be acquired. Among them, the parameters in group 3 are the optimal values (Table 16.1).

Table 16.1 The control values of process parameter optimization

Group	Process parameter1	Process parameter2	Process parameter3	Process parameter4
1	27.5102	80.6410	76.2715	101.6618
2	26.0396	78.1867	71.2622	104.1408
3	**26.0080**	**79.1592**	**76.1718**	**98.5151**
4	26.5196	80.3038	76.3107	98.9732
5	25.9413	78.4241	75.9898	100.9592

16.3.3 The Application of Near-Infrared Spectroscopy in the Quality Control of Dantonic™ Production

Shunnan Zhang, Junquan Wang, Xuesong Liu, Haiou Dong and Haibin Qu

Near-infrared (NIR) spectroscopy technology has many advantages, such as speed, nondestructive testing, simple sample preparation, environmental friendliness, real-time analysis, etc. It has been widely used in the pharmaceutical industry. In this section, we introduce the application of this method in the quality control of Dantonic™ production.

16.3.3.1 Fast PEG Quality Assay by NIR Spectroscopy

1. Principle and Methods
   (a) Devices and Materials: Antaris Fourier Transform NIR Spectrometer (Thermo Nicolet Inc., USA), equipped with Integrating Sphere for reflectance analyses, RESULT™ software and TQ Analyst™ software. Fifty nine PEG specimens from different manufacturers and with different average molecular weights; the first 42 samples were qualified PEG 6000 samples from the products used in regular production, and four samples were prepared by mixing PEG with different molecular weights. Detailed information about the samples can be seen in Table 16.2.
   (b) NIR Spectroscopy: crush the PEG samples and pass through a 60 mesh sieve, fill a quartz bottle with the PEG powder until half full, and collect NIR data with the integrated sphere diffuse reflection detector. Scanning conditions: scanning time was 32, resolution was 8 cm^{-1}, and the spectrum range was 4,000–10,000 cm^{-1}. Collect three spectra from each sample, and use the average values for data processing. The quartz bottles were shaken before each scanning. The original NIR spectra of PEG samples are shown in Fig. 16.8, which illustrates that PEG samples with different average molecular weights had obviously different spectra; whereas samples with same average MW but from different manufacturers also had certain differences in their NIR spectra due to the differences in product quality.
   (c) Pretreatment of the Spectra: Fig. 16.9 is the NIR spectra after pretreatment. It shows that treatment of the spectra sequentially with the multiplicative scatter correction (MSC) and the second derivative (13-point Savitzky-Golay Smoothing) could not only well remove the baseline drift, but also reflect the detailed differences among different samples.
   (d) Selection of Spectral Wavelength Range: The region shorter than 4,500 cm^{-1} had greater noise; the signals in the region longer than 8,000 cm^{-1} were mainly straight lines after pretreatment, showing almost no differences between each other. After comprehensive reviewing of the spectral information and NIR absorption region, the spectral wavelength for the PEG sample analysis was set between 4,560 and 7,320 cm^{-1}.

Table 16.2 The sources and molecular weights of PEG samples

Sample no.	Source	Average molecular weight	Sample no.	Source	Average molecular weight
1	Tasly	6,000	31	Tasly	6,000
2	Tasly	6,000	32	Tasly	6,000
3	Tasly	6,000	33	Tasly	6,000
4	Tasly	6,000	34	Tasly	6,000
5	Tasly	6,000	35	Tasly	6,000
6	Tasly	6,000	36	Tasly	6,000
7	Tasly	6,000	37	Tasly	6,000
8	Tasly	6,000	38	Tasly	6,000
9	Tasly	6,000	39	Tasly	6,000
10	Tasly	6,000	40	Tasly	6,000
11	Tasly	6,000	41	Tasly	6,000
12	Tasly	6,000	42	Tasly	6,000
13	Tasly	6,000	43	Nanjing	6,000
14	Tasly	6,000	44	Nanjing	6,000
15	Tasly	6,000	45	Nanjing	6,000
16	Tasly	6,000	46	Tasly	4,000
17	Tasly	6,000	47[a]	Tasly	5,000
18	Tasly	6,000	48[a]	Tasly	6,000
19	Tasly	6,000	49[a]	Tasly	6,000
20	Tasly	6,000	50[a]	Tasly	7,000
21	Tasly	6,000	51	Tasly	8,000
22	Tasly	6,000	52	Tasly	10,000
23	Tasly	6,000	53	Tasly	15,000
24	Tasly	6,000	54	Tasly	20,000
25	Tasly	6,000	55	Shanghai TCM chemical reagents inc.	2,000
26	Tasly	6,000	56	Shanghai pudong gaonan chemical plant	4,000
27	Tasly	6,000	57	Shanghai pudong gaonan chemical plant	6,000
28	Tasly	6,000	58	Shanghai TCM chemical reagents inc.	10,000
29	Tasly	6,000	59	Shanghai TCM chemical reagents inc.	20,000
30	Tasly	6,000			

[a] The samples were prepared by mixing PEG samples with different average molecular weights

2. Results and Discussion
   (a) Principal Component Analysis: The score of the first principal component (PC1, x-axis) was plotted against the score of the second principal constituent (PC2, y-axis), and the result showed that the PEG 6000 samples used in regular production clustered in a limited area, and had significant differences in distribution from the other samples (Fig. 16.10). Thus, a

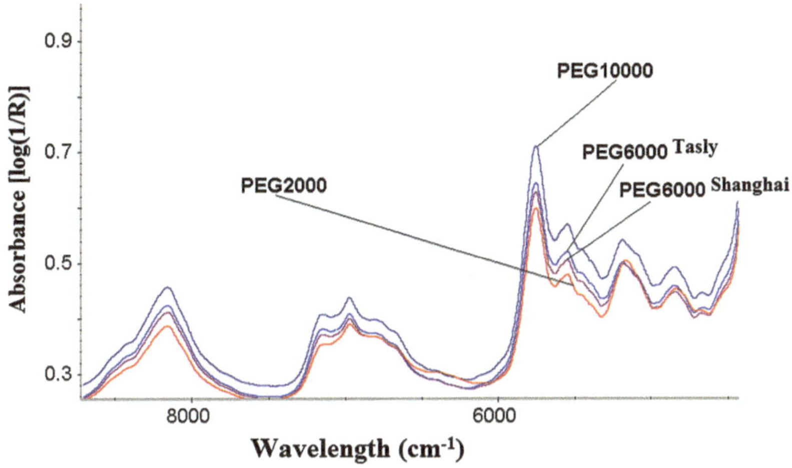

Fig. 16.8 The original PEG NIR spectra

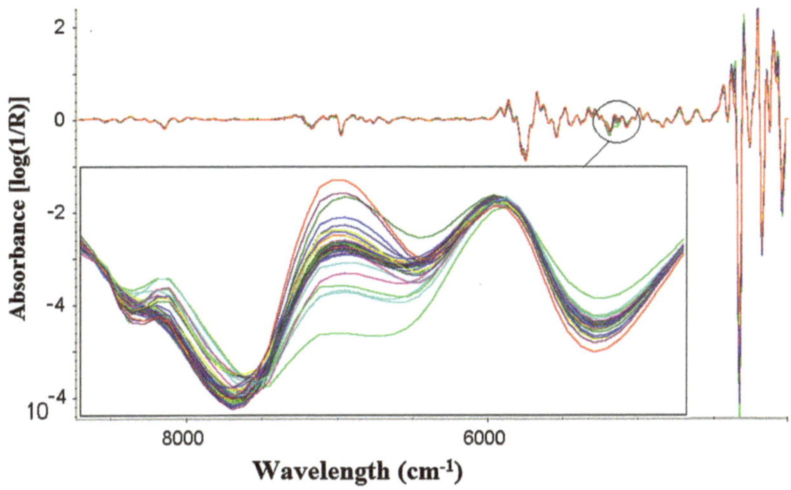

Fig. 16.9 The PEG NIR spectra after pretreatment

sample's quality can be rapidly identified by the pattern recognition method using its NIR spectral data.

(b) Cluster Analysis of NIR Spectral Data: Hierarchical cluster analysis (HCA) is a kind of unsupervised learning method which classifies spectral deviations into multiple levels of complexity. The HCA results are shown in Fig. 16.11. In the figure, the left side of the red arrow represent PEG 6000 samples used in production, and only sample No. 28 was allocated in the wrong cluster. It can be seen that different classes of PEG can be clearly separated by the method.

(c) Mahalanobis Distance: All samples were divided into three groups; one was of PEG 6000 whose quality meets the production standards, another was of the PEG samples whose quality is unqualified for production, and the rest of the samples were divided into three groups with the cross-validation method, i.e., two groups were taken as the calibration sets for modeling at each time, and the third group was used for prediction. The cycle was repeated. Sample No. 28 was classified into the wrong group in every classification, and when it was deleted, the performance of the model improved

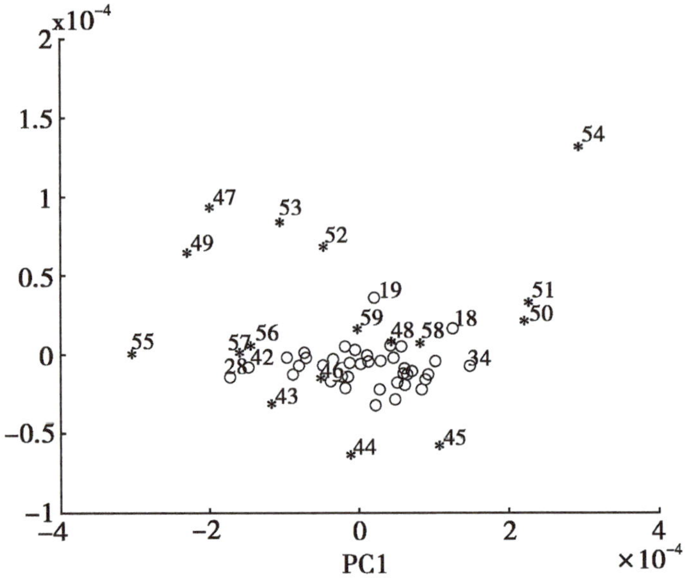

Fig. 16.10 The principle component analysis of NIR spectral data. *Circle* samples used in drug production; *asterisk* other samples

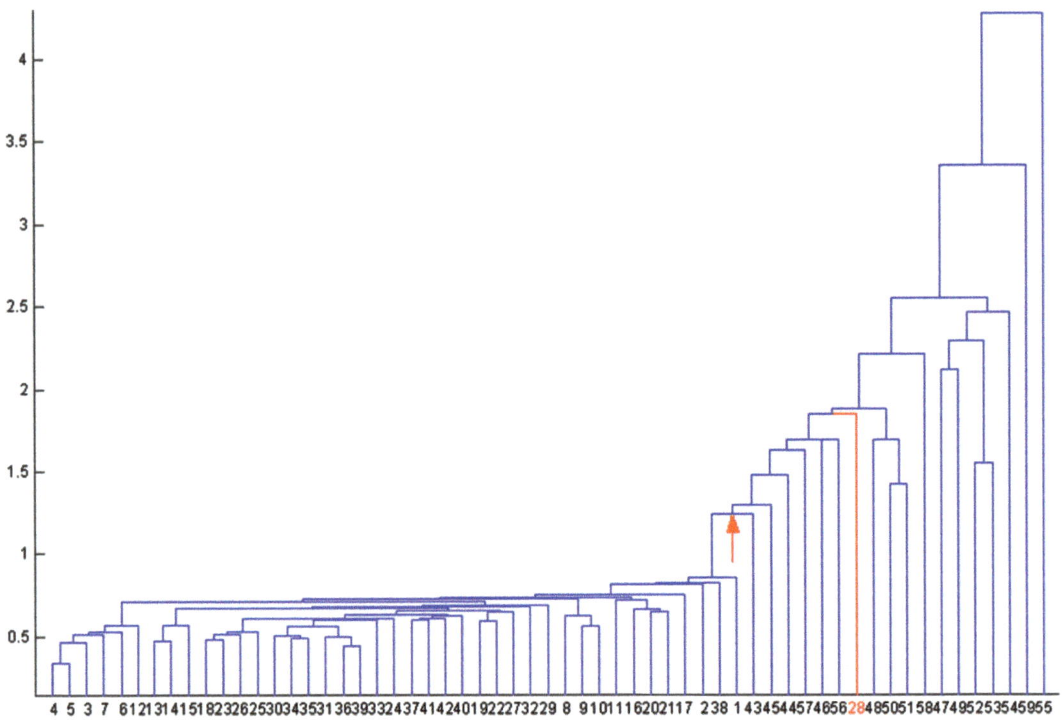

Fig. 16.11 Cluster analysis diagram of PEG samples based on the NIR spectral data

remarkably. Therefore, the sample was regarded as a singular point and deleted. The results of triple modeling are shown in Table 16.3 and Fig. 16.12, which show that the rate of correct cross-validation was 100 %.

Table 16.3 Cross-validation results of Mahalanobis distance modeling

Samples		Average distance to class 1	Average distance to class 2	Correct rate (%)
First	Production samples	0.670	1.858	100
	Rest samples	2.405	1.510	100
Second	Production samples	0.658	2.139	100
	Rest samples	2.607	1.678	100
Third	Production samples	0.733	1.989	100
	Rest samples	2.443	1.449	100

Class 1 qualified products, *Class* 2 unqualified products

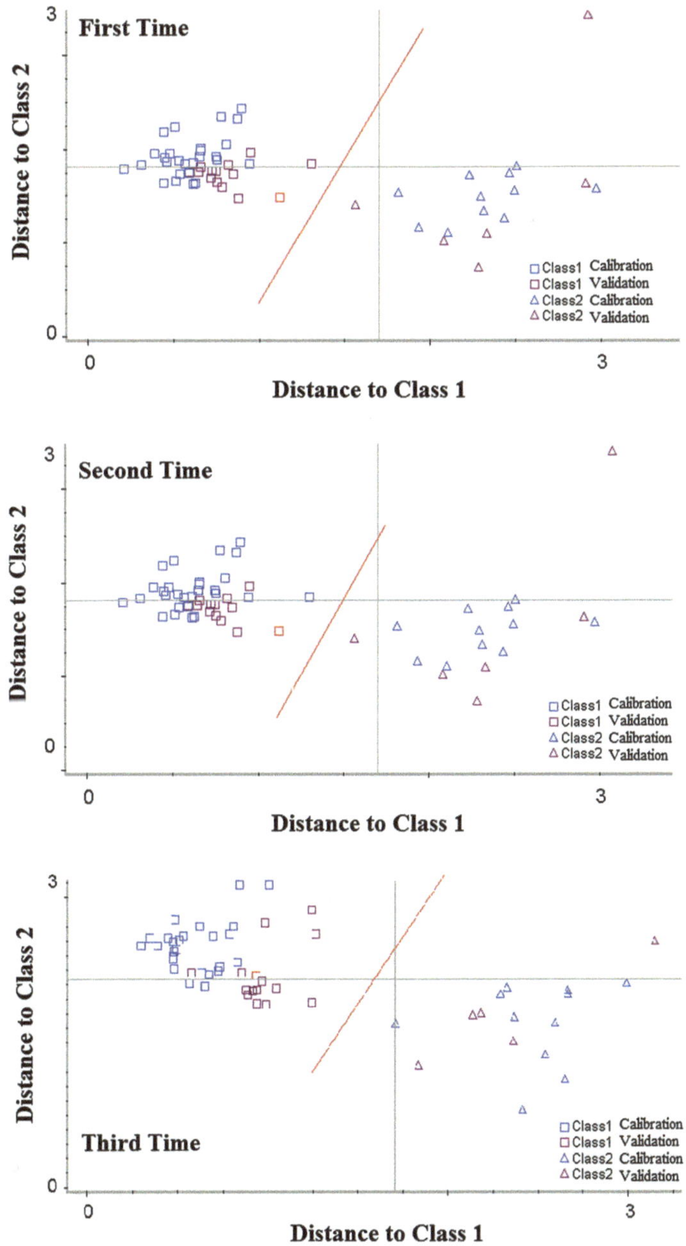

Fig. 16.12 Cross-validation results of Mahalanobis distance modeling

3. Summary
The rapid PEG quality identification method based on NIR diffuse reflectance spectroscopy can be used in practice to ensure the quality of Dantonic™.

16.3.3.2 Rapid Identification of the Quality of Condensation Agent by NIR Spectroscopy

1. Principles and Methods
   (a) Device and Material: Antaris Fourier Transform NIR Spectrometer (Thermo Nicolet Inc., USA), equipped with SabIR™ fiber optic reflectance probe, RESULT™ software and TQ Analyst™ software. Type TD-01 Digital Dripping Pill machine (Tasly).
   There were 189 liquid paraffin samples used for normal drug production; among them, 109 were used for modeling, and 80 were used for prediction. There were 121 liquid paraffin samples prepared from two batches of waste liquid paraffin provided by the workshop after dripping; 76 of them were used for modeling, and the other samples were used for prediction.
   (b) NIR spectra acquisition of samples: attach the transmission and reflection accessories on the SabIR™ optical fiber probe, fix an optical path of 4 mm, then immerse it into liquid paraffin to a specific depth to acquire the spectral data. Format of absorbance data: log 1/R; Range of spectrum: 4,000–10,000 cm^{-1}; Number of scans: 32; Resolution: 8 cm^{-1}, without attenuation. Three spectra were collected from each sample, and the average spectral value was used for data processing. The liquid paraffin was mixed well before each scanning. When the spectrum of each sample was acquired, the probe and accessories were cleaned. The original NIR spectra of the liquid paraffin samples are shown in Fig. 16.13.
   (c) Pretreatment of Spectra: Fig. 16.14 shows the NIR spectra after pretreatment, and the zoomed section is shown in Fig. 16.15. The figures show that after sequential treatment by second derivative and 15-point Savitzky-Golay Smoothing, the baseline drift was removed, and the figure reflects the more detailed difference among the samples.
   (d) Selection of Wavelength Range: After the pretreatment for the spectra, the wavelength range for qualitative identification was selected. The above figures show that liquid paraffin has a saturation peak near 4,300 cm^{-1}. Excluding the saturation peak zone and the zone without absorption, the remaining wavelength range, 8,400–4,500 cm^{-1}, was selected.
2. Results and Discussion
The samples were divided into two groups, one of those meeting the production requirements, the other of the waste samples which could not be used for Dantonic™ production.

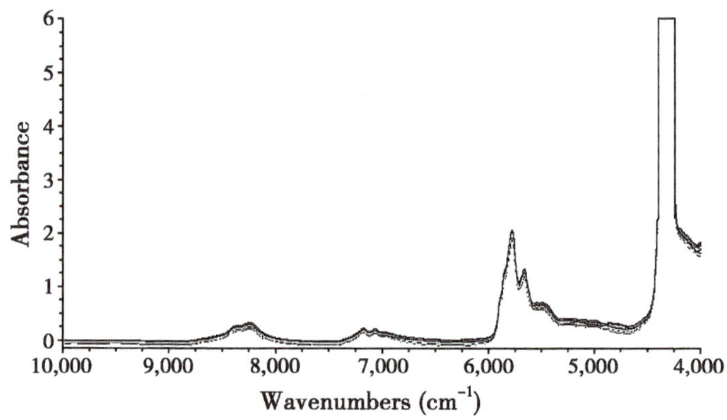

Fig. 16.13 The original NIR spectra of liquid paraffin samples

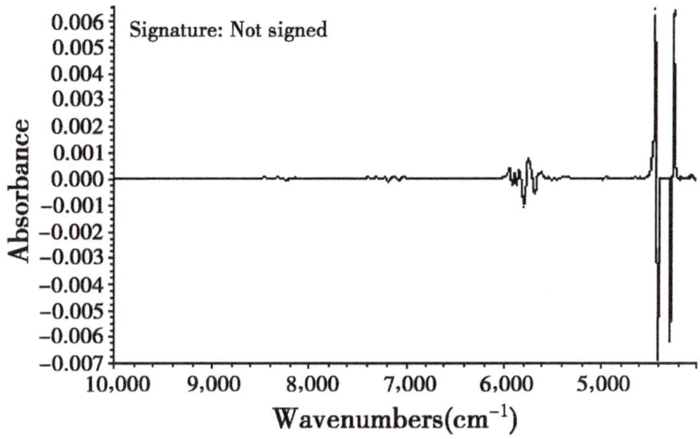

Fig. 16.14 NIR spectra of liquid paraffin after pretreatment

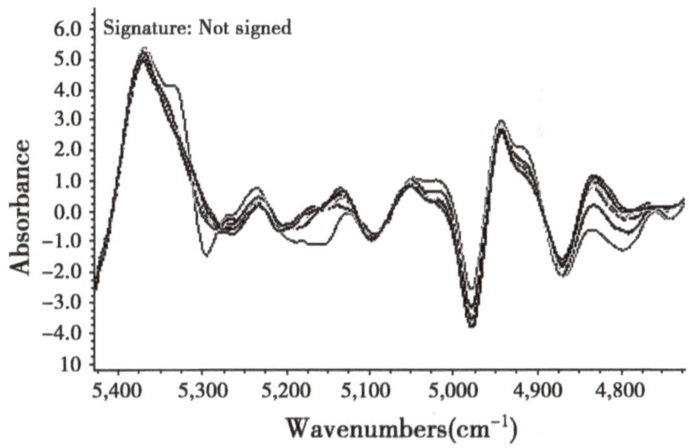

Fig. 16.15 Partial enlarged figure of NIR spectra after pretreatment

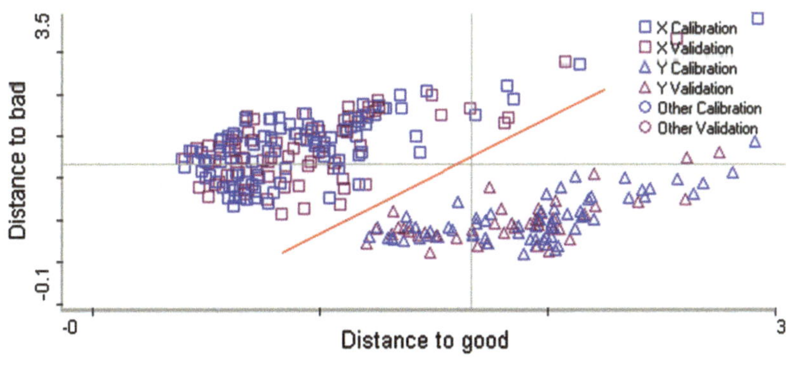

Fig. 16.16 Mahalanobis distance modeling results. *Square* normal liquid paraffin; *triangle* waste liquid paraffin

The Mahalanobis distance modeling results are shown in Fig. 16.16, with the two groups of liquid paraffin clustered on the two sides of the straight line, respectively, and the correct rate was 100 %.

3. Summary
   The NIR transflective spectroscopy-based rapid assay for liquid paraffin quality is capable of determining a sample's quality whether it can be used for Dantonic™ production.

16.3.3.3 Rapid Determination of the Effective Constituents of Dantonic™ by NIR Spectroscopy

1. Principles and Methods
   (a) Instruments and Reagents
   Instruments: Agilent 1100 HPLC equipped with quaternary pump, online degasser, automatic sampler, DAD, column temperature chamber, and ChemStation work station. Antaris Fourier Transformation NIR Spectrometer (Thermo Nicolet Inc., USA), equipped with the integrating sphere diffuse reflection detector, and RESULT spectral acquisition, and TQ Analyst data management software.
   Reagents: acetonitriles (chromatographic pure, Fisher Inc., USA), phosphoric acid (AR), acetic acid (superior grade pure), Wa–Ha–Ha purified water (commercial product). The reference substances of Danshen, such as danshensu, protocatechuic aldehyde, and salvianolic acid B were purchased from the National Institutes for Food and Drug Control.
   (b) HPLC Methods
   (i) Preparation of the test samples: Weigh 10 pills of Dantonic™ from each batch and place in a 10 ml volumetric flask, respectively, add an appropriate amount of distilled water, resolve with ultrasound and bring the volume to the scale, mix well, and centrifugate; the supernatant solutions were used for the sample assay.
   (ii) HPLC analysis conditions: Agilent SB-C_{18} analytical column (4.6 mm × 250 mm, Zorbax SB USCL010304); Mobile phase: phase A: 0.02 % (v/v) phosphoric acid solution, phase B: acetonitrile solution with 0.02 % (v/v) phosphoric acid; gradient elution, elution program is shown in Table 16.4. Flow rate: 1 ml/min, detection wavelength: 280 nm, column temperature 30 °C, sample size 10 μl. The HPLC chromatogram is shown in Fig. 16.17, wherein the identities of danshensu, protocatechuic aldehyde and salvianolic acid B were confirmed by the control tests and mass spectroscopy.
   (c) NIR Spectra Acquisition
   (i) Sample collection: the contents and distribution of modeling samples have great influence on the performance of the corrected model. Except for the Dantonic™ samples provided by the manufacturer, some Dantonic™ samples were prepared in the laboratory, simulating the manufacturing conditions of a workshop and using raw materials of different grades, from different production regions, and in different amounts. The total sample size was 60, of which 21 were samples prepared in the laboratory.
   (ii) Analysis conditions: The integrating sphere diffuse reflectance detection system; scan wave range: 4,000–10,000 cm^{-1}; number of scans: 32; resolution: 4 cm^{-1}, reference: internal background. Three samples were taken from each batch, each sample was tested three times, and the average values were used for data processing. The original NIR spectra are shown in Fig. 16.18.
   (d) Establishing the Calibration Model
   Based on appropriate pretreatment of the spectra, the partial least squares (PLS) method was used to construct a NIR spectral multivariant calibration

Table 16.4 HPLC gradient elution program

Time (min)	A % (v/v)	B % (v/v)
0	92	8
8	82	15
15	79	21
40	66	34

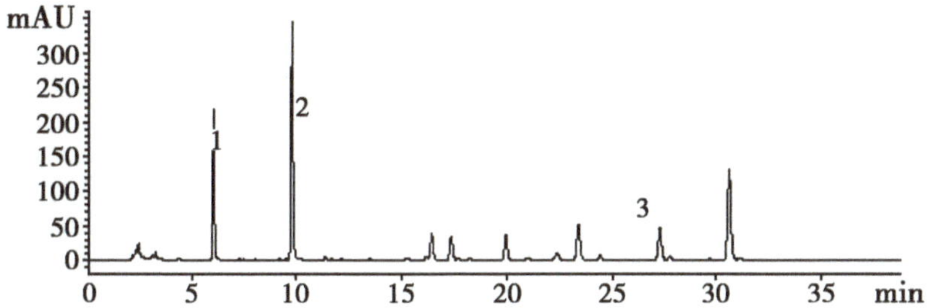

Fig. 16.17 The HPLC chromatogram of phenolic acids in Dantonic™. *1* danshensu; *2* protocatechuic aldehyde; *3* salvianolic acid B

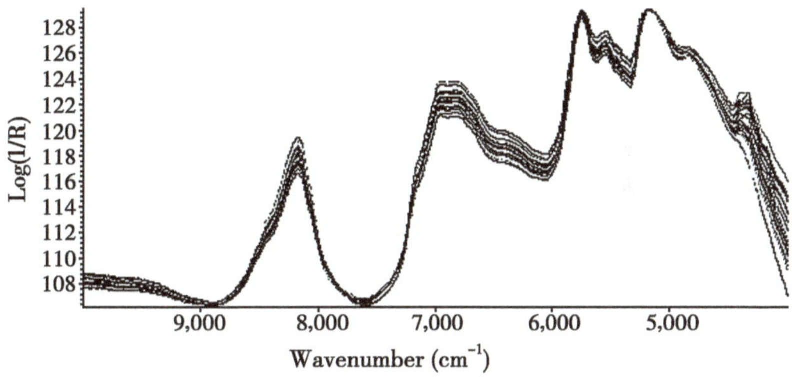

Fig. 16.18 The NIR diffuse reflectance spectra of Dantonic™ samples

model. With the root mean square error of cross-validation (RMSECV) as an indicator, the leave-one-out cross-validation (LOOCV) method was used to determine the optimal number of PLS major factors. The prediction performance of the model on unknown samples was examined by root mean square error of prediction (RMSEP), and the relative prediction deviation values of the calibration set and verification set of the model were verified with relative standard error (RSE).

2. Results and Discussion
   (a) Pretreatment of spectra: the MSC technique and derivative spectral analysis were combined to pretreat the NIR spectra in order to remove the scattering effect and baseline drift. Taking danshensu content as an example, the prediction performance of the model under different spectral pretreatments is shown in Fig. 16.19, in which HPLC analytical values were on the horizontal axis and NIR spectral prediction values were on the vertical coordinate. It shows that pretreatment by a combination of MSC and second derivative spectra gave better results. In establishing the calibration model for salvianolic acid B, it was found that the results of first derivative treatment were better than those of second derivative treatment. Therefore, the final established calibration model for salvianolic acid B was pretreated with the combination of MSC and first derivative spectra. First, remove the water absorbance band, then select the optimal modeling waveband according to RMSEP values. The results are shown in Table 16.5.
   (b) Determining the optimal number of modeling factors: the optimal numbers of three-component modeling PLS major factors were 7, 7, and 5, respectively.
   (c) Investigating the prediction performance of PLSR model: an independent

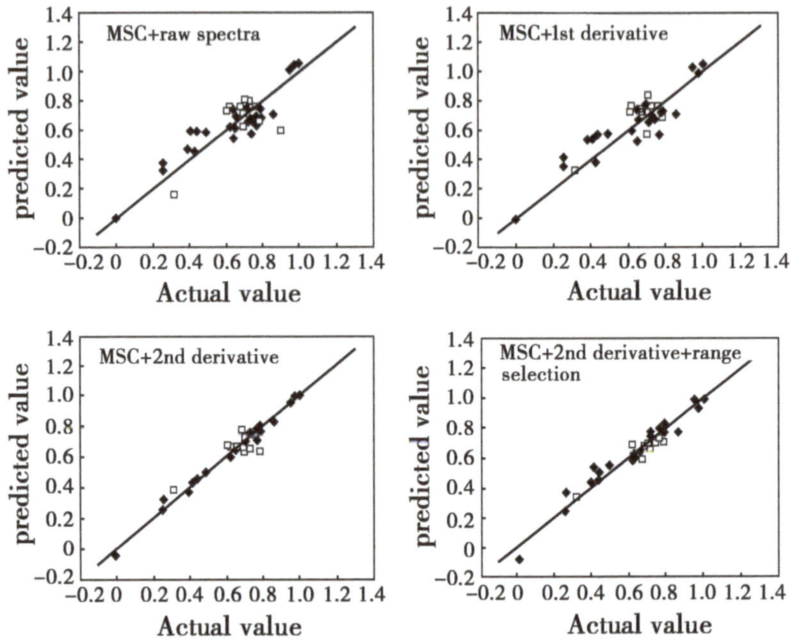

Fig. 16.19 The prediction results of danshensu calibration model with different spectral pretreatments. *Black diamond* calibration set; *square* verification set

Table 16.5 The NIR spectral modeling wavebands

	Danshensu	Protocatechuic aldehyde	Salvianolic acid B
Wavelength range (cm^{-1})	8,700.00–7,305.00	9,770.00–7,155.00	9,852.00–7,057.00
	6,600.00–6,350.00	6,950.00–5,150.00	6,728.00–5,143.00
	4,950.00–4,150.00	4,700.00–4,500.00	4,935.00–4,449.00

validation sample set was set up to verify the calibration model; the validation set was composed of the samples from 16 batches of real Dantonic™. The results showed that the correlationship between NIR spectral prediction values and HPLC analytical values was good, especially for the models for danshensu and protocatechuic aldehyde; the correlation coefficients were greater than 0.99, and RSEP was less than 10 %. The correlation coefficient of salvianolic acid B model was greater than 0.97 (Fig. 16.20 and Table 16.6).

3. Summary

The contents of several phenolic acids in Dantonic™ were simultaneously determined with the NIR spectral method, and the results were satisfactory. This method can be used for the fast determination of the effective constituents in other compound TCM preparations.

16.3.3.4 Rapid Determination of the Appearance Yield of Dantonic™ by NIR Spectroscopy

1. Principles and Methods
   (a) Instruments and reagents: Antaris Fourier Transformation NIR spectrometer (Thermo Nicolet Inc., USA); accessory configurations: integrating sphere diffuse reflection detector, rotary sample cup diver, quartz sample cups (internal diameter, 29 mm). Software packages: RESULTS software used for spectral acquisition, TQ Analyst software used

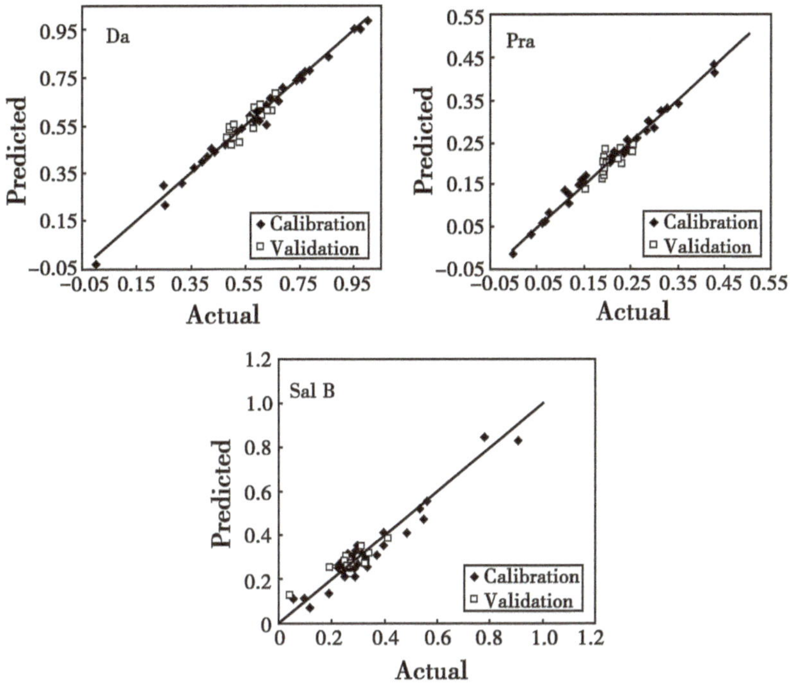

Fig. 16.20 The diagrams of correlations between HPLC analytical values and NIR spectral prediction values

Table 16.6 The prediction performance of NIR spectral calibration models

	R	RMSEC	RMSEP	RSEC (%)	RSEP (%)
Danshensu	0.9942	0.0210	0.0326	3.4037	5.8161
Protocatechuic aldehyde	0.9965	0.0082	0.0193	3.5743	9.2917
Salvianolic acid B	0.9714	0.0402	0.0363	11.7347	12.4232

for spectra treatment and calculation. Dantonic™ samples: qualified products, large pill defectives, and rough surface defectives.

(b) Preparation of samples: take one batch of qualified Dantonic™ products (lot number 20020116), mix with large pill defectives in the proportions listed in Table 16.7, and prepare 18 samples (group 1). Each sample contains 200 pills.
Take one batch of qualified Dantonic™ products (lot number 20020116) and mix with rough surface defectives in different proportions to prepare 20 samples (group 2), each sample containing 200 pills. The defective rates in each sample were similar to those in Table 16.7, with the addition of two more samples with defective rates of 85 and 95 %.
In the actual normal production process, the number of defective products in each batch is generally small; thus, it is more meaningful that NIR spectroscopy can detect low defective rates. Therefore, group 3 and group 4 samples were prepared as follows: group 3 contained the same defective products as group 1, but with an even gradient of defective rates from 0 to 10 % for a total of 11 samples; group 4 contained the same defective products as group 2, and the defective rates and sample size were the same as in group 3.

Table 16.7 Defective rates of samples in group 1

Serial no.	Defective rate %	Serial no.	Defective rate %	Serial no.	Defective rate %	Serial no.	Defective rate %
1	0	6	25	11	50	16	80
2	5	7	30	12	55	17	90
3	10	8	35	13	60	18	100
4	15	9	40	14	65		
5	20	10	45	15	70		

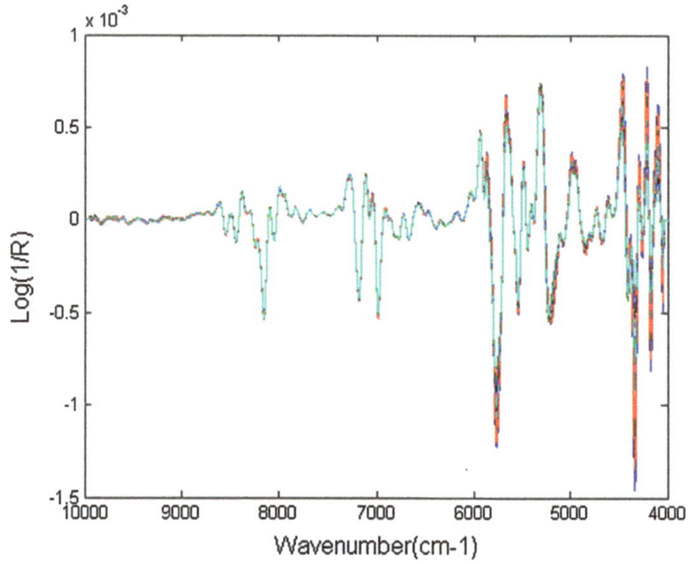

Fig. 16.21 Second NIR derivative spectra after Norris filtering

(c) Acquisition of NIR spectra: mix samples well and scan the samples with a NIR spectrometer using the integrating sphere diffuse reflection detector. Before scanning the samples, the internal background was scanned.

(i) Test conditions: scan frequency: 32; resolution: 4 cm^{-1}, scan range: 10,000–4,000 cm^{-1}, repeat 5 times and take the average values.

(ii) Pretreatment of spectra: to remove baseline drift, the original spectra were processed by a differential, smoothed through Norris derivative filter, and the second derivative spectra are shown in Fig. 16.21.

2. Results and Discussion

(a) Optimal number of PLS major factors: the selection of PLS major factors was determined by LOOCV method. The results showed that the optimal numbers of major factors were 4, 1, 6, and 4, respectively (Fig. 16.22).

(b) Calibration model performance evaluation: The correlation coefficient of NIR calibration model for Dantonic™ appearance was greater than 0.98, indicating good performance. Figure 16.23 shows the results of model correction, in which the abscissa is the true value of defective rates, whereas the coordinate is the calculated values. It is shown that the calibration model has a remarkable linear relationship, and both calibration errors and prediction errors are small, indicating high accuracy of the model. The good results were obtained under conditions of no waveband selection and

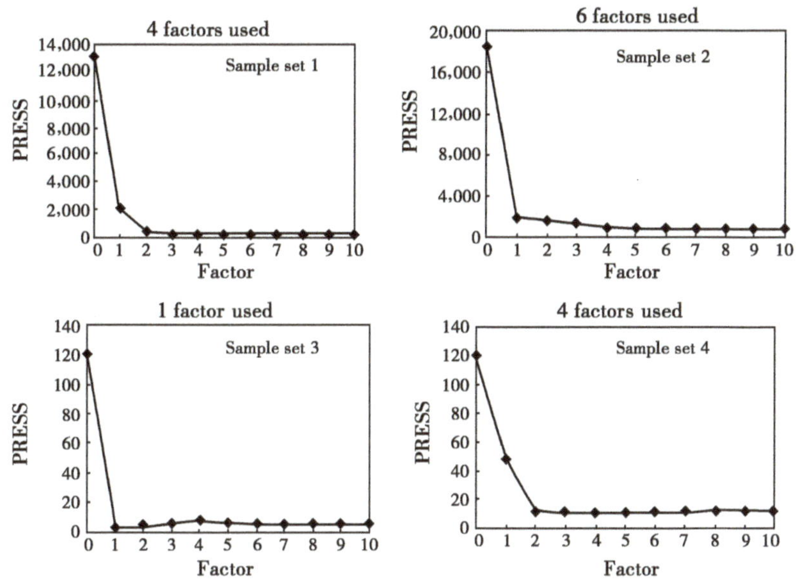

Fig. 16.22 Determination of PLS major factor number

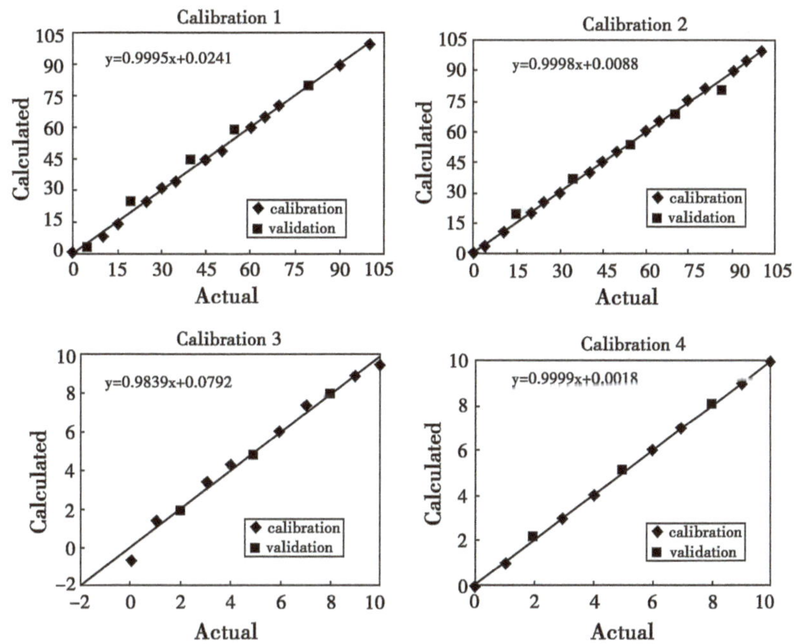

Fig. 16.23 Actual defective rates and the defective rates predicted by NIR spectroscopy

linear modeling, suggesting that the NIR diffuse reflectance spectral method is suitable for the fast determination of pill appearance quality (Table 16.8).

3. Summary

NIR spectroscopy was used to determine the quality of pill appearance. The results showed that the defective rate could be detected through NIR diffuse reflectance spectra based on a PLS multivariant calibration model. This method is suitable for the fast assay of the appearance of dripping pills and related preparations.

Table 16.8 The performance indices of calibration models

Sample set	R^2	RMSEC	RMSEP	RSEC (%)	RMSECV	RSEP (%)
1	0.9995	0.6250	3.300	1.150	4.190	6.900
2	0.9998	0.4190	3.030	0.7140	7.470	5.260
3	0.9844	0.4240	0.1570	7.010	0.6430	2.820
4	0.9999	0.01590	0.1030	0.2630	1.180	1.850

16.3.3.5 Rapid Determination of the Moisture Content in Dantonic™ by NIR Spectroscopy

1. Principles and Methods
   (a) Instruments and material: Antaris Fourier Transformation NIR Spectrometer (Thermo Nicolet Inc., USA); accessory configuration: integrating sphere diffuse reflection detector, rotary sample cup diver, quartz sample cup (internal diameter 29 mm); software packages: RESULT software used for spectral acquisition, TQ Analyst software used for the treatment and calculation of spectra.

   To obtain samples with high moisture contents, samples were sealed in a desiccator with saturated KNO_3 solution for 4–5 days at 25℃. The samples were taken out at different times. The moisture contents of the samples thus obtained were 4.8–8.2 %, and the total number of samples was 25. When moistening, most of the water molecules attach to the surface of the pills, and the pills' surfaces resolve as the water content increases. Thus, it is difficult to increase too much the moisture contents of DANTONIC™.

   (b) NIR spectral acquisition: Put two layers of sample pills in the rotary cup to cover the bottom, and collect spectral data by the integrating sphere diffuse reflectance accessory. Absorbance data format: log 1/R; range of spectral acquisition: 4,000–10,000 cm^{-1}; number of scans: 64, 8 × plus; resolution: 8 cm^{-1}. Three spectra were acquired from each sample, and their average value was used for data processing.

   Reference method: after NIR spectroscopy analysis, transfer the pills in the rotary cup to a weighing cup to weigh, dry for 5 h at 105 °C, and weigh after cooling down to room temperature in a desiccator. Repeat the above process until a constant weight is reached.

2. Results and Discussion
   (a) Selection of Wavelength Range: There were two spectral regions (4,800–5,240 and 6,720–7,130 cm^{-1}) in which the correlation values were good. These regions overlap with the region where water has significant NIR absorption peaks (5,120 and 6,850 cm^{-1}), suggesting they are suitable for the quantitation of moisture content.

   (b) Results of PLSR Modeling: In the region between 4,800–5,240 and 6,720 –7,130 cm^{-1}, the results pretreated with the first derivative were the best. A total of 16 samples were used for modeling, and 9 for prediction. The optimal PLS calibration model is shown in Fig. 16.24.

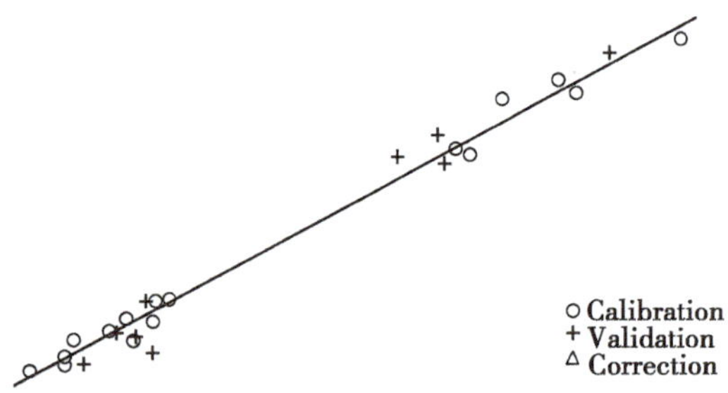

Fig. 16.24 The correlation diagram between NIR prediction values and actual values

The correlation coefficient of the model was 0.9953, root mean square error of calibration (RMSEC) was 0.110, and RMSEP (RMSEP) was 0.199.

3. Summary

The results showed that NIR spectroscopy could be used for the determination of the moisture content in Dantonic™. Compared with traditional methods such as HPLC, this method requires only simple sample treatment, and can save much analysis time and cost.

16.3.3.6 Rapid Determination of the Coating Thickness of Dantonic™ by NIR Spectroscopy

1. Principles and Methods
   (a) Instruments and Reagents: Antaris Fourier Transformation NIR spectrometer (Thermo Nicolet Inc., USA); accessory configuration: integrating sphere diffuse reflection detector, quartz sample cup; software packages: RESULT software used for spectral acquisition, TQ Analyst software used for the treatment and calculation of spectra; BS 200S-WEI electronic balance (Sartorius). Dantonic™ coating samples, a total of 54 from six different batches, were provided by Tianjin Tasly Pharmaceutical Co., Ltd. Each batch contained one sample of naked pills (uncoated) as the control.
   (b) NIR spectral Acquisition: The samples of Dantonic™ were put into the refrigerator to freeze at a constant temperature, and then transferred to a sample cup to gather NIR spectra through the integrating sphere diffuse reflection detector. Spectral acquisition conditions: with the internal background of instrument as the reference, the wavelength range was 10,000–4,000 cm^{-1}, number of scans was 32, and resolution was 8.0 cm^{-1}. Three spectra were collected from each sample, with the quartz cup rotated 120° after each spectral acquisition, and the average values were used for data processing. The original NIR spectra of the samples are shown in Fig. 16.25.
   (c) Determination of sample weight: take exactly 100 pills from each group and weigh. Weigh the samples obtained at different time points of the coating process and compare with naked pills of the same batch to calculate the weight gain rate, namely, sample weight minus the naked pill weight, divided by naked pill weight. (The pills gain less than 6 % weight after coating, and the changes in diameter are smaller than 2 %. It is believed that weight gain rate is directly proportional to the coating thickness.)
2. Results and Discussion
   (a) Selection of Wavelength Range: The major component of Dantonic™ coating

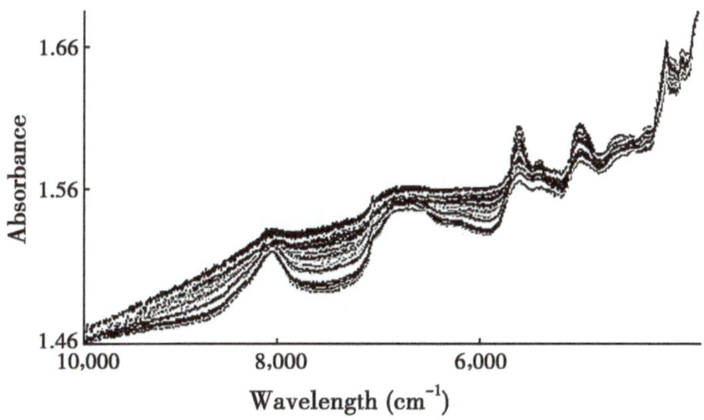

Fig. 16.25 The original NIR diffuse reflectance spectra of the samples

Table 16.9 Influence of different spectral pretreatments on PLSR modeling

Pretreatment method	Correlation coefficient R	RMSEC	RMSEP
Absorbance (source spectrum)	0.944	0.333	0.423
MSC	0.780	0.132e3	0.132e3
MSC + S–G	0.780	0.132e3	0.132e3
MSC + first derivative	0.980	0.198	0.434
MSC + second derivative	0.879	0.480	1.16
MSC + S–G + first derivative	0.974	0.228	0.338
MSC + S–G + second derivative	0.924	0.385	0.832
SNV	0.967	0.256	0.355
SNV + S–G	0.962	0.276	0.350
SNV + first derivative	0.901	0.437	0.321
SNV + second derivative	0.901	0.437	0.321
SNV + S–G + first derivative	0.901	0.437	0.321
SNV + S–G + second derivative	0.901	0.437	0.321
MSC + Norris + first derivative	0.980	0.200	0.369
MSC + Norris + second derivative	0.870	0.497	0.454
SNV + Norris + first derivative	0.901	0.437	0.321

material is cellulose. Its hydroxyl group has strong absorption in the 6,900–6,000 cm^{-1} and 4,900–4,500 cm^{-1} regions, which could be the characteristic wave bands for the determination of coating thickness.

(b) Spectral Pretreatment: Various kinds of spectral pretreatment methods were combined and used for PLSR modeling. The results are shown in Table 16.9.

(c) Selection of PLSR Major Factor Number and Cross-Validation Results: The results are shown in Fig. 16.26, and the optimal number of major factors was 2.

(d) Results of PLSR Modeling: The established quantitation calibration model was used to predict the average coating thickness using the prediction set, and the results are shown in Fig. 16.27. The correlation coefficient of the calibration model was 0.974, the RMSEC was 0.228, and RMSEP was 0.338, suggesting that the calibration model could accurately predict the average coating thickness.

3. Summary

NIR diffuse reflectance spectroscopy was used for the determination of the coating thickness of a TCM drug. The predicted results from the model matched well with the actual average coating thickness, thus the method could be used to determine the average coating thickness in the production process.

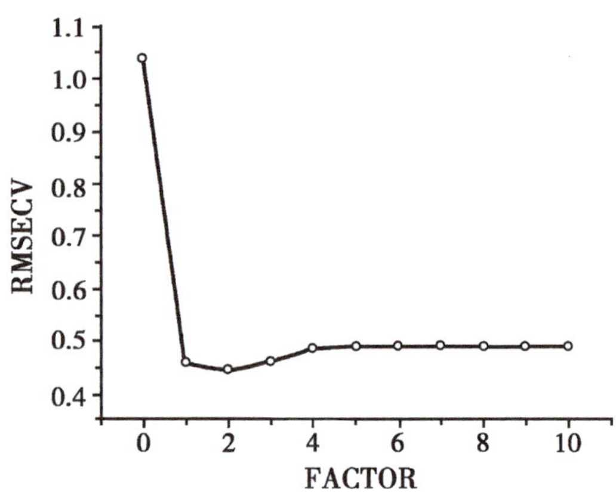

Fig. 16.26 The optimal major factor number and cross-validation results

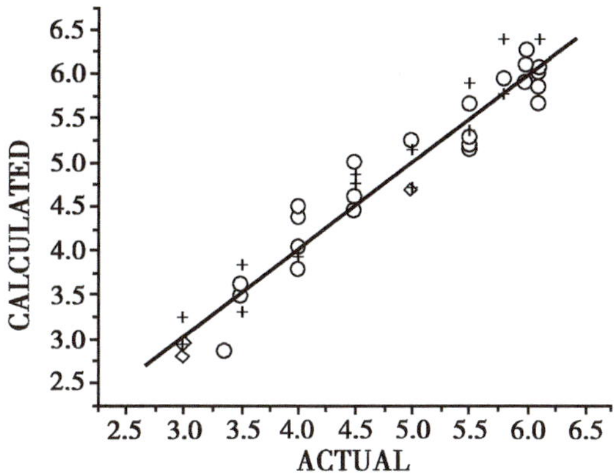

Fig. 16.27 The correlation between the predicted NIR spectral values of the prediction sample set and the average coating thickness

16.3.3.7 On-Line Monitoring of the Homogeneity of the Material Melting Process in Dantonic™ Production by NIR Spectroscopy

1. Instruments and Reagents: Antaris Fourier Transformation NIR Spectrometer (Thermo Nicolet Inc., USA), SabIR optical fiber probe, transmission reflectance accessories with software packages, RESULT software used for spectral acquisition, TQ Analyst software used for spectral treatment and calculation; WATERS 2695 HPLC, Waters 2487 UV detector; AE240 electrical balance (Mettler), TD-01 digital dripping machine (Tasly).
   Dantonic™ extractum, PEG 6000, and borneol were provided by Tianjin Tasly Pharmaceutical Co., Ltd. The reference products, sodium danshensu and protocatechuic aldehyde, were purchased from the National Institutes for Food and Drug Control. Methanol was chromatographic pure (Merck), purified water, and other agents were all analytical pure.
   Preparation of the stocks of effective components: add Dantonic™ extractum, PEG 6000, and borneol in a beaker, melt the mixture, stir well, and scan the liquid to acquire the NIR spectra. Cool the liquid down, and test the stock with HPLC. A total of four batches of Dantonic™ extractum were used, and 78 samples were obtained.
2. NIR Spectra Acquisition from the Stocks
   Attach the transmission and reflection accessories on the SabIR™ optical fiber probe, fix the optical path, then immerse it into the well-mixed stock to a specific depth to gather the spectra. The format of absorbance data was log 1/R, the range of spectral gathering was 4,000–10,000 cm^{-1}, the number of scans was 32, and the resolution was 8 cm^{-1} without attenuation. Three spectra were acquired from each sample, and the average values were used for data processing. The stock was stirred before each scan. The results are shown in Fig. 16.28.

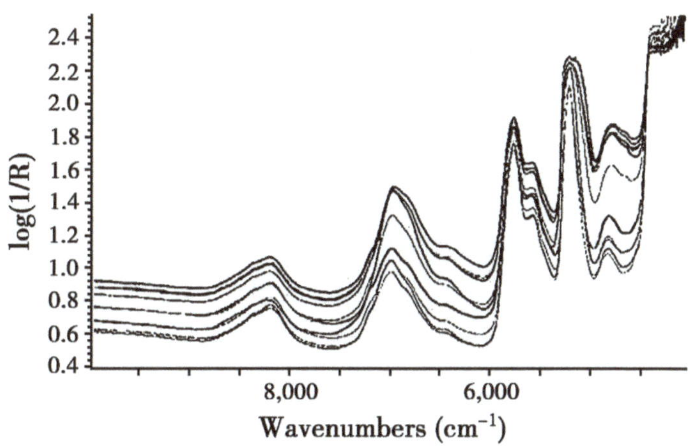

Fig. 16.28 The NIR transmission reflection spectra of Dantonic™ stock samples

Table 16.10 The content distribution of the effective constituents in the calibration and validation sample sets

Sample	Number of samples	Concentration range (mg g^{-1})		Average (mg g^{-1})	
		Danshensu	Protocatechuic aldehyde	Danshensu	Protocatechuic aldehyde
Calibration set	46	4.47–13.23	1.05–4.51	7.52	2.45
Validation set	30	4.89–11.63	1.21–3.74	7.86	2.22

The consistency of the material melting process: combine melted PEG 6000 together with extractum and borneol in the material melting tank of the dripping machine. Insert the SabIR™ optical fiber probe with attached transmission and reflection accessories into the tank, and begin stirring for 60 min, acquiring a spectrum every minute. The acquired spectra would be processed by the PLSR quantitative calibration model in real time to determine the contents of danshensu and protocatechuic aldehyde, which would be used to evaluate whether the stock was well mixed. This methodological validation test was conducted three times.

3. Determination of the contents of the effective constituents by HPLC: Agilent SB C_{18} analytical column (4.6 mm × 250 mm, 5 μm) was used; the mobile phase was methanol-water-glacial acetic acid (8:91:1). The sample injection volume was 10 μl, flow rate was 1 ml/min, detection wavelength was 281 nm, and the column temperature was 30 °C.
   The quality was based on the retention time, and quantitation was based on the peak area, and the external standard method was used to calculate. The content distribution of the effective constituents in Dantonic™ stock is shown in Table 16.10. In the results, the HPLC analytical values of danshensu and protocatechuic aldehyde in the calibration sample set and validation sample set showed even distributions from high to low with good linearity.
4. Data Processing
   Assign the samples into the calibration set (46) and validation set (30) randomly, making sure that the concentration range in the calibration set is larger than the range of the validation set. Two samples were considered singularities due to their large variance and were deleted. The samples in the calibration set were used to establish a calibration model. Select the spectral region for modeling, compare different spectral pretreatment methods, and use the PLSR method to establish a NIR spectral quantitative calibration model. Using RMSECV as an index to optimize model construction by selecting the modeling parameters, inspect the prediction performance of the model with RMSEP. All data processing was performed by TQ Analyst software.
5. Results and Discussion
   (a) Selection of Modeling Region: The content information of the samples is expressed as the absorbance in the NIR spectrum by C–H, N–H, and O–H groups. Therefore, the spectral region can be selected based on the structural features of the substance. Correlation spectroscopy generates relevant spectra by calculating correlation coefficients at each wavelength, which show the degree of correlation between the spectral information of different wavelengths and constituent contents [6]. We used correlation spectroscopy to select the spectral regions and processed the original spectral data with effective pretreatment methods. The selected spectral regions for danshensu were near C–H combined frequency and first-order frequency multiplication, 5,152–4,420 cm^{-1} and 6,850–5,940 cm^{-1}; the selected spectral region for protocatechuic aldehyde was near C–H combined frequency and first-order and second-order frequency

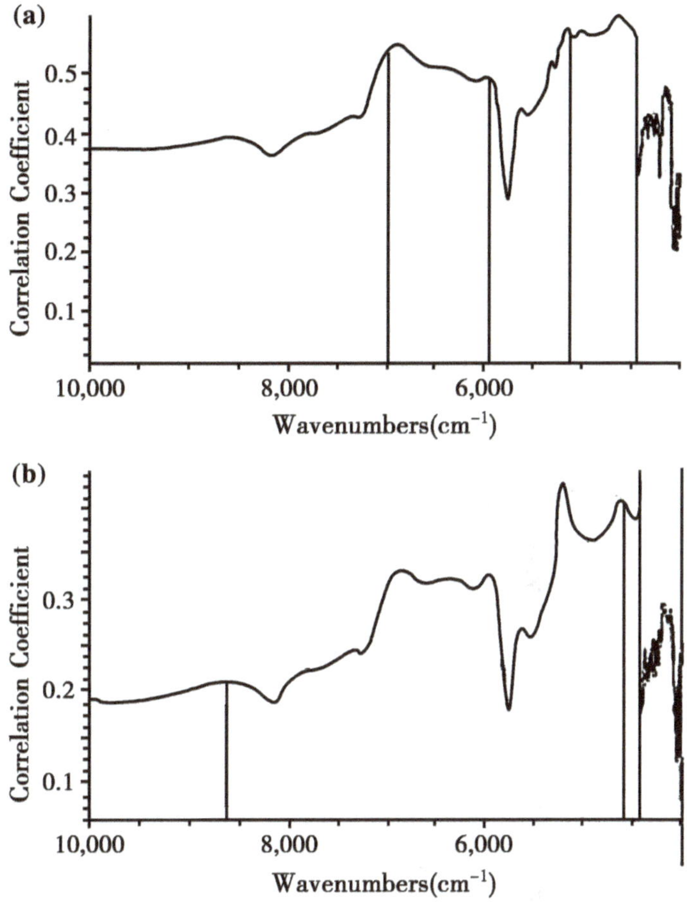

Fig. 16.29 The relevant spectra of the effective constituents in the stock of Dantonic™. **a** Danshensu; **b** protocatechuic aldehyde

multiplication, 8,659–4,624 cm^{-1}. The relevant spectra of the samples are shown in Fig. 16.29.

(b) Selection of Spectral Pretreatment Methods: The differences among samples in physical properties can cause changes in the spectral baseline and the slope. Therefore, in general, these influences should be eliminated by pretreatment with chemometric methods before establishing the calibration model. MSC is used to eliminate multi-spectral deviation, first-order and second-order derivative methods can maximally decrease the drift and shift of spectral peaks, and the Savitzky-Golay method is generally used for smoothing derivative spectra. The standard normal variate (SNV) technique can effectively diminish the useless spectral contribution to facilitate the resolution of analytical signals, meanwhile producing no noise signals. These pretreatment methods can be used together. We compared the different pretreatment methods and their combinations (MSC, first-order and second-order derivatives, SNV, MSC plus first-order derivative, MSC plus second-order derivative) to determine the optimal method. Five spectra were chosen for cross-validation at each time, and the correlation coefficient (R) and RMSECV were used to evaluate the performance of the calibration models. Table 16.11 shows that the most optimal pretreatment method for both danshensu and protocatechuic aldehyde was MSC plus first-order derivative.

Table 16.11 The effects of different pretreatments on the calibration models

Pretreatment method	Danshensu		Protocatechuic aldehyde	
	R	RMSECV	R	RMSECV
None	0.9065	0.687	0.8876	0.414
First-order derivative	0.9317	0.592	0.9672	0.233
Second-order derivative	0.9219	0.633	0.9326	0.341
MSC	0.9294	0.601	0.9399	0.302
SNV	0.9379	0.564	0.9344	0.316
MSC + first-order derivative	0.9664	0.419	0.9849	0.153
MSC + second-order derivative	0.9122	0.668	0.9288	0.348

(c) Determination of Model Major Factor Number: When establishing a quantitative calibration model by PLSR, the model's prediction performance has great variance due to different major factor numbers. When sample number in the calibration set is constant, if fewer factor numbers are used, the modeling information will be insufficient and the predictive ability of the model will be too low. On the other hand, if more factor numbers are used, the model will be too complex and cause an overfitting phenomenon during the training. Thus, it should make a reasonable selection of the number of major factors. By using cross-validation methods, we examined the influence of major factor number on RMSECV value. Figure 16.30 shows the change of RMSECV with major factor number, and the arrows in the figures show the optimal value. The optimal major factor numbers of danshensu and protocatechuic aldehyde in calibration model were 9 and 8, respectively.

(d) Establishment of Calibration Model: The optimal calibration models for danshensu and protocatechuic aldehyde were established by PLSR in the regions of 5,152–4,420 cm^{-1} and 6,850–5,940 cm^{-1}, and 8,659–4,624 cm^{-1}, respectively, after pretreating the spectral data by MSC plus first-order derivative. The correlation coefficients of the models were 0.9929 and 0.9937, respectively; root mean square errors of calibration (RMSEC) were 0.193 and 0.098, respectively; and RMSEP were 0.368 and 0.159, respectively. The correlation between the predicted values and the analytical values was high, indicating that the performance of the models was good (Fig. 16.31).

(e) On-line Monitoring of the Homogeneity of Material Melting Process: The

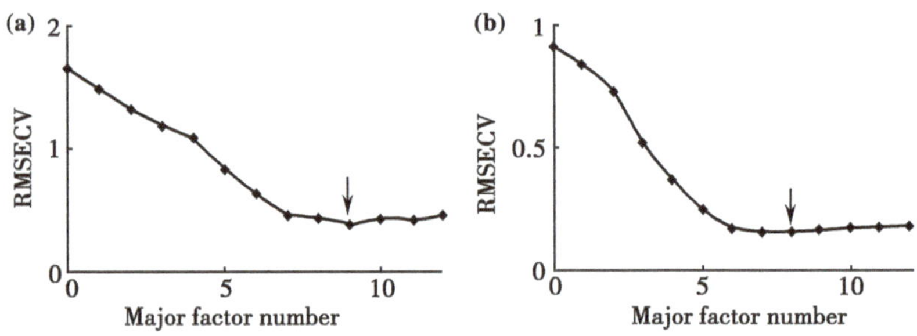

Fig. 16.30 The changes of RMSECV with PLSR major factor numbers. **a** Danshensu; **b** protocatechuic aldehyde

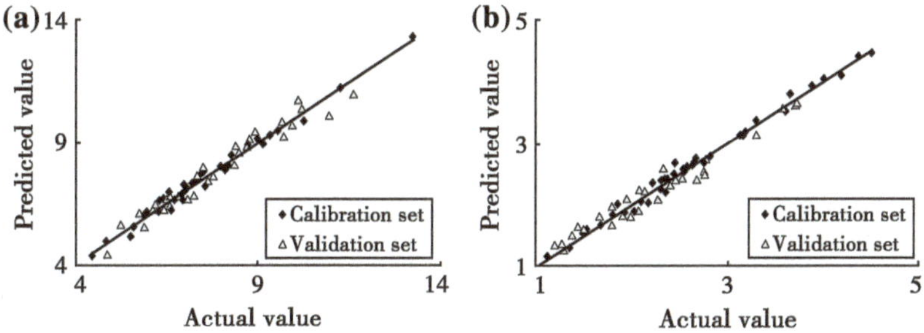

Fig. 16.31 The correlation between the values predicted by NIR calibration models and the values actually measured. **a** Danshensu; **b** protocatechuic aldehyde

established PLSR quantitation model was used to monitor the real-time contents of danshensu and protocatechuic aldehyde in the stock in the dripping machine to evaluate the homogeneity of the material melting process and determine the endpoint of the process.

The results of three series of tests (Fig. 16.32) showed that NIR spectroscopy could be used for real-time monitoring of changes in the component contents at the sampling points. When the stock reached homogeneity, the values detected by NIR spectroscopy were comparable with those determined by HPLC, indicating that the results of NIR spectroscopy can correctly reflect the mixing process of material melting: at the beginning, the contents of danshensu and protocatechuic aldehyde begin to increase and change rapidly; after 25 min, the contents of the two components reach a stable phase (small fluctuation within the range of predictive errors). At this time, the stock was mixed evenly and could be used for dripping. During the test, a standard spectral comparison method was used to evaluate the material melting homogeneity (see Fig. 16.33), i. e., when HPLC results indicated that the stock had reached homogeneity, its standard spectra were collected and compared with the real-time spectra, calculating the similarities of material melting homogeneity. Once the similarity of material melting homogeneity reached a certain value, the dripping process would begin. In practice, either content evaluation method or the similarity evaluation method could be used, and of course real-time monitoring the material melting process could also be used

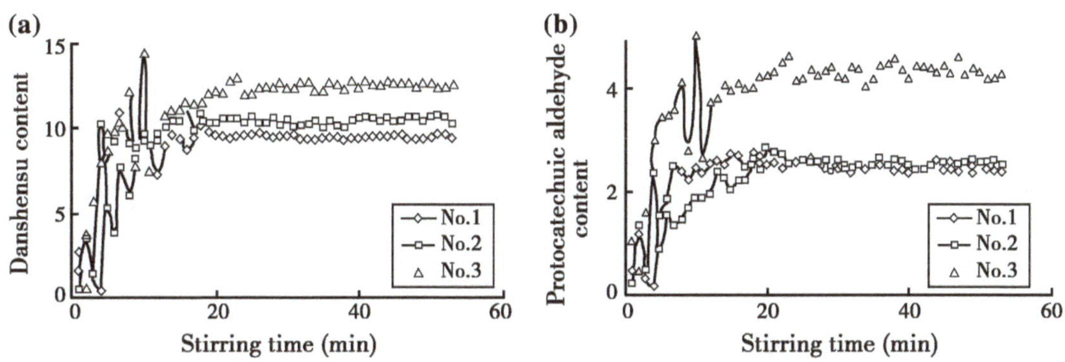

Fig. 16.32 Online monitoring of the homogeneity of material melting in a dripping machine

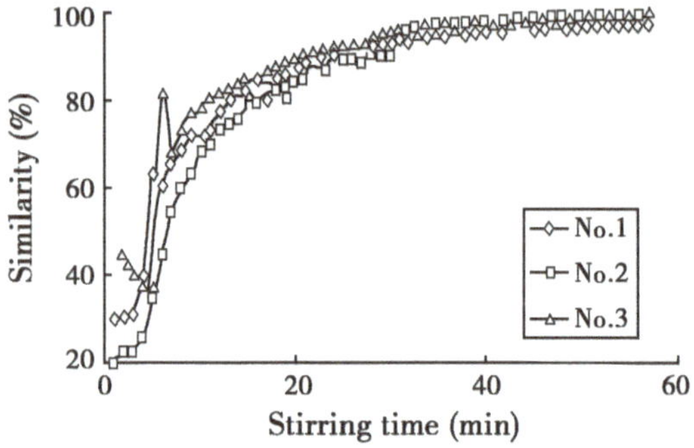

Fig. 16.33 The material melting homogeneity monitored by the standard spectral comparison method

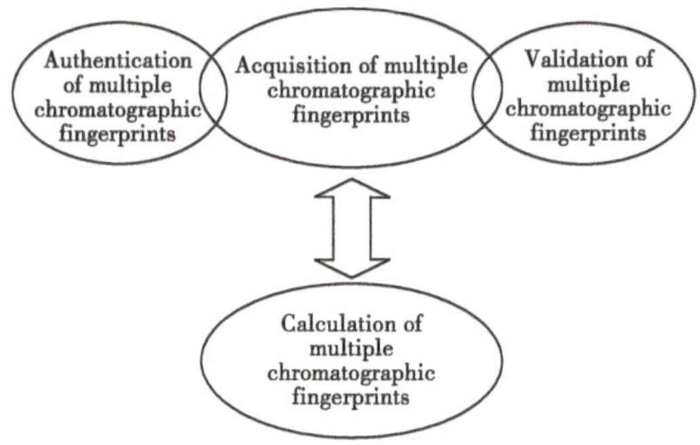

Fig. 16.34 Schematic diagram of multiple chromatographic fingerprinting technology

simultaneously, depending on the demand of different material melting processes.

5. Summary

   By utilizing the NIR spectral analytical technique, it is possible to promptly analyze the contents of the effective constituents in the stock of Dantonic™, and to quickly determine stock homogeneity in the material melting process. The method employs transmission and reflection optical fiber technology, does not require sample preparation, and the calibration models perform well.

16.4 Dantonic™ Quality Control Technique Based on Multivariant HPLC Fingerprinting

Xiaohui Fan

As mentioned above, a single HPLC chromatogram has difficulty in simultaneously showing the major effective constituents of Dantonic™, namely, the salvianolic acids (or known as water-soluble components) and notoginseng saponins. While undertaking a project from the Fifteenth National Research Programs—The Application and Demonstration of Fingerprinting Technology, we developed a multiple chemical fingerprinting method, i.e., using two chromatographic fingerprints, one for the salvianolic acids from Danshen and the other for the saponins from notoginseng, to show the two major effective constituents of Dantonic™. It is possible to show the overall chemical composition characteristics of Dantonic™ with this method [5–7].

The chromatographic fingerprinting of salvianolic acids in Dantonic™ was introduced in Chap. 15, so in this section, we introduce the chromatographic fingerprinting of saponins from notoginseng in Dantonic™ and the method of multicomponent chemical fingerprinting calculation (Fig. 16.34). The practical application of the technology in the quality control of Dantonic™ is also discussed.

16.4.1 Acquisition of Multiple Chromatographic Fingerprints of Dantonic™

16.4.1.1 Acquisition of HPLC Fingerprints of Salvianolic AcidS from Danshen in Dantonic™

Please see Chap. 15 for details.

16.4.1.2 Acquisition of HPLC Fingerprints of Saponinse from Notoginseng in Dantonic™

1. Instruments and Reagents

   Instrument: Agilent 1100 HPLC, TGL-16C

desk centrifuge (1.6×10^4 rpm) (Shanghai Anting Scientific Instrument Factory), HENGAO T&G filters (HENGGAO T&G), synthetic fiber filter membrane (0.45 μm) (Shanghai Xingya Purification Materials Factory).
Reagents: acetonitrile (chromatographic pure, Fisher Inc., USA), phosphoric acid (analytical pure), Wa-Ha-Ha purified water.
Dantonic™ (final product, intermediate product, and raw materials): provided by Tianjin Tasly Pharmaceutical Co., Ltd.

2. Preparation of Test Samples
Dantonic™ final product test sample preparation: weigh 1.0 g Dantonic™ from each batch, dissolve in 10 ml 4 % ammonia water, treat with ultrasound, filter the solution through a 0.45 μm membrane, load 5 ml of filtrate onto a C_{18} column (SUPELCO, USA), wash with 10 ml of methanol-4 % ammonia water (20:80) solution, rinse with 10 ml water, elute with methanol, bring the volume to 5 ml, centrifuge, and discard the pellet. Prepare two test samples from each batch. Injection volume: 20 μl.
Dantonic™ intermediate product test sample preparation: same as above, except beginning with 0.2 g of Dantonic™ intermediate.
Notoginseng herb test sample preparation: weigh 2.5 g of the herb, put into a round bottom flask, add 50 ml of distilled water, reflux for 1.5 h, filter, add 50 ml distilled water to the filter residue, reflux for 1 h, filter, combine the filtrates, bring the volume to 100 ml, centrifuge, discard the pellet, and prepare for determination. Prepare two test samples from each batch. Injection volume: 20 μl.
3. Determination of HPLC Conditions
After comprehensive study of the detection wavelength, mobile phase composition and elution gradient, column temperature, and the test sample preparation, the HPLC conditions were set as the following: Agilent SB-C_{18} analytical column (4.6×250 mm); mobile phase: phase A was 0.01 % acetic acid solution, phase B was acetonitrile in 0.01 % acetic acid; the program for gradient elution was 0–15 min, phase B increases from 20 to 35 %; 15–25 min, phase B remains at 35 %; 25–40 min, phase B increases from 35 to 43 %; 40–50 min, phase B remains at 43 %; 50–65 min, phase B increases from 43 to 58 %; 65–75 min, phase B increases from 58 to 75 %; flow rate was 0.8 ml/min; detection wavelength was 203 nm; column temperature was 30 °C; injection volume was 20 μl.

16.4.2 Authentication of Dantonic™ Multiple Chromatographic Fingerprints

The studies on Dantonic™ chemical composition have demonstrated that its major components are the water-soluble components of Danshen and soponins from notoginseng. Therefore, if a chromatograph can completely show these components with good correlation between the raw materials, intermediates, and final product, it can be considered the fingerprint of Dantonic™.

16.4.2.1 Authentication of HPLC Fingerprints of Salvianolic Acids

Please see Chap. 15 for details.

16.4.2.2 Authentication of HPLC Fingerprints of Saponins

The HPLC chromatographs obtained by the methods described above could well reflect the correlations among the raw materials, the intermediates, and the final product of Dantonic™ (Fig. 16.35).

Also, the major peaks in the chromatograph were identified [8] and the results are shown in Table 16.12. The HPLC chromatograph obviously reflects the notoginseng saponins of Dantonic™.

In summary, the notoginseng saponins and salvianolic acids in Dantonic™ were well represented in the HPLC fingerprints, which also showed good correlation among the raw

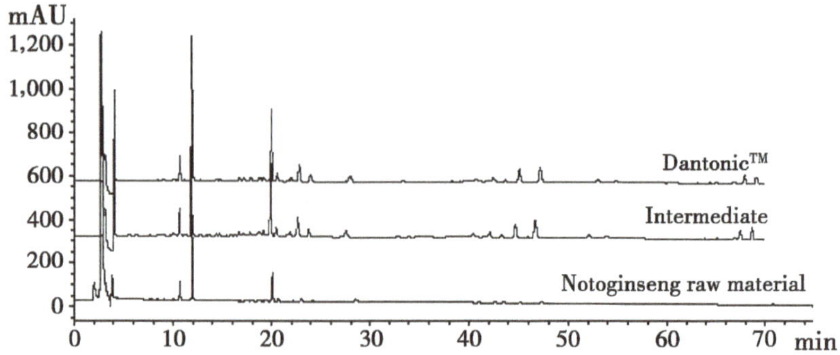

Fig. 16.35 The HPLC fingerprints of notoginseng saponin components of Dantonic™

Table 16.12 Identification of the major peaks in the chromatograph of Dantonic™

Peak no.	Identity	Peak no.	Identity
1	Notoginseng saponin R1	10	–
2	Ginsenoside Re	11	–
2	Ginsenoside Rg1	12	Ginsenoside Rga/f1
3	Ginsenoside Rb	13	Ginsenoside Rk_3/Rh_4
4	Notoginseng saponin R2	14	Ginsenoside Rk_3/Rh_4
5	Ginsenoside R_{g2}	15	Ginsenoside 20(R)$R9_3$
6	Ginsenoside R_{h1}	16	Ginsenoside 20(S)R_{g2}
7	Ginsenoside Rh1 iso.	17	Ginsenoside Rk_l/Rg_5
8	Ginsenoside Rd	18	Ginsenoside Rk_l/Rg_5
9	Ginsenoside Rd iso.		

materials, intermediates, and final product, so they could be used as the chemical fingerprints of Dantonic™.

16.4.3 Validation of Dantonic™ Multiple Chromatographic Fingerprints

16.4.3.1 Inspection of the Instrument Precision of the Method

Take a Dantonic™ test sample and continuously test six times; use danshensu and notoginseng saponin R1 as the reference substances, and calculate the relative retention time and peak area ratio of the multiple fingerprints. The results showed that the RSD of the relative retention time of each chromatographic peak was less than 0.2 %, and the RSD of the peak area ratio of the peaks with area greater than 10 % of the total peak area was less than 0.7 %.

16.4.3.2 Inspection of the Reproducibility of the Method

Take six samples from the same batch of Dantonic™, prepare the test samples, and perform multiple HPLC fingerprinting according to the method described above. The results showed that the RSD of the relative retention time of each chromatographic peak was less than 0.2 %, and the RSD of the peak area ratio of the peaks with area greater than 10 % of the total peak area was less than 0.5 %.

16.4.3.3 Inspection of the Stability of the Method

Take one DantonicTM test solution and determine its multiple HPLC fingerprints at 0, 2, 4, 6, 8, and 10 h. The results showed that the RSD of the relative retention time of each chromatographic peak was less than 0.2 %, and the RSD of the peak area ratio of the peaks with area greater than 10 % of the total peak area was less than 0.7 %.

16.4.3.4 Inspection of the Tolerance of the Method

1. The Tolerance of HPLC Fingerprinting for Salvianolic AcidS in DantonicTM
   Please see Chap. 15 for details.
2. The Tolerance of HPLC Fingerprinting for Notoginseng Saponins in DantonicTM
   (a) The Influence of Different Chromatographic Columns: Five batches of DantonicTM test samples were analyzed by the same HPLC instrument with different chromatographic columns to investigate the influence of chromatographic column on HPLC fingerprinting of notoginseng saponins in DantonicTM. During the course, a total of four columns were tested, and their information is listed in Table 16.13.
   A certain DantonicTM sample was tested on four different chromatographic columns, and the fingerprints of notoginseng chemical components are shown in Fig. 16.36. The results showed that except for column 1 (Extend C_{18}), which had a small shift in retention times, the chromatographs of other columns were basically the same. The fingerprint obtained from the fourth chromatographic column (the one used routinely) was designated as the standard fingerprint, and the similarities calculated from each sample's fingerprint were all greater than 0.98.
   (b) Influence of Different Instruments: Five batches of DantonicTM test solutions were tested under the same conditions except for the use of different instruments to evaluate the influence of instruments on the HPLC fingerprints of notoginseng saponins. Figure 16.37 shows the HPLC fingerprints of a certain sample determined by 4 HPLC instruments (3 Agilent 1100 and 1 Waters 2695). The fingerprint obtained from Agilent 1 (the instrument used in practice) was designated as the standard fingerprint, and the similarities calculated from each sample's fingerprint were all greater than 0.99, demonstrating that there was little influence from the Agilent and Waters instruments on the fingerprints of notoginseng saponins, when the other experimental conditions were the same.
   The above results demonstrated that the HPLC fingerprinting of notoginseng saponins has good tolerance.

16.4.4 Multiple Chromatographic Fingerprinting Calculation Based on Information Fusion

The method currently used for calculating the similarity of TCM fingerprints is suitable only for the one-to-one calculation between the sample fingerprint and the standard fingerprint, but not suitable for the many-to-many comparison of multiple fingerprints. Taking the overall similarity evaluation of multiple HPLC fingerprints of DantonicTM, for example, if comparing the HPLC

Table 16.13 Chromatographic columns used in the test

No.	Size	Filler	Manufacturer	Lot.
1	250 mm × 4.6 mm	Extend C_{18}	Agilent Zorbax	USHR002173
2	250 mm × 4.6 mm	Lichrospher C_{18}	Hanbang	01111604
3	250 mm × 4.6 mm	SB C_{18}	Agilent Zorbax	USCL013109
4	250 mm × 4.6 mm	SB C_{18}	Agilent Zorbax	USCL010304

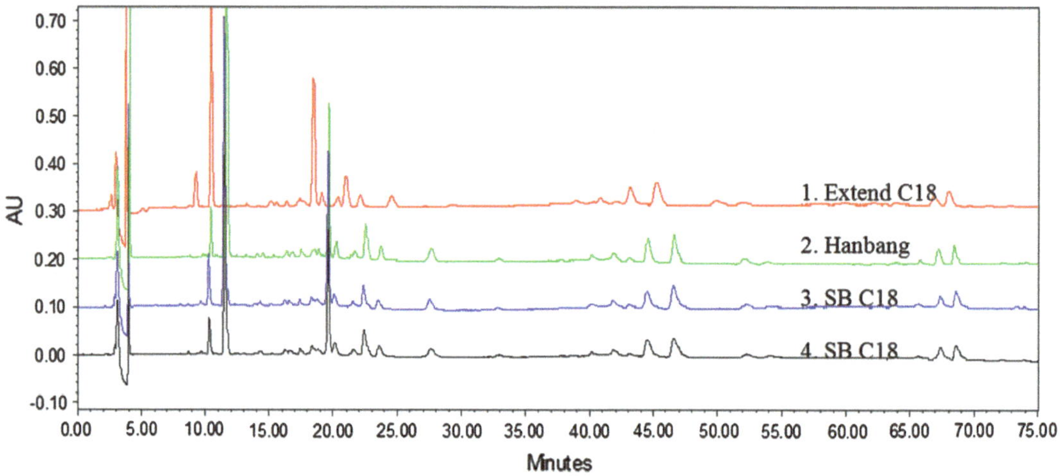

Fig. 16.36 The HPLC fingerprints of notoginseng saponins in Dantonic™: results from different chromatographic columns

Fig. 16.37 Comparison of HPLC fingerprints of notoginseng saponins of Dantonic™: results from different instruments

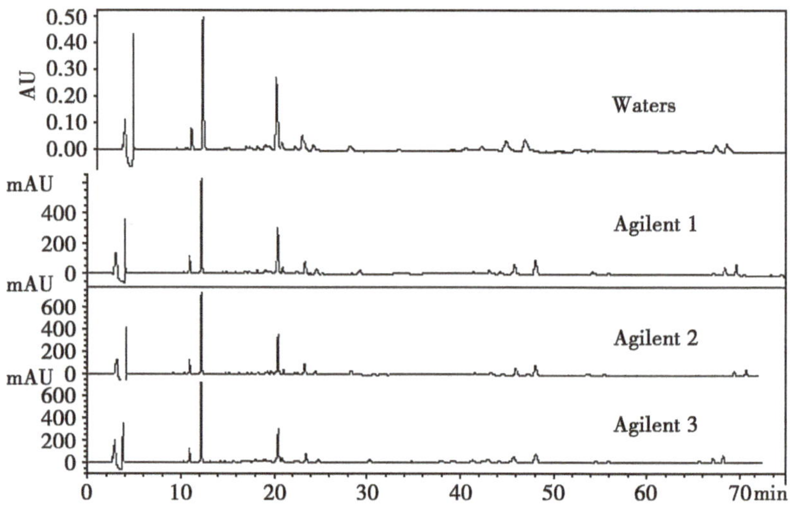

fingerprint similarity of salvianolic acids and notoginseng saponins separately, different similarities will be obtained, making it difficult to determine the overall similarity of multiple fingerprints. Thus, it is necessary to establish a set of calculation methods for multiple fingerprinting.

With respect to its mathematical nature, the calculation of multiple fingerprints is a problem concerned with multi-information fusion. So, we introduced a multiple fingerprinting calculation method based on information fusion: first, conduct information fusion of each subfingerprint at different levels, and then carry out the similarity calculation or pattern classification according to the fusion results. The application of the method in simulated data and in Dantonic™ samples demonstrated that the TCM multiple fingerprint similarity calculation method based on information fusion could evaluate the similarity of multiple fingerprints, and quantitatively represent the quality fluctuation of TCM drugs among different batches.

16.4.4.1 Principles

1. Introduction
Information fusion is the process simulation of the human brain to comprehensively process information. It has been extensively used in areas such as medical image processing, biological sensor information processing, etc. Its fundamental idea is to integrate diverse observational information according to some criteria so as to obtain a reasonable interpretation or description of the tested object. According to the level of the object, information fusion can be classified into three levels: pixel/data level fusion, feature level fusion, and decision level fusion [9].
Based on different vector combination strategies, information fusion technology can also be divided into serial and parallel fusions. The serial fusion strategy is to make a head-to-tail ligation of the vectors in a tandem way to form a new vector, with its dimension number equal to the sum of the dimensions of all original vectors. The principle of parallel strategy is to reasonably express the fusion vector by a complex vector. Taking the situation of two sets of original variables (α and β) for an example, the fusion vector can be expressed as a complex vector $\gamma = \alpha + i\beta$ (i is the imaginary unit). When the vector dimensions of α and β are not equal, the dimension could be made up by adding a zero value to the lower dimension vector. For example, for $\alpha = (\alpha_1, \alpha_2, \alpha_3)^T$, $\beta = (\beta_1, \beta_2)^T$, the combination feature is $\gamma = (\alpha_1 + i\beta_1, \alpha_2 + i\beta_2, \alpha_3 + i^*0)^T$. Thereby, the fusion vector is expressed as an n-dimensional complex vector, where n is the dimension of the bigger of vectors α and β.
2. Method for Pixel-Level Information Fusion of Multiple Chromatographic Fingerprints
According to the principle described above, the method for serial and parallel pixel-level information fusion of two fingerprints of multicomponent chemical fingerprinting serial and parallel is:
Suppose vector $\alpha_1 = [\alpha_{11}, \alpha_{12}, \ldots, \alpha_{1m}]$ and $\alpha_2 = (\alpha_{21}, \alpha_{22}, \ldots, \alpha_{2n})$ represent multiple fingerprints I and II (where α_{1i} and α_{2i} are sample data or chromatographic peak areas). When conducting serial data level fusion, the fusion vector $\phi = [\alpha_1, \theta\alpha_2]$; whereas in parallel fusion $\phi = [a_1; \theta\alpha_2]$ (when the dimensions of α_1 and α_2 are not equal, the smaller dimension would be made up by adding a zero value). θ is the combination coefficient, and its value could be determined according to specific situations.
If the multiple fingerprint is composed of three fingerprints, the serial fusion result is $\phi = [\alpha_1, \theta_1\alpha_2, \theta_2\alpha_3]$, and the parallel fusion result is $\phi = [\alpha_1; \theta_1\alpha_2; \theta_2\alpha_3]$; the rest may be deduced by analogy. For the multiple fingerprinting map composed of n fingerprints, their serial fusion vector is $\phi = [\alpha_1, \theta_1\alpha_2, \ldots, \theta_{n-1}\alpha_n]$ and their parallel fusion vector is $\phi = [\alpha_1; \theta_1\alpha_2, \ldots, \theta_{n-1}\alpha_n]$.
In addition, due to the differences in sample pretreatment methods and analytical instruments, it is possible that the vector of different fingerprints could have great variance in absolute values. Thus, it is necessary to carry out a normalization treatment for the original vectors first,

$$\alpha' = \alpha/\text{SUM}(\alpha) \tag{16.1}$$

where $\text{SUM}(\alpha)$ is the sum of variables in vector α.

3. Method for Calculating Multiple Chromatographic Fingerprint Similarity
Because the results of the above-mentioned chemical information fusion reflect the overall information of a multiple fingerprint, the stability of product quality among different batches can be evaluated by calculating the similarity of multiple fingerprints based on the fusion vector. Here, we use angle cosine measurements to calculate the similarity of multiple fingerprints among products from different batches.

$$S(\phi_i, \phi_0) = \frac{\sum_{k=1}^{m} \phi_{ik} \cdot \phi_{0k}}{\sum_{k=1}^{m} \phi_{ik}^2 \sum_{k=1}^{m} \phi_{0k}^2} \tag{16.2}$$

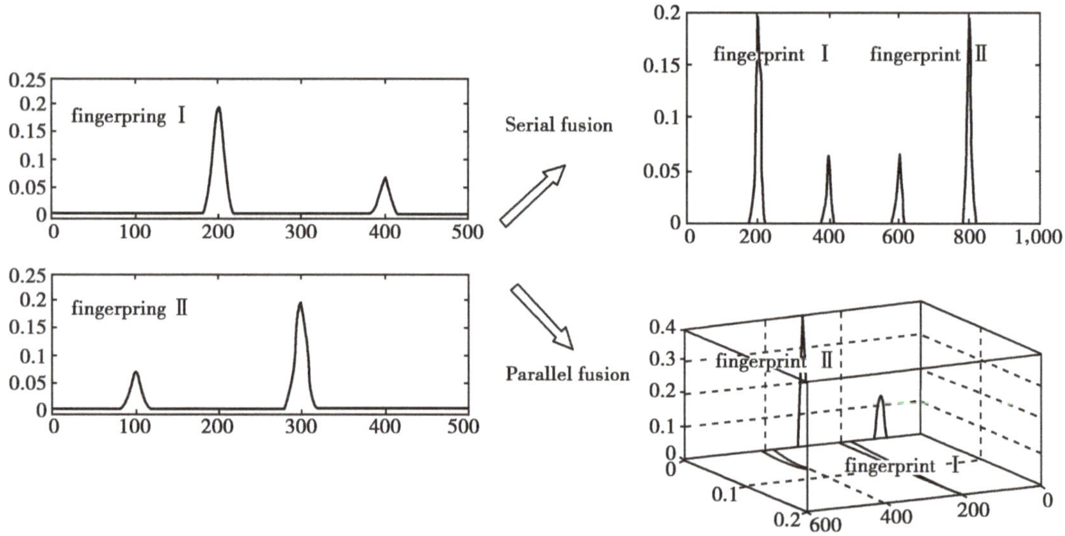

Fig. 16.38 Schematic diagram of the method for data level information fusion of binary fingerprints

where ϕ_i is the information fusion result from the sample multiple fingerprint; ϕ_0 is the information fusion result from the reference multiple fingerprint.

16.4.4.2 Computer Simulation Studies

To prove the feasibility of this technology, a computer simulation approach was used to investigate the data level information fusion of multiple fingerprints. Simulation fingerprints were generated by a Gaussian model; its mathematical model expression is:

$$\alpha_i(t) = \sum_{i=1}^{N} \frac{A_i}{\sqrt{2\pi\rho}} \exp\left(-\frac{(t-t_i)}{2\sigma^2}\right) \quad (16.3)$$

where N is the total peak number, t_i is the retention time of peak i, A_i is the peak area of peak i, and σ is the standard deviation of the Gaussian function.

For simplicity, assume a multiple fingerprint is composed of two fingerprints. Designate fingerprint I as α_1 = (1, 3); and fingerprint II as α_2 = (1, 3). Substitute the parameters into Formula (16.3), generating a multiple fingerprint composed of two fingerprints. If θ_1 = 1, according to the method shown in Fig. 16.38, the result of serial fusion is $\phi = [3, 1, 1, 3]$, and parallel fusion gives $\phi = \begin{bmatrix} 3, 1 \\ 1, 3 \end{bmatrix}$.

Change the peak areas in fingerprints I and II, and calculate the similarities of individual fingerprints and multiple fingerprints. A portion of the simulation results is shown in Table 16.14, which demonstrates that the similarities obtained using the two fusion strategies are consistent, and both can reflect the variation in the unit fingerprints. By the same reason, if a multiple fingerprint is composed of more than 2 fingerprints, similar results can be obtained.

Table 16.14 The calculation results of simulated fingerprints under different fusion strategies

	Similarity			
	Fingerprint I	Fingerprint II	Serial fusion	Parallel fusion
1	1.0	0.6	0.8	0.8
2	0.6	1.0	0.8	0.8
3	0.6	0.6	0.6	0.6

16.4.4.3 Case Study

The practicality of the method was investigated by using the similarity evaluation of multiple HPLC fingerprints of Dantonic™ as an example. Dantonic™ samples were provided by Tianjin Tasly Pharmaceutical Co., Ltd. Among the samples, 13 batches were qualified products, numbered from 1 to 13; and 7 batches were with abnormal process parameters, numbered 14–20. The multiple HPLC fingerprints were determined according to the method described above, and the integral similarity was calculated to evaluate the lot-to-lot consistency of the product.

1. Determination of Combination Coefficient θ
   In the traditional application fields of information fusion such as medical image processing, human face recognition, etc., the combination coefficient θ is generally set as 1 or as the ratio of the characteristic dimensions of each group. The θ of TCM multiple fingerprints should be set according to the fingerprinting peaks of the effective constituents. The HPLC multiple fingerprints of Dantonic™ include two major effective constituent groups: fingerprint I represents the chemical components of salvianolic acids, and fingerprint II represents the chemical components of notoginseng saponins. Obviously, if setting $\theta = 1$, it seem unreasonable to treat these two groups equally because their contents differ too much. So, θ is set as the content ratio of chemical components represented by fingerprint II and chemical components represented by fingerprint I. Further chemical study shows that the salvianolic acids represented by fingerprint I account for 5 % of the effective constituents of Dantonic™; whereas notoginseng saponins represented by fingerprint II account for 11 % of the effective constituents of Dantonic™. Based on this fact, we set θ at 2.2. In fact, how to precisely estimate the combination coefficient to obtain optimal effects is still an issue to be studied.
2. Multiple Fingerprint Similarity Calculation Based on the Data Level Information Fusion
   The areas of 10 common peaks in HPLC fingerprint I of salvianolic acids of Dantonic™ and 18 peaks in HPLC fingerprint II of notoginseng saponins of Dantonic™ were measured and normalized using Formula (16.1). The information was then fused in serial and parallel manners. The similarity values were calculated using the mean values of the qualified products (No. 1–13) as the reference/standard fingerprints. The results are shown in Fig. 16.39. The content of danshensu is a key indicator of Dantonic™ quality control. For comparison, the peak areas of danshensu from the same batch are shown in Fig. 16.40. Figure 16.40 shows that there was no significant pattern difference in danshensu peak areas between the two groups of samples

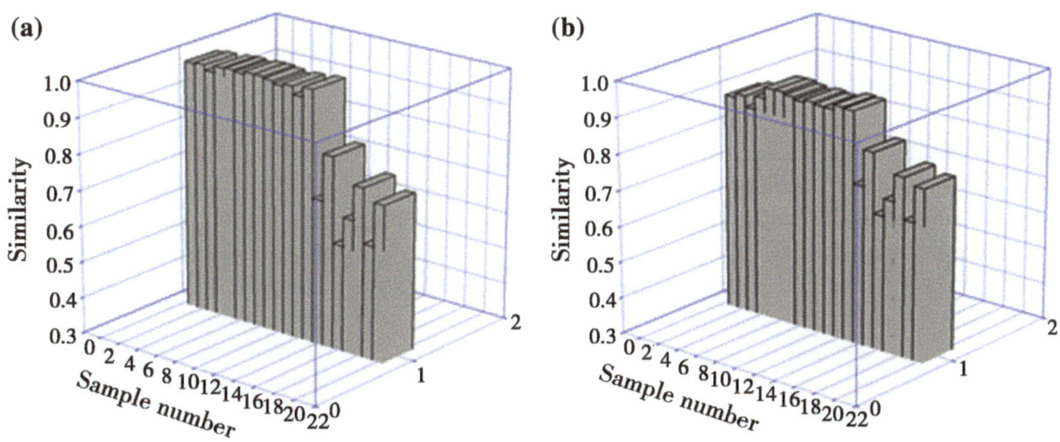

Fig. 16.39 The calculated integral similarity of Dantonic™ samples from different batches. **a** Serial fusion; **b** parallel fusion

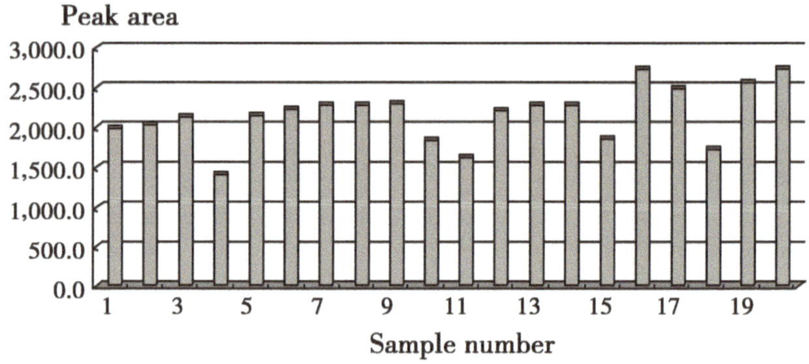

Fig. 16.40 Histogram of danshensu peak areas in Dantonic™ samples from different batches

Meanwhile, in spite of adding a zero value to make up the dimension, the multiple fingerprint similarity results calculated by parallel fusion at the raw data level were still not as good as those calculated by serial data level information fusion. However, both results showed similar trends: the similarities of the qualified products were relatively higher (greater than 0.9), whereas the similarities of the products with abnormal process parameters were lower (less than 0.8), indicating that calculation methods based on either serial or parallel information fusion could reflect the quality difference among different batches of Dantonic™, and thus could distinguish unqualified products from qualified products.

16.5 Method for Determining the Quality Uniformity of Compound Danshen Extract

Haibin Qu

Danshen extract is the aqueous extract of Danshen raw material, and Danshen compound extract is the aqueous extract of Danshen and notoginseng raw materials; both are the intermediate products of Danshen preparations. During the extraction process, the variation of the manufacturing conditions and the experience of the workers could cause nonconformity of the extract quality, thereby influencing the stability and homogeneity of the final product. In this section, we introduce the mixing uniformity method which has been used to control the quality of Danshen extract and compound Danshen extract to improve the stability of their quality.

16.5.1 Mixing Uniformity Method

16.5.1.1 The Basic Idea

TCM extract mixing uniformity method is a method to reduce lot-to-lot variations by mixing different batches of extracts in certain proportions to obtain an extract mixture which is as close to the quality of standard extract as possible. The mixing ratio used in this method can be deduced with an appropriate optimal algorithm according to the established mathematical model.

16.5.1.2 Mathematical Model

1. Explanation of Symbols
   The symbols involved in the modeling are listed in Table 16.15.
2. Linear Mixing Model
   To establish a mixing uniformity model, it is necessary to know the changes of peak areas in the fingerprints before and after mixing. We can construct a mixed model to calculate the changes. For simplicity, the following two assumptions can be made:
   Assumption 1.1 The peak areas in the fingerprints are linearly correlated to the corresponding contents of the ingredients in the TCM extract.

Table 16.15 Explanation of symbols used in mixing uniformity model

Symbol	Explanation
i	Serial batch number of the TCM extract used for mixing uniformity
j	Serial peak number in the fingerprint
k_i	Serial number of key component peak in the fingerprint
m	Total numbers of batches used for mixing uniformity
n	Total of peaks in the fingerprint
l	Number of indicator components in the extract
x_i	Fingerprint of TCM extract from batch i
x_s	Fingerprint of standard extract
x_b	Fingerprint of mixed extract
b_i	Mixing percentage of the extract i
x_{ij}	Peak j in x_i
x_{sj}	Peak j in x_s
x_{bj}	Peak j in x_b

Assumption 1.2 There are no interactions among the different ingredients in the extract. Based on the two assumptions and the laws of conservation of mass, the peak areas after mixing should satisfy the following linear model:

$$x_{bj} = \sum_{l=1}^{m} x_{ij} \cdot (b_i/100) j = 1, 2, \Lambda, n$$

To most mixing processes of TCM extracts, the above assumptions are generally true, thus, the linear mixing model can be used to calculate peak areas after mixing.

3. The Indicators for TCM Drug Quality Evaluation

   The similarity (S) of TCM fingerprinting and the relative deviation of concentrations (RDC) of indicator ingredients are used as the quality evaluation indicators. The formulas for calculating S and RDC_{ki} are as follows:

$$S = \frac{\sum_{i=1}^{n} (x_{bj} - \overline{x}_b)(x_{sj} - \overline{x}_s)}{\sqrt{\sum_{i=1}^{n} (x_{bj} - \overline{x}_b)^2} \sqrt{\sum_{i=1}^{n} (x_{sj} - \overline{x}_s)^2}}$$

$$RDC_{ki} = |x_{bki} - x_{ski}| / x_{ski}, \quad i = 1, 2, \ldots, l$$

where RDC_{ki} is the indicator ingredient RDC corresponding to peaks k_i in the mixed extract and the standard extract; X_{bki} and X_{ski} are the areas of peak k_i of the mixed extract and standard extract, respectively.

4. Modeling
   (a) Objective Function: In the TCM mixing uniformity problem, to minimize the variance between the mixing sample and standard sample, it is required that the RDC between the two samples be as small as possible and the similarity S between the two be as large as possible. Thus, the object function can be defined as:

$$\min \frac{1}{l}(RDC_{k_1} + RDC_{k_2} + \cdots + RDC_{k_l}) + (1 - S)$$

   (b) Restrictive Conditions: In TCM production practice, it is generally required that the RDC of every indicator ingredient be less than the predetermined threshold RDC_O (the extract with higher value than the threshold is considered unqualified), and the similarity be larger than the predetermined threshold S_O (the extract with lower value than the threshold is considered unqualified). Therefore, there are the following restrictive conditions:

$$RDC_{k_t} \le RDC_0 \quad t = 1, 2, \ldots, l$$
$$S \ge S_0$$

In addition, the sum of the percentages of each batch extract in the mixing extract should be 100 %, and the percentage of each batch extract should be not less than 0, i.e.:

$$\sum_{i=1}^{m} b_i = 100 \qquad b_i \ge 0 \quad i = 1, 2, \Lambda, m$$

(c) Mixing Uniformity Model: Mixing uniformity method can be described as the following optimization problem in mathematics:

Model 5.1
$$\min \frac{1}{l}(RDC_{k_1} + RDC_{k_2} + \cdots + RDC_{k_l}) + (1 - S)$$
$$\text{s.t.} \quad RDC_{k_j} \leq RDC_0 \quad 1 \leq j \leq l$$
$$S \geq S_0$$
$$\sum_{i=1}^{m} b_i = 100$$
$$b_i \geq 0 i = 1, 2, \cdots, m$$

The optimization model is a nonlinear programming problem with constraints, and its solution is the optimal mixing ratio b_i. The problem can be solved with sequential quadratic programming (SQP).

16.5.2 Mixing Uniformity Method Based on Multiple Fingerprinting

When the sample material system is complex (such as with TCM compound preparations), it is usually difficult to completely reflect the characteristics of the sample's chemical composition with a single chemical fingerprint. The multiple fingerprinting technique introduced in Sect. 16.4 of this chapter, however, is capable of doing that. Obviously, if a mixing uniformity method based on multiple fingerprinting is established, it would better control the quality of compound TCM.

The key to the construction of a multiple fingerprinting-based mixing uniformity model is to determine its objective function. When the objective function is determined, the other parts of the model can follow the corresponding parts in model 5.1.

After fusing the fingerprinting vectors of each component in a multiple fingerprint by information fusion technology, the definition of objective function has two possibilities, objective function 1 and objective function 2, as shown below:

$$\min \frac{1}{l}[RDC_{k_1} + RDC_{k_2} + \cdots + RDC_{k_l}) + (1 - S)]$$
$$\min \left\{ \frac{1}{l}(RDC_{k_1} + RDC_{k_2} + \cdots + RDC_{k_l}) + \frac{1}{l}[(1 - S_{k1}) + (1 - S_{k2}) + \cdots + (1 - S_{ld})] \right\}$$

In the above formulas, RDC_{k_i} is the RDC of the indicator ingredient corresponding to the fingerprint peak k_i between the mixing extract and the standard extract, S_{kl} is the fingerprint similarity of component l in the compound extract, and S is the overall similarity of the compound extract.

To objective function 1, even though the overall similarity reaches the required threshold, there is still the possibility that the similarity of a single component is below the threshold. On the other hand, objective function 2 has a set single component threshold, so it is impossible that a single component cannot reach the threshold; thus it is more suitable as the objective function of a mixing uniformity model based on multiple fingerprinting.

16.5.3 Experiment Part

16.5.3.1 Danshen Extract

1. Instruments and Reagents
   Instruments: Agilent 1100 HPLC, equipped with a quaternary pump, a vacuum degasser, an autosampler, a column heater–cooler, a diode-array detector, an UV detector, and Agilent 1100 ChemStation system.
   Reagents: acetonitrile (chromatographic pure, Merck Inc.), phosphoric acid (analytical pure), miniQ water.
   Samples: a total of four Danshen extract samples (labeled a–d) were provided by Tianjin Tasly pharmaceutical Co., Ltd.
2. Chromatographic Conditions
   Hanbang C_{18} column, mobile phase: A (0.02 % phosphoric acid-acetonitrile)—B (0.02 phosphoric acid-water) system, elution gradients: see Table 16.16. Flow rate was

Table 16.16 Danshen HPLC fingerprinting gradient elution program

Time (min)	A: 0.02 % $H_3PO_4 - CH_3CN$ (%)	B: 0.02 % $H_3PO_4 - H_2O$ (%)
0	8	92
8	18	82
15	21	79
35	31.5	68.5
36	80	20

1.0 ml/min, injection volume was 5 μl, column temperature was 30 °C, and detection wavelength was 280 nm.

3. Sample Preparation Method
   The samples provide by the manufacturer were diluted 10 fold and centrifuged before injection.
4. Model Calculation
   All calculations were carried out with MATLAB 6.5. The similarity calculation method is by the Euclidean distance method; S threshold was set at 0.90, and indicator ingredients RDC threshold was set at 0.10.

16.5.3.2 Compound Danshen Extract

1. Instruments and Reagents
   Instruments: Agilent 1100 HPLC, equipped with a quaternary pump, a vacuum degasser, an autosampler, a column heater–cooler, a diode-array detector, an UV detector, and Agilent 1100 ChemStation system.
   Reagents: acetonitrile (chromatographic pure, Merck Inc.), phosphoric acid (analytical pure), miniQ water.
   Samples: a total of 4 compound Danshen extract samples (labeled a–d) were provides by Tianjin Tasly pharmaceutical Co., Ltd.
2. Acquisition of Fingerprints
   See Sect. 16.4 of this chapter.
3. Sample Preparation Method
   Sample prepared method for multiple fingerprint I: place 0.1 g of compound Danshen extract in a 10 ml volumetric flask, add an appropriate amount of distilled water, treat with ultrasound for 15 min, bring volume to scale, and centrifuge before injection.
   Sample prepared method for multiple fingerprint II: dissolve 0.2 g of compound Danshen extract in 10 ml 4 % ammonia water, treat with ultrasound, pass through a 0.45 μm filter membrane, filter 5 ml of the filtrate through a C_{18} column, wash with 10 ml methanol-4 % ammonia water (20:80), rinse with water to remove solvent, elute with methanol, bring volume to 5 ml, and centrifuge before injection.
4. Model Calculation
   All the calculations were carried out with MATLAB 6.5. The similarity calculation method is the Euclidean distance method, the S threshold was set at 0.90, and the indicator ingredients RDC threshold was set at 0.10.

16.5.4 Results and Discussion

Haibin Qu

16.5.4.1 Mixing Uniformity Results of Danshen Extract

1. The Fingerprinting of Danshen Extract
   A total of 10 peaks were selected as fingerprint peaks, of which peak 1 was danshensu and peak 2 was protocatechuic aldehyde; both are commonly used indicator components for Danshen quality control. Other peaks were salvianolic acids, rosmarinic acid and lithospermic acid (Fig. 16.41).
2. The Fingerprints of Danshen Extract Samples from Different Batches
   The results are shown in Fig. 16.42.
3. The Results of Mixing Uniformity
   Using sample a as the standard extract, input the chromatographic fingerprinting data of the standard extract and the other three samples into the mixing uniformity model. The calculated optimal mixing ratios of batches b, c, and d were 11.86, 49.86, and 38.28 %, respectively. After rounding, the mixing ratios were 12, 50, and 38 %, respectively.
   Mixing samples b, c and d according to the above ratios, obtain the fingerprints of mixture e and sample a, as shown in Fig. 16.43.
4. The Quality Parameters of Danshen Extract
   Using sample a as the standard extract, and

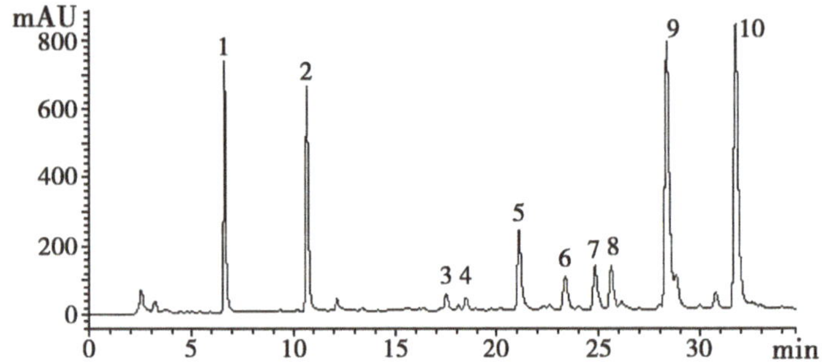

Fig. 16.41 The fingerprint of Danshen extract

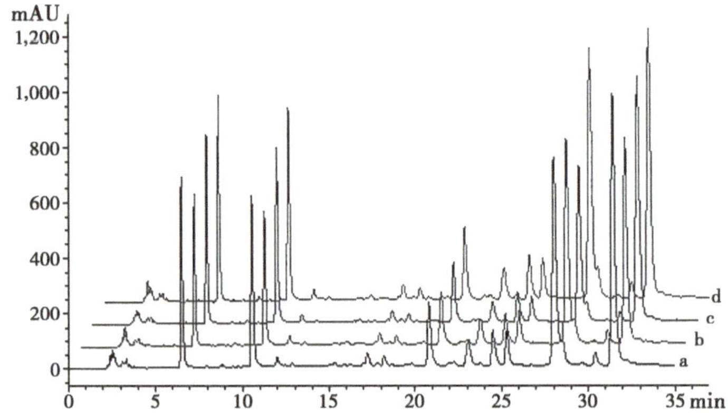

Fig. 16.42 The fingerprints of Danshen extract samples from different batches

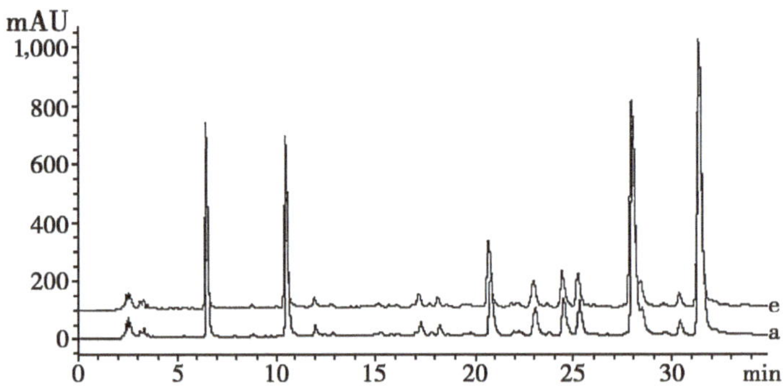

Fig. 16.43 The fingerprints of standard extract *a* and extract mixture *e*

Table 16.17 Quality parameters of Danshen extracts from different batches

Batch	b	c	d	Average	e
S	0.7936	0.8122	0.8748	0.8269	0.9434
RDC1	0.3114	0.0565	0.0429	0.1369	0.0409
RDC2	0.3124	0.0112	0.0831	0.1356	0.0069

similarity S and indicator components RDC as Danshen extract quality evaluation indicators, the quality parameters of each batch are shown in Table 16.17, where RDC1 is the RDC of danshensu and RDC2 is the RDC of protocatechuic aldehyde.

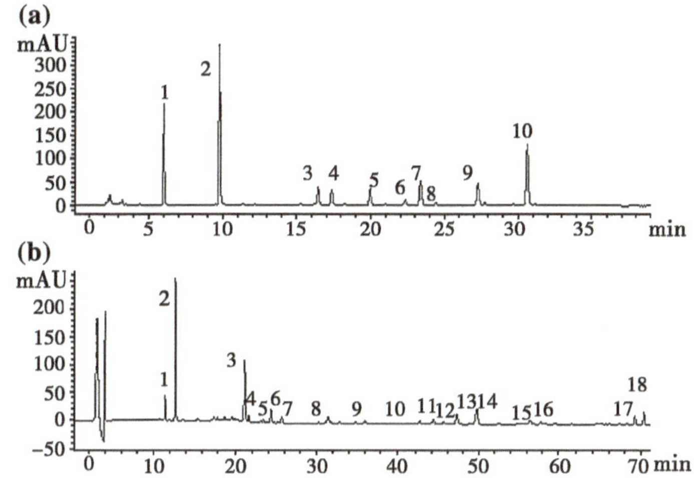

Fig. 16.44 The multiple HPLC fingerprints of compound Danshen extract

16.5.4.2 Mixing Uniformity Results of Compound Danshen Extract

1. The Fingerprinting of Compound Danshen Extract
   The fingerprints of compound Danshen extract are shown in Fig. 16.44, in which A is fingerprint I and B is fingerprint II. Select 10 peaks in fingerprint I as fingerprint peaks, of which peak 1 and 2 are danshensu and protocatechuic aldehyde, respectively. Select 18 peaks in fingerprint II as fingerprint peaks, of which peak 1 is notoginseng saponin R1, the characteristic chemical constituent of notoginseng.
2. The Fingerprints of Compound Danshen Extract Samples from Different Batches
   The results are shown in Fig. 16.45.
3. The Results of Mixing Uniformity
   Using sample a as the standard extract, conduct serial fusion of fingerprint I and fingerprint II by the information fusion technique, set the fusion coefficient at 1, and obtain the fusion fingerprinting data. Input the fused chromatographic fingerprinting data of the standard extract and the other 3 batches of extracts into the mixing uniformity model. The calculated optimal mixing ratios of batches b, c and d were 16.97, 55.60, and 27.43 %, respectively. After rounding, the ratios of the three batches were 17, 56, and 27 %, respectively. Mixing samples b, c, and d according to the ratios, the fingerprints of the mixed sample e and sample a are shown in Fig. 16.46.
4. The Quality Parameters of Compound Danshen Extract
   Using sample a as the standard extract and similarity S and RDC of the indicator components as the quality evaluation indicators for compound Danshen extract, the quality parameters of each batch are shown in Table 16.18, where *S*1 is the similarity of fingerprint I, *S*2 is the similarity of fingerprint II, RDC1 is the RDC of danshensu, RDC2 is the RDC of protocatechuic aldehyde, and RDC3 is the RDC of notoginseng saponin R1.

16.5.4.3 Discussion

1. The Fluctuation of Extract Quality
   In industrial production, due to the variations in operators and manufacturing conditions, the extract quality fluctuates to some extent. In the examples provided in this section, the lowest similarity of Danshen extract was 0.7936 and the highest RDC of the indicator components was 0.3124. For compound Danshen extract, the lowest similarity was 0.7301 and the highest RDC of the indicator components was 0.3362. If an extract with

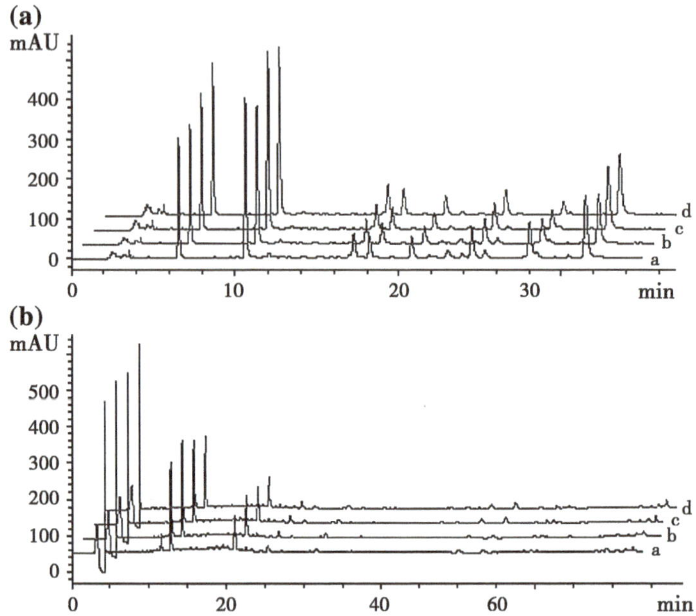

Fig. 16.45 The multiple HPLC fingerprints of compound Danshen extract from different batches. **a** Fingerprint I; **b** Fingerprint II

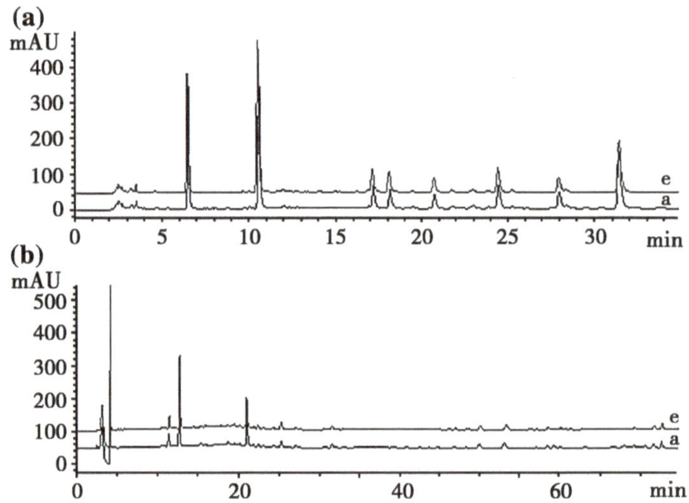

Fig. 16.46 The multiple fingerprints of standard extract *a* and mixed extract *e*. **a** Fingerprint I; **b** Fingerprint II

Table 16.18 Quality parameters of extracts from different batches and the mixed extract

Batch	b	c	d	Average	e
S1	0.8384	0.8146	0.7907	0.8146	0.9564
S2	0.8583	0.7778	0.7301	0.7887	0.9354
RDC1	0.1373	0.0583	0.1571	0.1176	0.0097
RDC2	0.1124	0.2336	0.2661	0.2040	0.0243
RDC3	0.1118	0.2131	0.3362	0.2203	0.0100

quality fluctuation were used for subsequent production, it would affect the quality stability of the final product.

2. Mixing Uniformity Method Can Effectively Improve Extract Quality
   The change in Danshen extract quality: using sample a as the standard, the fingerprint similarity of the mixed sample increased from 0.8269 premixing to 0.9434, the RDC of danshensu decreased from an average of 0.1369 premixing to 0.0409, and the RDC of protocatechuic aldehyde decreased from an average of 0.1356 premixing to 0.0069.
   The change in compound Danshen extract quality: using sample a as the standard, the fingerprint similarities between the mixed extract and standard extract were 0.9564 and 0.9354, far higher than the premixing averages of 0.8146 and 0.7887; the RDC of danshensu decreased from the premixing average of 0.1176–0.0097, the RDC of protocatechuic aldehyde decreased from the premixing average of 0.2040–0.0243, and RDC of notoginseng saponin R1 decreased from the premixing average of 0.2203–0.0100.
   These results show that the quality control parameters of mixed extract are all superior to those of the extracts from batches b, c, and d, and meet the predetermined specifications (RDC of indicator ingredients below 0.1, and similarity above 0.90), whereas none of the original batches, b, c, and d, met the quality requirements. It demonstrates that extract quality is improved after mixing each batch according to calculated ratios based on a mixing uniformity method.

References

1. Yang TT. J Pharm Sci. 1986;75(1):53.
2. Zen DH. Manufacture and theory investigation of dripping pills. Beijing: Chinese Medicine Science and Technology Press; 1994.
3. Zou WQ, et al. Physical chemistry experiments and techniques. Shanghai: East China Chemical Institute Press, 1990; 115:40–2.
4. Studies of modeling, optimization and application for TCM dripping pills. Tianjin Tasly Pharmaceutical Co., Ltd, 2005.
5. Fan XH, Cheng YY, Ye ZL, et al. Anal Chim Acta. 2006;555(2):217–24.
6. Fan XH, Ye ZL, Cheng JY. Chem J Chin Univ. 2006;27(I):26–9.
7. Fan XH, Ye ZL, Cheng JY. Chin J Pharm Anal. 2005;26(9):7–10.
8. Zhang MJ. Identification method studies of TCM material basis based on HPLC-electrospray MS multiple techniques. Zhejiang University, 2004.
9. Yang J, Yang JY, Zhang D, et al. Pattern Recogn. 2003;36(6):1369–81.
10. Cheng YY, Chen MJ, Wu YJ. Acta Chim Sinica. 2002;60(II):2017–21.
11. Ye ZL, Yu K, Cheng JY. Chem J Chin Univ. 2007;28(3):441–4.
12. Gong YF, Liu XS, Zhang SN, et al. Chin Pharm J. 2007;42(7):509–11.
13. Qu HB, Ye YY, Cheng YY. Chin J Pharm Anal. 2006;26(10):1413–6.
14. Yu K, Cheng YY. Chin J Anal Chem. 2006;34(4):561–4.
15. Ke BK, Liu XS, Chen Y, et al. Chin Tradit Herbal Drugs. 2006;37(5):685–7.
16. Yu K, Hu CC, Cheng YY. Chin Pharm J. 2006;41(3):226–9.
17. Qu HB, Liu XX, Cheng YY. Chem J Chin Univ. 2004;25(1):39–43.
18. Shi CS, Liu XS, Chen Y, et al. Chin Pharm J. 2006;41(23):1771–3.
19. Luo XF, Mo BQ, Ye YY, et al. Chin Tradit and Herbal Drugs. 2007;38(12):14–6.

In Vivo Metabolism of Danshen and Its Preparations

17

Aihua Liu, Dean Guo, Jinlan Zhang and Jianghao Sun

Danshen and its preparations are widely used for the treatment of cardiovascular and cerebrovascular diseases, with remarkable curative effects. However, there are few research reports on their in vivo metabolism. Studies in this area can elucidate the mechanisms of action of Danshen's active components, and provide a scientific basis for the clinical application of Danshen and Danshen preparations.

17.1 Pharmacokinetics and In Vivo Metabolism of Total Salvianolic Acids

17.1.1 Pharmacokinetics Study of Total Salvianolic Acids

Aihua Liu and Dean Guo

The major active compound in total Salvianolic Acids is Salvianolic Acid B (SalB), which accounts for 2.8 % of total Salvianolic acid s. In vivo pharmacokinetic studies of SalB in rats can elucidate the kinetic characteristics of absorption, distribution, and excretion, and provide a scientific basis for new preparation design, SalB-containing drug development, and clinical rational drug use. It can also establish the foundation for elucidating the material basis of the efficacy provided by Danshen.

17.1.1.1 Instruments and Surgical Devices

Instruments: Agilent 1100 series HPLC, Zorbax Extend C_{18} column (4.6 mm × 250 mm, 5 μm); Sardoris electronic balance, Eyela rotary evaporator, centrifuge (Shanghai Medical Centrifuge Factory), and homogenizer.

Surgical devices: surgical scissors, ophthalmic scissors, dressing forceps, ophthalmic forceps, hemostatic forceps, syringe, and intragastric administration needles for rats.

17.1.1.2 Reagents and Drugs

Methanol (GR), hydrochloride (AR), acetic ester (AR), and ether (AR) were the products of Beijing Chemical Plant; heparin sodium (biochemical reagent) was produced by Shanghai Biochemical Reagent Factory; medical normal saline and acetonitrile were produced by Caledon, Canada; total salvianolic acids: extracted from Danshen which was obtained from the Sichuan Zhongjiang Base; vanillic acid was purchased from the National Institutes for Food and Drug Control; SalB was self-prepared with purity >97 % (determined by HPLC).

A. Liu · D. Guo (✉) · J. Zhang · J. Sun
Shanghai Institute of Materia Medica, Shanghai, China
e-mail: daguo@mail.shcnc.ac.cn

X. Yan (ed.), *Dan Shen (Salvia miltiorrhiza) in Medicine*,
DOI 10.1007/978-94-017-9463-3_17,

Table 17.1 Gradient elution program

Time (min)	Acetonitrile (%)	0.026 % phosphorous acid water solution (%)
0	6	94
30	37	63
35	6	94

17.1.1.3 Laboratory Animals and Collection of Biological Samples

Spraque-Dawley male rats with body weights of 200 ± 20 g were provided by the Laboratory Animal Center of Peking University Health Science Center.

Prior to administration, fast the rats for 24 h in a metabolic cage. Administer total salvianolic acids by intragastric gavage at a dose of 200 mg/kg. Collect blood samples prior to and after administration at 15, 30, 45, 60, 75, 90, 120, 150, and 180 min. At each time point, administer ether anesthesia to six rats, draw blood from the abdominal aorta, use heparin sodium for anticoagulation, and centrifuge at 3,500 rpm for 5 min; collect the supernatants and store at −20 °C for further treatment and analysis.

17.1.1.4 Pretreatment of Blood Samples

Measure 1 ml of blank rat plasma and 1 ml of postadministration rat plasma, respectively; adjust pH with 10 % HCl to 2.5; extract with equal volumes of acetic ester or three times; pool the acetic ester layers; concentrate and evaporate; dissolve the residue in 0.5 ml of methanol; treat with ultrasound; filter through a 0.45 μm microporous membrane; analyze the filtrate by HPLC.

17.1.1.5 HPLC-UV Analysis Conditions

Mobile phase: acetonitrile–0.026 % phosphorous acid water system. Perform gradient elution according to the program shown in Table 17.1.

Flow rate: 0.8 ml/min; column temperature: 30 °C; detection wavelength: 288 nm; sample volume: 10 μl.

The pretreatment method for biological samples and the conditions for HPLC analysis are simple and fast, and there are no endogenous impurities in the subject to be analyzed or around the internal standard, so the accuracy of the analysis is ensured. A representative chromatogram is shown in Fig. 17.1.

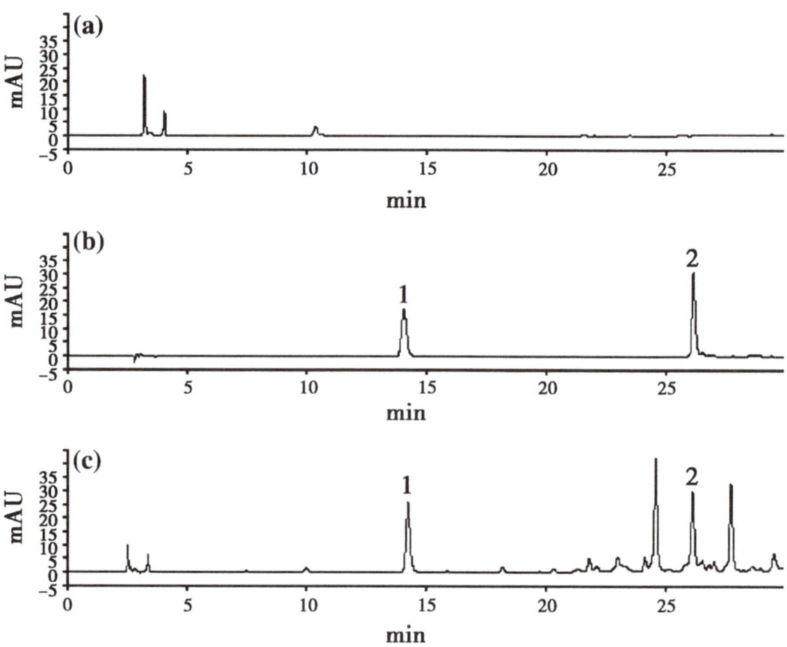

Fig. 17.1 HPLC chromatogram of blank and postadministration rat plasma. **a** Blank plasma; **b** Blank plasma with 34.6 μg/ml of SalB and 48.0 μg/ml of internal standard; **c** Rat plasma obtained 30 min postadministration

17.1.1.6 Standard Curve and Linear Range

Precisely weigh 4.3 mg of SalB and 3.2 mg of internal standard vanillic acid; transfer to a 2 ml flask; dissolve in ethanol and dilute to volume, thus obtaining the stock solutions of SalB and the internal standard, respectively. Precisely transfer different amounts of the stock SalB solution and 15 μl of the internal standard solution to a 10 ml tube; remove solvent under reduced pressure; add 1 ml of blank plasma; treat according to the pretreatment method for samples; and finally, obtain the standard working curve solutions of SalB with concentrations of 10.8, 21.6, 34.5, 64.9, 129.7, and 259.4 μg/ml, as well as of the internal standard solution with a concentration of 48.0 μg/ml. Inject 10 μl of the sample, then perform HPLC analysis. Draw the standard curve by plotting the SalB internal standard peak area ratio versus concentration. The linear regression equation was $Y = 0.0198X \pm 0.000346$, $r = 0.9911 \pm 0.0015$ ($n = 3$). With this method, results showed good linearity in the range of 10.8–259.4 μg/ml. The lower limit of detection was 2.7 μg/ml and the lower limit of quantification was 10.8 μg/ml.

17.1.1.7 Method Validation

1. Intraday and Interday Precision of the Method
   Measure 1 ml of blank plasma and prepare quality control (QC) samples with low, medium, and high concentrations of SalB (10.8, 34.5, and 129.7 μg/ml, respectively) according to the procedure described in the "standard curve and linear range." Six samples from each concentration level were analyzed for three consecutive days to test the precision of the method. The results are shown in Table 17.2. These results showed that intraday RSD < 5.1 % ($n = 18$) and interday RSD < 6.4 % ($n = 18$) when this method was used for analyzing SalB. The accuracy was between −4.9 and 3.2 %, indicating that this method can be used to determine the concentration of SalB in rat plasma.
2. Recovery
   Measure 1 ml of blank plasma and add standard SalB serial solutions to prepare plasma samples with 10.8, 34.5, and 129.7 μg/ml of SalB, respectively. For each concentration, prepare six samples. Test the samples by HPLC, calculate the peak area ratio of SalB to the corresponding internal standard, then compare with the peak area obtained from the standard solution and calculate the absolute recovery of SalB. The average recovery of SalB was 84 % ($n = 6$) with RSD < 5.6 %.

17.1.1.8 Pharmacokinetic Studies of SalB

After oral administration of total salvianolic acids to rats, collect the plasma samples at different time points. Analyze the samples and perform data processing to produce a concentration–time curve (Fig. 17.2); Pharmacokinetics parameters were obtained after

Table 17.2 Intraday and interday test precision of samples ($n = 6$)

Theoretical concentration (μg/ml)	Observed concentration (μg/ml)	Accuracy (%)	Precision (RSD %)
Intraday tests			
10.8	10.2 ± 0.52	95.1	5.1
34.5	33.9 ± 1.64	98.3	4.9
129.7	131.3 ± 3.72	101.2	2.8
Interday tests			
10.8	11.15 ± 0.71	103.2	6.4
34.5	35.1 ± 2.70	101.7	5.9
129.7	132.7 ± 6.24	102.3	4.7
Average		100.3	3.0

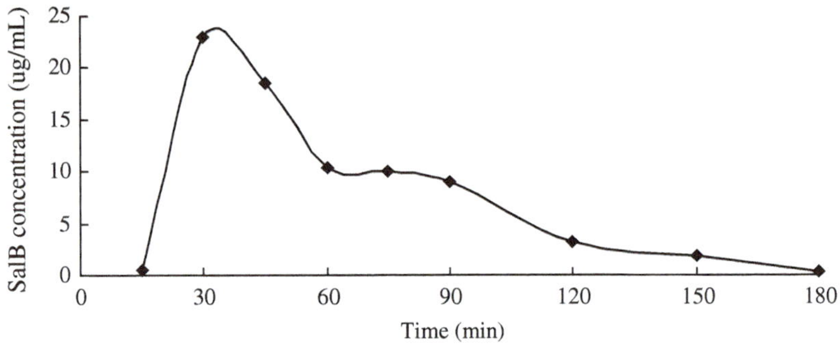

Fig. 17.2 The average concentration–time curve of SalB in rats after oral administration of total salvianolic acids ($n = 6$)

Table 17.3 The major Pharmacokinetic parameters after oral administration of total salvianolic acids to rats ($n = 6$)

Parameter	Unit	Value
A	(μg/ml)	190.98
α	min^{-1}	0.05
B	(μg/ml)	139.15
β	min^{-1}	0.04
k_a	min^{-1}	0.05
lag time	min	14.77
V/F	mg/kg/(μg/ml)	1.33
$t_{1/2\alpha}$	min	13.83
$t_{1/2\beta}$	min	19.16
$t_{1/2ka}$	min	13.66
k_{21}	min^{-1}	0.05
k_{10}	min^{-1}	0.04
k_{12}	min^{-1}	0.0002
AUC	(mg/ml) × min	1150.04
CL	mg/kg/min/(μg/ml)	0.05

processing the data with software 3p97 (Table 17.3).

The concentration of SalB in rat plasma follows a two-compartment model, belonging to a first-order absorption and elimination process. The absorption of SalB is prompt and reaches peak absorption within 0.5 h. The elimination of SalB from circulation is also prompt, and a large V_c indicates that SalB distributes quickly into extravascular organs and tissues.

17.1.2 The Metabolites of Total Salvianolic Acids from Danshen in Rats

Jinlan Zhang and Dean Guo

17.1.2.1 Instruments and Surgical Devices

Instruments: Sardoris electronic balance; Eyela rotary evaporator; centrifuge produced by Shanghai Medical Centrifuge Factory; homogenizer; Agilent 1100 series HPLC system equipped with DAD; Finigan Advantage ion trap mass spectrometer; Zorbax Extend-C_{18} chromatographic column (4.6 mm × 250 mm, 5 μm).

Surgical devices: surgical scissors; ophthalmic scissors; dressing forceps; ophthalmic forceps; hemostatic forceps; syringe; intragastric administration needles for rats.

17.1.2.2 Reagents and Drugs

Methanol (GR), hydrochloride (AR), acetic ester (AR), and ether (AR) were produced by Beijing Chemical Plant; heparin sodium (biochemical reagent) was produced by Shanghai Biochemical Reagent Factory; medical normal saline, acetonitrile were produced by Caledon, Canada; total salvianolic acids were extracted from Danshen produced at the Sichuan Zhongjiang Base; danshensu, protocatechuic aldehyde, caffeic acid, and ferulic acid purchased from The National Institutes for Food and Drug Control.

17.1.2.3 Laboratory Animals and Collection of Biological Samples

Sprague-Dawley male rats with body weights of 200 ± 20 g were provided by the Laboratory Animal Center of Peking University Health Science Center.

Fast rats for 12 h in a metabolic cage. During the fasting period, collect blank urine and feces samples, then feed rats total salvianolic acids by intragastric gavage at a dose of 200 mg/kg. Collect urine and feces samples during 0–12 h and 12–24 h postadministration. Store the samples at −20 °C for further treatment and analysis.

17.1.2.4 Establishment of Pretreatment Method for Samples

1. Exploration of Pretreatment Methods for Biological Samples
   To effectively extract and isolate the metabolites and effective components of the injection or total phenolic acids from the biological matrix, and to remove the endogenous substances as much as possible, the following pretreatment methods were tested (before the pretreatment, centrifuge the urine sample and save the supernatant for treatment and analysis).
   (a) Liquid–liquid extraction: Measure 10 ml of urine, adjust pH with 10 % HCl to 2.5; extract with threefold volume of acetic ester; pool acetic ester layers; concentrate and evaporate; dissolve the residue in 0.3 ml of methanol; treat with ultrasound; filter through a microporous membrane; use the filtrate for HPLC analysis.o
   (b) Solid-phase extraction: Measure 10 ml of urine; pass through an Alhech 200 mg ODS column; wash with two volumes of water, elute with equal volume of 20 % methanol, 50 % methanol and methanol; collect the eluants.
   Measure 10 ml of urine; flow through a 500 mg D201 column; wash with 2 volumes of water, elute with equal volume of 20 % methanol, 50 % methanol and methanol; collect the eluants, concentrate and evaporate; dissolve the residue in methanol; filter through a 0.45 μm-diameter microporous membrane; use the filtrate for HPLC analysis.
2. HPLC-DAD Analysis of the Pretreated Biological Samples
   (a) HPLC Analysis Conditions: Mobile Phase: Acetonitrile–0.1 % trifluoroacetic acid system. Perform gradient elution according to the program shown in Table 17.4.
   Flow rate: 0.8 ml/min; temperature of column: 30 °C; detection wavelength: 288 nm, sample volume: 20 μl.
   (b) HPLC Analysis Results
   The liquid–liquid extraction pretreatment method could extract the metabolites from urine samples. However, interference from endogenous substances (especially from those with weak polarity) was severe, while endogenous substances with strong polarity caused little trouble. When the urine samples were pretreated with the ODS solid-phase extraction method, the metabolites appeared in the flow-through solution rather than being absorbed to the ODS column, while endogenous substances with weak polarity were bound to the ODS column and those with strong polarity were not removed. When the urine samples were pretreated with the D201 solid-phase extraction method, the metabolites appeared in the equal volume washing water and some endogenous substances with strong polarity were removed, but there was still a large amount of endogenous substances with weak polarity.
   (c) Selecting a pretreatment method: based on the above results, a comprehensive treatment method was adopted: measure 10 ml of urine; pass through an Alhech

Table 17.4 The gradient elution program for HPLC analysis of urine samples

Time (min)	Acetonitrile (%)	0.1 %TFA (%)
0	2	98
60	33	67
70	2	98

200 mg ODS column; let the flow-through pass through a D201 column, wash with an equal volume of water, collect the eluant; adjust pH with 10 % HCl to 2.5; extract with 3 volumes of acetic ester; pool the acetic ester layers; concentrate and evaporate; dissolve the residue in 0.3 ml of methanol; treat with ultrasound; filter through a microporous membrane; use the filtrate for HPLC analysis.

17.1.2.5 HPLC-DAD Analysis of the Biological Samples

Treat blank and 0–24 h postadministration urine and feces samples according to the established pretreatment method, then perform HPLC analysis under the above-mentioned conditions. The results are shown in Figs. 17.3 and 17.4. These results demonstrate: (1) Pretreatment could effectively extract the metabolites from the samples and there was little interference caused by endogenous impurities. (2) No polyphenolic acids, such as SalB and SalA, were found in rat urine or feces samples collected after oral administration of total salvianolic acids, indicating that metabolic transformation had already occurred and the metabolic rate was high. (3) The metabolites were eliminated in urine within 0–12 h and few metabolites appeared in feces. By comparing with the blank control and comparing UV spectra with those of phenolic acid ingredients, the preliminary conclusion was that there were five metabolites, and they were named M1–M5 according to the peak order. The retention time and UV spectrum of M1 were consistent with those of danshensu, confirming M1 to be danshensu. The identities of the other metabolites remain to be confirmed.

17.1.2.6 HPLC-MS Analysis of the Biological Samples

The rat urine samples collected 0–12 h after oral administration of total phenolic acids were analyzed by HPLC-MS. Mass spectrometry conditions were as follows: ion source: Turbo

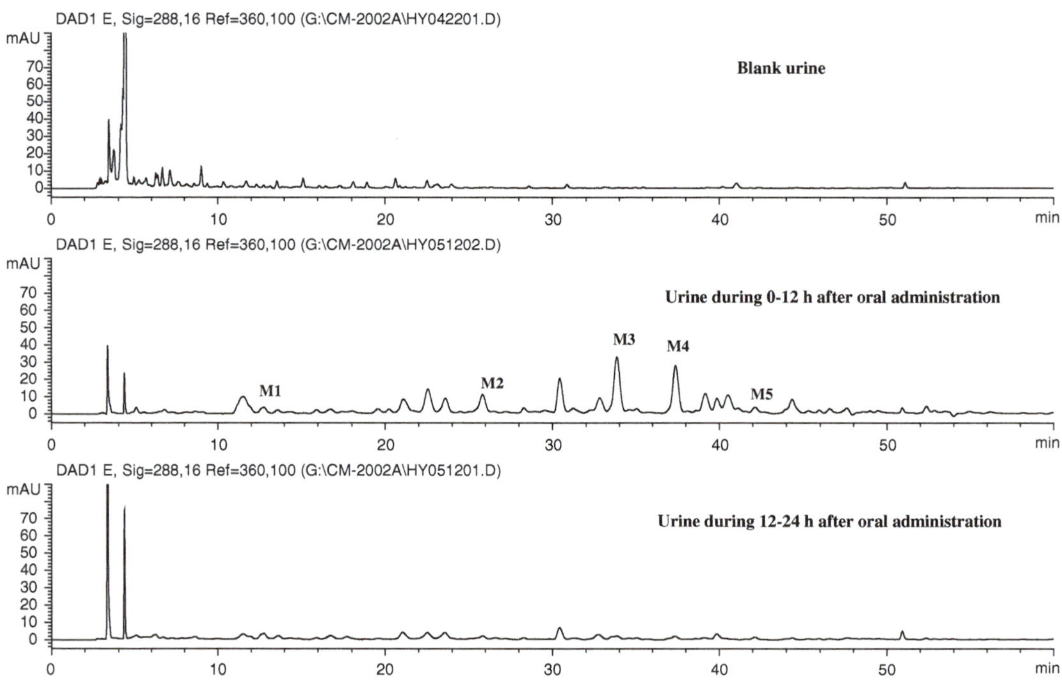

Fig. 17.3 The HPLC chromatograms of blank urine and urine collected 0–24 h after oral administration of total phenolic acids to rats

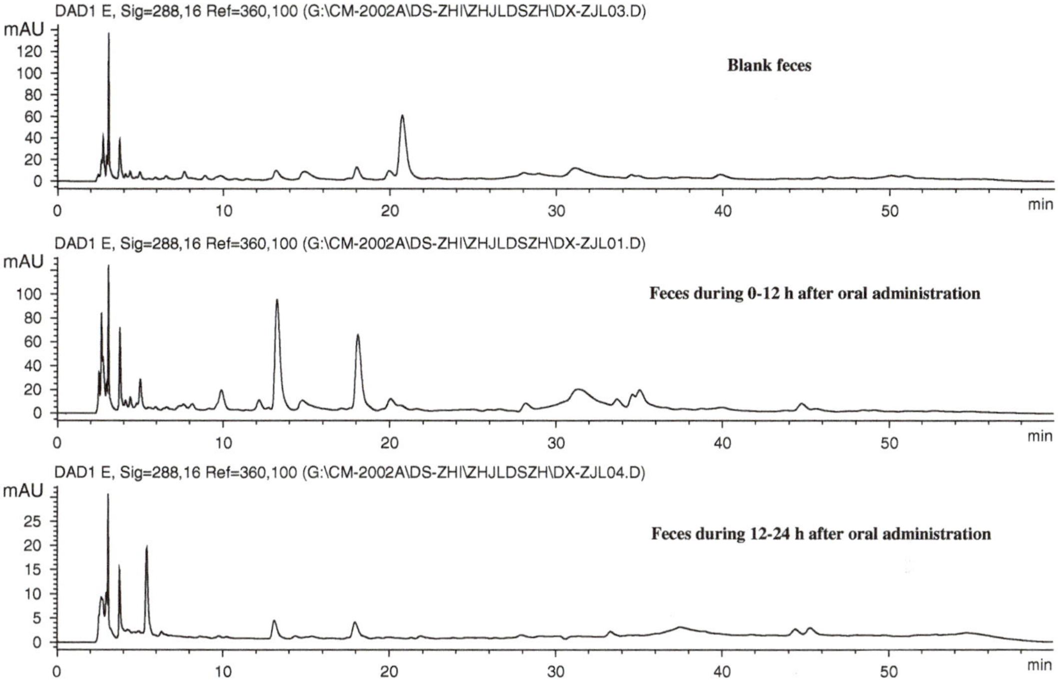

Fig. 17.4 The HPLC chromatograms of blank feces and feces collected 0–24 h after oral administration of total phenolic acids to rats

Ionspray; ionization mode: ESI positive ion mode; ionization voltage: 5.0 kV; MS range: 100–1,500 m/z amu; sheath gas flow rate: 60 arb; auxiliary gas flow rate: 5 arb; temperature of ion source: 350 °C. The selected ion recording (SIR) spectra are shown in Fig. 17.5 and the MS spectra of metabolites M2–M5 are shown in Figs. 17.6, 17.7, 17.8, and 17.9.

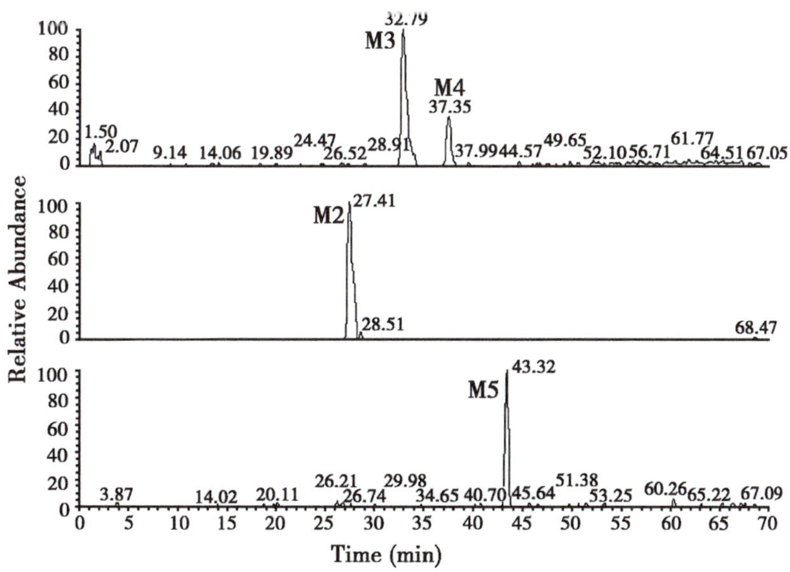

Fig. 17.5 The SIR graph of rat urine samples collected 0–12 h after oral administration of total phenolic acids

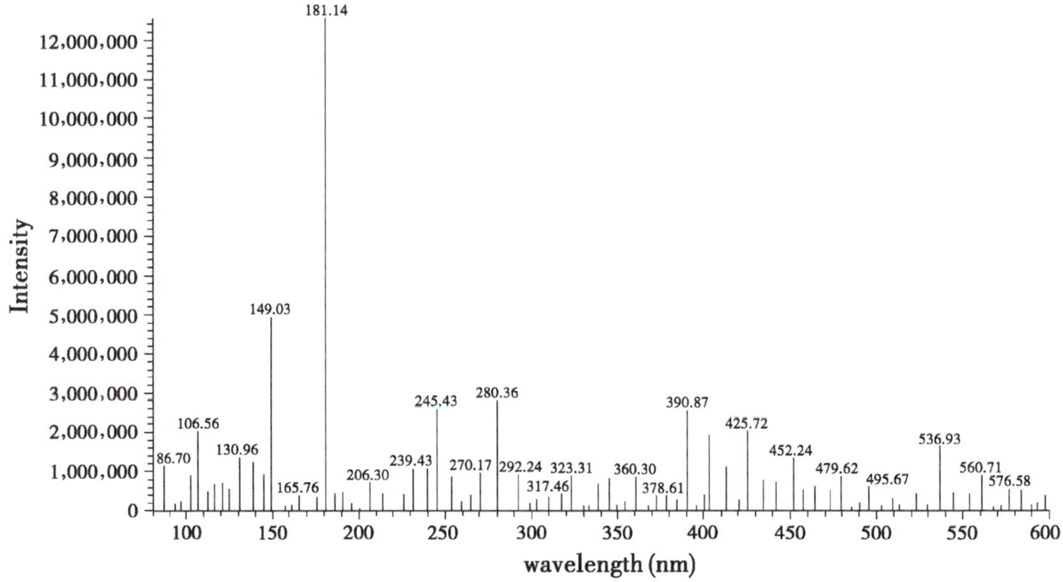

Fig. 17.6 The MS spectrum of M2

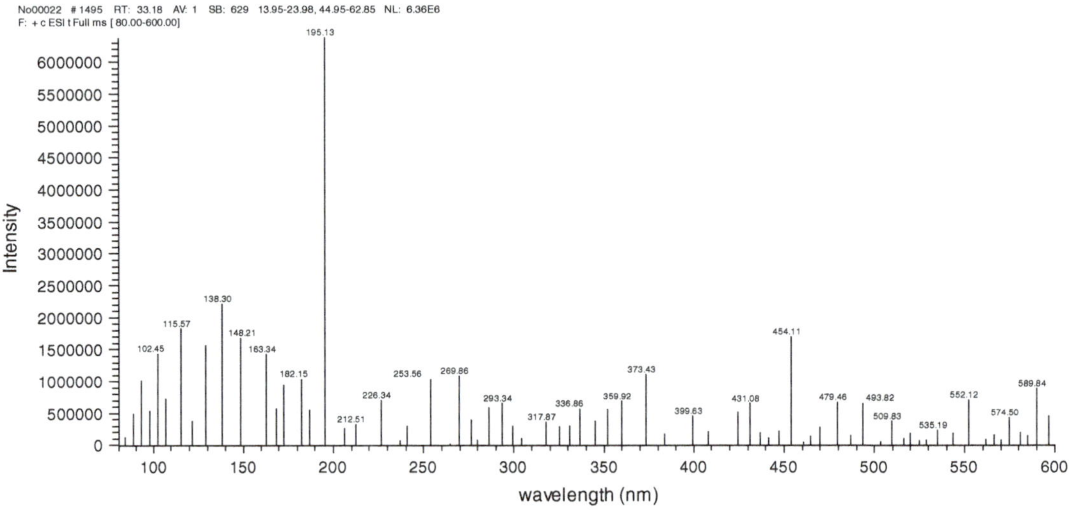

Fig. 17.7 The MS spectrum of M3

The HPLC-MS analysis results confirmed that the quasi-molecular ion $[M + 1]^+$ of M2 was 181 and its molecular weight was 180, inferring that M2 is caffeic acid; the quasi-molecular ions $[M + 1]^+$ of both M3 and M4 were 195 and its molecular weight was 194, inferring that M3 and M4 are ferulic acid or isoferulic acid produced by methylation of caffeic acid; the quasi-molecular ion $[M + 1]^+$ of M5 was 209 and its molecular weight was 208, inferring that M5 is methylated ferulic acid produced by further methylation of ferulic acid. The reference solutions of danshensu, protocatechuic aldehyde, caffeic acid, and ferulic acid were analyzed by HPLC-DAD under the same conditions as were the metabolites, to further confirm the identities of the metabolites. The chromatogram is shown in Fig. 17.10. The results further confirm that

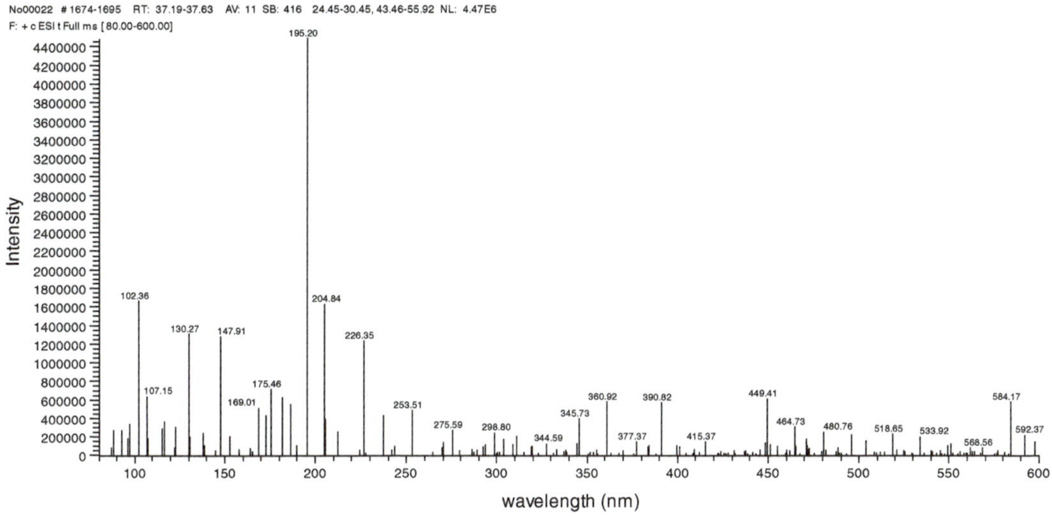

Fig. 17.8 The MS spectrum of M4

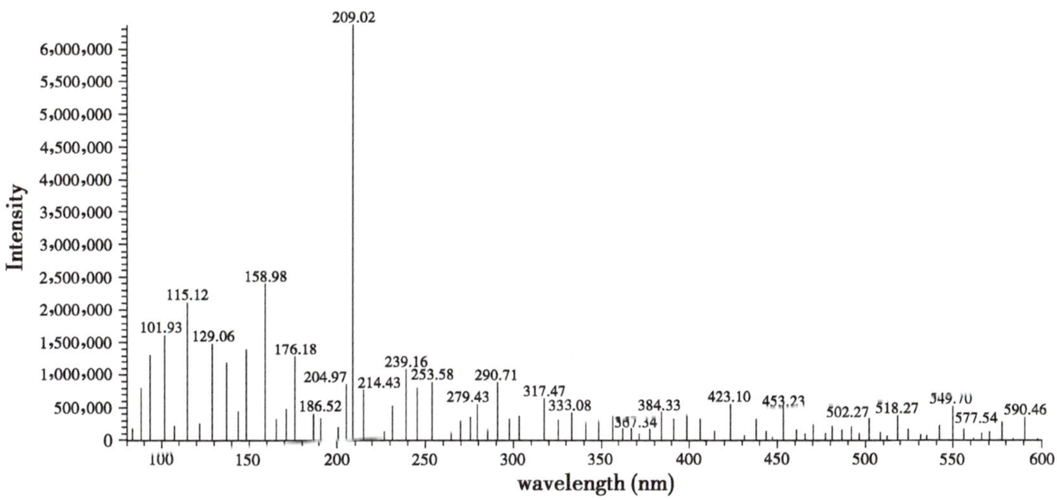

Fig. 17.9 The MS spectrum of M5

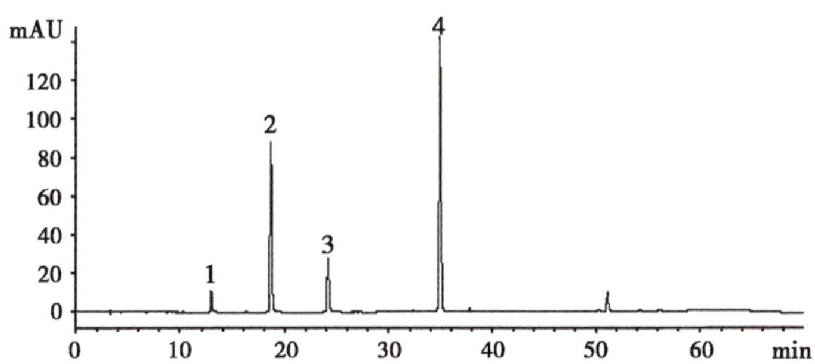

Fig. 17.10 The HPLC chromatogram of the reference substances. *1* danshensu, *2* protocatechuic aldehyde, *3* caffeic acid, *4* ferulic acid

M1 is danshensu, M2 is caffeic acid, M3 is ferulic acid, M4 is isoferulic acid, and M5 is methylized ferulic acid.

17.1.2.7 The Possible Metabolic Pathway

Based on the above results, it is proposed that after oral administration of total phenolic acids to rats, the polyphenolic acids are initially metabolized into danshensu (M1) and caffeic acid (M2), then caffeic acid is methylized into ferulic acid (M3) and isoferulic acid (M4), M3 and M4 are further transformed into methylated ferulic acid (M5). The proposed metabolism pathway is shown in Fig. 17.11.

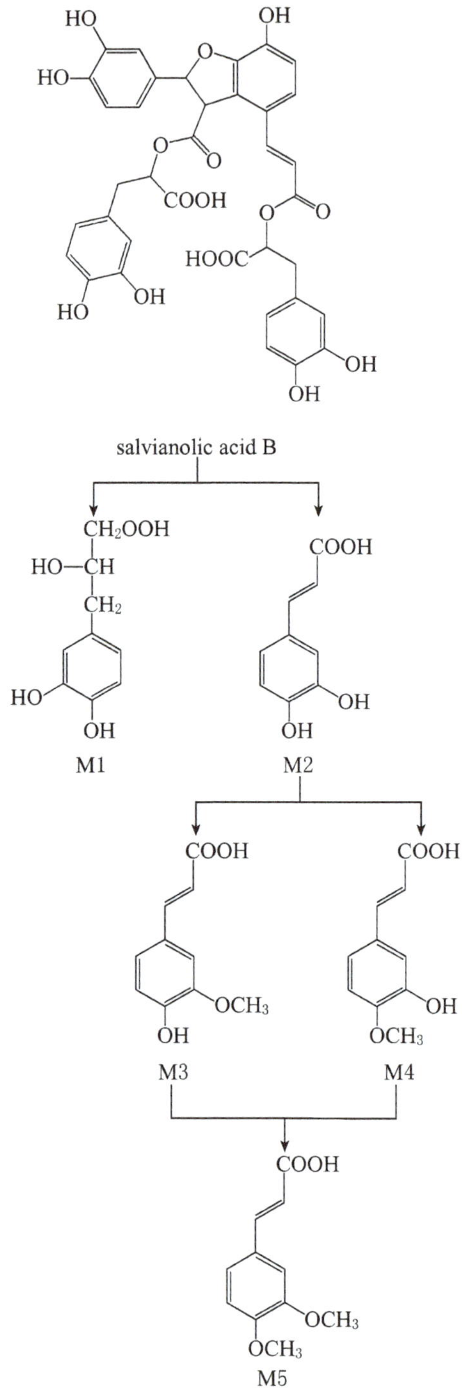

Fig. 17.11 The proposed metabolic pathway in rats orally administered with total phenolic acids

17.2 Pharmacokinetics and In Vivo Metabolism of Danshen Preparations

17.2.1 Preliminary Pharmacokinetic Study of Compound Danshen Tablet

Jinlan Zhang and Dean Guo

The metabolism pathways of Compound Danshen Tablet (CDT) were studied by analyzing the ingredients in the plasma, tissues, urine, and feces of rats after intragastric administration of the tablets.

17.2.1.1 Instruments

Agilent 1100 series HPLC system (Agilent, US), equipped with DAD, quaternary pump, autosampler, and Chemstation 10.02A; N-1000V-W rotary evaporator (EYEIA, Japan); Satorious electronic analytical balance (Satorious, Germany); TGL-16G-A high-speed refrigerated centrifuge (Beijing Anting Scientific Instrument Factory); XW-80 vortex mixer (Shanghai First Medical Instrument Factory).

17.2.1.2 Reagents and Drugs

CDT was produced by Guangzhou Baiyunshan Pharmaceutical Co., Ltd. Acetonitrile (chromatographic pure) and methanol were produced by Burdick & Jackson, MI, USA; phosphoric acid (chromatographic pure) was from Tedia, USA; hydrochloric acid was produced by Beijing Chemical Plant.

17.2.1.3 Laboratory Animals and Biological Sample Collection

1. Laboratory Animals
Spraque-Dawley male rats with body weights of 250 ± 20 g were provided by the Laboratory Animal Center of Shanghai Institute of Materia Medica, Chinese Academy of Sciences.
2. Drug administration
Crush CDT; sift the crushed tablets through a 100 mesh sieve; dissolve the powder with 0.5 % sodium carboxyl methyl cellulose (CMC-Na) solution to make a 1.8 g/ml suspension for intragastric administration. After grouping, the rats were fed in metabolic cages for 3 days. Fast the rats for 24 h before administering CDT by intragastric administration at a dose of 1.8 g/100 g.
3. Sample Collection
Plasma collection: blood samples were collected from the portal vein at 10, 30, 60, 90, 120, 180, 240, 360, 720 min post-administration in centrifuge tubes previously soaked in heparin sodium. Let the tubes stand for 2 h, then centrifuge at 3,500 rpm for 20 min; separate the plasma, and store the samples at −80 °C.
Urine collection: place the rats in metabolic cages after administration; collect urine samples at 12 and 24 h; centrifuge the samples at 3,000 rpm for 10 min; store the supernatant at −80 °C.
Feces collection: place the rats after in metabolic cages after administration; collect feces samples at 12 and 24 h; store the samples at −80 °C.
Tissue collection: after administration, sacrifice the rats by decapitation at 10, 30, 60, 120, 180, 360, 600, 720 min; dissect the rats, take out heart, liver, spleen, lungs, and kidneys, wash the tissues with normal saline; store the tissues at −80 °C.

17.2.1.4 Preparation of Biological Samples and HPLC Chromatographic Conditions

1. Preparation of Biological Samples
Preparation of blood samples: transfer 1 ml of rat plasma to a 10 ml centrifuge tube with stopper; add 150 μl of 10 % HCl and 2 ml of acetic ester; shake for 5 min in a vortex shaker; centrifuge at 4,500 rpm for 20 min; separate and extract the upper-layer organic phase; extract the lower layer water phase twice with acetic ester; pool the acetic ester layers; evaporate acetic ester with nitrogen gas at 29 °C; add 150 μl of acetonitrile–0.026 % phosphorous acid (50:50, V/V) to the residue; vortex for 1 min, dissolve with ultrasound for 1 min, then centrifuge at 6,000 rpm for 5 min; filter through a 0.22 μm membrane; inject 10 μl of the sample into an HPLC system to analyze.
Preparation of urine samples: transfer 2 ml of rat urine to a 10 ml centrifuge tube with stopper; add 300 μl of 10 % HCl and then 4 ml of acetic ester; shake at 1,900 rpm for 5 min in a vortex shaker; centrifuge at 4,500 rpm for 20 min; separate and extract the upper layer organic phase; extract the lower layer water phase twice with acetic ester; pool the acetic ester layers; evaporate with nitrogen gas at 29 °C; add 150 μl of acetonitrile–0.026 % phosphorous acid (50:50, V/V) to the residue; vortex for 1 min, dissolve with ultrasound for 1 min, then centrifuge at 6,000 rpm for 5 min; filter through a 0.22 μm membrane; inject 10 μl of sample into an HPLC system to analyze.
Preparation of feces samples: weigh the feces sample, add 4 volumes of water to the sample; precisely weigh the sample and homogenize; add 20 ml of acetonitrile and 1.5 ml of 10 % HCl to the homogenized solution containing 3.0 g of sample and then transfer to a conical flask with stopper; treat with ultrasound for

30 min; centrifuge at 3,000 rpm for 15 min; separate the supernatant and evaporate at ambient temperature and under reduced pressure. Add 150 μl of acetonitrile–0.026 % phosphorous acid (50:50, V/V) to the residue; shake in a vortex for 1 min, dissolve with ultrasound for 1 min, then centrifuge at 6,000 rpm for 5 min; filter through a 0.20 μm membrane; inject 10 μl of sample into an HPLC system to analyze.

Preparation of tissue samples: take intact heart, liver, spleen, lung, and kidney, precisely weigh, add four volumes of water to the sample; then precisely weigh the sample and homogenize; add 20 ml of acetonitrile and 1.5 ml of 10 % HCl to the homogenized solution containing 3.0 g of sample and then transfer to a conical flask with stopper; treat with ultrasound for 30 min; centrifuge at 3,000 rpm for 15 min; separate the supernatant and evaporate at ambient temperature and under reduced pressure. Add 150 μl of acetonitrile–0.026 % phosphorous acid (50:50, V/V) to the residue; shake in a vortex for 1 min, dissolve with ultrasound for 1 min, then centrifuge at 6,000 rpm for 5 min; filter through a 0.20 μm membrane; inject 10 μl of sample into an HPLC system to analyze.

2. Chromatography Conditions
   Zorbax Extend-C_{18} column (4.6 mm × 250 mm, 5 μm); Zorbax Extend-C_{18} precolumn (20 mm × 4 mm, 5 μm); mobile phase: (A) acetonitrile, (B) 0.026 % phosphorous acid aqueous solution. Perform gradient elution. Flow rate: 0.8 ml/min, detection wavelength: 280 nm; full-spectrum scanning: 20–400 nm; sample volume: 15 μl; temperature of column: 20 °C.

17.2.1.5 Results

1. The Results of Blood Samples
   The blood sample results are shown in Fig. 17.12. These results showed that the ingredients of Danshen were hardly detectable in rat plasma after intragastric administration of CDT.
2. The Results of Tissue Samples
   The test results of the tissue samples collected after intragastric administration of CDT to rats are shown in Figs. 17.13, 17.14, 17.15, 17.16, and 17.17. These results showed that the ingredients from Danshen were hardly detectable in rat tissues after administration.
3. The Results of Feces Samples
   The results of feces samples collected after intragastric administration of CDT to rats are shown in Fig. 17.18. These results showed that the phenolic acid ingredients of Danshen were hardly detectable in rat feces after CDT administration. However, tanshinones were detected.

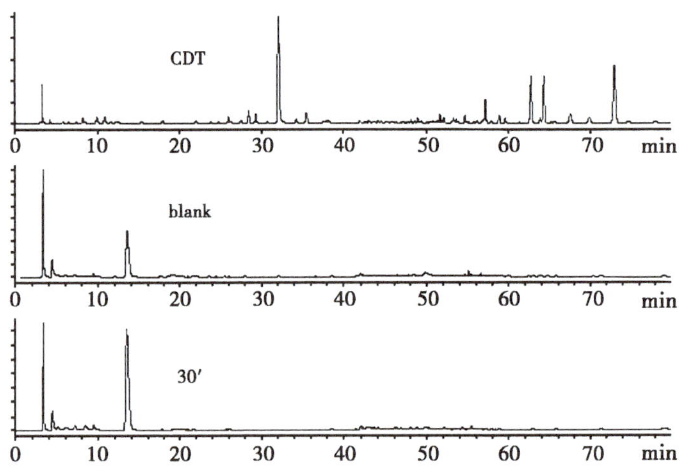

Fig. 17.12 The HPLC chromatograms of CDT sample, blank rat plasma sample, and rat plasma sample collected 30 min after CDT administration

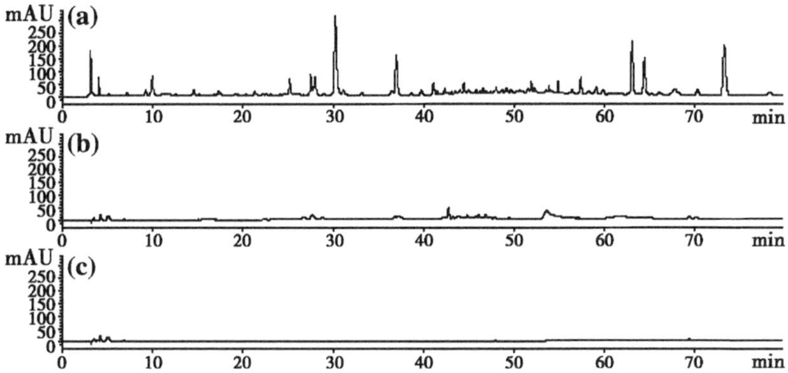

Fig. 17.13 The HPLC chromatogram of CDT sample (**a**), blank rat heart sample (**b**) and rat heart sample collected 90 min after CDT administration (**c**)

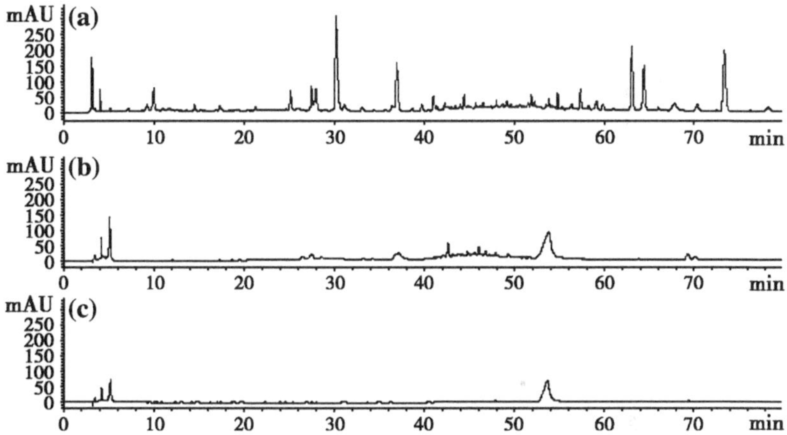

Fig. 17.14 The HPLC chromatogram of CDT sample (**a**), blank rat liver sample (**b**) and rat liver sample collected 90 min after CDT administration (**c**)

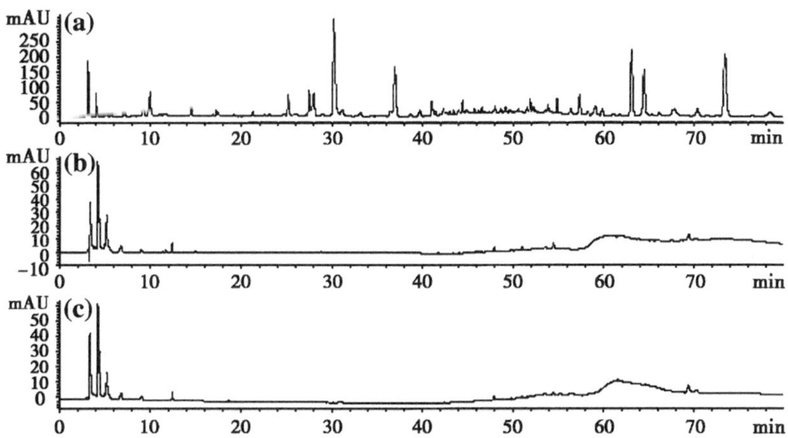

Fig. 17.15 The HPLC chromatogram of CDT sample (**a**), blank rat spleen sample (**b**) and rat spleen sample collected 90 min after CDT administration (**c**)

4. The Results of Urine Samples

The test results of urine samples collected after intragastric administration of CDT to rats are shown in Fig. 17.19. These results showed that the phenolic acids of Danshen were hardly detectable in rat urine after CDT administration; however, metabolites with strong polarities were found in the urine.

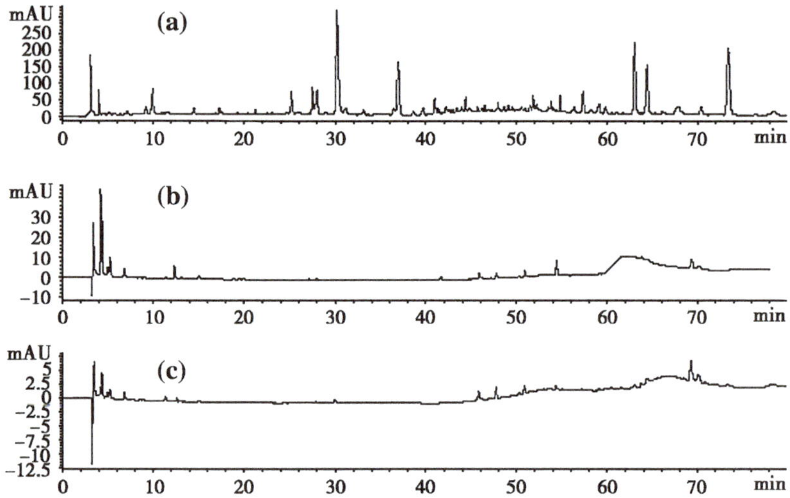

Fig. 17.16 The HPLC chromatogram of CDT sample (**a**), blank rat lung sample (**b**) and lung sample collected 90 min after CDT administration (**c**)

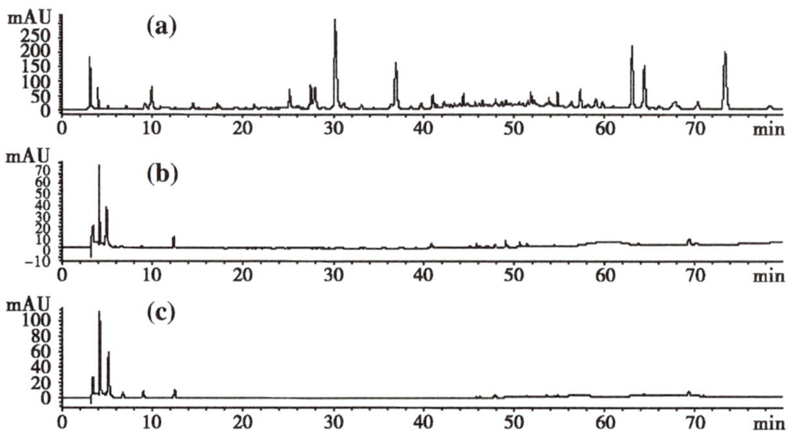

Fig. 17.17 The HPLC chromatogram of CDT sample (**a**), blank rat kidney sample (**b**) and kidney sample collected 90 min after CDT administration (**c**)

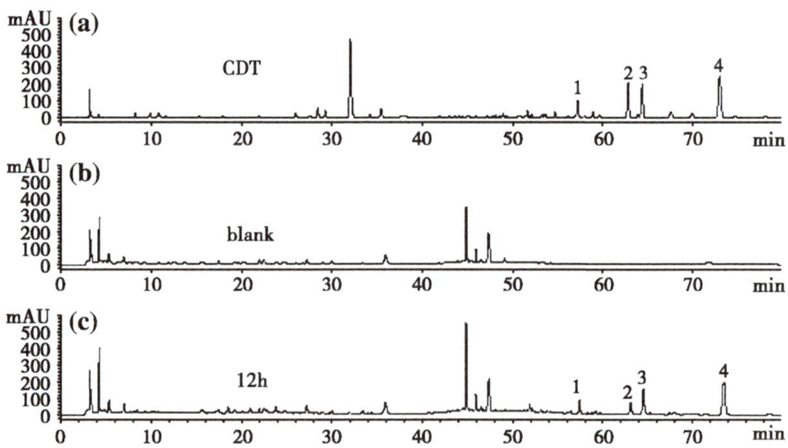

Fig. 17.18 The HPLC chromatogram of CDT sample, blank rat feces sample, and rat feces sample collected 12 h after CDT administration. *1* dihydrotanshinone I; *2* cryptotanshinone; *3* tanshinone I; *4* tanshinone II_A

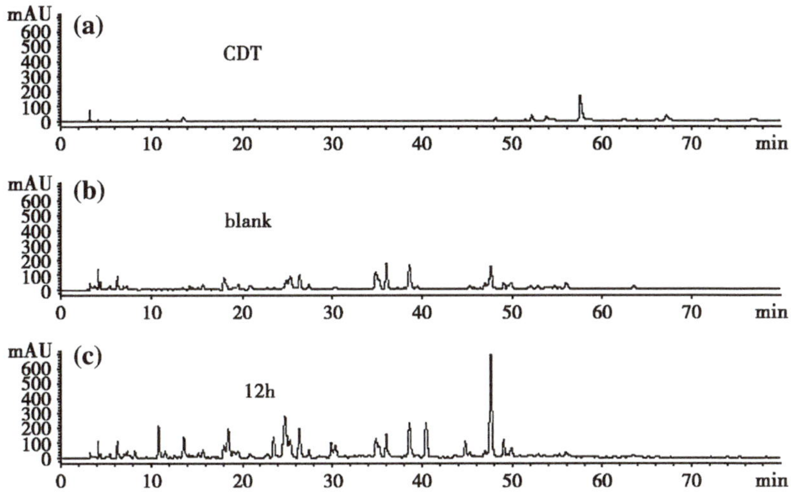

Fig. 17.19 The HPLC chromatogram of CDT sample, blank rat urine sample, and rat urine sample collected 12 h after CDT administration

17.2.1.6 Result Analysis

This study provided preliminary results on the absorption, distribution, metabolism, and elimination of Danshen ingredients in rats after oral administration of CDT. The original phenolic acid ingredients or tanshinone ingredients can hardly be detected in rat plasma after administration, suggesting that phenolic acids or tanshinone ingredients of CDT were poorly absorbed or completely metabolized. Further tests on feces indicated that a large amount of tanshinone ingredients were detected in rat feces after the administration, and the contents of major tanshinones (dihydrotanshinone I, cryptotanshinone, tanshinone I, and tanshinone II_A) in feces were similar to those in CDT, suggesting that the tanshinone ingredients of CDT were poorly absorbed following oral administration and the majority was excreted in its original form in the feces. Meanwhile, phenolic acid ingredients were undetectable in feces, indicating that these ingredients were possibly transformed to metabolites. A large amount of metabolites was detected in the urine samples, and the high polarity of these metabolites further confirms that the phenolic acids of CDT were possibly transformed into these metabolites. The original ingredients of Danshen were not detected in rat heart, liver, spleen, lung, or kidney following administration.

17.2.2 Metabolic Fingerprinting of Danshen Injection

Jinlan Zhang and Dean Guo

In order to better understand the material basis of the efficacy and mechanism of action of Danshen Injection, we investigated the metabolic fingerprints of rat plasma at different time points after intravenous administration of Danshen Injection.

17.2.2.1 Instruments and Surgical Devices

Sardoris electronic balance; Eyela rotary evaporator; centrifuge produced by Shanghai Medical Centrifuge Factory; homogenizer; Agilent 1100 series HPLC system equipped with diode-array detector; PE SCIEX QSTAR MASS instrument; Zorbax Extend-C_{18} chromatographic column (4.6 mm × 250 mm, 5 μm).

Surgical devices: surgical scissors, ophthalmic scissors, dressing forceps, ophthalmic forceps, hemostatic forceps, syringe, intragastric administration needles for rats.

17.2.2.2 Reagents and Drugs

Methanol (GR), hydrochloride (AR), acetic ester (AR), and ether (AR) were produced by Beijing Chemical Plant; heparin sodium (biochemical reagent) was produced by Shanghai Biochemical Reagent Factory; medical normal saline, acetonitrile were produced by Caledon, Canada; Danshen Injection samples were produced by Ya'an 999 Pharmaceutical Co., Ltd. (Lot No. 020603), size: 10 ml; content of Danshen: 1 g/ml; total salvianolic acids: extracted from Danshen produced at Sichuan Zhongjiang Base; danshensu, protocatechuic aldehyde, caffeic acid, and ferulic acid were purchased from The National Institutes for Food and Drug Control; SalB and SalC were self-prepared with purity >97 % (calculated by HPLC peak area normalization method).

17.2.2.3 Laboratory Animals and Collection of Biological Samples

SD male rats with body weights of 200 ± 20 g were provided by the Laboratory Animal Center of Peking University Health Science Center.

Fast rats for 12 h in a metabolic cage before administering Danshen Injection by tail vein injection at a dose of 1 g/kg. Collect blood samples 5, 10, and 30 min later. At each time point, administer ether anesthesia to three rats, draw blood from the abdominal aorta and use heparin sodium for anticoagulation, and centrifuge at 3,500 rpm for 5 min; collect supernatants and store at −20 °C for further treatment and analysis.

17.2.2.4 Pretreatment of Blood Samples

Take three blank plasma samples and three rat plasma samples obtained 10 min post-administration, 2 ml each; extract one sample with 3 volumes of methanol and adjust the pH of the other two samples with 10 % HCl to 2.5; extract one of them with an equal volume of acetic ester three times, pool acetic ester layers; extract the other with an equal volume of methanol for three times, pool the methanol layers; concentrate and evaporate; dissolve the residue in 0.3 ml of methanol; treat with ultrasound; filter through a 0.45 μm microporous membrane; analyze the samples on HPLC. The results showed that the samples acidified and extracted with acetic ester were associated with little interference caused by impurities and contained relatively complete water-soluble ingredients from the Injection.

17.2.2.5 HPLC-DAD and HPLC-MS Analysis of the Blood Samples

1. HPLC-DAD and HPLC-MS Analysis Conditions
   The conditions were the same as those for fingerprint analysis of CDT.
2. HPLC-DAD Analysis Results
   The HPLC fingerprint of Danshen Injection is shown in Fig. 17.20. The metabolic fingerprints of pretreated blank and postadministration rat plasma samples are shown in Fig. 17.21. These results showed that: (1) The established pretreatment method for blood samples could be used in the analysis of blood samples, which had little interference from the impurities with HPLC analysis. The method could effectively extract major ingredients in the Injection; (2) By comparing the fingerprint of the injection with

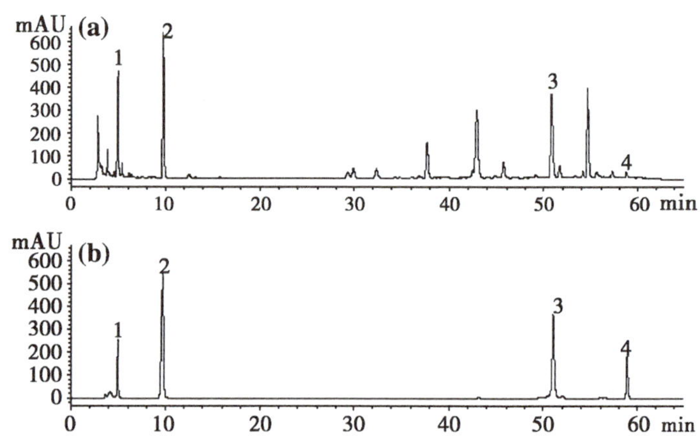

Fig. 17.20 The HPLC fingerprints of water-soluble ingredients in Danshen Injection (**a**) and the reference substances (**b**). *1* danshensu, *2* protocatechuic aldehyde, *3* SalB; and *4* SalC

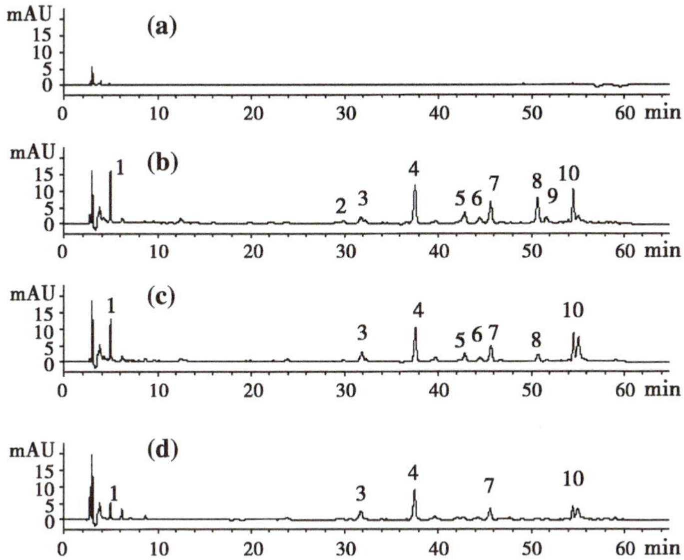

Fig. 17.21 The HPLC metabolic fingerprints of rat plasma after i.v. Danshen Injection administration. **a** blank plasma; **b** plasma 5 min postadministration; **c** plasma 10 min postadministration; and **d** plasma 30 min postadministration (For the peak identities in the figures, See Table 17.5)

those of the postadministration plasma samples, it was clear that the major active ingredients in the injection all appeared in the postadministration plasma samples; however, the ratio of these ingredients had changed significantly, which might have been caused by their different binding rates to carrying proteins in the plasma and their metabolic rates in vivo. (3) Danshensu, protocatechuic aldehyde, SalB and SalA were the major ingredients of the Injection, while the major components in postadministration plasma were danshensu, SalD, lithospermic acid, SalB and SalA. Danshensu and SalB were promptly eliminated from the plasma, while the elimination of SalD, SalA, and lithospermic acid was slow. The pharmacological pharmacodynamics of these components need to be further investigated. These results could fully address the material basis for the efficacy and mechanism of action of Danshen Injection.

3. Identification of Phenolic Acid Components in the Metabolic Fingerprint

In order to understand more about the structures of components in the HPLC metabolic fingerprints of Danshen Injection, we performed HPLC-MS analysis of rat plasma collected 10 min postadministration. The analysis conditions were the same as those for CDT. The structural information obtained from the analysis is shown in Table 17.5.

Table 17.5 Identification of phenolic acid peaks in HPLC metabolic fingerprints

Peak No.	Retention time (min)	$(M-1)^-$	M.W.	Identification
1	4.9	197	198	Danshensu
2	27.40	537	538	Salvianolic acids H and I
3	34.88	339	340	Salvianolic acid G
4	35.78	417	418	Salvianolic acid D
5	41.03	359	360	Rosmarinic acid
6	43.90	493	494	Unknown
7	45.97	537	538	Lithospermic acid
8	49.18	717	718	Salvianolic acid B
9	53.28	717	718	Salvianolic acid E
10	53.55	493	494	Salvianolic acid A

17.2.3 The Metabolites of Danshen Injection in Rats After Intravenous Administration

Jinlan Zhang and Dean Guo

17.2.3.1 Instruments and Surgical Devices

Instruments: Sardoris electronic balance; Eyela rotary evaporator; centrifuge produced by Shanghai Medical Centrifuge Factory; homogenizer; Agilent 1100 series HPLC system equipped with diode-array detector; Finigan Advantage ion trap mass spectrometer; Zorbax Extend-C_{18} chromatographic column (4.6 mm × 250 mm, 5 μm).

Surgical devices: surgical scissors, ophthalmic scissors, dressing forceps, ophthalmic forceps, hemostatic forceps, syringe, intragastric administration needles for rats.

17.2.3.2 Reagents and Drugs

Methanol (GR), hydrochloride (AR), acetic ester (AR), and ether (AR) were produced by Beijing Chemical Plant; heparin sodium (biochemical reagent) was produced by Shanghai Biochemical Reagent Factory; medical normal saline and acetonitrile were produced by Caledon, Canada; total salvianolic acids were isolated from Danshen produced at Sichuan Zhongjiang Base; danshensu, protocatechuic aldehyde, caffeic acid, and ferulic acid were purchased from The National Institutes for Food and Drug Control.

17.2.3.3 Laboratory Animals and Collection of Biological Samples

SD male rats with body weight of 200 ± 20 g were provided by the Laboratory Animal Center of Peking University Health Science Center.

Fast rats for 12 h in a metabolic cage. During the fasting period, collect blank urine and feces samples. Then administer Danshen Injection by tail vein injection at a dose of 1 ml/kg (1 ml of the Injection was equivalent to 1 g of Danshen herb). Collect urine and feces samples during 0–12 h and 12–24 h postadministration. Store samples at −20 °C for further treatment and analysis.

17.2.3.4 Pretreatment of Samples and HPLC-DAD Analysis

The methods for the pretreatment of samples and HPLC-DAD analysis were the same as those for oral administration of total salvianolic acids. The HPLC-DAD results are shown in Fig. 17.22. The results showed that: (1) The metabolites were eliminated within 0–12 h after tail vein injection of Danshen Injection to rats. By comparing with the blank control and UV spectra of phenolic acid ingredients, the preliminary conclusion was that there were four metabolites of Danshen Injection, which were named M1–M4 according to the peak order. The retention times and UV spectra of M1 and M2 were consistent with those of danshensu and protocatechuic aldehyde, respectively, therefore they were identified accordingly. The identities of the other two metabolites remained to be confirmed. (2) Other polyphenolic acid ingredients (SalB, SalA, rosmarinic acid and lithospermic acid) in the injection were not found in the urine or feces samples, indicating these ingredients had already been metabolized, and that their metabolic rates were rather high.

17.2.3.5 HPLC-MS Analysis of the Biological Samples

The analysis method and conditions were the same as those described in Sect. 17.1, and the results are shown in Figs. 17.23 and 17.24. These results showed that the quasi-molecular ion $[M + 1]^+$ of M3 was 181 and its molecular weight was 180, indicating that M3 is caffeic acid; quasi-molecular ion $[M + 1]^+$ of M4 was 195 and its molecular weight was 194, indicating that M4 is ferulic acid or isoferulic acid produced by methylation of caffeic acid. Analyzing the reference solutions of danshensu, protocatechuic aldehyde, caffeic acid, and ferulic acid under the same HPLC-DAD analysis conditions as these for the metabolites further confirmed the identities of the metabolites: M1 is danshensu, M2 is protocatechuic aldehyde, M3 is caffeic acid, and M4 is ferulic acid.

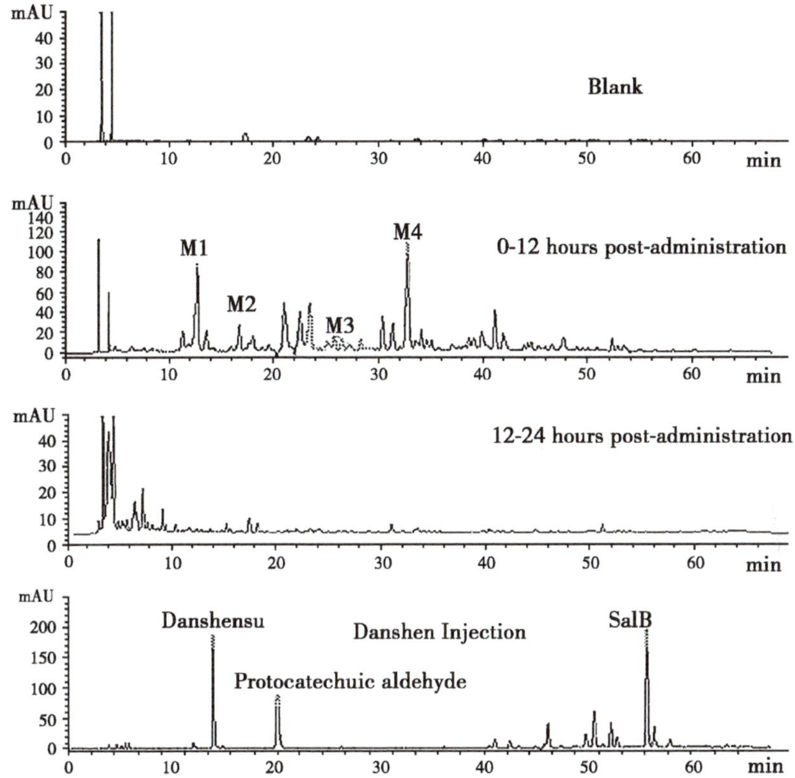

Fig. 17.22 The HPLC-DAD chromatograms of rat urine samples collected before (*blank*) or after administration of Danshen Injection. The bottom panel is the HPLC-DAD chromatogram of Danshen Injection

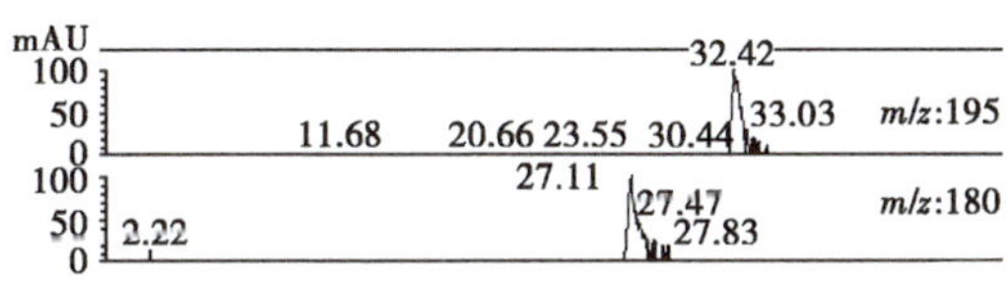

Fig. 17.23 SIR spectra of rat urine samples collected during 0–12 h postadministration of Danshen Injection

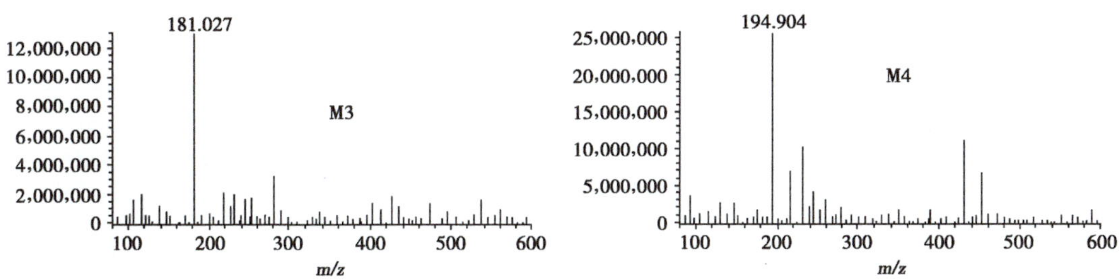

Fig. 17.24 The MS spectra of M3 and M4

17.2.3.6 The Possible Metabolic Pathway

After tail vein administration of Danshen Injection to rats, the major ingredients, danshensu and protocatechuic aldehyde, are excreted unchanged in urine; polyphenolic acid ingredients are transformed into danshensu and caffeic acid, with the latter further metabolized to ferulic acid. Polyphenolic acid ingredients have a high metabolic rate and are excreted mainly as metabolites in urine within 0–12 h following administration.

17.3 The Metabolism of Monomer Tanshinone Components

Dean Guo and Jianghao Sun

17.3.1 In Vivo Pharmacokinetics of Tanshinone II_A in Rats

As one of the main lipid-soluble components in total tanshinone extracts, tanshinone II_A has potential efficacy for treating cardiovascular diseases, tumors, and leukemia, and has attracted much attention from investigators around the world. Recent pharmacological studies have demonstrated that tanshinone II_A may inhibit left ventricular hypertrophy, protect the myocardium against hypoxia; protect the liver against fibrosis and hepatic carcinoma; prevent the occurrence of local cerebral ischemia-reperfusion injury; inhibit the growth of various tumor cells; provide protection against promyelocytic leukemia; induce mesenchymal stem cells in adults to differentiate into neuron-like cells; inhibit the proliferation of scar fibroblasts; provide antibacterial and bacteria-inhibiting effects; and protect against ischemia-reperfusion injury [1–12].

By using HPLC, we investigated the absorption, distribution, metabolism, and elimination processes of tanshinone II_A after a single dose and multiple doses, in order to provide evidence for the reasonable clinical application of tanshinone II_A.

17.3.1.1 Experiment

1. Instruments and Reagents/Drugs
   TSQ Quantum triple quadrupole mass spectrometer, equipped with Thermo Surveyor HPLC system, PDA detector, column heater (Thermo Finnigan, USA); Aglient 1100 series liquid chromatography system (Aglient, US), including DAD, quaternary pump, autosampler, and Chemstation 10.02A; N-1000V-W rotary evaporator (Eyela, Japan); Satorious electronic analytical balance (Satorious, Germany); Sigma 2–16 K low-temperature centrifuge (Sigma, Germany); XW-80 vortex mixer (Shanghai First Medical Instrument Factory); Branson SB 5200 ultrasonic extractor (Shanghai Branson Co., Ltd.).
   Tanshinone II_A and tanshinone I (internal standards) were purchased from The National Institutes for Food and Drug Control, with purity >98 % (determined by HPLC); chromatographic pure acetonitrile and methanol (Burdick, USA); Oasis solid-phase extraction columns (Waters, USA). Other reagents were analytically pure, produced by Shanghai First Reagent Works.
2. Laboratory Animals
   SD male rats with body weights of 250 ± 20 g were provided by the Laboratory Animal Center of Shanghai Institute of Materia Medica, Chinese Academy of Sciences.
3. Study Design
   (a) Pharmacokinetics Study Following Single i.v. Administration of Tanshinone II_A: prepare 2 mg/ml tanshinone II_A propylene glycol solution and administer by tail vein injection at a dose of 4 mg/kg. Draw blood from the orbital veniplex at 5, 7, 10, 15, 20, 30, 60, 90, and 120 min postadministration. At each time point, draw 0.5 ml of blood. Centrifuge at 4,000 rpm for 10 min, and then take 0.2 ml of plasma for further use.
   (b) Absorption by and Tissue Distribution in Rats After Single Oral Administration of

Tanshinone II_A: prepare 10 mg/ml tanshinone II_A suspension with 1 % Tween-80, shake well before intragastric gavage. Decapitate rats at 0, 0.25, 0.5, 1, 2, 4, 6, 8, 12, and 24 h postadministration, respectively. Collect blood samples in heparinized centrifuge tubes, then immediately dissect and take the heart, lung, liver, stomach, spleen, kidney, brain, fat, intestine, and testicles. Flush away superficial blood from the organs with normal saline. Weigh the organs. Collect the contents of the stomach and intestine. Centrifuge blood samples and separate the plasma. All samples were stored at −80 °C before treatment.

(c) Bile Excretion Test in Rats: 12 SD rats were randomized into a low-dose group and a high-dose group, each group with 6 rats. Draw blank bile from another two rats. Administer by intragastric gavage at a dose of 40 and 80 mg/kg to the two groups, respectively. Collect bile samples 0–4, 4–8, 8–12, 12–16, 16–20, and 20–24 h postadministration. Record the volume of collected bile. All samples were stored at −80 °C before treatment.

(d) Excretion Test in Rats: administer by intragastric gavage at a dose of 40 mg/kg to 6 rats, then place the rats in metabolic cages. Animals may drink water freely. Collect urine and feces samples 0–8, 8–16, and 16–24 h postadministration. All samples were stored at −80 °C before treatment.

(e) Qualitative Study Following Multiple Oral Administrations: give the drug to four rats at a dose of 40 mg/kg twice daily (9:00, 21:00) for 3 days. Another rat served as a blank control. During the drug administration period, give 1 ml of 10 % sucrose solution to the rats by intragastric gavage every day. Collect feces and urine samples during this period. Sacrifice animals 1 h after the seventh dose and draw blood. Then immediately dissect and take the heart, lung, liver, stomach, spleen, kidney, brain, fat, intestine, and testicles. Flush away superficial blood from the organs with normal saline. Weigh the organs. Collect the contents of the stomach and intestine. Centrifuge blood samples and separate the plasma. All samples were stored at −80 °C before treatment.

(f) Preparation of Stock Solutions, Working Solutions, and Internal Standard Solutions: precisely weigh certain amounts of tanshinone II_A and tanshinone I, respectively; transfer to a volumetric flask and add methanol to produce 1.0 and 0.1 mg/ml stock solutions, respectively. To prepare standard solutions, dilute stock solution with methanol to a series of working solutions. Store all solutions at −20 °C before use. These solutions can be kept stable for at least 1 month.

(g) Pretreatment of Biological Samples:

Pretreatment of plasma samples: activate a Waters HLB solid-phase extraction (SPE) column (5 ml, 200 mg of 40 μm-diameter reverse-phase silica gel) with 4 ml of methanol; then equilibrate with 5 ml of water. Add 100 μl of internal standard solution to the SPE column; elute with 5 ml of water; add 2 ml (for intragastric administration) or 200 μl (for intravenous administration) of plasma; shake in a vortex shaker for 5 min; load the samples to a SPE column. Elute by gravity: initially with 5 ml of water, then with 12 ml of 60 % methanol, finally with 12 ml of 100 % methanol. Collect the pure methanol eluant. Evaporate the methanol at ambient temperature and under reduced pressure; dissolve the residue in 200 μl of methanol; filter the solution through a 0.22 μm microporous membrane. Injection volume: 25 μl.

Pretreatment of tissue samples and gastrointestinal contents: add normal saline to tissue samples in a ratio of 1:3 (W/V) to prepare homogenized tissue. Add 100 μl of internal standard solution to a centrifuge tube, evaporate with nitrogen gas at 30 °C; then add 1 ml of homogenized tissue; mix well in a vortex mixer; add 5 ml of acetic ester; mix in a vortex mixer for 10 min; centrifuge at 4,024 g for 10 min; separate the acetic ester layer and measure 4 ml of separated acetic ester, evaporate with nitrogen gas; then add 500 μl of methanol to redissolve. Injection volume: 25 μl. Wash gastrointestinal contents with normal saline, then transfer the contents to a 50 ml volumetric flask; measure 1 ml of contents and treat in the same way as for tissue samples.

Pretreatment of bile samples: add 200 μl of bile sample to a centrifuge tube in which the internal standard solution had been blow dried; add 1 ml of acetic ester to the tube; vortex mix for 10 min; centrifuge at 4,024 g for 10 min; separate acetic ester layer and measure 0.8 ml to blow-dry with nitrogen gas; then add 200 μl of methanol to redissolve; filter through a 0.22 μm microporous membrane. Injection volume: 25 μl.

Pretreatment of urine samples: pool all urine samples, then centrifuge and filter; add three volumes of acetic ester to extract; then extract the organic phase with an equal volume of water; discard the water phase; evaporate the organic phase at 30 °C to dryness; add 500 μl of methanol to redissolve; filter through a 0.22 μm microporous membrane. Injection volume: 25 μl.

Pretreatment of feces samples: air-dry the feces samples; grind feces in a mortar to powder; transfer the powder to a conical flask with stopper; add 50 ml of methanol to the flask, then measure the weight; extract with ultrasound for 1 h; add methanol to compensate for lost weight. Separate the supernatant and filter through a 0.22 μm microporous membrane. Injection volume: 25 μl. Directly quantify feces samples by the external standard method.

(h) Preparation of QC samples: precisely pipette 100 μl of the reference solution and internal standard solution at different concentrations to SPE columns, and prepare plasma QC samples at low, medium, and high concentrations according to the method of Pretreatment of Plasma Samples.

Precisely pipette 100 μl of reference and internal standard solutions at different concentrations to a centrifuge tube, add 1 ml of blank homogenized liver tissue to the tube after evaporating the solution with nitrogen gas, then prepare tissue QC samples at low, medium, and high concentrations according to the method of Pretreatment of Tissue Samples.

Precisely pipette 100 μl of reference and internal standard solutions at different concentrations to a centrifuge tube, add 200 μl of blank bile to the tube after evaporating the solution with nitrogen gas, then prepare bile QC samples at low, medium, and high concentrations according to the method described under Pretreatment of Bile Samples.

4. Chromatography Conditions

   Column: Agilent Zorbax XDB-C_{18} (250 mm × 4.6 mm, 5 μm) connected with Agilent Zorbax XDB-C_{18} precolumn (20 mm × 4 mm, 5 μm). Quantitative analysis chromatographic conditions: acetonitrile–water (75:25) constant gradient elution. Qualitative analysis of chromatographic conditions: 0–20 min, acetonitrile increases linearly from 20 to 30 %; 20–30 min, acetonitrile increases linearly from 30 to 70 %; 30–40 min, acetonitrile increases linearly from 70 to 80 %; 40–50 min, elute with 80 % acetonitrile for 10 min. Flow rate: 0.8 ml/min.

5. Method Validation

   The quantitative method's specificity, linear range, lower limit of quantification, precision, accuracy, and stability were tested. Also tested was the extraction recovery of tanshinone II_A and internal standards from various biological materials.

6. Pharmacokinetics Study

   Determine the concentration of tanshinone II_A in each biological sample by the established analysis method, analyze the plasma concentration by 3P87, and calculate pharmacokinetic parameters. The study also investigated the tissue distribution and elimination processes of tanshinone II_A after oral administration.

17.3.1.2 Results and Discussion

1. Results of Method Validation
   (a) The Specificity of the Method: since tanshinone II_A is a compound with high lipid, solubility and endogenous interference is mainly caused by compounds with strong polarity, a mobile system containing a high-percentage organic phase is chosen to perform constant gradient elution. Under selected chromatography conditions, the endogenous substances in plasma and tissues do not interfere with the determination of tanshinone II_A and internal standards. The representative chromatograms are shown in Fig. 17.25.
   (b) Method for Pretreatment of Biological Samples: after trying several methods for sample pretreatment, including protein precipitation for plasma samples, acetic ester extraction, dichloromethane extraction, chloroform–methanol extraction, and solid-phase extraction, we found that solid-phase extraction had the highest recovery. Therefore, solid-phase extraction was chosen for the pretreatment of

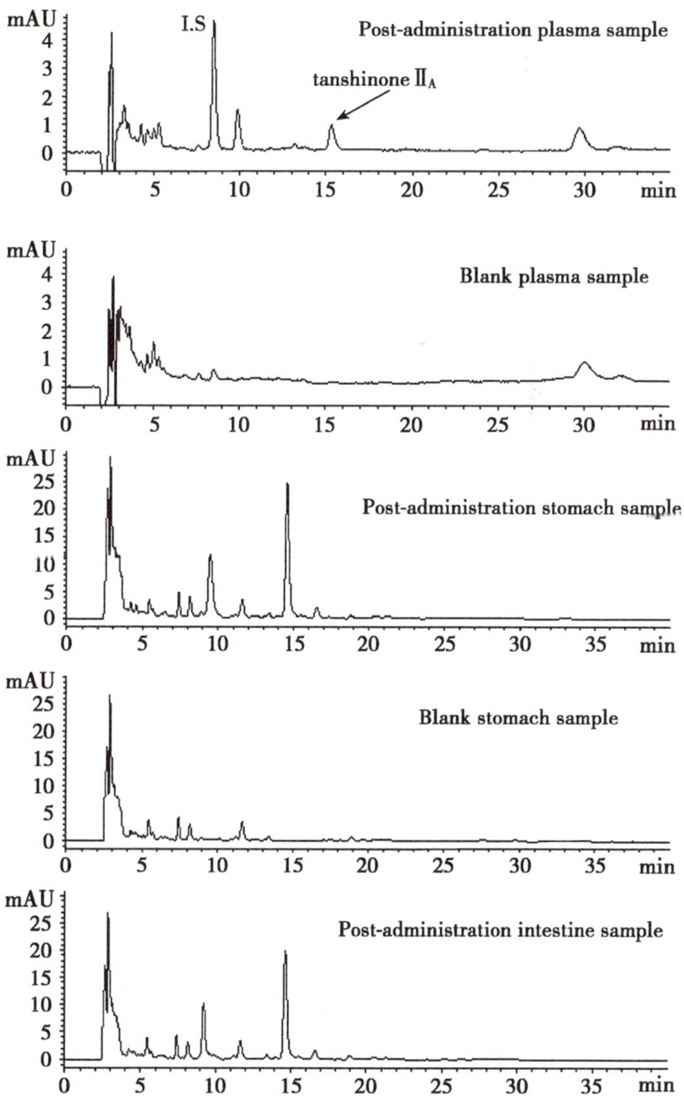

Fig. 17.25 The representative chromatograms of rat samples before and after tanshinone II_A administration

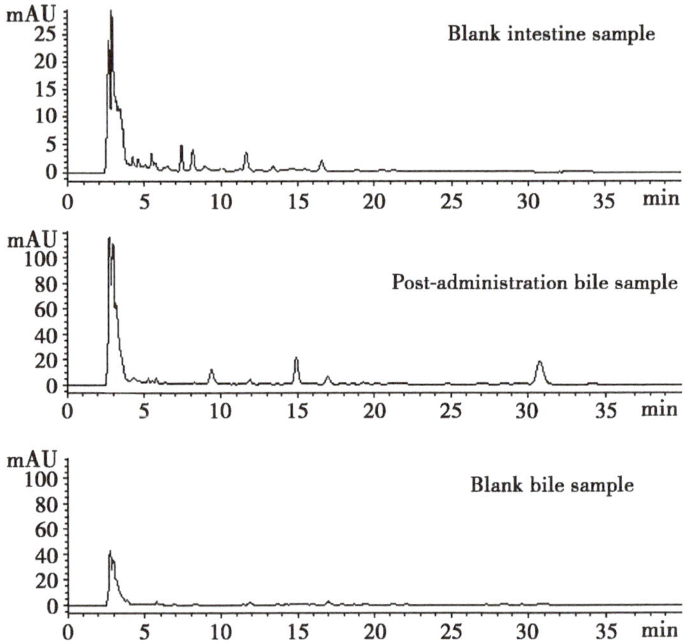

Fig. 17.25 (continued)

plasma samples. Meanwhile, we compared two solid-phase extraction columns: Alltech C_{18} and Waters OASIS HLB. On Alltech C_{18} columns, 60 % water–methanol could elute tanshinone II_A, while on Waters OASIS HLB columns, only when the concentration of water–methanol exceeded 80 % could tanshinone II_A be eluted. Since tanshinone II_A is better retained on Waters columns, eluting with 60 % water–methanol to remove endogenous impurities in plasma followed by eluting with 100 % methanol may simplify subsequent treatments and better purify the plasma samples.

When deciding the optimal solid-phase extraction method, the initially selected method was to add internal standard solution to a centrifuge tube, evaporate the solvent, then add blank plasma to mix well, and subsequently flow through a SPE column. However, tanshinones were almost completely adsorbed on the tube after solvent evaporation due to their poor water solubility, making it difficult to transfer these components to a SPE column and consequently producing uneven results. Therefore, it was finally decided to add internal standard solution to the solid-phase extraction column, then elute with water in order to replace trace methanol on the solid-phase extraction column, thus preventing protein precipitation in plasma caused by methanol from occluding the column.

To the tissue samples and bile samples, general liquid–liquid extraction was good enough to meet the requirements for analysis, so acetic ester extraction was chosen to treat the samples.

(c) Selection of Internal Standard: in order to ensure the parallelism of sample processing, an internal standard method was chosen to determine the contents. It was finally decided to use tanshinone I, which has a similar structure to tanshinone II_A and is stable, as the internal standard.

(d) Linear Range and Lower Limit of Quantification: the standard curves of tanshinone II_A in the plasma, liver, intestine, stomach, gastrointestinal content, and feces under selected test conditions are shown in Table 17.6. These results showed that tanshinone II_A had good

Table 17.6 The standard curves of tanshinone II_A in various tissue samples

Biological sample	Intercept ($\times10^{-2}$)	Slope	*r*	Linear range (μg/ml)	Lower limit of quantification (ng/ml)
Plasma (p.o)	0.074	4.22	0.9984	0.0251–0.502	25.1
Plasma (i.v.)	0.032	1.57	0.99	0.0251–2.51	25.1
Liver	−0.16	3.14	0.9978	0.204–1.25	20.4
Stomach	−0.057	2.80	0.9936	0.204–2.55	20.4
Intestine	−0.047	2.82	0.9932	0.204–2.55	20.4
Bile	−0.049	0.56	0.9946	1.02–12.75	25.0
Gastrointestinal contents	−0.15	3.16	0.9985	0.204–3.06	20.4
Feces	8.75	128.99	0.9999	10.3–102.3	103

linearity in various tissue samples over selected concentration ranges.

(e) Precision and Accuracy: the precision and accuracy of the method were estimated by analyzing the QC samples of plasma, tissue, and bile at low, medium, and high concentrations. The results are listed in Table 17.7. These results showed that both intraday and interday tests precision and accuracy of the method meet the requirements defined in national and international guidelines [13–16].

(f) Stability: the stability test results are listed in Table 17.8. The results showed that the samples were stable after storing at room temperature for 24 h or for three freeze-thaw cycles.

(g) Extraction Recovery: The recovery of tanshinone II_A from blank plasma was 92.5 ± 2.7 %, tanshinone I was 89.4 ± 3.4 %; the recovery of tanshinone II_A from blank liver was 88.3 ± 4.5 %, tanshinone I was 90.7 ± 3.7 %; the recovery of tanshinone II_A from bile was 89.2 ± 4.2 %, tanshinone I was 88.1 ± 4.7 %.

2. Pharmacokinetics of Tanshinone II_A in Rats

(a) Pharmacokinetics of Tanshinone II_A Following a Single Intravenous Dose: Tanshinone II_A was very quickly eliminated from plasma following intravenous administration to rats; its plasma concentration 5 min postadministration was only 1,370 ng/ml, and that at 2 h postadministration was below the lower limit of quantification. The concentration–time curve of tanshinone II_A is shown in Fig. 17.26.

3P97 software was used to calculate the pharmacokinetic parameters based on plasma concentration. The calculated results showed that for a single dose of

Table 17.7 The precision and accuracy of the method

QC sample	Concentration (μg/ml)	Intraday test RSD (%)	Interday RSD (%)	Relative error (%)
Plasma	0.03	4.5	4.7	7.2
	0.05	3.8	7.3	3.4
	0.10	2.7	3.9	1.1
Tissue	0.20	4.4	10.0	3.5
	0.41	2.6	4.5	−2.6
	1.02	7.3	6.4	5.9
Bile	2.0	2.2	4.9	2.2
	4.1	3.0	3.9	1.4
	8.2	1.5	7.1	14.2

Table 17.8 The stability test results of the QC samples

QC sample	Concentration (μg/ml)	Relative error (%)	
		After 24 h at room temperature	After 3 freeze-thaw cycles
Plasma	0.025	12.2	10.9
	0.050	10.5	8.6
	0.201	9.6	−7.8
Tissue	0.20	5.4	2.1
	0.41	−3.2	3.2
	1.02	2.6	−3.3
Bile	2.0	5.7	−0.2
	4.1	2.1	−0.8
	8.2	6.3	7.0

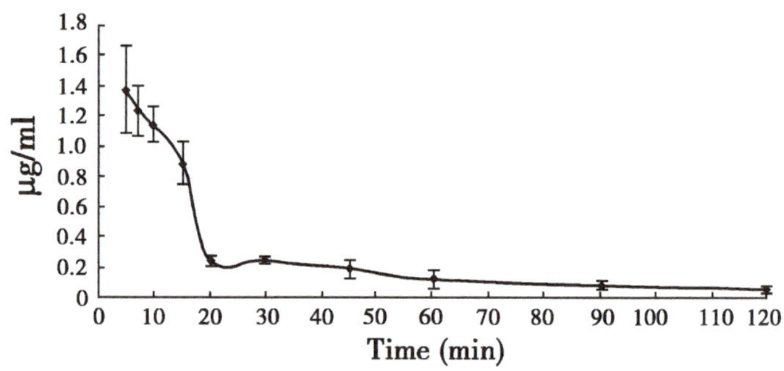

Fig. 17.26 The average plasma concentration–time curve of tanshinone II_A in rats after single intravenous dose (n = 3)

2 mg/kg, the three-compartment model fitted the concentration–time curve well when the weight coefficient was 1/C. The major pharmacokinetic parameters are listed in Table 17.9. The V_d (apparent distribution volume) was 845.74 ± 182.25 ml/kg, well above the total blood volume of a rat, indicating that tanshinone II_A is widely distributed in the body.

(b) Pharmacokinetics of Tanshinone II_A Following a Single Oral Dose: The plasma concentration of tanshinone II_A was extremely low after a single oral dose in rats. Since the volume of the plasma sample obtained by rat orbital bleeding was too small and tanshinone II_A was undetectable in the 200 μl of plasma sample after pretreatment, the volume of plasma sample was increased by drawing blood after decapitation and 2 ml of plasma sample was treated each time. The tanshinone II_A in plasma was enriched by solid-phase extraction to meet analysis requirements.

The results showed that the absorption of tanshinone II_A was irregular following a single oral dose, and no suitable compartment model could fit the time course of plasma concentration. Therefore, a noncompartment model was adopted to calculate its pharmacokinetic parameters, and the observed values were used for C_{max} and T_{max}. The plasma concentration–time curve is shown in Fig. 17.27. The observed C_{max} and T_{max} were 55.1 ± 6.8 ng/ml and 60 min, respectively. When calculating the parameters by statistics moments of a noncompartment model, the results were: MRT_{0-t} 671.6 min, AUC_{0-t} 56,442.23 (ng/ml) min (Table 17.10). When calculating oral bioavailability based on the following equation,

Table 17.9 The pharmacokinetic parameters of tanshinone II_A after a single intravenous dose in rats (2 mg/kg, $n = 5$)

Parameter	Unit	Relative value					Mean	Standard deviation
		1	2	3	4	5		
V_d	ml/kg	568.70	1006.12	805.19	833.85	1014.86	845.74	182.25
$t_{1/2\alpha}$	min	51.09	45.61	87.52	205.84	44.89	86.99	68.74
$t_{1/2\beta}$	min	5.24	10.39	6.48	113.12	7.78	28.60	47.29
$t_{1/2\gamma}$	min	4.72	10.27	6.06	6.50	6.66	6.84	2.06
K_{12}	l/min	0.00023	0.0000020	0.00011	0.050	0.000011	0.010	0.022
K_{21}	l/min	0.14	0.067	0.11	0.012	0.089	0.084	0.048
K_{13}	l/min	0.043	0.00089	0.048	0.0029	0.030	0.025	0.022
K_{31}	l/min	0.022	0.015	0.016	0.0038	0.025	0.016	0.0082
K_{10}	l/min	0.084	0.066	0.056	0.047	0.065	0.064	0.014
AUC	(ng/ml) min	42060.41	30222.78	44649.95	51112.12	30283.37	39665.73	9203.19
CL	ml/kg/ min	47.55	66.18	44.79	39.13	66.04	52.74	12.58
$MRT_{0-\infty}$	min	30.70	39.52	43.03	57.43	27.39	39.61	11.81
MRT_{0-t}	min	20.92	22.24	26.06	27.4	21.95	23.71	2.84

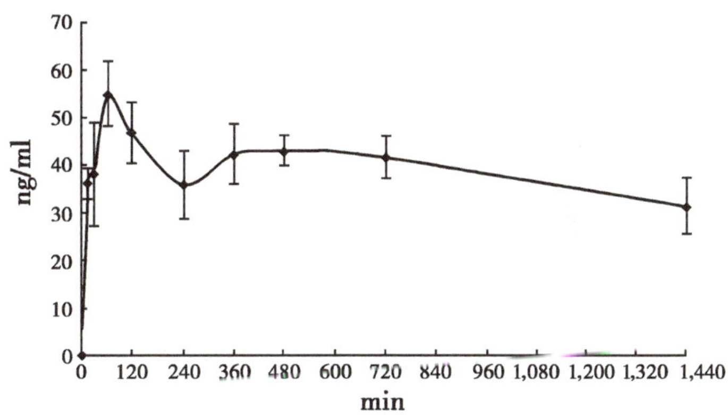

Fig. 17.27 The average plasma concentration–time curve of tanshinone II_A in rat following a single oral dose ($n = 3$)

Table 17.10 The pharmacokinetic parameters of tanshinone II_A following a single oral dose in rats (40 mg/kg)

Parameter	Unit	Value
AUC_{0-t}	(ng/ml) min	56442.23
MRT_{0-t}	min	671.603760
$AUC_{0-\infty}$	(ng/ml) min	162366.88
$MRT_{0-\infty}$	min	3372.21
C_{max}	ng/ml	55.1 ± 6.8
T_{max}	min	60
F_{po}		7.11

$$F_{po} = \left(AUC_{po} \times Dose_{iv} / AUC_{iv} \times Dose_{po}\right) \times 100\,\%$$

The calculated F_{po} was 7.11 %, suggesting that tanshinone II_A has low oral bioavailability.

3. Tissue Distribution Following a Single Oral Dose

After a single oral administration of tanshinone II_A to rats, the heart, spleen, kidney, fat, testicle, and brain were homogenized, extracted, and treated. However, no tanshinone II_A was detected in any of these samples. As shown in Fig. 17.28, tanshinone II_A could only be detected in the digestive organs such as the stomach, intestine, and liver. Considering the low plasma concentration of tanshinone II_A, it suggests that tanshinone II_A might have a strong first-pass effect in the liver, and a large amount of unabsorbed tanshinone II_A is recovered from gastrointestinal contents. 15 and 30 min postadministration, more than 90 % tanshinone II_A was recovered from gastrointestinal tract; 24 h postadministration, about 30 % tanshinone II_A was recovered from gastrointestinal tract; while the concentration of tanshinone II_A in rat liver remained at a low level (about 5 μg/g).

Although tanshinone II_A has high lipid solubility, it was not detected in brain tissue, suggesting that it is not easy for tanshinone II_A to pass the blood–brain barrier. Tanshinone II_A was undetectable in fat tissues, suggesting that it was not distributed in these tissues. Tanshinone II_A was undetectable in kidneys, suggesting that it was not excreted by the kidney following a single oral dose.

4. Bile Excretion Results

After oral administration of tanshinone II_A at a high dose of 80 mg/kg and a low dose of 40 mg/kg to rats, bile excretion did not reach a plateau after 24 h postadministration, and cumulative bile excretion appeared to increase in an obvious linear manner (Fig. 17.29). Retention of tanshinone II_A in the gastrointestinal tract for a long period seems to be an important reason explaining why bile excretion did not reach a plateau 24 h postadministration. At that time point, cumulative excretion of tanshinone II_A in bile accounts for 1.70 % (high dose) and 1.45 % (low dose) of the administered dose, respectively. Bile excretion has no significant difference between high dose and low dose, indicating that bile excretion may be a saturated process.

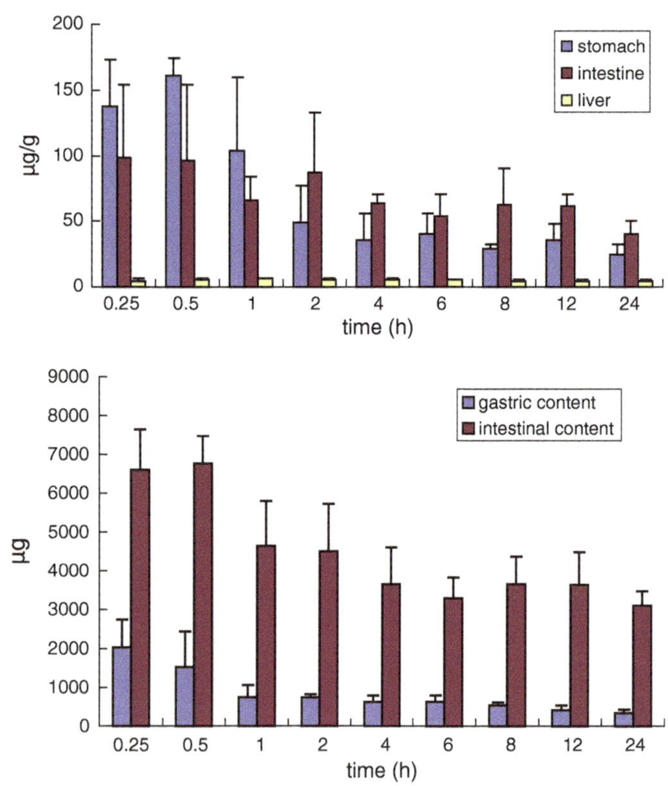

Fig. 17.28 Tissue distribution of tanshinone II_A in rats following a single oral dose (n = 3)

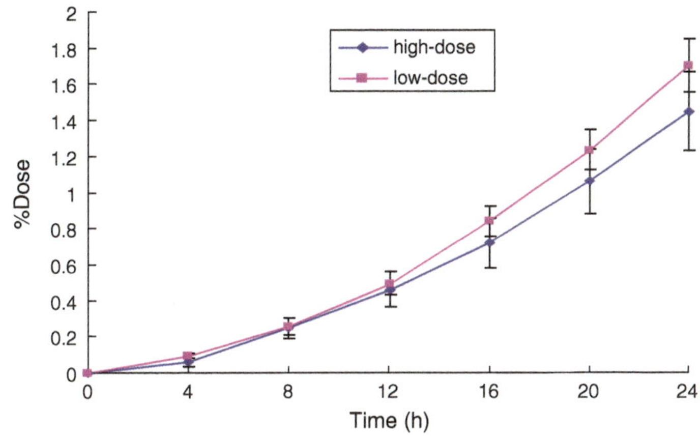

Fig. 17.29 The cumulative excretion of tanshinone II_A in rats ($n = 6$)

5. Feces and Urine Excretion Results
   Filter the rat urine samples collected during a 24 h period, centrifuge at 4,024 g, and pretreat the samples according to the method described above. No tanshinone II_A was detected in the samples, which further confirms the results that tanshinone II_A was hardly excreted through the kidneys following a single dose. It should be noted that because a large amount of tanshinone II_A exists in feces and the metabolic cages used in the experiment have certain defects in design, which resulted in inevitable urine flow through feces and consequent transfer of tanshinone II_A from feces to urine, pretreatment of urine must therefore include filtration and high-speed centrifugation in order to minimize interference from the contamination. A large amount of unabsorbed tanshinone II_A was recovered in feces; during the first 24 h post-administration, 62.07 ± 7.03 % of the administered dose was recovered in the feces (Table 17.11).
6. Tissue Distribution Following Multiple Oral Doses in Rats
   The concentration of tanshinone II_A in various rat organs following seven consecutive oral administrations at 40 mg/kg was significantly higher than that following a single dose. In those organs where tanshinone II_A was undetectable following a single dose, such as the heart, lung, spleen, kidneys and testicles, tanshinone II_A was detected after multiple doses. Danshen is a conventional medicine for cardiovascular diseases, and a multiple-dose study has shown that lipid-soluble ingredients, such as tanshinone II_A, could reach the target organs only after long-term administration (Fig. 17.30).

Table 17.11 The cumulative excretion of tanshinone II_A in rat feces

Time (h)	Dose (%)	Time (h)	Dose (%)
0–8	43.80 ± 10.93	16–24	5.83 ± 3.45
8–16	12.43 ± 2.63	0–24	62.07 ± 7.29

7. The Metabolites in Rat Intestinal Contents and Feces After Multiple Oral Doses of Tanshinone II_A
   Following multiple doses of tanshinone II_A to rats, five metabolites were found in the intestinal contents and feces, among which IM1, IM2, and IM3 all have a protonated molecule of 311; by comparing with reference substances and the literature, these three metabolites were identified to be tanshinone II_B, 3α-hydroxyl tanshinone II_A and przewaquinone A. The UV absorption of IM4 and IM5 also showed a characteristic tanshinone II_A backbone (Fig. 17.31). In order to better identify the structures, IM4 and IM5 were semi-prepared from intestinal contents and subjected directly to mass spectrometry analysis.
   Under selected mass spectrometry conditions, a full-scan spectrum of IM4 displays ions with base peaks of *m/z* 370, *m/z* 370 and a very weak ion of *m/z* 253, as well as a characteristic ion with a tanshinone II_A backbone, *m/z* 293. A

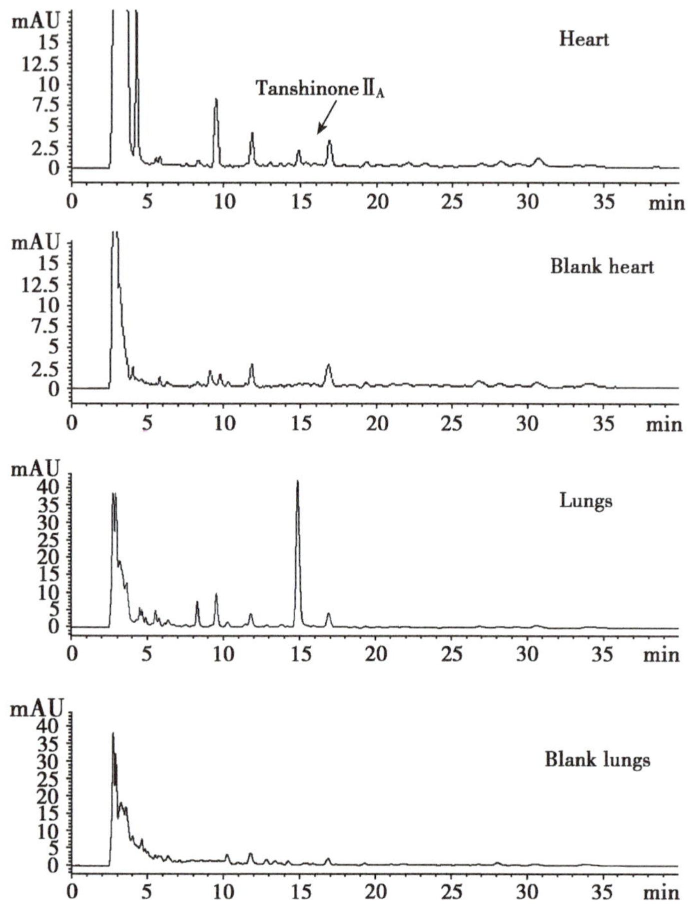

Fig. 17.30 The HPLC chromatograms of tanshinone II_A in various rat blank and postadministration organ samples following multiple doses

second-stage mass spectrum of *m/z* 375 showed that it was very difficult for this ion to trigger secondary fragmentation and a higher collision energy (35 %) was required; with a collision energy of 60 %, relatively complete fragmentation occurred, but it produced fragments with a low abundance, mainly *m/z* 293. The *m/z* 370 ion tended to break up easily: a collision energy of 10 % could induce near complete fragmentation and *m/z* 353 and *m/z* 293 ions were found during secondary fragmentation. Similarly, a second-stage mass spectrum of *m/z* 353 also displayed an ion with base peak of *m/z* 293. Therefore, it suggests that *m/z* 375, *m/z* 370 and *m/z* 353 are a group of associated ions and based on their mass, they are most likely $[M + Na]^+$, $[M + NH_4]^+$, and $[M + H]^+$. To validate the speculation, the precursor ion *m/z* 293 was scanned and it was found that *m/z* 293 origins from *m/z* 353, *m/z* 370 and *m/z* 375, thus confirming the speculation. Therefore, the molecular weight of metabolite IM4 should be 352.

The protonated molecule *m/z* 353 produced by IM4 lost 60 ($M + H–CH_3COOH$) in its second-stage mass spectrum to form *m/z* 293 ion. Its fragmentation pathway was similar to that of przewaquinone A (see Fig. 17.32).

To identify this metabolite, tanshinone II_B, 3α-tanshinone II_A and przewaquinone A were in vitro acetylated in acetic anhydride-pyridine. It was found that the retention time of the acetylated product derived from tanshinone II_B was closest to that of IM4, an in vivo metabolite, but not exactly the same as the latter, with a difference of 1 min. The MS fragmentation pathway of IM was similar to that of przewaquinone A, suggesting that IM is possibly a

Fig. 17.30 (continued)

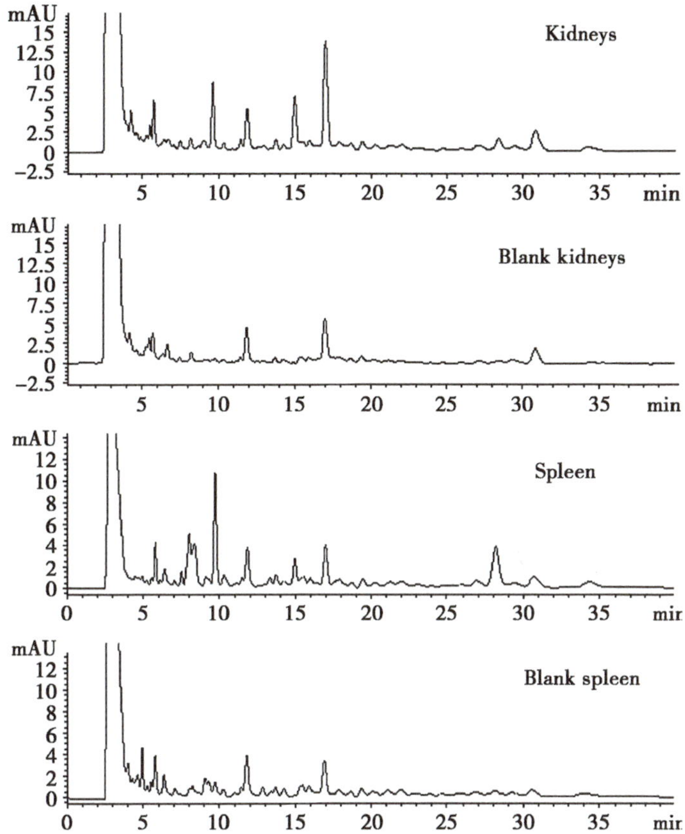

chiral isomer of an acetylated product derived from przewaquinone A through an in vitro reaction (Fig. 17.33).

Similarly, the analysis method for IM4 is also applied to IM5. As shown in the full-scan primary MS spectrum, IM5 produced a pair of ions, *m/z* 347 and *m/z* 325, and it was very difficult for ion *m/z* 347 to break up further. Therefore, it was proposed that the molecular weight of IM5 was 324 and that *m/z*347 was the $[M + Na]^+$ adduct ion. Further fragmentation of *m/z* 325 ion indicated that the ion produced mainly *m/z* 293 by losing a methanol molecule and *m/z* 310 ion was also produced by losing a methyl group. Therefore, it was concluded that this metabolite was a methylated product of hydroxylated tanshinone II_A (Fig. 17.34).

IM4 and IM5 were only found in rat feces and in intestinal contents, and were not found in any other biological samples, suggesting that these two metabolites were produced by intestinal bacteria.

8. The Metabolites in Rat Urine After Multiple Oral Doses of Tanshinone II_A

As shown in Fig. 17.35, tanshinone II_A and several trace metabolites were also found in rat urine following multiple doses. The urine samples have a strong ion-inhibitory effect; to alleviate the effect, the urine samples were extracted initially with acetic ester, which was then back extracted with water to further remove inorganic salts and other water-soluble impurities in urine samples to enhance the response of tanshinones in urine samples in mass spectrometry analysis.

In addition to tanshinone II_A, several other metabolites were also detected in rat urine following multiple doses.

The protonated molecules of UM1 and UM2 were 311 and these two metabolites showed

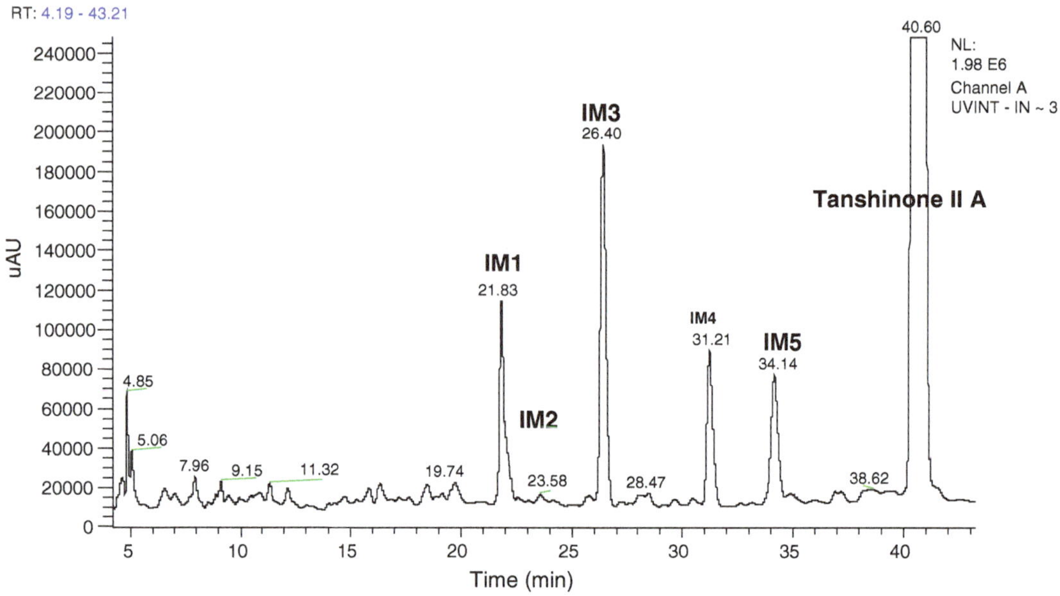

Fig. 17.31 The HPLC chromatograms of metabolites in rat intestinal contents and feces following multiple oral doses of tanshinone II_A

UV-absorption characteristics of tanshinone II_A, and had similar retention times to those of tanshinone II_B and 3α-hydroxyl tanshinone II_A, respectively. Therefore, they were identified as tanshinone II_B and 3α-hydroxyl tanshinone II_A.

UM3 and UM4, two trace metabolites, were only found in the extracted ion current chromatogram. As shown in the mass spectrum, their protonated molecules were 293, consistent with that of M8 and M9 in bile samples as identified above. It was concluded that UM3

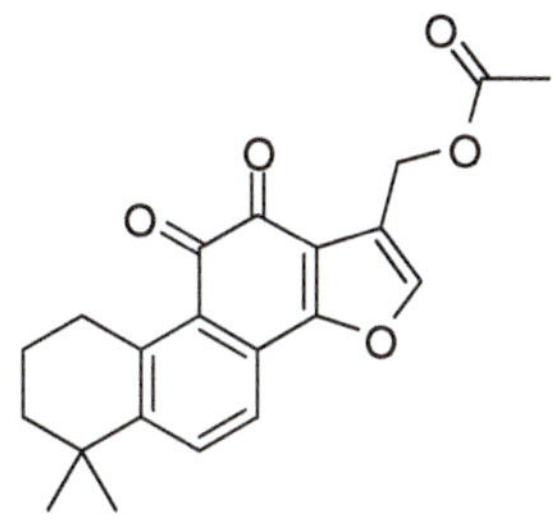

Proposed chemical structure of IM4

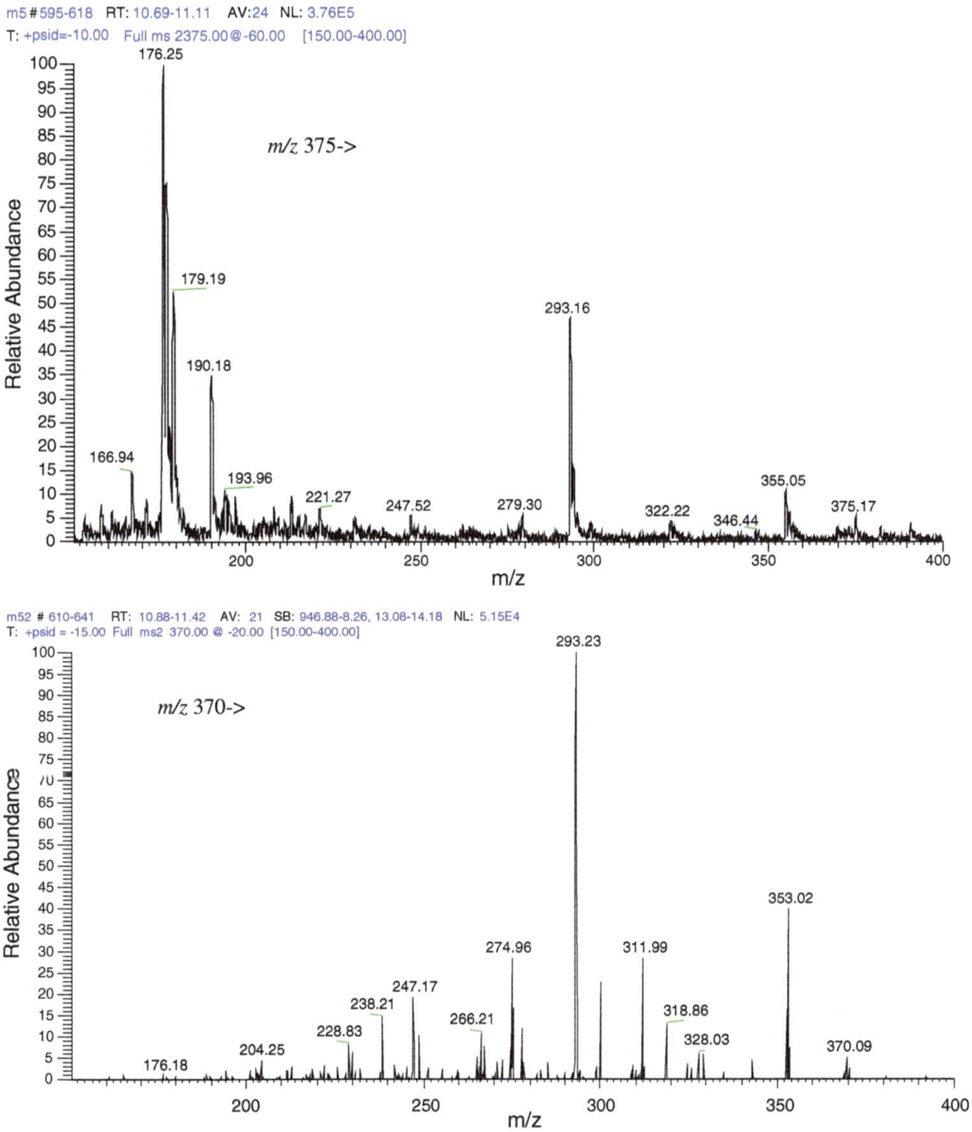

Fig. 17.32 The MS spectra of IM4

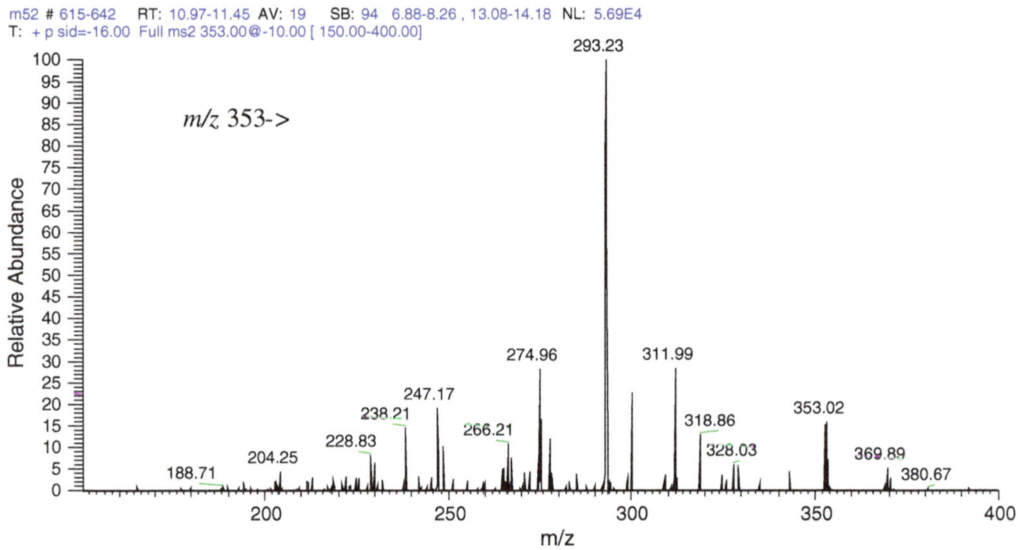

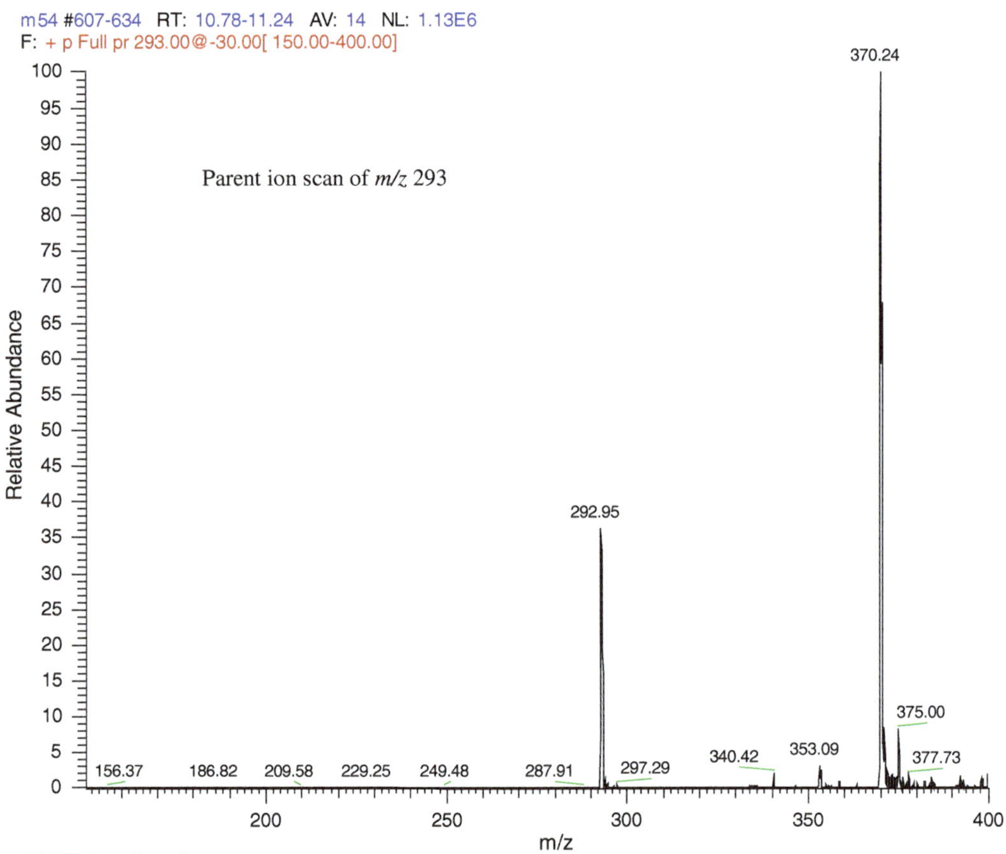

Fig. 17.32 (continued)

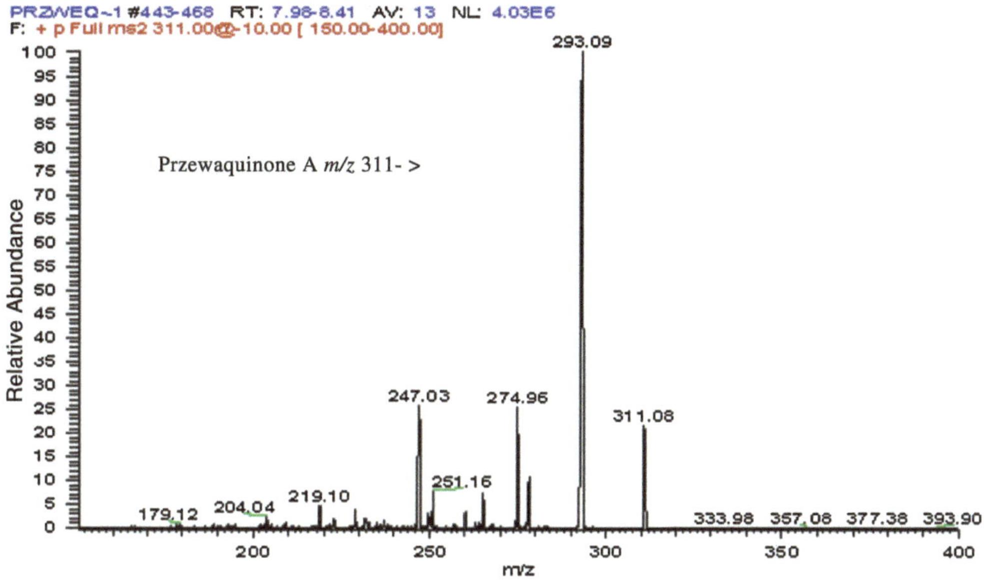

Fig. 17.32 (continued)

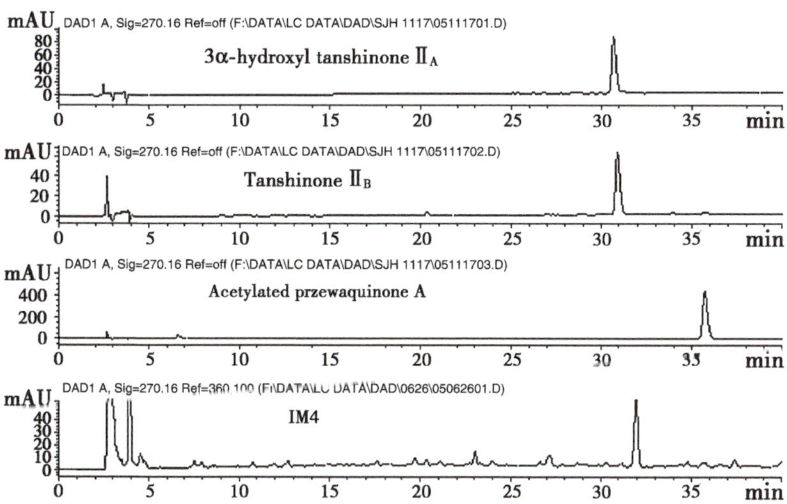

Fig. 17.33 The HPLC chromatograms of 3α-hydroxyl tanshinone II_A, tanshinone II_B, acetylated przewaquinone A and IM4

and UM4 were the metabolites produced by the dehydrogenation of tanshinone II_A, and they were named 1,2-dehydrotanshinone II_A or 2,3-dehydrotanshinone II_A (Fig. 17.36).

The mass spectrum of UM5 showed that the molecular weight of UM5 was 460. As shown in second-stage mass spectrum, the characteristic diagnostic ion *m/z* 293 was produced after an obvious neutral loss of 167u. Further fragmentation of ion *m/z* 293 was similar to that of przewaquinone A and no przewaquinone A was found in urine, suggesting that UM5 was possibly a product produced by conjugation of przewaquinone A and endogenous nitrogen-containing compounds in urine. The multistage mass spectra of UM5 are shown in Fig. 17.37.

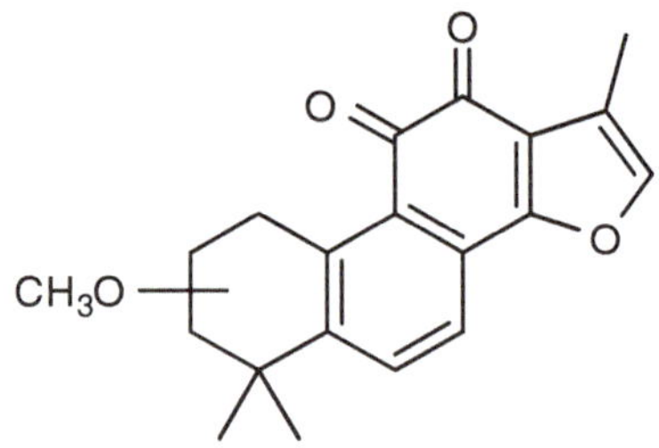

Proposed chemical structure of IM5

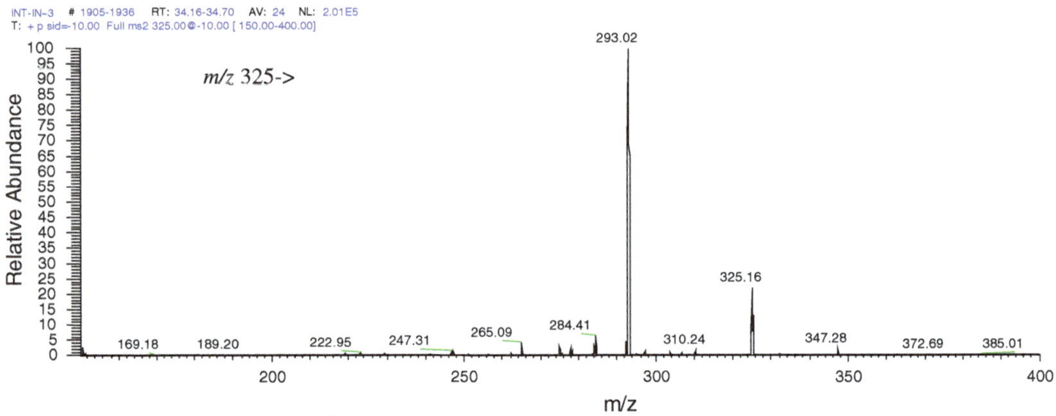

Fig. 17.34 The MS^2 spectrum of IM5.

17.3.1.3 Summary

After oral administration of tanshinone II_A, the plasma concentration of the compound in rats was kept at a low level. The reasons for the low oral bioavailability of tanshinone II_A are explained as follows.

1. Poor Water Solubility: A moderate lipid/water partition coefficient (log *P*) is required for a drug to be absorbed in the gastrointestinal tract and enter into the circulation system. Searching databases (CA, SciFinder) found that log P of tanshinone is 3.74, suggesting that it is a drug with high lipid solubility and is practically insoluble in water (solubility of 2.1×10^{-6} mol/L), thus severely reducing its absorption.
2. It is reported in literature that cryptotanshinone is a substrate of P-glycoprotein and its low bioavailability is related to P-glycoprotein mediated efflux [17]. Studies have shown that a majority of tanshinone II_A in the body is excreted unchanged in bile and that a strong first-pass effect exists in the liver. P-glycoprotein is mainly found in the intestine and liver and tanshinone II_A is also a potential substrate of P-glycoprotein, thus leading to an obvious efflux and consequently an extremely low oral bioavailability. Some investigators demonstrate in the rat intestinal perfusion model that absorption of tanshinone may increase when co-administered with verapamil (*P*-glycoprotein inhibitor) in rat [18].
3. High-plasma protein binding. No suitable solvent can be used to perform in vitro plasma protein binding assay due to the high lipid solubility of tanshinone II_A. There was an indirect estimate: assay the postadministration plasma directly by ultrafiltration and the results showed that in vivo plasma protein binding of tanshinone II_A was 99.2 %, mainly binding to lipid albumin with a binding rate of 75.5 % [19]. This is also an important factor contributing to the extremely low plasma concentration of tanshinone II_A following either an intravenous or oral dose.
4. Tanshinone II_A was mainly distributed in the organs of the digestive system including the

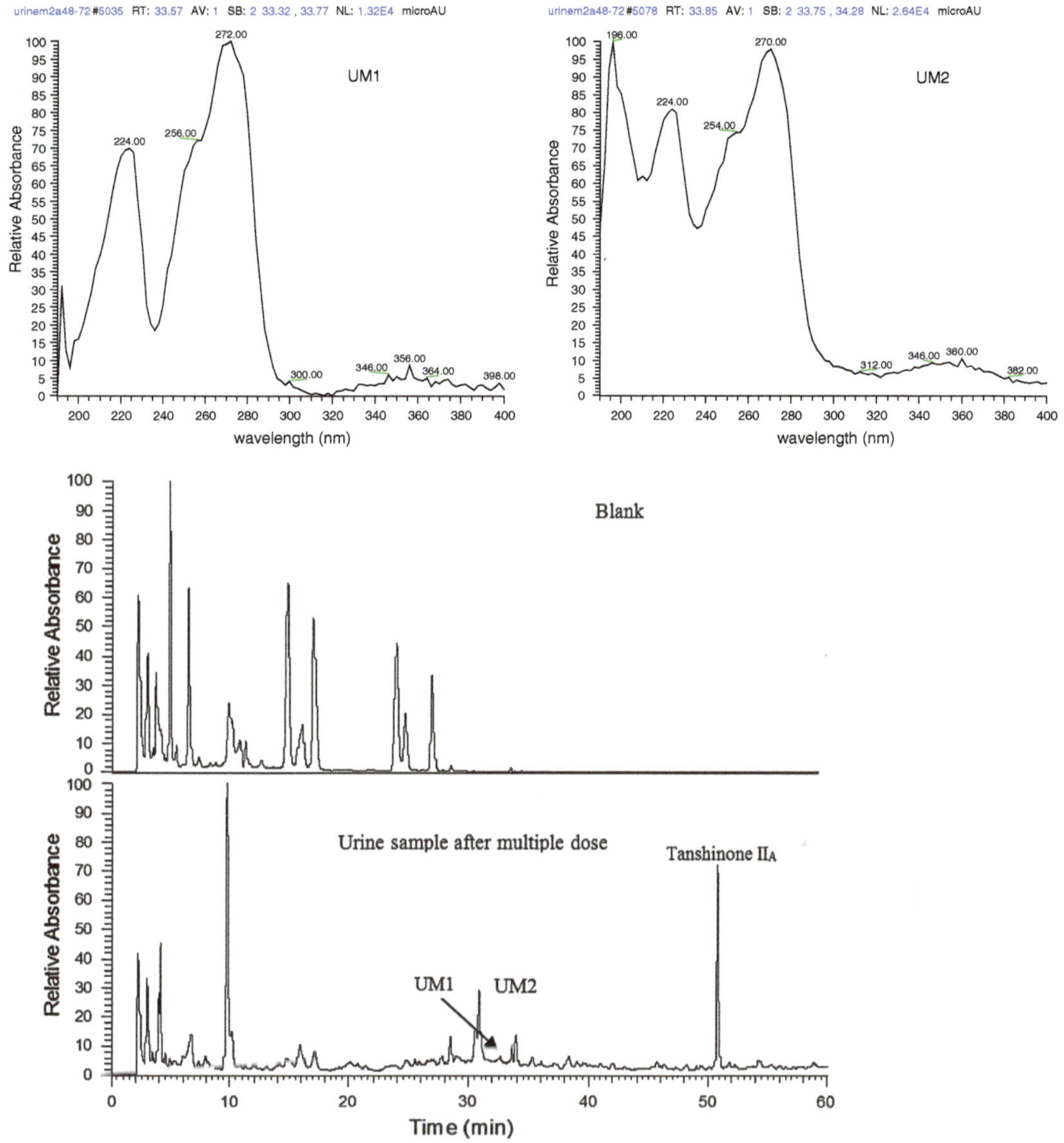

Fig. 17.35 The UV spectra of UM1 and UM2 and HPLC chromatograms of blank and postadministration rat urine samples following multiple doses

liver, intestine, and stomach following a single dose; a large amount of tanshinone II_A is excreted unchanged in the gastrointestinal tract, while a small amount of tanshinone II_A which enters systemic circulation is mainly excreted in the bile. Plasma concentration remains at a low level, and renal excretion is not a major elimination route for tanshinone II_A since it is not detected in urine. On the other hand, tanshinone II_A can be seen in the organs of the circulatory system, such as the heart, lungs, spleen, and kidneys, suggesting that multiple doses of tanshinone preparation are required to maintain certain concentration in the plasma and target organs and provide clinical efficacy. After multiple doses, tanshinone II_A was also detected in rat urine in an unchanged form, or as hydroxylated and dehydrogenated metabolites. In multiple-dose studies, two new metabolites were found

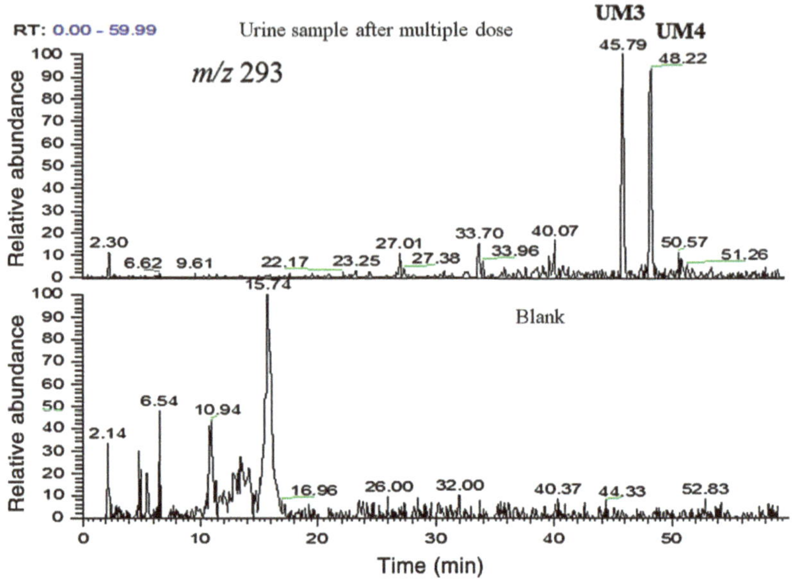

Fig. 17.36 The EIC chromatogram of UM3 and UM4 with a *m/z* of 293 in blank and postadministration rat urine following multiple doses

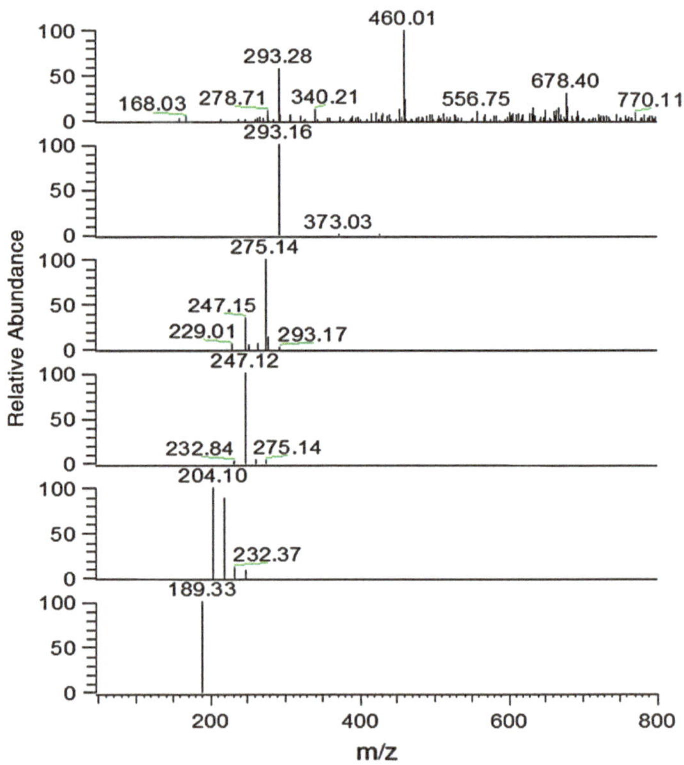

Fig. 17.37 The MS^n spectra of UM5 in rat urine following multiple doses

during the long-time retention of tanshinone II_A in gastrointestinal tract; these two metabolites have never been found in the biological fluid of rats or reported in the literature. The results of several mass spectrometry scan modes have confirmed that the molecular weights of these two metabolites are 352 and 325, respectively, and suggested they are possibly acetylated and methylated products produced by intestinal bacteria when tanshinone II_A is excreted from the bile to the intestine following hydroxylation.

17.3.2 The Metabolites in Rat Bile After Intravenous Administration of 7 Tanshinones

A drug's absorption, distribution, and elimination in vivo are accompanied by its changes in chemical structure, i.e., its metabolism, which is a process of biotransformation of the drug molecules to different extents. Therefore, a drug's metabolism is an important part of the interaction between the drug and the organism, and is an essential factor affecting the disposition and action of the drug in organisms. The metabolic reactions of a drug mainly occur in the liver and they are generally classified into phase I and phase II reactions. The phase I reactions are also called functional group reactions, which involve the introduction of functional groups into the structure of the drug by oxidation, reduction, hydrolysis, and isomerization so as to increase the polarity of the drug and consequently change the drug into a form more easily excreted, and also to a potential substrate for a phase II reaction. The phase II reactions are also known as conjugation reactions, which involve further conjugation of the drug or its metabolites with endogenous glucuronic acid, sulfonates, and amino acids, thus increasing the water solubility and polarity of the drug to facilitate its elimination. Drugs may be metabolized by various routes, and drug metabolism studies look for answers to questions like whether the metabolites produced by various routes have pharmacological activities or toxicities. Most drugs lose or reduce their pharmacological activities after metabolism, a process called pharmacological deactivation.

A number of literatures and pilot studies have demonstrated that tanshinones have extremely low oral bioavailability and their plasma concentrations are at the nanogram level [20–23], while their concentrations in bile are much higher than in plasma and tissues. The concentrations of tanshinone II_A and cryptotanshinone in bile are two orders of magnitude higher than those in plasma; therefore we chose bile as our study subject.

Since tanshinones are mainly excreted in their unchanged forms and as metabolites in bile, and most metabolites exist in trace amounts, it is impossible to separate and identify these metabolites by conventional chemical methods after cumulating biological samples (feces or urine). Furthermore, bile contains a large amount of endogenous substances, such as bile acids, phospholipids, cholesterols, bilirubin, proteins, and lots of organic salts, which pose a great challenge to analysis methods. In order to elucidate the in vivo metabolic process of tanshinones, we used liquid chromatography–electrospray ion trap mass spectrometry to analyze rat bile samples collected following intravenous administration of seven tanshinones with different structures. By analyzing the retention times of parent compounds and their metabolites, multistage mass spectral fragmentation data, and UV absorption spectra, we deduced the structures of 35 metabolites and preliminarily addressed the relationship between the structures of tanshinones and their metabolic pathways.

17.3.2.1 Experiment

1. Instruments and Reagents
   Bruker 400 MHz NMR Spectrometer; Alltech 426 HPLC equipped with UV-Vis200 detector; Agilent 1100 HPLC equipped with quaternary pump, autosampler and DAD; Finnigan ICQ Advantage ion trap mass spectrometer; ESI ionization source, Xcalibur 1.4 data management system; EYEIA N-1000V-W rotary evaporator; TGL-16G-A refrigerated centrifuge (Shanghai Anting Medical Analytical Instrument Factory); HGC-12 nitrogen evaporator (Shanghai Hegong Scientific Instrument Factory); chromatographic pure methanol and acetonitrile (Fisher, USA); double-distilled water, dichloromethane, acetic ester, and other reagents were analytical pure (Beijing Analytical Reagent Works).

The seven tanshinone reference substances are tanshinone II_A, sodium tanshinone II_A sulfonate (STS), cryptotanshinone, 15,16-dihydrotanshinone I, tanshinone II_B, przewaquinone A, and tanshinone I. Among them, STS was purchased from Guangzhou Meichen Pharmaceutical Co., Ltd., the others and tanshinaldehyde, 3α-hydroxyl tanshinone were all self-prepared with purity >99 % as determined by HPLC.

2. Extraction and Isolation of Tanshinone Components
   Load 20 g of tanshinone extract onto 600 g silica gel chromatographic column; perform gradient elution with dichloromethane-acetic ester (1:0, 40:1, 30:1, 20:1, 15:1, 10:1); collect a sample in 400 ml aliquots; evaporate the solvent in a rotary evaporator under reduced pressure; perform thin-layer chromatography to detect the compound, and combine the positive aliquots. Recrystallize fractions 2–7 with dichloromethane and obtain 5 g of compound 1; run 3.2 g of material from fractions 29–35 through a silica gel chromatographic column, elute with dichloromethane and obtain 1.2 g of compound 2; let fractions 61–69 go through the silica gel chromatographic column, elute with dichloromethane and recrystalize with chloroform to obtain compounds 3–4; purify fractions 124–134 with preparative liquid chromatography, elute with methanol–water (80:20) and obtain compounds 5, 6, and 7.
3. Structural Identification
   Based on the results of NMR and MS spectral analysis and published reports [24–29], these compounds were identified as tanshinone II_A (1), tanshinone I (2), 15,16-dihydrotanshinone I (3), cryptotanshinone (4), przewaquinone A (5), tanshinone II_B (6), and tanshindiol B (7). The structures of these compounds are shown in Fig. 17.38 and spectral data are shown in Table 17.12.

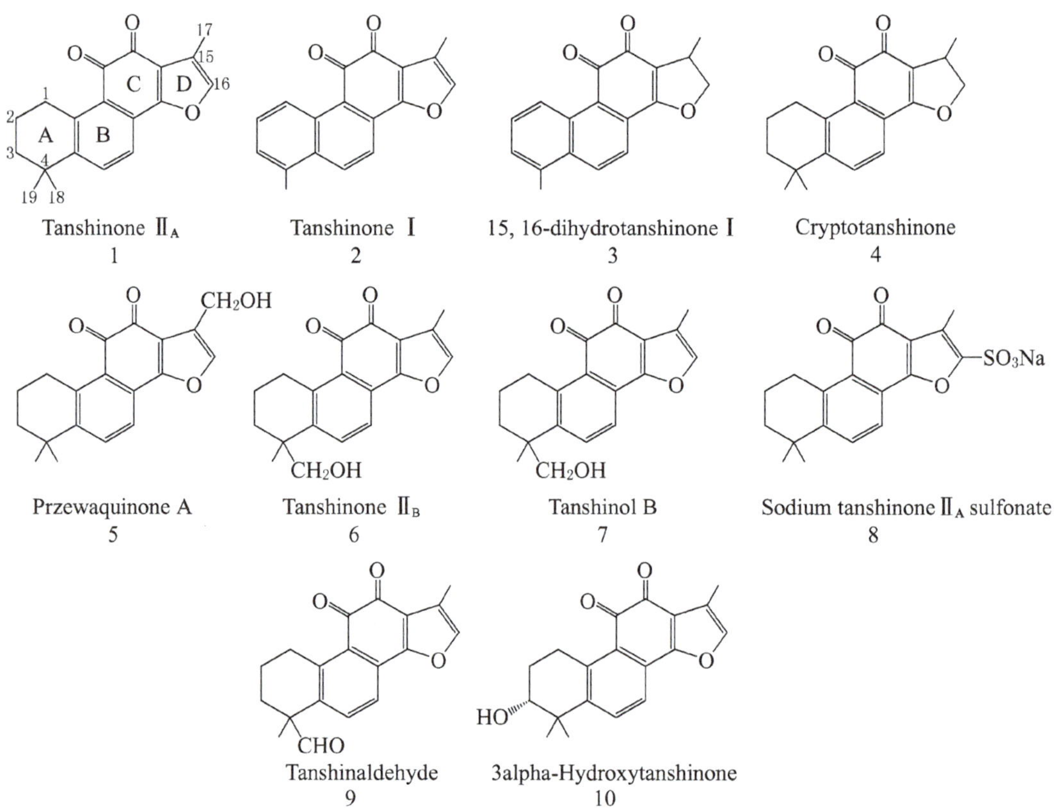

Fig. 17.38 The structural formulas of compounds 1–7 and STS

Table 17.12 ^{1}H-NMR spectral data of compounds 1–7 (400 MHz, $CDC1_3$, δ, ppm, Hz)

H	1	2	3	4	5	6	7[a]
1	3.18 (2H,dd. J = 12.8,12.8)	9.21 (1H,d, J = 8.7)	9.1 1 (1H,d,J = 8.9)	3.19 (2H,dd,J = 12.8,12.8)	3.19 (2H,dd, J = 6.4,6.4)	3.16 (2H,m)	3.00 (2H,m)
2	1.79 (2H,m)	7.52 (1H,dd, J = 8.7,8.7)	7.45 (1H,dd,J = 8.9,7.1)	1.79 (2H,m)	1.81 (2H,m)	1.99 (1H,m),1.77 (1H,m)	1.89 (1H,m) 1.79 (1H,m)
3	1.66 (2H,m)	7.33 (1H,d, J = 8.7)	7.28 (1H,d,J = 7.1)	1.65 (2H,m)	1.67 (2H,m)	1.87 (1H,m),1.58 (1H,m)	1.79 (1H,m) 1.67 (1H,m)
6	7.63 (1H,d, J = 8.1)	8.26 (1H,d, J = 8.6)	8.03 (1H,d,J = 8.7)	7.63 (1H,d,J = 8.1)	7.66 (1H,d, J = 8.1)	7.65 (1H,d,J = 8.1)	7.94 (1H,d, J = 8.1)
7	7.54 (1H,d, J = 8.1)	7.76 (1H,d, J = 8.6)	7.50 (1H,d,J = 8.7)	7.47 (1H,d,J = 8.1)	7.58 (1H,d, J = 8.1)	7.51 (1H,d,J = 8.1)	7.61 (1H,d, J = 8.1)
15			3.59 (1H,ddq, J = 9.5,6.3,6.8)	3.59 (1H,m)			
16	7.22 (1H,q, J = 1.3)	7.27 (1H,q, J = 1.4)	4.94 (1H,d,J = 9.3.9.5)4.38 (1H,d,J = 9,3,6.3)	4.90 (1H,dd,J = 9.3,9.5)4.37 (1H,dd,J = 9.3,6.1)	7.39 (1H,s)	7.20 (1H,q,J = 1.2)	7.75 (1H,d, J = 1.3)
17	2.27 (3H,d, J = 1.3)	2.28 (3H,d, J = 1.4)	1.39 (3H,d,6.8)	1.37 (3H,d,J = 7.1)	4.67 (2H,s)	2.26 (3H,d,J = 1.2)	2.16 (3H,d, J = 1.0)
18	1.31 (3H,S)	2.67 (3H,s)	2.57 (3H,s)	1.31 (3H,s)	1.32 (3H,s)	3.80 (1H,d,J = 11.1)3.63 (1H,d,J = 11.1)	
19	1.31 (3H,S)			1.31 (3H,s)	1.32 (3H,s)	1.29 (3H,s)	1.39 (3H,S)

[a] The solvent is d_6-DMSO

4. MS Conditions for Direct Sample Injection of Monomer Compounds
   MS conditions for sodium tanshinone II_A sulfonate: electrospray negative ion scan; spray voltage, 4.5 kV; sheath gas, 30 units; capillary temperature, 340 °C; capillary voltage, −4 V; lens compensation voltage, −10 V; scanning range: 60–800. The sample was directly injected through an injection pump at a flow rate of 12.5 μl/min.
   Mass spectrum conditions for other tanshinone components: electrospray positive ion scan; spray voltage, 4.5 kV; sheath gas, 30 units; capillary temperature, 340 °C; capillary voltage, 30 V; scanning range: 100–500. Samples were directly injected through an injection pump at a flow rate of 12.5 μl/min.
5. LC and LC-MS Conditions
   Chromatography column: Agilent Zorbax Extend-C_{18} reverse-phase column (5 μm, 250 mm × 5 mm), Agilent Zorbax SB-C_{18} guard column (5 μm, 20 mm × 4 mm). Detection wavelength was set at the maximal absorption wavelength of each compound: tanshinone II_A, tanshinone II_B, and przewaquinone A, 270 nm; STS, 242 nm; cryptotanshinone, 264 nm; 15,16-dihyrotanshinone, 242 nm. Flow rate: 1 ml/min. Two different mobile phase systems were used in analyzing STS and other tanshinone components, respectively. The chromatographic conditions are shown in Table 17.13 and mass spectral conditions are shown in Table 17.14.

Table 17.13 The HPLC chromatographic conditions

Gradient	Time (min)	Acetonitrile: 0.03 % formic acid (v/v)
Tanshinone compounds	0	20:80
	20	30:70
	30	50:50
	40	40:60
	50	80:20
	60	80:20
STS	Time (min)	Acetonitrile: 0.1 % ammonia water (v/v)
	0	15:85
	15	50:50

6. Setting the Data-Dependent Scanning Mode
   A data-dependent scanning mode was used for HPLC/MS: the two ions with the highest abundance in each stage would be the parent ions for MS analysis in the next stage. Relative collision energy: STS, 60 %; other components, 45 %. At the time of setting the data-dependent scanning mode, dynamic exclusion was also set to prevent ions with high abundance from disturbing the determination of target ions. Width of ion was set to be 1 Th. All data were processed by Xcalibur 1.4 software.
7. Laboratory Animals
   SD male rats with body weights of 250 ± 20 g were provided by Laboratory Animal Center of Peking University Health Science Center. The animals were fasted for 12 h before the experiment, but with no limits on drinking water.
8. The Biliary Tract Intubation and Drug Administration in Rats
   16 SD rats were randomized to 8 groups (8 × 2). Give the same drug to both animals in the same group. Fix rats on their back following 20 % urethane anesthesia (1 g/kg). Make a longitudinal incision of 2–3 cm (V1) below the xiphoid; find the bile tract in the mucous membrane of the upper duodenum, dissect the bile tract and make a small incision at the distal bile tract, insert a 0.8 mm-diameter PE tube toward the liver, ligate, and the fix bile tract. When bile successfully flows out, close the wound. Then inject different tanshinones into the tail vein at a dose of 4 mg/kg and collect bile 0–24 h postadministration.
9. Treatment of Rat Bile Samples
   Collect bile samples following intravenous administration of STS, concentrate the bile, filter through a 0.22 μm microporous membrane, and inject 15 μl for LC-MS analysis. For bile samples collected after the administration of other tanshinones, vortex to mix and extract with 3 volumes of acetic ester 3 times, each time lasting for 10 min. Pool the acetic ester layers, evaporate with nitrogen gas at 30 °C, add 1 ml of methanol to redissolve, and inject 15 μl for LC-MS analysis.

Table 17.14 The LC-MS^n conditions

Compound	1	4	3	5	6	3	STS
Spray voltage (kV)	4.5	4.5	4.5	4.5	4.5	4.5	4.5
Sheath gas (arbitrary unit)	50	50	50	50	50	50	50
Auxiliary gas (arb unit)	10	10	10	10	10	10	10
Capillary temperature (°C)	340	340	340	340	300	350	350
Capillary voltage (V)	20	−4	18	25	3	14	23
Tube lens offset (V)	15	−25	16	14	20	20	10
Source CID (V)	0	10	0	0	0	0	0

17.3.2.2 Results and Discussion

Because drug metabolic reactions are generally the structural modification of parent compounds, the metabolites usually have structures very much similar to those of their parent compounds. Therefore, we initially conduct studies on the MS fragmentation pathways of the original compounds so as to understand their fragmentation characteristics or to obtain diagnostic ions for identifying the metabolites and speculating on their possible structures.

1. MS Fragmentation Pattern of STS

   It has been reported that tanshinone components had good MS responses in electrospray positive ion mode, while STS had a poor response in the same mode and the abundance of the $[M + H]^+$ ion was extremely low; however, STS showed a good response in negative ion mode [30–32]. Compared to other tanshinones, STS requires higher collision energy to be broken up. As shown in Fig. 17.39, in negative ion full-scan first-stage mass spectrometry, the peak of its *m/z* 373 ion, $[M–Na]^-$, was the base peak, so this ion was chosen to perform the second-stage scan. When 60 % collision energy was used, *m/z* 309 and *m/z* 293 ions were obvious in the second-stage spectrum. These two ions were derived from the *m/z* 373 ion by losing a molecule of SO_2 and SO_3, respectively. Since the sulfur-containing groups are the characteristic groups of STS, this fragmentation pathway can be used to identify the metabolites of STS. In addition, *m/z* 358 ion produced by losing a methyl free radical was also very obvious in the second-stage spectrum. Further fragmentation of *m/z* 309 ion could produce *m/z* 281, *m/z* 294, and *m/z* 265 by losing CO, CH_3 and CO_2, respectively.

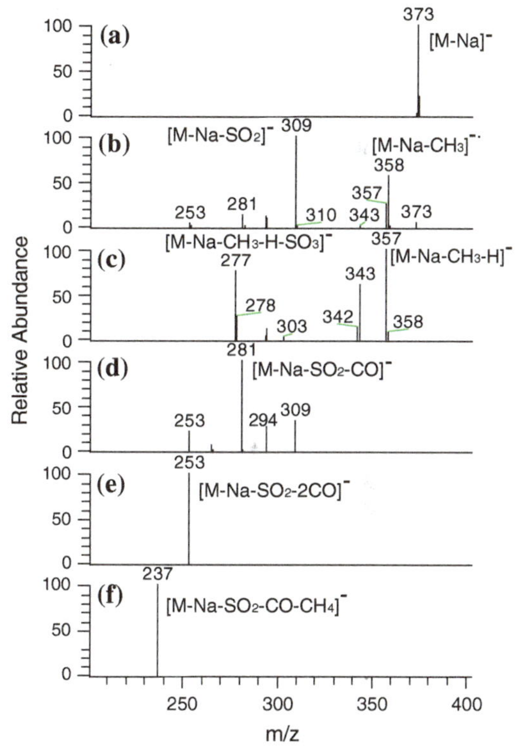

Fig. 17.39 The ESI (−) MS and MS^n spectra of STS. **a** ESI MS spectrum; **b** MS^2 spectrum of $[M–Na]^-$ ion (*m/z* 373); **c** MS^3 spectrum of $[M–Na–CH_3]^-$ ion (*m/z* 358); **d** MS^3 spectrum of $[M–Na–SO_2]^-$ ion (*m/z* 309); **e** MS^4 spectrum of $[M–Na–SO_2–CO]^-$ ion (*m/z* 281); and **f** MS^5 spectrum of $[M–Na–SO_2–CO]^-$ ion (*m/z* 253)

The *m/z* 358 ion tends to lose a H free radical and produce *m/z* 357; these two ions can be further broken up to lose SO_2 and SO_3 to produce two pairs of ions: *m/z* 278 and *m/z* 277, and *m/z* 243 and *m/z* 242. Its possible fragmentation mechanism is shown in Fig. 17.40.

2. The Fragmentation Patterns of Other Tanshinone Compounds
   As reported in the literature [33], the fragmentation patterns of tanshinone II_A, cryptotanshinone, 15,16-dihydrotanshinone I, tanshinone II_B, przewaquinone A and tanshinone I are as follows: these compounds mainly lose H_2O and CO, accompanied by the loss of fragments such as methyl and propylene. The fragment-losing sequence and the abundance of ions produced vary with conjugated systems. For example, 15,16-dihydrotanshinone I loses a water molecule first; while tanshinone I, which has a large conjugated system, loses a CO molecule first. The chance of losing a propylene fragment from cryptotanshinone is significantly greater than that from tanshinone II_A. These patterns can be directly used to identify the metabolites of tanshinones in rat bile.

17.3.2.3 Identification of the Metabolites Following Intravenous Administration of Different Tanshinones to Rats

1. The Metabolites of STS in Rat Bile
   STS is the sulfonate of tanshinone II_A and exists as ions in bile. When chromatographed on a common C_{18} reverse-phase column, the peak showed severe tailing. According to one report [34], ion-pair reversed-phase HPLC was used for the analysis of STS. However, LC-MS requires that the mobile phase must have the following features: moderate conductivity to support the formation of ions, and low surface tension and low solvation energy to facilitate evaporation of solvents and deadsorption of sample ions. It was found that an acetonitrile–2 % triethylamine system could significantly improve the peak's heavy tailing, but triethylamine was not suitable for LC-MS analysis. Therefore, triethylamine was replaced by 0.1 % ammonia water. Although ammonia water could not improve the peak shape as much as triethylamine did, it could significantly increase the ionization efficiency of STS in negative ion mode. Therefore,

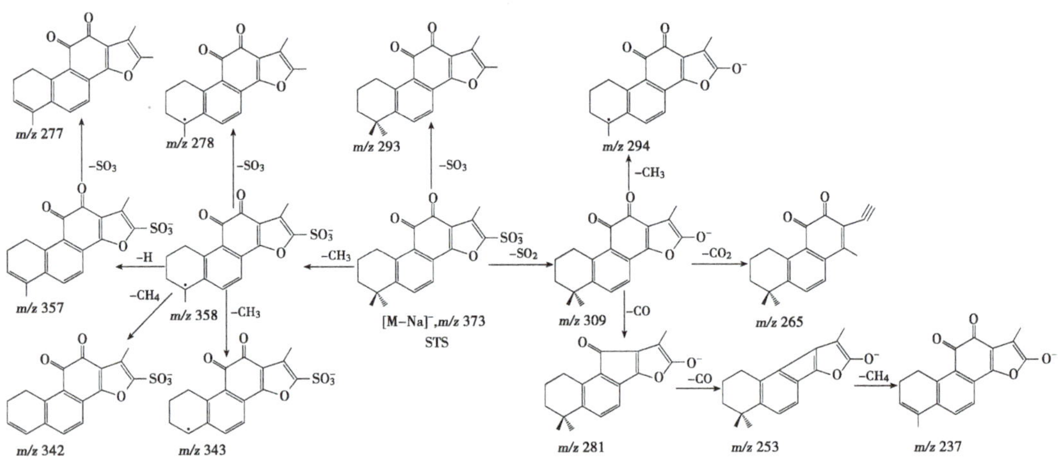

Fig. 17.40 The possible fragmentation pathway of STS

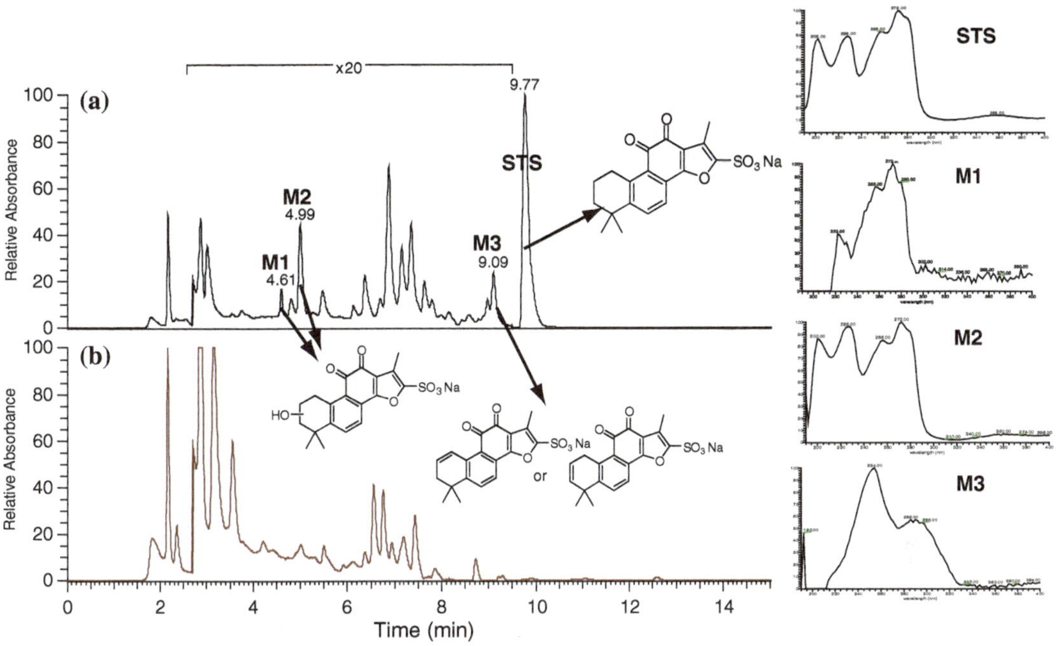

Fig. 17.41 The enlarged HPLC chromatograms of rat bile sample collected after intravenous administration of STS (**a**) and the blank bile sample (**b**) (272 nm), and the UV spectra of STS and its metabolites

acetonitrile–0.1 % ammonia water was chosen as the mobile phase in this experiment.

As shown in Fig. 17.41, the three peaks on the enlarged chromatograms with retention times of 4.6, 5.0, and 9.1 min, respectively, were possibly the metabolites based on a comparison with blank bile samples; the second-stage chromatograms of these three peaks all showed obvious $[M–Na–SO_2]^-$ and $[M–Na–SO_3]^-$ fragments. Furthermore, the UV absorptions of M1 and M2 were very similar to that of STS. As shown in the full-scan first-stage MS spectrum, the quasi-molecule $[M–Na]^-$ of both M1 and M2 was *m/z* 389, heavier than their parent compound by 16 Da, suggesting that these two compounds might be the hydroxylated STS. By comparing the second-stage MS spectra of M1 and M2 (Fig. 17.42), it can be seen that M2 did not produce a dehydration peak of *m/z* 371 $[M–Na–H_2O]^-$, but instead produced a demethanolized peak with high abundance, $[M–Na–CH_3OH]^-$. In negative scan mode, the only reasonable explanation for this phenomenon is that hydroxylation occurred on the methyl group of C-17 or C-18/19 and formed a hydroxymethyl structure which tends to produce a demethanolized peak instead of a dehydrated peak. On the other hand, the existence of a dehydrated peak proves that hydroxy-substitution occurred on the ring A in case of M1.

The quasi-molecule ion peak of M3 was 371, lower than STS by 2 Da, suggesting that it may be a dehydrogenated metabolite. Considering that the UV absorption spectrum of M3 was much different from that of the parent drug, it indicates dehydrogenation on ring A is impossible to occur at site 1, 3; instead, only site 1, 2 or 2, 3 could be the location for dehydrogenation to occur. Therefore, M3 is tentatively identified as 1,2-Dehydro-STS or 2,3-Dehydro-STS.

2. Metabolites of Tanshinone II_A in Rat Bile

After intravenous administration of tanshinone II_A to rats, several metabolites were found in rat bile, including three hydroxylated metabolites and several trace metabolites (Fig. 17.43). By comparing with reference substances, these three hydroxylated metabolites were identified as tanshinone II_B (M4), 3α-hydroxytanshinone II_A (M5), and przewaquinone A (M6). M5,

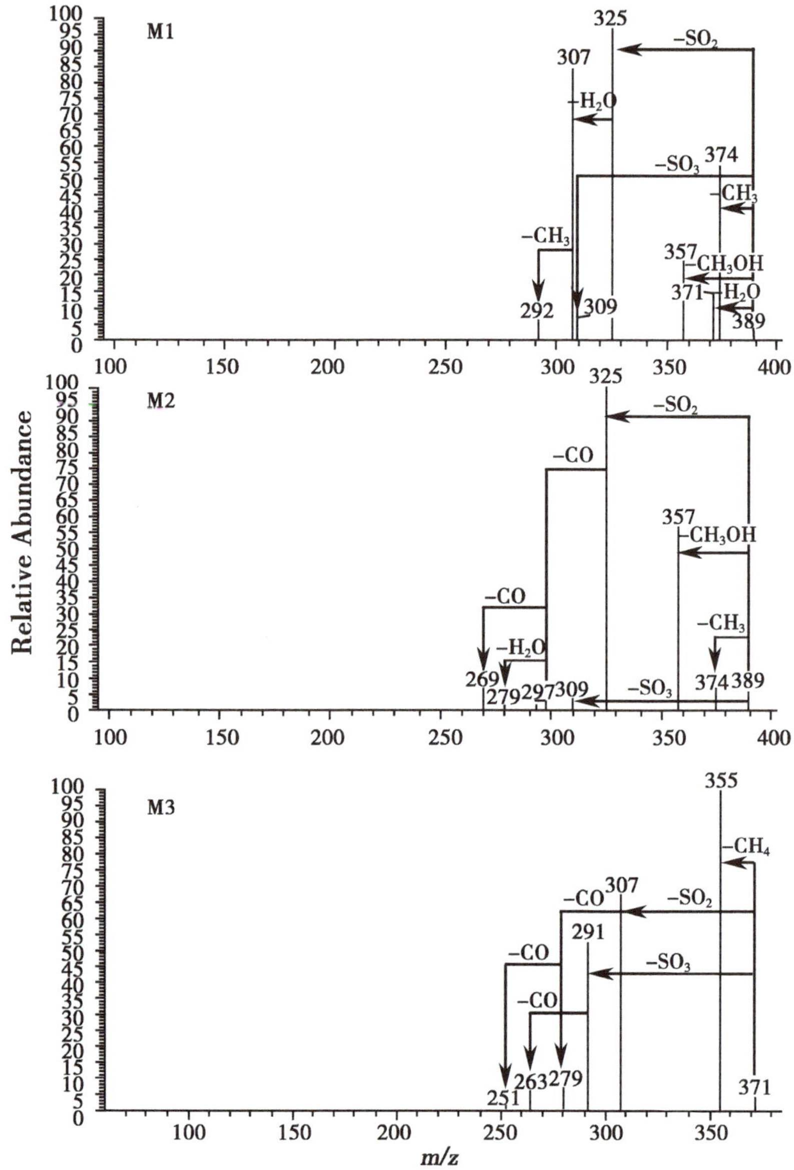

Fig. 17.42 The MS^2 spectra of $[M–Na–H]^-$ ions derived from metabolites of STS

which was previously identified as the chiral isomer of tanshinone II_B [35], was reidentified as 3α-hydroxytanshinone II_A by comparing with the reference substance.

At retention times of 44.9 min (M7), 47.3 min (M8), and 49.8 min (M9), a $[M + H]^+$ quasi-molecule ion peak of m/z 293 was produced. Among these metabolites, the UV spectrum of M4 was consistent with that of tanshinone II_A, indicating that M4 retains the conjugated system of the tanshinone II_A backbone. Furthermore, the molecular weight of M4 was lower than that of tanshinone II_A by 2 Da, and only 1,3-dihydrotanshinone II_A has this structural feature; therefore M4 was identified as 1,3-dihydrotanshinone II_A. The UV spectra of M5 and M6 indicated that the conjugated system in the backbone of tanshinone II_A had changed. Since double-bond substitution only occurs on ring A, M5 and M6 were identified as 1,2-dihydrotanshinone II_A or 2,3-dihydrotanshinone II_A. The multiple-stage MS spectra of

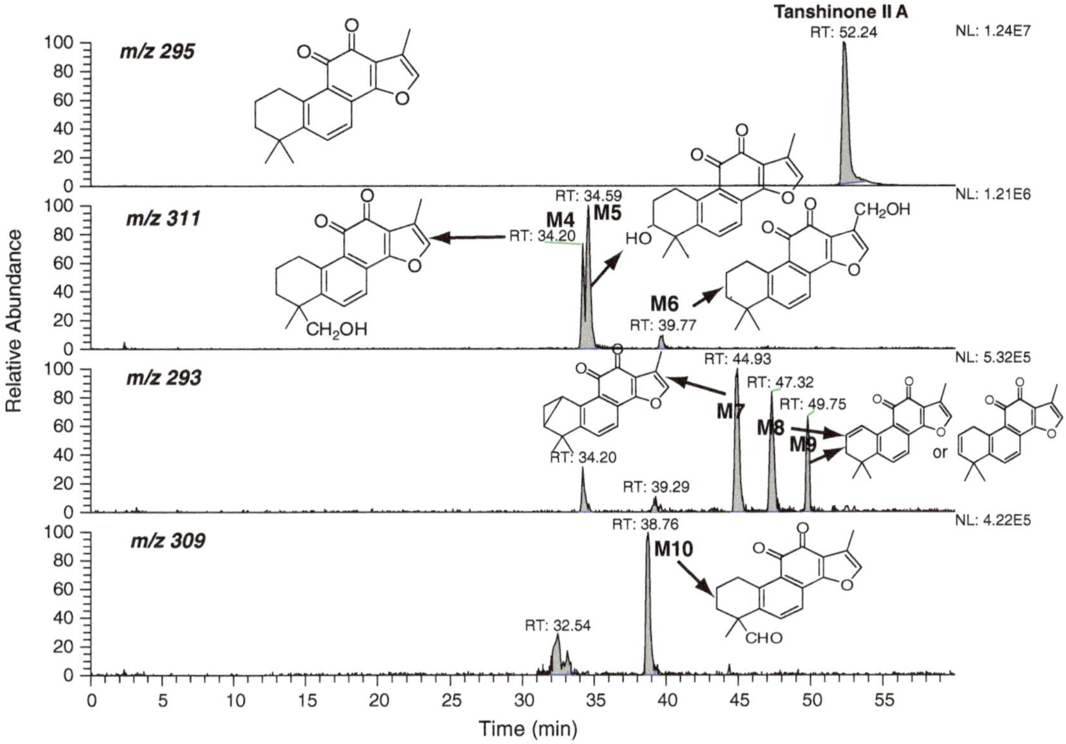

Fig. 17.43 The $[M + H]^+$ ion chromatograms of tanshinone II_A metabolites in rat bile 24 h after intravenous administration

these three metabolites were quite similar to each other, as shown in Fig. 17.44.

The first-stage MS spectrum of the peak (M10) with retention time of 38.8 min showed that its protonated molecule was *m/z* 309 and the multiple-stage MS spectra showed strong ions, *m/z* 281 $[M + H\text{–}CO]^+$, *m/z* 263 $[M\text{–}LH\text{–}CO\text{–}H_2O]^+$, *m/z* 252 $[M + H\text{–}CO\text{–}29]^+$, and *m/z* 235 $[M\text{–}H\text{–}CO\text{–}2H_20]^+$. The retention time and the fragmentation pattern of M10 were consistent with that of tanshinaldehyde, so M10 was identified as such (Fig. 17.45).

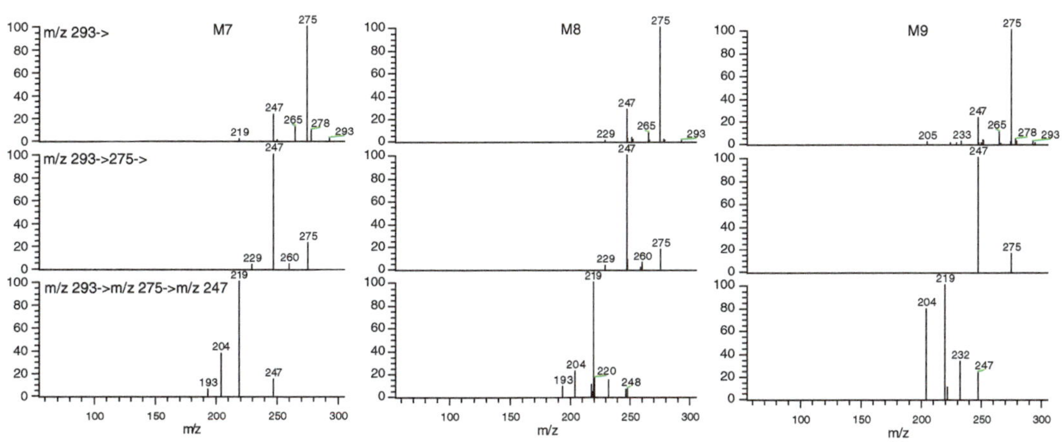

Fig. 17.44 The MS^n spectra of M7, M8, and M9

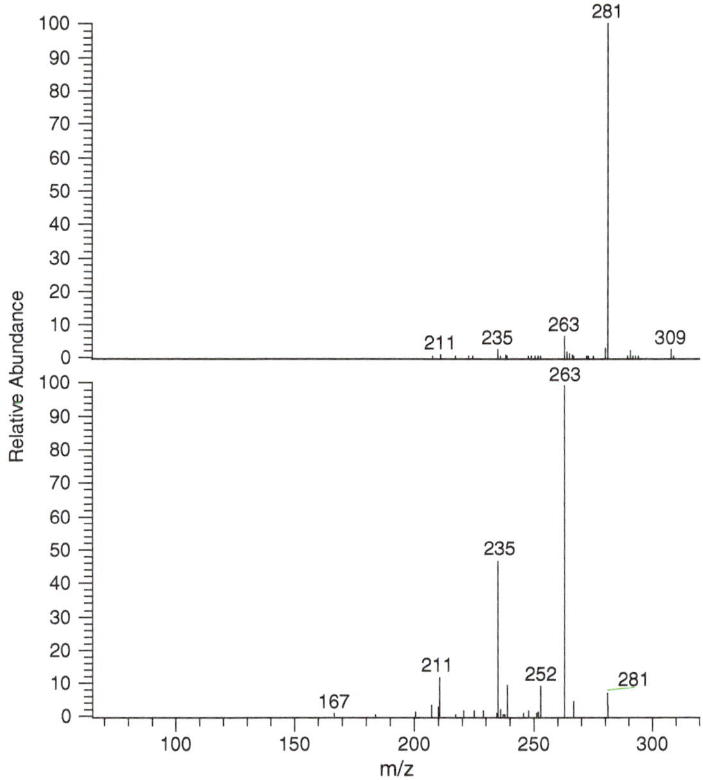

Fig. 17.45 The MS^n spectra of M10

The metabolites of tanshinone II_A in rat bile following intravenous administration were previously reported [35]. Comparing with the report, we did not find double-hydroxyl metabolites or phase II metabolites in this experiment.

3. The Metabolites of Cryptotanshinone in Rat Bile

   As shown in Fig. 17.46, cryptotanshinone tends to be metabolized more easily than other tanshinones because it has two saturated lipid rings, A and D. The major metabolite of cryptotanshinone is the dehydrogenated metabolite tanshinone II_A (M19). In quantitative analysis, described in the next section, about 46 % cryptotanshinone could be transformed to tanshinone II_A. In addition to tanshinone II_A, several trace metabolites were also found.

   M11, M12, M13, M14, and M15 all produced an ion of *m/z* 313. Comparing with cryptotanshinone, the mass number was increased by 16, suggesting that these metabolites were the hydroxylated products. In second-stage MS spectra, we found very obvious *m/z* 295 ions derived from *m/z* 313 by losing a molecule of water. Therefore, these metabolites were identified as hydroxyl cryptotanshinone. Their hydroxy-substitution sites were preliminarily affirmed by referring to the fragmentation patterns of tanshinone II_B, 3α-hydroxytanshinone II_A, and przewaquinone A.

   As shown in Fig. 17.47, 3α-hydroxytanshinone II_A with direct hydroxy-substitution on the lipid ring tends to break up more easily and produce more fragment ions. Furthermore, *m/z* 275 ion losing two H_2O molecules and *m/z* 265 ion losing one H_2O molecule and then one molecule of CO were very obvious in second-stage spectra, while tanshinone II_B and przewaquinone A with hydroxy-substitutions on 17,18-methyl groups produced few fragments except for a dehydrated base peak. Among these five metabolites of cryptotanshinone with a single hydroxy-substitution, M12 and M13 had second- and third-stage MS spectra similar to those

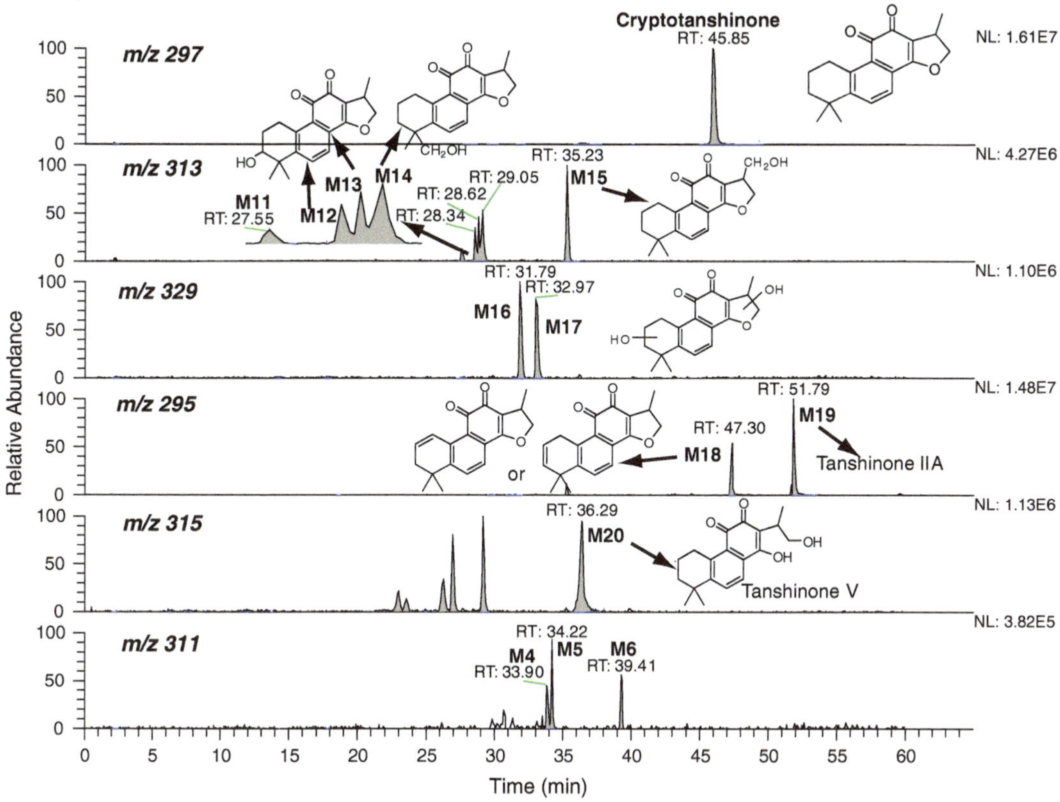

Fig. 17.46 The extracted $[M + H]^+$ ion chromatogram of the metabolites in rat bile 24 h after intravenous administration of cryptotanshinone

of 3α-hydroxytanshinone II_A and produced more obvious fragmentation with 45 % collision energy than M11, M14, and M15. Furthermore, M12 and M13 produced almost the same second- and third-stage MS spectra with the same collision energy, indicating that they were a pair of chiral isomers.

The obvious difference between tanshinone II_B and przewaquinone A, or between tanshinone II_B and 3α-hydroxytanshinone II_A is that under 45 % collision energy, an obvious *m/z* 251 ion was produced by tanshinone II_B with a relative abundance of about 50 % in the third-stage MS spectrum; among the single-hydroxyl metabolites of cryptotanshinone, M14 had a multistage fragmentation pattern very much similar to that of tanshinone II_B and produced a very obvious *m/z* 253 ion in the third-stage spectrum. Therefore, M14 was tentatively identified as 18-hydroxycryptotanshinone (Fig. 17.48). Based on the similarity of fragmentation patterns between M15 and przewaquinone A, M15 was identified as 17-hydroxycryptotanshinone (Fig. 17.49).

The quasi-molecule ion $[M + H]^+$ peak produced by M16 and M17 was 329, and there were consecutive dehydration peaks in the second- and third-stage mass spectra, suggesting that these two metabolites were double-hydroxylated cryptotanshinone (Fig. 17.50).

M18 was another dehydrogenated metabolite in addition to tanshinone II_A. Its quasi-molecule ion $[M + H]^+$ peak was 295 and M18 had a fragmentation pattern similar to that of tanshinone II_A. Therefore, M18 was tentatively identified as 1,2-dehydrocryptotanshinone or 2,3-dehydrocryptotanshinone (Fig. 17.51).

The quasi-molecule ion $[M + H]^+$ peak

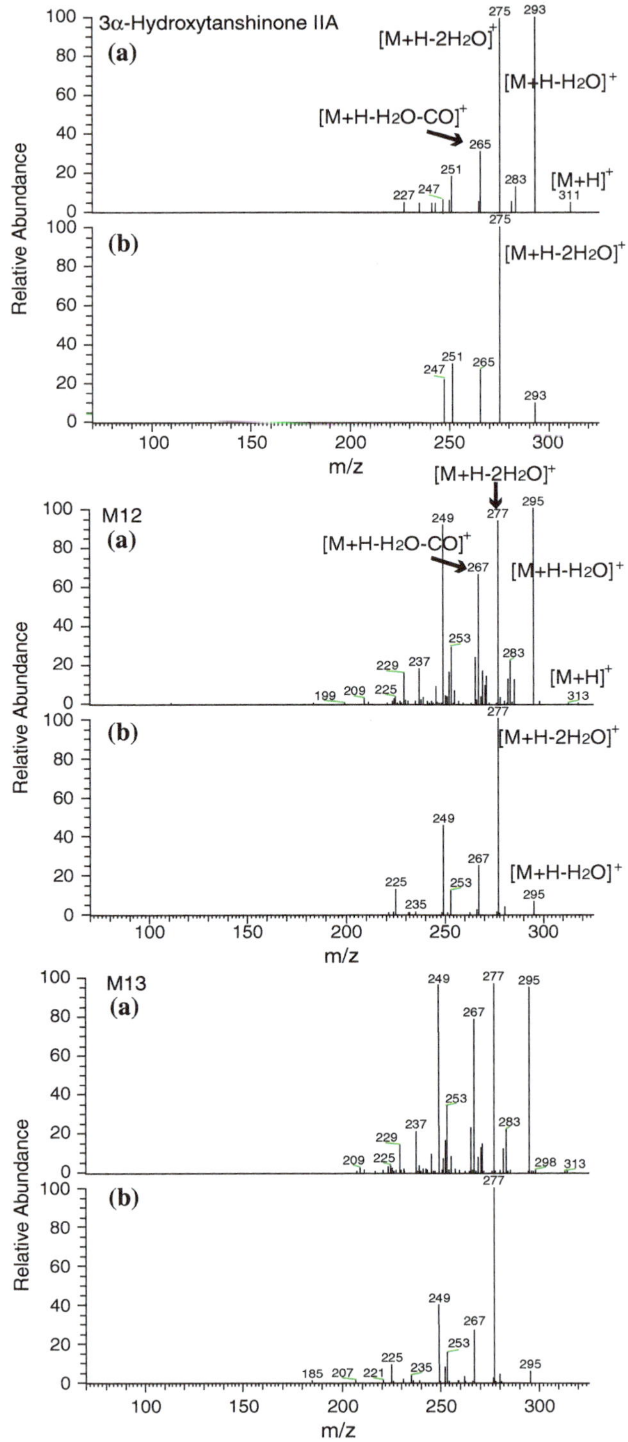

Fig. 17.47 The MS^2 (**a**) and MS^3 (**b**) spectra of 3α-hydroxytanshinone II$_A$, M12, and M13

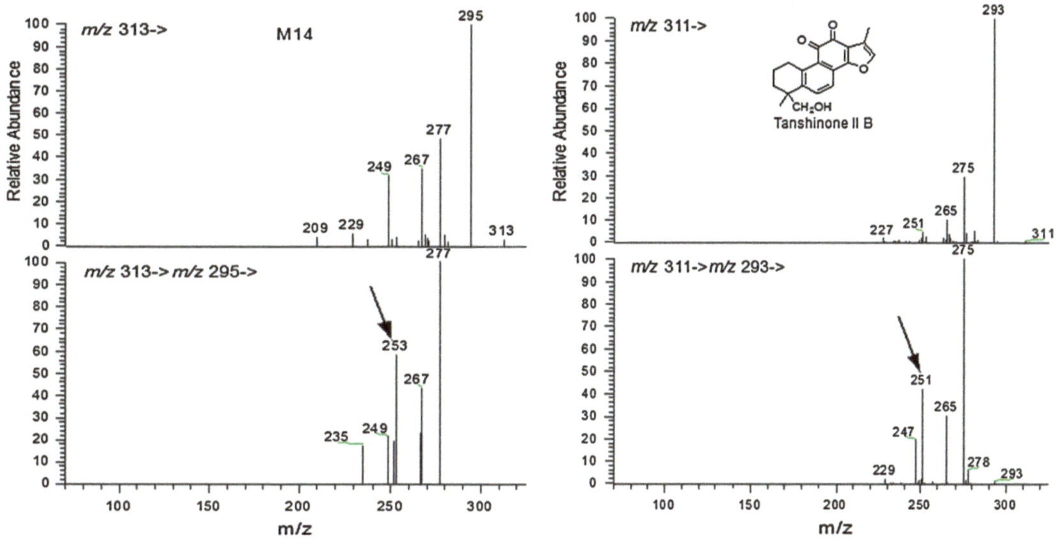

Fig. 17.48 The MS^n spectra of M14 and tanshinone II_B

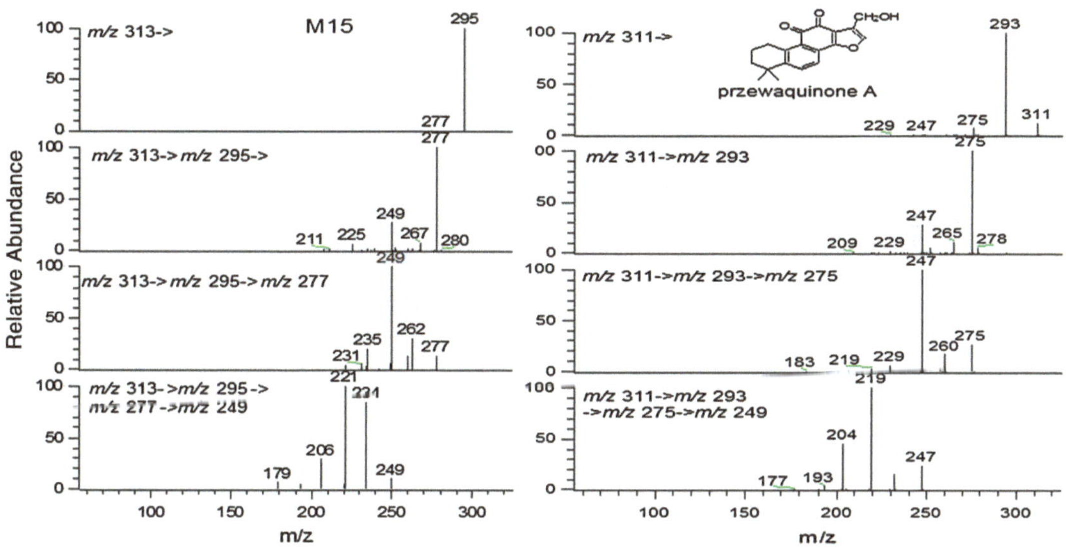

Fig. 17.49 The MS^n spectra of M15 and przewaquinone A

produced by M20 was 315 and there were consecutive dehydration peaks in the second- and third-stage mass spectra, suggesting that two hydroxyl groups exist in its structure. By comparing with the literature [36], M20 was identified as tanshinone V, produced by opening of ring D and hydrolysis of cryptotanshinone (Fig. 17.52).

4. The Metabolites of 15,16-Dihydrotanshinone I in Rat Bile

The structure of 15,16-dihydrotanshinone is similar to that of cryptotanshinone in that both have a saturated bond between positions 15 and 16. Their difference, though, is that ring A of 15,16-dihydrotanshinone I is an unsaturated aromatic ring. The level of hydroxylated

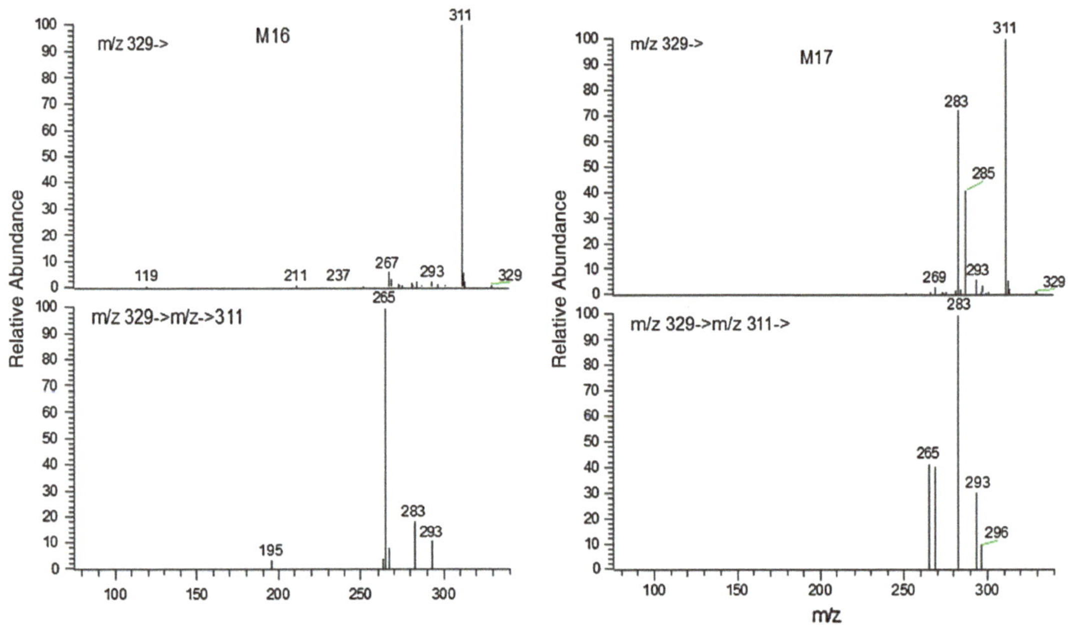

Fig. 17.50 The MS^n spectra of M16 and M17

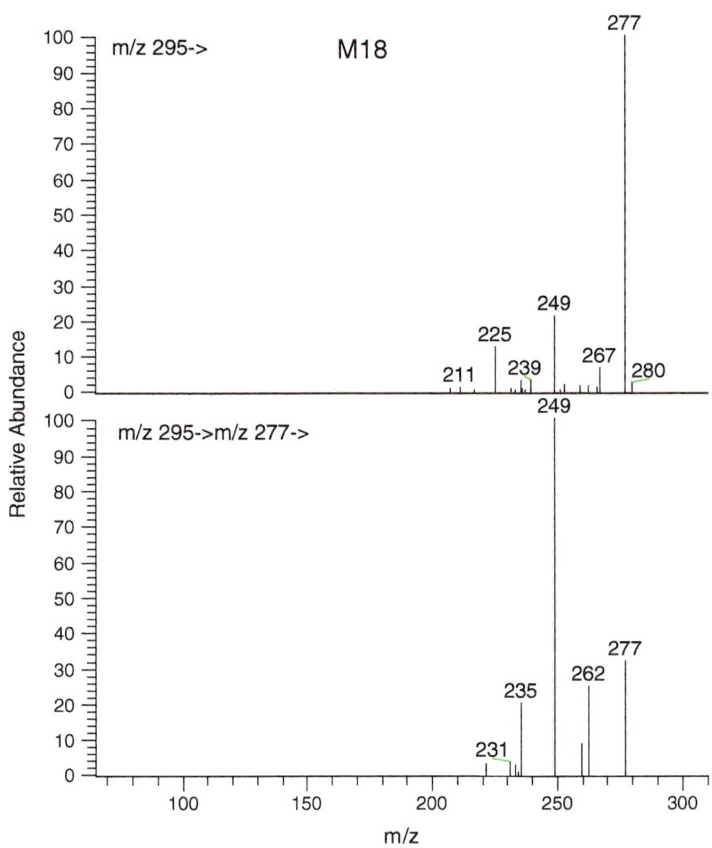

Fig. 17.51 The MS^n spectra of M18

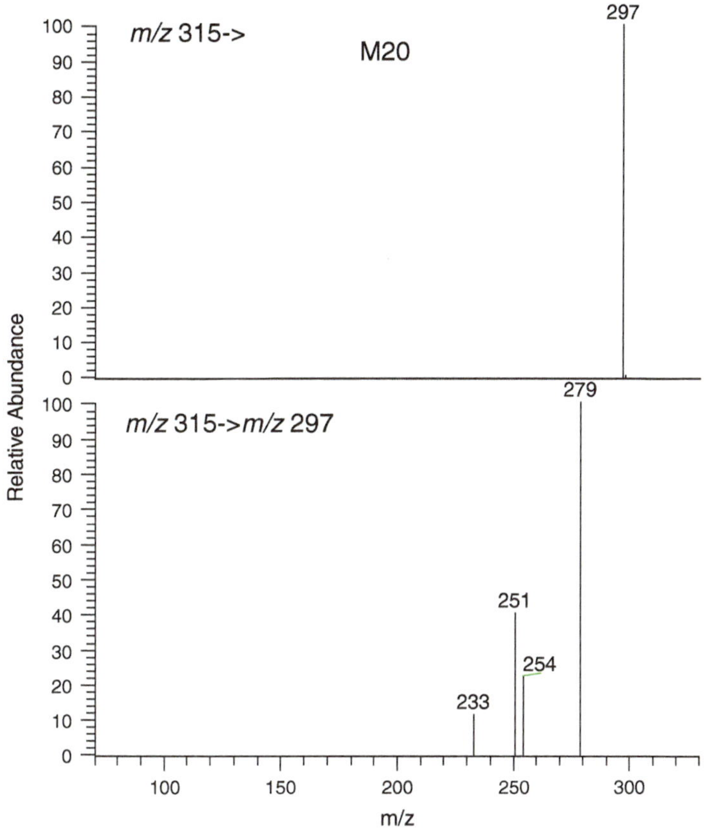

Fig. 17.52 The MS^n spectra of M20

metabolites of 15,16-dihydrotanshinone I produced in rats was extremely low, but it was still possible to detect these compounds by extracting ions (M21, *m/z* 295, Fig. 17.53). The metabolites that could be identified in a LC chromatogram were M22 and M23, with retention times of 24.0 and 36.4 min, respectively. Full-scan first-stage mass spectra showed that the quasi-molecular ion peak $[M + H]^+$ of these two metabolites was 297, heavier than 15,16-hydrotanshinone I by 18 Da. The fragmentation pathways of these two compounds after losing a water molecule were the same as those of 15,16-dihydrotanshinone I (Fig. 17.54). By comparisons in the literature [33], M23 was identified as tanshinone VI, a metabolite produced by the hydrolysis and breaking up of ring D in 15,16-dihydrotanshinone I, and M22 was possibly produced by M23 via keto-enol tautomerization.

The retention time of M24 was 47.6 min and its MS fragmentation pattern was consistent with that of tanshinone I (Fig. 17.55). Therefore, this metabolite was identified as tanshinone I. As reported in the literature [37], tanshinone I was the major metabolite of 15,16-dihydrotanshinone I in an in vitro homogenized liver incubation experiment. However, only trace amounts of tanshinone I were found in an in vivo experiment.

5. The Metabolites of Tanshinone II_B and Przewaquinone A in Rat Bile

Among the seven tanshinones, both tanshinone II_B and przewaquinone A have hydroxy-substitutions in their structure. The metabolism of these two compounds is quite similar. Tanshinone II_B and przewaquinone A were excreted extensively unchanged in bile (Fig. 17.56).

After intravenous administering of tanshinone II_B to rats, there were three metabolites with

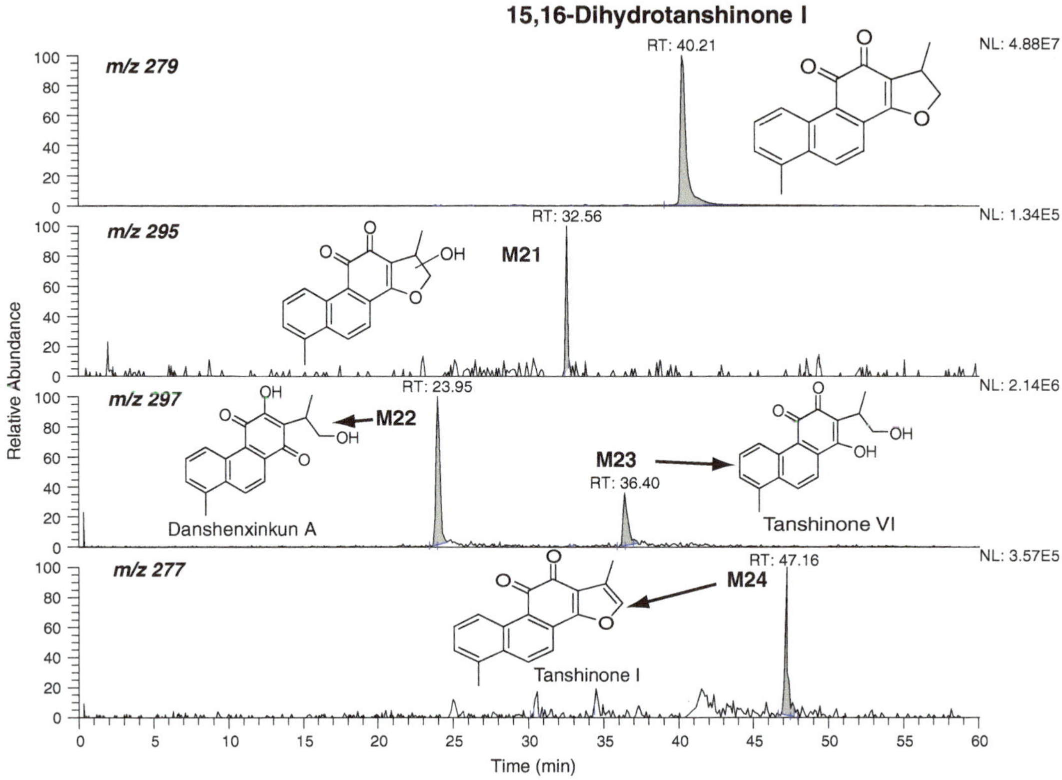

Fig. 17.53 The extracted $[M + H]^+$ ion chromatograms of metabolites in rat bile 24 h after intravenous administration of 15,16-dihydrotanshinone I

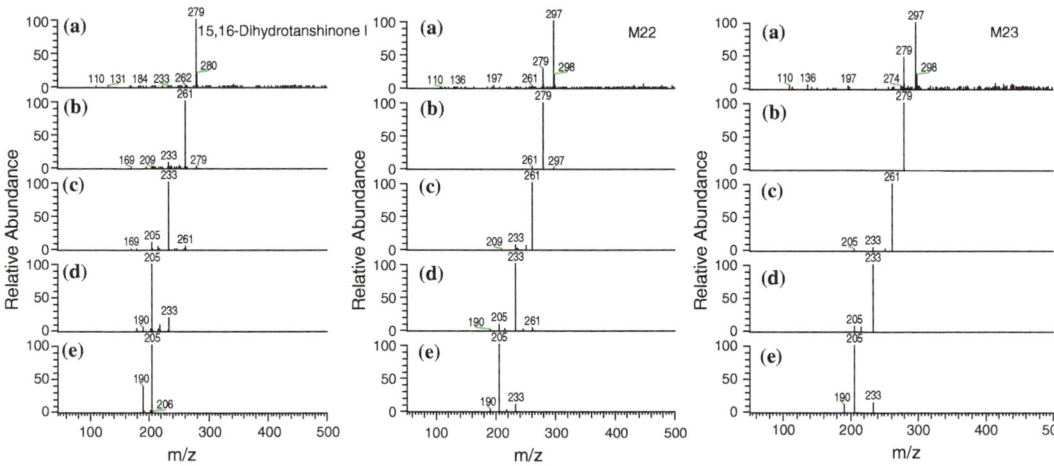

Fig. 17.54 The full-scan mass (**a**) and Ms^2-MS^5 (**b–e**) spectra of 15,16-dihydrotanshinone I, M22, and M23

retention times of 19.9, 21.7, and 22.0 min showing UV absorption features similar to those of tanshinone II_B, and they all produced ions of $[M + H]^+$ *m/z* 327, which was 16 Da heavier than tanshinone II_B. Furthermore, the second-stage mass spectra of these three

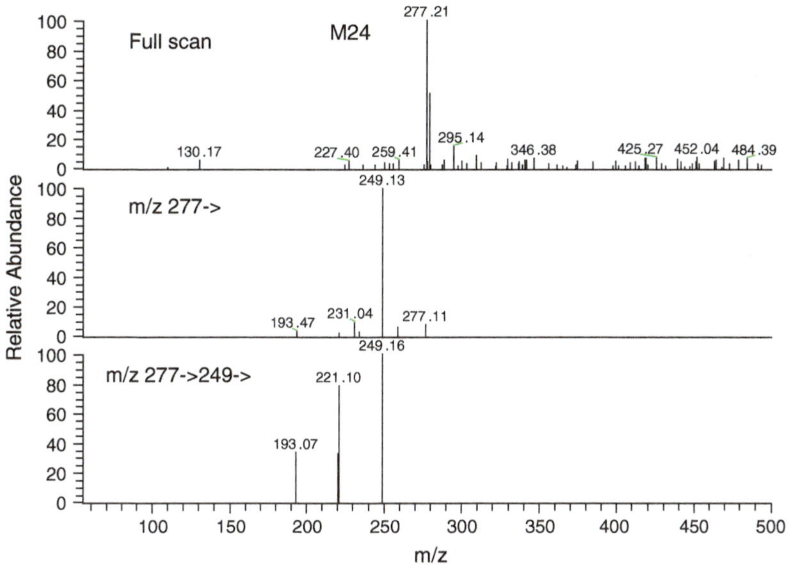

Fig. 17.55 The full-scan MS and MS^n spectra of M24

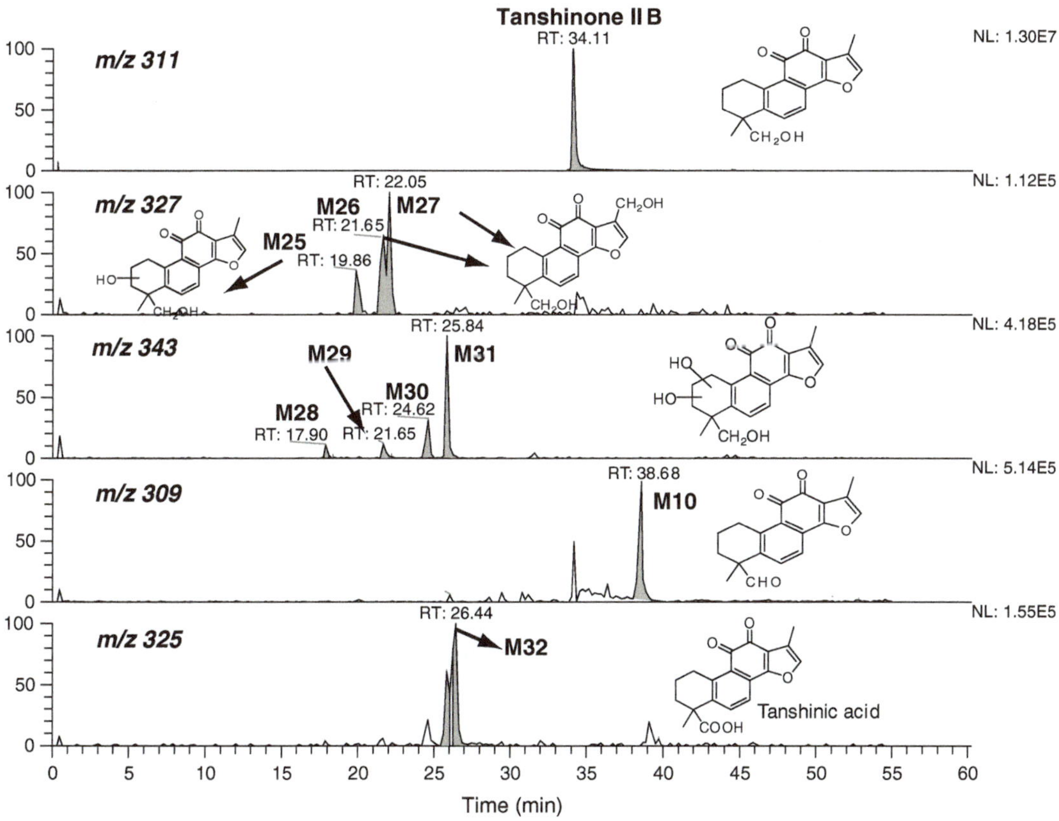

Fig. 17.56 The extracted $[M + H]^+$ ion chromatograms of the metabolites of tanshinone II_B in rat bile 24 h after intravenous administration

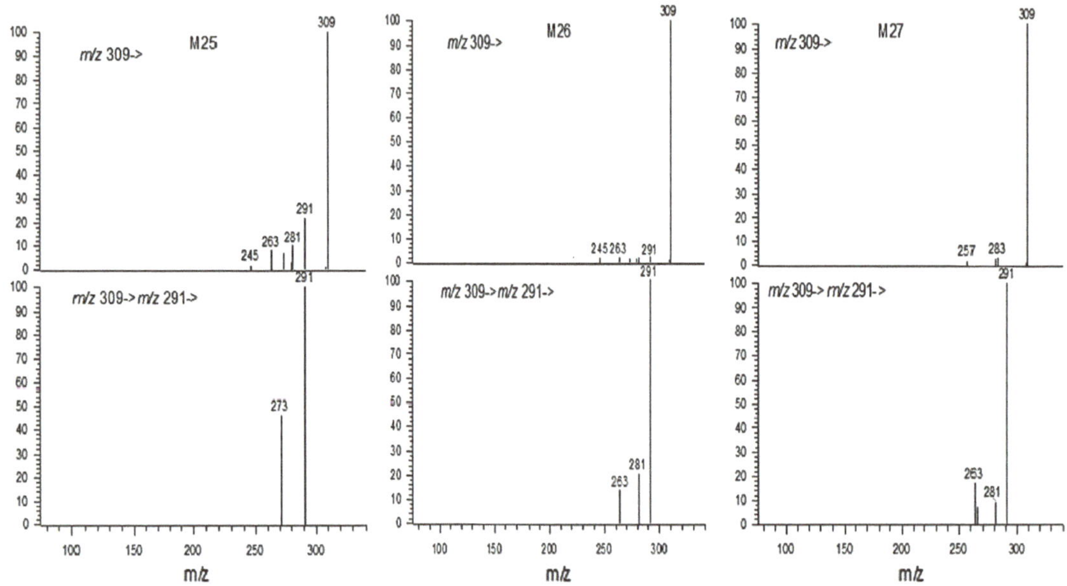

Fig. 17.57 The Ms^n spectra of M25, M26, and M27

compounds all showed very obvious dehydrated peaks. Therefore, these three metabolites were identified as single-hydroxylated tanshinone II_B (Fig. 17.57).

The quasi-molecule ion peak of M10 was 309, and its retention time was consistent with that of above-mentioned tanshinaldehyde. Therefore, M10 was identified as tanshinaldehyde. The detection of M10 in rat bile samples obtained after administration of tanshinone II_B further verified the existence of the metabolic reactions of tanshinones on the C-4 side chain, a metabolism pathway which was found for the first time. The quasi-molecule ion peak of M32 was 325 and obvious *m/z* 281 ion losing a CO_2 molecule could be observed in the second-stage mass spectrum, suggesting the possible presence of a hydroxyl group in the structure of M32. Therefore, M32 was tentatively identified as tanshinic acid (Fig. 17.58).

Przewaquinone A has a 17-hydroxy group in its structure and its metabolism is similar to that of tanshinone II_B (Fig. 17.59). Three single-hydroxylated metabolites were found in the postadministration bile samples, among which two with retention times of 21.5 and 21.9 min showed MS fragmentation patterns consistent with those of M26 and M27 in bile samples following administration of tanshinone II_B. Therefore, these two metabolites were identified as 17,18-dihydroxytanshinone II_A and its isomer. M33 was another single-hydroxyl metabolite (Fig. 17.60).

Similar to tanshinone II_B, the protonated molecules of M34 and M35 were found to be *m/z* 309 and *m/z* 325, respectively. It was speculated that M34 and M35 were the metabolites produced by oxidation of the hydroxymethyl group on the C-15 side chain, so they were tentatively named przewaquinaldehyde and prze-waquinic acid, respectively, (Fig. 17.61).

Four double-hydroxyl metabolites (M28–M31) were detected following administration of tanshinone II_B to rats; however, no double-hydroxyl metabolites were detected in bile samples following administration of przewaquinone A (Fig. 17.62).

6. The Metabolites of Tanshinone I in Rat Bile

Tanshinone I has unsaturated rings A and D as a fully conjugated planar structure. No metabolites were detected following intravenous administration of tanshinone I to rats,

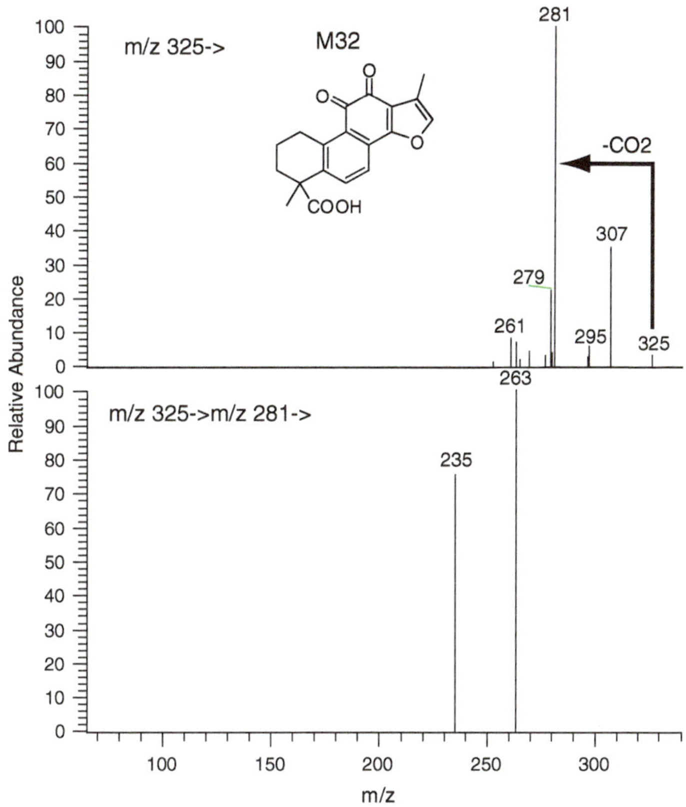

Fig. 17.58 The MS^n spectra of M32

and the abundance of the parent compound was extremely low, suggesting that it was difficult to make structural modifications to tanshinone I.

17.3.2.4 Discussion

In order to increase the sensitivity of ion trap mass spectrometry and the throughput of an analysis cycle, and to save biological samples, this study sets the following parameters in data-dependent scans: in the "parent ion list," input ion mass as M + H, M + 16, M − 2, M + 32, etc., which are the possible parent compounds or quasi-molecule ions of the metabolite. Also, select "No parent ion Most intense ion" to ensure that the data-dependent scan would be activated with priority to ions with input mass numbers. If no target ion was found, a data-dependent scan would be activated by ions with the highest abundance to avoid missing unknown metabolites. Meanwhile, fill the mass numbers of endogenous substances in the bile into the exclusion ion list, such as 464, 460, 496, and 220, to avoid activating a data-dependent scan by ions derived from endogenous substances and consequently lowering MS sensitivity.

HPLC/MS^n technology has great advantages for the identification of drug metabolites, including high throughput, promptness, and high sensitivity. HPLC can satisfactorily separate multiple components in complex biological media, including compounds' geometric isomers. For those components which are poorly separated by chromatography, mass spectrometry can differentiate by its high selectivity as long as they have different molecular weights. Using ESI-MS^n ion trap technology, information about the molecular weights of the analytes and multistage MS fragment ions could be obtained under gentle conditions. This technology has shown good reproducibility. By comparing the fragmentation patterns of diagnostic ions with those of

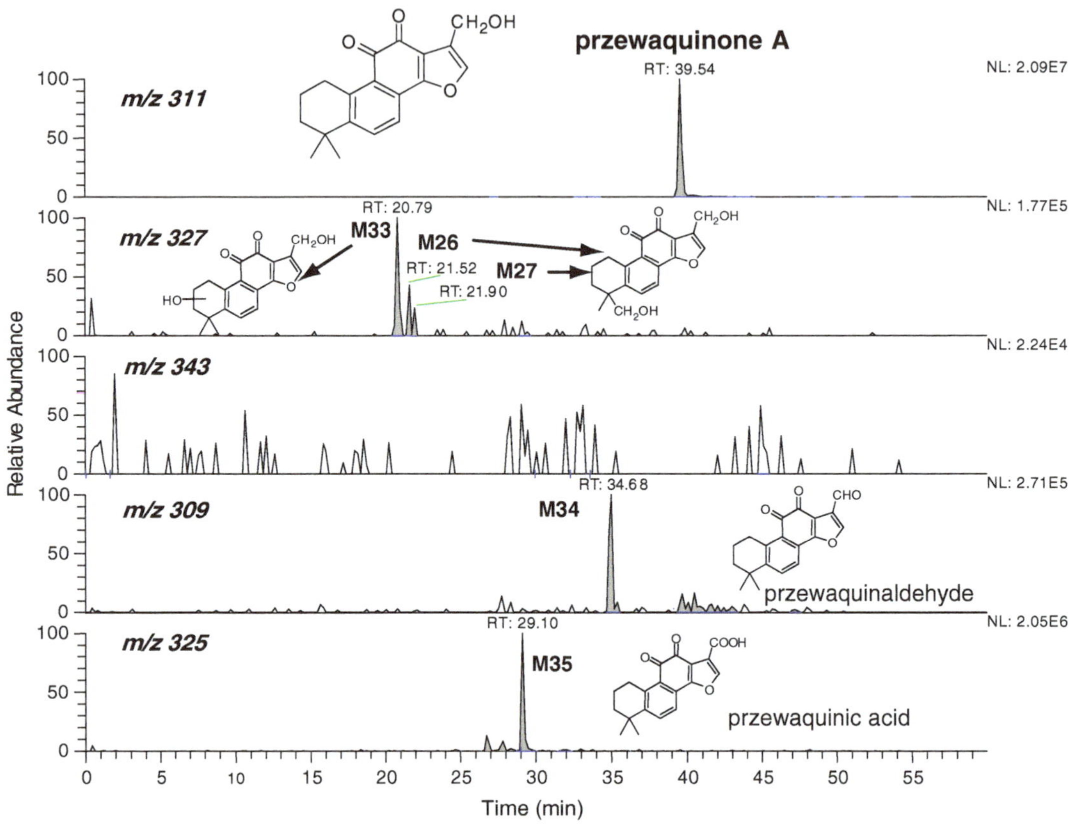

Fig. 17.59 The extracted $[M + H]^+$ ion chromatograms of the metabolites of przewaquinone A in rat bile 24 h after intravenous administration

reference substances, the structures of unknown compounds could be reasonably and accurately deducted even without standard substances.

The HPLC/MS^n analysis method established by our institute used data-dependent scanning and dynamic exclusion technology in a reasonable manner and identified 35 metabolites in rat bile following intravenous administration of seven tanshinones with different structures. This method is especially suitable for the identification of trace metabolites in a complex system.

It was found in this study that tanshinones with different structures produced different metabolites. The extent of ring conjugation and the type and position of the substituting groups affect the metabolism of these tanshinones. Cryptotanshinone with 2 saturated rings tends to undergo dehydrogenation reactions in its metabolism very easily. Furthermore, the metabolic reactions of cryptotanshinone were most diverse, including dehydrogenation, monohydroxylation, double hydroxylation and breaking up of the D ring, etc. On the other hand, it is difficult for tanshinone I, which also has two saturated rings, to be metabolized, and even the parent drug excreted in bile was at a very low level. For 15,16-dihydrotanshinone I, although it was reported in the literature that this compound would be transformed to tanshinone I very easily in a homogenized liver incubation experiment, only trace amounts of tanshinone I could be detected in bile samples following administration of dihydrotanshinone I. When incubating cryptotanshinone and tanshinone II_A in blank bile samples in vitro, the results showed that these compounds tended to easily lose the saturated bond between positions 15 and 16 to form a double bond, as shown in Fig. 17.63.

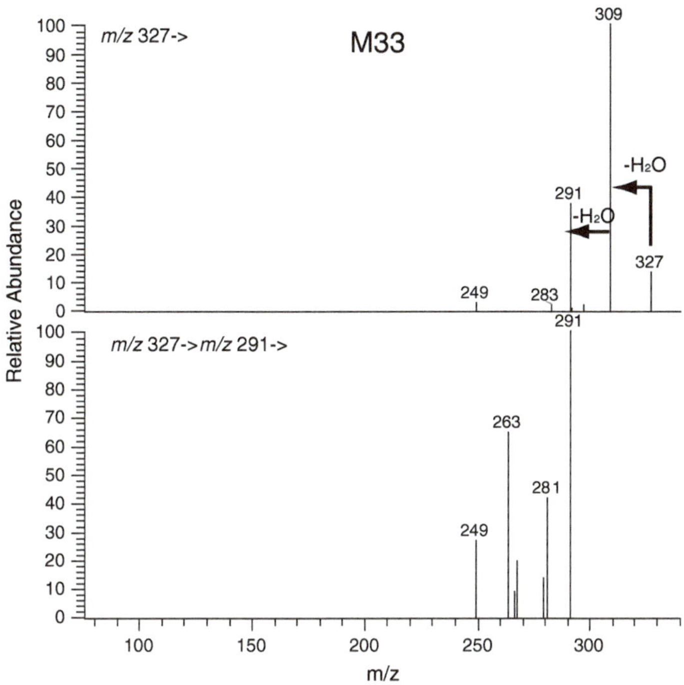

Fig. 17.60 The MS[n] spectra of M33

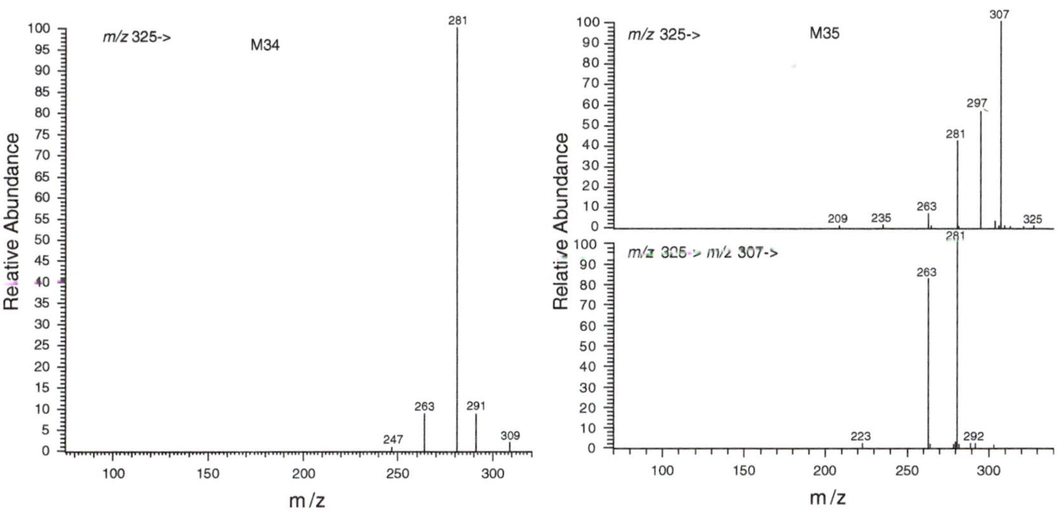

Fig. 17.61 The MS[n] spectra of M34 and M35

Tanshinone II_A, STS, tanshinone II_B, and przewaquinone A have the same conjugation systems and similar metabolic pathways. Hydroxylation was the major metabolic pathway for all these compounds. However, for those compounds with substitutions in their backbones, such as tanshinone II_B, STS, and przewaquinone A, hydroxylation reactions had more difficulty occurring than in tanshinone II_A, which has no substitutions. The metabolic reactions involving the transformation of a methyl group to a hydroxymethyl group, aldehyde group, and

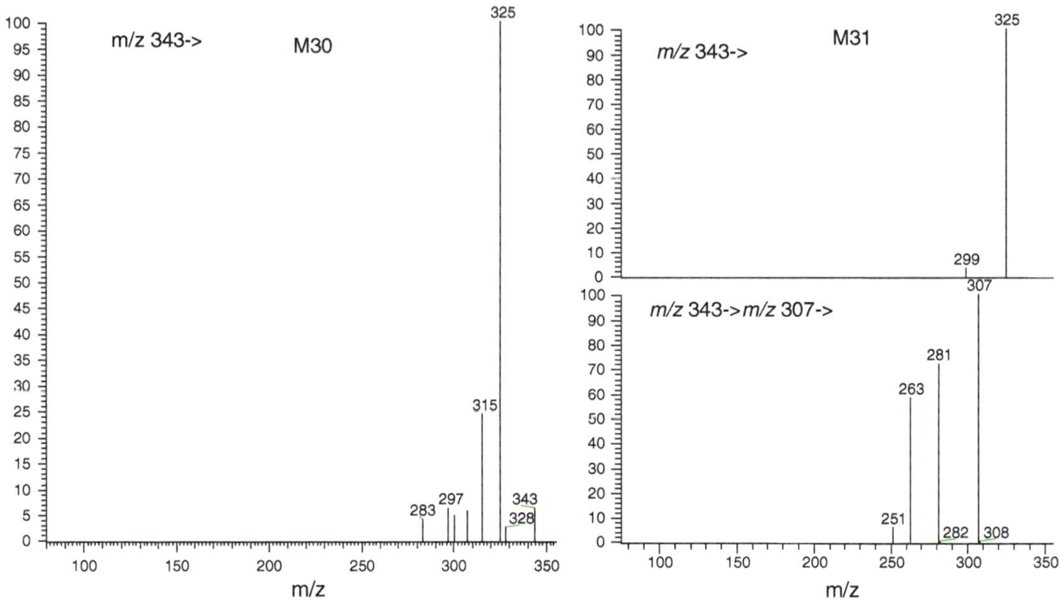

Fig. 17.62 The MS^n spectra of M30 and M31

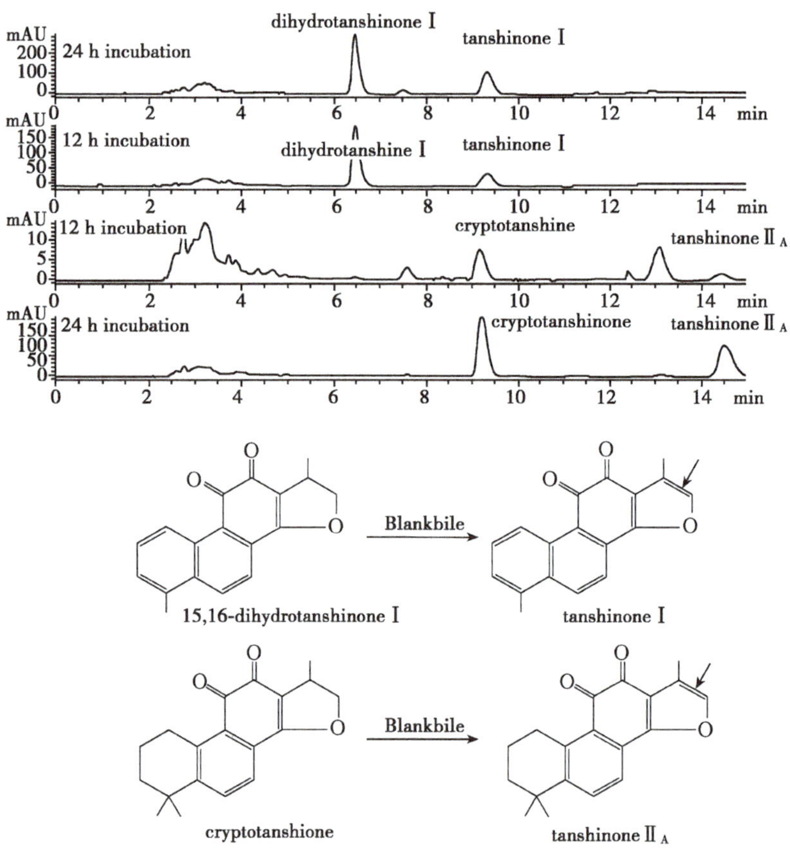

Fig. 17.63 In vitro incubation of cryptotanshinone and 15,16-dihydrotanshinone I in blank rat bile samples

Table 17.15 The characteristics of metabolic reactions of 7 tanshinones

Compounds	Metabolic reactions				
	Mono-hydroxylation	Dihydroxylation	Dehydrogenation	Oxidation in side chain	D-ring cleavage
STS	+	–	+	–	–
Tanshinone II_A	+	–	+	+	–
Cryptotanshione	+	+	+	–	+
15,16-Dihydrotanshine I	+	–	+	–	+
Tanshinone II_B	+	+	–	+	–
Przewaquinone A	+	–	–	+	–
Tanshinone I	–	–	–	–	–

\+ positive, − negative

carboxyl group on the C-4 and C-15 side chains were first found in tanshinones. The metabolic reactions of the seven tanshinones are summarized in Table 17.15. The retention times, maximum UV absorption wavelengths, and multistage MS fragmentation data of the 35 metabolites are shown in Table 17.16.

17.3.3 Quantitative Analysis of Bile Excretion Following Intravenous Administration of 7 Tanshinones to Rats

In this section, we describe the quantitative analysis by HPLC of the cumulative excretion in bile 24 h after the administration of seven tanshinones (tanshinone II_A, STS, cryptotanshinone, 15,16-dihydrotanshinone I, tanshinone II_B, przewaquinone, and tanshinone I). The relationship between the structures of tanshinones and the rates of excretion and cumulative excretion was investigated.

17.3.3.1 Experiment

1. Instruments, Reagents, and Drugs
   Agilent 1100 HPLC system equipped with a quaternary pump, autosampler, and DAD; TGL-16G-A high-speed refrigerated centrifuge; XW-80 vortex mixer; EYELA N-1000V-W rotary evaporator; HGC-12 nitrogen evaporator; 7 tanshinones as described in Sect. 17.2; 4-chlorobiphenyl (Shanghai First Reagent Works).
2. Laboratory Animals
   SD male rats with body weights of 250 ± 20 g were provided by the Laboratory Animal Center of Peking University Health Science Center.
3. Drug Administration and Bile Sample Collection
   Precisely weigh 10 mg of tanshinone II_A, cryptotanshinone, 15,16-dihydrotanshinone I, tanshinone II_B, przewaquinone A, and tanshinone I, respectively. Prepare 2 mg/ml solutions with 1,2-propylene glycol. Precisely weigh 10 mg of STS and prepare a 2 mg/ml solution with normal saline.
   Randomize 32 rats into 8 groups, one of which was set as the blank, to which the equivalent volume of propylene glycol or normal saline was injected. To the other seven groups, the drugs were administered by tail vein injection at a dose of 4 mg/kg. Collect bile samples from the animals receiving STS at 0–0.25, 0.25–0.5, 0.5–0.75, 0.75–1, 1–1.5, 1.5–2, 2–2.5, 2.5–3, 3–3.5, 3.5–4, 4–5, 5–6, 6–8, 8–12, and 12–24 h postadministration in graduated tubes and record the volume of bile. Collect bile samples from the animals in other groups at 0–1, 1–2, 2–3, 3–4, 4–6, 6–8, 8–10, 10–12, 12–16, 16–20, and 20–24 h postadministration in graduated tubes and record the volume of bile. Store all bile samples at −80 °C.

Table 17.16 Identification of the metabolites in rat bile following intravenous administration of different tanshinones

Parent compound	Metabolites	Parention	UV λ_{max} (nm)	Retention time (min)	Data-dependent MS^n data (% base peak)	Identify	Metabolic reaction
STS	M1	$[M-Na]^-$ 389	272	4.6	MS^2[389]: 389(4), 374(55), 371(14), 357(26), 325(100), 309(7), 307(88), , 252(8)	Hydroxylated STS	Mono-hydroxylation
	M2	$[M-Na]^-$ 389	272	5.0	MS^2[389]: 389(7), 374(7), 357(58), 325(100), 309(3), 297(2), 293(2), 269(9) MS^3[389 → 325]: 297(100), 269(43)	Hydroxylated STS	Mono-hydroxylation
	M3	$[M-Na]^-$ 371	252	9.1	MS^2[371]: 371(5), 355(100), 307(64), 291(55), 279(5), 251(2)	Sodium 1,2-dehydrotanshinone II$_A$ sulfonate or Sodium 2,3-dehydrotanshinone II$_A$ sulfonate	Dehydrogenation
Tanshinone II$_A$	[a]M4	$[M+H]^+$ 311	270	34.2	MS^2[311]: 293(100), 275(6), 251(2) MS^3[311 → 293]: 293(3), 278(17), 275(100), 265(15), 251(82), 249(4), 247(6), 229(2) MS^4[311 → 293 → 275]: 275(100), 247(98)	Tanshinone II$_B$	Mono-hydroxylation
	[a]M5	$[M+H]^+$ 311	270	34.6	MS^2[311]: 293(100), 283(13), 281(5), 275(99), 265(36), 251(18), 250(6), 247(7), 243(4), 241(4), 235(4), 227(5) MS^3[311 → 293]: 293(10), 275(100), 265(27), 251(30), 247(16)	3α-Hydroxytanshinone II$_A$	Mono-hydroxylation
	[a]M6	$[M+H]^+$ 311	270	39.8	MS^2[311]: 293(100), 275(5) MS^3[311 → 293]: 275(100), 265(9), 251(247(31) MS^4[311 → 293 → 275]: 275(12),260(46), 247(100), 219(8)	Przewaquinone A	Mono-hydroxylation
	M7	$[M+H]^+$ 293	270	45.0	MS^2[293]: 278(7), 275(100), 265(16), 251(6), 247(32) MS^3[293 → 275]: 275(10), 247(100) MS^4[293 → 275 → 247]: 247(63), 232(57), 219(100), 204(32), 203(22)	1,3-Dehydrotanshinone II$_A$	Dehydrogenation
	M8	$[M+H]^+$ 293	–	47.3	MS^2[293]: 275(100), 265(8), 247(28) MS^3[293 → 275]: 275(17), 260(6), 247(100) MS^4[293 → 275 → 247]: 248(7), 247(7), 232(14), 219 (100), 218(4), 217(11), 204(22), 193(9)	1,2-Dehydrotanshinone II$_A$ or 2,3-Dehydrotanshinone II$_A$	Dehydrogenation
	M9	$[M+H]^+$ 293	–	49.8	MS^2[293]: 293(2), 275(100), 265(10), 247(23) MS^3[293 → 275]: 275(16), 247(100) MS^4[293 → 275 → 247]: 232(33), 221(10), 219(100), 204(79)	1,2-Dehydrotanshinone II$_A$ or 2,3-Dehydrotanshinone II$_A$	Dehydrogenation
	[a]M10	$[M+H]^+$ 309	270	38.8	MS^2[309]: 291(2), 281(100), 263(5), 235(2) MS^3[309 → 281]: 281(56), 266(10), 263(64), 239(26), 235(5), 211(100)	Tanshinaldehyde	Oxidation of C-18
Cryptotanshinone	M11	$[M+H]^+$ 313	–	27.6	MS^2[313]: 295(100), 283(25), 277(73), 270(24), 267(74), 265(24), 252(13), 249(70), 237(40) MS^3[313 → 295]: 277(77), 253(63), 249(100), 220(63)	Hydroxyl cryptotanshinone	Mono-hydroxylation
	M12	$[M+H]^+$ 313	–	28.3	MS^2[313]: 313(3), 295(100), 283(22), 277(94), 271(14), 270(9), 267(66), 265(4), 253(30), 252(16), 249(92), 237 (18), 229(16) MS^3[313 → 295]: 295(6), 277(100), 267(25), 253(12), 252(10), 249(46), 225(12)	3-Hydroxycryptotanshinone	Mono-hydroxylation

(continued)

Table 17.16 (continued)

Parent compound	Metabolites	Parention	UV λ_{max} (nm)	Retention time (min)	Data-dependent MS^n data (% base peak)	Identify	Metabolic reaction
	M13	$[M + H]^+$ 313	–	28.6	MS^2[313]: 313(2), 295(94), 283(22), 282(12), 277(97), 271(14), 270(12), 269(7), 267(78), 265(23), 255(8), 253 (21), 252(16), 251(7), 249(98), 245(9), 237(21), 229(14) MS^3[313 → 295]: 295(5), 280(4), 277(100), 267(26), 253(11), 252(8), 249(57), 225(9)	3-Hydroxycryptotanshinone	Mono-hydroxylation
	M14	$[M + H]^+$ 313	–	29.1	MS^2[313]: 295(100), 280(4), 277(33), 267(33), 265(2), 254(9),249(15) MS^3[313 → 295]: 295(14), 280(28), 277(100), 267(40), 253(20), 249(17)	18-Hydroxycryptotanshinone	Mono-hydroxylation
	M15	$[M + H]^+$ 313	264	35.2	MS^2[313]: 295(100) MS^3[313 → 295]: 277(100), 249(12), 239(7), 225(8) MS^4[313 → 295 → 277]: 262(46), 249(100), 235(16), 217(12), 207(13) MS^5[313 → 295 → 277 → 249]: 234(100), 207(48), 205 (45)	17-Hydroxycryptotanshinone	Mono-hydroxylation
	M16	$[M + H]^+$ 329	–	31.8	MS^2[329]: 311(100), 293(3), 267(7) MS^3[329 → 311]: 293(15), 283(19), 267(3), 265(100), 195(2)	Dihydroxyl cryptotanshinone	Di-hydroxylation
	M17	$[M + H]^+$ 329	–	33.0	MS^2[329]:329(2), 311(100), 297(3), 293(5), 285(42), 283 (77), MS^3[329 → 311]: 296(15), 293(28), 283(100), 269 (42), 267(41)	Di-hydroxyl cryptotanshinone	Di-hydroxylation
	M18	$[M + H]^+$ 295	–	47.3	MS^2[295]: 280(2), 277(100), 267(4), 253(5), 249(12), 235(2), 225(9) MS^3[295 → 277]: 277(9), 262(12), 249(100), 197(11)	1,2-Dehydrocryptotanshinone or 2,3-Dehydrocryptotanshinone	Dehydrogenation
	[a]M19	$[M + H]^+$ 295	270	51.8	MS^2[295]: 277(100), 259(2), 253(5), 249(13), 235(4) MS^3[295 → 277]: 277(26), 262(24), 259(5), 249(100), 248(6), 235(5), 234(6), 231(8), 221(5) MS^4[295 → 277 → 249]: 249(30), 234(64), 231(9), 221 (100), 220(7), 206(37), 193(8) MS^5[295 → 277 → 249 → 221]: 221(68), 206(100), 193 (28) 179(5)	Tanshinone II_A	Dehydrogenation
	M20	$[M + H]^+$ 315	248	36.3	MS^2[315]: 297(100) MS^3[315 → 297]: 279(100), 264(20), 254(17), 251(46), 233(11) MS^4[315 → 297 → 279]: 279(21), 251(53), 237(100), 233(50)	Tanshinone V	D-ring hydrolization
	[a]M4	$[M + H]^+$ 311	–	33.9	MS^2[311]: 293(100), 275(5) MS^3[311 → 293]: 293(4), 278(17), 275(100), 265(16), 251(92)	Tanshinone II_B	Mono-hydroxylation of M19
	[a]M5	$[M + H]^+$ 311	–	34.2	MS^2[311]: 293(72), 281(19), 275(100), 269(8), 267(36), 265(9), 263(11), 253(8), 251(13), 250(6), 227(8)	3α-Hydroxytanshinone II_A	Mono-hydroxylation of M19
	[a]M6	$[M + H]^+$ 311	–	39.4	MS^2[311]: 311(4), 293(100), 275(3) MS^3[311 → 293]: 278(5), 275(100), 265(6), 247(30)	Przewaquinone A	Mono-hydroxylation of M19

(continued)

Table 17.16 (continued)

Parent compound	Metabolites	Parention	UV λ_{max} (nm)	Retention time (min)	Data-dependent MS^n data (% base peak)	Identify	Metabolic reaction
Dihydrotanshinone I	M21	$[M + H]^+$ 295	242	32.6	MS^2[295]: 277(100), 249(20) MS^3[295 → 277]: 262(39), 249(100), 231(11), 221(10) MS^4[295 → 277 → 249]: 231(63), 221(100)	Hydroxyl dihydrotanshinone I	Mono-hydroxylation
	M22	$[M + H]^+$ 297		24.0	MS^2[297]: 297(2), 279(100), MS^3[297 → 279]: 261(100), 251(6), 237(1), 235(2), 233 (6) MS^4[297 → 279 → 261]: 261(3), 233(100), 215(2), 205 (9) MS^5[297 → 279 → 261 → 233]: 233(10), 205(100), 190 (3)	Danshenxinkun A	D-ring hydrolysis
	M23	$[M + H]^+$ 297		36.4	MS^2[297]: 279(100) MS^3[297 → 279]: 261(100), 251(3), 233(4),209(2) MS^4[297 → 279 → 261]: 233(100), 205(7) MS^5[297 → 279 → 261 → 233]: 233(15), 205(100), 190 (12)	Tanshinone VI	D-ring hydrolysis
	[a]M24	$[M + H]^+$ 277	–	47.2	MS^2[277]: 277(7), 259(5), 249(100), 234(2), 231(9), 221 (2), 194(3) MS^3[277 → 249]: 249(100), 221(78), 220(33), 193(33)	Tanshinone I	Dehydrogenation
Tanshinone II_B	M25	$[M + H]^+$ 327	270	19.9	MS^2[327]: 309(100), 291(23), 281(12), 279(3), 273(6), 263(7), 245(3) MS^3[327 → 309]: 291(100), 273(47)	Hydroxyl tanshinone II_B	Mono-hydroxylation
	M26	$[M + H]^+$ 327	270	21.7	MS^2[327]: 327(3), 309(100), 299(3), 291(3), 283(24), 265(8), 223(2)	Hydroxyl tanshinone II_B	Mono-hydroxylation
	M27	$[M + H]^+$ 327	270	22.0	MS^2[327]: 309(100), 283(3), 281(2), 257(2) MS^3[327 → 309]: 291(100), 281(9), 265(7), 263(17)	Hydroxyl tanshinone II_B	Mono-hydroxylation
	M28	$[M + H]^+$ 343	–	17.9	–	Hydroxyl tanshinone II_B	Di-hydroxylation
	M29	$[M + H]^+$ 343	–	21.7	–	Hydroxyl tanshinone II_B	Di-hydroxylation
	M30	$[M + H]^+$ 343	–	24.4	MS^2[343]: 328(3), 325(100), 315(24), 307(6), 300(5), 297(6), 283(4)	Hydroxyl tanshinone II_B	Di-hydroxylation
	M31	$[M + H]^+$ 343	–	25.9	MS^2[343]: 325(100), 299(3) MS^3[343 → 325]: 307(100), 281(72), 263(58), 251(6)	Hydroxyl tanshinone II_B	Di-hydroxylation
	M10	$[M + H]^+$ 309	–	39.2	MS^2[309]: 291(9), 281(100), 263(7), 245(2), 221(2)	Tanshinaldehyde	Oxidation of C-18
	M32	$[M + H]^+$ 325	–	26.3	MS^2[325]: 307(33), 295(8), 281(100), 279(25), 263(13), 261(4) MS^3[325 → 281]: 263(100), 235(76) MS^3[325 → 307]: 289(2), 263(100)	Tanshinic acid	Oxidation of C-18

(continued)

Table 17.16 (continued)

Parent compound	Metabolites	Parention	UV λ_{max} (nm)	Retention time (min)	Data-dependent MS^n data (% base peak)	Identify	Metabolic reaction
Przewaquinone A	M33	$[M + H]^+$ 327	270	20.8	MS^2[327]: 327(10), 309(100), 291(18) MS^3[327 → 309]: 291(100), 267(16), 266(23), 263(50) MS^4[327 → 309 → 291]: 291(100), 263(98)	Hydroxylated przewaquinone A	Mono-hydroxylation
	M26	$[M + H]^+$ 327	270	21.5	MS^2[327]: 309(100) MS^3[327 → 309]: 291(100), 281(15), 263(17)	Hydroxylated przewaquinone A	Mono-hydroxylation
	M27	$[M + H]^+$ 327	270	21.9	MS^2[327]: 327(13), 309(100), 291(37)	Hydroxylated przewaquinone A	Mono-hydroxylation
	M34	$[M + H]^+$ 309	–	34.7	MS^2[309]: 291(11), 281(100), 263(11)	Przwaquinaldehyde[b]	Oxidation of C-15 side chain
	M35	$[M + H]^+$ 325	–	29.6	MS^2[325]: 325(2), 307(100), 297(53), 281(37), 263(5) MS^3[325 → 307]: 289(3), 279(3), 263(100)	Przwaquinic acid[b]	Oxidation of C-15 side chain
Tanshinone I	–	–	–	–	–	–	No metabolite found

[a] Identified with the reference compounds
[b] Compound named by author
– Not found

4. Preparation of Standard Solutions
   Precisely weigh seven reference substances; add methanol to make the following stock solutions: 400 μg/ml tanshinone II_A, 400 μg/ml cryptotanshinone, 400 μg/ml 15,16-dihydrotanshinone I, 500 μg/ml tanshinone II_B, 650 μg/ml przewaquinone A, 21 μg/ml tanshinone I, and 2,000 μg/ml STS, respectively. Make serial dilutions with methanol to obtain solutions for standard curve drawing. Precisely weigh a certain amount of 4-chlorobiphenyl and prepare a 500 μg/ml internal standard solution with methanol. Store all stock solutions at −20 °C. These solutions were stable for 1 month.
5. Pretreatment of Bile Samples
   Measure 1 ml of bile collected after intravenous administration of STS (add double-distilled water to 1 ml if the volume of bile was less than 1 ml); filter through a 0.22 μm microporous membrane, then directly inject 20 μl of the sample into HPLC for analysis. Record the chromatograms.
   Transfer bile samples collected during each interval postadministration to graduated tubes containing pre-evaporated internal standard solution; dilute to 3 ml; add 8 ml of acetic ester to extract; vortex to mix for 10 min; centrifuge at 4,000 rpm for 5 min; measure 7 ml of the solution, blow to dryness with nitrogen gas; precisely add 500 μl of methanol to redissolve; filter through a 0.22 μm microporous membrane, then directly inject 20 μl of the samples into HPLC for analysis. Record the chromatograms.
6. Preparation of QC Samples
   Precisely pipette 100 μl of STS reference solution to a centrifuge tube; blow to dryness with nitrogen gas; precisely add 1 ml of blank bile; vortex mix for 5 min to dissolve.
   Precisely pipette 100 μl of STS reference solution to a centrifuge tube; blow to dryness with nitrogen gas; precisely add certain volume of blank bile; vortex mix for 5 min to dissolve, then treat the samples according to the "pretreatment of bile samples" method described above.
   Prepare each tanshinone QC samples with low, medium, and high concentrations. The QC samples were used for tests of the accuracy, precision, and stability of the analysis methods.
7. HPLC Conditions for Quantitative Analysis
   Chromatographic column: Agilent Zorbax Extend-C_{18} reversed-phase column (250 mm × 5 mm, 5 μm); Agilent Zorbax SB-C $_{18}$ guard column (20 mm × 4 mm, 5 μm).
   Detection wavelength was set to the maximal absorption wavelength of each compound: tanshinone II_A, tanshinone II_B, and przewaquinone: 270 nm; STS: 242 nm; cryptotanshinone: 264 nm; 15,16-dihydrotanshinone I: 242 nm.
   Mobile phase 1: used to separate STS: acetonitrile–0.1 % ammonia water gradient elution; 0–15 min, acetonitrile concentration increases linearly from 15 to 85 %.
   Mobile phase 2: used to separate tanshinone II_A and tanshinone I: acetonitrile–0.03 % formic acid solution (75:25).
   Mobile phase 3: used to separate cryptotanshinone and 15,16-dihydrotanshinone I: acetonitrile–0.03 % formic acid solution (75:23).
   Mobile phase 4: used to separate tanshinone II_B and przewaquinone A: acetonitrile–0.03 % formic acid solution. 0–25 min, acetonitrile concentration increases linearly from 50 to 100 %.
   Flow rate: 1 ml/min.
   Temperature of the columns: 30 °C.

17.3.3.2 Method Validation

1. System Suitability
   The requirements were that the theoretical plate number of each of the seven tanshinones in bile samples be more than 7,000, the separation between any adjacent peaks be larger than 1.2, and the symmetry factor be less than 1.1.
2. Method Specificity
   Optimize the chromatographic conditions to avoid the disturbance caused by endogenous substances in bile to the test substances and internal standard. The peak shape should be good.

3. Extraction Recovery
The peak areas of the samples, which were made by blending the blank bile with the reference solutions and pretreated according to the protocol of "pretreatment of bile samples," were compared with those of the standard solutions. Their ratios were then multiplied by the corresponding dilution factors.
4. Standard Working Curve
Precisely pipette 20 μl of tanshinone reference serially diluted solutions and the internal standard solution, respectively; blow dry with nitrogen gas; precisely add certain volume of blank bile; vortex mix for 1 min; pretreat these samples according to the method described in "pretreatment of bile samples" section; prepare the following solutions: tanshinone II_A with concentrations of 0.26, 0.51, 1.02, 2.04, 4.08, 10.2, 20.4, and 40.8 μg/ml, respectively; cryptotanshinone with concentrations of 0.25, 0.50, 1.00, 2.00, 4.08, 10.2, 20.4, and 40.8 μg/ml, respectively; 15,16-dihydrotanshinone I with concentrations of 0.25, 0.50, 1.00, 2.00, 4.08, 10.2, 20.4, and 40.8 μg/ml, respectively; tanshinone II_B with concentrations of 1.00, 2.50, 5.00, 10.0, 20.0, 40.0, 50.0, and 100.0 μg/ml, respectively; przewaquinone A with concentrations of 0.65, 1.30, 3.25, 6.50, 13.0, 26.0, 52.0, 65.0, and 130.0 μg/ml, respectively; tanshinone I with concentrations of 0.20, 0.51, 1.02, 2.04 and 4.08 μg/ml, respectively; STS with concentrations of 1.25, 2.50, 5.00, 6.25, 12.5, 25.0, 50.0, 125.0, 250.0 and 500.0 μg/ml, respectively. Record the peak area ratio of the test substance to the internal standard substance on the longitudinal axis (Y) and the concentration of the test substance on the abscissa axis (X), and conduct a weighted least square regression with a weight coefficient of $1/Y^2$ to obtain the standard curves of the seven tanshinones in bile.
5. Intraday Precision, Interday Precision, and Accuracy
Take the QC samples with high, medium, and low concentrations to perform analysis. For each sample, analyze six times to determine the intraday precision. Meanwhile, take the QC samples with high, medium, and low concentrations and analyze for three consecutive days to determine the interday precision. This test was performed simultaneously with standard curve drawing.
The accuracy was evaluated by the recovery results. Calculate the ratio between the determined concentrations of the QC samples with the high, medium, and low concentrations and their theoretical concentrations of the reference substances. The required recoveries were in the range of 85–115 %.
6. Lower Limit of Quantification
Serially dilute the solutions with concentrations at the lower ends of standard curves and obtain a series of samples. For each sample, determine its concentration 5 times. When the results showed a precision of RSD <20 % and an accuracy (theoretical concentration deviates from the true concentration) <±20 %, the lowest concentration of the test substance in the sample is the LOQ.
7. Stability Studies
Investigate the stability of the QC samples after three freeze-thaw cycles or storage at room temperature for 24 h.
8. Bile Excretion Studies
Use the above-mentioned methods to determine the concentration of tanshinones in bile during each interval following intravenous administration, and calculate the cumulative excretion (percentage of administered dose) in bile in 24 h postadministration.

17.3.3.3 Results and Discussion

1. HPLC Conditions
To ensure the complete separation of all test substances in bile samples from endogenous impurities, we compared two mobile phase systems: acetonitrile–water system and methanol–water system. The results showed that the former was more suitable for separating tanshinones and produced peaks with better shapes. For cryptotanshinone and 15,16-dihydrotanshinone I, adding 0.03 % formic acid to the mobile phase could produce more symmetric peaks and reduce tailing. The

tanshinone compounds have extremely low polarity, so a higher percentage of organic phase could shorten the analysis time. Because ammonia water was needed for separating STS and the pH of the mobile phase was about 10.5, Zorbax Extend C_{18} was found to be more suitable. When the column temperature was at 30 °C, the endogenous substances in bile did not disturb the determination. Under the selected chromatographic conditions, every tanshinone in bile was well separated, as shown in Fig. 17.64a, b.

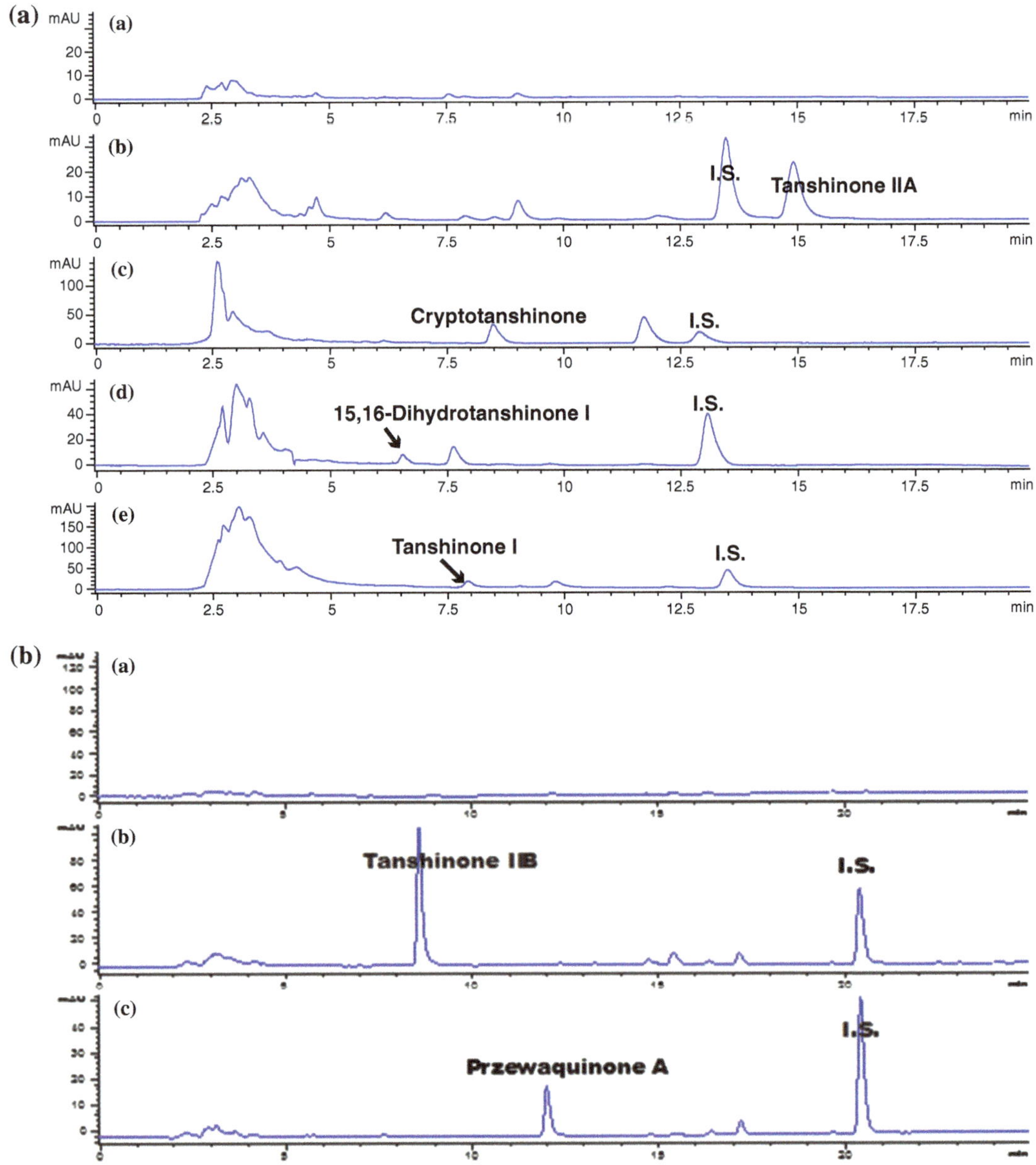

Fig. 17.64 **a** The HPLC chromatograms of blank bile (*a*), and bile samples following intravenous administration of tanshinone II_A (*b*), cryptotanshinone (*c*), 15,16-dihydrotanshinoneI (*d*), and tanshinoneI (*e*). *I.S.* internal standard. **b** The HPLC chromatograms of blank bile (*a*), bile samples following intravenous administration of tanshinone II_B (*b*), and przewaquinone (*c*)

2. Selection of Pretreatment Method for Bile Samples
   Because STS is a sulfonate of tanshinone II_A with good solubility in bile, its samples were therefore directly used for analysis after dilution and filtration. Since the pretreatment processes like extraction and evaporation were omitted, the external standard method was chosen for quantitation. Although the endogenous substances in untreated bile samples generated many peaks, they did not interfere with the determination of STS.
   For the bile samples of other tanshinones, acetic ester extraction was chosen in this experiment for sample pretreatment due to their high lipid-solubility. This method was convenient and had an extraction recovery higher than 80 %.
3. Selection of Internal Standard Substances
   Internal standard method is a good option to ensure the parallelism and consistence of the extraction process. An internal standard substance should be a substance similar in structure and physicochemical properties to the test compound. However, the structures of tanshinones could transform into one another through metabolism; for example, cryptotanshinone could be metabolized to tanshinone II_A, 15,16-dihydrotanshinone I could be metabolized to tanshinone, tanshinone II_A could be metabolized to tanshinone II_B, 3α-hydroxytanshinone II_A, and przewaquinone A, etc. Therefore, 4-chlorobiphenyl was chosen to be the internal standard as reported in literature. Under the selected chromatographic conditions, the internal standard peak and the peaks of the 6 tanshinones could be separated very well.

1. Results of Method Validation
   (a) Extraction Recovery: the extraction recoveries of tanshinone II_A, cryptotanshinone, 15,16-dihydrotanshinone I, tanshinone II_B, przewaquinone, tanshinone I and internal standard substance from blank bile were 90.5 ± 1.9 %; 83.4 ± 3.5 %; 81.3 ± 3.2 %; 87.5 ± 4.6 %; 87.6 ± 2.2 %; 75.3 ± 3.7 %; 70.4 ± 2.3 %, respectively (n = 5).
   (b) Linear Range and Lower Limit of Quantification: the linear range and the lower limit of quantification of tanshinone II_A, cryptotanshinone, 15,16-dihydrotanshinone I, tanshinone II_B, przewaquinone A, tanshinone I and STS under the selected test conditions are shown in Table 17.17.
   (c) Intraday Precision, Interday Precision, and Accuracy: the results are shown in Table 17.18. These results demonstrated that both the precision and accuracy of the method met the requirements defined in national and international guidelines [13–16].
   (d) Stability: after storing at room temperature for 24 h and undergoing 3 freeze-thaw cycles, the relative errors of variation in drug concentrations of the treated QC samples were less than ±11.8 %, indicating that the samples were stable for at least 24 h at room temperature. The results are shown in Table 17.19.
   (e) Bile Excretion: the cumulative excretion of the 7 tanshinones during 24 h post-administration were: tanshinone II_A, 2.42 ± 0.18 %; cryptotanshinone, 4.16 ± 0.28 %; tanshinone II_A excreted in cryptotanshinone

Table 17.17 The linear range and lower limit of quantification of the seven tanshinones in rat bile

Analytes	Intercept ($\times 10^{-2}$)	Slope	r	Linear range (μg/ml)	LOQ (ng/ml)
Tanshinone II_A	−0.111	0.215	0.9976	0.26–40.8	26
cryptotanshinone	−0.153	0.125	0.9993	0.25–40.8	25
15,16-dihydrotanshinone I	−0.186	0.188	0.9986	0.25–40.8	25
Tanshinone II_B	2.069	0.171	0.9996	1.00–100.0	50
Przewaquinone A	1.425	0.156	0.9972	0.65–130.0	65
Tanshinone I	0.0880	0.126	0.9999	0.20–4.08	50
STS	3857.312	43.425	0.9983	1.25–500.0	50

Table 17.18 The precision and accuracy of the method

Analytes	Concentration (μg/ml)	Intraday RSD (%)	Interday RSD (%)	Relative error (%)
Tanshinone II_A	0.5	2.7	4.0	5.9
	4.1	1.7	3.1	0.6
	40.8	3.5	9.0	2.2
Cryptotanshinone	0.5	8.4	5.1	7.0
	4.1	1.3	3.0	1.7
	40.8	3.4	2.7	1.2
15,16-dihydrotanshinone I	0.3	10.3	6.8	−4.9
	2.0	5.4	3.0	−2.9
	20.4	4.1	5.7	11.6
Tanshinone II_B	2.5	6.4	5.0	−3.9
	10.0	6.0	4.9	−4.6
	50.0	3.9	5.9	−4.0
Przewaquinone A	1.3	5.3	4.1	−2.3
	6.5	4.3	5.1	−0.7
	65.0	3.3	3.2	11.4
Tanshinone I	0.2	3.6	3.6	−2.2
	1.0	2.9	2.9	0.8
	4.1	2.4	2.4	4.8
STS	2.5	8.7	4.7	−14.2
	12.5	1.2	1.3	3.6
	125.0	1.4	7.6	−2.9

bile samples was calculated as cryptotanshinone, 3.58 ± 0.24 %; 15,16-dihydrotanshinone I, 3.75 ± 0.39 %; tanshinone II_B, 9.99 ± 0.82 %; przewatanshinone A, 8.02 ± 0.80 %; tanshinone I, 0.18 ± 0.08 %; STS, 94.34 ± 4.03 %.

As shown in Fig. 17.65, bile excretion of most tanshinones reached a plateau period about 4 h after administration; the bile excretion rate and cumulative excretion of tanshintones varied greatly with their structures. Analyzing the excretion rate using tanshinone II_A as a reference, it can be seen that tanshinones tend to excrete in bile more easily, and excretion rate and cumulative excretion will increase if a hydroxyl substitution is introduced into the structure, such as in tanshinone II_B and przewaquinone A.

When a sulfonic group was introduced into the structure, such as in STS, the bile excretion rate would increase and about 80 % of the administered dose would be excreted in bile within 1 h. It was found in the experiment that red bile (normal bile is golden yellow and STS solution is dark red) flew out immediately after intravenous administration of STS to rats, and bile samples collected during the first two intervals were all an obvious red color, indicating that STS was excreted in bile very quickly. Therefore, sampling intervals were set to be very dense in the experimental design. Experimental results demonstrated that 94.34 % of administered STS was excreted in bile within 24 h.

When the conjugated system of a tanshinone structure was weakened (such as in

Table 17.19 The stability of the seven tanshinone QC samples

Analytes	Concentration (μg/ml)	Relative error (%)	
		After 24 h at room temperature	After three freeze-thaw cycles
Cryptotanshinone	0.5	−11.8	−8.5
	4.1	−4.2	−1.3
	40.8	2.6	4.1
15,16-dihydrotanshinone I	0.5	4.7	−6.7
	4.1	−3.0	−2.8
	40.8	−1.6	−2.0
Tanshinone II_B	0.3	−5.0	−6.3
	2.0	−1.6	2.7
	20.4	−3.5	−1.8
Przewaquinone A	2.5	−3.4	−6.6
	10.0	−2.5	−1.4
	50.0	2.1	1.7
Tanshinone I	1.3	−1.0	−3.1
	6.5	−0.2	0.9
	65.0	−0.6	−0.8
STS	0.2	−3.8	−4.4
	1.0	−0.6	0.5
	4.1	−2.3	−3.4

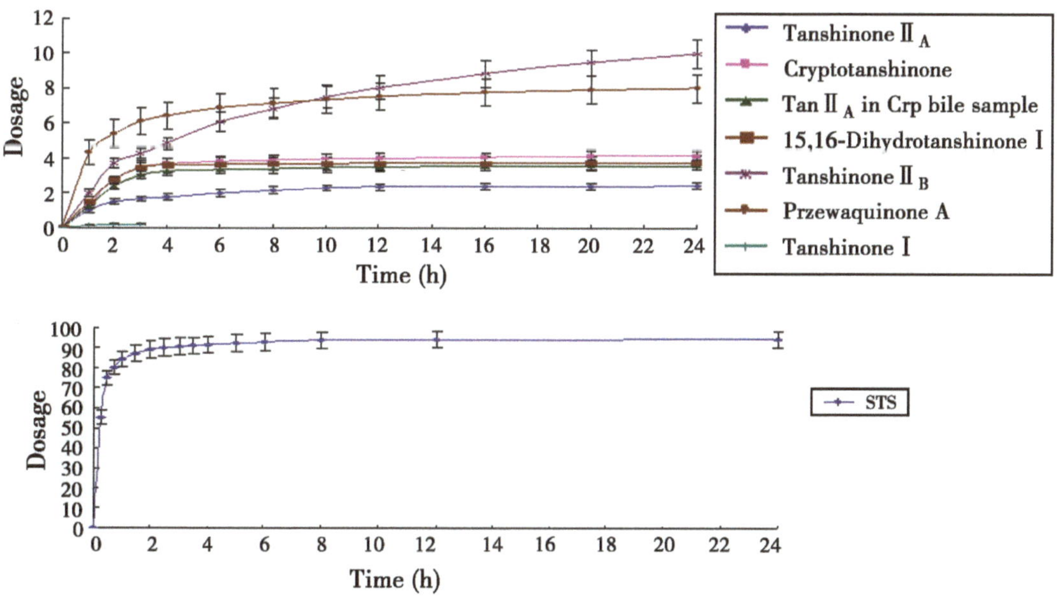

Fig. 17.65 The cumulative excretion of seven tanshinones during 24 h after administration

cryptotanshinone) and both ring A and ring D were present in the structure, bile excretion rate and cumulative excretion were also higher than those of tanshinone II_A; about 46 % of cryptotanshinone excreted in bile was excreted as metabolites, tanshinone II_A. When the conjugated system of a tanshinone structure was weakened (such as in cryptotanshinone) and both ring A and ring D were present in the structure, bile excretion rate and cumulative excretion were also higher than those of tanshinone II_A; about 46 % of cryptotanshinone excreted in bile was excreted as metabolites, tanshinone II_A. When the conjugated system of a tanshinone structure was weakened (such as in cryptotanshinone) and both ring A and ring D were present in the structure, bile excretion rate and cumulative excretion were also higher than those of tanshinone II_A; about 46 % of cryptotanshinone excreted in bile was excreted as metabolites, tanshinone II_A.
In summary, it can be concluded that the extent of conjugation and the substituting groups on the parent nucleus of tanshinones affect the bile excretion rate and cumulative excretion. The higher the conjugation extent of the parent nucleus, the lower the bile excretion rate and the smaller the cumulative excretion. Introducing a hydroxyl group or sulfonic acid group into the parent nucleus will considerably increase the bile excretion rate and cumulative excretion of tanshinones. This finding will help the structural modification and reformation of existing tanshinones and facilitate the discovery of lead compounds with better pharmacokinetic parameters.

References

1. Sun LP, Xiong W, Zheng Z, et al. West China J Pharm Sci. 2004;19:24–8.
2. Hei MY, Yin P, Kuang SJ. J Clin Res. 2004;21:876–80.
3. Chen LJ, Li T, Li C. J Chin Med Mater. 2003;26:504–6.
4. Tang Z, Tang Y, Fu LB. J First Mil Med Univ. 2003;23:595–8.
5. Ye LB, Xi T, Chen F, et al. J Chin Pharm Univ. 2004;35:267–70.
6. Yuan SL, Song Y, Wang XJ, et al. West China J Pharm Sci. 2003;18:327–31.
7. Qin P, Qu Y. J West China Med Univ. 2002;33:370–97.
8. Xia WJ, Xiang P, Zhang LR, et al. Chin J Pathophysiol. 2003;19:865–8.
9. Zuo Q, Zhang PH, Luo SJ, et al. Guangdong Med. 2004;25:636–9.
10. Zhu JR, Luo HW. J Chin Pharm Univ. 2004;35:368–72.
11. Hu MX, Zhou MS, Hu XM, et al. Chin Pharmacol Bull. 2006;22:436–40.
12. Sun XG, Jia JH, Chen YY. Shandong J Chin Tradit Med. 2000;19:358–60.
13. Drug Evaluation Center of SFDA. Guideline on Preclinical Pharmacokinetics studies of chemical drugs. 2001.
14. Xiao C, Chen JX. Chin Pharm J. 1993;24:425–6.
15. Shah VP, Midha KK, Dighe S, et al. J Pharm Sci. 1992;81:309–12.
16. Karnes HT, March C. Pharm Ires. 1993;10:1420–2.
17. Zhang J, Huang M, Guan S, et al. J Pharmacol Exp Ther. 2006;317:1285–94.
18. Yuan Y, Jiang XH, Zhou J, et al. West China J Pharm Sci. 2002;17:246–8.
19. Hao H, Wang G, Cui N, et al. Planta Ivied. 2006;72:1311–7.
20. Hao H, Wang G, Li P, et al. J Pharm Biomed Anal. 2006;40:382–8.
21. Song M, Hang TJ, Zhang ZhX, et al. J Chromatogr B. 2005;827(205):209.
22. Xue M, Cui Y, Wang HQ, et al. J Pharm Biomed Anal. 1999;21:207–13.
23. Bi HC, Law FC, Zhong GP, et al. Biomed Chromatogr, 2007 Mar 14. (in press).
24. Okumura Y, Kakisawa H, Kato M, et al. Bull Chem Soc Jpn. 1961;34:895–7.
25. Baillie AC, Thomson RH. J Chem Soc. 1968;12:48–52.
26. Takium K, Koizumi K. Chem Pharm Bull. 1962;10:112–6.
27. Romanova AS, Patudin AV, Pervykh LN, et al. Khimba Prirodnykh Soedinenii. 1978;4:515–6.
28. Qian Mk, Yang PC, Chin KC, et al. Yaoxue Tongbao 1980; 15:1–5.
29. Ryu SY, No Z, Kim SH, et al. Planta Med. 1997;63:44–6.
30. Liu AH, Lin YH, Yang M, et al. J Chromatogr B. 2007;846:32–41.
31. Hu P, Luo GA, Zhao ZZ, et al. Chem Pharm Bull. 2005;53:705–9.

32. Pan JY, Cheng YY. J Pharm Biomed Anal. 2006;42:565–72.
33. Yang M, Liu AH, Guan SH, et al. Rapid Commun Mass Spectrom. 2006;20:1266–80.
34. Mao SJ, Hou SX, Liang Z, et al. J Chromatogr B. 2006;831:163–8.
35. Li P, Wang GJ, Li J, et al. J Mass Spectrom. 2006;41:670–84.
36. He Q, Hu XJ, Cheng YY. J Pharrn Biomed Anal. 2006;41:485–92.
37. Luo HW, Sheng LS, Zhang SQ, et al. Acta Pharm Sinica. 1983;18:1–6.

Postscript

Prospects of Pharmacological Research on Danshen

Danshen can be considered as a typical example of modern pharmacological research on a traditional Chinese medicine. After reviewing the research data on Danshen, we have found that not only is the data abundant, but the scope of the research is also broad. The research material covers fields from plants and ecology to medical preparations and products; from decoction pieces and decoctum to chemical components; from basic research to clinical trials; from single chemical components, active fractions, and single herbal drugs to compound preparations. It is really a pharmacological monograph with abundant content.

Research on Danshen has received great attention, possibly for the following reasons. The first is the curative effect of Danshen. As a commonly used drug "to activate circulation and to dissipate blood stasis" in traditional Chinese medicine, Danshen has played an important role in clinical practice for a long time, and has received great attention. The second reason is the theory of "activating circulation to dissipate blood stasis." As an important theory in traditional Chinese medicine, it has also received great attention. Although symptoms of blood stasis cannot be explained completely by modern medicine theory, the understanding of the diseases and the curative effects have been confirmed in clinical practice. The third reason is the confirmation of Danshen's effects through research, which establishes the basis for understanding the pharmacological effects of Danshen. Because Danshen can withstand the validation of modern pharmacological examinations, research on Danshen has been greatly promoted.

The research studies on Danshen have the following characteristics:

1. **The studies on the chemical constituents are comprehensive**

The identification and acquisition of chemical constituents is the basis for research on natural products, and for TCM, it is particularly so. Whether it is a simple extract or a single chemical constituent, no pharmacological research can be performed without this basis. In the research on Danshen, great efforts have been made to study the extracts, fractions, and constituents of Danshen, which establish the basis of pharmacological research. Furthermore, the chemical constituents of Danshen are well understood: we not only know a lot about the easily obtainable liposoluble constituents, but have also acquired a great number of hydrosoluble constituents with strong pharmacological activities. These researches on the chemical constituents establish the material basis of pharmacological research on Danshen.

2. **The understanding of the pharmacological effects is extensive**

There are four primary types of research on the pharmacological effects of Danshen:

(1) Using modern pharmacological methods to study and demonstrate the effects and functions recorded in ancient literature. Many confirmative results have been obtained in these studies. For example, research on the

X. Yan (ed.), *Dan Shen (Salvia miltiorrhiza) in Medicine*,
DOI 10.1007/978-94-017-9463-3,

curative effects of Danshen on myocardial ischemia and coronary artery disease, research on antibacterial activities, and research on antiplatelet activity are all based on the understanding of the ancient classics. These studies have demonstrated the connectivity and compatibility of TCM and modern medicine. Although no theories have been established on the connectivity and compatibility, many facts have proved the scientific nature of TCM.

(2) Discovering the pharmacological effects of Danshen and its constituents on the basis of the understanding of current applications and findings in related research, and of theoretical reasoning and analysis. For example, Danshen was mainly used for the treatment of cardiovascular diseases in TCM. Although multiple functions were recorded in ancient literature, clinical applications in other areas were limited. With the advancement of related research, the clinical application of Danshen extended into the treatment of nephropathy and respiratory diseases. Of course, the curative effects of Danshen in these extended applications are still to be validated by more research.

(3) Inferring the possible functions and activities of Danshen based on theories of modern medicine. For example, research has confirmed that Danshen had good effect on the treatment of cardiovascular diseases, and had protective effects on ischemia injury and ischemia reperfusion injury, which suggests that its antioxidation activity might be involved. Combined with chemistry research, it is proven that the salvianolic acid compounds of Danshen are one of the strongest natural antioxidants.

(4) Confirming the pharmacological effects deduced from the results of modern research by experiments. After a great amount of research on the chemical constituents of Danshen, some researchers thought it could have effects on not only the cardiovascular system but also on the cerebrovascular system. Research has proved that salvianolic acids are an ideal type of drug to treat cerebral ischemia diseases. Preliminary clinical trials confirmed the good curative effect of Danshen.

Pharmacological research on Danshen and its chemical constituents has reflected perfectly the researchers' understanding of modern medicine and traditional Chinese medicine and their creativity. Creative research further promotes the understanding of Danshen's multiple pharmacological effects.

3. The relationship between research and application is intertwined

Until now, no other traditional Chinese medicines have established such a close tie between research and clinical application from the very beginning, which has resulted in a great number of modern pharmaceutical preparations for clinical application. Some of these preparations were the chemical constituents of Danshen, such as Danshen Tablet, Tanshinol Tablet, Danshen Injection, Tanshinone Tablet, and Tanshinone Capsule, and some were the compound preparations of Danshen. These preparations were produced in many factories with a huge sales volume, and new preparations are coming out continuously. It has been a hotspot in medical research. The reasons for this phenomenon include two parts. First, Danshen is a good drug with good curative effects. Second, the research on Danshen is thorough. Scientific research has played an important supportive role in the application of Danshen.

4. The combination of traditional Chinese medicine and modern medicine

The research on Danshen has shown the close integration of TCM and modern medicine. Although no proper entry point has been found for merging of these two medical theories, the integration is totally reflected in the research on Danshen nevertheless. Research on the chemical constituents and molecular mechanisms have enhanced the application of Danshen under the guidance of modern medicine theory, while pharmacological research further proved the rationality and scientific nature of TCM. The application of modern preparations under the guidance of TCM theory is a unique feature in the clinical practice.

5. **The scientific basis of traditional understanding is explained by modern research**

As we all know, Danshen has multiple effects which are the result of different constituents in Danshen with different pharmacological effects. The existence of these constituents and the manifestation of different effects have established the basis of understanding the curative effects of Danshen in traditional application. Great respect should be given to our antecessors for their precise and painstaking work on the recognition and application of the drugs. It is amazing that they were able to manage the exact effect of Danshen in different diseases. In contrast, with plenty of information and high-performance techniques, our understanding of Danshen is not more accurate than that of our predecessors, which requires careful reflection on our part.

Of course, as the author mentioned in the preface to this part, there is plenty of information on Danshen with little or limited scientific value. The occurrence of this phenomenon is not accidental, but an inevitable result in the development of science. However, the existence of the phenomenon requires us to treat these results carefully, verify and understand them with more rigorous methods, and continue to improve our understanding of Danshen.

Therefore, research on Danshen is still a long-term task. With the advancement and development of science, the research will explore deeper and broader, which will effectively enhance the application of Danshen in the clinical practice reasonably and scientifically, and produce better results in the prevention and treatment of diseases. The research on Danshen will focus on the following aspects:

(1) **The chemical constituents and their pharmacological effects**

Although hundreds of constituents in Danshen have been identified so far, our understanding of these chemical constituents is still not complete. Plenty of trace constituents may exist in Danshen, which would produce certain pharmacological effects. Recent research has suggested that some constituents of Danshen in low content could produce good effects. These constituents would be important in the clinical application of Danshen.

(2) **The mechanism for the functions of the effective constituents**

There is no end to the understanding of the mechanisms of action. With the advancement of science, this understanding will go deeper and deeper. As for the research on Danshen, there is a more important meaning with the constant advancement, and that is the combination of the theories of traditional Chinese medicine and modern medicine.

(3) **Scientific evaluation of effective constituents of Danshen**

At present, plenty of chemical constituents of Danshen have been identified. But there are still many theoretical difficulties and confusions in the criteria for defining effective constituents. Therefore, after the development of the first drug with a single chemical compound, the development of the second, third, and fourth single chemical compounds will follow, which may not be scientific or feasible. For this reason, an important topic we have to face will be how to scientifically evaluate these effective constituents.

(4) **Development of ideal preparations of danshen**

The objective of drug research is to prevent and treat diseases. Research on the Danshen will spur the development and research of new drugs, which will produce ideal drugs with better curative effects and benefit humankind.

As a medicinal herb used in traditional Chinese medicine for thousands of years, Danshen is not only the object of scientific research that needs more effort, but is also an important material that will bring benefit to man. Its contents are inexhaustible, its effects are complex, and the materials are plain, which are as marvelous as traditional Chinese medicine theories.

Elimination of interference in the research and impurity in the results is a long-term task for scientific researchers.

This part is a simple summary of the pharmacological effects of Danshen. Due to the deficiency of in-depth literature analysis and experimental confirmation, controversial results are unavoidable. Even so, it still has certain reference value for scientific researchers in understanding the current status of research on Danshen. The authors wish this part to provide basic information for scientific researchers and people in the related areas, and to facilitative the scientific research, clinical application, and industrial development of Danshen.

Guanhua Du, Juntian Zhang

Index

X. Yan (ed.), *Dan Shen (Salvia miltiorrhiza) in Medicine*,
DOI 10.1007/978-94-017-9463-3,